PREFACE

The International Association for Cryptologic Research (IACR) organizes two international conferences every year, one in Europe and one in the United States. EUROCRYPT '89 was the seventh European conference and was held in Houthalen, Belgium on April 10-13, 1989. With close to 300 participants, it was perhaps the largest open conference on cryptography ever held.

The field of cryptography is expanding not only because of the increased vulnerability of computer systems and networks to an increasing range of threats, but also because of the rapid progress in cryptographic methods, that the readers can witness by reading the book.

The present proceedings contain nearly all contributions which were presented including the talks at the rump session. The chapters correspond to the sessions at the conference. It was the first time that a rump session was organized on a Eurocrypt conference. Sixteen impromptu talks were given, and the authors were invited to submit short abstracts of their presentations. Because of the special character of this session, the editors have taken the liberty to shorten some of these.

We are grateful to all authors for the careful preparation of their contributions. It is a pleasure to thank the members of the Program Committee for having made the conference such an interesting and stimulating meeting. In particular, we were very pleased with the interesting rump session organized by J. Gordon and the animated open problem session organized by E. Brickell. We are indebted to the sponsors for their generous donations and to the members of the Organization Committee for the smooth organization of the meeting.

Louvain-la-Neuve, Belgium   J.-J.Q.
Louvain, Belgium   J.V.
July 1990

Contents

Section 1: Public-key cryptosystems

Section 2: Theory

Section 3: Zero-knowledge protocols

Section 4: Applications

Section 5: Signature and untraceability

Section 6: Cryptanalysis

Section 7: Sharing and authentication schemes

Section 8: Sequences

Section 9: Algorithms

Section 10: Old problems

Section 11: Rump Session (impromptu talks)

Section 1

Public-key cryptosystems

THE ADOLESCENCE OF PUBLIC-KEY CRYPTOGRAPHY

Whitfield Diffie

Northern Telecom, 685A E. Middlefield Road,
94043 Mountain View, CA, USA

Abstract

©1988 IEEE. Reprinted, with permission, from Proceedings IEEE, Vol. 76, No. 5, pp. 560-577, May 1988.

Public-key cryptosystems separate the capacities for encryption and decryption so that 1) many people can encrypt messages in such a way that only one person can read them, or 2) one person can encrypt messages in such a way that many people can read them. This separation allows important improvements in the management of cryptographic keys and makes it possible to 'sign' a purely digital message.

Public-key cryptography was discovered in the Spring of 1975 and has followed a surprising course. Although diverse systems were proposed early on, the ones that appear both practical and secure today are all very closely related and the search for new and different ones has met with little success. Despite this reliance on a limited mathematical foundation public-key cryptography is revolutionizing communication security by making possible secure communication networks with hundreds of thousands of subscribers.

Equally important is the impact of public-key cryptography on the theoretical side of communication security. It has given cryptographers a systematic means of addressing a broad range of security objectives and pointed the way toward a more theoretical approach that allows the development of cryptographic protocols with proven security characteristics.

A Secure Public-Key Authentication Scheme

Zvi Galil
Tel Aviv University
and
Columbia University
Department of Computer Science
New York, N.Y. 10027

Stuart Haber
Bellcore
445 South Street
Morristown, N.J. 07960

Moti Yung
IBM Research Division
T.J. Watson Research Center
P.O. Box 218
Yorktown Heights, N.Y. 10598

Abstract

We propose interactive probabilistic public-key encryption schemes so that:

(1) the sender and the receiver of a message, as well as the message itself, can be authenticated;

(2) the scheme is secure against any feasible attack by a participant, including chosen-ciphertext attack.

Our suggested protocols can use any one-way trapdoor functions. In order to formulate and prove the properties of our procedures, we propose several new complexity-theoretic definitions of different levels of cryptographic security for systems which allow interaction. Chosen-ciphertext security is achieved using the techniques of minimum-knowledge interactive proofs, and requires only a constant number of message exchanges at the system's initiation stage.

1 Introduction

To authenticate something is to prove that it is genuine or to establish its validity. There are several sorts of on-line authentication problem that face the designer of a cryptosystem. For example, he may wish to authenticate either a physical or a virtual channel between certain pairs of users; he may wish to allow users to establish their identities; or he may wish to enable the authentication of certain characteristics of specific messages.

Among the directions that researchers have taken in studying authentication problems are the "algebraic" approach of [14, 4, 5], the "authentication channel" of [15], the "digital signature" approach initiated by Diffie and Hellman [3] (which actually deals with a different problem that we do not study here), and the "identification scheme" of [6] (whose purpose is to identify a user without regard to specific messages).

In this paper we will assume that the communications media are not physically or otherwise authenticated. Our setting is that of the original Diffie-Hellman public-key model [3]: each user has an encryption key that is "public," or available to all other users (as well as to any adversary), and a corresponding private decryption key; all cryptographic tasks, including encryption and authentication of messages, must employ these keys. Our procedures are not based on the intractability of a specific computational problem, but are based instead on the more general assumption that a one-way trapdoor function exists.

In this setting, we distinguish several authentication problems that arise when A sends a message to B. There is the problem of *sender authentication*, which is to convince B that it was indeed A who sent the message, and the complementary problem of *receiver authentication*, which is to convince A that it is indeed B who received the message. There is also the problem of *message authentication*, which is to convince B that the message he receives is indeed the message that A intended to send.

In [7] we suggested an interactive procedure for the authentication both of the sender and of the receiver of a probabilistically encoded message. In this paper we extend our scheme so as to enable message authentication; and we show how to use interactive proofs of knowledge [16, 6] in order to refine our procedures so that they are secure against chosen-ciphertext attack. A major contribution of this paper is the formulation of definitions of security and privacy, using the language of computational complexity. In proving the properties of our constructions, we adapt the formal definitions of cryptographic security that were proposed by [17, 10], generalizing them to the setting of interacting users of a cryptosystem. Our definitions capture the intuitive notions that have often been used by cryptographers without careful definition, while allowing us to give precise proofs of the strength of the cryptographic procedures that we propose.

After describing our assumptions and notational conventions, we give our definitions of several different degrees of cryptographic security in §2. We give our authentication procedures and state their properties in §3. Finally, in §4, we formalize chosen-ciphertext attack and give our procedures for achieving security against such attack. We give our new definitions in some detail in this abstract, but due to lack of space we omit the proofs in §§3-4; for these, see the full paper [9].

2 Definitions of cryptographic security

2.1 Preliminaries

We model the users of a cryptosystem as *interactive Turing machines*, as defined in [11, 8].

Our constructions assume the existence of a family of one-way functions with an associated family of hard-bit predicates, as follows [1, 17]. Let I be an infinite family of strings, and for each positive integer k let $I_k = \{i \in I \mid |i| = k\}$ denote the set of strings in I of length k. A *hard-bit family* is a family $\mathcal{F} = \{(f_i, b_i) \mid i \in I\}$, where for each i of length k, $f_i : D_i \to D_i$ is a function defined on a domain D_i consisting of k-bit strings, and $b_i : D_i \to \{0,1\}^{l(k)}$ is an $l(k)$-bit "predicate". We require that these sets be *accessible*: there is a probabilistic algorithm, running in expected polynomial time, that, given k, chooses $i \in I_k$ uniformly and at random; and there is another such algorithm that, given i, chooses $x \in D_i$ uniformly and at random. Both $(i, x) \mapsto f_i(x)$ and $(i, x) \mapsto b_i(x)$ are easy computations (i.e. of cost polynomial in k); the functions f_i are *one-way*, and furthermore it is computationally intractable to predict the bits of $b_i(x)$, given only the value $f_i(x) \in D_i$.

$$x \quad \to \quad f_i(x)$$
$$\downarrow \quad \diagup \quad \text{hard bits}$$
$$b_i(x)$$

(Typically, one is given functions f_i that are assumed to be one-way, and then one proves that the "hard bits" $b_i \circ f_i^{-1}$ polynomially reduce to f_i^{-1}. By a recent result of Impagliazzo, Levin, and Luby [12], it suffices merely to assume the existence of a family $\{f_i\}$ of one-way functions.)

For our applications, we assume in addition that $\mathcal{F} = \{(f_i, b_i) \mid i \in I\}$ is a *trapdoor hard-bit family*: for each $i \in I$ there is a string d_i—the *trapdoor information* or the *secret key* associated with i—that enables the inversion of f_i, so that the computation $(i, y, d_i) \mapsto x \in f_i^{-1}(y)$ is easy, for any $y \in f_i(D_i)$. Furthermore, there is a probabilistic algorithm that, on input k, chooses a pair (i, d_i), with i uniformly distributed in the set I_k.

For the sake of greater efficiency in one of our constructions in §4 below, we assume that (as is the case with all the suggested examples of hard-bit families $\mathcal{F}$) the sets D_i are groups, and that each function f_i is a group automorphism.

The following presentation is adapted from that of [10, 13, 2]. We will say that a *public-key cryptosystem* consists of a *security parameter* k, a sequence of *message spaces*

$\{M_k\}$ (probability distributions on strings of polynomial length $l(k)$), a key-generator, possibly a public-key encryption-decryption algorithm pair, and one or more message-sending protocols:

- The *key-generator* is a probabilistic algorithm G that, on input 1^k, halts in expected time polynomial in k and writes out a pair of strings (e, d): a *public key* and a *secret key*. For example, for any trapdoor hard-bit family $\mathcal{F}$, there exists a corresponding key-generator $G_{\mathcal{F}}$ which, on input k, chooses $i \in I_k$ at random and writes out the pair (i, d_i).

- A *public-key encryption-decryption algorithm pair* consists of a pair of (possibly probabilistic) algorithms (E, D). The encryption algorithm E takes as input a message m chosen according to the message space M_k and a public key e with security parameter k, and produces a ciphertext c; when E is probabilistic, there may be many possible ciphertexts for each input pair (m, e). The decryption algorithm D takes as input a ciphertext c and a private key d, and produces the cleartext message m.

- A *message-sending protocol* is a pair of interacting probabilistic Turing machines (S, R), a *sender* and a *receiver*. Each machine takes as input a pair of public keys (e_S, e_R) (belonging to S and R, respectively) as well as its own private key d_S or d_R (respectively). The sender receives as additional input a message chosen according to the message space M_k. The machines proceed with their computation. At any point, either party may *reject* the execution, writing a special **reject** symbol on its private output tape. Otherwise, when they halt the receiver writes out, as a private output, the cleartext message m' that it has *accepted*. We also require that S write a message m (presumably, the one it has just tried to send) on its private output tape. The protocol is *correct* if $m' = m$ with overwhelming probability.

2.2 Security against cryptanalysis

In this section we repeat the definition first proposed by Goldwasser and Micali for the "polynomial security" of a probabilistic public-key encryption algorithm [10], and then we adapt the definition in order to handle more general public-key message-sending protocols. In our definitions, attacks are deemed to be successful if they result in very weak sorts of cryptanalytic ability; hence any procedure that satisfies one of our definitions is all the stronger against attacks that attempt to achieve more.

Suppose that we are given a public-key encryption-decryption algorithm pair (E, D) (along with a sequence of message spaces $\{M_k\}$ and a key-generator G). A *message-finder* is a family of probabilistic circuits $F = \{F_k\}$, as follows: on input a specific public key with security parameter k, the circuit F_k produces as output two messages $m_0, m_1 \in M_k$. A *message-distinguisher* is a family of probabilistic circuits $D = \{D_k\}$, each of which takes four input strings and produces one output bit: the first input to D_k is a public key e with security parameter k; its second and third inputs are strings of length $l(k)$, for example a pair of messages chosen from M_k; and its fourth input is a string of length $l'(k)$, such as an encryption produced by E. (In [10] "message-distinguishers" were called "line-tappers.") Note that both F_k and D_k may have information about G, about M_k, and about E "hard-wired" into them.

Consider the following experiment. Run the algorithm G on input 1^k to produce a pair of keys (e, d); give the public key e as input to F_k to obtain a pair of messages $m_0, m_1 \in M_k$; choose one of these at random, m_b say, and give (m_b, e) to E to produce an encryption α; finally, give (e, m_0, m_1, α) as input to D_k to obtain the bit b'. (This bit b' may depend, of course, on the sequence of random bits used by these probabilistic circuits and probabilistic Turing machines.) The experiment is a success if $b = b'$. We say that the encryption-decryption algorithm pair (E, D) is *polynomially secure against cryptanalysis* if for any message-finder F, for any message-distinguisher D, and for any constant c, the probability of success is less than $\frac{1}{2} + \frac{1}{k^c}$ for sufficiently large k.

Suppose now that we are given a message-sending protocol (S, R) (along with $\{M_k\}$ and G). To analyze the security of such a protocol we define a *message-finder* to be a family of probabilistic circuits $F = \{F_k\}$ whose kth member takes as input a pair of public keys with security parameter k, and produces as output two messages $m_0, m_1 \in M_k$. Similarly, we define a *message-distinguisher* $D = \{D_k\}$ as before, except that D_k's inputs include a pair of public keys with security parameter k instead of just a single public key, and the transcript of a protocol execution instead of an output of the encryption algorithm. In addition, we give D_k another input string consisting of the random-bit string that was F_k's random input; this is a technical consideration that could conceivably increase the power of the attacking circuits. (Once again, both F_k and D_k may have information about M_k, about S, and about R "hard-wired" into them.)

As with a conventional public-key encryption algorithm, we consider the following experiment. Run the algorithm G twice on input 1^k to produce two pairs of keys (e_S, d_S) and (e_R, d_R), and give the public keys e_S, e_R (along with a random bit-sequence ρ) as input to F_k to obtain a pair of messages $m_0, m_1 \in M_k$. Choose one of these at random, m_b say, and use S and R to perform the given protocol to send the message m_b; let τ be the transcript of strings written by S and R on their communication tapes during the protocol execution. Finally, give $(e_S, e_R, \rho, m_0, m_1, \tau)$ as input to D_k to obtain the bit b'—its guess as to whether the message sent was m_0 or m_1. The experiment is a success if $b = b'$. We say that the message-sending protocol is *polynomially secure against cryptanalysis* if for any message-finder F, for any message-distinguisher D, and for any constant d, the probability of success is less than $\frac{1}{2} + \frac{1}{k^d}$ for sufficiently large k.

We remark that a conventional public-key encryption-decryption algorithm pair (E, D) can be regarded as a special kind of interactive message-sending protocol, one in which there is only one round of interaction. For such a message-sending protocol, the definition of security just proposed reduces to the Goldwasser-Micali definition described above.

Next we deal with the provision of authentication (for each message sent) in the context of the probabilistic public-key model. In §3 below we present our public-key solution to the problems of sender, receiver, and message authentication. In the rest of §2 we give our formal definitions of authentication security.

2.3 Security against sender impersonation

We formalize the property of sender authentication by asking what advantage an adversary can gain by impersonating the legitimate sender of a message. Thus we define a *sender-impersonator* to be an interactive Turing machine $\tilde{S}$ that takes as input a pair of public keys (e_S, e_R). Note that $\tilde{S}$ is not given either of the corresponding secret keys d_S, d_R.

Consider the following experiment. Run the algorithm G twice on input 1^k to produce two pairs of keys (e_S, d_S) and (e_R, d_R). Then perform the given protocol, with $\tilde{S}$ (instead of S) acting as the sender, and R performing the role of the receiver (as specified); let m be the private output of $\tilde{S}$. The experiment is successful if R accepts the same message m. We say that the protocol is *polynomially secure against sender impersonation* if for any sender-impersonator $\tilde{S}$ and for any constant d, the probability that the experiment is successful is less than $\frac{1}{2^{l(k)}} + \frac{1}{k^d}$ for sufficiently large k.

2.4 Security against receiver impersonation

We formalize the property of receiver authentication by asking what advantage an adversary can gain by impersonating the legitimate receiver of a message. Thus we define a *receiver-impersonator* to be an interactive Turing machine $\tilde{R}$ that takes as input a pair of public keys (e_S, e_R) and a pair of messages m_0, m_1. We define a *message-distinguisher* $D = \{D_k\}$ as above, except that the last input of D_k is $\tilde{R}$'s view of a protocol execution with security parameter k (in other words, a string of, say, $l'(k)$ bits). Note that D_k may have information about G, about M_k, about (S, R), and about $\tilde{R}$ "hard-wired" into it. (However, neither $\tilde{R}$ nor D_k is given either of the secret keys d_S, d_R corresponding to e_S, e_R.)

Let $F = \{F_k\}$ be a message-finder, and consider the following experiment. Run the algorithm G twice on input 1^k to produce two pairs of keys (e_S, d_S) and (e_R, d_R), and give the public keys e_S, e_R as input to F_k to obtain a pair of messages $m_0, m_1 \in M_k$. Choose one of these at random, m_b say, and perform the given protocol to send the message m_b, with S performing the role of the sender (as specified), and $\tilde{R}$ acting as the receiver. Finally, give $\tilde{R}$'s view of the protocol execution to D_k to obtain a bit b' (its guess as to whether the message sent was m_0 or m_1). Call the experiment a success if $b = b'$. We say that the protocol is *polynomially secure against receiver impersonation* if for any message-finder F, for any receiver-impersonator $\tilde{R}$, for any message-distinguisher D, and for any constant d, the probability of success is less than $\frac{1}{2} + \frac{1}{k^d}$ for sufficiently large k.

2.5 Security against random-message attack

In order to address the problem of message authentication formally, we challenge an adversary, without knowing a sender's private key, to "force" a legitimate receiver to accept some legal message. Consider, therefore, the following experiment. Let $\tilde{S}$ be a sender-impersonator. Run the algorithm G twice on input 1^k to produce two pairs of keys (e_S, d_S) and (e_R, d_R). Then perform the given protocol, with $\tilde{S}$ (instead of S) acting as the sender, and R performing the role of the receiver (as specified). The experiment is successful if R accepts a message $m \in M_k$—any message at all. We say that the protocol is *polynomially secure against random-message attack* if for any sender-impersonator $\tilde{S}$ and for any constant d, the probability that the experiment is successful is less than $\frac{1}{k^d}$ for sufficiently large k.

Observe that a message-sending protocol which is polynomially secure against random-message attack is clearly also polynomially secure against sender impersonation.

3 Authentication schemes

Assume that we are given a trapdoor hard-bit family $\mathcal{F} = \{(f_i, b_i) \mid i \in I\}$, where b_i is an $l(k)$-bit predicate for $i \in I$ of length k. The following is a protocol for S to send an authenticated $l(k)$-bit message m to R. The sender's inputs include its public and secret keys (i_S, d_{i_S}), the receiver's public key i_R, and the message m; the receiver's inputs include its public and secret keys (i_R, d_{i_R}) and the sender's public key i_S. **Protocol 1**

1. $S \rightarrow R$: "Hi, this is S sending a message to R."

2. R chooses $x_S \in D_{i_S}$ at random, and computes $p_S := b_{i_S}(x_S)$ and $y_S := f_{i_S}(x_S)$.
   $R \rightarrow S$: y_S

3. S computes $x_S := f_{i_S}^{-1}(y_S)$ and $p_S := b_{i_S}(x_S)$, chooses $x_R \in D_{i_R}$ at random, and computes $p_R := b_{i_R}(x_R)$, $y_R := f_{i_R}(x_R)$, $c := m \oplus p_S \oplus p_R$.
   $S \rightarrow R$: $[c, y_R]$

4. R computes $x_R := f_{i_R}^{-1}(y_R)$, $p_R := b_{i_R}(x_R)$, $c \oplus p_S \oplus p_R = m$; and then R accepts and writes as output this message m.

   end-protocol

Theorem 1 *Protocol 1, based on the trapdoor hard-bit family $\mathcal{F} = \{(f_i, b_i) \mid i \in I\}$, is a correct message-sending protocol that is polynomially secure against cryptanalysis, against sender impersonation, and against receiver impersonation.*

Protocol 1 is not secure against random-message attack. A cheating user S', not knowing S's trapdoor d_{i_S}, can still succeed in sending *some* message to R simply by choosing a string c at random in step 3. In order to foil such an attack, we adapt the protocol, as follows. In step 2, R chooses not just one but k random elements x_S, computing the corresponding values y_S and p_S for each one; similarly, S runs k versions of step 3 in parallel. Finally, R executes k versions of the computation $m' := c \oplus p_S \oplus p_R$, and only accepts a message if they all give the same value m'. Call the resulting procedure Protocol 2.

Theorem 2 *Protocol 2, based on the trapdoor hard-bit family $\mathcal{F} = \{(f_i, b_i) \mid i \in I\}$, is a correct message-sending protocol that has all the security properties of Protocol 1 and is also polynomially secure against random-message attack.*

4 Chosen-ciphertext security

The designer of a multi-user cryptosystem must be concerned not only with passive attacks or unauthenticated messages sent by impersonators but also with attacks carried on by active participants. One or more malicious users may take advantage of their user privileges, sending and receiving specially computed plaintext and ciphertext messages, or deviating from the system's message-sending protocols in some other manner, in order to attack a legitimate user's security. In this section we deal with such attacks.

4.1 Definition of chosen-ciphertext security

In a chosen-ciphertext attack on a standard encryption algorthm, the adversary is allowed to choose several ciphertext messages, and then is given the corresponding plaintext messages (if they exist). Generalizing to the context of message-sending protocols, we may regard such an attack as consisting of two stages: a participation stage during which the attackers take part in protocol executions and interact with other users, and an extraction stage during which the attackers try to infer additional information about legitimate users' messages or private keys. As in the definitions above, we will call the attack successful if the extraction stage results in a very weak sort of cryptanalytic ability.

As in §2.2 above, it may be helpful formally to define a chosen-ciphertext attack on a public-key encryption-decryption algorithm pair (E, D) before defining the attack on a more general message-sending protocol. A *chosen-ciphertext attack* on (E, D) consists of an interactive Turing machine $\mathcal{A}$, a message-finder $F = \{F_k\}$, and a message-distinguisher

$D = \{D_k\}$, as follows. The input to $\mathcal{A}$ consists of a security parameter k and a public key e; this key is the target of $\mathcal{A}$'s attack. Several times during the course of its computation, $\mathcal{A}$ interacts with D by requesting that a ciphertext string c of its choice be decrypted using the private key d that corresponds to e. (More precisely, $\mathcal{A}$ writes c on one of its communication tapes, and then reads $D(c, d)$ from its other communication tape.) The message-finder circuit F_k takes as input the target key e along with the contents of $\mathcal{A}$'s history tape, and computes as output two messages $m_0, m_1 \in M_k$. The message-distinguisher D_k takes as input a public key, the contents of an interactive Turing machine's history tape, a pair of messages in M_k, and an encryption produced by E.

Consider the following experiment. Run G on 1^k to get (e, d). Run $\mathcal{A}$ with input e, using D (and the private key d) to decrypt the requested strings; this computation is the participation stage of the attack. Next, in the extraction stage, give e and $\mathcal{A}$'s history tape h to F_k to obtain a pair of messages $m_0, m_1 \in M_k$. Choose one of these at random, m_b say, and give (m_b, e) to E to produce an encryption α; finally, give (e, h, m_0, m_1, α) as input to D_k to obtain the bit b'. The experiment is a success if $b = b'$. We say that (E, D) is *polynomially secure against chosen-ciphertext attack* if for any interactive Turing machine $\mathcal{A}$, for any message-finder F, for any message-distinguisher T, and for any constant d, the probability of success is less than $\frac{1}{2} + \frac{1}{k^d}$ for sufficiently large k.

Next we generalize this definition to deal with the more general setting of interactive protocols. Suppose that we are given a message-sending protocol (S, R) (along with $\{M_k\}$ and G). A *chosen-ciphertext attack* on the protocol consists of an interactive Turing machine $\mathcal{A}$, a message-finder $F = \{F_k\}$, and a message-distinguisher $D = \{D_k\}$, as follows. The input to $\mathcal{A}$ consists of a security parameter k and two public keys e_0 and e_1; these keys are the targets of $\mathcal{A}$'s attack. After performing several executions of the message-sending protocol, as described below, $\mathcal{A}$ writes out its view of these executions. The kth message-finding circuit F_k takes as input the target keys e_0, e_1 along with $\mathcal{A}$'s execution views, and computes as output two messages $m_0, m_1 \in M_k$ as well as a *choice* of either 0 or 1. The message-distinguisher is as above, except that D_k's inputs include a pair of public keys with security parameter k instead of a single public key, the machine $\mathcal{A}$'s history string, and the transcript of a protocol execution instead of an output of the encryption algorithm.

The attacking machine $\mathcal{A}$ is meant to model a coalition of several malicious users; it operates as follows. Several (polynomially many) times, $\mathcal{A}$ chooses to participate in the given message-sending protocol, each time choosing either to send or to receive a message, as well as choosing one of the two keys e_0 or e_1 to be used by the "legitimate" receiver or sender.

- If the choice is to send, then $\mathcal{A}$ takes the role of the sender in a protocol execution with R. In this execution, $\mathcal{A}$ may "send" a message from the message space M_k or a message chosen according to some other computation. R uses either (e_0, d_0) or (e_1, d_1) according to $\mathcal{A}$'s choice. At the end of the execution, $\mathcal{A}$ is given R's private output (the message that R accepted). For convenience, we may refer to $\mathcal{A}$'s computation during this execution as a "subroutine" $\tilde{S}$.

- If the choice is to receive, then $\mathcal{A}$ takes the role of the receiver in a protocol execution with S, in which S attempts to send a message chosen according to the distribution M_k. S uses either (e_0, d_0) or (e_1, d_1) according to $\mathcal{A}$'s choice. At the end of the execution, $\mathcal{A}$ is given S's private output (the message that S tried to send). Once again, we may refer to $\mathcal{A}$'s "subroutine" $\tilde{R}$.

In each execution, $\mathcal{A}$ may request a new public-key, private-key pair generated by G (with security parameter k), or may use a pair requested earlier. These pairs are generated independently of the target keys.

$\mathcal{A}$'s attack may be "adaptive": any of its computation steps may depend on previous steps. For instance, the sending (or receiving) subroutine $\tilde{S}$ (or $\tilde{R}$) invoked by $\mathcal{A}$ during a particular execution of the protocol may be different from the subroutine used by $\mathcal{A}$ in an earlier execution. Of course, the computation of R (or of S)—modelling the actions of a legitimate receiver (or sender)—is not adaptive; each step is an independent execution with a new sequence of random coin-flips. Without loss of generality we may require that the concatenation of $\mathcal{A}$'s execution views be a string v of length at most $l(k)$, where l is a polynomial.

Consider the following experiment. Run G twice on input 1^k to obtain (e_0, d_0) and (e_1, d_1). Run $\mathcal{A}$ with input (e_0, e_1) as described above to produce the execution history v; this computation is the participation stage of the attack. Next, in the extraction stage, give (e_0, e_1) and v to F_k to produce $m_0, m_1 \in M_k$ and a "choice" of 0 or 1; this is a choice as to whether to use the target keys e_0 and e_1 as the receiver's key and the sender's key or vice versa in the next run of the protocol. Next, choose one of the messages m_0, m_1 at random, m_b say, and run (S, R) to send m_b, using as keys either (e_0, d_0) and (e_1, d_1) or (e_1, d_1) and (e_0, d_0) according to F_k's choice of 0 or 1, respectively; let τ be the ("public") transcript of this run. Finally, give the public keys e_0 and e_1, the pair of messages (m_0, m_1), the transcript τ, and the history string v as input to D_k to obtain the bit b'—its guess as to whether the message sent was m_0 or m_1. The experiment is a success if $b = b'$. We say that the message-sending protocol is *polynomially secure against chosen-ciphertext attack* if for any interactive Turing machine $\mathcal{A}$, for any message-finder F, for any message-distinguisher D, and for any constant d, the probability of success is less than $\frac{1}{2} + \frac{1}{k^d}$ for sufficiently large k.

4.2 Achieving chosen-ciphertext security

The main tool we use in order to achieve chosen-ciphertext security is the *zero-knowledge* (or *minimum-knowledge*) *interactive proof of knowledge* that was formalized by [16, 6]. In the form that we need it, this is a procedure whereby one party (the "prover") can prove to another party (the "verifier") that he "knows" or can compute a quantity without revealing to the verifier any computational knowledge about the value of that quantity. For example, if f_i is a one-way function and y is known beforehand to both parties, the prover can convince the verifier that he knows a pre-image of y, i.e. an element x that satisfies the relation $f_i(x) = y$, without revealing anything about the bits of x. In the case that each function f_i is a group automorphism of its domain D_i (thus providing an example of a "random self-reducible problem"), this interactive proof can be carried out especially efficiently.

We specify Protocol 3 by refining Protocol 1 in the following way. After sending a value y_S in step 2, R proves to S that he knows a preimage x_S. Similarly, after sending a value y_R in step 3, S proves to R that she knows a corresponding preimage x_R. At any point, if one of these interactive proofs is not successful—i.e. if the verifier is not "convinced"—then the unconvinced verifier rejects the attempted message-sending and halts the protocol.

The complete protocol is as follows. As in §3, this is a protocol for S to send an authenticated $l(k)$-bit message m to R. The sender's inputs include its public and secret keys (i_S, d_{i_S}), the receiver's public key i_R, and the message m; the receiver's inputs include its public and secret keys (i_R, d_{i_R}) and the sender's public key i_S.

Protocol 3

1. $S \to R$: "Hi, this is S sending a message to R."

2. R chooses $x_S \in D_{i_S}$ at random, and computes $p_S := b_{i_S}(x_S)$ and $y_S := f_{i_S}(x_S)$.
   $R \to S$: y_S

   R proves to S that it can compute an element of $f_{i_S}^{-1}(y_S)$; if the verifier would reject the proof, then S halts the protocol.

3. otherwise S computes $x_S := f_{i_S}^{-1}(y_S)$ and $p_S := b_{i_S}(x_S)$, chooses $x_R \in D_{i_R}$ at random, and computes $p_R := b_{i_R}(x_R)$, $y_R := f_{i_R}(x_R)$, $c := m \oplus p_S \oplus p_R$.
   $S \to R$: $[c, y_R]$

   S proves to R that it can compute an element of $f_{i_R}^{-1}(y_R)$; if the verifier would reject the proof, then R halts the protocol.

4. otherwise R computes $x_R := f_{i_R}^{-1}(y_R)$, $p_R := b_{i_R}(x_R)$, $c \oplus p_S \oplus p_R = m$; R accepts this message.

end-protocol

Theorem 3 *Protocol 3, based on the trapdoor hard-bit family* $\mathcal{F} = \{(f_i, b_i) \mid i \in I\}$*, is a correct message-sending protocol that has all the security properties of Protocol 1 and is also polynomially secure against chosen-ciphertext attack.*

We can refine Protocol 2 in a similar manner, so that the resulting protocol is polynomially secure against random-message attack and against chosen-ciphertext attack.

Finally, we briefly describe an adaptation of our protocols so as to provide a message-sending protocol that is polynomially secure against chosen-ciphertext attack, and requires overhead whose amortized cost (per bit of plaintext message sent) can be arbitrarily small.

In this protocol one of the two parties chooses a short random bit-string r, and then the two parties use the refined version of Protocol 2 so that he can send r to the other party. They then use r as a seed—known only to them—for a pseudo-random bit-generator so that they share a (simulated) one-time pad of length polynomial in the security parameter; different pieces of this one-time pad may be used in order to encode messages sent from either one of the two parties to the other. This enables the exchange of polynomially many messages, in such a way that the system as a whole is secure against chosen-ciphertext attack. This method minimizes the cryptographic tools required; it uses only the parties' public keys, with no additional keys produced in order to send additional messages.

References

[1] M. Blum and S. Micali. How to generate cryptographically strong sequences of pseudo-random bits. *SIAM J. Comp.*, 13(4):850–864, Nov. 1984.

[2] G. Brassard. *Modern Cryptology: A Tutorial*, volume 325 of *Lecture Notes in Computer Science*. Springer-Verlag, 1988.

[3] W. Diffie and M.E. Hellman. New directions in cryptography. *IEEE Trans. on Inform. Theory*, IT-22:644–654, November 1976.

[4] D. Dolev and A.C. Yao. On the security of public-key protocols. *IEEE Trans. on Inform. Theory*, IT-29:198–208, 1983.

[5] S. Even and O. Goldreich. On the security of multi-party ping-pong protocols. In *Proc. 24th FOCS*, pages 34–39, 1983.

[6] U. Feige, A. Fiat, and A. Shamir. Zero knowledge proofs of identity. *J. of Cryptology*, 1(2):77–94, 1988.

[7] Z. Galil, S. Haber, and M. Yung. Symmetric public-key encryption. In *Crypto '85*. Springer-Verlag, 1986.

[8] Z. Galil, S. Haber, and M. Yung. Minimum-knowledge interactive proofs for decision problems. *SIAM J. Comput.*, 18(4):711–739, 1989.

[9] Z. Galil, S. Haber, and M. Yung. Symmetric public-key cryptosystems. Submitted for publication, 1989.

[10] S. Goldwasser and S. Micali. Probabilistic encryption. *JCSS*, 28:270–299, April 1984.

[11] S. Goldwasser, S. Micali, and C. Rackoff. The knowledge complexity of interactive proof systems. *SIAM J. Comput.*, 18(1):186–208, 1989.

[12] R. Impagliazzo, L.A. Levin, and M. Luby. Pseudo-random generation from one-way functions. In *Proc. 21st STOC*, pages 12–24. ACM, 1989.

[13] S. Micali, C. Rackoff, and B. Sloan. The notion of security for probabilistic cryptosystems. *SIAM J. Comput.*, 17(2):412–426, 1988.

[14] R.M. Needham and M.D. Schroeder. Using encryption for authentication in large networks of computers. *Communications of the ACM*, 21(12):993–999, 1978.

[15] G.J. Simmons. A survey of information authentication. *Proceedings of the IEEE*, 76(5):603–620, May 1988.

[16] M. Tompa and H. Woll. Random self-reducibility and zero knowledge interactive proofs of possession of information. In *Proc. 28th FOCS*, pages 472–482. IEEE, 1987.

[17] A.C. Yao. Theory and applications of trapdoor functions. In *Proc. 23rd FOCS*, pages 80–91. IEEE, 1982.

How to improve signature schemes

Gilles BRASSARD †

Département IRO
Université de Montréal
C.P. 6128, Succursale "A"
Montréal (Québec)
CANADA H3C 3J7

ABSTRACT

Bellare and Micali have shown how to build strong signature schemes from the mere assumption that trapdoor permutation generators exist. Subsequently, Naor and Yung have shown how to weaken the assumption under which a strong signature scheme can be built: it is enough to start from permutations that are one-way rather than trapdoor. In this paper, which is independent from and orthogonal to the work of Naor and Yung, we weaken in a different way the assumption under which a strong signature scheme can be built: it is enough to start from what we call a weak signature scheme (defined below). Weak signature schemes are trapdoor in nature, but they need not be based on permutations. As an application, the Guillou–Quisquater–Simmons signature scheme (a variant on Williams' and Rabin's schemes, also defined below) can be used to build a strong signature scheme, whereas it is not clear that it gives rise directly to an efficient trapdoor (or even one-way) permutation generator.

1. Introduction

In a very nice paper [BM], Bellare and Micali have shown how to build a strong signature scheme from the mere assumption that trapdoor *permutation* generators exist (trapdoor *functions* are not shown to suffice, despite the title of [BM]). Here, "strong" means "non existentially forgeable under an adaptive chosen message attack". Refer to [GMR] for a precise definition of this concept. This was a significant improvement over [GMR], which needed the (possibly) stronger assumption that claw-free pairs exist in order to build strong signature schemes.

This result of Bellare and Micali can be extended in several directions. One such extension was worked out by Naor and Yung, who showed that it is enough to start from permutations that are one-way rather than trapdoor [NY].

† Supported in part by Canada NSERC grant A4107.

In this paper, which is independent from the work of Naor and Yung, we extend the result of Bellare and Micali in an orthogonal direction: in order to build strong signature schemes, it suffices to start with a signature scheme that is "randomly simulatable", but "non randomly forgeable under a key-only attack", which we shall refer to as a "weak signature scheme" (a precise definition is given in Section 2). This generalization is not achieved through a more clever construction, but rather through the observation that the original construction of Bellare and Micali works just as well under our weaker assumption. Following [GMR], a "key-only attack" is an attack in which the enemy knows only the legitimate signer's public key. A "random forgery" is the ability to forge a signature for a message selected at random, with non-negligible probability of success. (This is somewhere between existential and universal forgery, in the terminology of [GMR].) In contrast, the scheme is "randomly simulatable" if knowledge of the public key suffices to produce pairs (m, s) of signed messages whose probability distribution is the same as if message m had been chosen randomly and signature s had been provided by the legitimate signer. The difference between these notions is best grasped if one thinks of the RSA signature scheme [RSA], which is randomly simulatable and conjectured not to be randomly forgeable.

It is clear that a trapdoor permutation generator such as those used as building block in [BM] can be used to obtain a weak signature scheme, but the converse may not hold. In order to build a weak signature scheme from a trapdoor permutation generator, generate a pair (x, y) such that $E(x, \bullet)$ and $I(y, \bullet)$ are permutations that are inverses of each other (refer to [BM] for the notation, not needed for the remainder of this paper), and write down x in the public directory. In order to sign message m, use the secret information y to compute signature $s = I(y, m)$. In order to verify that signature s is valid for message m, use the public information x to verify that $E(x, s) = m$. Moreover, this signature scheme is randomly simulatable provided that knowledge of x is sufficient to draw randomly and uniformly in the domain of $E(x, \bullet)$.

Thus, it has been common practice since Diffie and Hellman [DH] to think of the signing process as computing a trapdoor permutation in the hard direction (the direction that requires knowledge of the trapdoor), whereas the verification process corresponds to computing the trapdoor permutation in the easy direction. However, this perspective is unnecessarily restrictive for the following reasons:

1) The public verification procedure is given both the message and its purported signature. In general, it could compute on both of them in order to decide whether the signature is valid — rather than computing on the signature alone and then using the message merely for the purpose of a final comparison. In fact, it would make perfect sense to have a signature scheme such that, given a signature, no one — perhaps not even the legitimate signer — could figure out which message is (or which messages are) actually signed by this signature, yet given a message and its signature, everyone could verify the validity of the signature.

2) A signature scheme could be secure even if some (or all) messages had more than one valid signature. In terms of trapdoor permutations, this would translate into allowing the function $E(x, \bullet)$ not to be one-one, thus the "function" $I(y, \bullet)$ would be multi-valued (hence not a function).

3) A signature scheme could be secure even if some (or all) signatures were valid for more than one message. In terms of trapdoor permutations, this would translate into allowing the function $I(y, \bullet)$ not to be one-one, thus the "function" $E(x, \bullet)$ would be multi-valued (hence not a function).

4) The space of messages and the space of signatures need not be the same, and they could even have different cardinalities (which is clearly not allowed in the trapdoor *permutation* setting). Moreover, the set of signatures could be different from one instance of a signature scheme to another even if the set of messages to be signed were the same. (As pointed out in [BM], a cross product construction due to Yao [Ya] can be used to bypass this difficulty. Nevertheless, from a practical point of view, it is preferable if Yao's construction can be avoided.)

After defining formally the notion of weak signature scheme in Section 2 and sketching how to transform any one of them into a strong signature scheme in Section 3, we conclude this paper in Section 4 with a discussion of a signature scheme due independently to Guillou and Quisquater [GQ] and to Simmons [S], which we shall refer to as the GQS signature scheme. This signature scheme is based on the scheme of Williams [Wi] — and thus also similar to Rabin's [Ra]. Notice that all these schemes, including the GQS scheme, are totally broken under a directed chosen-message attack. Nevertheless, the GQS signature scheme fits our definition of a weak signature scheme, hence it can be used directly to build a strong signature scheme. Transforming the GQS signature scheme into an efficient trapdoor (or even one-way) permutation generator, on the other hand, would be difficult because of problems (2), (3) and (4) above.

2. Definition of a weak signature scheme

Let X be a finite set. Denote by $ps[X]$ the set of functions f from X to the real interval $[0, 1]$ such that $\sum_{x \in X} f(x) = 1$. (Think of "$ps[X]$" as the set of all probability distributions over X.)

Let k be an integer parameter. Consider the set $M = \{0, 1\}^k$ of length k messages, a set S of signatures (arbitrary for now), and two functions $sig: M \to ps[S]$ and $ver: M \times S \to \{true, false\}$ such that for all $m \in M$ and $s \in S$, $ver(m, s) = true$ if and only if $(sig(m))(s) > 0$. Intuitively, this means that the verification function should accept s as a valid signature for m precisely when the probability that the signing process on m would produce s is nonzero. A *weak signature pair* (with parameter k) is a pair of (possibly probabilistic) efficient algorithms *SIG* and *VER* such that *VER* computes the function *ver* and such that the probability that *SIG* on input m returns s is precisely $(sig(m))(s)$ for all m and s. Furthermore, we require that:

- *The signature scheme is non randomly forgeable under a key-only attack*: given a randomly chosen $m \in M$, knowledge of the algorithm *VER* does not enable one (in feasible time) to find even one $s \in S$ such that $ver(m,s) = true$ (except with negligible probability).

- *The signature scheme is randomly simulatable*: knowledge of the algorithm *VER* does enable one to come up efficiently with pairs (m,s) such that

  - $ver(m,s) = true$;

  - the marginal probability on the m thus generated is uniform on M ; and

  - no matter which m is generated, the conditional probability on s is given by $sig(m)$.

In other words, knowledge of VER does not enable one to forge signatures for randomly chosen messages, but enables one to forge pairs (m,s) that look just like what the legitimate signer would produce where she to choose a random message and sign it.

A *weak signature scheme* is a generator of weak signature pairs. More precisely, it is a probabilistic algorithm G that outputs such a pair on input 1^k. The description of algorithms *SIG* and *VER* produced by G must be of a length polynomially related to k. The set S may be different from pair to pair, but the set M must always be $\{0,1\}^k$. We require that every pair $< SIG, VER >$ thus generated be randomly simulatable. However, we only require that it be non randomly forgeable in a probabilistic and uniform sense: given any (possibly probabilistic) polynomial-time algorithm A, any polynomial p, and any sufficiently large integer k, the probability that $VER(m, A(k, VER, m)) = true$ is less than $1/p(k)$, where *VER* is obtained by a call on $G(1^k)$ and m is a random element of $\{0,1\}^k$. (The probabilities are taken over all random choices of G and A, and over the random choice of m.)

3. How to improve weak signature schemes

Assume the existence of a weak signature scheme. A strong signature scheme can be obtained with the techniques of Bellare and Micali [BM]. The only modification is that algorithms for trapdoor permutations are replaced by the *SIG* algorithms and, similarly, the inverse of the trapdoor permutations are replaced in the obvious way by the application of the *VER* algorithms. For the sake of completeness, here is a brief sketch of the construction of Bellare and Micali, adapted for our purpose.

Let k be an integer safety parameter and let l be the maximum length of the description of *VER* that can be produced by a call on $G(1^k)$. In order to set up a strong signature capability, each user obtains one weak signature pair $< SIG, VER >$ by a call on $G(1^k)$. The user also chooses $l+1$ pairs (x_0, y_0), (x_1, y_1), . . . , (x_l, y_l) of elements drawn uniformly at random among $\{0,1\}^k$. These $l+1$ pairs are made public, together with the description of algorithm *VER*. In order to sign a first bit b, the user exhibits $SIG(x_0)$ if $b=0$ or $SIG(y_0)$ if $b=1$. Furthermore, the user calls $G(1^k)$ again, thus producing a new pair $<SIG_1, VER_1>$. The pairs $(x_1, y_1), \ldots, (x_l, y_l)$ are used to sign the description of VER_1 bit-by-bit, much the same way that (x_0, y_0) had been used to sign bit b. At this point, a second bit can be signed by producing either $SIG_1(x_0)$ or $SIG_1(y_0)$, depending on the value of the bit to be signed. This process is continued in order to sign an arbitrary (polynomial in k) number of bits.

The process by which such a signature can be verified should be clear. The reader is referred to [BM] for more detail, in particular for the various ways in which more than one message can be signed. The proof that this scheme is non existentially forgeable under an adaptive chosen message attack [GMR] (assuming that the underlying signature scheme is weakly secure) follows the lines of the proof given in [BM] and is not repeated here.

4. The GQS signature scheme and how to use it

Our main motivation for this work was to be able to use a simple, elegant and natural variant on Williams' signature scheme as basis for the construction of Bellare and Micali [BM]. This work was necessary since this signature scheme does not yield a trapdoor permutation generator, because of difficulties (2) and (3) mentioned in Section 1, and because the removal of difficulty (4) through Yao's construction would be expensive in practice

Williams' key observation [Wi] is that if $n=pq$ where p and q are primes congruent to 3 and 7 modulo 8, respectively, then -1 is a quadratic non-residue modulo n with Jacobi symbol $+1$, whereas 2 has Jacobi symbol -1. Such an integer is called a Williams' integer. In his paper, Williams uses this property to remove a difficulty found in Rabin's previously proposed scheme [Ra]: without using the secret factorization of n, Williams transforms any element of $\mathbb{Z}_n^*$ between 1 and $n/8$ (or "any odd number between 1 and $n/4$" [Wi]) into an element of $\mathbb{Z}_n^*$ that can be signed directly and deterministically (with knowledge of the factors of n) by a scheme as hard to break as it is to factor n (under a key-only attack). (Recall that $\mathbb{Z}_n$ denotes the set of integers modulo n, whereas $\mathbb{Z}_n^*$ denotes the subset of $\mathbb{Z}_n$ consisting of those integers relatively prime with n. For simplicity, we confuse a residue class with its smallest non-negative representative, so that it makes sense to talk about an odd element of $\mathbb{Z}_n^*$.)

In the opinion of the current author, Williams' observation could have been used in a much simpler way: if n is of the form proposed by Williams and if $x \in \mathbb{Z}_n^*$, then exactly one among $\{x, -x, 2x, -2x\}$ (modulo n) is a quadratic residue, and whichever it is can be signed by providing one of its square roots modulo n. This idea was discovered independently by Guillou and Quisquater [GQ] and by Simmons [S] (see also [SP]). (In a personal communication, Quisquater has given credit to Goldwasser for the observation that there exists exactly one quadratic residue among $\{x, -x, 2x, -2x\}$ modulo a Williams' integer.) We refer to the resulting scheme and its immediate variants as the Guillou–Quisquater–Simmons (or GQS) signature scheme. We now describe in more detail a version of this scheme that is well suited for our purpose.

Let k be an integer parameter. In order to build a weak signature pair, randomly choose two distinct primes p and q of binary length $1 + \lfloor k/2 \rfloor$ and $1 + \lceil k/2 \rceil$, respectively, such that $p \equiv 3 \pmod 8$ and $q \equiv 7 \pmod 8$. Compute $n = pq$. Note that $2^k < n < 4 \times 2^k$. For any $x \in \mathbb{Z}_n^*$, let $\sqrt{x}$ denote the (possibly empty) set $\{y \in \mathbb{Z}_n^* \mid x \equiv y^2 \pmod n\}$, and let $car(x)$ stand for the unique element of $\{x, -x, 2x, -2x\}$ (modulo n) that is a quadratic residue. If X is a non-empty finite set, let $unif(X)$ denote an element of X chosen randomly with uniform distribution.

The weak signature pair corresponding to n is defined by the following algorithms, where $M = \{0, 1\}^k$ will be confused with the set of integers between 0 and $2^k - 1$, and $S = \mathbb{Z}_n$:

- $SIG(m) \quad = \quad \begin{cases} m & \text{if } \gcd(m, n) \neq 1 \\ unif(\sqrt{car(m)}) & \text{otherwise,} \end{cases}$

- $VER(m, s) \quad = \quad \begin{cases} true & \text{if } \gcd(m, n) \neq 1 \text{ and } m = s \\ true & \text{if } \gcd(m, n) = 1 \text{ and } s^2 \in \{m, -m, 2m, -2m\} \pmod n \\ false & \text{otherwise.} \end{cases}$

The signing process is well-defined because $n > 2^k$, hence $M \subseteq \mathbb{Z}_n$. The reader can verify that all the desired properties of a weak signature scheme are fulfilled under the conjecture that factoring Williams' integers is hard. (It is easy to see that it is non randomly forgeable under a key-only attack assuming the factoring conjecture; it is a bit more subtle to show that it is randomly simulatable — and the fact that $n \in O(2^k)$ is important here.) Therefore, this weak signature scheme can serve as basis for the construction of Bellare and Micali in order to obtain a strong signature scheme.

It should be pointed out that the version of the GQS signature scheme described above should *not* be used directly. Not only is it totally breakable under a chosen message attack, but it could even be broken under a known message attack [GMR] if the same message is ever signed twice by the legitimate signer! For this reason, Simmons [S] suggests that only one of the four elements of $\sqrt{car(m)}$ should be returned for any given m (for instance, we suggest that it be the one that is simultaneously odd and whose Jacobi symbol is +1 — this defines a unique signature because both prime factors of n are congruent to 3 modulo 4). Nevertheless, this safeguard is not necessary if the basic scheme is used only as building block in the construction of Bellare and Micali in order to obtain a strong signature scheme.

ACKNOWLEDGEMENTS

The author is very grateful to Gus Simmons, who presented this paper on his behalf at EUROCRYPT '89 when his newborn daughter Alice was too young for him to travel.

BIBLIOGRAPHY

[BM] Bellare, M. and Micali, M., "How to sign given any trapdoor function", *Proceedings of the 20th ACM Symposium on Theory of Computing*, 1988, pp. 32–42. (Also presented at *CRYPTO '88*.)

[DH] Diffie, W. and Hellman, M. E., "New directions in cryptography", *IEEE Transactions on Information Theory*, vol. IT-22, 1976, pp. 644–654.

[GMR] Goldwasser, S., Micali, S. and Rivest, R. L., "A digital signature scheme secure against adaptive chosen-message attacks", *SIAM Journal on Computing*, vol. 17, no. 2, April 1988, pp. 281–308.

[GQ] Guillou, L. and Quisquater, J.–J., "Efficient digital public-key signatures with shadow", *Advances in Cryptology — CRYPTO '87 Proceedings*, Springer-Verlag, 1988, p. 223.

[NY] Naor, M. and Yung, M., "Universal one-way hash functions and their cryptographic applications", *Proceedings of the 21st ACM Symposium on Theory of Computing*, 1989, pp. 33–43.

[Ra] Rabin, M. O., "Digital signatures and public-key functions as intractable as factorization", Technical Report MIT/LCS/TR-212, M.I.T., 1979.

[RSA] Rivest, R. L., Shamir, A. and Adleman, L. M., "A method for obtaining digital signatures and public-key cryptosystems", *Communications of the ACM*, vol. 21, 1978, pp. 120–126.

[S] Simmons, G. J., "A protocol to provide verifiable proof of identity and unforgeable transaction receipts", *IEEE Journal on Selected Areas of Communications*, vol. 7, no. 4, May 1989, pp. 435–447.

[SP] Simmons, G. J. and Purdy, G. B., "Zero-knowledge proofs of identity and veracity of transactions receipts", *Advances in Cryptology — EUROCRYPT '88 Proceedings*, Springer-Verlag, 1988, pp. 35–49.

[Wi] Williams, H. C., "A modification of the RSA public-key encryption procedure", *IEEE Transactions on Information Theory*, vol. IT-26, 1980, pp. 726–729.

[Ya] Yao, A. C.-C., "Theory and applications of trapdoor functions", *Proceedings of the 23rd IEEE Symposium on Foundations of Computer Science*, 1982, pp. 80–91.

A Generalization of El Gamal's Public Key Cryptosystem
W. J. Jaburek, GABE Vienna

The general scheme

El Gamal's Public Key Cryptosystem (El Gamal 1985) can be generalized as follows (compare Shamir 1980) giving a public key exchange system:

The potential receiver of encrypted messages chooses a function f and publishes his public keys

    s, k where k=f(s), f remains secret
    G a set of functions commutative to f

The sender of message m chooses g ∈ G and computes

    k' =g(k)=g(f(s))

He uses k' as a key for a symmetric Cryptosystem such as DES or even simpler computes

    m' = m xor k'

and sends m', g(s) to the receiver. The latter computes

    f(g(s)) = g(f(s)) = k'
and
    m' xor k' = m

and has received m in a safe way.

Associative Operations

(Modular) Multiplication per se does not offer a secure way of encryption. But multiplying an integer m n times by itself gives a very popular encrypting function, modular exponentiation, which has been used by El Gamal (El Gamal 1985) and by Rivest-Shamir-Adleman (Rivest 1978) as well.

The advantage modular exponentiation gives the friend against the foe is the possibility to compute f(x) in ld n steps (cf Knuth 1981, p 441) whereas the enemy nearly has to go through $O(n)$ steps to get n by trial and error. This advantage is caused by the associativity of (modular) multiplication.

Associativity of the basic operation causes commutativity of the exponentiation, too.

$$a^{x^y} = a^{y^x}$$

Generalisation of exponentiation

Multiplication cannot be the only possible associative operation
in that respect. Perhaps there are other operations that are
easier to compute and more secure in a cryptographic sense. That
would imply that the resulting pseudo-exponentiation is more
easily applied to real life cryptography without special hardware. Rueppel (Rueppel 1988) is following the track of considering function composition as a basis for pseudo-exponentiation. In
this paper binary operations are considered.

The "pseudo-exponentiation" defined as follows - at least to the
author - sounds very promising in the light of fast computation:

Let
x ... bitstring
f(x) = pa(pa(... pa(pa(x,x),x) ...),x)

the pseudoaddition as defined below applied n times to x, n
integer

```
function pseudoaddition(x,y)
(x,y,acc1,acc2,carry: bitstrings of length l)
acc1: =x
acc2: =y
while acc2<>0
   carry: = acc1 and acc2
   acc1: = acc1 xor acc2
   (* Transformation of carry into acc2 *)
   acc2: = 0
   for i: =1 to l do
     if (Bit i in carry equal to 1)
     then acc2: =acc2 or tabelle[i]
   end_for
end_while
```

Tabelle[i] is a bitstring of length l, the i-th bit being zero
and none or another few bits being one. For all bits in
tabelle[i], i=1 .. l, the j-th bit (j=1 .. l) only once has value
1 because otherwise the or-function in the above pseudo-code must
be replaced by a recursive call of pseudoaddition in order that
pseudoaddition remains associative.

In general there exist l^l possible values for tabelle, as tabelle
describes the mapping of l source bits into l bits, where each
source bit may be used zero to l times (Variations with repetition).

The while-loop must terminate, because after the and-operation
any bit of the carry has value one with probability p=0.25 and
after the xor-operation any bit of acc1 has value one with p=0.5
with both probabilities clearly being smaller than one. The or-
operation with a tabelle satisfying the above given conditions
does not change the number of one-bits.

Example for tabelle with l=4

```
tabelle[1] = 0010
tabelle[2] = 0000
tabelle[3] = 0001
tabelle[4] = 1100
```

Note that tabelle[4] results in the urgently needed non-linearity!

Remark: Tabelle with values

```
                0010
                0100
                1000
                0001
```

describes binary addition modulo 15.

Lemma
Pseudoaddition is associative.

Proof
Can easily be verified by considering the similarity with addition.

Lemma
By repeating pseudoaddition a pseudo-exponentiation can be defined. Pseudo-exponentiation takes ld n (n being the number of times pseudoaddition is repeated in the trivial way of computation) pseudoadditions.

Proof
Just take the square-and-multiply-algorithm for exponentiation and substitute pseudoaddition for multiplication (cf Knuth 1981, p 441).

Example
Pseudoaddition using the above given tabelle, (2,0,1,12) when representing the bitstrings as decimal numbers, applied to 0011 or 3 gives values when repeated: 3, 13, 1, 14, 2, 12, 15, 3, ... a sequence that cannot be matched with the modular powers of 3 with any integer modulus.

Computational Complexity

Pseudoaddition takes n bit-operations in the for-loop times the number of times the while-loop is taken. The latter depends on the effect of carry-propagation. By applying the idea of a Carry Save Adder (Vgl Brickell 1982 and the literature given there) the while-loop ceases to exist (except in the case of normalizing the result of the whole operation). By using special hardware operating on all n bits at once, Pseudoaddition only takes $O(1)$ step. Pseudoexponentiation therfore takes $O(ld\ n)$ steps, which is faster than modular exponentiation by a factor of n, as the latter takes $O(n \cdot ld(n))$ steps in good hardware.

Remark

By implementing tabelle in hardware n^n basic functions can be chosen, adding even more security against possible attacks. In order to prevent easy reading of the chip, it should not respond to requests with low exponents.

Security Assessment

Up to now the author did not do a concise exploration of the properties of the resulting set of binary operations. By using the following 3-bit-pseudoaddition some properties of the operations are discussed.

```
tabelle[1]=     010
tabelle[2]=     101
tabelle[3]=     000
```

results in the Cayley table for pseudoaddition

f	000	001	010	011	100	101	110	111
000	000	001	010	011	100	101	110	111
001	001	010	011	101	101	110	111	001
010	010	011	101	110	110	111	001	010
011	011	101	110	111	111	001	010	011
100	100	101	110	111	000	001	010	011
101	101	110	111	001	001	010	011	101
110	110	111	001	010	010	011	101	110
111	111	001	010	011	011	101	110	111

Some properties can be deduced:

* There is an Identity Element: 000 and a sort of dual representation of it: 111 (Compare to addition with negative numbers
represented as one's complement). The latter 11..111 could be
called pseudo-identity.

* For each possible operand x there exists a value y so that
$pa(x, y) = 111...1111$, the pseudo-identity. The Pseudo-Inverse of
each bitstring can be calculated by applying the NOT-operation.
* Operations in general are not commutative as can be shown by
using a second operation g described by tabelle 110, 000, 001 or
the Cayley table:

g	000	001	010	011	100	101	110	111
000	000	001	010	011	100	101	110	111
001	001	110	011	100	101	011	111	001
010	010	011	000	001	110	111	100	101
011	011	100	001	110	111	001	101	011
100	100	101	110	111	001	110	011	100
101	101	011	111	001	110	111	100	101
110	110	111	100	101	011	100	001	110
111	111	001	101	011	100	101	110	111

For example $f(g(010,110),101) = f(100,101) = 001$ and $g(f(010,110),101) = g(001,101) = 011$.

* Operations f with any tabelle[i] that includes two one-bits, applied ø times to one value x result in the value x itself. ø is Euler's Totient Function of the largest prime-number in the value range used. In the example given above ø = p-1 = 6 as 7 is the largest prime number representable with 3 bits.

Potential Weaknesses:

1. Pseudo-exponentiation could be represented as multiplication and therefore easily inverted, as one of the pseudoadditions is addition. Examples chosen at will show that pseudoexponentiation cannot be represented neither as addition nor as multiplications, except in the linear case of addition or permutations of addition's carry-tabelle.

2. By applying pseudoaddition repeatedly the identity-element may be produced. That is the same problem with modular exponentiation and therefore does not seem to be critical.

3. For some values in the example-f
   pseudoaddition(a,b) = pseudoaddition(a+1, b-1)

The author did not find a way how to exploit that potential weakness.

If other security threats to the system should become known it seems to be possible to expand the algorithm for pseudo-addition in a number of ways - e. g. by using a more complex transformation of carry into acc2 - without changing the run-time complexitiy of the algorithm.

Conclusion

The above given idea of creating pseudoadditions has two advantages over El Gamal's scheme:

* The basic function only takes O(1) step. El Gamal's takes O(n).

* No large primes have to be calculated for initializing the system.

The new operation pseudo-exponentiation can be applied to all cryptographic procedures using modular exponentiation as a one-way-function, e.g. the Pohlig-Hellmann Public Key Distribution System (Pohlig 1978) or the one-way encipherment of passwords in computer-systems. Thus a new set of operations worth studying for cryptographic purposes seems to emerge.

References

Brickell 1982
Brickell, E.F., A Fast Modular Multiplication Algorithm With Applications to Two Key Cryptography, in: Chaum et al(Eds), Advances in Cryptology - Proceedings of Crypto 82, Plenum 1983, 51-60

El Gamal 1985
El Gamal, A public key cryptosystem and a signature scheme based on discrete logarithms, IEEE Trans. Inf. Theory, IT-31 (1985), 469-472

Knuth 1981
Knuth, D.E., The Art of Computer Programming², Vol 2: Semi-numerical Algorithms, Addison-Wesley 1981

Pohlig 1978
Pohlig - Hellmann, An Improved Algorithm for Computing Logarithms Over GF(p) and its Cryptographic Significance, IEEE Trans. Inf. Theory, IT-24 (1978), 106-110

Rivest 1978
Rivest - Shamir - Adleman, A Method of Obtaining Digital Signatures and Public Key Cryptosystems, Communications of the ACM 1978, 120 - 126

Rueppel 1988
Rueppel, R., Key Agreements Based on Function Composition, Proc. EUROCRYPT' 88, LNCS 330, Springer 1988, 3-10

Shamir 1980
Shamir, A., On the Power of Commutativity in Cryptography, in: Automata, languages, and programming, Proc ICALP 1980, Lecture Notes in Computer Science 85, Springer 1980, 582-595

AN IDENTITY-BASED KEY-EXCHANGE PROTOCOL

Christoph G. Günther
Asea Brown Boveri
Corporate Research
CH-5405 Baden, Switzerland

ABSTRACT

The distribution of cryptographic keys has always been a major problem in applications with many users. Solutions were found for closed user groups and small open systems. These are, however, not efficient for large networks. We propose an identity-based approach to that problem which is simple and applicable to networks of arbitrary size. With the solution proposed, the user group can, furthermore, be extended at will. Each new user needs only to visit a key authentication center (KAC) once and is from then on able to exchange authenticated keys with each other user of the network. We expect this type of approach, which was originally conceived for authentication and signatures, to play an increasing role in the solution of all types of key distribution problems.

I. INTRODUCTION

The transmission of data at medium to high rate requires the use of symmetric encryption algorithms. The key distribution problem implied in this mode is frequently solved by using the Diffie-Hellman key-exchange algorithm [1] or some of its variants. A major concern in large networks is then the authentication of the public keys used in the algorithm. A local storage of this list requires a large storage capacity and is, in addition, inflexible with respect to network extensions. A centralised storage, on the other side, implies a communication complexity comparable to the communication complexity of classical key distribution protocols and therefore ruins the advantage of the scheme. The same situation occurs with the El-Gamal signature protocol [2].

Rivest, Shamir and Adleman [3] have indicated a solution to the problem of authenticating public keys: A key authentication center (KAC) signs the public key of each user and thereby guarantees its authenticity. The implementation of this solution for the authentication of the public keys used in the Diffie-Hellman scheme is typically not very practical. Fiat [4] has proposed an interesting approach to identification and signatures. In this approach the user of some communication facility only needs to know the "name" of his communication partner and the public key of the KAC. Such schemes are correspondingly called identity-based.

We adapt this approach for the construction of an identity-based key-exchange scheme (section III). In this protocol the two parties construct keys which agree if they are both legitimate and do both conform to the protocol. The actual authentication is established when the decryption of the message sent by the other party is meaningful. It is obvious that such a protocol cannot be zero-knowledge in the sense of Goldwasser, Micali and Rackoff [5] or Feige, Fiat and Shamir [6], since no simulator can construct the key in polynomial time if the encryption scheme is reasonable. Nevertheless, the protocol has some kind of zero-knowledge property, which will be discussed elsewhere. In section III we shall make some further remarks.

In the following, we assume that p is prime and we use the definition $\mathbf{Z}_m := \{0, 1, \ldots, m-1\}$ and $\mathrm{GF}^*(p) := $ *the multiplicative group of* $\mathrm{GF}(p)$. Finally, $t \in_R \mathbf{Z}_{p-1}$ means t is chosen at random from $\mathbf{Z}_{p-1}$.

II. IDENTITY-BASED PROTOCOLS

Identity-based protocols were mainly considered for authentication and signature. Examples are given by Fiat and Shamir [7], Beth [8] and Guillou and Quisquater [9]. Identity-based protocols run in three phases: a set-up phase, a preauthentication phase and an authentication phase. The first two phases involve a key authentication center (KAC), which is trusted by all parties. The essence of the protocol can be summarised as follows: The set-up phase is used by the KAC to lay down all the system parameters. In the preauthentication phase all those who wish to join the network visit the KAC and identify themselves. Let Alice be such a user, then after verification of her identity the KAC forwards her the signature of her name and the system parameters. *The central property of identity-based protocols is that after completion of this preauthentication phase Alice is able to authenticate herself (authentication phase) to any other user without further communication with the KAC and without uncovering the secret signature of her name.* We would like to use such a protocol in order to authenticate the public key r^s used in the Diffie-Hellman scheme. The El-Gamal signature scheme [2] is, as we shall see, very well adapted to solve this and other authentication problems. The steps read as follows:

Set-up:
The KAC chooses a one-way function f, a finite field $GF(p)$ in which it is difficult to compute discrete logarithms, a primitive element $\alpha \in GF^*(p)$ and at random some number $x \in \mathbf{Z}_{p-1}$ which is not divisible by the largest prime factor of $p - 1$. The number x is the KAC's private key. It is used to compute the public key $y = \alpha^x$.

Preauthentication:
Alice visits the KAC and identifies herself. If the KAC accepts her, it provides her with f, $GF(p)$, α and y. Furthermore, it computes the El-Gamal signature (r, s) of $ID = f(\text{description of Alice})$, gives it to Alice and keeps it secret otherwise. The "description of Alice," D, may include Alice's name, birthday, physical description or whatever is suitable for the application intended. The one-way function f is used in order to increase the redundancy of D if the inherent redundancy is either too small or difficult to use. (A certain amount of redundancy is needed in order to avoid El-Gamal's attack 5 [2], *i.e.*, in order to avoid the generation of valid triples (ID, r, s).) The computation of the signature (r, s) runs as follows [2]: the KAC chooses at random $k \in \mathbf{Z}_{p-1}$, with $\gcd(k, p-1) = 1$, computes $r := \alpha^k$, and solves the equation $ID = xr + ks \mod (p-1)$ for s. We note that no k should be used repeatedly, since this would uncover the secret key x. We also note that, due to the assumption $\gcd(k, p-1) = 1$, the element r is primitive.

Authentication:

The verification equation for the signature reads:

$$\alpha^{ID} = y^r r^s, \tag{1}$$

and can be rewritten in the form

$$r^s = \alpha^{ID} y^{-r}. \tag{2}$$

This equation leads to the following reinterpretation of the El-Gamal scheme: *It is a scheme for the computation of the discrete logarithm s to a primitive basis r of an expression that only depends on publicly known quantities and on the base r to which the logarithm is taken.*

We note that making r public does not compromise the secret key s, since determining a pair (r, s) which signs the message ID is at least as difficult if r is prescribed as it is when r can be chosen freely. The base r does also not need to be authenticated since the determination of a pair (r, s) is precisely breaking an instance of the El-Gamal signature scheme.

If Alice now wishes to authenticate herself, she uses the Chaum, Evertse, van de Graaf protocol [10] in order to "prove" in zero-knowledge that she knows s. This is Beth's identity-based zero-knowledge proof of identity [8].

In section III we shall consider a corresponding protocol for key-exchange. Here, we conclude by noting that the reinterpretation of equation (2) also leads to an identity-based El-Gamal signature scheme. If Alice wishes to sign a message $m \in \mathbf{Z}_{p-1}$, she chooses $\kappa \in \mathbf{Z}_{p-1}$ with $\gcd(\kappa, p-1) = 1$, determines $\rho = r^\kappa$ and computes σ by solving the equation $f(m) = s\rho + \kappa\sigma \bmod (p-1)$. The signature (ρ, σ) then satisfies the verification equation:

$$\begin{aligned} r^{f(m)} &= (r^s)^\rho \rho^\sigma \\ &= (\alpha^{ID} y^{-r})^\rho \rho^\sigma. \end{aligned} \tag{3}$$

In particular, Alice can act as a KAC for another user David, if she chooses $m = ID' = ID_{David}, \rho = r', \sigma = s'$. In this way whole hierarchies of KAC's can be constructed. Proving the security of this scheme seems to be outside the scope of todays methods. It is closely related to the security of the El-Gamal scheme itself.

III. AUTHENTICATED KEY EXCHANGE

In section II we have seen how to authenticate a number r^s for which the KAC can compute the discrete logarithm s to the base r. It is natural to use this scheme to authenticate the public keys in the Diffie-Hellman scheme. There is, however, one additional step to do: The basis r and r' of any two parties must be different, since else two of them could coalesce and, by sharing their secret keys s and s', determine the KAC's secret key. The Diffie-Hellman algorithm must, therefore, be adapted to accommodate different basis for the parties. Incidentally, this adaption has the advantage to generate a different key at each session. The resulting protocol reads

Alice		Bob

Step 0:

$$y \in \mathrm{GF}^*(p) \qquad\qquad y \in \mathrm{GF}^*(p)$$

$$r \in \mathrm{GF}^*(p) \qquad\qquad r' \in \mathrm{GF}^*(p)$$

$$s \in \mathbf{Z}_{p-1} \qquad\qquad s' \in \mathbf{Z}_{p-1}$$

Step 1:

$$\xrightarrow{\quad D,\, r \quad}$$

$$\xleftarrow{\quad D',\, r' \quad}$$

$$ID' = f(D') \qquad\qquad ID = f(D)$$

$$r'^{s'} := \alpha^{ID'} y^{-r'} \qquad\qquad r^s := \alpha^{ID} y^{-r}$$

Step 2:

$$t \in_R \mathbf{Z}_{p-1} \qquad\qquad t' \in_R \mathbf{Z}_{p-1}$$

$$u := r'^t \qquad\qquad u' := r^{t'}$$

$$\xrightarrow{\quad u \quad}$$

$$\xleftarrow{\quad u' \quad}$$

Step 3:

$$z = u'^s \qquad\qquad z = (r^s)^{t'}$$

$$z' = (r'^{s'})^t \qquad\qquad z' = u^{s'}$$

$$\zeta = zz' \qquad\qquad \zeta = zz'$$

As stated in the introduction, this protocol cannot be zero-knowledge in the traditional sense. Let us, however, make two remarks to show that the protocol does not disclose much information on Alice's secret s and correspondingly on Bob's secret s'. Alice only sends two quantities to Bob. These are r and u:

- sending r gives no useful information to Bob or any other party. The reason is as follows: Bob or the other party can either break the El-Gamal scheme in which case they do not need to receive r, or they cannot break that scheme. In the latter case, they can, however, not determine s, since this would precisely mean to break an instance of the El-Gamal scheme with a prescribed r. (We note that not even the KAC is able to determine an s associated with an r of which it does not know the discrete logarithm.)

- u is uncorrelated to s.

In order to discuss the soundness, we assume that Clair wants to impersonate Alice. Then she has to determine ζ without knowing s. She can certainly determine z'. Her problem is to compute $z = r^{t's}$ from $r^{t'}$ and $r^s = \alpha^{ID}y^{-r}$. We expect this to be of a comparable difficulty as breaking the Diffie-Hellman scheme. Due to the lack of further results on the El-Gamal and the Diffie-Hellman scheme this can, however, not yet be proved.

Let us finally consider the following slight modification of the steps 2 and 3 of the protocol:

Step 2:

$$t \in_R \mathbf{Z}_{p-1},\ u := r'^t \qquad\qquad t' \in_R \mathbf{Z}_{p-1},\ u' := r^{t'}$$
$$w \in_R \mathbf{Z}_{p-1},\ v := \alpha^w \qquad\qquad w' \in_R \mathbf{Z}_{p-1},\ v' := \alpha^{w'}$$

$$\xrightarrow{\ \ u,\ v\ \ }$$

$$\xleftarrow{\ \ u',\ v'\ \ }$$

Step 3:

$$z = u'^s \qquad\qquad\qquad z = (r^s)^{t'}$$
$$z' = (r'^{s'})^t \qquad\qquad\qquad z' = u^{s'}$$
$$\tilde{z} = v'^w \qquad\qquad\qquad \tilde{z} = v^{w'}$$
$$\zeta = zz'\tilde{z} \qquad\qquad\qquad \zeta = zz'\tilde{z}$$

This modification restores a property of the Diffie-Hellman scheme, which we could call *perfect forward secrecy:* If Alice and Bob are not impersonated, when the protocol is run, finding the key ζ is as difficult as breaking the Diffie-Hellman scheme for *every* third party. We note that even the KAC could be the third party. This has the important consequence that if by accident the KAC's secret key becomes known, the confidentiality of past message would not be compromised. Only the authenticity in the future would be lost.

IV. RELATED WORK

Bauspieß and Knobloch [11] have obtained a key-exchange scheme very similar to the identity-based protocol of section III. In their protocol Alice and Bob first run Beth's zero-knowledge identification scheme once in each direction. Alice and Bob then use the commitments of the respective verifiers in these protocols (which are authenticated if the protocols end successfully) as inputs to two Diffie-Hellman key-exchanges. They thus end up with two keys, one authenticated by Bob and the other one by Alice, which they could then suitably combine. Their protocol has the property of perfect forward secrecy from the beginning but is somewhat more involved than ours. The great advantage of the approach chosen by Bauspieß and Knobloch is, however, that the soundness of their protocol only depends on the security of the El-Gamal and of the Diffie-Hellman schemes, taken separately.

At the conference, we learned about a result of Okamoto and Tanaka, which has appeared in the mean time [12]. Okamoto and Tanaka transform the Diffie-Hellman key-exchange scheme into an identity-based one by using the RSA-scheme [3] as a trap-door function for the computation of the discrete logarithm of the ID-number. Their scheme uses only one data-exchange and is very attractive, due to its low communication complexity. It is, however, not perfectly forward secure. The introduction of that property would require to leave the ring $\mathbf{Z}_n$ $(n = p \cdot q)$ with some part of the protocol. Unfortunately, the security of the protocol seems also difficult to assess.

V. CONCLUSION

The authenticated key-exchange algorithm described in section III is simple and only needs few data exchanges. The security level only depends on the length of the words exchanged and not on the number of exchanges. The operations involved in the protocol are identical to those involved in a Diffie-Hellman key-exchange. The security could not be assessed within the current terminology, but some arguments were given why the scheme should be secure. We would thus expect that this type of protocols, including those of Okamoto and Tanaka [12], and Bauspieß and Knobloch [11], will play an increasing role for the security in large data systems.

REFERENCES

[1] W. Diffie, M.E. Hellman, "New Directions in Cryptography," *IEEE Trans. on Inform. Theory*, vol. IT-22, pp. 644-654, Nov. 1976.

[2] T. El-Gamal, "A Public Key Cryptosystem and a Signature Scheme Based on Discrete Logarithms," *IEEE Trans. on Inform. Theory*, vol. IT-31, pp. 469-472, July 1985.

[3] R.L. Rivest, A. Shamir, L. Adleman, "A Method for Obtaining Digital Signatures and Public-Key Cryptosystems," *Comm. ACM*, vol. 21, pp. 120-126, Feb. 1978.

[4] A. Shamir, "Identity-Based Cryptosystems and Signature Schemes," *Advances in Cryptology - CRYPTO'84*, Lect. Notes in Computer Science, vol. 196, pp. 47-53, Springer-Verlag (1985).

[5] S. Goldwasser, S. Micali, C. Rackoff, "The Knowledge Complexity of Interactive Proof Systems," *SIAM J. Comput.*, vol. 18, pp. 186-208, Feb. 1989.

[6] U. Feige, A. Fiat, A. Shamir, "Zero-Knowledge Proofs of Identity," *J. of Cryptology*, vol. 1, pp. 77-94, 1988.

[7] A. Fiat, A. Shamir, "How to Prove Yourself: Practical Solutions to Identification and Signature Problems," *Advances in Cryptology - CRYPTO'86*, Lect. Notes in Computer Science, vol. 263, pp. 186-194, Springer-Verlag (1987).

[8] T. Beth, "Efficient Zero-Knowledge Identification Scheme for Smart Cards," *Advances in Cryptology - EUROCRYPT'88*, Lect. Notes in Computer Science, vol. 330, pp. 77-84, Springer-Verlag (1988).

[9] L.C. Guillou, J.-J. Quisquater, " A Practical Zero-Knowledge Protocol Fitted to Security Microprocessor Minimizing Both Transmission and Memory," *Advances in Cryptology - EUROCRYPT'88*, Lect. Notes in Computer Science, vol. 330, pp. 123-128, Springer-Verlag (1988).

[10] D. Chaum, J.-H. Evertse, J. van de Graaf, "An Improved Protocol for Demonstrating Possession of Discrete Logarithms and Some Generalizations," *Advances in Cryptology - EUROCRYPT'87*, Lect. Notes in Computer Science, vol. 304, pp. 127-141, Springer-Verlag (1988).

[11] F. Bauspieß, H.-J. Knobloch, "How to Keep Authenticity Alive in a Computer Network," *Advances in Cryptology - EUROCRYPT'89*, Lect. Notes in Computer Science, this issue, Springer Verlag.

[12] E. Okamoto, K. Tanaka, "Key Distribution System Based on Identification Information," *IEEE J. Select. Areas Commun.*, vol. SAC-7, pp. 481-485, May 1989.

How to Keep Authenticity Alive in a Computer Network

Fritz Bauspieß

Hans-Joachim Knobloch

Universität Karlsruhe
Europäisches Institut für Systemsicherheit
European Institute for System Security

Kaiserstr. 8
D-7500 Karlsruhe 1

Abstract

In this paper we present a cryptograhic scheme that allows to ensure the ongoing authenticity and security of connections in a computer network. This is achieved by combining a zero-knowledge authentication and a public key exchange protocol. It is noteworthy that due to the combination both protocols gain additional security against attacks that would otherwise be successful. The scheme is applicable to both local area networks and internetworks.

1 Introduction

In recent computer networks there are two developments demanding for the use of new methods of user authentication. On one hand advanced workstations largely increased the number of machines in a network and internetworking LAN-LAN connections or remote access telecommunication lines increased the connectivity of these machines, thus making the network much more vulnerable against unauthorized manipulation. On the other hand through the widespread use of computers more and more people gain the knowledge, the wish and the possibility to carry out such en-vogue manipulations like hacking or programming computer viruses.

A network administrator wants methods to prevent these manipulations or at least to be able to sue an individual user for damages if manipulations occured, thus deterring other possible ab-users of his network. Likewise network users want methods to prevent them from being accused of manipulations other people made pretending their authentication.

Even if encryption is used to protect an interactive session there is a fundamental problem whenever key exchange for the session encryption and authentication of the session are two separate protocols.

The possible intruder who is capable of suppressing and forging messages will let the authentication data pass unchanged. But he will try to intercept a session key instead, so that his requests encrypted under this key appear authentic.

If e.g. the Diffie-Hellman key exchange protocol [3] were used, the intruder may perform the well-known switching-in attack resulting in two keys, one between user and intruder and the other between intruder and host. The key shared with the user might be used to ask him for authentication data as the pretended host. The same attack would also work against the key exchange using function composition [5].

So it appears to be essential that key exchange and authentication are inseparably combined.

2 Authentication by key exchange

Suppose all users are known to the host before they attempt to establish a session. Then secret information common between the user and the host can be used to generate a session key. The user is under this approach authenticated by using the correct key and thereby showing his knowledge of the secret information he was given in advance.

For example the shared secret information may be a master key to encrypt and decrypt the session key generated by the host. There are numerous slight modifications of this scheme.

Another possibility is to use a public key exchange protocol where an essential part of information is not transmitted publicly but kept as common secret. Using the Diffie-Hellman protocol the steps could be:

- Host B chooses a prime power q and a primitive element ω of $GF(q)$

- Upon registration user A chooses his secret $a \in \mathbf{Z}_{q-1}$ and passes $\omega^a \in GF(q)$ to B (along with his identification)

- Every time a session is established, B generates a random number $s \in \mathbf{Z}_{q-1}$ and sends $\omega^s \in GF(q)$ to A. Only A and B are then capable to compute the session key ω^{as}

One major disadvantage of this protocol is that B has to keep a datafile of all the possible users A_i and their ω^{a_i}. This will be a large and often changing file. Moreover if ω^{a_i} is used to ensure the user that he is connected to the correct host (as proposed in SELANE [1]) this file contains sensitive data und thus is an additional breakpoint into the system.
SELANE therefore proposes a CSC (Central Security Controller) to hold these data. Updating and securing this file is then made easy. In addition this CSC can be used to note all connections made in the network (for later reconstruction) - no one can bypass it!

3 Key exchange by authentication

An alternative possibility to combine authentication and key exchange is to use data exchanged during the authentication protocol, and therefore authentic data, to construct the session key. Under this approach the use of a zero-knowledge authentication protocol provides additional advantages, since then the host does not have to know all possible users but only the secure key issuing authority (SKIA). All sensitive data is either offline within the SKIA or protected by the zero-knowledge scheme. Also concerning internetworking this solution has advantages over the one mentioned before, because there are not many SKIAs within a whole internetwork nor is there a large fluctuation of their public data as compared to user data.

A zero-knowledge identification scheme that is very well suited for our purposes is the one Beth introduced at the Eurocrypt '88 [2] which we will recall here for the reader's convenience.

The scheme consists of three phases. In the first phase the SKIA chooses some constants that are common to all participants of the scheme. In the next phase the SKIA computes data that serve as the individual user's credentials and issues them in a secure token device. The last phase is the authentication itself.

Initiation:

SKIA chooses a finite field $GF(q)$ with primitive element ω,
$\qquad$ random $x_1, \ldots, x_m \in \mathbf{Z}_{q-1}$ and
$\qquad$ a oneway function $f : \mathbf{Z} \times \mathbf{Z}_m \rightarrow \mathbf{Z}_{q-1}$

$\quad$ computes $y_j := \omega^{x_j}$ in $GF(q)$ for $j = 1, \ldots, m$
$\quad$ q, w, $y_1, \ldots, y_m$ and f are published, $x_1, \ldots, x_m$ are kept secret

Registration of user A:

SKIA checks A and gives her a $name_A \in \mathbf{Z}$
$\qquad$ computes $I_{A,j} := f(name_A, j)$ for $j = 1, \ldots, m$
$\qquad$ chooses a random $k_A \in \mathbf{Z}_{q-1}$ and computes $r_A := \omega^{k_A}$ in $GF(q)$
$\qquad$ determines solutions $s_{A,j}$ of $x_j r_A + k_A s_{A,j} \equiv I_{A,j} \bmod (q-1)$ for $j = 1, \ldots, m$
$\qquad$ issues a token device to A containing $name_A$, r_A, $s_{A,1}, \ldots, s_{A,m}$

Authentication of user A versus host B:

A $\hfill$ B

$$name_A, r_A$$

$$\longrightarrow$$

$\qquad\qquad\qquad\qquad$ for $j = 1, \ldots, m$ computes
$\qquad\qquad\qquad\qquad$ $I_{A,j} = f(name_A, j)$

$$\text{for } i = 1, \ldots, h \text{ do}$$

- -

chooses random $t_{A,i} \in \mathbf{Z}_{q-1}$

computes $z_{A,i} := r_A^{-t_{A,i}}$ in $\mathrm{GF}(q)$

$$z_{A,i}$$

$$\longrightarrow$$

chooses random
$$(b_{A,1,i}, \ldots, b_{A,m,i}) \in \mathbf{Z}_{q-1}^m$$

$$(b_{A,j})_i$$

$$\longleftarrow$$

computes

$$u_{A,i} := t_{A,i} + \sum_j b_{A,j,i}\, s_{A,j} \bmod (q-1)$$

$$u_{A,i}$$

$$\longrightarrow$$

computes (in $\mathrm{GF}(q)$)

$$\gamma_{A,i} := \left(\prod_j y^{r_A b_{A,j,i}}\right) r_A^{u_{A,i}}\, z_{A,i} - \omega^{\sum_j b_{A,j,i} I_{A,j}}$$

- -

$$\text{od}$$

accepts the authentication if
$\gamma_{A,i} = 0$ for all $i = 1, \ldots, h$

Now consider the case $m = 1$ and $h = 1$.

Observation: There is the value $z_A = r^{-t_A}$ the form of which resembles the public parts of the key in the Diffie-Hellman protocol.

Idea: Use z_A as part of the Diffie-Hellman key. B chooses a random $d_B \in \mathbf{Z}_{q-1}$ to generate and exchange the other part $e_B := r^{d_B}$ of the key. A computes $e_B^{-t_A}$, B computes $z_A^{d_B}$, both of which are the same key $\omega^{-k_A t_A d_B}$.

Notes:

1. The value z_A is an integral part of the authentication protocol and therefore it cannot be altered by an intruder without causing the authentication to fail. Using it as a part of the session key ia an effective way to combine authentication and key exchange as demanded above.

2. SKIA must take care that the multiplicative order of $r_A = \omega^{k_A}$ is as high as possible, which can be guaranteed if $GF(q)$ is a Fermat-Field.

3. Neither A nor B need the knowledge of k_A. On the other hand they must not have this knowledge, otherwise x could be reconstructed from r_A, s_A, k_A and I_A.

4. The trivial cases $t_A = 1$ or $b_A = 0$ have to be avoided.

5. $m > 1$ or $h > 1$ significantly decrease the probability that (ab-)user C pretending to be A can successfully guess the matrix $((b_{A,j,i}))$ and "tune" his $z_{A,i}$ and $u_{A,i}$ accordingly.

6. If $m > 1$, the $s_{A,j}$ must be stored in secure memory. Otherwise a number of conspiring users would be able to pool their information to create new valid IDs without participation of SKIA.

7. $h > 1$ leads to the problem of having to combine several $z_{A,j}$ during the key exchange. Otherwise only the security of one pass of the authentication protocol would apply to the key.

8. Even if, for $m, h = 1$ a forger C could guess the right challenge b_A and would send $z'_A := \omega^{b_A I_A} y^{-r_A b_A} r_A^{-u_C} = (r_A^{s_A})^{b_A} r_A^{-u_C}$ instead of z_A and $u'_A := u_C$ instead of u_A for a chosen u_C he could only gain a correct authentication. There seems to be no effective way for him to compute the discrete logarithm of his certain z'_A which he needs to compute the correct session key. So he cannot make use of the authenticated session. Thus, the security of the combined protocol seems to be independant from the difficulty of guessing $((b_{A,j,i}))$ and $m, h = 1$ may be used without decreasing the overall security.

9. In Beth's original publication the choice of $((b_{A,j,i}))$ is limited to a "suitably chosen subset" $R^{m \times h}$ of $\mathbf{Z}_{q-1}^{m \times h}$ for proof technical reasons. Increasing the choice space for a practical implementation seems to result in higher, not in lower security.

10. As the amount of data needed for the protocol is small and these data are rarely changing and not sensitive, there is no direct demand for a CSC as mentioned in the previous section. Even if a CSC were used for logging established connections, its failure would not crash the whole cryptographic system. So its availability is less important than it was in SELANE.

The protocol in this form ensures B that A is authentic but not vice versa. This is acceptable if B is a host and A a user. However it is desirable that both communicating parties are sure about each other's identity, especially in a network with equivalent nodes that are clients as well as servers to other nodes.

So we propose a symmetric version of the above scheme (for $m, h = 1$):

A   B

$$name_A, r_A$$

$$name_B, r_B$$

computes   computes
$I_B = f(name_B, 1)$ $I_A = f(name_A, 1)$
chooses random $t_A \in \mathbf{Z}_{q-1}$ chooses random $t_B \in \mathbf{Z}_{q-1}$

computes $z_A := r_A^{-t_A}$ computes $z_B := r_B^{-t_B}$

$$z_A$$

$$z_B$$

chooses random   chooses random
$b_B \in \mathbf{Z}_{q-1}$ $b_A \in \mathbf{Z}_{q-1}$

$$b_B$$

$$b_A$$

computes   computes
$u_A := t_A + b_A s_A \quad mod(q-1)$ $u_B := t_B + b_B s_B \quad mod(q-1)$

$$u_A$$

$$u_B$$

verifies   verifies
$y^{r_B b_B} r_B^{u_B} z_B =^? \omega^{b_B I_B}$ $y^{r_A b_A} r_A^{u_A} z_A =^? \omega^{b_A I_A}$
chooses random $d_A \in \mathbf{Z}_{q-1}$ chooses random $d_B \in \mathbf{Z}_{q-1}$
computes $e_A := r_B^{-d_A}$ computes $e_B := r_A^{-d_B}$

$$e_A$$

$$e_B$$

computes   computes
$c_1 = z_B^{d_A} = \omega^{-k_B t_B d_A}$ $c_1 = e_A^{t_B} = \omega^{-k_B t_B d_A}$
$c_2 = e_B^{t_A} = \omega^{-k_A t_A d_B}$ $c_2 = z_A^{d_B} = \omega^{-k_A t_A d_B}$

Notes:

1. The protocol provides two keys. z_A and z_B cannot be used to construct one key as they are powers of different bases r_A and r_B and their logarithms k_A and k_B are unknown to A and B.

2. A is sure that z_B is authentic so she knows that she shares key c_1 with B. But she is not sure who sent e_B and thus with whom she shares key c_2. The same holds for B concerning z_A and e_A and respectively keys c_2 and c_1.
   However A can be sure that if she would share key $c_2' := e_C^{t_A} = \omega^{-k_A t_A d_C}$ with an intruder C then C cannot share another key $c_2'' := z_C^{d_B}$ with B, because B uses the authentic z_A to generate his key c_2 and C does not know the corresponding t_A. Again the same holds for B and key c_1.

To overcome the problem mentioned in the note above consider the following extension of the protocol:

- A chooses a random string g and sends it to B encrypted under key c_2.

- B receives and decrypts g under key c_2, encrypts it under key c_1 and returns it to A.

- If A receives g encrypted under key c_1, then key c_2 is considered valid and will be used as the session key.

Notes:

1. If A receives g encrypted under c_1, she is sure that it was sent to her by B. So B must have received it from her under key c_2 as mentioned above in note 2.

2. Nobody else besides A and B can share key c_2 unless the Diffie-Hellman protocol is broken.

3. Repeat the same protocol starting with B and a different random string h to verify key c_1. The second key may e.g. be used for synchronisation or as replacement of c_2 during the session. If the involved cipher is a stream cipher or equivalent, c_1 can be verified even with only one additional transmission.

4. An implementation must provide a suitable timeout when A waits for the return of g (and of course also for any other transmission).

4   Acknowledgements

The authors would like to thank Th. Beth, M. Clausen, D. Gollmann, H.-P. Rieß and S. Stempel for interesting discussions about the ideas presented in this paper.

It is noteworthy that Christoph Günther devised a surprisingly similar scheme for combined authentication and key exchange under a different approach [4]. We wish to thank him for the time he spent working with us on the differences and similarities of the two approaches.

References

[1] F. Bauspieß: *SELANE - SEcure Local Area Network Environment*, Studienarbeit, Universität Karlsruhe, 1988

[2] Th. Beth: *Efficient Zero-Knowledge Identification Scheme for Smart Cards*, Proc. of Eurocrypt '88, Springer LNCS 330, pp. 77-84, 1988

[3] W. Diffie, M. Hellman: *New Directions in Cryptography*, IEEE Trans. on Information Theory, IT-22, pp. 644-654, 1976

[4] Ch. Günther: *Diffie-Hellman and El-Gamal protocols with one single authentication key*, Proc. of Eurocrypt '89

[5] R. Rueppel: *Key Agreements based on Function Composition*, Proc. of Eurocrypt '88, Springer LNCS 330, pp. 3-10, 1988

The Use of Fractions in Public-Key Cryptosystems

Hartmut Isselhorst

Gesellschaft für elektronische Informationsverarbeitung
Oxfordstr. 12-16, D-5300 Bonn 1
West Germany

Abstract. This paper discusses an asymmetric cryptosystem based on fractions, the R^k-system, which can be implemented fast using only additions and multiplications. Also it is very simple to initialize the system and to generate new keys. The R^k-system makes use of the difficulty to compute the numerator and the denumerator of a fraction only knowing the rounded floating point representation. It is also based on the difficulty of a simultaneous diophantine approximation with many parameters and only a little error bound.

INTRODUCTION

Many known public-key cryptosystems deal with integer problems like factorization, discrete logarithms or knapsacks. Searching for another foundation of security we allow the use of real numbers, especially fractions.

Everyone knows that it is easy to choose two primes p and q and to compute the product $n \cdot p \cdot q$. But up to now it is difficult to calculate the factors p and q only knowing n, if n is greater than 10^{200}. But knowing n one has enough information to compute p and q, because factorization is deterministic. To avoid this one can try the following: Allowing real numbers it is possible to replace the multiplication by the division. To be more precisely, we pose the

Problem

Let a and p be integers with $1 < a < p < 10^{1000}$ and $gcd(a, p) = 1$. Denote

$$x_n \cdot 10^{-n} \cdot \lfloor 10^n \cdot a/p \rfloor, \quad n \in \mathbb{N}.$$

1. Is it possible to compute a and p from x_n with a suitable parameter n?

2. Is it possible to choose the parameter n in a way such that it is impossible to calculate a and p from x_n?

The following theorem solves the problem.

Theorem 1

Let a, p, k be integers with $10^{k-1} < p < 10^k$, $1 < a < p$, $gcd(a,p) = 1$.

1. Only knowing x_{2k} it is easy to compute a and p.

2. One cannot calculate a and p from x_n, if $0 < n < 2k-50$ and p is a prime.

Proof:

1. Let $0 < s < t < 1$ and $s = /s_1,...,s_r/$, $t = /t_1,...,t_m/$ (the continued fractions of s and t). Put formally $s_i = \infty$ for all $i > r$ and $t_i = \infty$ for all $i > m$. Then find j with $s_i \cdot t_i$ for all $i \in [1:j-1]$ and $s_j \neq t_j$. Define

$$q = \begin{cases} s_j + 1 & j \in 2\mathbb{N}, \ j \geq r \\ s_j & j \in 2\mathbb{N}, \ j \geq r \\ t_j + 1 & j \in 2\mathbb{N}+1, \ j < m \\ t_j & j \in 2\mathbb{N}+1, \ j \geq m. \end{cases}$$

Then $v = /s_1,...,s_{j-1},q/$ is the irreducible fraction in [s,t] with the lowest denominator ([Knuth 81, p.606]).

If a/b and c/d are consecutive fractions in the Farey-sequence F_n, $n \geq 2$, it holds

$$\left| a/b - c/d \right| = \frac{1}{bd} \geq \frac{1}{n(n-1)}$$

([Niven and Zuckerman 76, p.186])

Hence $a/p \in [x_{2k}, x_{2k} + 10^{-2k}] \cap F_{10^k} = \{ a/p \}$.

So the algorithm above computes a and p from the input $s = x_{2k}$, $t = x_{2k} + 10^{-2k}$.

2. If a/p, c/d are consecutive fractions in F_p we have

$$[x_n, x_n + 10^{-n}] \cap F_p = \{a/p\} \Leftrightarrow 10^{-n} \le \left| \frac{a}{p} - \frac{c}{d} \right| = \frac{1}{dp} \Leftrightarrow n \ge \log_{10}(p \cdot d)$$

From [Horster and Isselhorst 89, p.101] we have

$$\frac{1}{|F_p|} \cdot \sum_{\substack{\frac{a}{b}, \frac{c}{d} \in F_p \\ \text{consecutiv}}} \log_{10}(b \cdot d) \approx 2 \cdot \log_{10}(p+1) - \frac{1}{\log_e(10)} \cdot$$

so to compute a and p one needs to know x_n with $n = 2 \cdot \log_{10}(p)$ almost everytime. Knowing only $2 \cdot \log_{10}(p) - 50$ digits, one has to guess 50 sequential digits following x_n or approximately 50 partial quotients of a/p ([Isselhorst 88, p.104]).

Here one should observe, that the probability of a/p having a short period is nearly zero, because there are only a few primes q having a short period in 1/q ([Horster and Isselhorst 89, p. 89]).

Remark

Now it is possible to use the fraction a/p with $2 \cdot \log_{10}(p) - 50$ digits as a public key, because it is impossible to compute a and p having not enough information about a/p.

THE PROPOSED PUBLIC-KEY CRYPTOSYSTEM

Knowing the results about fractions we look for a way to use them for building a public-key cryptosystem. One possibility to do this is based on the computation with a real modulus:

$$a \equiv b \pmod{c} \Leftrightarrow (a-b)/c \in \mathbb{N}, \; a,b,c \in \mathbb{R}^+$$
$$a \text{ MOD } b := a - \lfloor a/b \rfloor \cdot b$$

The following lemma combines the results about fractions with a real modulus.

Lemma

Let p be a prime, $t > 0$, a, $a^* \in [1:p-1]$ with $a \cdot a^* \equiv 1 \pmod{p}$ and denote

$$c := t \cdot a / p$$
$$E(x) := (c \cdot x) \text{ MOD } t, \; x \in [0:p-1]$$

$$D(y) := (y/t \cdot p \cdot a^*) \text{ MOD } p$$

then $\quad D(E(x)) = x \quad$ for all $x \in [0:p-1]$.

Proof:

Since $a/r \text{ MOD } b/r = a/r - \lfloor (a/r)/(b/r) \rfloor \cdot b/r = (a \text{ MOD } b)/r$ for all $r > 0$ we get

$$
\begin{aligned}
D(E(x)) &= ((\ t \cdot a/p \cdot x \text{ MOD } t\ )\ /t \cdot p \cdot a^*\ ) \text{ MOD } p \\
&= ((\ t/p \cdot (a \cdot x \text{ MOD } p))\ /t \cdot p \cdot a^*) \text{ MOD } p \\
&= (a \cdot x \text{ MOD } p) \cdot a^* \text{ MOD } p = x. \ \square
\end{aligned}
$$

This can be interpreted as a model of a cryptosystem, which uses the fraction a/p in the encryption function $E(x)$, but uses the integer $p \cdot a^*$ in the decryption function $D(y)$. Here it is important to see, that the integer $p \cdot a^*$ is not the same as the fraction p/a.

However it works only if one uses exact arithmetic, it is possible to get a and p knowing c. But when the system is made fault tolerant with rounded numbers, it can be secured and implemented using the results above.

So the lemma can be improved to a public-key cryptosystem, which will be discussed here:

The R^k-System

Assumptions: Let
- p be a prime, $p > 10^{250}$, $k \in \mathbb{N}+2$
- $A = (a_{i,j}) \in \mathbb{Z}_p^{k \times k}$, $\det(A) \neq 0 \pmod{p}$
- $A^* = (a_{i,j}^*) \in \mathbb{Z}_p^{k \times k}$ with $A \cdot A^* \equiv I \pmod{p}$
- $t \in (0,p)$ (for example $t = 1$)
- $z = \lceil \log_{10}(4 \cdot p/t) \rceil$
- $c_{i,j}^n = 10^{-n} \lfloor 10^n \cdot t \cdot a_{i,j} /p \rfloor$, $C_n = (c_{i,j}^n) \in \mathbb{R}^{k \times k}$ with $n = \lceil 2 \cdot \log_{10}(p) - 50 - \log_{10}(t) \rceil$

Plaintext:
- $X \in \mathbb{Z}_m^k$ with $m = \lfloor \dfrac{p}{10^{50} \cdot 4k} \rfloor$

Encryption function: $\quad E(X) = 10^{-z} \lfloor 10^z \cdot \{ (C_n \cdot X) \text{ MOD } t \} \rfloor$

Public keys: $\quad$ - C_n, t, z

Decryption function: $\quad$ $D(Y) = (A^* \cdot \lfloor Y \cdot p / t + 1/2 \rfloor) \bmod p$

Secret keys: $\quad$ - p, A^*

Theorem 2

The R^k-system holds $D(E(X)) = X$ for all $X \in \mathbb{Z}_m^k$.

Proof (sketch):

The central step is

$$0 \le [(\sum_{j=1}^{k} c_{i,j}^n \cdot x_j) \bmod t] \cdot \frac{p}{t} - [10^{-z} \lfloor 10^z (\sum_{j=1}^{k} c_{i,j}^n \cdot x_j) \bmod t \rfloor] \cdot \frac{p}{t} \le \frac{1}{4}.$$

With [Isselhorst 88, p. 131-134] we also have

$$| [(\sum_{j=1}^{k} t \cdot a_{i,j} / p \cdot x_j) \bmod t] \cdot p/t - [(\sum_{j=1}^{k} c_{i,j}^n \cdot x_j) \bmod t] \cdot \frac{p}{t} | =$$

$$| (\sum_{j=1}^{k} a_{i,j} \cdot x_j) \bmod p - [(\sum_{j=1}^{k} c_{i,j}^n \cdot x_j) \bmod t] \cdot \frac{p}{t} | < 1/4.$$

Taking both inequalities together implies

$$| (\sum_{j=1}^{k} a_{i,j} \cdot x_j) \bmod p - [10^{-z} \lfloor 10^z (\sum_{j=1}^{k} c_{i,j}^n \cdot x_j) \bmod t \rfloor] \cdot \frac{p}{t} | <$$

$$| (\sum_{j=1}^{k} a_{i,j} \cdot x_j) \bmod p - [(\sum_{j=1}^{k} c_{i,j}^n \cdot x_j) \bmod t] \cdot \frac{p}{t} | +$$

$$| [(\sum_{j=1}^{k} c_{i,j}^n \cdot x_j) \bmod t] \cdot \frac{p}{t} - [10^{-z} \lfloor 10^z (\sum_{j=1}^{k} c_{i,j}^n \cdot x_j) \bmod t \rfloor] \cdot \frac{p}{t} | < \frac{1}{4} + \frac{1}{4} = \frac{1}{2}.$$

and finally $(\sum_{j=1}^{k} a_{i,j} \cdot x_j) \bmod p = \lfloor [10^{-z} \lfloor 10^z (\sum_{j=1}^{k} c_{i,j}^n \cdot x_j) \bmod t \rfloor] \cdot \frac{p}{t} + \frac{1}{2} \rfloor$. $\square$

DISCUSSION

a) SECURITY

i) The R^k-system with parameter k=1 is not secure. It is easy to approximate the number $c_n/t \approx e/f$ by continued fractions. Then one can simulate the original

R^1-system with e/f instead of a/p. So one can break a R^1-system without knowing the secret keys a and p.

ii) With $k \geq 2$ one can try the same attack: one looks for an approximation of $c_{i,j}^n/t$ $\approx e_{i,j}/f$, which is a simultaneous diophantine approximation.

But note the following facts:

- the number of simultaneous diophantine approximations increases quadratically with k
- the error bound is always very small ($\sim 10^{50}/p^2$, [Horster and Isselhorst 89])
- the common denominator f has to be bounded: $f \leq 10 \cdot p$.

Furthermore the best algorithm to solve simultaneous diophantine approximation problems of this kind would in my opinion Lagarias' algorithm [Lagarias 85], which uses $O(k^{12} \cdot (k^2 \cdot \log_2(10^n) + \log_2(p))^4)$ bit-operations to find *some* approximation. It is not guaranteed that solutions found by this procedure will work.

b) ADVANTAGE

i) The advantage of the R^k-system is, that it works fast. To encrypt and decrypt k integers out of [0:m] ($m \approx p \cdot 10^{-50}$) there are ($2k^2+10k$) operations like addition, multiplication and reduction. (With k=10, t=1 there are 9 additions, 10 multiplications and 1 reduction to encrypt or decrypt one number). It is possible to choose t=1, so that the reduction mod t is very simple.

ii) It is easy to initialize the R^k-system and to generate new keys, because one needs only one prime p and an invertible matrix A with the $\mathbb{Z}_p$-invers A^*.

iii) To strengthen the system one can select a higher dimension k without the need to use larger numbers as in the RSA-scheme.

iv) The R^k-system provides another way to build a public-key cipher without using the well known arithmetical problems like factorization or knapsacks.

c) DISADVANTAGE

i) The security of the R^k-system is not proved.

ii) The size of the public and the secret key might be regarded as a disadvantage. But unlike knapsack-schemes within the R^k-system one encrypts $\log_2(m)/k >> 1$ bits with every component of the key.

FURTHER RESEARCH

i) Look for other attacks for the R^k-system: One is to try to get the prime p with a simultaneous diophantine approximation with only a few components of the key matrix C_n.

ii) Examine if the security of the R^k-system holds when p is an arbitrary integer and not necessarily a prime, and $\det(A) \neq 0 \pmod{p}$. So the initialization becomes easier.

iii) Examine if one can select a small number $k \in [2:10]$, such that the R^k-system is very fast. This should be used for messages which have to be secret only for a short time (like one hour or one day e. g. in military use).

CONCLUSION

The paper shows how to use fractions in a public-key cryptosystem, which is based on the problem of a simultaneous diophantine approximation with many parameters. The new R^k-system can be implemented in a fast way using only addition and multiplication with only one reduction. Also new keys can be produced very simply, so that one can use a different pair of keys in every communication.

REFERENCES

[Horster and Isselhorst 89]:
P. Horster, H. Isselhorst, Approximative Public-Key-Kryptosysteme, Informatik-Fachberichte 206 (Heidelberg: Springer, 1989)

[Isselhorst 88]:

H. Isselhorst, Ein Beitrag zur Verwendung rationaler Zahlen in Public-Key Krypto-systemen (Heidelberg: Hüthig, 1988)

[Knuth 81]:

D. E. Knuth, The Art of Computer Programming Vol.2, Seminumerical Algorithms, 2ed (Reading: Addison Wesley, 1981)

[Lagarias 85]:

J. C. Lagarias, "The Computational Complexity of Simultaneous Diophantine Approximation Problems," SIAM J. Comput. Vol.14 No1 (1985), p.196-209

[Niven and Zuckerman 76]:

I. Niven, H. S. Zuckerman, Einführung in die Zahlentheorie I (Mannheim: Bibliographisches Institut, 1976)

SYMBOLS

$\mathbb{N} = \{0, 1, 2, 3, ...\}$ $\mathbb{N}+i = \{i, i+1, i+2, i+3, ...\}$

$\mathbb{Z} = \{ ..., -2, -1, 0, 1, 2, ...\}$ $\mathbb{Z}_p = \{0, 1, 2, ..., p-2, p-1\}$

$\mathbb{R}$ = real numbers

$\lfloor x \rfloor$ = greatest integer less than x $\lceil x \rceil$ = lowest integer greater than x

gcd(a,b) = greatest common divisor of a and b

$[a:b] = \{a, a+1, a+2, ..., b-1, b\}$

I = unit matrix

SMALL EXAMPLE

The prime: $p = 64301$

The dimension $k = 2$

The invertible matrix $A = \begin{pmatrix} 5387 & 2993 \\ 7461 & 4001 \end{pmatrix}$

The inverse matrix $A^* = \begin{pmatrix} 14109 & 19322 \\ 59703 & 52039 \end{pmatrix}$

the modulus	$t = 1$
The constants	$n = 9, z = 6$

The key-matrix

$$C_n = \begin{pmatrix} 0.083777857 & 0.0465467 \\ 0.116032410 & 0.06222298 \end{pmatrix}$$

Plaintext

$$\begin{pmatrix} x_1 \\ x_2 \end{pmatrix} \in \mathbb{Z}_{1000}^2 , \; m = 1000$$

Encryption

$$E\begin{pmatrix} x_1 \\ x_2 \end{pmatrix} = 10^{-6} \cdot \lfloor 10^6 \cdot \{ (C_9 \cdot \begin{pmatrix} x_1 \\ x_2 \end{pmatrix}) \bmod 1 \} \rfloor$$

Decryption:

$$D\begin{pmatrix} y_1 \\ y_2 \end{pmatrix} = A^{-1} \cdot \lfloor \begin{pmatrix} y_1 \\ y_2 \end{pmatrix} \cdot 64301 + 1/2 \rfloor \bmod 64301$$

$$E\begin{pmatrix} 500 \\ 501 \end{pmatrix} = \begin{pmatrix} 0.20883 \\ 0.189918 \end{pmatrix}, \quad D\begin{pmatrix} 0.208830 \\ 0.189918 \end{pmatrix} = \begin{pmatrix} 500 \\ 501 \end{pmatrix}$$

A Practical Protocol for Large Group Oriented Networks

Yair Frankel
Electrical Engineering and Computer Science Department
University of Wisconsin-Milwaukee
Milwaukee, WI 53201

1   Introduction

It is infeasible in large networks for every individual to have his own key. In a group oriented society, public keys are needed for communications between one organization and another. The organization might also want the supervisors to read some of the messages received by the employees. In the case of urgent messages, the organization may want any member to be able to read it [3]. This paper proposes a method in which individuals at separate organizations can communicate without the advance coordination of keys between the individuals. Thus, reducing the number of keys needed to communicate between two organizations. Also, the destination company can create its own policy on who can read messages and the type of public, or conventional, key system used within organization, without any involvement with the sending organization. This system will solve one of the many problems presented in [3].

In this system, an individual sends parts of messages to clerks(devices) who proceed to transmit messages to destination organizations using public keys. The clerks need to know key(s) for the source and the destination organization(s), not the individuals within the organizations. These clerks, when not acting in collusion, cannot recover the message but can only determine the destinations.

At the destination organization, clerks do some calculation to the messages using their private keys and send the result to the destination(s). The individuals at the destinations combine the messages received from their clerks to recover the original message. The clerks at the destination must also act in collusion to recover the original message.

Public key cryptosystems [5] can be used in large communication networks. However as the size of the network increases, the quantity of keys increases to the point where key management becomes a major problem. A system that does not require that a message sender know the key(s) of the intended recipients offers a considerable advantage.

Threshold schemes [1, 8] prevent a (small) number of individuals from acting in collusion to view a message. then the threshold. These schemes are used in this paper to provide protection against collusion on the part of clerks and against communication failures.

Tamper proof devices have been used to show relationships between classes of cryptosystems [4] and to simplify implementations of protocols [2]. In the following, tamperproof devices can be used to replace the clerks, therefore further reducing the collusion problem.

2 Basic System

The general concept is that the sender breaks a message into separate parts. Half of the message pieces will be encrypted and go to one of the clerks and the key will go to the *other* clerk. The same will be done with the other half of the message. The clerks will not be able to understand the pieces of the message that they receive since the other clerk will have the key needed to decrypt the message that he/she receives. Each clerk will then transmit the message using the destination company's public key. Since keys are only needed between the companies rather than the individuals in the company, the number of keys is drasticly reduced.

Both of the clerks at the destination organization will receive the message and do some calculations on it. For example, they will both receive an encrypted version of one of the keys sent by the source organization's clerk. They, however, will not be able to know the key after they do the calculation. The destination clerks will both give the result of their calculation to the destination. When the individual(s) receive the message from the clerks, he/she will multiply the two results together to get the key. The rest of the message pieces will be received in the same manner, but the destination will use the key that he/she received earlier to decrypt the message.

Four distinct tasks are necessary for the protocol's operation. These tasks are described in greater detail in the following.

2.1 Source

The source partitions the message stream into sequence M_i $i = 0, 1, 2 \ldots$ of non-overlapping packets. The source will then generate two keys K_1 and K_2 in which K_1 will be used to encrypt the odd numbered messages and K_2 the even numbered messages using a conventional cryptosystem of required strength.

$$C(M_{2i+1}, K_1) = C_{2i+1}$$

and
$$C(M_{2i}, K_2) = C_{2i} \text{ where } i = 0, 1, 2, \ldots.$$

It will become clear that the keys can be selected as often as needed and the source and eventual destination(s) do not need to choose the keys in advance, only the cryptosystem.

Next K_1 and all the even number C_i's are signed and then encrypted using the key of one of the clerks, and transmitted to the clerk. Similarly K_2 and the odd number C_i's are signed, encrypted using key of *other* clerk, and then transmitted to the *other* clerk.

The cryptosystem protecting the channel between source and clerk is selected at the discretion of the company's security manager. There is no necessity for coordination between companies in the selection of this system.

2.2   Clerks at Source Company

When a source clerk receives a message, he/she decrypts it using his/her key and recovers K_2 and C_{2i+1}'s, or K_1 and the C_{2i}'s.

At this point the K_i's are exposed. *Note however that each clerk sees the key of the other clerk's C's, not the key necessary to decrypt the cipher text in his possession.* To compromise this system the clerks must act together or both clerks must be bugged.

Then the clerk authenticates the sender and encrypts the result using the public key of the destination company.

$$\hat{C}_{K_1} = (K_1)^\alpha \bmod n, \quad \hat{C}_{M_{2i+1}} = (M_{2i+1})^\alpha \bmod n$$

and
$$\hat{C}_{K_2} = (K_2)^\alpha \bmod n, \quad \hat{C}_{M_{2i}} = (M_{2i})^\alpha \bmod n \text{ where } i = 0, 1, 2, \ldots.$$

Each clerk broadcasts his messages to the destination company.

In this system, the transform $\hat{C}$ is the RSA cryptosystem [7] with modulus n. The α and n form the public key of the destination organization and are generated by a trusted key distribution center or the destination organization.

2.3   Clerks at Destination Company

Each clerk receives and performs a calculation (described below) on their copy of of every packet received by the organization. It does not matter which clerk sent the message from the sending organization.

To do the calculations, each clerk is given a private key which will not decrypt the packet "completely". That is, the clerks will not be able to read the packet since they did not use the key, but they have done part of the calculations needed to decrypt the message without getting any information about the message or key. The calculation is the same as if one would decrypt a message using RSA with his private keys.

$$\dot{C}_p^t = (\hat{C}_p)^t \bmod n, \quad \dot{C}_p^s = (\hat{C}_p)^s \bmod n,$$

where p is a packet received by the destination company.

If the private keys of the clerks (at destination) are s and t then the following relation holds:

$$\alpha t + \alpha s \equiv 1 \bmod \phi(n).$$

Barring collusion between the destination clerks, not even K_1 or K_2 are exposed. Each clerk transmits the packet he did calculations on (i.e. $\dot{C}_p^t$), encrypts, signs and transmit it to the destination. If the organizations policy states that certain individual must read the message also, the clerk will also send message to them. If the clerks want to collude, they must transmit both parts to an unauthorized individual or the other clerk.

The public key system protecting the channel between the clerks and the destination is selected at the discretion of the destination company's security manager. There is no necessity of cooperation between companies in the selection of this system.

2.4 Destination

The destination decrypts the incoming packets $(\dot{C}_p^s, \dot{C}_p^t)$ which are the result of the computation done by destination clerks, and multiplies the packets together to recover the K_1, K_2 and the C_i's. This is possible since,

$$
\begin{aligned}
\dot{C}_p^t * \dot{C}_p^s &= (\hat{C}_p)^{\alpha s} * (\hat{C}_p)^{\alpha t} \bmod n \\
&= (\hat{C}_p)^{\alpha t + \alpha s} \bmod n \\
&= (\hat{C}_p)^1 = M_p \text{or} K_p.
\end{aligned}
$$

The destination then can easily recover the M_i's using K_1 and K_2.

3 Proofs of Equivalence to Existing Cryptosystems

It is easy to show that this cryptosystem is equivalent to RSA. The approach is similar to the method of Kranakis [6], but is slightly different. However, our system is more secure than [6]. Since if there exists public keys e_i, e_j in the Krankis method such that $\gcd(e_i, e_j) = 1$, it is trivialy breakable.

Since $t + s = \alpha^{-1} \bmod \phi(n)$ and t is chosen randomly, finding s is equivalent to calculating α^{-1}.

4   Extending the Basic System

The total dependence of the trustworthyness of the source clerks is a weak point of the basic system. To remove this deficiency and to increase the reliability, multiple source clerks can be employed. The cryptosystem C above is expanded to require M keys, and the number of source clerks is increased to N. Each clerk has sufficient numbers of keys and the copies of the keys are distributed in such a fashion that any set of K clerks have all N keys.

$$C(M_i, K_j) = C_i \text{ where } i = 0, 1, 2, \ldots \text{ and } j = 0, 1, 2, \ldots, M.$$

Since the above key distribution will operate in the face of N minus K inoperative clerks, the partitioned message stream can distributed to the clerks such that every message block is sent to exactly $N - K + 1$ clerks. If the M keys are randomly generated and are exclusive or'd to form a single key as part of the encipherment C the the knowledge of any M minus 1 keys given an enemy no knowledge of the actual key. This method becomes impractical due to message expansion.

In the scheme presented in section 2, collusion on the part of the two destination clerks, in the basic system results in compromising the message. To prevent this more clerks can be employed but as the number grows the reliability of the system degrades. Each destination clerk j posses k keys such that

$$\alpha \sum_{i=1}^{k} x_{j,i} \equiv 1 \bmod \phi(n) \quad \text{where} j = 1, \ldots, n.$$

No clerk possess more than one $x_{j,i}$ from the same congruence. When a clerk receives a message he generates k messages by raising the message to the power of each key mod n. The destination can choose any k messages that resulted from the same message (that the destination clerks *received*) provided that the keys are from the same congruence.

Tamper proof devices can be used to replace the clerks by performing all their duties. These devices must have at their disposal an authenticated routing table and public/private key database.

5   Conclusion

This protocol is a method in which one can use RSA in a large network. Since our society is organized into groups, this system is not only practical but represents a robust method since the techniques allow a single sender or receiver to communicate with a company rather than require companies to commincate only.

6 Acknowledgements

The author wishes to extend his thanks to Professor Yvo Desmedt for his suggestions and references. The author wishes to thank Brian Matt for his many helpful discussions and suggestions.

REFERENCES

[1] G. R. Blakley. Safeguarding cryptographic keys. In *Proc. Nat. Computer Conf. AFIPS Conf. Proc.*, pages 313–317, 1979. vol.48.

[2] G.I. Davida and B.J. Matt. Arbitration in tamper proof systems. In C. Pomerance, editor, *Advances in Cryptology, Proc. of Crypto'87 (Lecture Notes in Computer Science 293)*, pages 216–223. Springer–Verlag, 1988. Santa Barbara, California, U.S.A., August 16–20.

[3] Y. Desmedt. Society and group oriented cryptography : a new concept. In C. Pomerance, editor, *Advances in Cryptology, Proc. of Crypto'87 (Lecture Notes in Computer Science 293)*, pages 120–127. Springer–Verlag, 1988. Santa Barbara, California, U.S.A., August 16–20.

[4] Y. Desmedt and J.-J. Quisquater. Public key systems based on the difficulty of tampering (Is there a difference between DES and RSA?). In A. Odlyzko, editor, *Advances in Cryptology, Proc. of Crypto'86 (Lecture Notes in Computer Science 263)*, pages 111–117. Springer–Verlag, 1987. Santa Barbara, California, U.S.A., August 11–15.

[5] W. Diffie and M. E. Hellman. New directions in cryptography. *IEEE Trans. Inform. Theory*, IT–22(6):644–654, November 1976.

[6] E. Kranakis. A class of cryptosystems equivalent to RSA. *Dept. of Computer Science, Yale Univ. tech report 315*, April 1984.

[7] R. L. Rivest, A. Shamir, and L. Adleman. A method for obtaining digital signatures and public key cryptosystems. *Commun. ACM*, 21:294 – 299, April 1978.

[8] A. Shamir. How to share a secret. *Commun. ACM*, 22:612 – 613, November 1979.

Section 2

Theory

COUNTING FUNCTIONS SATISFYING A HIGHER ORDER STRICT AVALANCHE CRITERION

Sheelagh Lloyd

Hewlett-Packard Laboratories
Filton Road, Stoke Gifford
Bristol, ENGLAND
BS12 6QZ

I . INTRODUCTION

The strict avalanche criterion was introduced by Webster and Tavares [3] in order to combine the ideas of completeness and the avalanche effect. A cryptographic transformation is complete if each output bit depends on all the input bits, and it exhibits the avalanche effect if an average of one half of the output bits change whenever a single input bit is complemented. To fulfil the strict avalanche criterion, each output bit should change with probability one half whenever a single input bit is complemented. This means, in particular, that there is no good lower order (fewer bits) approximation to the function. This is clearly a desirable cryptographic property since such an approximation would enable a corresponding reduction in the amount of work needed for an exhaustive search.

The notion of strict avalanche criterion was recently extended by Forré to consider subfunctions obtained from the original function by keeping one or more input bits constant. This is also important cryptographically because, in a chosen plaintext attack, the cryptanalyst could arrange for certain input bits to be kept constant. Forré defined the strict avalanche criterion of order m, with order 0 being the original strict avalanche criterion, and made a conjecture, supported by experimental evidence, concerning the number of functions satisfying a higher order strict avalanche criterion.

In this paper, we shall first present unified definitions of the three concepts of completeness, the avalanche effect and the strict avalanche criterion. We shall then show how Forré's definition of the higher order strict avalanche criterion may be simplified, and then use this simplified form to prove her conjecture.

II . DEFINITIONS

In this section we shall discuss more fully the ideas of completeness, the avalanche effect and the strict avalanche criterion. We shall present these three criteria in a unified framework, in order to highlight the connections between them.

Let $f : \mathbf{Z}_2^n \to \mathbf{Z}_2^m$ ($n \geq m$) be a cryptographic transformation. Then f is said to be complete [2] if and only if for each pair (i, j), $1 \leq i \leq n$, $1 \leq j \leq m$, there exists a pair of n-bit vectors $\underline{x}$ and $\underline{x}'$ such that $\underline{x}$ and $\underline{x}'$ differ only in bit i, and $f(\underline{x})$ and $f(\underline{x}')$ differ in at least bit j. This property ensures that each output bit depends on all the input bits. If some output bits depended on only a few input bits, then, by observing a significant number of input-output pairs, a cryptanalyst might be able to detect these relations and use this information to aid the search for the key.

Webster and Tavares [3] pointed out that this condition can be restated as follows. Let us fix i, $1 \leq i \leq n$, and write $\underline{c}_i$ for the n-bit vector with a 1 in the ith position and 0 elsewhere. Now consider the set of m-bit vectors $f(\underline{x}) \oplus f(\underline{x} \oplus \underline{c}_i)$ as $\underline{x}$ ranges over $\mathbf{Z}_2^n$. (These are called "avalanche vectors" in [3]). For each j, $1 \leq j \leq m$, at least one of these vectors has a 1 in the jth position. So, if we add these vectors together (as elements of $\mathbf{Z}^m$ rather than elements of $\mathbf{Z}_2^m$), all components should be greater than 0. Thus we have the following definition.

Definition 2.1 Let $f : \mathbf{Z}_2^n \to \mathbf{Z}_2^m$ be a cryptographic transformation. Then f is complete if and only if

$$\sum_{\underline{x} \in \mathbf{Z}_2^n} f(\underline{x}) \oplus f(\underline{x} \oplus \underline{c}_i) > (0, .., 0) \qquad \text{for all } i, \, 1 \le i \le n$$

where both the summation and the greater-than are component-wise over $\mathbf{Z}^n$.

Forré considers only the case $m = 1$, and, in this case, we see that $f : \mathbf{Z}_2^n \to \mathbf{Z}_2$ is complete if and only if

$$\sum_{\underline{x} \in \mathbf{Z}_2^n} f(\underline{x}) \oplus f(\underline{x} \oplus \underline{c}_i) > 0 \qquad \text{for all } i, \, 1 \le i \le n.$$

We consider now the avalanche effect. A function exhibits the avalanche effect if and only if an average of one half of the output bits change whenever a single input bit is changed. This may be formalised [3] by considering again the "avalanche vectors" $f(\underline{x}) \oplus f(\underline{x} \oplus \underline{c}_i)$ as $\underline{x}$ varies over $\mathbf{Z}_2^n$. The bits of these vectors are referred to as "avalanche variables". Then f is said to exhibit the avalanche effect if and only if, for each i, $1 \le i \le n$, one half of the avalanche variables are equal to 1.

We shall write this condition as follows. For each i, there are 2^n avalanche vectors, and hence $m2^n$ avalanche variables. If exactly half of them are 1, then their sum must be $m2^{n-1}$. Thus we have the definition below.

Definition 2.2 Let $f : \mathbf{Z}_2^n \to \mathbf{Z}_2^m$ be a cryptographic transformation. Let w denote the Hamming weight function. Then f exhibits the avalanche effect if and only if

$$\sum_{\underline{x} \in \mathbf{Z}_2^n} w(f(\underline{x}) \oplus f(\underline{x} \oplus \underline{c}_i)) = m2^{n-1} \qquad \text{for all } i, \, 1 \le i \le n.$$

In the case $m = 1$, w is essentially the identity, so we see that $f : \mathbf{Z}_2^n \to \mathbf{Z}_2$ exhibits the avalanche effect if and only if

$$\sum_{\underline{x} \in \mathbf{Z}_2^n} f(\underline{x}) \oplus f(\underline{x} \oplus \underline{c}_i) = 2^{n-1} \qquad \text{for all } i,\, 1 \leq i \leq n.$$

Note that, in this case, if f exhibits the avalanche effect, then f must automatically be complete.

Finally, we consider the strict avalanche criterion. For a function to satisfy this, each output bit should change with probability one half whenever a single input bit changes. This can be written as below (see, e.g. [3]).

Definition 2.3 Let $f : \mathbf{Z}_2^n \to \mathbf{Z}_2^m$ be a cryptographic transformation. Then f satisfies the strict avalanche criterion if and only if

$$\sum_{\underline{x} \in \mathbf{Z}_2^n} f(\underline{x}) \oplus f(\underline{x} \oplus \underline{c}_i) = (2^{n-1}, .., 2^{n-1}) \qquad \text{for all } i,\, 1 \leq i \leq n.$$

In the case $m = 1$, we see that the strict avalanche criterion is exactly the same as the avalanche criterion.

It is clear that a necessary and sufficient condition for a function $f : \mathbf{Z}_2^n \to \mathbf{Z}_2^m$ to satisfy the strict avalanche criterion is that the m functions which specify the behaviour of each bit should all satisfy the strict avalanche criterion. We are, therefore, justified in considering only the case $m = 1$ in what follows.

III . PRELIMINARIES

Let f be a function from $\mathbf{Z}_2^n$ to $\mathbf{Z}_2$. It turns out to be convenient to consider instead the function $\hat{f}$ defined by $\hat{f}(\underline{x}) = (-1)^{f(\underline{x})}$ which has the same domain as f, but takes values in $\{1, -1\}$ rather than in $\mathbf{Z}_2$. We shall use the following characterisation of functions satisfying the strict avalanche criterion (SAC) due to Forré.

Theorem 3.1 [1] A function $f : Z_2^n \rightarrow Z_2$ satisfies the SAC if and only if

$$\sum_{\underline{x} \in Z_2^n} \hat{f}(\underline{x}) \hat{f}(\underline{x} \oplus \underline{c}) = 0$$

for all $\underline{c} \in Z_2^n$ with Hamming weight 1, where $\hat{f}(\underline{x}) = (-1)^{f(\underline{x})}$.

The definition of the higher order SAC is as follows:

Definition 3.2 [1] A function $f : Z_2^n \rightarrow Z_2$ satisfies the SAC of order m, where $1 \leq m \leq (n-2)$ if and only if

(a) any function obtained from f by keeping m of its input bits constant satisfies the SAC (for any choice of the positions and of the values of the constant bits)

and

(b) f satisfies the SAC of order $m - 1$.

Note that it would be impossible for a function to satisfy the SAC of order $(n-1)$, since this would be equivalent to finding a function $g : Z_2 \rightarrow Z_2$ satisfying the SAC. Using Theorem 3.1, this would mean that $\hat{g}(0)\hat{g}(1) = 0$ which is impossible, since both $\hat{g}(0)$ and $\hat{g}(1)$ have values in $\{1, -1\}$.

IV . SIMPLIFICATION

In this section we shall show that condition (b) of Definition 3.2 is not necessary. For ease of notation, we introduce the following terminology.

Definition 4.1 A function $f : Z_2^n \rightarrow Z_2$ satisfies the partial strict avalanche criterion (PSAC) of order m, where $0 \leq m \leq (n-2)$ if and only if any function obtained from f by keeping m of its input bits constant satisfies the SAC (for any choice of the positions and of the values of the constant bits).

Note that the PSAC of order 0 is exactly the same as the SAC (of order 0) and that a function satisfies the SAC of order m if and only if it satisfies both the PSAC of order m and the SAC of order $(m-1)$.

We shall show that the PSAC of order m alone is sufficient to ensure the SAC of order m. We shall first prove a result concerning the PSAC.

Lemma 4.2 Let f be a function from $\mathbf{Z}_2^n$ to $\mathbf{Z}_2$ satisfying the PSAC of order m $(1 \leq m \leq (n-2))$. Then f satisfies the PSAC of order $(m-1)$.

Proof

Let g be a function obtained from f by keeping $(m-1)$ input bits fixed. We must show that g satisfies the SAC. By Theorem 3.1, we need to show that

$$S = \sum_{\underline{x} \in \mathbf{Z}_2^{n-m+1}} \hat{g}(\underline{x})\hat{g}(\underline{x} \oplus \underline{c}) = 0$$

for all $\underline{c} \in \mathbf{Z}_2^{n-k}$ with Hamming weight 1. Without loss of generality, we may assume that $\underline{c} = (0, .., 0, 1)$. Now, since $(n - m + 1) > 1$, we know that $\underline{x}$ and $\underline{x} \oplus \underline{c}$ agree on the first bit, so we may split the sum up into those terms where the first bit of $\underline{x}$ is 0, and those where it is 1.

$$S = \sum_{\underline{y} \in \mathbf{Z}_2^{n-m}} \hat{g}_0(\underline{y})\hat{g}_0(\underline{y} \oplus \underline{c}') + \sum_{\underline{y} \in \mathbf{Z}_2^{n-m}} \hat{g}_1(\underline{y})\hat{g}_1(\underline{y} \oplus \underline{c}')$$

where g_0, g_1 denote the functions obtained from g by setting the first input bit to 0, 1 respectively, and $\underline{c}'$ denotes the vector of length $(n - m)$ obtained from $\underline{c}$ by deleting the first bit. Now both g_0 and g_1 are obtained from g by fixing one bit, and hence from f by fixing m bits, so by our assumption, they satisfy the SAC. Hence by Theorem 3.1, both the sums above are zero, and so $S = 0$ as required. We have therefore proved the result.

We are now able to prove the following theorem.

Theorem 4.3 A function $f : \mathbf{Z}_2^n \rightarrow \mathbf{Z}_2$ satisfies the Strict Avalanche Criterion (SAC) of order m if and only if any function obtained from f by keeping m of its input bits constant satisfies the SAC (for any choice of the positions and of the values of the constant bits)

Proof

In other words, we must prove that f satisfies the SAC of order m if and only if it satisfies the PSAC of order m. Clearly, by the definitions of the SAC and PSAC, we know that if f satisfies the SAC of order m, then it satisfies the PSAC of order m. Hence we need only prove that if f satisfies the PSAC of order m, then it satisfies the SAC of order m.

The proof is by induction. The base step is trivial, since, as we have already remarked, the PSAC of order 0 is identical to the SAC (of order 0). So let us assume the result for $m = k$, and try to prove it for $m = k+1$. Suppose that f satisfies the PSAC of order $(k+1)$. We must show that f satisfies the SAC of order $(k+1)$. By Lemma 4.2, f satisfies the PSAC of order k. By the inductive hypothesis, therefore, f satisfies the SAC of order k. Hence, by the definition of the SAC, f satisfies the SAC of order $(k+1)$.

V . COUNTING FUNCTIONS

In this section, we shall prove a result conjectured by Forré [1] in the light of experimental results.

Theorem 5.1 Let $n \in \mathbf{Z}$ be such that $n \geq 2$. Then the number of functions $f : \mathbf{Z}_2^n \to \mathbf{Z}_2$ satisfying the SAC of order $(n-2)$ is 2^{n+1}.

We shall prove this by giving an explicit form for the functions satisfying the SAC of order $(n-2)$.

Lemma 5.2 Suppose that $n \in \mathbf{Z}$, $n \geq 2$ and $f : \mathbf{Z}_2^n \to \mathbf{Z}_2$. Then f satisfies the SAC of order $(n-2)$ if and only if for all $S \subseteq \{1, 2, .., n\}$,

$$\hat{f}(\underline{e}_S) = (-1)^{\frac{|S|(|S|-1)}{2}} (\hat{f}(\underline{0}))^{(|S|+1)} \prod_{r \in S} \hat{f}(\underline{e}_{\{r\}})$$

where $\underline{e}_S$ denotes the element of $\mathbf{Z}_2^n$ which satisfies $e_i = 1 \iff i \in S$ and as before, $\hat{f}(\underline{x}) = (-1)^{f(\underline{x})}$.

Proof

By Theorem 4.3 and Theorem 3.1, f satisfies the SAC of order $(n-2)$ if and only if

$$\sum_{\underline{x} \in \mathbf{Z}_2^2} \hat{g}(\underline{x})\hat{g}(\underline{x} \oplus \underline{c}) = 0$$

for all $\underline{c} \in \mathbf{Z}_2^2$ with Hamming weight 1, and all functions g obtained from f by fixing $(n-2)$ bits. There are two choices for $\underline{c}$, namely $\underline{c} = (1, 0)$ or $\underline{c} = (0, 1)$.

In fact, these two give rise to the same equation as follows.

$$\sum_{\underline{x} \in \mathbb{Z}_2^2} \hat{g}(\underline{x}) \hat{g}(\underline{x} \oplus (1,0)) = 2(\hat{g}(0,0)\hat{g}(1,0) + \hat{g}(0,1)\hat{g}(1,1)) = 0$$

and

$$\sum_{\underline{x} \in \mathbb{Z}_2^2} \hat{g}(\underline{x}) \hat{g}(\underline{x} \oplus (0,1)) = 2(\hat{g}(0,0)\hat{g}(0,1) + \hat{g}(1,0)\hat{g}(1,1)) = 0.$$

But all $\hat{g}(\underline{x})$ have the value ± 1, so we may multiply the second equation through by $\hat{g}(0,1)\hat{g}(1,0)$ to obtain the first equation. Hence we have essentially one equation for each function g. This equation may be written as

$$\hat{g}(1,1) = -\hat{g}(0,0)\hat{g}(0,1)\hat{g}(1,0).$$

For example, if g were obtained from f by setting the last $(n-2)$ bits to 0, then this equation becomes, in terms of values of f

$$\hat{f}(1,1,0,..,0) = -\hat{f}(0,..,0)\hat{f}(0,1,0,..,0)\hat{f}(1,0,..,0).$$

In general, the equation defines the value of $\hat{f}$ at a point $\underline{x}$ of Hamming weight $(w+2)$ in terms of the values of $\hat{f}$ at points with Hamming weight $(w+1)$ and w. This means that we can express the value of f at all points of Hamming weight greater than or equal to 1 in terms of the values of f at points with Hamming weight 0 or 1. Of course, in general, there will be more than one equation defining the value of $f(\underline{x})$, so we will need to check that these equations are consistent. Each equation corresponds to a set $S \subseteq \{1, 2, .., n\}$ with $|S| \geq 2$ together with a pair $i, j \in S$, $i \neq j$. This corresponds to the function obtained from f by setting all but the ith and jth bits to agree with $\underline{e}_S$. Then the equation obtained is (writing T for $S \setminus \{i, j\}$)

$$\hat{f}(\underline{e}_S) = -\hat{f}(\underline{e}_T)\hat{f}(\underline{e}_{T \cup \{i\}})\hat{f}(\underline{e}_{T \cup \{j\}})$$

We shall prove the result by induction on the size of the set S. The base step consists of the cases in which $|S| \leq 1$. We may subdivide these cases into $S = \emptyset$ and $S = \{r\}$ for some $r \in \{1, .., n\}$. In the first case, we want to show that

$$\hat{f}(\underline{0}) = (-1)^0 (\hat{f}(\underline{0}))^1$$

which is clearly true, and in the second case that

$$\hat{f}(\underline{e}_{\{r\}}) = (-1)^0 (\hat{f}(\underline{0}))^2 \hat{f}(\underline{e}_{\{r\}})$$

which is also clearly true.

Now let us assume that the result is true for all T with $|T| \leq k$ ($k \geq 1$) and let $|S| = k + 1$. Then we obtain a set of equations

$$\hat{f}(\underline{e}_S) = -\hat{f}(\underline{e}_T)\hat{f}(\underline{e}_{T \cup \{i\}})\hat{f}(\underline{e}_{T \cup \{j\}})$$

for each distinct pair $i, j \in S$, where $T = S \setminus \{i, j\}$. Now since $|T| = k - 1$ and $|T \cup \{i\}| = |T \cup \{j\}| = k$, we may apply the induction hypothesis and obtain

$$\hat{f}(\underline{e}_T) = (-1)^{\frac{(k-1)(k-2)}{2}} (\hat{f}(\underline{0}))^k \prod_{r \in T} \hat{f}(\underline{e}_{\{r\}})$$

$$\hat{f}(\underline{e}_{T \cup \{i\}}) = (-1)^{\frac{k(k-1)}{2}} (\hat{f}(\underline{0}))^{(k+1)} \prod_{r \in T \cup \{i\}} \hat{f}(\underline{e}_{\{r\}})$$

$$\hat{f}(\underline{e}_{T \cup \{j\}}) = (-1)^{\frac{k(k-1)}{2}} (\hat{f}(\underline{0}))^{(k+1)} \prod_{r \in T \cup \{j\}} \hat{f}(\underline{e}_{\{r\}})$$

Now for each $r \in T$, the term $\hat{f}(\underline{e}_{\{r\}})$ will occur once in each of those expressions, and so exactly three times in the expression for $\hat{f}(\underline{e}_S)$, while the terms $\hat{f}(\underline{e}_{\{i\}})$ and $\hat{f}(\underline{e}_{\{j\}})$ occur exactly once each in the expression for $\hat{f}(\underline{e}_S)$. Furthermore, the terms $(-1)^{\frac{k(k-1)}{2}}$ and $(\hat{f}(\underline{0}))^{k+1}$ both occur twice in the expression and so cancel out, leaving

$$\hat{f}(\underline{e}_S) = (-1)^{1 + \frac{(k-1)(k-2)}{2}} (\hat{f}(\underline{0}))^k \prod_{r \in S} \hat{f}(\underline{e}_{\{r\}})$$

Now

$$1 + \frac{(k-1)(k-2)}{2} = \frac{k^2 - 3k + 4}{2} \equiv \frac{k(k+1)}{2} (\mathrm{mod}2)$$

and $k \equiv (k+2)(\mathrm{mod}2)$, so

$$\hat{f}(\underline{e}_S) = (-1)^{\frac{k(k+1)}{2}} (\hat{f}(\underline{0}))^{(k+2)} \prod_{r \in S} \hat{f}(\underline{e}_{\{r\}})$$

as required.

Example

Let us consider the simple case of $n = 3$. We want to discover which functions f satisfy the SAC of order $(n-2) = 1$. Then as at the beginning of the proof of the lemma, we must have

$$\hat{g}(1,1) = -\hat{g}(0,0)\hat{g}(0,1)\hat{g}(1,0)$$

for all g obtained from f by fixing one bit. Now there are six such functions g, giving us the following equations.

$$\hat{f}(0,1,1) = -\hat{f}(0,0,0)\hat{f}(0,0,1)\hat{f}(0,1,0)$$
$$\hat{f}(1,1,1) = -\hat{f}(1,0,0)\hat{f}(1,0,1)\hat{f}(1,1,0)$$
$$\hat{f}(1,0,1) = -\hat{f}(0,0,0)\hat{f}(0,0,1)\hat{f}(1,0,0)$$
$$\hat{f}(1,1,1) = -\hat{f}(0,1,0)\hat{f}(0,1,1)\hat{f}(1,1,0)$$
$$\hat{f}(1,1,0) = -\hat{f}(0,0,0)\hat{f}(0,1,0)\hat{f}(1,0,0)$$
$$\hat{f}(1,1,1) = -\hat{f}(0,0,1)\hat{f}(0,1,1)\hat{f}(1,0,1)$$

Rearranging these, we obtain

$$\hat{f}(0,1,1) = -\hat{f}(0,0,0)\hat{f}(0,0,1)\hat{f}(0,1,0)$$
$$\hat{f}(1,0,1) = -\hat{f}(0,0,0)\hat{f}(0,0,1)\hat{f}(1,0,0)$$
$$\hat{f}(1,1,0) = -\hat{f}(0,0,0)\hat{f}(0,1,0)\hat{f}(1,0,0)$$
$$\hat{f}(1,1,1) = -\hat{f}(0,0,1)\hat{f}(0,1,0)\hat{f}(1,0,0).$$

So we see that the 16 functions $f_1, .. f_{16} : \mathbf{Z}_2^3 \to \mathbf{Z}_2$ satisfying the SAC of order 1 are the following.

$\underline{x}$	000	001	010	100	011	101	110	111
$f_1(\underline{x})$	0	0	0	0	1	1	1	1
$f_2(\underline{x})$	0	0	0	1	1	0	0	0
$f_3(\underline{x})$	0	0	1	0	0	1	0	0
$f_4(\underline{x})$	0	0	1	1	0	0	1	1
$f_5(\underline{x})$	0	1	0	0	0	0	1	0
$f_6(\underline{x})$	0	1	0	1	0	1	0	1
$f_7(\underline{x})$	0	1	1	0	1	0	0	1
$f_8(\underline{x})$	0	1	1	1	1	1	1	0
$f_9(\underline{x})$	1	0	0	0	0	0	0	1
$f_{10}(\underline{x})$	1	0	0	1	0	1	1	0
$f_{11}(\underline{x})$	1	0	1	0	1	0	1	0
$f_{12}(\underline{x})$	1	0	1	1	1	1	0	1
$f_{13}(\underline{x})$	1	1	0	0	1	1	0	0
$f_{14}(\underline{x})$	1	1	0	1	1	0	1	1
$f_{15}(\underline{x})$	1	1	1	0	0	1	1	1
$f_{16}(\underline{x})$	1	1	1	1	0	0	0	0

Proof of Theorem 5.1

By Lemma 5.2, the set of functions $f : \mathbf{Z}_2^n \to \mathbf{Z}_2$ satisfying the SAC of order $(n-2)$ is the same as the set of functions $f : \mathbf{Z}_2^n \to \mathbf{Z}_2$ satisfying the set of equations

$$\hat{f}(\underline{e}_S) = (-1)^{\frac{|S|(|S|-1)}{2}} (\hat{f}(\underline{0}))^{(|S|+1)} \prod_{r \in S} \hat{f}(\underline{e}_{\{r\}})$$

where the notation is as in the statement of Lemma 5.2. Now, since any element in $\mathbf{Z}_2^n$ can be written as $\underline{e}_S$ for exactly one set $S \subseteq \{1, 2, .., n\}$, this determines the value of $g(\underline{x})$ for all values of $\underline{x}$ with Hamming weight greater than 1 in terms of the values of $g(\underline{x})$ for values of $\underline{x}$ with Hamming weight less than or equal to 1. In other words, if we choose values for $g(\underline{0})$ and for $g(\underline{e}_{\{r\}})$ for all $r \in \{1, .., n\}$, then g is completely determined on the whole of $\mathbf{Z}_2^n$. Thus there are 2^{n+1} ways to choose such a function, and so the size of the set of these functions is 2^{n+1}.

We have the following immediate corollary.

Corollary 5.3 The proportion of functions $f : \mathbf{Z}_2^n \to \mathbf{Z}_2$ satisfying the SAC of order $(n-2)$ is $2^{n+1}/2^{(2^n)}$.

VI . CONCLUSIONS

We have presented a simplified definition of the higher order strict avalanche criterion and showed its equivalence with the original. We have then used this to calculate the number of n-bit binary functions which satisfy the strict avalanche criterion of order $(n-2)$.

REFERENCES

[1] R. Forré, *The Strict Avalanche Criterion : Spectral Properties of Boolean Functions and an Extended Definition*, Abstracts CRYPTO88.

[2] J.B. Kam and G.I. Davida, *A structured design of substitution-permutation encryption networks*, IEEE Trans. on Computers, Vol. 28, No. 10 (1979).

[3] A.F. Webster and S.E. Tavares, *On the design of S-boxes*, Advances in Cryptology, Proceedings CRYPTO85, Springer Verlag, Heidelberg, 1986

A Key Distribution System Based On Any One-Way Function

(Extended Abstract)

George Davida   Yvo Desmedt   René Peralta

Dept. EE & CS
Univ. of Wisconsin – Milwaukee
P.O. Box 784
WI 53201 Milwaukee
U.S.A.

1   Introduction

Assuming the existence of one-way functions, we describe a simple protocol to exchange secret keys through an insecure (but authenticated) channel. If no pre-computation is allowed, our scheme uses $O(n)$ time for agreement on a number in the range $1..n^2$. An intruder takes time $O(n^2)$ to obtain the secret key. Thus, the number of steps necessary to cryptanalyze is the square of the number of steps in the protocol. If pre-computation is allowed to one of the parties in the key-exchange and also to the enemy, then this performance can be improved significantly. The assumptions necessary about the one-way-function are weaker than the assumptions in [Mer78] and in [DH76]. [1] The potential applications of our protocol also are more general than those of Merkle's protocol.

2   The assumptions

Let F_α be a family of bijections parametrized by α and with domain $\{1...K\}$. We suppose F_α is implemented by a specific circuit. In our protocol, players A and B will use F_α to exchange a secret key k over an open channel. Player E (the eavesdropper) will have access to the whole communication. Player E's goal is to compute k given A and B's communication. We make the following assumptions:

- $|\alpha| < \sqrt{K}$.

---

[1] The set of functions usable in Merkle's protocol is a subset of the set of one-way trap-door functions. The set of functions usable in our protocol is a superset of the set of one-way functions.

- the fastest algorithm to compute k given α and $F_\alpha(k)$ uses exhaustive search on a set of expected size $O(K)$.

- We assume the existence of an authenticated channel.

- We assume that E's technology is comparable to A and B's technology (E, however, may spend much more resources computing k than A and B).

Note that the first and second assumptions do not appear to imply that one-way functions exist.

3 The protocol

In the following protocol, players A and B will agree on a common secret key $k \in \{1...K\}$.

step 1 Player A chooses α at random.

step 2 Player A computes and stores $(r_i, F_\alpha(r_i))$ for n distinct randomly chosen $r_i \in \{1...K\}$.

step 3 Player A sends α to player B.

step 4 Player B chooses a random $k \in \{1...K\}$ and sends $F_\alpha(k)$ to player A.

step 5 Player A checks whether $k = r_j$ for some j by checking $F_\alpha(k)$ against the values computed at step 2. If this is the case, then A sends B a 1 (meaning k is the agreed-upon key). Otherwise A sends B a 0.

step 6 Steps 4-5 are repeated until an agreement is achieved.

Alternatively, B may send, at step 4, sufficiently many random $F_\alpha(k)$'s so that the probability of at least one $F_\alpha(k)$ being in A's table is high. In this case, A would tell B, at step 5, which $F_\alpha(k)$ was in its table.

4 Analysis

The running time of the protocol depends on n, the size of A's precomputed table. Assuming no memory or communication constraints for either player, the choice of n which minimizes the running time is $\sqrt{K}$. With this choice of n, the expected number of iterations of steps 4-5 is $\sqrt{K}$, since the experiment is a sequence of independent Bernoulli trials with $p = 1/\sqrt{K}$. The probability of no agreement after cn iterations of steps 4-5 is $(1 - 1/n)^{cn} \approx e^{-c}$. The eavesdropper will find the key, using exhaustive search, in (expected) $K/2$ steps. Therefore the resources needed to break the protocol are proportional to the square of the resources invested in the protocol.[2]

Note that the open channel is used *after* player A has computed a large table with known encryptions. If we assume that player A has more resources (e.g. time, memory, security) than player B, then we may choose n differently. For example, player A could compute a table of size $K^{3/4}$. Then, following the protocol, the open channel is used to agree on a key in expected number of messages equal to $O(K^{1/4})$. But the eavesdropper's time is still $O(K)$. Thus, the number of steps necessary to cryptanalyze would be the fourth power of the number of steps in the protocol (not counting precomputation by A).

In the next section we discuss the performance of our protocol under different constraints regarding channel capacity, time, and memory.

5   Choosing parameters under constraints

The optimal choice of parameters K and n depends on the rate R at which $F_\alpha(t)$ can be computed (in mappings per second). Since R can be varied (see, for example, [QDD86]), we will treat it as a parameter. Note that the expected number of messages in our protocol is K/n.

One constraint on the choice of parameters K, R and n is the maximum number C of $F_\alpha(r_i)$ messages which can be communicated. A second constraint is the maximum size M of A's table, in number of $(r_i, F_\alpha(r_i))$ pairs stored. A third constraint is given by a maximum time T, in seconds, allowed for the pre-computation of A's table. The expected time necessary to obtain the secret key by exhaustive search is $K/(2 \cdot R)$. Thus, we must choose K, n, and R such that K/R is maximized subject to the constraining inequalities

$$\frac{K}{n} \le C \quad ; \quad n \le M \quad ; \quad \frac{n}{R} \le T.$$

It is clear that one should choose $n = M$ and $K = C \cdot M$. This implies $K/R = C \cdot M/R$, and therefore R should be minimized subject to $n/R = M/R \le T$. Thus, the optimal value of R is M/T. The time to find the secret key by exhaustive search is $K/(2 \cdot R) = C \cdot T/2$ seconds.

In order to see how the protocol performs in practice, we substitute typical values for T, M, and C.

Let

$$M = 10^8 \quad ; \quad T = 10^5 \quad ; \quad C = 10^6.$$

Then the optimal R is $M/T = 10^3$. Thus, a function or chip should be chosen such that $F_\alpha(t)$ can be computed in one millisecond (*and no faster*). There will

---

[2]we have chosen to ignore the fact that it takes slightly longer than n trials to generate n distinct random elements from a set of size $K > n$.

be $K = C \cdot M = 10^{14}$ possible keys. Precomputation time will be $M/R = 10^5$ seconds. The time to obtain the secret key by exhaustive search is $K/(2 \cdot R) = 10^{11}/2$ seconds (approximately 1.5 thousand years). If we allow $C = 10^7$ messages communicated, then a secret key is obtained which takes 15 thousand years to find by exhaustive search.

Thus, we have shown that the protocol can be used in practice.

5.1 The effect on security of increased resources

It is useful to analyze the protocol's security when resources available to all parties are increased by a factor ω. Let

$$M' = \omega \cdot M \quad ; \quad T' = \omega \cdot T \quad ; \quad C' = \omega \cdot C$$

be the new memory, precomputation time, and communication constraints. Suppose also that the eavesdropper has ω chips for the computation of F_α. That is, the enemy can compute many $F_\alpha(t)$ at a time, even though he cannot compute a particular $F_\alpha(t)$ any faster than A and B can. The reader can verify that, under the new constraints, the rate R remains the same but the size of the key space in increased by ω^2. Thus, the eavesdropper's time to find the secret key is increased by ω, even though he can now search ω possible keys at a time.

6 Conclusions and future research

There is a generalized feeling in the cryptographic community that modern cryptography strongly depends on assumptions about the asymptotic complexity of certain functions and their inverses. In particular, the fact that there is an odd possibility that $P = NP$ seems to make cryptographers very nervous. The usual definition of protocol security goes somewhat like this

- a protocol is secure if, when the legitimate parties use resources in an amount N, then the resources necessary to break the protocol is an exponential function of N.

We have exhibited a protocol which can be broken by an amount of resources which is only the square of the resources used by the protocol itself. The fact that this protocol seems secure suggests that the above definition of security may be stronger than necessary.

We would like to point out the following problems suggested by our research:

- Our protocol takes n steps to agree on a secret key which can be found by the eavesdropper in $O(n^2)$ steps. Is there a protocol which achieves security

$O(n^3)$ or $O(n^4)$, under the same assumptions? [3]

- Can the assumptions be weakened? In particular, can we trade keys without one-way functions?

- Given a chip which computes a mapping F at a rate R of mappings per second, it is possible to define a function which (apparently) can only be computed at a much slower rate. For example, we could simply define the function $G(t) = F^{(i)}(t)$, where i is an integer and $F^{(i)}$ is the composition of F with itself i times. Can we guarantee that G is not computable at a faster rate than R/i? If not, is there a provably secure way to decrease the rate R?

REFERENCES

[DH76] W. Diffie and M. E. Hellman. New directions in cryptography. *IEEE Trans. Inform. Theory*, IT–22(6):644–654, November 1976.

[IR89] R. Impagliazzo and S. Rudich. Limits on the provable consequences of one-way permutations. *Proceedings of the 21th Annual ACM Symposium on the Theory of Computing*, pages 44–61, 1989.

[Mer78] Ralph Merkle. Secure communications over insecure channels. *Communications of the ACM*, 21(4):294 – 299, 1978.

[QDD86] J.-J. Quisquater, Y. Desmedt, and M. Davio. The importance of 'good' key scheduling schemes (how to make a secure DES scheme with $\leq$ 48 bit keys?). In Hugh C. Williams, editor, *Advances in Cryptology. Proc. of Crypto'85 (Lecture Notes in Computer Science 218)*, pages 537–542. Springer–Verlag, 1986. Santa Barbara, California, U.S.A., August 18–20.

---

[3]This, however, may prove to be a very hard problem, as suggested by recent research by Impagliazzo and Rudich [IR89]

Non-linearity of Exponent Permutations*

Josef P. Pieprzyk
Department of Computer Science
University College
University of New South Wales
ADFA
Canberra, ACT 2600, AUSTRALIA

Abstract

The paper deals with an examination of exponent permutations with respect to their non-linearity. The first part gives the necessary background to be able to determine permutation non-linearity. The second examines the interrelation between non-linearity and Walsh transform. The next part summarizes results gathered while experimenting with different binary fields. In the last part of the work, we discuss the results obtained and questions which are still open.

---

*Support for this research was provided in part by the Australian Research Council under reference number A48830241

1 Introduction

In [PF88] the authors analysed the non-linearity of permutations as one of the indicators of their quality for cryptographic use. They have given the upper boundary on non-linearities. Of course an application of permutations of the maximum non-linearity does not guarantee that an encryption algorithm based on them generates a "strong" cipher. For example the well-known DES algorithm is built using 32 permutations (each S-box consists of four permutations) and none of them attains the maximum non-linearity. Of course the selection of permutations in the DES has been made using a collection of properties (some of which may still remain unidentified). For the most complete list of such properties see [Bro88].

It has been shown in [PF88] that there are permutations whose non-linearity attains the upper bound for Galois fields $GF(2^3)$, $GF(2^4)$, and $GF(2^5)$. Unfortunately, it is not known if such permutations exist for larger binary fields (for $n > 5$). Experiments we have done point out that such permutations exist and can be easily obtained using exponentiation.

2 Background

Consider a Boolean function $f \in \mathcal{F}_n$ where $\mathcal{F}_n$ is the set of all Boolean functions of n variables. Its non-linearity N_f is defined as the Hamming distance between the function f and the set of all linear functions $\mathcal{L}_n$ existing in $\mathcal{F}_n$ i.e.

$$N_f = d(f, \mathcal{L}_n) = \min_{\alpha \in \mathcal{L}_n} d(f, \alpha) \tag{1}$$

For a given permutation $\mathbf{f} \in \mathcal{P}_n$, where $\mathcal{P}_n$ is the set of all permutations over $GF(2^n)$, we define its non-linearity as

$$N_{\mathbf{f}} = \min_i (N_{f_i}, N_{f_i^{-1}}) \tag{2}$$

where $\mathbf{f} = (f_1, f_2, ..., f_n)$ and $\mathbf{f}^{-1} = (f_1^{-1}, f_2^{-1}, ..., f_n^{-1})$ are coordinates of the original and the inverse permutation, respectively.

In [PF88], it has been shown that there is a bound on non-linearity which can be attained by permutations for a given dimension n. The dimension n also can be seen as the number of binary inputs of a permutation block. The bound N_n is expressable by the following formula

$$N_n = \begin{cases} \sum_{i=1}^{n-3}{}_{/2(n-3)} \, 2^{i+1} & \text{if } n = 3, 5, 7, ... \\ \sum_{i=1}^{n-4}{}_{/2(n-4)} \, 2^{i+2} & \text{if } n = 4, 6, 8, ... \end{cases} \tag{3}$$

We note that all permutations in $\mathcal{P}_2$ are linear so their non-linearities are equal to zero.

When designing cryptographic algorithms, one looks for those permutations which are both easy to implement and demonstrate a satisfactorily high non-linearity. Although the definition given in (2) is good to characterize non-linearity, we can also determine non-linearity of a permutation $\mathbf{f} \in \mathcal{P}_n$ as follows:

$$\mathcal{N}_{\mathbf{f}} = \frac{\sum_{i=1}^{n} N_{f_i} + \sum_{i=1}^{n} N_{f_i^{-1}}}{2n} \tag{4}$$

For example (see [PF88]) all permutations applied in the DES have the same non-linearity measured by the formula (2) and equal to 2 which is a half of the possible maximum. If we take the definition (4), then they have different non-linearities which vary from 3.73 (there are 6 of them; one in S_1 and S_8 and two in S_2 and S_6) to 3.00 (there are 4 of them; single permutations in S_2, S_6, S_7, and S_8). The average non-linearity of permutations in the DES is equal to 3.408 and is close to the maximum that is 4. For details see [PF88].

In fact in DES, all the permutations are used in their original form (their inverses are never applied during either enciphering and deciphering process). So if we count non-linearities of coordinates of the original permutations only, we find only 14 out of 32 that reach the maximum non-linearity.

The non-linearity of permutations given in (2) can be calculated differently as we can apply the following theorem.

Theorem 2.1 *Given a permutation* $\mathbf{f}$, *its non-linearity can be calculated as*

$$
\begin{aligned}
N_f &= \min_{\alpha \in \mathcal{L}_n^n} \min_{i=1,\ldots,n} N_{(\mathbf{f} \star \alpha)_i} \\
&= \min_{\alpha \in \mathcal{L}_n^n} \min_{i=1,\ldots,n} N_{(\mathbf{f}^{-1} \star \alpha)_i}
\end{aligned}
\tag{5}
$$

where $\mathcal{L}_n^n$ *is a set of all linear permutations, and* $(\mathbf{f} \star \alpha)_i$ *stands for i-th coordinate of the composite permutation* $\mathbf{f} \star \alpha$.

Proof: Clearly the set of all coordinates of all linear permutations creates the set $\mathcal{L}_n - \{0, 1\}$. For our permutation $\mathbf{f}$, we can define the following set

$$
\mathcal{L}_{\mathbf{f}} = \{g \in \mathcal{F}_n; g = a_0 \oplus a_1 f_1 \oplus \ldots \oplus a_n f_n\}
\tag{6}
$$

where a_i are binary elements and the set

$$
\mathcal{L}_{\mathbf{f}^{-1}} = \{g \in \mathcal{F}_n; g = a_0 \oplus a_1 f_1^{-1} \oplus \ldots \oplus a_n f_n^{-1}\}
\tag{7}
$$

The both sets consist of all Boolean functions which are coordinates of compositions $\mathbf{f} \star \alpha$ and $\mathbf{f}^{-1} \star \alpha$, respectively. Notice that as α is a linear permutation, non-linearities of $\mathbf{f}$ and $\mathbf{f} \star \alpha$ coordinates are the same.

Observe that the relation between two sets $(\mathcal{L}_n, \mathcal{L}_{\mathbf{f}})$ and $(\mathcal{L}_n, \mathcal{L}_{\mathbf{f}^{-1}})$ is symmetric. This is to say that if we create a composition of $\mathbf{f} \star \mathbf{f}^{-1}$ the set $\mathcal{L}_{\mathbf{f}}$ plays the same role to the inverse as the set $\mathcal{L}_n$ to the original. If we take the second possibility $\mathbf{f}^{-1} \star \mathbf{f}$, we can draw the same conclusion for sets $\mathcal{L}_{\mathbf{f}^{-1}}$ and $\mathcal{L}_n$. In other words, the non-linearity of a permutation can be expressed as a distance between two sets $\mathcal{L}_n$ and $\mathcal{L}_{\mathbf{f}}$ or equivalently between $\mathcal{L}_{\mathbf{f}^{-1}}$ and $\mathcal{L}_n$.

□

We can conclude from the theorem that if there is at least one permutation coordinate that may be expressed by linear combination of the rest of the coordinates, this permutation has "0" non-linearity. It means that the inverse has at least one linear coordinate.

It has been shown [PF88] that it is easy to generate permutations of the maximum non-linearity at random for $GF(2^3)$ and $GF(2^4)$. Unfortunately, the generation of such permutations becomes more and more difficult as the dimension of Galois field grows.

Ideally, we would like to have a method that would generate permutations of the maximum non-linearity or at least close enough to it. We are going to examine an exponential function and its application to the generation of non-linear permutations for different dimensions n of a Galois field.

3   Walsh transform and non-linearity

Spectral tests are useful tools for detecting non-randomness in binary strings. Originally the first such test was proposed by Gait [Gai77] who used the discrete Fourier transform to examine binary strings generated by the DES. At the same time Yuen [Yue77] suggested another test based on the Walsh transform which was later improved by Feldman [Fel87]. As Forre [For88] has shown the Walsh transform can also be used to estimate the strict avalanche criterion of Boolean functions.

Before we show the relation between the Walsh transform and non-linearities of Boolean functions, we give the necessary definitions and notions. Assume we have a Boolean function $f \in \mathcal{F}_n$ of n variables $\vec{x}=(x_1,...,x_n)$. As the Walsh transform may be applied to real-valued functions only, we treat $f(\vec{x})$ as such a function. It is taken as 0 if it is false and 1 if it is true. The Walsh transform $F(\vec{\omega})$ of $f(\vec{x})$ is defined as (see [Bea75]):

$$F(\vec{\omega}) = \sum_{\vec{x} \in \mathcal{Z}_2^n} f(\vec{x})(-1)^{\vec{\omega} \cdot \vec{x}} \tag{8}$$

where $\mathcal{Z}_2^n$ is a space of all binary sequences of length n, $f(\vec{x})$ is the transformed function (considered as a real-valued function), and $\vec{\omega} \cdot \vec{x}$ stands for the dot-product of $\vec{\omega}$ and $\vec{x}$:

$$\vec{\omega} \cdot \vec{x} = \omega_1 x_1 \oplus \omega_2 x_2 \oplus \cdots \oplus \omega_n x_n. \tag{9}$$

Having $F(\vec{\omega})$, we can recreate the function $f(\vec{x})$ using the inverse Walsh transform that is:

$$f(\vec{x}) = 2^{-n} \sum_{\vec{\omega} \in \mathcal{Z}_2^n} F(\vec{\omega})(-1)^{\vec{\omega} \cdot \vec{x}}. \tag{10}$$

First notice that the Walsh spectrum of a linear Boolean function has a specific form which is described by the following theorem.

Theorem 3.1 *The Walsh transform $F(\vec{\omega})$ of a linear Boolean function has two non-zero components only.*

Proof: Clearly, the first non-zero component is

$$F(0) = \sum_{\vec{x} \in \mathcal{Z}_2^n} f(\vec{x}) \tag{11}$$

as $\vec{\omega} \cdot \vec{x}=0$ for $\vec{\omega}=0$ and therefore $(-1)^{\vec{\omega} \cdot \vec{x}}=1$. In other words $F(0)$ gives the number of 1's (the number of TRUE values) in the function f.

Notice that for a fixed $\vec{\omega}$, its dot product $\vec{\omega} \cdot \vec{x}$ indicates the linear function

$$l_\omega(\vec{x}) = \omega_1 x_1 \oplus ... \oplus \omega_n x_n.$$

Clearly, $\vec{\omega}$ generates a half of all linear functions of n Boolean variables (recall that we assign value "0" to the FALSE and value "1" to the TRUE). The rest may be obtained using negation

$$\overline{l_\omega(\vec{x})} = \omega_1 x_1 \oplus ... \oplus \omega_n x_n \oplus 1.$$

So if we have a linear function f, it is either of the form

$$\omega'_1 x_1 \oplus ... \oplus \omega'_n x_n$$

or

$$\omega'_1 x_1 \oplus \ldots \oplus \omega'_n x_n \oplus 1$$

for a suitable sequence of $\vec{\omega}' = (\omega'_1, \ldots, \omega'_n)$. There are three possible cases:

- f is different from a linear function generated by the dot product given by $\vec{\omega}''$. In that case the Hamming distance between f and the linear function $l_{\omega''}(\vec{x})$ is equal to $2^n - 1$. It means that

$$F(\vec{\omega}'') = \sum_{\vec{x} \in Z_2^n} f(\vec{x})(-1)^{\vec{\omega}'' \cdot \vec{x}} = 0;$$

- f is equal to a linear function $l_{\omega''}(\vec{x})$ generated by the dot product then the Hamming distance between these two is equal to 0. It implies that $f=1$ while $\vec{\omega}'' \cdot \vec{x} = 1$ so $F(\vec{\omega}'') = -2^{n-1}$;

- f is a complement of $l_{\omega''}(\vec{x})$ generated by the dot product, then $d(f, l) = 2^n$ and $f = 1$ while $l_{\omega''}(\vec{x}) = 0$ so $F(\vec{\omega}'') = +2^{n-1}$;

One of the last two gives the second non-zero component of the Walsh spectrum.

□

The next theorem explains the interrelation between non-linearity of a Boolean function and its Walsh spectrum.

Theorem 3.2 *If a given Boolean function $f(x_1, \ldots, x_n)$ has its non-linearity equal to η, then*

$$\eta = \min(\min_{\vec{\omega} \neq 0}(2^{n-1} - |F(\vec{\omega})|), 2^n - |F(0)|, |F(0)|). \tag{12}$$

Proof: As we know the dot product generates a half of all linear functions. If the non-linearity of f is equal to η, then there is a linear function α such that

$$\eta = d(f, \alpha). \tag{13}$$

Therefore we consider two cases.

- Case I - the closest linear function α can be expressed by an element of the dot-product set, that is

$$\alpha(\vec{x}) = \vec{\omega}' \cdot \vec{x}. \tag{14}$$

So

$$\begin{aligned}
F(\vec{\omega}') &= \sum_{\vec{x} \in Z_2^n} f(\vec{x})(-1)^{\vec{\omega}' \cdot \vec{x}} \\
&= \sum_{\vec{x} \in Z_2^n} \alpha(\vec{x})(-1)^{\vec{\omega}' \cdot \vec{x}} \\
&\quad - \sum_{\vec{x}; f(\vec{x}) \neq \alpha(\vec{x}); \vec{\omega}' \cdot \vec{x} = 1} (-1)^1 + \sum_{\vec{x}; f(\vec{x}) \neq \alpha(\vec{x}); \vec{\omega}' \cdot \vec{x} = 0} (-1)^0.
\end{aligned} \tag{15}$$

As the first part of the formula

$$\sum_{\vec{x} \in Z_2^n} \alpha(\vec{x})(-1)^{\vec{\omega}' \cdot \vec{x}} = -2^{n-1} \tag{16}$$

and the rest is equal to η, we get

$$F(\vec{\omega}') = -2^{n-1} + \eta. \tag{17}$$

- case II - the closest linear function can be expressed by a negation of an element of the dot-product set. It means that

$$\eta = d(f, \overline{\alpha}),$$ (18)

and

$$
\begin{aligned}
F(\vec{\omega}') &= \sum_{\vec{x} \in \mathcal{Z}_2^n} f(\vec{x})(-1)^{\vec{\omega}' \cdot \vec{x}} \\
&= \sum_{\vec{x} \in \mathcal{Z}_2^n} \overline{\alpha(\vec{x})}(-1)^{\vec{\omega}' \cdot \vec{x}} \\
&+ \sum_{\vec{x}; f(\vec{x}) \neq \alpha(\vec{x}); \vec{\omega}' \cdot \vec{x}=1} (-1)^1 - \sum_{\vec{x}; f(\vec{x}) \neq \alpha(\vec{x}); \vec{\omega}' \cdot \vec{x}=0} (-1)^0.
\end{aligned}
$$ (19)

The first part of the formula is given by

$$\sum_{\vec{x} \in \mathcal{Z}_2^n} \overline{\alpha(\vec{x})}(-1)^{\vec{\omega}' \cdot \vec{x}} = 2^{n-1}$$ (20)

and the rest is equal to η. So, we get

$$F(\vec{\omega}') = 2^{n-1} - \eta.$$ (21)

Considering both cases, we can write that

$$\eta = 2^{n-1} - |F(\vec{\omega}')|.$$ (22)

Finally, if we consider the case for $\vec{\omega}=0$, we obtain the result.
□

Theorem 3.1 says that any linear Boolean function is easily identified by looking at its Walsh spectrum. In general any linear permutation over $GF(2^n)$ has a specific Walsh spectrum. Before we present the relation between linear permutations and their Walsh spectra, let us consider the following theorem which expresses the dependence between Walsh spectra of permutations and their coordinates.

Theorem 3.3 *Given a permutation* $f \in \mathcal{P}_n$. *If* $F_i(\vec{\omega})$ *is the Walsh transform of i-th coordinate of the permutation (i=1,...,n), then the Walsh transform of* f *is expressable as*

$$F(\vec{\omega}) = \sum_{i=1}^n 2^{i-1} F_i(\omega)$$ (23)

where $f = (f_1, ..., f_n)$ *and* $F_i \leftrightarrow f_i$.

<u>Proof:</u> For any value of $\vec{x}$, $f(\vec{x}) = f_1 + 2f_2(\vec{x}) + ... + 2^{n-1} f_n(\vec{x})$. Using the definition of the Walsh transform (8), we get the final result.
□

The above theorem gives us, the following conclusion.

Collorary 3.1 *If a permutation* $\mathbf{f} \in \mathcal{P}_n$ *is linear, then its Walsh transform has* $n-1$ *non-zero components. The first component is given by*

$$F(0) = 2^{n-1} \sum_{i=1}^{n} 2^{i-1} \tag{24}$$

and the other n *components create a permutation of* n *integers* $(\pm 2^{n-1}, \pm 2^n, ..., \pm 2^{2(n-1)})$

Example. Consider the identity permutation over $GF(2^3)$. The permutation $\mathbf{f} = (f_1, f_2, f_3)$ is shown in the table below. Its Walsh spectrum is given in the same table where F_1, F_2, F_3 are Walsh spectra of the Boolean functions f_1, f_2, f_3, respectively, and F is the spectrum of the permutation.

$\bar{x}/\bar{\omega}$	f_1	f_2	f_3	f	F_1	F_2	F_3	F
000 (0)	0	0	0	0	4	4	4	28
001 (1)	0	0	1	1	-4	0	0	-4
010 (2)	0	1	0	2	0	-4	0	-8
011 (3)	0	1	1	3	0	0	0	0
100 (4)	1	0	0	4	0	0	-4	-16
101 (5)	1	0	1	5	0	0	0	0
110 (6)	1	1	0	6	0	0	0	0
111 (7)	1	1	1	7	0	0	0	0

4 Exponent permutations

In [PF88] it has been shown that there are permutations of the maximum non-linearity for n=3,4,5 and 6. For $n \geq 7$, it is very hard to find such permutations. It would be interesting if we could find such permutations using a simple method. Experiments have shown exponentiation generates permutations whose non-linearities tend to be close to the maximum.

In general, we have dealt with permutations of the form:

$$\mathbf{f}(x) = (h(x))^k \bmod g(x) \tag{25}$$

where $g(x)$ is an irreducible polynomial of degree n which generates a Galois field $GF(2^n)$ and $h(x)$ represents elements of $GF(2^n)$.

Example. Consider $GF(2^3)$ generated by $g(x)=x^3 + x^2 + 1$ and two permutations: the first

$$\mathbf{f}_1(x) = (h(x))^2 (\bmod g(x)) \tag{26}$$

and the second

$$\mathbf{f}_2(x) = (h(x))^3 (\bmod g(x)) \tag{27}$$

The permutations along with their Walsh transforms are presented in the tables below:

$\bar{x}/\bar{\omega}$	f	F_1	F_2	F_3	F
0	0	4	4	4	28
1	1	-4	0	0	-4
2	4	0	0	0	0
3	5	0	0	0	0
4	7	0	-4	0	-8
5	6	0	0	0	0
6	3	0	0	-4	-16
7	2	0	0	0	0

and

$\bar{x}/\bar{\omega}$	f	F_1	F_2	F_3	F
0	0	4	4	4	28
1	1	2	0	0	2
2	5	0	-2	0	-4
3	2	-2	-2	0	-6
4	6	0	-2	-2	-12
5	4	-2	2	-2	-6
6	7	0	0	-2	-8
7	3	-2	0	2	6

The first permutation is linear. The second, however, attains the maximum non-linearity for all its coordinates. The Walsh transforms F of the whole permutation can be easily computed from Walsh transforms of its coordinates (F_1, F_2, F_3) according to $F(\bar{\omega}) = F_1 + 2F_2 + 4F_3$.

The first permutation in the example illustrates the general property of exponent permutation that can be expressed as

Collorary 4.1 *Any permutation*

$$(h(x))^k \, mod \, g(x) \tag{28}$$

where $g(x)$ is a generator of $GF(2^n)$, is a linear permutation for $k=1,2,4,...,2^{n-1}$

The collorary results from the well-known fact that squaring in any binary field $GF(2^n)$ is a linear operation - for details see [Ber68].

The second permutation has a very regular pattern of integers in its Walsh transforms. It is no coincidence. In other words (see [PF88]), the space of all linear functions $\mathcal{L}_n$ may be divided into two sets with respect to any coordinate. The first subset consists of all linear functions whose Hamming distances to the given coordinate are equal 2^{n-1} (corresponding Walsh transforms are equal 0). The second subset comprises all linear functions whose distances to the given coordinate are either $2^{n-1} - 2^{(n-1)/2}$ or $2^{n-1} + 2^{(n-1)/2}$. It means that the corresponding Walsh components are equal $\pm 2^{(n-1)/2}$ and therefore their non-linearities are equal to $2^{(n-1)} - 2^{(n-1)/2}$. This observation is valid for n odd and greater than 1.

Concatenation of exponential blocks gives the identity permutation if and only if the product of all exponents is equal to 1 modulo $(2^n - 1)$. Consider squaring. Exponents after using subsequent squaring are as follows:

$$2^2, \, 2^3, \, ... \, 2^n \tag{29}$$

Note that the last element $2^n = 1 \, mod \, (2^n - 1)$ and it corresponds to concatenation of n blocks. In general, if we deal with an exponential permutation described by an exponent 2^i where $i = 2, 3, ..., n - 1$, then we may construct the following sequence of exponents:

$$2^i, \, 2^{2i}, \, ... \, , 2^{ni} = 1 \, (mod \, 2^n - 1) \tag{30}$$

This leads us to the next conclusion (see [Ber68]).

Collorary 4.2 *If n is prime, then all linear exponents $(2, 4, 8, 16, ...)$ have the same minimal length of concatenation n after which we have got the identity permutation.*

Collorary 4.3 *If n is not prime, then for any linear permutation described by its exponent 2^i, we can find such an integer k ($k < n$) that $2^{ki} = 2^n$, where k and i are factors of n. It means that we can use a smaller number of exponential blocks to create the identity permutation.*

Consider an exponent permutation $\mathbf{f}(\alpha) = \alpha^a$ in $GF(2^n)$ given by its exponent a. Any concatenation of the same $\phi(2^n - 1)$ exponential blocks (permutations) gives the identity permutation, where $\phi(N)$ is the Euler ϕ-function or the Euler totient function - see [Ber68]. So all exponents can be divided into several disjoined classes depending upon their orders modulo $2^n - 1$ that have to be factors of $\phi(2^n - 1)$.

If $2^n - 1$ is a prime, then $\phi(2^n - 1) = 2^n - 2$ and there is always an exponent e that generates, by exponentiation, all other nonzero exponents. It is called a primitive element of the cyclic group CG created by all nonzero elements of $GF(2^n)$ and with multiplication as the group operation. In other words the sequence of exponents

$$e, \, e^2, \, e^3, \, ... \, e^{2^n - 2} \tag{31}$$

generates all possible exponential permutations. Thus we have got the following conclusion.

Collorary 4.4 *If $2^n - 1$ is a prime, then concatenation of $\frac{2^n - 2}{n}$ exponential blocks defined by a primitive element e of CG generates a linear permutation.*

Take an exponent e of order m modulo $2^n - 1$ (e has order m modulo $2^n - 1$ if $e^m = 1 \, mod \, 2^n - 1$) and m is different from 2, then we can create concatenation of two blocks: the first expressed by e, and the second by 2^i; $i = 1, ..., n - 1$ and its order is $n \cdot m$ but its non-linearity stays the same.

We have experimented extensively, starting the search with $n=3$. The set of all exponent permutations is modest and it splits into two subsets. Elements of $\{1,2,4\}$ give linear permutations. The set $\{3,5,6\}$ consists of all non-linear permutations and corresponding permutations share the same, maximum non-linearity which is equal to 2. In fact, we use the only independent non-linear exponent $e = 3$. Exponent 5 is the inverse of 3 so $5=3^{-1}$, but $6=6^{-1}=2 \cdot 3$. Therefore all three exponent must have the same non-linearity.

In space $GF(2^4)$, we can identify the set of linear exponents $\{1,2,4,8\}$ and the set of non-linear ones $\{7,11,13,14\}$. The rest of exponents do not give permutations - they are factors of

$2^4 - 1 = 15$ and they do not have their inverses. Non-linear exponent permutations have the same maximum non-linearity equal to 4.

For $n=5$, $2^5 - 1$ is a prime and $\phi(31)=30$, so there are four basic subsets of exponents (elements of CG). Each basic set contains elements of the same order modulo $2^n - 1$. The basic sets are as follows:

$$
\begin{aligned}
Z_1 &= \{1\} \\
Z_2 &= \{30\} \\
Z_3 &= \{5,25\} \\
Z_5 &= \{2,4,8,16\}
\end{aligned}
$$

and they correspond to the factorization of the integer $\phi(31)$. Sets Z_5 and Z_1 contain all linear exponents. Z_2 has the only element which gives a permutation of non-linearity 10. The set Z_3 consists of two permutations of the maximum non-linearity (equal to 12). The rest of exponents (permutations) can be seen as a product of exponents from the basic sets Z_2, Z_3, and Z_5. If we create them using Z_2 and Z_3, we have $Z_6 = Z_2 \times Z_3 = \{14,26\}$ - both elements have non-linearity 12, where $Z_2 \times Z_3$ means the set whose elements are of the form $e = e' \times e"$; $e' \in Z_2$, $e" \in Z_3$. Finally, we obtain

$$
\begin{aligned}
Z_{10} = Z_2 \times Z_5 &= \{15,23,27,29\} \\
Z_{15} = Z_3 \times Z_5 &= \{7,9,10,14,18,19,20,28\} \\
Z_{30} = Z_2 \times Z_3 \times Z_5 &= \{3,11,12,13,17,21,22,24\}
\end{aligned}
$$

Non-linearities are 10,12,12 for Z_{10}, Z_{15}, Z_{30}, respectively.

The case of $n=6$ is especially interesting. As $2^6 - 1 = 63$ and $\phi(63)=36=6 \cdot 6$, all orders of exponents must be factors of 6. Therefore we get

$$
\begin{aligned}
Z_1 &= \{1\} \\
Z_2 &= \{8,55,62\} \\
Z_3 &= \{4,16,22,25,37,43,46,58\} \\
Z_6 &= \{2,5,10,11,13,17,19,20,23,26,29,31, \\
&\qquad 32,34,38,40,41,44,47,50,52,53,59,61\}
\end{aligned}
$$

The set Z_2 consists of one linear exponent and two of the maximum non-linearity (equal to 24). All non-linear elements of Z_3 share the same non-linearity 20. The set Z_6 has a mixture of exponents of different non-linearities (0, 20, 24).

Consider $GF(2^7)$. In this field $2^7 - 1 = 127$ is prime and $\phi(127)=126=2 \cdot 3^2 \cdot 7$. The set of all exponents splits up into following subsets:

$$
\begin{aligned}
Z_1 &= \{1\} - non-linearity\ 0 \\
Z_2 &= \{126\} - non-linearity\ 54 \\
Z_3 &= \{19,107\} - non-linearity\ 44 \\
Z_7 &= \{2,4,8,16,32,64\} - non-linearity\ 0 \\
Z_9 &= \{22,37,52,68,99,103\} - non-linearities\ 44\ and\ 56 \\
Z_6 &= Z_2 \times Z_3 - non-linearity\ 56
\end{aligned}
$$

$$Z_{14} = Z_2 \times Z_7 - non - linearity\ 54$$
$$Z_{18} = Z_2 \times Z_9 - non - linearities\ 44\ and\ 56$$
$$Z_{21} = Z_3 \times Z_7 - non - linearity\ 44$$
$$Z_{42} = Z_6 \times Z_7 - non - linearity\ 56$$
$$Z_{63} = Z_9 \times Z_7 - non - linearities\ 44\ and\ 56$$
$$Z_{126} = Z_2 \times Z_{63} - non - linearities\ 44\ and\ 56$$

For example exponents 3 and 7 belong to Z_{126} (they are primitive elements of CG) but their non-linearities are different. The exponent $e=3$ yields the permutation of the maximum non-linearity (which is 56) while $e=7$ produces the permutation of non-linearity 44.

Our experiments, let us draw the following conclusions:

- non-linearity of exponent permutations does not depend on the primitive polynomial that generates $GF(2^n)$,

- non-linearity of exponent permutations does depend upon the internal structure of CG,

- non-linearity of the permutation for a given exponent that is a primitive element of CG does not necessarily attain the maximum,

- concatenation of $\frac{\phi(2^n-1)}{n}$ the same exponent permutations generates a linear permutation.

5 Conclusions

Boolean functions can be characterized by their non-linearities. We have assumed a definition of non-linearity as the minimum distance between a given function and the set of all linear Boolean functions. Non-linearity of a function can be determined by looking at its Walsh transform. In general, the Walsh transform of a Boolean function can be seen as a projection of the function into the vector basis created by linear functions. Functions of the maximum non-linearity have a specific pattern of their Walsh spectra. It means that they are pretty rare in the space of all Boolean functions of a given dimension n. The probability of choosing such a function at random diminishes as the space dimension grows.

Walsh transforms can be applied to generate Boolean permutations of the maximum non-linearity. Clearly, a permutation is a collection of n Boolean functions (coordinates). Any such a function has its Walsh spectrum which consists of $2^{n-1} - 1$ zeros, 2^{n-1} elements are equal $\pm(2^{n-1} - \eta)$ where η is the maximum non-linearity, and the last element $F(0)$ is always equal to 2^{n-1}. Among themselves, Walsh spectra of coordinates have to fulfil the same conditions as Boolean functions do. So, for any pair of coordinates, a pair of suitable nonzero Walsh spectrum elements $F(\vec{\omega})$ ($\vec{\omega} \neq 0$) must overlap for 2^{n-2} elements (signs do not matter).

Exponentiation is an convenient way to produce permutations of the maximum or close to maximum non-linearity. The production of non-linear Boolean function or non-linear permutations in $GF(2^n)$ is required in many applications for example while designing cryptographic algorithms, pseudorandom generators, etc. Squaring is always a linear operation in $GF(2^n)$ but cubing provides a permutation of the maximum non-linearity for $n=3,5,7,9$, note that for $n=4,6,8,\ldots$, it does not generate a permutation as 3 divides $2^n - 1$. It seems that any generator of a multiplicative group CG should share the same non-linearity as each generator produces by concatenation all possible exponent permutations. We have found that this statement is not true in general. There are, however, many still open questions about exponent permutations and their non-linearities. Some of them are listed below.

1. What is the relation between the value of an exponent and its permutation non-linearity ?

2. What is the dependence between the order modulo $2^n - 1$ of an exponent and its permutation non-linearity ?

3. What is the non-linearity spectrum of all exponents which produce permutations ?

4. What is the non-linearity spectrum of primitive generators of CG ?

5. Do permutations for $e = 3$ attain the maximum for all $GF(2^n)$, where n is odd ?

6. Does non-linearity of exponent permutations depend upon a field generator for larger n or not ?

There is a common consensus that non-linearity is a desirable cryptographic feature. However, in practice, while choosing non-linear permutations for a DES-like cryptographic algorithm, we face the question of whether permutations of the maximum non-linearity are "good" from a cryptographic point of view or permutations of the average non-linearity are better. Although exponent permutations are sucessfully being used in both the Rivest-Shamir-Adleman and the Diffie-Hellman cryptosystems and exponentiation itself is difficult to invert if you are dealing with large enough instance (large enough $GF(2^n)$), it is difficult to say if exponent permutations for small parameters n could generate a "strong" enciphering algorithm.

Considering the fact that concatenation of any $\frac{\phi(2^n - 1)}{n}$ exponent permutations generates a linear one, we notice that a well known iteration attack (see [SN77] or [SP88]) on exponential cryptosystems is more efficient in $GF(2^n)$. The iteration attack has always worked well if there is a small number of concatenations of exponent permutations (the exponent permutation is given by a public key) that give the identity permutation. In $GF(2^n)$, however, it is sufficient to create a concatenation of exponent permutations which produces a linear permutation. As the set of all linear permutations is closed according to the inversion operation, we can easily find the inverse linear permutation and generate the identity permutation.

ACKNOWLEDGMENT

I would like to thank Professor Jennifer Seberry for her critical comments and presenting this paper on my behalf during EUROCRYPT'89. It is also a pleasure to thank Sophie Richards and Patrick Tang for their assistance during preparation of this work.

References

[Bea75] K. G. Beauchamp. *Walsh Functions and Their Applications.* Academic Press, New York, 1975.

[Ber68] E. R. Berlekamp. *Algebraic Coding Theory.* MacGraw-Hill Book Company, New York, 1968.

[Bro88]   L. Brown. A proposed design for an extended DES. IFIP TC11, International Confer-
          ence on Computer Security : IFIP/SEC'88, W.J. Caelli (ed), Elsevier (to appear), May
          1988. Abstracts, Gold Coast, Queensland, Australia.

[Fel87]   F. A. Feldman. Fast spectral tests for measuring nonrandomness and the DES. Papers
          and Abstracts, CRYPTO'87, 1987.

[For88]   R. Forre. The strict avalanche criterion: spectral properties of Boolean functions and
          an extended definition. Papers and Abstracts, CRYPTO'88, August 1988.

[Gai77]   J. Gait. A new nonlinear pseudorandom number generator. *IEEE Transactions on
          Software Engineering*, SE-3(5):359–363, September 1977.

[PF88]   J. P. Pieprzyk and G. Finkelstein. Towards an effective non-linear cryptosystem design.
          IFIP TC11, International Conference on Computer Security : IFIP/SEC'88, W.J. Caelli
          (ed), Elsevier (to appear), May 1988. Abstracts, Gold Coast, Queensland, Australia.

[SN77]   G. J. Simmons and M. J. Norris. Preliminary comments on the MIT public-key cryp-
          tosystem. *Cryptologia*, Vol.1:406–414, 1977.

[SP88]   J. Seberry and J. Pieprzyk. *Cryptography: An Introduction to Computer Security*.
          Prentice Hall, Inc., Englewood Cliffs, New Jersey, 1988.

[Yue77]   C. Yuen. Testing random number generators by Walsh transform. *IEEE Transactions
          on Computers*, C-26(4):329–333, April 1977.

INFORMATIONAL DIVERGENCE BOUNDS FOR AUTHENTICATION CODES

Andrea Sgarro

Department of Mathematics and Computer Science
University of Udine
I-33100 Udine, Italy
Department of Mathematical Sciences
University of Trieste
I-34100 Trieste, Italy

Abstract

We give an easy derivation of Simmons' lower bound for impersonation games which is based on the non-negativity of the informational divergence. We show that substitution games can be reduced to ancillary impersonation games. We use this fact to extend Simmons' bound to substitution games: the lower bound we obtain performs quite well against those available in the literature.

1 Introduction

As already observed by the author at Eurocrypt 88 (corridor discussion; cf also /1/), a basic lower bound for impersonation games due to Simmons (cf e.g. /2/ or /3/) is a straightforward consequence of the non-negativity of the *informational divergence*:

$$D(P,Q) = \sum_i p_i \log \frac{p_i}{q_i} \geq 0 \ , \ \ D(P,Q) = 0 \ \text{iff} \ P = Q \qquad (1)$$

($D(P,Q)$ is called also: discrimination, cross entropy, Kullback-Leibler's number; above P and Q are two probability distributions with the same number of components; logs are to a base greater than 1, e.g. 2 or e; for $a > 0$ set $0 \log \frac{0}{a} = 0 \log \frac{0}{0} = 0$, $a \log \frac{a}{0} = +\infty$). The proof of (1) is itself quite short: one has $\ln x \leq x - 1$ with equality iff $x = 1$ ($\ln$ is strictly concave and lies below its tangent at $x = 1$); set $x = \frac{q_i}{p_i}$, multiply by -1 and plug the result into the definition of $D(P, Q)$.

A remarkable divergence is the *mutual information* $I(Y, Z)$; here P is the joint distribution of the random couple YZ and Q is the joint distribution obtained by multiplication of the marginal probabilities "as if" Y and Z were independent; (1) implies that the mutual information is non-negative, and is zero iff Y and Z are independent.

Actually, we need conditional divergences: if YZ and YW are two random couples over the same product-space the *conditional divergence* $D(Z, W \mid Y)$ is defined as the weighted average $\sum_y \Pr(Y = y) D(P_y, Q_y)$, where P_y (Q_y, respectively) is the conditional distribution of Z (of W, respectively) given $Y = y$. Of course, by (1):

$$D(Z, W \mid Y) \geq 0 \text{ , with equality iff } YZ \simeq YW \qquad (2)$$

("$\simeq$" denotes equality of distributions)

Below we review the easy derivation of Simmons' bound and show that it can be extended to substitution games.

2 Impersonation games

Let YZ be the random couple cryptogram-key. In the impersonation game the opponent chooses a cryptogram y hoping it to be taken as legal. The probability of success is

$$\Pr(Z \in A_y) \text{ , with} \qquad (3)$$

$$A_y = \{z : \Pr(Y = y, Z = z) \neq 0\} \tag{4}$$

(A_y is the set of keys which authenticate y). The best strategy for the opponent is to choose y as to maximize the probability in (3). Let P_I be this maximal probability of success in the impersonation game:

$$P_I = \max_y \Pr(Z \in A_y)$$

Set $\Pr(W = z \mid Y = y) = \Pr(Z = z \mid Z \in A_y)$. Plug this into (2) and recall the definition of the mutual information $I(Y, Z)$ to obtain

$$E_Y \log \Pr(Z \in A_Y) \geq -I(Y, Z)$$

(E_Y denotes expectation.) Noting that P_I is the maximal value for the random probability $\Pr(Z \in A_Y)$ one has Simmons' bound (we take binary logs):

$$P_I \geq 2^{-I(Y,Z)} \tag{5}$$

The conditions for equality are soon found: YZ and YW must have the same distribution, and $\Pr(Z \in A_y)$ must be constant in y. An authentication system for which (5) holds with equality will be called, in the spirit of Simmons' definition, *perfect* (actually, perfection as defined by Simmons is more ambitious and covers also the substitution game: ours is perfection with respect to the impersonation game only).

We note that the original derivation of (5) is 'tedious and long', and even the streamlined derivation given in /3/ is 'somewhat lengthy' (we are quoting /2/ and /3/, respectively).

3 Substitution games

Assume cryptogram c has been legally sent (and assume that there are at least two cryptograms). The opponent deletes c and sends instead $y \neq c$;

his probability of success is now:

$$\Pr(Z \in A_y \mid Y = c) = \Pr(Z \in A_y \cap A_c \mid Y = c) \tag{6}$$

The best strategy is to maximize (6) with respect to $y \neq c$. Let $P_S(c)$ be the corresponding maximal probability of success:

$$P_S(c) = \max_{y \neq c} \Pr(Z \in A_y \cap A_c \mid Y = c)$$

Set $a_c(z) = \sum_{y \neq c} \Pr(Y = y \mid Z = z)$. $a_c(z) = 0$ means that z authenticates *only* c (z is a "one-cryptogram key"). Assume for the moment that $a_c(z) \neq 0$ for all z in A_c, and define the random couple $Y_c Z_c$ as follows:

$$\Pr(Z_c = z) = \Pr(Z = z \mid Y = c) \ , \ z \in A_c \tag{7}$$

$$\Pr(Y_c = y \mid Z_c = z) = a_c(z)^{-1} \Pr(Y = y \mid Z = z) \ , \ y \neq c \tag{8}$$

Note that in $Y_c Z_c$ the conditional probabilities of the cryptograms have been "pumped up" so as to sum to 1 (c is no more there). Let us play the impersonation game for $Y_c Z_c$: one has (5) with $Y_c Z_c$ instead of YZ. If $P_I(c)$ denotes the probability of success for the ancillary impersonation game, one has:

$$P_I(c) = \max_{y \neq c} \Pr(Z_c \in A_y^c) = \max_{y \neq c} \Pr(Z \in A_y^c \mid Y = c)$$

$$= max_{y \neq c} \Pr(Z \in A_y^c \cap A_c \mid Y = c)$$

where A_y^c is defined as in (4) with $Y_c Z_c$ instead of YZ. However, as one easily checks, the set of keys which authenticate y in the ancillary impersonation game is the same as the set of keys which authenticate y in the original substitution game:

$$A_y^c \cap A_c = A_y \cap A_c$$

(to see this, recall definition (4) and observe that $\Pr(Z_c = z, Y_c = y)\Pr(Z = z, Y = c) \neq 0$ iff $\Pr(Z = z, Y = y)\Pr(Z = z, Y = c) \neq 0$). Recalling the definition of $P_S(c)$, we obtain the reduction formula we had announced:

$$P_I(c) = P_S(c) \tag{9}$$

Therefore, the lower bound for $P_I(c)$ is also a lower bound for $P_S(c)$:

$$P_S(c) \geq 2^{-I(Y_c, Z_c)} \ , \ \ Y_c Z_c \text{ defined as in (7),(8)} \tag{10}$$

By averaging with respect to $\Pr(Y = c)$ one obtains a lower bound for $P_S = \sum_c \Pr(Y = c) P_S(c)$, the overall probability of a successful substitution:

$$P_S \geq \sum_c \Pr(Y = c) \ 2^{-I(Y_c, Z_c)} \tag{11}$$

Equality holds in (10) iff the system $Y_c Z_c$ is perfect, in (11) iff the systems $Y_c Z_c$ are perfect for all cryptograms c's.

(10) and (11) have been derived under the assumption $a_c(z) \neq 0$. We shall show below (section 4) that this assumption always holds in cases of interest. For reasons of mathematical completeness, however, we shall cover also the case when some of the $a_c(z)$ are zero. Then, an extra step is needed. Actually, it is enough to observe that, keys with $a_c(z) = 0$ being hopeless for the opponent whatever his strategy, he can as well devise his optimal strategy under the assumption that the key Z does not belong to the set B of one-cryptogram keys:

$$B = \{z : \exists d : \Pr(Y = d \mid Z = Z) = 1\}$$

More formally, choose $y \neq c$ and write the corresponding probability of success as:

$$\Pr(Z \in A_y \cap B \mid Y = c) + \Pr(Z \in A_y - B \mid Y = c)$$

The first term in the sum is zero and $\Pr(Z \notin B \mid Y = c)$ does not depend on y. Assume the probability just written is not zero (else the whole thing is trivial: substitution attacks on c are unfeasible and so $P_S(c) = 0$); then the purpose of the opponent is to maximize in y, $y \neq c$:

$$\Pr(z \in A_y \mid Y = c, Z \notin B)$$

However, this is the same problem as above (cf (6)) once the conditioning $Z \notin B$ is imposed. One obtains, in full generality:

$$P_S(c) \geq \Pr(Z \notin B \mid Y = c) \, 2^{-I(Y_c, Z_c)}$$

Here Y_c and Z_c are defined as in (7) and (8), only adding the conditioning $Z \notin B$ in the definition of Z_c. Equality holds iff either the system $Y_c Z_c$ is perfect, or the lower bound is zero (keys which authenticate c authenticate *only* c; then, to be fastidious, $I(Y_c, Z_c)$ is undefined, so that we have to make the convention that zero times an undefined term is zero). By averaging with respect to $\Pr(Y = c)$ one obtains a bound for P_S, with conditions for equality.

4 A toy example

An authentication code is more accurately described as a random triple XYZ rather than as a random couple YZ (cf e.g. /2/ or /3/): the term added is the random clearmessage. One assumes Y to be a deterministic function of the couple XZ and X to be a deterministic function of the couple YZ: these functions correspond to enciphering and (legal) deciphering, respectively; the random key and the random clearmessage are assumed to be independent random variables. By requiring enciphering to be deterministic we rule out *splitting*; formally, our results hold true also in the case with splitting: however, substitution games are meaningful only when substitution of the legal cryptogram implies substitution of the correct clearmessage, that is only when enciphering is deterministic. (Note that often, e.g. in /2/ and /4/, the cryptographic-flavoured terms clearmessage, cryptogram and key as used here are replaced by the terms source-state, message (helas!) and encoding rule, respectively; a third alternative, more in keeping with the usage of non-secret coding theory, would be: message, codeword and encoding rule.) A nice way to represent the enciphering and deciphering functions for an authentication code XYZ is to draw up a matrix whose row-headers and column-headers are the keys and the clearmessages, respectively, and whose entries are the corresponding cryptograms. Of course the XYZ model is a special case of the YZ model, and so the lower bounds given above still hold. Actually, since the same cryptogram cannot appear more than once in the same row of the matrix (deciphering is deterministic) one-cryptogram keys are ruled out and only bounds (10) and (11) are needed (we are assuming that the matrix has at least two columns, which is an obvious assumption to make when one plays substitution).

Let us work out a toy example with 2 keys, 2 clearmessages and 3 cryptograms:

*	1	2
1	1	2
2	3	1

We assume equiprobability both for the key and the clearmessage; then $\Pr(Y = 1) = \frac{1}{2}$, $\Pr(Y = 2) = \Pr(Y = 3) = \frac{1}{4}$. Clearly, the (optimal-strategy) probability of successful substitution is $\frac{1}{2}$ if cryptogram 1 is substituted, else is 1. So the overall probability of a successful substitution is $P_S = \frac{3}{4}$. Since this value is returned also by bound (11), as one easily checks, our bound for P_S holds here with equality.

In /4/ four lower bounds are given for P_S (theorems 2.6, 2.9, 2.10, 2.14); however they return values strictly smaller than $\frac{3}{4}$ for our toy example. A computer search has been performed by A. Mereu, a student at Udine, on a large sample of small-order examples: in no case the bound we have proposed has been strictly beaten.

5 Final remark

Formula (9) allows one to reduce substitution games to impersonation games. There is a snag, however. Even if we start by a "complete" authentication code XYZ, the ancillary codes we were able to construct are random couples $Y_c Z_c$ and *not* random triples $X_c Y_c Z_c$. We could convert Simmons' bound for impersonation into a bound for substitution only because the former does not involve the random clearmessage in any way.

References

/1/ A. Sgarro, *Three-types of perfection in Shannon-theoretic cryptography*, Proceedings of a Workshop on Sequences, Positano, June 6-11, 1988

/2/ G.J. Simmons, *A survey of information authentication*, Proceedings of the IEEE, May 1988, pp. 603-620

/3/ J.L. Massey, *An introduction to contemporary cryptology*, Proceedings of the IEEE, May 1988, pp. 533-549

/4/ D.R. Stinson, *Some constructions and bounds for authentication codes* Journal of Cryptology, 1,1 (1988) pp. 37-52

2n-BIT HASH-FUNCTIONS USING n-BIT SYMMETRIC BLOCK CIPHER ALGORITHMS

Jean-Jacques Quisquater[1]   Marc Girault[2]

[1]Philips Research Laboratory
Avenue Van Becelaere, 2
B-1170 Brussels, Belgium

[2]Service d'Etudes communes des Postes et Télécommunications
42 rue des Coutures
BP6243, 14066 Caen, France

ABSTRACT

We present a new hash-function, which provides 2n-bit hash-results, using any n-bit symmetric block cipher algorithm. This hash-function can be considered as a extension of an already known one, which only provided n-bit hash-results. The difference is crucial, because a lot of symmetric block cipher algorithms use 64-bit blocks and recent works have shown that a 64-bit hash-result is greatly insufficient.

1. INTRODUCTION

Public-key systems provide methods of producing digital signatures of messages. Nonetheless, if these systems are used according to their primitive description, the signature produced is at least as long as the message itself and production time may be very high. This is the reason why one generally prefers to efficiently reduce the message to a short imprint, prior to applying the secret function of the public-key system. Such a reduction must be carefully designed, so that it does not introduce any weakness in the resulting digital signature scheme.

The functions which achieve this sort of reduction are often called hash-functions and may be defined as cryptographically secure methods of computing a fixed-length imprint of a message. A signature of this message can thereafter be generated by applying the secret function of a digital signature scheme to the imprint instead of applying it to the whole message.

More precisely, a hash-function is said secure if it is collision-free, i.e. if it is computationally infeasible to construct distinct messages which hash to the same imprint. Generally speaking, the collision-free property requires that the size h of the imprint be at least about 100 bits (say 128 bits, to preserve a safety margin). Indeed if it is much smaller (for example 64 bits), an attack exists which allows to efficiently construct distinct messages with the same imprint.

This attack, due to Yuval [Yu], consists in preparing two sets of $2^{h/2}$ messages. Each of these sets can be easily built by creating a few (h/2) variations of a unique message and by combining them together. It can be shown that the probability of finding a message M_1 from the first set and a message M_2 from the second set which have the same imprint is greater than 1/2. Now, such twin messages can be found by sorting the imprints of the first set and matching them with each imprint of the second set.

As best known (sequential) techniques of sorting are of time complexity O(NlogN), where N is the size of the list, it appears that h/2 should be greater than 32 in order to make this type of attack computationnally infeasible. This leads to a convenient length of more than 64 bits for the imprint. Moreover, new techniques due to Quisquater and Delescaille [QD] allow both to avoid sorting step and to use very few memory space, so that the so-called twin messages can be found in a much more efficient manner. Therefore, it appears reasonable to require the imprint to be (say) 128-bit long.

This size can however be reduced in some cases, due to the following reasons:

- the above mentioned attack is no longer effective if a random number is systematically inserted in the messages to be hashed, or if the initializing vector I is randomly chosen; nonetheless, it must be pointed out that the attack remains effective for the signer himself, since I is chosen by him.

- the collision-free property is not required in some applications, when it appears that the opponent has no practical way to profit from the collisions he found. Even in this case, however, the hash-function has still to be one-way in the following strong sense: given a message M and its imprint H, it must be computationally infeasible to find another message M' with the same imprint H; for that reason, a mimimum size of 64 bits is required.

Nevertheless, a size of 128 bits or more appears (nowadays) to be secure for all types of applications. Various authors have, in the past, also recommended such a size (e.g. [Ju]).

2. THE "SINGLE-LENGTH" HASH-FUNCTIONS

Much attention has been paid to hash-functions based on a symmetric block-cipher algorithm, generally DES. But until now, only schemes providing imprints of length equal to n (the block-length of the cipher algorithm) have been proposed. As n is often equal to 64, it results from the discussion of section 1 that such schemes are not secure enough from a general point of view.

For example, the following scheme (attributed sometimes to Davies, sometimes to Meyer), which we will call DM, is a good example of a "single-length" hash-function using DES [DP]: first, the message M is split into 56-bit blocks M_1, M_2, ..., M_r. Then, the imprint H is calculated in the following iterative way (where (+) stands for bitwise Exclusive-OR and $DES_K(X)$ denotes the encipherment of X with key K):

$$H_0 = I \quad \text{(initializing value)}$$

then: $$H_i = DES_{Mi}(H_{i-1}) \ (+) \ H_{i-1} \quad \text{for i from 1 to r}$$

The imprint is $H = H_r$.

This hash-function is as good as possible (apart from the fact that weak and semi-weak keys should be avoided; see [MS]). In particular, it seems to be resistant to a "meet-in-the-middle" attack ([DP] or [Co]). This sort of attack must be considered with care, especially after very recent results [QD2] which show its efficiency when implemented with a right time-memory trade-off.

In fact, the only default of DM-scheme is to provide too short imprints. So the question is: can we specify an efficient and secure scheme which provides twice as long imprints, still using a 64-bit cipher algorithm?

3. OUR PROPOSAL: A "DOUBLE-LENGTH" HASH-FUNCTION

3.1 General

We propose here a secure scheme which provides 2n-bit imprints using n-bit block cipher algorithms. Moreover, computation time of this scheme is almost the same as for DM (or similar) scheme, contrary to some other ones [MS]. In that way, we can answer "yes" to the question raised at the end of the previous section.

This scheme is the "good" generalization of DM scheme, in that it uses also feedforward techniques to avoid meet-in-the-middle attacks, and is specified in such a way that all the possible attacks (exhaustive or birthday ones) require a number of steps which is the square of the number required in the DM scheme.

As in DM scheme, the number of encipherments is equal to the number of blocks of the message to be hashed. The other operations are only bitwise Exclusive-Or and addition modulo 2^k-1, so that the scheme is almost as efficient as DM scheme, while it offers a much greater security.

This scheme is general, in that it can a priori use any symmetric block cipher algorithm. But it must be kept in memory that any weakness in this algorithm will probably induce a weakness in the scheme itself (e.g. weak or semi-weak keys of DES).

3.2 Informal description

The major problem to solve, which is inherent to the goal we wish to achieve, is the following one: since the basic operation is an n-bit encipherment, it is a priori still possible to perform a "local" attack at this n-bit level. This is particularly true after

Quisquater and Delescaille results already mentioned [QD], which show that one or two hours may suffice to find a DES collision, i.e., given a value I, two distinct keys K and K' such that $DES_K(I) = DES_{K'}(I)$. This attack only requires one very fast DES chip (or ten moderately fast DES chips!) and a personal computer to pilot this chip.

This statement has the following incidence: each block of the message should appear at least twice along the hashing operation, in a form or another. Meyer and Schilling [MS] propose that each block be involved in two encipherments, but this leads to a computation time which is twice as long as computation time of DM scheme.

We rather suggest to introduce two supplementary blocks at the end of the message; each of these blocks is dependent on all the preceding significant blocks, calculated by very simple (but as "independent" as possible) functions.

The basic step of our hash-function is composed of two encipherments (with blocks M_{2i+1} and M_{2i+2}), followed by Exclusive-Or operation with the hash-result which was obtained at the end of the previous step ($H_{2i-1}||H_{2i}$) to provide the new current hash-result ($H_{2i+1}||H_{2i+2}$), where $||$ is the symbol for concatenation. This feedforward connection is the analogue of the feedforward connection of DM scheme. Its role is to prevent from going backwards in the hash-function, in order to defeat meet-in-the middle attacks.

3.3 Formal description

Let e be a symmetric block-cipher algorithm, whose block-length is n and key-length is k (for example, n=64 and k=56 if e is DES). We denote the encipherment of input X under key K by eK(X). Let I and J be two n-bit initializing values, preferably chosen at random. Then, the imprint H of a binary message M is calculated in four steps.

Step 1 (splitting):
M is split into k-bit blocks $M_1, M_2,...$

Step 2 (first completion):
If the number of blocks is even, a supplementary block filled with '0's is added. Let n=2m be the number of blocks at the end of this step.

Step 3 (second completion):
Two supplementary blocks are added to the message. The first one, M_{n+1}, is equal to the Exclusive-Or of all the preceding blocks:

$M_{n+1} = M_1 (+) M_2 (+) ... (+) M_n$

The second one, M_{n+2}, is equal to the addition modulo 2^k-1 of the same blocks, seen as integers expressed in base 2:

$M_{n+2} = M_1 + M_2 +... + M_n \quad$ modulo 2^k-1

Step 4 (iteration): The output values $H_1, H_2,...,H_{n+1}, H_{n+2}$ are calculated in the following iterative way (see figure 1) :

$H_{-1} = I$ $\qquad\qquad\qquad\qquad\qquad\qquad H_0 = J$

$H'_{2i+1} = eM_{2i+1}(H_{2i-1}) (+) H_{2i}$ $\qquad\qquad H'_{2i+2} = eM_{2i+2}(H'_{2i+1}) (+) H_{2i-1}$

$H_{2i+1} = H'_{2i+2} (+) H_{2i}$ $\qquad\qquad\qquad H_{2i+2} = H'_{2i+1} (+) H_{2i-1}$

for i from 0 to m.

The imprint is $H = H_{n+1} \| H_{n+2}$.

3.4 Remark

Another iteration step had been suggested at the time of Eurocrypt'89 conference, but it was later shown to be weak by D. Coppersmith [Co2].

4. CONCLUSION

We have described a hash-function providing 2n-bit imprints, using a n-bit symmetric block cipher algorithm. In that way, Yuval's attack requires 2^n calculations (instead of only $2^{n/2}$), a prohitive number if, e.g., n=64 (there are more than 500,000 years in 2^{64} microseconds). Meet-in-the-middle attacks are made impossible, because of a feedforward connection similar to the one of Davies-Meyer scheme. Other attacks are also rendered unpractical, because of two supplementary blocks which are introduced at the end of the message

5. BIBLIOGRAPHY

[Co] D. Coppersmith, "Another birthday attack", Advances in Cryptology, Proc. of Crypto '85, LNCS, Vol. 218, Springer-Verlag, 1986, pp. 14-17.

[Co2] D. Coppersmith, private communication.

[DP] D.W. Davies and W.L. Price, "Security for computer networks", ed. J. Wiley & Sons, 1984.

[Ju] R.R. Jueneman, "A high speed Manipulation Detection Code", Advances in Cryptology, Proc. of Crypto '86, LNCS, Vol. 263, Springer-Verlag, 1987, pp. 327-346.

[MS] C.H. Meyer and M. Schilling, "Secure program load with Manipulation Detection Code", Securicom 88, pp. 111-130.

[QD] J.J. Quisquater and J.P.Delescaille, "How easy is collision search? Application to DES", these proceedings.

[QD] J.J. Quisquater and J.P.Delescaille, "How easy is collision search? New results and applications to DES", Proc. of Crypto '89, to appear.

[Yu] G. Yuval, "How to swindle Rabin", Cryptologia, Vol. 3, Jul. 1979, pp. 187-189.

Figure 1

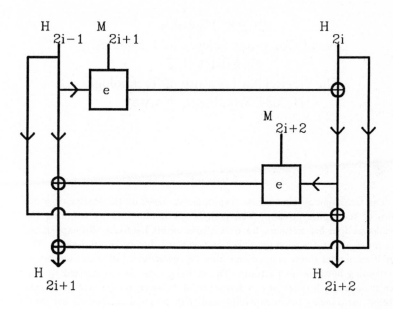

H_{-1}, H_0 : initializing values   e : block−cipher algorithm

$M_1, M_2, \ldots$: message blocks

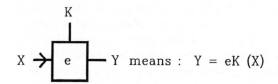

X $\rightarrow$ e $-$ Y means : $Y = eK\,(X)$

A Simple Technique for Diffusing Cryptoperiods

Stig F. Mjølsnes
Div. of Computer Science and Telematics
ELAB-RUNIT
The Norwegian Institute of Technology
N-7034 Trondheim, Norway

Abstract

The technique obtains diffuse cryptoperiods based on the stochastic properties of the cipher stream. The periods are randomized by scanning the pseudo-random bit sequence for occurrences of bit patterns. No explicit information about the change of key is necessary during transmission. The statistical model shows a deviation from the geometrical distribution due to overlapping between bit patterns. The technique can be generalized to randomize and synchronize any common event between sender and recipients without introducing extra signalling and with minimal computational overhead under the assumption of a reliable communication channel.

1 Introduction

In several key-management schemes proposed for conventional cryptosystems, the concept of a session key is prevailing [NS78, MM82, DP84]. Similar concepts are labeled data-encrypting key, primary communication key or file key. The keys are dynamically generated as needed and the approach is mostly to allocate one key to each communication session. For every new session, a new encryption key is generated, either by a common trusted party or by the communicators themselves. One good reason for doing this is to limit the amount of text to be encrypted under a specific key. Nevertheless, a session may be short or long depending on what type of communication is going on. This indicates that the amount of text enciphered under a single key can vary considerably. In other words, the "cryptoperiods" do not have similar length. As a result, the keys used in the same cryptosystem do not have a uniform security load.

There are various ways of handling this problem. Changing the key in the middle of a transmission can be carried out by introducing extra signalling, for example by adding another type of data packet. However, to modify the formats and functioning of an existing communication protocol is most likely not a viable solution. Another possibility is to modify the initial phase of connection. The protocol can negotiate a period length parameter during the set up procedure, which is then used throughout the entire connection.

This paper proposes to change the encryption key as a function of the bits communicated. If this can be carried out in a randomized manner, the cryptanalyst can now no longer be confident that all ciphertext intercepted from one session is encrypted with the same key. Choosing a suitable mode of operation, the only way to discover the cryptoperiods seems to be to break the specific encryption algorithm applied. By selecting an appropriate value of some parameter, the amount of text enciphered by each key can even be reduced to the unicity distance with some probability, if enough keybits can be spared, without reducing effective bandwidth of the channel.

A direct way to diffuse the duration of the cryptoperiods is to choose the lengths independently and according to some suitable probability distribution, either prior to, or during the transmission phase. An acceptable solution could be to select uniformly between a lower and an upper length limit. Such a method appears to be applicable to diverse classes of communication protocols that involve cryptographic functions where the secret keys must be changed in a randomized mode.

The special solution presented herein randomize the cryptoperiods based on the occurrence of patterns in a pseudo-random bit sequence. Moreover, I suggest to make use of a pseudo-random stream that is readily available in the cryptosystem itself; the cipher stream.

Some advantages of the scheme to be presented are:

- The exact time of key-change is kept secret (if the cryptoalgorithm is not broken); accordingly, no guarantee exists for a wiretapper that a single key is used for all ciphertext intercepted during a communication session.

- No extra signalling needed in the communication protocol,which makes it easy to add the technique to existing communication protocols.

- Data transfer is not interrupted regardless of session duration because key-change can be done at network speed.

2 Description of the technique

A strong cryptosystem can be characterized by many properties. Among them is the property of generating a strong pseudo-random bit sequence. Such a sequence should be indistinguishable from a random source of bits, in order to thwart statistical attacks of any kind. For the purpose of randomizing, it is sufficient to assume that the stream is indistinguishable from a random source with respect to the specific statistical test of matching patterns. However, for security reasons it must not be possible to distinguish between cipher streams encrypted under different keys.

The reliability and the authenticity of the bits received are taken for granted, for example by using the services of some link or transport protocol. Exploiting the pseudo-randomness of the cipher stream and its integrity, the cipher stream can be used to signal when the common event of key-change occurs. Both sender and receiver are scanning the cipherbit stream for a bit pattern or *flag* to appear, and when it appears the key will be replaced by some predetermined algorithm.

Neither the receiver nor any eavesdropper will know in advance when the next key-change is going to take place. However, the receiver will be able to detect the event by using some shared secret with the sender: the flag. The eavesdropper seems to be no better off than guessing the exact time of key-change, lacking information about which flag to look for and being unable to distinguish between streams enciphered under different keys.

If the receiver is able to distinguish between meaningful and meaningless decipherments, then he can decide with probability close to unity at which point the cryptoperiod terminates. If the next key to be used is known or can be computed at this stage, there is no need for explicit knowledge of the flag. The disadvantages of this method of decryption is the extra amount of computation and back tracking that has to be carried out whenever a period ends. The knowledge of the flag patterns make the decryption process simpler and faster for the legitimate receiver, and is accordingly assumed in the model.

If a reliable and authenticated multicast protocol can be provided, it is possible to extend the technique to include multiple receivers [Mjo89].

A special case will be to keep the flag pattern fixed. Now it becomes possible to resynchronize the encryption-decryption process at any time during a session, assuming a method for determining the new key is agreed upon in advance. Henceforth it is possible to relax the reliability demands of the channel. Bit synchronization can be obtained throughout the communication session with granularity as a function of the length of the flag.

2.1   The model

A set W contains all possible words of length l over the binary alphabet $\{0,1\}$. A set of words $\{w_i\}_{i=1}^n$ is picked at random from W. Similarly, a set of encryption keys $\{k_i\}_{i=1}^n$ is chosen at random from the set of possible keys K. The set of ordered pairs $\{(w_i, k_i)\}_{i=1}^n$ represents the shared secret between sender and receiver. A cipher stream is generated by an encryption algorithm $C = E_{\{k\}}(M)$ by input of a message stream M. Both sender and receiver carry out the following algorithm in order to switch to a new encryption key:

Init: Set $key \leftarrow k_1$; $flag \leftarrow w_1$; $i \leftarrow 1$
and start sending/receiving cipherbit stream C.

Matching: Scan stream C until match with $flag$.

Newperiod: Set $i \leftarrow i + 1$; $flag \leftarrow w_i$; $key \leftarrow k_i$
and continue with "Matching".

This is a very high level algorithm description that hides much of the details. One concern might be traffic analysis of a possible delay in transmission caused by changing the keys. A fast matching detector and use of buffering can prevent this. More considerations for implementation are treated in section 4. The matching procedure itself can be performed differently. The details of this procedure affect the probabilistic model as shown in the next section. For this purpose it is useful to distinguish between *hopping* and *sliding*.

Hopping: Hopping is defined as increasing the position in the stream by l bits between every test for match. No bits in the stream are used in more than one comparison.

Sliding: Sliding consists of increasing the position by one before next comparison. Except for the case when $l = 1$, succeeding comparisons are dependent because $l - 1$ bits are common between the two comparisons.

A variation of the scanning procedure is to make a hop before the first comparison after a match has taken place, and then proceed as for normal scanning until next match. This is called *sliding with initialization*.

3   The statistics of matches

This section presents two probability models of the scanning algorithm; the first model is a slightly modified geometric distribution, the other does not fit any standard model, and is partly based on results by Blom and Thorburn [BT82].

3.1   Assumptions and definitions

If the analysis that follows shall be valid to the scheme proposed, a basic assumption about the stochastic properties of the cipher stream must be made. For a strong cipher it is reasonable to assume that it is not possible to apply any statistical test and be able to infer useful information about the cleartext from the ciphertext, other than that it appears to be random. For the present analysis a weaker assumption is sufficient, namely that the cipherstream is indistinguishable from a random stream with respect to one specific statistical test: The flag matching process.

Assumption 1 (Randomness) *The flag matching process cannot distinguish between C and a sequence of independent and identical distributed random variables.*

Let $C = [c_1, c_2, \cdots c_t, \cdots)$ be the semi-infinite sequence of cipher bits. A subsequence of length l at position t is denoted $[c_{t-l+1}, \cdots, c_{t-1}, c_t]$. A flag of length l is one of the 2^l different bit patterns possible.

Definition 1 (Match) *There exists a match between the flag w of length l and the sequence C at position t if $w = [c_{t-l+1}, \cdots, c_{t-1}, c_t]$.*

It follows from this definition that a match can only occur at position $t \geq l$. The distance between two matches at positions t and t' is $|t - t'|$. The distance normally considered is the distance between succeeding occurrences.

Definition 2 (First match) *A match at position t in C is the first match of flag w if there exists no match of w for positions $l \leq r < t$.*

The figure below depicts position of matches and distances to and between them.

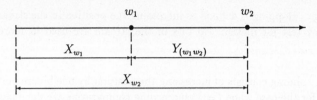

Let X_w be the waiting time or distance to first match of w when starting without any preconditions. Let $Y_{(w_1 w_2)}$ be the distance between consecutive matches of w_1 and w_2, in that order.

Note that the flag length itself is included in the distance until first match. The distance between two matches is equal to the number of *new* bits examined.

In order to be consistent on the words "left" and "right", think of the scanning process as normal reading. All patterns are read from left to right. For an arbitrary pattern w, let the left r bits be denoted $w_{[r]}$, and the right r bits be denoted $w^{[r]}$. The notion of pattern overlapping is defined as:

Definition 3 (Overlap) *If the right r digits of w_1 are equal to the left r digits of w_2, then w_2 overlaps w_1 with length r. This is denoted by the indicator variable:*

$$\varepsilon_r(w_1, w_2) = \begin{cases} 1 & \text{if } (w_1^{[r]} = w_{2[r]}) \\ 0 & \text{otherwise} \end{cases}$$

Note that the operator "overlaps" is not commutative in general. It is also convenient to define a notation for the total number of overlaps there are between two patterns:

$$\nu(w_1, w_2) = \sum_{r=1}^{l} \varepsilon_r(w_1, w_2)$$

If $\nu(w, w) > 1$ then w is *self-overlapping*. The longest overlapping subsequence between w_1 and w_2 is denoted $\kappa(w_1, w_2)$. If there exists no overlapping subsequence then $\kappa(w_1, w_2) = \lambda$, where λ is the empty string.

For convenience; $\varepsilon_r(w, w) = \varepsilon_r(w)$; $\nu(w, w) = \nu(w)$; $\kappa(w, w) = \kappa(w)$.

Example: If $\nu(w_1, w_2) = 1$, then $\kappa(w_1, w_2) = \lambda$. We have always that $\nu(w) \geq 1$ because $(w^{[r]} = w_{[r]})$ is always true for $r = l$.

3.2 Hopping

Recall that searching for the occurrence of a flag pattern in a random sequence by shifting l bits between successive comparisons is termed hopping, where l is the length of the flag. By assumption 1 and the fact that l new bits of the sequence are compared with the flag lead to independence between comparisons. The probability of getting a match in one comparison is $\Pr(\text{match}) = p = 2^{-l}$, by symmetry.

Let T be the number of comparisons until first match. From the conditions it follows that the random variable T is geometric distributed:

$$\Pr(T = t) = p(1 - p)^{t-1}$$

where t is the number of comparisons carried out.

The position in the sequence after t samples will be lt, and hence the expected distance to first match is:

$$E(X_w) = l \cdot E(T) = l/p = l \cdot 2^l \tag{1}$$

where $E(T)$ is the expectation of the geometric distribution.

The variance of the distance in bits to the first match is

$$Var(X_w) = l^2 \cdot Var(T) = l^2(1-p)/p^2 = \frac{l^2(1-2^{-l})}{2^{-2l}} \tag{2}$$

where $Var(T)$ is the variance of the geometric distribution.

Let $Y_{(ww)}$ be the number of comparisons between matches of a specific flag w. Then the distribution of $Y_{(ww)}$ is equal to the distribution of X_w because the conditions are the same. Similarly, the distribution of the number of comparisons between *any* pair of flags of length l is equal to the distribution of X_w. Therefore the equations 1 and 2 also apply to the distance *between* matches.

Lemma 1 *For all* $w, w_1, w_2 \in \{0,1\}^l : X_w$ *and* $Y_{(w_1 w_2)}$ *are identically distributed when the matching process is performed by hopping. The expectation of the distance between any pairs of flags is* $E(X_w) = E(Y_{(w_1 w_2)}) = l \cdot 2^l$.

In other words, the frequency of matches is a random variable with a distribution only dependent on the length of the flag. The choice of the length of the flag determines the expected frequency of key-changes to be $1/E(X_w)$.

Example: If the flag is 8 bits long, the expected distance between key-changes is 256 bytes (2048 bits). Increasing the length to 12 bits, the expected distance between key-changes becomes 6144 bytes (49152 bits). The standard deviation has approximately the same value as the expectation.

3.3 Sliding

Searching for the occurrence of a flag pattern in a random sequence by shifting only one bit between successive comparisons is termed sliding. Trivially, if the length of the pattern is only 1 bit, the stochastic conditions are still according to the geometric distribution. In general, however, interdependencies are introduced which complicates the model to some extent.

The expected distance between two matches of the same flag can be found by the following lemma:

Lemma 2 *For all* $w \in \{0,1\}^l$ *the expected distance between consecutive occurrences of* w *is* $E(Y_{(ww)}) = 2^l$.

The proof is based on the elementary renewal theorem. Note that there is no restrictions on the pattern of w.

The sliding procedure prescribes that the flag pattern shall be changed after each match. If this is neglected then the lemma can be used, but choosing another flag introduce dependencies, and the process seize to be a renewal process. These dependencies can be formalized by the definition of overlapping sequences. And there is an indirect way of finding the probability distribution by looking at distances to first match.

Let the set of flags be restricted to those that are not self-overlapping $\{w_n\}$, then the following lemma connects the two distances X_{w_n} and $Y_{(w_n w_n)}$:

Lemma 3 *For all $w_n \in \{0,1\}^l$; if $\nu(w_n) = 1$ then the expected distance to first match is $E(X_{w_n}) = E(Y_{(w_n,w_n)}) = 2^l$.*

Proof(sketch): Let w_n be a flag for which $\nu(w_n) = 1$, and let X_{w_n} be the distance to first match of this flag. By observing that X_{w_n} and $Y_{(w_n,w_n)}$ are identically distributed because the flag is not self-overlapping, the conclusion of the lemma follows. $\square$

It is now possible to generalize to self-overlapping flags where $\nu(w) > 1$.

Theorem 1 *Let X_w be the distance to first occurrence of the flag w. The expected distance to first match of $w \in \{0,1\}^l$ is*

$$E(X_w) = \sum_{r=1}^{l} \varepsilon_r(w) \cdot 2^r$$

Proof: The proof is based on applying lemma 2 and lemma 3.

The claim is that the distance to the first match of w is the sum of the distance to the first match of $\kappa(w)$ and the distance between successive matches of w:

$$X_w = X_{\kappa(w)} + Y_{(ww)} \tag{3}$$

where $X_\lambda = 0$.

Assume, without loss of generality, that the bitstring $\kappa(w)$ is at position t. Then the initial necessary and sufficient conditions for $Y_{(ww)}$ are exactly satisfied at position t. By recursively applying equation 3 $\nu(w)$ times, the distance X_w can be expressed as a sum like this:

$$X_w = Y_{(ww)} + Y_{(\kappa(w)\kappa(w))} + \dots Y_{(\kappa^{\nu-1}(w)\kappa^{\nu-1}(w))} + X_{\kappa^{\nu-1}(w)}$$

where κ^r denotes the function product $\kappa \circ \kappa \cdots \kappa$ where κ appears r times. $\kappa^{\nu-1}$ represents the shortest overlapping subsequence of w and hence is not a self-overlapping sequence itself. By lemma 3 $X_{\kappa^{\nu-1}(w)} = Y_{(\kappa^{\nu-1}(w)\kappa^{\nu-1}(w))}$. Definition 3 implies that

$$\varepsilon_r(w) = 1 \Leftrightarrow r = |\kappa^s(w)|, \ s = 0 \dots \nu(w) - 1$$

The theorem follows by using lemma 2 and asserting independence between the random variables. $\square$

The next theorem gives the expected distance between flag w_1 and flag w_2:

Theorem 2 *For all $w_1, w_2 \in \{0,1\}^l$; the expected distance between consecutive matches in the order w_1 and w_2 is:*

$$
\begin{aligned}
E(Y_{(w_1 w_2)}) &= E(X_{w_2}) - E(X_{\kappa(w_1,w_2)}) \\
&= \sum_{r=1}^{l} \varepsilon_r(w_2) 2^r - \sum_{r=1}^{|\kappa(w_1,w_2)|} \varepsilon_r(\kappa(w_1,w_2)) 2^r
\end{aligned}
$$

Proof: If $\kappa(w_1, w_2) = \lambda$ then $Y_{(w_1 w_2)} \geq l$, which is equivalent to theorem 1. More general, the lower bound on the distance is $Y_{(w_1 w_2)} \geq l - |\kappa(w_1, w_2)|$. Since $\kappa(w_1, w_2)$ is a subsequence of w_2, the distance to first match of w_2 is $X_{w_2} = X_{\kappa(w_1, w_2)} + Y_{(w_1 w_2)}$. By assumption 1 the variables $X_{\kappa(w_1, w_2)}$ and $Y_{(w_1 w_2)}$ are independent. $\square$

Now it is possible to state the following bounds from theorem 1 and 2.

$$2^l \leq E(X_w) \leq 2^{l+1} - 2$$

$$2 \leq E(Y_{(w_1 w_2)}) \leq 2^{l+1} - 2$$

If the flag pairs (w_1, w_2) are chosen at random the average expected distance between matches turns nicely out to be $E(Y_{()}) = 2^l$ for long streams.

Remark: It is possible to find the probability generating function for the distance to first match [BT82], and hence the probability distribution, the expectation and variance can be derived from that function [Mjo89].

Example: If the flag is 8 bits long, the expected distance between matches will vary from 2 bits to 510 bits. For 12 bits, the expected distance runs to 8190 bits, for 16 bits length the expected distance range 131070 bits.

3.4 Deciding cryptoperiods

The main classes of attacks (ciphertext-only, known-plaintext and chosen-plaintext) assume normally that the same key is used for all text involved. Changing keys more often at random places can make these attacks harder to achieve.

Keeping the flag constant gives 2^l possible partitions of a given cipher stream. This is independent of the length of the stream, so that it is not hard to test exhaustively for reasonable lengths of the flag.

Changing the flag by every match makes the number of possible partitions grow exponentially with the length of the cipher stream. Let $1/\delta$ be the matching intensity, and $|C|$ the length of the communication session, then the number of possible partitions is $O((2^l)^{|C|/\delta})$.

Selecting flags independent insures that detecting one cryptoperiod does not reveal anything about past or future events.

3.5 A simple cryptanalytic model

One possible start for a cryptanalytic attack is to include the stack of keys and flags as part of the cryptoalgorithm, and then go on and analyze the box as an autonomous bit generator. A basic algorithm might be a linear feedback shiftregister enhanced with a stack of states (the keys) and a bit pattern detector. This also indicates a possible method of making a linear generator non-linear, although a first analysis shows that quite strong restrictions on the feedback must be assumed [Ing88].

4 Implementation considerations

In practice a decision has to be made about when exactly the key change shall occur, which is highly dependent on the specific cryptosystem applied. Looking for a specific pattern match by scanning the cipher stream is easily implemented by hardware containing a register and some comparator circuitry. This allows for minimal delays such that the output stream will not be interrupted because of a key change.

A lower and an upper bound on the number of bits encrypted with the same key can easily be enforced with a counter, in order to avoid those rare occasions where a match does not take place after a reasonable number of bits.

Of course, it is of no use to set the expected distance between matches shorter than the blocksize of the cryptosystem. The clustering of events is still a problem that need careful attention. A possible solution is to use sliding with initialization, or even to skip the rest of the block. The scanning process will then be according to theorem 2 plus the amount of bits that are skipped.

4.0.1 Variations and extensions

Some ideas for extensions and variations are only briefly mentioned here:

- Pseudorandom generation of flags and keys by some strong cryptographic function.

- Computing the next flag to be used by some one-way function of variables like a masterkey, time, input and output traffic and other state information.

- Scanning for the r'th match.

- Scanning the stream for several flags combined by a Boolean function.

- Keeping a pool of keys, picking the next one to use according to some one-way function of where the flag appears.

- Adding memory to the flag detector.

- Multicast protocols.

5 Related work

Nicolai[Nic83] suggested a method of increasing the the linear complexity of a key bit generator by making random jumps in the cycle of the generator. He exemplified this by describing a generator constructed of several maximal length linear feedback shiftregister sub-generators that are XORed together. It is possible to jump from one point in the cycle to another by modifying the state of some of the subgenerators, for example by having the ability to step each sub-generator independently. The "code branch" should have the following properties:

1. The jump should be long to avoid reuse of key bits.

2. The jump is determined by a one-way function of the previous state of the generator.

The receiver can determine when a code branch has occurred by detecting error in decipherment, for example by relying on some integrity check. When this happens the decryption procedure backtracks until a correct decipherment appears.

Applying the scheme proposed in this paper, the code branching event will occur when the flag matches the ciphertext. If the flag patterns are chosen independent and at random then the code branching will be randomized. The backtracking procedure is not necessary if the flag sequence is predistributed. Still there is the need for some integrity check in order to keep synchronization. Using a one-way function of the previous state to determine the target state of the jump appears to a practical solution of key generation.

During the reviewing of this paper another related reference was brought to my attention. A Swiss patent of 1986 [KW86] describes a similar scheme, for the purpose of maintaining cryptographic synchronism between transmitter and receiver. The patent addresses the problem of cryptographic resynchronization in a stream cipher system where temporary errors such as bit slips are acceptable to a certain degree. Resynchronization is made possible by letting the flag pattern be fixed in advance. Whenever the flag appears in the cipherstream the key-bit generator is initialized as a function of a given number of bits following the predetermined bit pattern. This method is currently in widespread use [Sch89].

The fixed flag scheme appears to be the converse of what is presented in this paper. While the patent claims that diffuse cryptoperiods can be made distinct by flag matching, this paper focus on using flag matching to obtain diffuse cryptoperiods. Moreover, it is crucial that the flag pattern is changed continuously.

6 Generalizing the problem

The original idea was to diffuse the cryptoperiods of a cipher stream. A cryptoperiod can loosely be defined as that part of a message stream enciphered with the same key. However, synchronizing change of encryption keys is a special instance of any common event between sender and receiver. The only necessary property of the event is that it can be related to the bitstream in some unique way. The target of the action of the event need not be the same stream, like the cryptoperiod diffusion scheme presented here.

If the consequence of the event is purely internal on both sides, there is no method to detect this from the outside. In the special case of switching encryption keys, this can in principle be computed from the outside given sufficient resources and ciphertext. We may define an internal action *inobservable* if what is communicated to the outside is independent of the change of internal state. Following this, an internal action is *computationally inobservable* if it is infeasible by any computation to detect the change of state, even though sufficient information is released to the outside. A synchronized internal action is originated by an event synchronized between two or more parties. An example of a synchronized, computationally inobservable action is the key-switch proposal of this paper.

Matching flags can be seen as a very specific and limited statistical test of the stochastic properties of the cipher stream. Any suitable test function, with or without memory, might be used instead.

Acknowledgements

My thanks to Ingemar Ingemarsson for commenting on the main idea and adding another; To David Chaum for pointing out related work and reading of an early manuscript; To Bo Lindquist for discussion and finding a useful reference in the literature; And to Bjarne Helvik for numerous helpful discussions.

References

[BT82]   G. Blom and D. Thorburn. How many random digits are required until given sequences are obtained? *J. Appl. Prob.*, 19():518–531, 1982.

[DP84]   D. W. Davies and W. L. Price. *Security for Computer Networks*, chapter six. Wiley, 1984.

[Ing88] Ingemar Ingemarsson. Bitmönsterstyrda händelser. November 1988. Private communication.

[KW86]   H. Klemenz and W. R. Widmer. Verfahren und vorrichtung zur chiffrierten datenübermittlung. Swiss patent CH 658 759 A5, November 1986.

[Mjo89] S. F. Mjølsnes. *Some Issues in Cryptographic Protocols*. PhD thesis, The Norwegian Institute of Technology, Trondheim, 1989. In preparation.

[MM82]   Carl H. Meyer and Stephen M. Matyas. *Cryptography: A New Dimension in Computer Data Security*, chapter six and seven. Wiley, 1982.

[Nic83] Carl R. Nicolai. Nondeterministic cryptography. In D. Chaum, R. L. Rivest, and A. T. Sherman, editors, *Advances in Cryptology: Proceedings of Crypto'82*, pages 323–326, Plenum Press, New York, 1983.

[NS78]   R.M. Needham and M. Schroeder. Using encryption for authentication in large networks of computers. *Communications of the ACM*, 21(12):993–999, December 1978.

[Sch89] P. Schmid. Private communication. June 1989. Omnisec AG, Regensdorf, Switzerland.

Section 3

Zero-knowledge protocols

A GENERAL ZERO-KNOWLEDGE SCHEME *

Mike V. D. Burmester
Dept. of Mathematics
RHBNC - University of London
Egham, Surrey TW20 OEX
U.K.

Yvo Desmedt[†]
Dept. EE & CS
Univ. of Wisconsin – Milwaukee
P.O. Box 784
WI 53201 Milwaukee
U.S.A.

Fred Piper
Dept. of Mathematics
RHBNC - University of London
Egham, Surrey TW20 OEX
U.K.

Michael Walker
Racal Research Ltd.
Worton Grange Industrial Estate
Reading, Berks RG2 OSB
U.K.

Extended Abstract

Abstract

There is a great similarity between the Fiat-Shamir zero-knowledge scheme [8], the Chaum-Evertse-van de Graaf [4], the Beth [1] and the Guillou-Quisquater [12] schemes. The Feige-Fiat-Shamir [7] and the Desmedt [6] proofs of knowledge also look alike. This suggests that a generalization is overdue. We present a general zero-knowledge proof which encompasses all these schemes.

I.   Introduction

An interactive proof-system, or simply a proof, is an interactive protocol by which, on input I, a prover $A(lice)$ attempts to convince a verifier $B(ob)$ that either (a) $I \in \mathcal{L}$, $\mathcal{L}$ a language (proof of membership), or (b) that she "knows" a witness S for which (I, S) satisfies a polynomial-time predicate $P(\cdot, \cdot)$ (proof of knowledge). A proof is zero-knowledge if it reveals no more than is strictly necessary (for a formal definition of a proof of membership see [11]; for proofs of knowledge see [7]). Many zero-knowledge proofs have been described in the literature and various definitions of a proof-system have been suggested. The property of zero-knowledge has also been analyzed and refined (e.g., [7]). One might wonder why so many different zero-knowledge proofs have been proposed. One reason is that schemes which are

---

* *Some* of the results in this paper have been briefly announced at the rump session of Crypto'88.

[†] Research partly done when visiting RHBNC and sponsored by SERC.

based on zero-knowledge protocols must be easy to implement. Another is the complexity of protocols: practical considerations make it necessary to increase the speed of a protocol [8], to reduce its storage requirements [1,12] and to reduce the number of its iterations [2]. Finally the theoretical approach to zero-knowledge is closely related to the theory of computational complexity [11].

The purpose of this paper is to provide a general setting for these zero-knowledge protocols and to show that many known protocols fit into this setting. The advantages of having such a generalization are that:

- it illustrates the essential features of the protocol,

- it provides a proof that a general class of protocols are zero-knowledge, thereby establishing a straightforward set of criteria to determine whether or not a given protocol is zero-knowledge.

In this paper we consider an algebraic framework which includes the systems of Fiat-Shamir [8], Feige-Fiat-Shamir [7], Chaum-Evertse-van de Graaf [4], Beth [1], Desmedt [6] and Guillou-Quisquater [12]. We shall not discuss non-interactive zero-knowledge protocols [2].

The Fiat-Shamir scheme

To start with we briefly describe the set up of the Fiat-Shamir scheme [8]. This will help the reader to appreciate the setting for our scheme and to understand the details. In the Fiat-Shamir scheme we have:

- a set of secret numbers $S_1, S_2, \ldots, S_m$ which are chosen from the group of units Z_n^* of the ring of integers modulo n.

- a set of public numbers $I_1, I_2, \ldots, I_m \in QR_n$, the set of quadratic residues.

- a predicate $P(I, S) \equiv (I = S^2 (\bmod n))$, satisfied by all the pairs (I_j, S_j).

The protocol repeats $t = O(|n|)$ times:

Step 1 A, the prover, selects a random integer X modulo n and sends B, the verifier, the number $Z = X^2 (\bmod n)$.

Step 2 B sends A the *random* bits $q_1, q_2, \ldots, q_m$ as a query.

Step 3 A sends B: $Y = X \cdot \prod_j S_j^{q_j} (\bmod n)$, when all $q_i \in \{0, 1\}$.

Step 4 B verifies that $Y \in Z_n^*$ and that $Y^2 = Z \cdot \prod_j I_j^{q_j} (\bmod n)$.

B accepts A's proof only if for all t iterations the verifications in Step 4 are successful.

Remark: If $Y \notin Z_n^*$ were allowed (as in the Fiat-Shamir protocol) then a crooked prover A' could convince the verifier B (who must adhere to the protocol) that some quadratic non-residues $\bar{I}$ belong to QR_n. E.g., if A' chooses $X \equiv 0 \,(\bmod\, n)$, then B will always accept.[1]

We will describe a protocol which generalizes this scheme and we will show that all the protocols in [1,4,6,7,8,12] are particular cases of this protocol. In Section III. we will prove that our protocol is a zero-knowledge proof of membership or a zero-knowledge proof of knowledge, depending on the setting.

II. A framework for a zero-knowledge proof

In our general scheme the "public numbers" $I_1, I_2, \ldots, I_m$ are taken from a set $\mathcal{H}$ and the "secret numbers" belong to a set $\mathcal{G}$. These numbers are related by a predicate $P(\cdot, \cdot)$, that is $P(I_j, S_j)$ for all j. We assume that $\mathcal{H}, \mathcal{G}$ have some algebraic structure and we take $P(I, S)$ to be the predicate $(I = f(S))$, where f is a homomorphism. Such predicates are a common feature of all the protocols we consider. We remark that the notion of group homomorphisms has also been used in [13] but in a different context. In our protocol we use the following:

- a monoid $\mathcal{G}''$, with subsets $\mathcal{G}, \mathcal{G}'$ such that $\mathcal{G} \subset \mathcal{G}' \subset \mathcal{G}''$. All the secret numbers S_i belong to $\mathcal{G}$. $\mathcal{G}'$ contains the identity and all the elements of $\mathcal{G}$ are units (it means invertible elements).

- a semigroup $\mathcal{H}''$, with subsets $\mathcal{H}, \mathcal{H}'$ such that $\mathcal{H} \subset \mathcal{H}' \subset \mathcal{H}''$. $\mathcal{H}'$ has an identity and its elements are units.

- a (possibly one-way) homomorphism $f : \mathcal{G}'' \to \mathcal{H}''$ with $f(\mathcal{G}) = \mathcal{H}$.

The security parameter is $|n| = O(\log n)$, where $n = |\mathcal{H}|$. We shall regard this framework as being a particular instance of a general framework which is defined for all (sufficiently large) integers n. We therefore are tacitly assuming that $\mathcal{G} = \mathcal{G}_n$, $\mathcal{H} = \mathcal{H}_n$, etc. In this setting we have a framework for (a) a proof of membership for the language $\mathcal{L} = \bigcup_n \mathcal{H}_n$: the prover wants to prove that all the public numbers I_j belong to $\mathcal{L}$; (b) a proof of knowledge for the predicate $P(I, S)$: the prover wants to prove that she "knows" secret numbers S_j such that $P(I_j, S_j)$ for all j. Let us now describe the protocol.

---

[1]An interesting case occurs when I_1 is a quadratic non-residue of p, $I_1 \equiv 1 \,(\bmod\, q)$, $n = pq$, and $m = 1$. If A' sends $Z = p^2$ in Step 1 and $Y = p$ in Step 2 then B will always accept ($p = 5$, $q = 7$, $I_1 = 8$ is worth exploring).

Protocol

First the verifier checks that all the $I_j \in \mathcal{H}'$. Then the protocol starts. Repeat t times:

Step 1 A selects a random $X \in \mathcal{G}''$ and sends B: $Z = f(X)$ (A's cover).

Step 2 B sends A a random $\mathbf{q} = (q_1, \ldots, q_m) \in Q^m$ (B's query).

Step 3 When all $q_i \in Q$, A sends B: $Y = X \cdot \prod_j S_j^{q_i}$ (A's answer).

Step 4 B verifies that $Y \in \mathcal{G}'$ and that $f(Y) = Z \cdot \prod_j I_j^{q_i}$ (B's verification).

If the precondition is satisfied, and if for all iterations the conditions in Step 4 are satisfied then B accepts A's proof.

Remark: An important feature of this protocol is the inbuilt probability ($(|(\mathcal{G}'' \setminus \mathcal{G}')|/|\mathcal{G}''|)$) that an honest prover fails to convince the verifier.

II.1. A group based framework

We now state conditions that make the protocol a zero-knowledge proof. First consider the case when $\mathcal{G} = \mathcal{G}' = \mathcal{G}''$ is a group. We assume that:

1. *Conditions for computational boundedness of B:*

   1.a) We can check if $I \in \mathcal{H}'$ in polynomial time.

   1.b) We can check if $Y \in \mathcal{G}'$ in polynomial time.

   1.c) Multiplication in $\mathcal{H}''$ can be executed in polynomial time.

   1.d) f is a polynomial time mapping.

2. *Completeness condition:*   none.

3. *Soundness conditions:*

   3.a) The set of exponents is Q is $\{0, 1\}$.

4. *Zero-knowledge condition:*

   4.a) We can choose at random with uniform distribution an element $X \in \mathcal{G}''$.

   4.b) m is $O(\log |n|)$.

5. *Conditions for Proofs of knowledge:*

   5.a) $\mathcal{H}' = \mathcal{H}$.

5.b) Multiplication in $\mathcal{G}'$ and taking inverses in $\mathcal{G}'$ are polynomial time operations.

We show in Section III. that the conditions above are sufficient to make the protocol a zero-knowledge proof. However these conditions are rather restrictive and we only get the Chaum-Evertse-van de Graaf protocols [4]. In the following section we relax these conditions and show that the [1,6,7,8,12] are also particular cases of our protocol.

The Chaum-Evertse-van de Graaf protocols

Many protocols related to the discrete logarithm problem in a general sense were presented by Chaum-Evertse-van de Graaf [4]. The first one, called the multiple discrete logarithm, proves existence (and knowledge) of S_j such that $\alpha^{S_j} = I_j$, where α is an element of a group $\mathcal{H}''$. Examples of $\mathcal{H}''$ are $Z_N^*(\cdot)$, where N is a prime or composite number. This is a particular case of our protocol for which

- $\mathcal{G} = Z_n(+)$, n is a multiple of the order of α,

- $\mathcal{H}'' = \mathcal{H}'$ is a group, $\mathcal{H} = \langle \alpha \rangle$ is the group generated by α,

- $Q = \{0,1\}$, $m = 1$, and f is the group homomorphism $f : Z_n \to \mathcal{H}; x \to \alpha^x$.

We assume that the verifier knows an upper bound for n. Let us check the above conditions. Conditions 1.b and 5.b are satisfied even if one does not know what n is. Conditions 1.a and 1.c must be satisfied by $\mathcal{H}'$, which is automatically the case when $\mathcal{H}' = Z_N^*$. All the other conditions are trivially satisfied.

Next let us consider the Chaum-Evertse-van de Graaf protocol for the relaxed discrete log and show that it is also a particular case. This proves existence (and knowledge) of $S = (s_1, s_2, \ldots, s_k)$ such that $\alpha_1^{s_1} \alpha_2^{s_2} \cdots \alpha_k^{s_k} = I$, where $\alpha_1, \alpha_2, \ldots, \alpha_k, I$ are elements of a group $\mathcal{H}''$. To relate this scheme to our protocol we use "direct product groups". We take:

- $\mathcal{G} = Z_{n_1}(+) \times Z_{n_2}(+) \times \cdots \times Z_{n_k}(+)$, where n_i is a multiple of the order of α_i $(1 \leq i \leq k)$,

- $\mathcal{H}'' = \mathcal{H}'$ is a group, $\mathcal{H} = \langle \alpha_1, \alpha_2, \ldots, \alpha_k \rangle$,

- $Q = \{0,1\}$, $f : \mathcal{G} \to \mathcal{H}; (x_1, x_2, \ldots, x_k) \to \alpha_1^{x_1} \alpha_2^{x_2} \cdots \alpha_k^{x_k}$.

As in Chaum-Evertse-van de Graaf, $\mathcal{H}''$ has to be commutative, ($\mathcal{G}$ is commutative). There is one difference between the Chaum-Evertse-van de Graaf scheme and our description of it. In the former, A sends $\alpha_1^{x_1}$, $\alpha_2^{x_2}$, $\ldots$, $\alpha_k^{x_k}$ in Step 1, whilst in ours A sends $f(X) = \alpha_1^{x_1}\alpha_2^{x_2}\cdots\alpha_k^{x_k}$. This means that the prover makes more multiplications, the verifier makes fewer multiplications, and less is communicated.

Chaum-Evertse-van de Graaf take m to be 1, which is not necessary. Indeed when $m > 1$ the protocol proves knowledge of the multiple relaxed discrete log. It proves knowledge of $S_1 = (s_{11}, \ldots, s_{1k})$, $S_2 = (s_{21}, \ldots, s_{2k})$, $\ldots$, $S_m = (s_{m1}, \ldots, s_{mk})$, such that $\alpha_1^{s_{11}}\cdots\alpha_k^{s_{1k}} = I_1$, $\alpha_1^{s_{21}}\cdots\alpha_k^{s_{2k}} = I_2$, $\ldots$, $\alpha_1^{s_{k1}}\cdots\alpha_k^{s_{kk}} = I_k$.

Chaum-Evertse-van de Graaf also discussed a protocol for the simultaneous discrete log. This proves knowledge of S such that $\alpha_1^S = I_1, \alpha_2^S = I_2, \ldots, \alpha_k^S = I_k$. For this protocol we have $\mathcal{G} = Z_n(+)$, $\mathcal{H} = \langle\alpha_1\rangle \times \langle\alpha_2\rangle \times \cdots \langle\alpha_k\rangle$, and $f : \mathcal{G} \to \mathcal{H}$; $x \to (\alpha_1^x, \alpha_2^x, \ldots, \alpha_k^x)$. The other sets an the remarks about the conditions are similar to those for the multiple discrete logarithm.

II.2. A monoid based framework

We relax the conditions of the group based framework by allowing the sets $\mathcal{G}, \mathcal{G}', \mathcal{G}''$ to be distinct, by taking the set of exponents Q to be any set of integers, and by introducing some new conditions and modifying others. We use the same numbering and list only those conditions which are new or modified.

2. *Completeness conditions:*

   2.a) $|\mathcal{G}'| / |\mathcal{G}''| \geq 1 - |n|^{-c}$, c any constant.

   2.b) $\mathcal{G}' \cdot \mathcal{G} \subset \mathcal{G}'$.

3. *Soundness conditions:*

   3.a) There is an a such that: (i) $|(Q \pm a) \cap Q| \geq \psi|Q|$, where $(Q \pm a) = (Q+a) \cup (Q-a)$ and $\psi \in (0,1]$ is a constant, and (ii) if $f(Y') = f(Y) \cdot I^a$ for some $Y, Y' \in \mathcal{G}'$ and $I \in \mathcal{H}$ then there exists an element $S \in \mathcal{G}$ such that $P(I, S)$.

4. *Zero-knowledge condition:*

   4.b) $m \log |Q|$ is $O(\log |n|)$.

5. *Condition for Proofs of knowledge:*

   5.b) (replaces 3.a (ii)) Given $Y, Y' \in \mathcal{G}'$ and $I \in \mathcal{H}'$ with $f(Y') = f(Y) \cdot I^a$, we can obtain in polynomial time an element $S \in \mathcal{G}$ such that $P(I, S)$.

Remark: In most cases Q is of the form $[0:m]$ or $[1:m]$, $a = 1$ and $\psi = 1$. If Y is a unit and $1 \in Q$ then Condition 3.a is trivially satisfied for $a = 1$ and $S = Y^{-1}Y'$.

The Fiat-Shamir scheme

This protocol was discussed earlier. We take, $\mathcal{G}'' = \mathcal{H}'' = Z_n(\cdot)$, n a product of two distinct primes, $\mathcal{G}' = \mathcal{G} = \mathcal{H}' = Z_n^*(\cdot)$, $\mathcal{H} = QR_n$, $Q = \{0,1\}$, $a = 1$ and $f : Z_n \to Z_n; x \to x^2$, which is a homomorphism of the monoid Z_n. The reader can easily check that all conditions of Section II.2. are satisfied.

The Feige-Fiat-Shamir scheme

For this scheme $I_j = \pm s_j^2$ [7] (to be consistent with our general presentation we have modified slightly the notation), so that the secrets S_j consists of two parts: the sign part and the s_j. To make the relation of the Feige-Fiat-Shamir scheme with our protocol we use direct products of monoids. Let $n = pq$, p, q distinct primes with $p \equiv q \equiv 3 \pmod 4$. Take

- $\mathcal{G}'' = \{-1, +1\}(\cdot) \times Z_n(\cdot)$, $\mathcal{G}' = \{-1, +1\} \times Z_n^0$, $Z_n^0 = Z_n \setminus \{0\}$, $\mathcal{G} = \{-1, +1\} \times Z_n^*$,

- $\mathcal{H} = \mathcal{H}' = Z_n(\cdot)$, $\mathcal{H} = Z_n^{+1} = \{y \in Z_n^* \mid (y \mid n) = 1\}$, where $(y \mid n)$ is the Jacobi symbol,

- $Q = \{0, 1\}$, $a = 1$ and $f : \{-1, 1\} \times Z_n \to Z_n; (g, x) \to gx^2$.

This scheme is essentially the same as the Feige-Fiat-Shamir scheme except that in Step 3 of the protocol the prover sends $Y = X \prod_j S_j^{q_j}$, where Y is a pair with a sign part $y_1 \in \{-1, 1\}$ and a number part $y_2 \in Z_n$, whereas in Feige-Fiat-Shamir only a number is sent. However in the latter the verifier must check if $Y^2 = Z \cdot \prod_j I_j^{q_j} \pmod n$ or if $Y^2 = -Z \cdot \prod_j I_j^{q_j} \pmod n$. By doing this he knows exactly what the sign y_1 is. Therefore, for us the prover sends one extra bit in Step 3 whereas in Feige-Fiat-Shamir the verifier has to check one more equation. The two schemes are essentially the same, only the actual implementation is slightly different. Observe that the remark about the Fiat-Shamir protocol in the introduction applies to this protocol as well: if $Y \notin Z_n^0$ were allowed then we do not have a proof system.

The Desmedt scheme

For this scheme [6] take the same parameters as we discussed for the Feige-Fiat-Shamir scheme, except that $f : \{-1, 1\} \times Z_n \to Z_n; (h, x) \to hx^{2^{|i|}}$. Take $I_j = R_j/g_i(1) \pmod n$, where $g_i(x) = g_{i_d}(g_{i_{d-1}}(\cdots(g_{i_1}(g_{i_0}(x)))\cdots))$, with $g_0(x) = x^2 \pmod n$ and $g_1(x) = 4x^2 \pmod n$.

The Guillou-Quisquater scheme

Take

- $\mathcal{G}'' = \mathcal{H}'' = Z_n(\cdot)$, n a product of two different primes, $\mathcal{G}' = \mathcal{G} = \mathcal{H}' = Z_n^*$,

- $\mathcal{H} = \{y \in Z_n^* \mid y = x^v, x \in Z_n^*\}$, v a prime, $Q = [0 : v\text{-}1]$, $a = 1$

- $f : Z_n \to Z_n; x \to x^v$.

For $m = 1$ we get the Guillou-Quisquater scheme [12]. We observe that:

1. When $v^{mt} = O(|n|^c)$, c a constant, this scheme is insecure (since then "guessing the query" is a convincing strategy). So we must have $mt \log v \succ \log |n|$.[2] In Section III. we shall see that this scheme is sound when $t \succ \log |n|$.

2. The zero-knowledge proof in Section III. requires that $tv^m = O(|n|^c)$, c a constant. This proof cannot be used when either $t \succ |n|^c$, or $v^m \succ |n|^c$.

The Beth scheme

In this scheme [1], a centre possesses the security numbers $x_1 \ldots x_m \in Z_{q-1}$ and makes public α, a primitive root of $GF(q)$ and the values $y_j = \alpha^{x_j}$ for all j. For each user the centre chooses a random $k \in Z_{q-1}$ and gives the user $r = \alpha^k$ as one part of her public number. The other part consists of the numbers $ID_1, \ldots, ID_m \in Z_{q-1}$. The centre determines the secret numbers $S_1, \ldots, S_m$ by solving the congruence

$$x_j r + k S_j \equiv ID_j \mod (q - 1), \qquad j = 1, \ldots, m. \tag{1}$$

In Step 1 of the protocol the prover sends $z = r^{-t}$ (t random in Z_{q-1}) to the verifier. In Step 2 the verifier replies with $\mathbf{b} = (b_1 \ldots b_m)$, $b_i \in Q \subset Z_{q-1}$, and finally in Step 3 the prover sends $u = t + \sum_j b_j S_j \in Z_{q-1}$. The verification is

$$\prod_j y_j^{r b_j} r^u z = \alpha^{\sum_j b_j ID_j}. \tag{2}$$

Let us now make the relation with our protocol. Take

- $\mathcal{G} = \mathcal{G}' = \mathcal{G}'' = Z_{q-1}(+)$, $Q \subset Z_{q-1}$, $\mathcal{H}'' = \mathcal{H}' = GF(q)^*(\cdot)$,

- $\mathcal{H} = \langle r \rangle$, $r \in GF(q)^*$, and $f : Z_{q-1} \to GF(q)^*; x \to r^x$.

---

[2]This means that $\log |n| (mt \log v)^{-1} \to 0$ as $|n| \to \infty$.

Clearly f is a homomorphism of $\mathcal{G}$ onto $\mathcal{H}$. This is a discrete logarithm proof which looks very similar to the Beth scheme, except for the relation between the public and secret keys of A and the consequences in Step 4. Let us discuss this difference. We have,

$$I_j = f(S_j) = r^{S_j} = \alpha^{kS_j} = \alpha^{ID_j}\alpha^{-x_j r} = \alpha^{ID_j} y_j^{-r},$$

using (1), so that we can rewrite (2) in the form

$$f(u) = r^u = z^{-1}\alpha^{\sum_j ID_j b_j}\prod_j y_j^{-rb_j} = z^{-1}\prod_j(\alpha^{ID_j} y_j^{-r})^{b_j} = z^{-1}\prod_j I_j^{b_j}.$$

This is the same as the verification in our protocol for $Y = u$, $Z = z^{-1}$ and $\mathbf{q} = \mathbf{b}$. So the Beth scheme is essentially a particular case of our protocol. Observe that the verifier can use the I_j's instead of the $\alpha^{ID_j} y_j^{-r}$, which simplifies the computations (if $0, 1 \in Q$ then the verifier can obtain I_j by sending the query $\mathbf{q} = q_1 \cdots q_m$ with all entries zero except the j-th entry which is 1). The difference between the Beth scheme and our scheme is that in the former it is hard for the user to make her own ID_j's, whereas in the latter it is trivial to make the I_j's. This is exactly the same difference as exists between the Fiat-Shamir versions in [8] and the Fiat-Shamir scheme of [7,9].

III.   Fundamentals of the scheme

Theorem 1 *If the conditions of Section II.1. are satisfied with* $\mathcal{G} = \mathcal{G}' = \mathcal{G}''$, *then the conditions in Section II.2. are also satisfied.*

Proof.   Trivial (take $a = 1$, $\psi = 1$ and $S = Y^{-1}Y'$). $\quad\square$

Theorem 2 *If the Conditions 1–4 of Section II.2. are satisfied, if* $m \log|Q| \preceq \log|n|$ *and if t is bounded by* $\log|n| \prec t \preceq |n|^c$, c *any constant, then the protocol in Section II. is a (perfect) zero-knowledge proof of membership for the language* $\mathcal{L} = \bigcup_n \mathcal{H}_n$. *If, furthermore, Conditions 5 are satisfied[3] then the protocol is a (perfect) zero-knowledge proof of knowledge for the predicate* $P(I, S)$.

Proof.   (sketch) We remark that we do not rely on unproven assumptions.

Completeness: *(If A is genuine then B accepts the proof of A with overwhelming probability)*
This is obvious since the mapping f is an operation preserving mapping.

---

[3]We can relax the condition $n = |\mathcal{H}|$ to $n = |\mathcal{G}|$ in this case.

Soundness: *(If A' is crooked then the probability that B accepts the proof of A' is negligible)*

The proof is an extension of the one in Feige-Fiat-Shamir [7]. Suppose that A' convinces B with non-negligible probability. We consider the *execution tree T* of (A', B): this is a truncated tree which describes the responses of A' to the requests of B. A vertex of T is *super heavy* if it has more than $\omega = 1 - \frac{1}{4}\psi$ sons (ψ is the constant in Condition 3.a of Section II.2.; in [7] we have heavy vertices with $\omega = \frac{1}{2}$). In the final paper we will show that the condition $\log|n| \prec t$ guarantees that T has at least one super heavy vertex. The following Lemma makes it possible to show that there exist S_j such that $P(I_j, S_j)$ for all j.

Lemma 1: *At a super heavy vertex, for each $j \in [1:m]$ there exists at least one pair of queries $\mathbf{q} = (q_i)$, $\mathbf{q}' = (q_i')$ with $q_i' = q_i$ for all $i \neq j$ and $q_j' = q_j + a$, which A' answers correctly.*

Proof: Will be given in the full paper.

Apply this Lemma to a super heavy vertex. For each pair of sons we have:

$$f(Y) = f(X) I_1^{q_1} \cdots I_{m-1}^{q_{m-1}} I_m^{q_m}$$
$$f(Y') = f(X') I_1^{q_1'} \cdots I_{m-1}^{q_{m-1}'} I_m^{q_m'}$$

with $f(X) = f(X')$. To find the S_j we use a recursive procedure: first we find S_m and then we use it to calculate S_{m-1} and continue in the same way until we find all the S_j. Suppose that $\mathbf{q}$ and $\mathbf{q}'$ differ in the last place. Since $I_m^{q_m}$ and $I_m^{q_m'} = I_m^{q_m+a}$ are units the equations above can be written in the form,

$$f(Y) I_m^{-q_m} = f(X) I_1^{q_1} \cdots I_{m-1}^{q_{m-1}}$$
$$f(Y') I_m^{-q_m-a} = f(X') I_1^{q_1} \cdots I_{m-1}^{q_{m-1}},$$

so that $f(Y') = f(Y) I_m^a$. Then using Condition 3.a we obtain an S_m such that $P(I_m, S_m)$. This solution is not necessarily *the* S_m, but it is a good substitute.

This procedure is repeated to find $S_{m-1}, S_{m-2}, \ldots, S_1$. This completes the proof, for proofs of membership. For proofs of knowledge we have to show that there exists a polynomial time Turing machine, the *interrogator M*, that will extract the secrets from A'. M is allowed to *reset* A' to any previous state: this means that it can "obtain" all the sons from a super heavy vertex and hence all the S_j in the manner described earlier, this time using Condition 5.b. It remains to show how the interrogator can find a super heavy vertex in polynomial time. In the extended proof we will show that:

Lemma 2: *At a suitable level i of the execution tree the fraction of super heavy vertices is at least γ, where $\gamma \in (0, 1]$ is a constant.*

Proof: Will be given in the full paper.

In the final paper we prove that M will find a super heavy vertex (with overwhelming probability) in polynomial time.

Zero-knowledge: *(For each B' there exists a probabilistic expected polynomial time Turing Machine $M_{B'}$ which can simulate the communication of A and B')* The simulator proceeds as follows:

Step 1 $M_{B'}$ chooses a random X from $\mathcal{G}''$ (using Condition 4.a) and a random vector $\mathbf{q}$ from Q^m and sends to B': $Z = f(X)(\prod_j I_j^{q_j})^{-1}$.

Step 2 $M_{B'}$ reads the answer of B', $\mathbf{q}'$. If $\mathbf{q}' = \mathbf{q}$ then it sends X to B'. If $\mathbf{q}' \neq \mathbf{q}$ then it rewinds B' to its configuration *at the beginning of the current iteration* and repeats Step 1 and Step 2 with *new* random choices.

When all the iterations are completed, $M_{B'}$ outputs its record. The expected number of probes for a complete run is $t|Q|^m = O(|n|^c)$. Observe that the probability distribution output by $M_{B'}$ is identical to that of the transcript set of (A, B'). So this scheme is a *perfect* zero-knowledge scheme [11]. $\square$

IV. Conclusion

In this paper we have shown that the schemes described in [1,4,6,7,8,9,12] are all particular cases of one protocol. This protocol has been further generalized to include the Goldreich-Micali-Wigderson graph isomorphism scheme [10], the Chaum-Evertse-van de Graaf-Peralta scheme [5], and schemes based on encryption functions, such as the Brassard-Chaum-Crepeau [3] scheme and the Goldreich-Micali-Wigderson proof of 3-colourability [10]. However this is not in the scope of the monoid based framework.

REFERENCES

[1] T. Beth. A Fiat-Shamir-like authentication protocol for the El-Gamal-scheme. In C. G. Günther, editor, *Advances in Cryptology, Proc. of Eurocrypt'88 (Lecture Notes in Computer Science 330)*, pp. 77–84. Springer-Verlag, May 1988. Davos, Switzerland.

[2] M. Blum, P. Feldman, and S. Micali. Non-interactive zero-knowledge and its applications. In *Proceedings of the twentieth ACM Symp. Theory of Computing, STOC*, pp. 103–112, May 2–4, 1988.

[3] G. Brassard, D. Chaum, and C. Crépeau. Minimum disclosure proofs of knowledge. *Journal of Computer and System Sciences*, 37(2), pp. 156–189, October 1988.

[4] D. Chaum, J.-H. Evertse, and J. van de Graaf. An improved protocol for demonstrating possession of discrete logarithms and some generalizations. In D. Chaum and W. L. Price, editors, *Advances in Cryptology — Eurocrypt'87 (Lecture Notes in Computer Science 304)*, pp. 127–141. Springer-Verlag, Berlin, 1988. Amsterdam, The Netherlands, April 13–15, 1987.

[5] D. Chaum, J.-H. Evertse, J. van de Graaf, and R. Peralta. Demonstrating possession of a discrete logarithm without revealing it. In A. Odlyzko, editor, *Advances in Cryptology. Proc. Crypto'86 (Lecture Notes in Computer Science 263)*, pp. 200–212. Springer-Verlag, 1987. Santa Barbara, California, U.S.A., August 11–15.

[6] Y. Desmedt. Subliminal-free authentication and signature. In C. G. Günther, editor, *Advances in Cryptology, Proc. of Eurocrypt'88 (Lecture Notes in Computer Science 330)*, pp. 23–33. Springer-Verlag, May 1988. Davos, Switzerland.

[7] U. Feige, A. Fiat, and A. Shamir. Zero knowledge proofs of identity. *Journal of Cryptology*, 1(2), pp. 77–94, 1988.

[8] A. Fiat and A. Shamir. How to prove yourself: Practical solutions to identification and signature problems. In A. Odlyzko, editor, *Advances in Cryptology, Proc. of Crypto'86 (Lecture Notes in Computer Science 263)*, pp. 186–194. Springer-Verlag, 1987. Santa Barbara, California, U. S. A., August 11–15.

[9] A. Fiat and A. Shamir. Unforgeable proofs of identity. In *Securicom 87*, pp. 147–153, March 4–6, 1987. Paris, France.

[10] O. Goldreich, S. Micali, and A. Wigderson. Proofs that yield nothing but their validity and a methodology of cryptographic protocol design. In *The Computer Society of IEEE, 27th Annual Symp. on Foundations of Computer Science (FOCS)*, pp. 174–187. IEEE Computer Society Press, 1986. Toronto, Ontario, Canada, October 27–29, 1986.

[11] S. Goldwasser, S. Micali, and C. Rackoff. The knowledge complexity of interactive proof systems. *Siam J. Comput.*, 18(1), pp. 186–208, February 1989.

[12] L.C. Guillou and J.-J. Quisquater. A practical zero-knowledge protocol fitted to security microprocessor minimizing both transmission and memory. In C. G. Günther, editor, *Advances in Cryptology, Proc. of Eurocrypt'88 (Lecture Notes in Computer Science 330)*, pp. 123–128. Springer-Verlag, May 1988. Davos, Switzerland.

[13] R. Impagliazzo and M. Yung. Direct minimum-knowledge computations. In C. Pomerance, editor, *Advances in Cryptology, Proc. of Crypto'87 (Lecture Notes in Computer Science 293)*, pp. 40–51. Springer-Verlag, 1988. Santa Barbara, California, U.S.A., August 16–20.

Divertible Zero Knowledge Interactive Proofs and Commutative Random Self-Reducibility

Tatsuaki Okamoto   *Kazuo Ohta*

NTT Communications and Information Processing Laboratories
Nippon Telegraph and Telephone Corporation
1-2356, Take, Yokosuka-shi, Kanagawa-ken, 238-03 Japan

Abstract

In this paper, a new class of zero knowledge interactive proofs, a *divertible* zero knowledge interactive proof, is presented. Informally speaking, we call (A,B,C), a triplet of Turing machines, a divertible zero knowledge interactive proof, if (A,B) and (B,C) are zero knowledge interactive proofs and B converts (A,B) into (B,C) such that any evidence regarding the relationship between (A,B) and (B,C) is concealed. It is shown that any *commutative random self-reducible* problem, which is a variant of the *random self-reducible* problem introduced by Angluin et al., has a divertible perfect zero knowledge interactive proof. We also show that a specific class of the commutative random self-reducible problems have *more practical* divertible perfect zero knowledge interactive proofs. This class of zero knowledge interactive proofs has two sides; one positive, the other negative. On the positive side, divertible zero knowledge interactive proofs can be used to protect privacy in networked and computerized environments. Electronic checking and secret electronic balloting are described in this paper to illustrate this side. On the negative side, identification systems based on these zero knowledge interactive proofs are vulnerable to an abuse, which is, however, for the most part common to all logical idenification schemes. This abuse and some measures to overcome it are also presented.

1. Introduction

In this paper, we consider the following question: Let (A, B) be a zero knowledge interactive proof (ZKIP) system regarding a problem, where A is a prover and B is a verifier. Can B prove this problem to another machine C in a zero knowledge manner under the condition that B does not leave any evidence of his utilizing A's power in order to prove it to C? In other words, this condition can be described as follows: B does not leave any evidence of the relationship between the A-B interactions and the B-C interactions. In many applications of this class of zero knowledge interactive proofs, this condition plays an essential role.

First, a new class of zero knowledge interactive proofs is defined, *divertible* zero knowledge interactive proofs, which satisfies the above-mentioned question. A new class of problems, *commutative random self-reducible (CRSR)* problems, are also defined. Basically, these are a variant of the *random self-reducible* problems introduced by Angluin et al., [AL] and Tompa et al., [TW]. We show that any CRSR problem has a divertible perfect zero knowledge interactive proof. We also show that a specific class of CRSR problems, *endomorphic CRSR (ECRSR)* problems, have *more practical (multi-keys and higher degree version)* divertible perfect zero knowledge interactive proofs.

Divertible zero knowledge interactive proofs have two sides; one positive, the other negative. On the positive side, this class of zero knowledge interactive proofs can be used to protect privacy in networked and computerized situations. For example, a blind digital signature scheme based on divertible zero knowledge interactive proofs can be constructed. The blind digital signature schemes based on the RSA scheme and the GMR scheme [GoMiRi] have been proposed for electronic check protocols and electronic secret ballot protocols [C1, C2, Oh2, Da]. However, the scheme based on the RSA is not provably secure against adaptive chosen message attacks and is not efficient. The scheme based on the GMR is provably secure against adaptive chosen message attacks under some reasonable assumptions but is not efficient. In contrast, our scheme, based on divertible zero knowledge interactive proofs, is provably secure against adaptive chosen message attacks under some assumptions [MS, S] and is efficient when some efficient problems (e.g., square root $\bmod N$) are used. In this paper, we show two applications of divertible zero knowledge interactive proofs: one for electronic checking, and the other for secret electronic balloting. As a different type of application, this class of zero-knowledge proofs can be used to construct subliminal-channel-free identification/signature systems based on zero-knowledge proofs in a manner similar to that shown by Desmedt et al., [DGB, De].

On the negative side, we show a new abuse of divertible zero knowledge interactive proofs, which is related to the *mafia fraud* described in [DGB]. These abuses, however, are for the most part common to all logical identification schemes, in which security depends only on secret information. In our abuse, Bob can pass himself off as Alice to anyone, when Alice proves her identity to Bob, and he conceals any evidence regarding the relationship between Bob's proof and Alice's proof. That is, although Bob proves himself to be Alice with Alice's help, he conceals any evidence that he used her help. To illuminate this situation, we discuss a number of measures for overcoming these abuses.

Note: The protocol described in [DGB] as a subliminal-channel-free identification system based on the Fiat-Shamir scheme satisfies the property of divertible zero knowledge interactive proofs. That is, although the notion of divertible zero-knowledge interactive proofs has not been proposed previously, an implementation of this class of zero knowledge interactive proofs based on the Fiat-Shamir scheme has been shown in [DGB], where this protocol corresponds to a protocol to be shown in Appendix B.

2. Divertible zero knowledge interactive proofs

There are two types of interactive proofs. One is an interactive proof *for membership in language L*, in which a membership of an instance in language L is demonstrated [GMR]. The other is an interactive proof *for possession of information*, in which a prover's possession of information is demonstrated [FFS, TW]. In this paper, we concentrate on the interactive proof for possession of information. The results in this paper can be applied to the interactive proof for language membership.

(A, B) is an interactive pair of Turing machines, where A is the prover, and B is the verifier [GMR, TM]. Let $T \in \{A, B\}$. $T(s)$ denotes T begun with s on its input work tape. $(A, B)(x)$ refers to the probability space that assigns to the string σ the probability that (A, B), on input x, outputs σ. $(\underline{A(s)}, B(t))(x)$, A's history, denotes the triplet (x, s, ρ, m), where ρ is the finite prefix of A's random tape that was read, and m is the final content of the communication channel tape on which B writes. Similarly, $(A(s), \underline{B(t)})(x)$, B's history, denotes the triplet (x, t, ρ', m'), where ρ' is the finite prefix of B's random tape that was read, and m' is the final content of the communication channel tape on which A writes. B^A means B with oracle A, where B^A's oracle tapes correspond to B's communication channel tapes with A. $R \subseteq X \times Y$ is a relation.

Definition 1. An interactive triple of Turing machines (A, B, C) is a *divertible (computational/perfect)* zero knowledge interactive proof that the prover can compute some y satisfying $(x, y) \in R$, if the following conditions hold.

(i) (A, B^C) is a (computational/perfect) zero knowledge interactive proof that the prover can compute some y satisfying $(x, y) \in R$ [TW].

(ii) (B^A, C) is a (computational/perfect) zero knowledge interactive proof that the prover can compute some y satisfying $(x, y) \in R$.

(iii) Only A can compute some y satisfying $(x, y) \in R$.

(iv) For any prover A^* accepted by a valid verifier C, any verifier C^*, any $(x, y) \in R$, and any strings s and t, $((\underline{A^*(y, s)}, B^{C^*(t)})(x)$, $(B^{A^*(y,s)}, \underline{C^*(t)})(x))$ and $((\underline{A^*(y, s)}, C)(x)$, $(A(y), \underline{C^*(t)})(x))$ are (polynomially indistinguishable/equivalent), where A is a valid prover.

3. Commutative random self-reducible problems and divertible perfect zero knowledge interactive proofs

3.1 Commutative random self-reducible

Definition 2. Let $\mathcal{N}$ be a countable infinite set. For any $N \in \mathcal{N}$, let $|N|$ denote the length of a suitable representation of N. For any $N \in \mathcal{N}$, let X_N, Y_N be finite sets, and

$R_N \subseteq X_N \times Y_N$ be a relation. Let

$$dom R_N = \{x \in X_N \mid (x, y) \in R_N \text{ for some } y \in Y_N\}$$

denote the *domain* of R_N,

$$R_N(x) = \{y \mid (x, y) \in R_N\}$$

the *image* of $x \in X_N$, and

$$R_N(X_N) = \{y \mid (x, y) \in R_N, x \in X_N\}$$

the *image* of R_N.

R is *commutative random self-reducible (CRSR)* if and only if there is a polynomial time algorithm A that, given any inputs $N \in \mathcal{N}$, $x \in dom R_N$, and $r \in R_N(X_N)$, outputs $x' = A(N, x, r) \in dom R_N$ satisfying the following five properties.

R1. If r is randomly and uniformly chosen on $R_N(X_N)$, then x' is uniformly distributed over $dom R_N$.

R2. There is a polynomial time algorithm that, given N, x, r, and any $y' \in R_N(x')$, outputs $y \in R_N(x)$.

R3. There is a polynomial time algorithm that, given N, x, r, and any $y \in R_N(x)$, outputs some $y' \in R_N(x')$. If, in addition, r is randomly and uniformly chosen on $R_N(X_N)$, then y' is uniformly distributed on $R_N(x')$.

R4. A law of composition $\bullet : R_N(X_N) \times R_N(X_N) \to R_N(X_N)$ is defined, and $(R_N(X_N), \bullet)$ is a commutative group. In addition, the following relation holds.

$$(x', y \bullet r) \in R_N$$

R5. There is a polynomial time algorithm that, given N, x, and x', outputs some $x^* \in dom R_N$ such that $(x^*, r^{-1}) \in R_N$.

In the conditions for *CRSR* in Definition 2, R1-R3 are the same as those for *random self-reducible (RSR)* shown in [TW], but R4 and R5 are added. In CRSR, $r \in R_N(X_N)$ replaces $r \in \{0, 1\}^\omega$ from RSR. Therefore, the set of CRSR relations is a subset of RSR relations.

Example 1. Let a function $f_N : Y \to X$ as follows:

(1) Laws of composition $* : X \times X \to X$ and $\circ : Y \times Y \to Y$ are defined, and $(X, *)$, $(Y, \circ)$ are commutative groups.

(2) f_N is homomorphic. That is,

$$f_N(y_1 \circ y_2) = f_N(y_1) * f_N(y_2),$$

where $y_1, y_2 \in Y$.

(3) f_N is regular. Here, f_N is regular [GKL] if there exists a function $m(\cdot)$ such that for every $x \in X$ the cardinality of $R_N(x)$ equals $m(|x|)$, where $R_N(x) = \{y \mid f_N(y) = x \in X, y \in Y\}$.

(4) For any $y \in Y$, if r is randomly and uniformly chosen on Y, then $y * r$ is uniformly distributed over Y.

(5) There are polynomial time algorithms to compute laws of composition $*$ and $\circ$, and to take inverses of these groups.

Then, the relation R is commutative random self-reducible, if

$$(x, f_N(x)) \in R_N,$$

$$\operatorname{dom} X_N = X,$$

$$R_N(X_N) = Y,$$

$$A(N, x, r) = x * f_N(r)$$

Example 2. The following three examples E1, E2, and E3 are random self-reducible. Among them, E1 and E2 are commutative random self-reducible, because they are included in Example 1. However, E3 is not commutative random self-reducible, because commutativity of a group for the condition R4 does not hold.

E1 (square roots $\operatorname{mod} N$).

$$(x = y^2 \bmod N, y) \in R_N,$$

$$A(N, x, r) = r^2 x \bmod N.$$

E2 (discrete logarithms).

$$(x = a^y \bmod p, y) \in R_{(p,a)},$$

$$A((p, a), x, r) = a^r x \bmod p.$$

E3 (graph isomorphism).

$$(G' = \pi(G), \pi) \in R_G,$$

$$A(G, G', \phi) = \phi(G'),$$

where G and G' are graphs, and $\pi : G \to G'$ and $\phi : G' \to G''$ are isomorphic transformations on graphs.

The following proposition is a collorary of Theorem 4 in [TW].

Proposition 1. On inputs N and x, there is a polynomial time perfect zero knowledge interactive proof that the prover can compute some y satisfying $(x, y) \in R_N$, if R satisfies the following conditions:

T0 R is CRSR.

T1 There is a probabilistic polynomial time algorithm that, given N, x', and y', determines whether $(x', y') \in R_N$.

T2 There is a probabilistic polynomial time algorithm that, given N, outputs random pairs $(x', y') \in R_N$ with x' uniformly distributed over $\operatorname{dom} R_N$ and y' uniformly distributed over $R_N(x')$.

Theorem 1. Let the relation R be CRSR and satisfy T1 and T2. Then, on inputs N and x, there is a polynomial time divertible perfect zero knowledge interactive proof (A, B, C) that the prover can compute some y satisfying $(x, y) \in R_N$.

Proof Sketch:

We start by describing a construction of divertible perfect zero knowledge interactive proof (A, B, C) that the prover can compute some y satisfying $(x, y) \in R_N$.

Construction: On inputs N and x, the (A, B, C) procedure is as follows.
The following procedure is repeated $t = O(|N|)$ times, where $x \in_R X$ denotes that x is uniformly and randomly chosen on X.
(Procedure)

A   B   C

$r \in_R R_N (X_N)$
$x' = A(N, x, r)$

$\xrightarrow{\quad x' \quad} >$

$e \in_R \{0, 1\}$
$u \in_R R_N (X_N)$
$x'' = A(N, x^*, u)$
$\begin{cases} x^* = x' & (\text{if } e = 0) \\ (x^*, r^{-1}) \in R_N & \\ & (\text{otherwise}) \end{cases}$

$\xrightarrow{\quad x'' \quad} >$

$< \xleftarrow{\quad \beta \quad}$ $\beta \in_R \{0, 1\}$

$\beta' = \beta \oplus e$

$< \xleftarrow{\quad \beta' \quad}$

$\begin{cases} z = r & (\text{if } \beta' = 0) \\ z = y \cdot r & (\text{otherwise}) \end{cases}$ $\xrightarrow{\quad z \quad} >$

$z' = u \cdot z^{1-2e}$

$\xrightarrow{\quad z' \quad} >$

$\begin{cases} x'' = \\ \quad A(N, x, z')? \\ \quad (\text{if } \beta = 0) \\ (x'', z') \in R_N ? \\ \quad (\text{otherwise}) \end{cases}$

(Relationship among variables when $e = 1$)

$$
\begin{array}{ccc}
x & \xrightarrow{\quad R \quad} & y \\
\downarrow & & \downarrow \cdot r \\
A(N, x, r) = x' & \xrightarrow{\quad R \quad} & y \cdot r \\
\downarrow & & \downarrow \cdot y^{-1} \cdot r^{-2} \quad \Big| \cdot (y \cdot r)^{-1} \cdot u \\
x^* & \xrightarrow{\quad R \quad} & r^{-1} \\
\downarrow & & \downarrow \cdot u \\
A(N, x^*, u) = x'' & \xrightarrow{\quad R \quad} & r^{-1} \cdot u
\end{array}
$$

Correctness: Clearly, this construction satisfies conditions (i), (ii), and (iii) of Definition 1. Thus, we will show that the construction satisfies condition (iv). First, since e is uniformly and randomly chosen on $\{0, 1\}$, then β' is uniformly distributed over $\{0, 1\}$ independent from β. Since u is uniformly and randomly chosen on $R_N(X_N)$, then, from condition $R1$ of Definition 2, x'' is uniformly distributed over $R_N(X_N)$ independent from x'. In addition, from condition $R3$ of Definition 2, z' is uniformly distributed over a set $Z' = \{z' \mid z'$ is validly verified by $C\}$. Thus, the construction satisfies condition (iv) of Definition 1. **QED**

The parallel version of the divertible zero knowledge interactive proof can be constructed in manners similar to those of the Fiat-Shamir scheme [FS, FFS]. (Here, its parallel version is not a zero knowledge interactive proof, but it has been proven to reveal no useful knowledge [FFS, OO, Ok].

3.2 Divertible zero knowledge interactive proof for digital signatures

An application to digital signatures of zero knowledge interactive proofs is shown in [FS, MS, GQ2, OO]. In this section, a blind digital signatures scheme is shown based on the divertible zero knowledge interactive proofs.

Definition 3. An interactive couple of Turing machines (A, B) is a *divertible (computational/perfect)* zero knowledge interactive proof for digital signatures, if the following conditions hold.

(i) (A, B) is a parallel version of (computational/perfect) zero knowledge interactive proof that the prover can compute some y satisfying $(x, y) \in R$.

(ii) B^A outputs a digital signature z of m based on the (computational/perfect) zero knowledge interactive proof with respect to $(x, y) \in R$, where m is a message chosen by B.

(iii) Only A can compute some y satisfying $(x, y) \in R$.

(iv) For any message m chosen by any party, any prover A^* accepted by a valid verifier C, and any string s, $((\underline{A^*(y, s)}, B(m))(x), Z(B^{A^*(y, s)})(m, x))$ and $((\underline{A^*(y, s)}, C)(x), Z(A(y))(m, x))$ are (polynomially indistinguishable/equivalent), where A is a valid generator of digital signatures based on the (computational/perfect) zero knowledge interactive proofs with respect to $(x, y) \in R$. $Z(T)(m, x)$ denotes the probability space that assigns to the signature z the probability that T outputs z, on input x and m.

Theorem 2. On inputs N and x, there is a polynomial time divertible perfect zero knowledge interactive proof for digital signatures (A, B) that the prover can compute some y satisfying $(x, y) \in R_N$, if the relation R is CRSR and satisfies T1 and T2'.

T2' There is a probabilistic polynomial time algorithm that, on input y', outputs x' satisfying $(x', y') \in R_N$. If, in addition, y' is randomly and uniformly chosen on $R_N(X_N)$, then x' is uniformly distributed on $dom R_N$.

Proof Sketch:

Construction. On inputs N and x, the procedure of (A, B) is as follows.

A · · · · · · · · · · · · · · B · · · · · · · · · · · · · · C

$$r_i \in_R R_N (X_N)$$
$$x_i' = A(N, x, r_i)$$

$$\xrightarrow{\quad x_1' \,..\, x_t' \quad}$$

$$e_i \in_R \{0, 1\}$$
$$u_i \in_R Y$$
$$x_i'' =$$
$$A(N, x_i{}^*, u_i)$$
$$\begin{cases} x_i{}^* = x' & (\text{if } e_i = 0) \\ (x_i{}^*, r_i^{-1}) \in R_N & \\ & (\text{otherwise}) \end{cases}$$
$$m : \text{Message}$$
$$\beta_i = h_i (m, x_1'', \cdots, x_t'')$$
$$\beta_i' = \beta_i \oplus e_i$$

$$\xleftarrow{\quad \beta_1' \,..\, \beta_t' \quad}$$

$$\begin{cases} z_i = r_i & \\ \quad (\text{if } \beta_i' = 0) & \\ z_i = y \cdot r_i & \\ \quad (\text{otherwise}) & \end{cases}$$

$$\xrightarrow{\quad z_1 \cdots z_t \quad}$$

$$z_i' = u_i \cdot z_i{}^{1-2e_i}$$

$$\xrightarrow{\quad m, \; z_1' \cdots z_t', \; \beta_1 \cdots \beta_t \quad}$$

$$\begin{cases} x_i'' = & \\ \quad A(N, x, z_i') & \\ \quad\quad (\text{if } \beta_i = 0) & \\ (x_i'', z_i') \in R_N & \\ \quad\quad (\text{otherwise}) & \end{cases}$$
$$\beta_i = h_i (m, x_1'' \cdots x_t'') \; ?$$

Correctness. It can be proven in a manner similar to the proof of Theorem 1 that this construction satisfies conditions (i)-(iv) of Definition 3. **QED**.

Note: When we replace β_i by x''_i; $(i = 1, \ldots, t)$ as a part of signature information of m, we do not need to replace condition $T2$ by $T2'$.

4. Practical implementation of divertible zero knowledge interactive proofs

Some practical protocols such as multi-key version and higher degree version [C3, FS, FFS, GQ, OO, Oh1] have been proposed based on a basic zero knowledge proof protocol for quadratic residuosity [GMR] or discrete logarithm problem.

In this section, we show that a specific class of CRSR problems, *endomorphic CRSR (ECRSR)* problems, has multi-keys and higher degree version divertible zero knowledge interactive proofs.

Definition 4. For any $N \in \mathcal{N}$, let X_N be a finite set, and $R_N \subseteq X_N \times X_N$ be a relation. R is *endomorphic commutative random self-reducible (ECRSR)* if

(1) A law of composition $\bullet : X_N \times X_N \to X_N$ is defined, and $(X, \bullet)$ is a commutative group.

(2) $(x = y^L, y) \in R_N$, where $x \in X_N$, $y \in X_N$, L is an integer, and

$$y^L = \overbrace{y \bullet \cdots \bullet y}^{L \text{ times}}.$$

(3) There exists a function $m(\cdot)$ such that for every $x \in dom R_N$ the cardinality of $R_N(x)$ equals $m(|N|)$.

(4) For any $y \in X_N$, if r is randomly and uniformly chosen on X_N, then $y \bullet r$ is uniformly distributed over X_N.

(5) There are polynomial time algorithms to compute the law of composition $\bullet$, and to take inverse of this group.

Here, using ECRSR relations, we show a protocol that is a divertible zero knowledge interactive proof with multi-keys and higher degree.

Protocol (multi-keys and higher degree version divertible zero knowledge interactive proof) On inputs N, L and $x_1, x_2, \ldots, x_k$, the following procedure is repeated $t = O(|N|)$ times.

Theorem 3. If the relation R is ECRSR, then this protocol is a polynomial time divertible perfect zero knowledge interactive proof that the prover can compute some y_i satisfying $(x_i, y_i) \in R_N$ for all $i \in \{1, 2, \ldots, k\}$.

Notes:
(1) Variations of this protocol are shown in Appendix B, which are based on the Fiat-Shamir type protocol (Appendix A).
(2) When $t = 1, k \cdot |L| = O(|N|)$, this protocol has not been proven to be zero knowledge, however, it has been proven to reveal no useful knowledge [FFS, OO, Ok].
(3) By combining the ideas shown in Section 3.2 and this section, we can easily construct multi-keys and higher degree version divertible perfect zero knowledge interactive proofs for digital signatures.

5. Applications
In this section, the positive properties of divertible zero knowledge interactive proofs are shown. They can be useful for electronic checking and secret electronic balloting. In these applications, divertible zero knowledge interactive proofs for digital signatures shown in Section 3.2 are used as follows:
(1) After authority A checks the identity of member B, A gives B his digital signature on a message made by B through the zero knowledge interactive proofs between A and B. However, A cannot see his own signature or the message that he signs. That is, A makes a blind signature.
(2) B presents A's signature on B's messsage to a verifier C. C checks whether the message was signed by A. However, C cannot determine who made the message.
(3) Even if A and C are colluding with each other, they cannot know who made B's message with A's signature. This is because there is no information that shows the relationship between the A-B interaction for the generation of A's signature and B's message with A's signature.

When the above-described digital signature protocol is used for an electronic checking protocol, A is a banker, B a customer and C the owner of a shop. A gives B a check, after A checks the identity of B. B uses the check at C's shop, where C checks the validity of the check. Even if A and C are in collusion with each other, they cannot know who used the check at C's shop. This protocol is useful for privacy protection regarding this customer's activities.

On the other hand, when this digital signature protocol is used for a secret ballot, A is a ballot publisher, B a voter, and C a ballot counter. After A checks the validity of B based on voter registration records, A signs (stamps) the outside of an unopened envelope that contains a ballot for B and a facing piece of carbon paper. B takes B's ballot with the carbon image of A's signature out of the envelope, and sends it to C. C counts it, after C checks the validity of the carbon image of the signature on the ballot. Even if A and C are in collusion with each other, they cannot know whose ballot it is. Therefore, the privacy of each voter is guaranteed.

The above check protocol and secret ballot protocol were proposed in [C1, C2, Oh2]. However, the digital signatures used in these protocols are the RSA scheme or an RSA-like scheme. Hence, these protocols are not provably secure and are not efficient. In contrast, when divertible zero knowledge interactive proofs for digital signatures are used for these protocols, they are provably secure under some conditions and are efficient if square root $\mod N$ is adopted as a proof problem.

6. Abuses and the protective measures

In this section, we turn to the negative side of divertible zero knowledge interactive proofs. We show some abuses and a number of measures to counter them.

6.1 Abuses

(1) Identification based on divertible zero knowledge interactive proofs

Let us explain our abuse by using an example similar to that shown in [DGB]. A(lice) identifies herself to B(ob). B impersonates A and claims to be A. Then, C(harlie) checks the identity of B who is claiming to be A. Even if A and C are aware of the abuse, they cannot obtain any evidence but the relationship between the time when A claimed to be A and that B claimed to be A. To make it easier understand, we assume B and C are the owners of a restaurant and a jewelry shop with electronics payment respectively, where customers can pay electronically. A is a customer of B's restaurant. At the moment that A is ready to pay and to prove her identity to B, B determines to buy an expensive thing at C's shop, and C is starting to check B's (in fact A's) identity. While C is checking the identity of B, B is checking the identity of A, where the interaction between A and B is affected by the interaction between B and C and vice-versa. In this abuse, B leaves no evidence that proves the relationship between the A-B interaction and the B-C interaction.

(2) Digital signatures based on divertible zero knowledge interactive proofs

By using the divertible zero knowledge interactive proof, we can construct an abuse of digital signatures, as described below.

A identifies herself to B. B tries to forge A's signature on any message made by B. Then, C checks the validity of the forged signature, which , B is claiming, was generated by A. Even if A and C are aware of the abuse, they cannot obtain evidence of it. For illustrative purposes, consider an example similar to that in (1). B is a shop owner, and C is a banker. A is a customer of B's shop. While B is checking the identity of A, B is forging A's signature on a promissory note to C's bank written by B. Here, B's interaction with A is determined according to the promissory note. In this abuse, B leaves no evidence which proves the relationship between the A-B interaction and the signature message forged by B.

6.2 Protective measures

Here, we show two types of measures to protect against the above-described abuses; operational measures and algorithmic measures. Note that in the applications shown in Section 5, only operational measures can be used to counter these abuses, because algorithmic measures cannot used without losing the positive properties of divertible zero knowledge interactive proofs.

(1) Operational measures

Regarding the abuse in identification, essentially there is no operational protective measure except using a unique physical description as mentioned in [DGB]. In order to protect against the abuse in digital signatures, using a key for digital signatures different from that for identification is effective. Then, even if a forged signature message from Alice is made by Bob through the abuse described in Section 6.1, she can claim that the signature is invalid, although it is valid with respect to her identification key.

(2) Algorithmic measures

For these divertible perfect zero knowledge interactive proofs, it is essential that a verifier can determine the values of random bits to be sent to a verifier. Therefore, there are algorithmic measures in which the values are not determined by only the verifier. Two measures are shown in the following.

(i) Measure 1

In the first measure, the values of random bits to be sent from a verifier to a prover are determined by the cooperation of the verifier and the prover. Here, the values cannot be controlled by either the prover alone or the verifier alone. A coin flipping protocol for two persons has been shown in [B, BL]. In this measure, the previous perfect zero knowledge interactive proofs are used, replacing the verifier's coin flips with two people's coin flips. The other procedures in the divertible perfect zero knowledge interactive proofs are the same.

(ii) Measure 2

Recently, non-interactive zero knowledge proofs have been proposed [BFM, DMP]. In these zero knowledge proofs, the prover and verifier share common random bits before the prover starts the proofs. Therefore, these proofs are algorithmic measures to protect against this abuse, because the common random bits are not determined only by the verifier.

7. Open problems

Many problems regarding the divertible zero knowledge interactive proofs remain open. Here, we introduce some typical ones:

(1) What class of relations has divertible zero knowledge interactive proofs except CRSR relations? (Do all NP relations have divertible zero knowledge interactive proofs?)

(2) What class of relations has divertible *perfect* zero knowledge interactive proofs except CRSR relations? (Do all RSR relations have divertible perfect zero knowledge interactive proofs?)

(3) What class of relations has *multi-keys or higher degree version* divertible zero knowledge interactive proofs except ECRSR relations? (Do all CRSR relations have multi-keys or higher degree version divertible zero knowledge interactive proofs?)

Acknowledgements: The authors would like to thank Prof. Adi Shamir for his valuable suggestions, especially on the formal definition of divertible zero knowledge interactive proofs. They would also like to thank Prof. Yvo Desmedt for informing them of the related protocols in [DGB].

References

[AFK] M.Abadi, J.Feigenbaum and J.Kilian, "On Hiding Information from an Oracle," STOC pp.195-203 (1987)

[AL] D.Angluin and D.Lichtenstein, "Provable Security of Cryptosystems: a Survey," Technical Report TR-288, Yale University (1983)

[B] M.Blum, "Coin Flipping by Telephone: A Protocol for Solving Impossible Problems," Compcon, pp133-137 (1982)

[BFM] M.Blum, P.Feldman and S.Micali, "Non-Interactive Zero-Knowledge and Its Applications," STOC, pp.103-112 (1988)

[BL] M.Ben-Or and N.Linial, "Collective Coin Flipping, Robust Voting Schemes and Minima of Banzhaf Values," FOCS, pp.408-416 (1985)

[C1] D.Chaum, "Security without Identification: Transaction Systems to Make Big Brother Obsolete," Comm. of the ACM, 28, 10, pp.1030-1044 (1985)

[C2] D.Chaum, "Blinding for Unanticipated Signatures," Eurocrypto'87 (1987)

[C3] D.Chaum, "An Improved Protocol for Demonstrating Possession of Discrete Logarithms and Some Generalizations," Eurocrypto'87 (1987)

[Da] I.B.Damgård, "Payment Systems and Credential Mechanisms with Provable Security against Abuse by Individuals," Crypto'88 (1988)

[De] Y.Desmedt, "Subliminal-Free Authentication and Signature," Eurocrypto'88 (1988)

[DGB] Y.Desmedt, C.Goutier and S.Bengio, "Special Uses and Abuses of the Fiat-Shamir Passport Protocol," Crypto'87 (1987)

[DMP] A.DeSantis, S.Micali and G.Persiano, "Non-Interactive Zero-Knowledge Proof Systems," Crypto'87 (1987)

[FFS] U.Feige, A.Fiat and A.Shamir, "Zero Knowledge Proofs of Identity," STOC, pp.210-217 (1987)

[FS] A.Fiat and A.Shamir, "How to Prove Yourself," Crypto'86 (1986)

[GKL] O.Goldreich, H.Krawczyk, and M.Luby, "On the Exixtence of Pseudorandom Generators," Crypto'88 (1988)

[GMR] S.Goldwasser, S.Micali, and C.Rackoff, "Knowledge Complexity of Interactive Proofs," STOC, pp291-304 (1985)

[GMW] O.Goldreich, S.Micali, and A.Wigderson, "Proofs that Yield Nothing But their Validity and a Methodology of Cryptographic Protocol Design," FOCS, pp.174-187 (1986)

[GoMiRi] S.Goldwasser, S.Micali, and R.Rivest, "A Paradoxical Solution to the Signature Problem," FOCS, pp.441-448 (1984)

[GQ1] L.C.Guillou, and J.J.Quisquater, "A Practical Zero-Knowledge Protocol Fitted to Security Microprocessors Minimizing Both Transmission and Memory," Eurocrypto'88 (1988)

[GQ2] L.C.Guillou, and J.J.Quisquater, "A "Paradoxical" Identity-Based Signature Scheme Resulting from Zero-Knowledge," Crypto'88 (1988)

[MS] S.Micali, and A.Shamir, "An Improvement of The Fiat-Shamir Identification and Signature Scheme," Crypto'88 (1988)

[Oh1] K.Ohta, "Efficient Identification and Signature Schemes," Electronics Letters, 24, 2, pp.115-116 (1988)

[Oh2] K.Ohta, "An Electrical Voting Scheme Using a Single Administrator" (in Japanese), Spring Conference of IEICE Japan, A-294 (1988)

[Ok] T.Okamoto "Proofs that Release No Use Knowledge and Their Applications," to appear

[OO] K.Ohta, and T.Okamoto "A Modification of the Fiat-Shamir Scheme," Crypto'88 (1988)

[S] A.Shamir, Private Communication (1988)

[TW] M.Tompa and H.Woll, "Random Self-Reducibility and Zero Knowledge Interactive Proofs of Possession of Information," FOCS, pp472-482 (1987)

Appendix A

In this appendix, we show two types of perfect zero knowledge interactive proofs based on the commmutative random self-reducible relation. One is the Tompa-Woll type [TW], and the other is the Fiat-Shamir type. Although all of the protocols shown in this paper are based on the Tompa-Woll type, we can construct similar protocols based on the Fiat-Shamir type, including the multi-keys and higher degree versions (Appendix B).

(Tompa-Woll type)

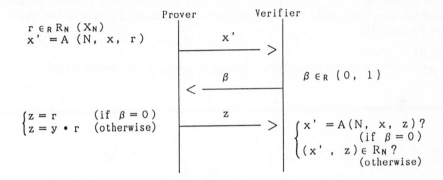

(Fiat-Shamir type)

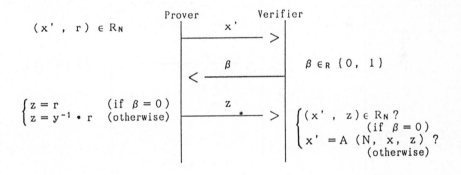

Appendix B

In this appendix, we show some protocols of divertible zero knowledge interactive proofs based on the Fiat-Shamir type (Appendix A), including the multi-keys and higher degree versions.

Protocol B1 is the Fiat-Shamir type of the protocol shown in the proof of Theorem 1 (basic version). In this protocol, we must replace condition R5 with R5' shown as follows:

R5'. There is a polynomial time algorithm that, given N, x, and x', outputs some $x^* \in dom R_N$ such that $(x^*, y \bullet r^{-1}) \in R_N$.

Protocol B2 is the Fiat-Shamir type of the protocol shown in Section 4 (multi-keys and higher degree version). The protocol described in [DGB] as a subliminal-channel-free identification system based on the Fiat-Shamir scheme corresponds to the quadratic version of Protocol B2.

(Protocol B1)

On inputs N and x, the following procedure is repeated $t = O(|N|)$ times.

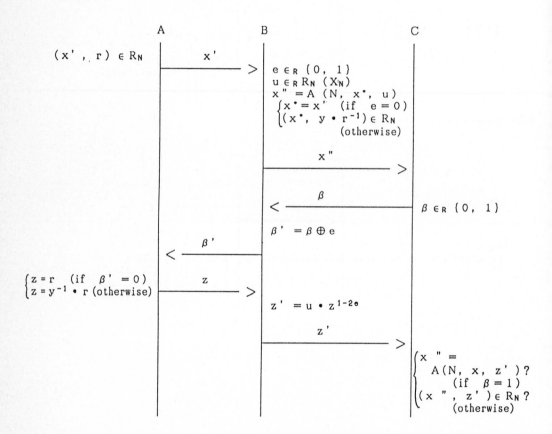

(Protocol B2)

On inputs N, L and $x_1, x_2, \ldots, x_k$, the following procedure is repeated t times.

Note: We can replace $x'' = u^L \bullet x' \bullet \prod_i x_i^{-e_i}$, $z = r \bullet \prod_i y_i^{-\beta'_i}$, $z' = u \bullet z \bullet \prod_i x_i^{-c_i}$, and $x'' = z'^L \bullet \prod_i x_i^{\beta_i}$ with $x'' = u^L \bullet x' \bullet \prod_i x_i^{e_i}$, $z = r \bullet \prod_i y_i^{\beta'_i}$, $z' = u \bullet z \bullet \prod_i x_i^{c_i}$, and $x'' = z'^L \bullet \prod_i x_i^{-\beta_i}$.

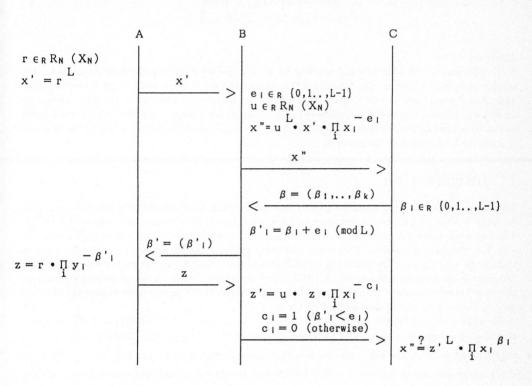

Verifiable Disclosure of Secrets and Applications
(Abstract)

Claude Crépeau *
MIT-Laboratory for Computer Science
545 Technology Square
Cambridge MA 02139 U.S.A.

Abstract

A $\binom{2}{1}$–Oblivious Bit Transfer protocol is a way for a party Rachel to get one bit from a pair b_0, b_1 that another party Sam offers her. The difficulty is that Sam should not find out which secret Rachel is getting while Rachel should not be able to get partial information about more than one of the bits. This paper shows a way to make "verifiable" this protocol (v-$\binom{2}{1}$–Oblivious Bit Transfer) and shows that it can be used to directly achieve oblivious circuit evaluation [Ki] and fair exchange of bits [MRL], assuming the existence of a non-verifiable version of the protocol.

1   Introduction

The study of disclosure protocols has greatly evolved recently. Oblivious transfer has now been used for quite a while as a standard primitive tool for construction of cryptographic protocols [Ra] [Bl]. The importance and extreme generality of this protocol was evidenced by the work of [BCR2], [Cr] and [Ki] who basically showed that every two-party protocol can be achieved using only oblivious transfer as a primitive. The results of [BCR2] and [Cr] are that a very general disclosure problem (all-or-nothing disclosure of secrets (ANDOS)) can be solved through a set of reductions from an oblivious transfer protocol. Kilian showed how to use the ANDOS primitive to implement the very general oblivious circuit evaluation (OCE) protocol.

In some sense this result is not very surprising. The earlier work of [GMW] and [CDG] hinted that any two party protocols could be achieved given a $\binom{2}{1}$–Oblivious Bit Transfer protocol. The only missing piece at the time was the fact that these protocols relied not only on $\binom{2}{1}$–Oblivious Bit Transfer but rather on a version of $\binom{2}{1}$–Oblivious Bit Transfer where the secrets are committed upon. This is what v-$\binom{2}{1}$–Oblivious Bit Transfer is about. We say that the secrets are verifiable because it is clear in Rachel's mind that the secret she eventually get is indeed one of the secret Sam had committed upon. The notion of $\binom{2}{1}$–Oblivious Bit Transfer and v-$\binom{2}{1}$–Oblivious Bit Transfer get generally confused in a setting with computational assumption, since commitments are used extensively in order to achieve $\binom{2}{1}$–Oblivious Bit Transfer. The result of [Ki] is a rather complex construction that achieves OCE through a brand new machinery and solves the problem from scratch. The current paper presents an easy way to extend any ANDOS protocol to a v-$\binom{2}{1}$–Oblivious Bit Transfer protocol, thus giving an alternate construction for the general OCE protocol.

2   Commitment Scheme

A Commitment Scheme is a way for Sam to commit himself to values that Rachel cannot determine but that are uniquely defined. We call a *blob* the piece of data used by Sam to commit to a value. By

*Research supported in part by an N.S.E.R.C. postgraduate scholarship. Some of this research was done while visiting Århus University.

opening a blob we mean to reveal the value represented by the blob in a verifiable way. We assume that the reader is familiar with this notion and that he knows the properties of such objects [BCC].

3 $\binom{2}{1}$–Oblivious Bit Transfer , ANDOS and v-$\binom{2}{1}$–Oblivious Bit Transfer

The All-or-Nothing Disclosure Of Secrets protocol is a way for a party Sam to reveal an element from a set of n strings of length t to another party Rachel, in a way that Sam does not learn which string Rachel gets and such that Rachel cannot get information about more than one of Sam's strings. A cryptographic solution to this problem can be found in [BCR1]. The well known $\binom{2}{1}$–Oblivious Bit Transfer protocol is simply a special case of this general protocol where $n = 2$ and $t = 1$. [BCR2] offers a set of reductions showing that a solution to the $\binom{2}{1}$–Oblivious Bit Transfer problem leads to an ANDOS protocol.

Although Sam cannot find out which string Rachel is getting he may nevertheless use some "garbage" secrets that are of no use to her. Because of this possibility, if the outcome of the protocol must be used by Rachel in some further interaction, Sam may, by offering "good" and "bad" secrets, determine from which set she chooses, depending on her ability to continue the protocol or not. In general, it might be necessary for Rachel to *verify* some properties of the secrets before getting one of them.

A Verifiable ANDOS (VANDOS) protocol is a way for Sam to commit to a set of n strings of length t such that Rachel can open exactly one set of blob corresponding to one of his strings. Sam should not learn which string Rachel is getting and Rachel should not learn information about more than one string. It is necessary also that Rachel is convinced that if she had chosen a different secret she would have been able to open it correctly as well (this is the verifiability property).

VANDOS can be achieved from v-$\binom{2}{1}$–Oblivious Bit Transfer (the verifiable version of $\binom{2}{1}$–Oblivious Bit Transfer) exactly like ANDOS can be obtained from $\binom{2}{1}$–Oblivious Bit Transfer . Therefore we focus only on this simpler case (v-$\binom{2}{1}$–Oblivious Bit Transfer) rather than the more complicated general problem.

Before going any further, we define exactly what properties we want our v-$\binom{2}{1}$–Oblivious Bit Transfer protocol to achieve: Assume Sam is committed to a pair of bits b_0, b_1 and Rachel is committed to a bit c. At the end of the interaction we want Rachel to get exactly bit b_c and to be committed to that result.

4 Building the blobs

Assume that Sam and Rachel have access to a protocol for ANDOS. Sam can commit to a bit b using the following technique. Sam generates s pairs of random bits $(l_1, r_1), (l_2, r_2), ..., (l_s, r_s)$ such that $b = l_i \oplus r_i$ for each i. Using the $\binom{2}{1}$–Oblivious Bit Transfer protocol Sam lets Rachel get one c_i from each pair (l_i, r_i). Obviously, since Rachel does not know any complete pair l_i, r_i she has no clue of what b might be.

To open the blob Sam must reveal each and every l_i, r_i. To be satisfied, Rachel should check that each pair XORes to the same bit ($l_i \oplus r_i = b$) and that each of the bits she got in the first place (c_i) is correct.

One can easily verify that Rachel has no way to cheat, while Sam can cheat only with probability 2^{-s} (corresponding to the event that he guesses Rachel's choices).

Notice that it is also possible to create *XOR*-blobs for the same amount of work: blobs for which the XOR relation ($c = b \oplus a$) can be proven without revealing their values. To do so, we need only to create all these blobs at once.

Let a, b and c be three bits. Sam can commit to these three bits in a way that he can convince Rachel that $c = b \oplus a$. Suppose that Sam chooses $6s$ random bits $A_1, ..., A_s, \alpha_1, ..., \alpha_s, B_1, ..., B_s,$ $\beta_1, ..., \beta_s, C_1, ..., C_s, \gamma_1, ..., \gamma_s$ such that $A_i \oplus \alpha_i = a$, $B_i \oplus \beta_i = b$ and $C_i \oplus \gamma_i = c$ for all $1 \le i \le s$. In order to commit to a, b, c Sam (using ANDOS) discloses to Rachel one triple out of (A_i, B_i, C_i) or $(\alpha_i, \beta_i, \gamma_i)$ for each i. Basically the commitment is the same as before except the Sam knows that Rachel is always reading the same entries for a, b and c. To open the blob a (resp. b or c) Sam reveals the A_i's and α_i's (resp. the B_i's and β_i's or C_i and γ_i). In order to show that $a \oplus b = c$ Sam reveals all the $A_i \oplus B_i \oplus C_i$ and $\alpha_i \oplus \beta_i \oplus \gamma_i$ to Rachel. She will accept this fact if the values of $A_i, B_i, C_i, \alpha_i, \beta_i, \gamma_i$ she had selected before agree with the later values of the XORs. This process can easily be extended to any number of blobs.

Rachel can commit to bits by exactly the same technique exchanging roles with Sam.

5 A bad v-$\binom{2}{1}$–Oblivious Bit Transfer protocol

Let b_0, b_1 be the secret bits of Sam. Assume that Sam commits to each of these bits using the commitment scheme of the previous section. For each b_i define $\mathcal{B}_i$ to be the $(2 \times s)$-bit-matrix used to commit to b_i.

First using ANDOS, Sam commits to each b_i (by revealing s entries of $\mathcal{B}_i$). Then using ANDOS again, Sam reveals to Rachel one of $(\mathcal{B}_0, \mathcal{B}_1)$ so that she can open the blob of one of the secret bits. Now Rachel can open the blob of the secret she chose and therefore make sure that she got exactly what Sam was committed to.

What is wrong with this protocol? The problem with it is that there is no validation performed on the ANDOS used for the n-tuples. Sam could have some "good" n-tuple that really open the corresponding blob and some "bad" ones that don't. This way Rachel would be able to open or not the blob depending on her choice. Which is a potential threat in a setting where the result must be used in some further interaction.

Indeed Rachel could complain about the fact that Sam gave her "garbage" matrices, but this would reveal him something about which secret she was after. The main idea of the solution is to make sure that Rachel can complain without revealing which secret she wants.

6 v-$\binom{2}{1}$–Oblivious Bit Transfer protocol

Let b_0 and b_1 be the two secret bits of Sam. He breaks each of them into pieces $b_0^1, b_0^2, ..., b_0^s$ and $b_1^1, b_1^2, ..., b_1^s$ such that $b_0 = \bigoplus_{i=1}^{s} b_0^i$ and $b_1 = b_0^i \oplus b_1 \oplus b_0$. Using the ANDOS protocol, Sam creates XOR blobs and commits to $b_0, b_0^1, ..., b_0^s, b_1, b_1^1, ..., b_1^s$. He proves to Rachel that they satisfy the appropriate XOR relation. Let $\mathcal{B}_0^i$ (resp. $\mathcal{B}_1^i$) be the $(2 \times s)$-bit-matrix used to commit to b_0^i (resp. b_1^i).

The trick we are setting up is that Rachel can get b_0 or b_1 in many different ways: to get b_0 (resp. b_1) she can use any sets of indices I, J such that $I \cap J = \emptyset$, $I \cup J = \{1, 2, ..., s\}$ and $\#J$ is even (resp. odd) since $b_0 (resp. b_1) = \bigoplus_{i \in I} b_0^i \oplus \bigoplus_{j \in J} b_1^j$ for all such I, J. Therefore the next step of our protocol is to let Rachel choose such sets I, J and get (using ANDOS) what she needs to open the corresponding b_0^i, $i \in I$ and b_1^j, $j \in J$. For this purpose Rachel gets one of $(\mathcal{B}_0^i, \mathcal{B}_1^i)$ so that she can open the corresponding b_0^i or b_1^i.

If Sam is dishonest and gives her "garbage" at any point, Rachel can complain and say that she got "bad" stuff (which does not enable her to open a blob). Because she could be reading a given b_0^i or b_1^i independently of the fact that she is trying to get b_0 or b_1, he does not learn anything about her intention. If Rachel is happy of all the b_0^i and b_1^i she is getting, she will be able to open exactly one of b_0 or b_1 at the end.

6.1 What about Rachel?

Now, we also want Rachel to be committed to her choice c and what she reads b_c. For that purpose, consider the above described protocol. Because it guaranties security from Rachel's point of view, we call it a R-v-$\binom{2}{1}$–Oblivious Bit Transfer . We use this protocol as a primitive to build the v-$\binom{2}{1}$–Oblivious Bit Transfer .

Let r_0, r_1 be two random bit chosen by Sam. Sam generates $4s$ random bits (r_0^1, r_1^1), (r_0^2, r_1^2),... (r_0^{2s}, r_1^{2s}) that he transfers to Rachel using the R-v-$\binom{2}{1}$–Oblivious Bit Transfer protocol from above. For each $1 \le i \le 2s$ Rachel randomly chooses c_i and get bit $r_{c_i}^i$. She commits to all the c_i and $r_{c_i}^i$. Sam picks a random subset H of $\{1, 2, ..., 2s\}$ of size s and asks Rachel to open the commitments to the c_i and $r_{c_i}^i$ for i not in H. If these commitments are correct, then Sam is guaranteed that with overwhelming probability, that the majority of the remaining $(c_i, r_{c_i}^i)$ pairs correspond to the values of c_i used and $r_{c_i}^i$ obtained. Rachel commits to her final choice c. She points out to Sam the subset C of H such that $c = c_i$ for $i \in C$ and prove this using the XOR property of the blobs. She also proves that $c \ne c_i$ for $i \in H - C$. Sam reveals to her the following two values after proving that they are properly built: $\hat{r}_0 = r_0 \oplus \bigoplus_{i \in C} r_0^i \oplus \bigoplus_{i \in H-C} r_1^i$ and $\hat{r}_1 = r_1 \oplus \bigoplus_{i \in C} r_1^i \oplus \bigoplus_{i \in H-C} r_0^i$.

Through this trick Sam increases his confidence that the committed value of c is likely to correspond to the value of r_c obtained by the protocol. Unfortunately this protocol gives little guarantee that the committed value of r_c corresponds to the value actually obtained. We solve this problem by repeating the above protocol s times with independent values of random bits (r_0^1, r_1^1), (r_0^2, r_1^2),... (r_0^{2s}, r_1^{2s}) but for the same r_0, r_1. Each time Rachel should come up with the same value for c and r_c. She proves this by proving equality of the values (c and r_c) used at each iteration of the protocol.

If all the cs and r_cs are the same then Sam is convinced that Rachel is actually committed to the c she used and the r_c she got. In this case Sam reveals $x_0 = r_0 \oplus b_0$ and $x_1 = r_1 \oplus b_1$ and proves the correctness of these values. Rachel produces a commitment to b_c by doing one of the following and showing the correctness of the relation: if $x_0 = x_1$ then Rachel computes $b_c = r_c \oplus x_0$, otherwise if $x_i = i$ then Rachel computes $b_c = r_c \oplus c$ and finally if $x_i = 1 - i$ then she computes $b_c = r_c \oplus \bar{c}$.

More details and a proof of correctness of this protocol will be provided in the final version of this paper.

7 Using v-$\binom{2}{1}$–Oblivious Bit Transfer to achieve OCE

Basically, we use the technique of [GV]; their result is based on a specific cryptographic assumption, but can be extended easily to any v-$\binom{2}{1}$–Oblivious Bit Transfer protocol. For each bit b in a computation, Sam owns x and Rachel y such that $b = x \oplus y$. The main step to perform is to manage to compute for instance the output of a NAND gate on two bits b_0, b_1 without revealing their value.

Assume Sam owns x_0, x_1 and Rachel owns y_0, y_1 such that $b_0 = x_0 \oplus y_0$ and $b_1 = x_1 \oplus y_1$ and that they wish to come up with secret values x_2 and y_2 such that $x_2 \oplus y_2 = b_2 = b_0 \overline{\wedge} b_1$. Sam offers the four following secrets from which Rachel get the appropriate one (corresponding to the cases $(y_0, y_1) = (0, 0); (0, 1); (1, 0); (1, 1).)$:

1. $(x_0 \wedge x_1) \oplus \overline{x_2}$

2. $(x_0 \wedge \overline{x_1}) \oplus \overline{x_2}$

3. $(\overline{x_0} \wedge x_1) \oplus \overline{x_2}$

4. $(\overline{x_0} \wedge \overline{x_1}) \oplus \overline{x_2}$

They both prove the correctness of their actions using the XOR-property of the blobs and the verifiability of v-$\binom{2}{1}$–Oblivious Bit Transfer . The outcome is y_2. By repeating this technique, all the gates of a circuit can be evaluated easily.

8 Fair OCE

A Fair OCE can be obtained by combining an OCE and a Fair Exchange of BIT protocol like the one of Micali, Rackoff and Luby [MRL]. Such a protocol can be achieved easily. Remember that the basic idea of [MRL]'s protocol is that Sam and Rachel who want to exchange bits b_0 and b_1, flip a coin secretly that is slightly biased toward $b_0 \oplus b_1$. Let $\epsilon = \frac{p}{q}$ be the bias they wish to obtain. Sam commits to q secrets from which p are b_0 and $q - p$ are $\overline{b_0}$. The correctness of this step can be proven using the XOR-property of the blobs. Rachel gets one of the bits (b) at random and gains this way a little bias toward the answer. To give this bias back to Sam she tells him the XOR of her result b with b_1 (and proves it using the XOR-property of the blobs).

9 Acknowledgments

I would like to acknowledge Gilles Brassard, Ernie Brickell, Ivan Damgård, Cynthia Dwork, Joe Kilian, and Silvio Micali for their valuable comments, ideas, and encouragement.

10 References

[Bl] Blum, Manuel. "Three applications of the oblivious transfer: Part I: Coin flipping by telephone; Part II: How to exchange secrets; Part III: How to send certified electronic mail", Department of EECS, University of California, Berkeley, CA, 1981.

[BCC] Brassard, Gilles, David Chaum, and Claude Crépeau. "Minimum disclosure proofs of knowledge" (revised version), *Technical Report PM-R8710*, Centre for Mathematics and Computer Science (CWI), Amsterdam, The Netherlands, 1987.

[BCR1] Brassard, Gilles, Claude Crépeau, and Jean-Marc Robert. "All-or-Nothing Disclosure of Secrets," *Proceedings Crypto 86*, Springer-Verlag, 1987.

[BCR2] Brassard, Gilles, Claude Crépeau, and Jean-Marc Robert. "Information Theoretic Reductions Among Disclosure Problems," *Proceedings of the 27th FOCS*, IEEE, 1986, 168–173.

[C] Crépeau Claude, "Equivalence Between Two Flavours of Oblivious Transfer", *Proceedings of Crypto 87*, 1988, Springer-Verlag.

[CDG] Chaum David, Ivan Damgård, and Jeroen Van de Graaf, "Multiparty computations ensuring privacy of each party's input and correctness of the result", *Advances in Cryptology CRYPTO '87 Proceedings*, Springer-Verlag, 1988, 87–119.

[GMW] Goldreich, Oded, Silvio Micali, and Avi Wigderson. "How to play any mental game, or: A completeness theorem for protocols with honest majority", *Proceedings of the 19th ACM Symposium on Theory of Computing*, 1987, 218–229.

[GV] Goldreich, Oded, Vainish. *Advances in Cryptology CRYPTO '87 Proceedings*, Springer-Verlag, 1988.

[Ki] Kilian, Joe, "On The Power of Oblivious Transfer," *Proceedings of the 20th STOC*, ACM, 1988.

[MRL] Micali, Silvio, Charles Rackoff, Mike Luby, "How to Simultaneously Exchange a Secret Bit by flipping Assymetrically Biased coins", *Proceedings of the 24th FOCS*, IEEE, 1983, 11–21.

[Ra] Rabin, Michael, "How to exchange secrets by oblivious transfer," Tech. Memo TR-81, Aiken Computation Laboratory, Harvard University, 1981.

Practical Zero-Knowledge Proofs: Giving Hints and Using Deficiencies

Joan Boyar, Katalin Friedl and Carsten Lund*

Computer Science Department

University of Chicago

Abstract

New practical zero-knowledge proofs are given for some number-theoretic problems. All of the problems are in NP, but the proofs given here are much more efficient than the previously known proofs. In addition, these proofs do not require the prover to be super-polynomial in power. A BPP prover with the appropriate trap-door knowledge is sufficient. The proofs are perfect or statistical zero-knowledge in all cases except one.

1 Introduction

Many researchers have studied zero-knowledge proofs and the classes of problems which have such zero-knowledge proofs. Little attention, however, has been paid to the practicality of these proofs. It is known, for example, that, under certain cryptographic assumptions, all problems in NP have zero-knowledge proofs [14], [7], [9]. But these proofs may involve a transformation to a circuit or to an NP-complete problem, so they are often quite inefficient. The first zero-knowledge proofs, those for quadratic residuosity and non-residuosity [17], were practical; they were efficient and the prover could be BPP if she[1] had the appropriate trap-door knowledge. Other efficient zero-knowledge proofs are given in [8], [10], [11], [13], [18], [25].

In this paper we present a practical zero-knowledge proof for a subproblem of primitivity. This protocol, which shows that an element of the multiplicative group modulo a prime is a generator, only requires that the prover know the complete factorization of $p - 1$. Note that the protocol given in [25] is not practical because the prover must be able to compute discrete logarithms. In order to avoid that problem in our protocol, we have the verifier give the prover "hints" which will help her find the discrete logarithms in question.

---

*This research was supported in part by NSA Grant No. MDA904-88-H-2006.

[1]In this paper, it will at times be convenient to think of the verifier as being named Vic, and the prover being named Peggy. This has the advantage that personal pronouns such as "he" and "she" can be used to unambiguously identify one of the parties.

Unfortunately, the portion of our protocol which shows that the element a is a primitive element of Z_p^* fails in some cases if $p - 1$ has large square factors. It fails, though, in such a well-defined manner that we can use its failure in a zero-knowledge proof that a number n is not square-free. This proof that a number is not square-free is zero-knowledge only under a certain reasonable cryptographic assumption and is thus only computational zero-knowledge rather than perfect or statistical zero-knowledge. The protocol does not, however, involve any bit encryption. All previous "natural" zero-knowledge proofs which are neither perfect nor statistical zero-knowledge have used bit encryptions. Furthermore, this zero-knowledge proof is efficient, assuming the Extended Riemann Hypothesis.

We also give practical zero-knowledge proofs for non-primitivity, and for membership and non-membership in $\{n| \ n$ and $\varphi(n)$ are relatively prime$\}$.

When we refer to a *practical zero-knoledge proof* we mean one in wich the prover is *BPP* and the proof is direct, i.e. it doesn't involve a transformation to a circuit or an NP-complete problem. Usually such a transformation would involve a very significant blowup in the size of the problem, greatly increasing the number of bits which must be communicated. For example, the circuit for proving that the element g is a primitive element of Z_p^* would presumably involve checking that a factorization of $p - 1$ is complete and checking for each prime factor q of $p - 1$, that g raised to the power $(p - 1)/q$ is not the identity. This circuit is not at all trivial; the protocol we give involves much less communication.

For a fixed zero-knowledge proof let $CC_k(N)$ be the number of bits communicated to achieve probability of error no more then $(1/2)^k$, where k is the security parameter and N the size of the input. In a typical zero-knowledge proof, one repeats a short protocol k times in order to obtain the required security.

2 The Zero-Knowledge Proofs

2.1 Primitivity

If we are allowing the prover to be all-powerful, it is easy to give a zero-knowledge proof that g is a generator of the multiplicative group modulo a prime p. In one such proof, the following would be repeated $k = \log_2 p$ times:

1. The verifier randomly and uniformly chooses $r \in Z_{p-1}^*$.

2. The verifier computes $h \equiv g^r \pmod{p}$ and sends it to the prover.

3. The verifier gives a proof of knowledge [13] of r [10].

4. The prover takes the discrete logarithm of h to get r.

5. The prover sends r back to the verifier who checks that it is correct.

This is slightly more complicated than the zero-knowledge proof in [25], and it still has the problem that the prover needs to be able to take discrete logarithms. We can eliminate this problem, however, by letting the verifier give the prover a hint which will help her to compute the discrete logarithm.

Let us assume that the prover initially has the complete factorization of $p-1$ on her private work tape. Fortunately, it is possible in expected polynomial time to create a random prime p with a given length, along with the complete factorization of $p-1$ [3], [1]. Now, we will modify the above zero-knowledge proof to include the following steps:

0. The verifier attempts to factor $p-1$ by trial division up to $c \log^c p$. For all nontrivial factors q found, the verifier checks that neither $g^q \pmod p$ nor $g^{(p-1)/q} \pmod p$ is the identity. Here c is an arbitrary constant.

1. The verifier randomly and uniformly chooses $r \in Z_{p-1}^*$.

2. The verifier computes $h \equiv g^r \pmod p$ and sends it to the prover.

$2\frac{1}{2}$. The verifier computes $x \equiv r^2 \pmod{p-1}$ and sends it to the prover.

3. The verifier gives a proof of knowledge [13] of the discrete logarithm of h [10].

4'. The prover takes the discrete logarithm of h to get r and checks that x has the correct form. If something fails, the prover terminates the protocol.

5. The prover sends r back to the verifier who checks that it is correct.

If we assume that $p-1$ is square-free, except possibly for powers of small primes less than $c \log^c p$, the above is a perfect zero-knowledge interactive proof system. Observe that in step 3 the verifier need not prove that x has the correct form because the prover can test this herself from the discrete logarithm. We do, however, have a protocol which proves directly that the form is correct (see Appendix A). It is not immediately obvious, though, that the above protocol is an interactive proof system (i.e. that the verifier should be convinced after the proof is completed), or that the prover can now compute the discrete logarithm (if $p-1$ has many factors).

Let us first show that it is, in fact, an interactive proof system. Suppose that g is not a generator. We will show that in this case the prover will fail to send back the correct r at least 50% of the time. If g is not a generator, then $g = f^k$ for some $f \in Z_p^*$, and $k = tq$ for some prime factor q of $p-1$. By assumption, q^2 does not divide $p-1$. Thus there is another square root r' of x modulo $p-1$ with $r' \equiv -r \pmod q$, but $r' \equiv r \pmod{\frac{p-1}{q}}$. This means that there exists an integer s such that $r' = r + s(\frac{p-1}{q})$ and $r' \not\equiv r \pmod{p-1}$. But $g^{r'} = f^{tq(r+s(\frac{p-1}{q}))} = f^{tqr} f^{ts(p-1)} = g^r$. Thus there are at least two distinct square roots of x which are discrete logarithms of h, so the prover has at best a 50-50 chance of guessing which one the verifier knows.

Now, we will look at the prover's algorithm for finding discrete logarithms given the hint x. The idea is to use the Chinese Remainder Theorem. For each of the prime power factors q of $p-1$, we find the two square roots r_1 and r_2 of x using [2], [6], [21] or [22]. In order to determine which is correct, we compute $(g^{r_1} \cdot h^{-1})^{\frac{p-1}{q}}$ and $(g^{r_2} \cdot h^{-1})^{\frac{p-1}{q}}$. Without loss of generality, suppose $r_1 \equiv r \pmod{q}$ and $r_2 \equiv -r \pmod{q}$. Then there exist k_1 and k_2 such that $r_1 = r + k_1 q$ and $r_2 = -r + k_2 q$. Then,

$$(g^{r_1} \cdot h^{-1})^{\frac{p-1}{q}} = (g^{r+k_1 q} \cdot g^{-r})^{\frac{p-1}{q}} = 1$$

and

$$(g^{r_2} \cdot h^{-1})^{\frac{p-1}{q}} = (g^{-r+k_2 q} \cdot g^{-r})^{\frac{p-1}{q}}$$
$$= g^{-2r(\frac{p-1}{q})} \neq 1$$

since r was chosen from Z_{p-1}^*. Thus, the prover can simply choose the square root which produces the identity in this formula, and then put all the square roots modulo the different prime power factors together, using the Chinese Remainder Theorem.

The proof that this protocol is in fact a perfect zero-knowledge proof system for primitivity when $p-1$ is "essentially" square-free, follows the lines of [15]. We will sketch some of the ideas for the construction of the simulator. The main idea is to use the verifier (here he can be any BPP-machine) and his proof that he knows r to find this r. In the verifier's proof, the following is done in parallel[2] for $1 \leq i \leq k = \log_2 p$.

1. The verifier randomly and uniformly chooses $r_i \in Z_{p-1}^*$.

2. The verifier computes $h_i \equiv g^{r_i} \pmod{p}$ and sends it to the prover.

3. The prover chooses $\beta_i \in \{0, 1\}$ randomly and sends β_i to the verifier.

4. If $\beta_i = 0$ then the verifier sets $\hat{r}_i = r_i$ otherwise he sets $\hat{r}_i = r_i + r$. Then he reveals $\hat{r}_i$

5. The prover checks that $h_i = g^{\hat{r}_i}/h^{\beta_i}$.

This subprotocol may no longer be zero-knowledge when run in parallel, but it does not help a cheating prover. If g is not a generator and $g^m \equiv 1 \pmod{p}$, then r and $r' = r + lm$ are both discrete logarithms of h. But the verifier could have chosen either r_i or $r_i - lm$. Thus $\hat{r}_i$ will never help the prover distinguish between these possibilities.

---

[2]This may be done sequentially. We are doing it in parallel to make it clearer that the entire primitivity protocol can be done in parallel.

Now the simulator for the primitivity protocol works as follows. It asks a question $(\beta_1, \beta_2, \ldots, \beta_k)$, and if it does not get a correct answer, it stops as the real prover would. If it gets a correct answer, it has to find the real r since this is what the real prover does. To do this, it resets the verifier to the point just before the question was asked and asks another random question $(\beta_1', \beta_2', \ldots, \beta_k')$. If it also gets a correct answer for this question it can find r, since if $\beta_i' \neq \beta_i$ we have $r = \pm(\hat{r}_i' - \hat{r}_i)$. If it does not get correct answer, then the simulator continues asking questions until it gets a correct answer or it has asked 2^k questions. In the second case it can continue to find r, using brute force. It can be shown that this simulator runs in expected polynomial time for all verifiers. See [15] for more details. Furthermore we can make this a bounded round protocol because this simulator works even if the protocol is run in parallel [15]. Hence we get a bounded round perfect zero-knowledge protocol. If the length of p is N, then the communication cost of the protocol is $O(k^2 N)$ because $O(kN)$ bits are communicated in step 3 to achieve error probability not greater then $(1/2)^k$. This gives

Theorem 1 *Let p be a prime and c be an arbitrary constant. Assume that if $q > c \log_2^c p$ is a prime then q^2 does not divide $p - 1$. Then there is a practical perfect zero-knowledge, bounded round, interactive proof system for*

$$\{g| < g >= Z_p^*\}.$$

The BPP prover's secret information is the complete factorization of $p - 1$. The communication cost of this protocol is $CC_k(N) = O(k^2 N)$.

The set of primes for which our proof works is of reasonable size since [24] proved that

$$\exists c > 0: \frac{\{p| \ p \leq x, p \text{ prime and } p - 1 \text{ squarefree}\}}{\{p| \ p \leq x \text{ and } p \text{ prime}\}} \geq c$$

for x sufficiently large.

Throughout this section, we have been looking at the multiplicative group Z_p^* of the integers modulo a prime p. It is easy, however, to generalize the proof system given above to many other cyclic group with known order. Consider, for example, the multiplicative group Z_q^* of the integers modulo $q = p^n$, where p is an odd prime and $n \geq 1$.[3] Almost all that is necessary is to substitute $\varphi(q) = p^{n-1}(p-1)$ in place of $p - 1$ throughout this exposition. (When q is prime, $\varphi(q)$, Euler's phi function, has the value $q - 1$.) Of course, $\varphi(q)$ is never square-free if $n > 2$. This is not a problem, however, because q is easy to factor, so the verifier and the simulator can find p and can check that $g^{p^{n-2}(p-1)} \not\equiv 1 \pmod{p}$. Thus, one can assume that $g \not\equiv h^{tp} \pmod{p}$ for any integer t, and one again only needs to worry about square factors of $p - 1$.

---

[3]Notice that this is even easier in this particular case because the problem of determining primitivity in the group $Z_{p^n}^*$ is efficiently reducible to that of determining primitivity in Z_p^*. This follows from the fact, that an element $g \in Z_{p^n}^*$ is primitive if and only if $g^{p^{n-2}(p-1)} \not\equiv 1 (\bmod p^n)$ and g is primitive when viewed as an element of the group Z_p^*.

2.2 Are n and $\varphi(n)$ relatively prime?

In the zero-knowledge proof system presented in the previous section, we had to assume that $p-1$ was square-free except for powers of small primes. This is unfortunate, particularly since there is no known efficient zero-knowledge proof for square-freeness. It is possible, however, to give an efficient proof that a number n and $\varphi(n)$, the number of elements in the multiplicative group modulo n, are relatively prime. This property implies that n is square-free. Thus, if $p-1 = 2^k r$, where r is odd, and if r and $\varphi(r)$ are relatively prime, the prover could prove that this is the case and afterwards she could prove primitivity. Unfortunately, it is possible to have r and $\varphi(r)$ not relatively prime even if $p-1$ is square-free, so this proof system will not work for quite as large a class as we would like. Combined with the proof system of the previous section, however, it gives a perfect zero-knowledge proof for

$$\{(p,g) \mid p \text{ is prime}, p-1 = 2^k r, \text{ where } r \text{ is odd}, \gcd(r, \varphi(r)) = 1, \text{ and } <g> = Z_p^*\}.$$

Suppose the prover knows $\varphi(n)$ for an odd integer n and wants to prove that n and $\varphi(n)$ are relatively prime. The prover and verifier can repeat the following $\log_2 n$ times.

1. The verifier randomly and uniformly chooses $x \in Z_n^*$ and sends it to the prover.

2. The prover chooses a random $r \in Z_n^*$ and sends the verifier $y \equiv r^n x \pmod{n}$.

3. The verifier chooses $\beta \in \{0,1\}$ randomly with equal probabilities and sends β to the prover.

4. If $\beta = 0$, the prover reveals r showing that y was formed correctly. If $\beta = 1$, the prover reveals an nth root of y, thus showing that x has an nth root modulo n.

To see that this works, suppose that n and $\varphi(n)$ are not relatively prime. Then, the $\gcd(n, \varphi(n)) = q$, where $1 < q < \varphi(n) < n$. Since there is some positive integer t such that, for every $g \in Z_n^*$, $g^n \equiv g^{tq} \pmod{n}$, every element which has n^{th} roots also has q^{th} roots. But no more than half of the elements of Z_n^* have q^{th} roots modulo n. If the verifier chooses an x which does not have an nth root, there is no more than a 50-50 chance that the prover will be able to answer the challenge chosen by the verifier. Thus, at each step, there is at least one chance in four that the prover will be caught, making the probability that the prover will succeed $\log_2 n$ times exponentially small. When n and $\varphi(n)$ are relatively prime $x \equiv (x^n)^k \pmod{n}$ where $k \equiv (n \pmod{\varphi(n)})^{-1} \pmod{\varphi(n)}$. Hence the prover can compute n^{th} roots of x and y.

This is clearly perfect zero-knowledge since the simulator has a 50-50 chance each time of guessing which β the verifier will choose. When the simulator guesses that $\beta = 0$, he computes y exactly as the prover would, so he has no problem revealing r. When the simulator guesses that $\beta = 1$, he sends $y \equiv r^n \pmod{n}$, so he can give an nth root of y. When he guesses incorrectly, he backs up the transcript tape and tries again. Thus the simulation will be in expected polynomial time. Since y is a random element of Z_n^* whether it is produced by the prover or the simulator, the transcripts produced by the simulator will have the same distribution as those produced by the true prover. Thus this protocol is perfect zero-knowledge. The protocol can furthermore be parallelized following the lines of [4]. The above discussion gives

Theorem 2 *There is a practical perfect zero-knowledge interactive proof system for*

$$\{n \mid \gcd(n, \varphi(n)) = 1\}$$

with communication cost $CC_k(N) = O(kN)$. The BPP prover's trapdoor information is the number $\varphi(n)$.

Theorem 3 *There is a practical perfect zero-knowledge interactive proof system for*

$$\{(p, g) \mid p \text{ prime}, p - 1 = 2^k r, \text{where } r \text{ is odd}, \gcd(r, \varphi(r)) = 1, \text{and} < g >= Z_p^*\}.$$

The BPP prover's secret information is the complete factorization of $p - 1$. The communication cost is $CC_k(N) = O(k^2 N)$.

If n and $\varphi(n)$ are not relatively prime, a prover who knows $\varphi(n)$ can give a practical zero-knowledge proof that they have a common factor, under certain assumptions. One such proof involves repeating the following $\log_2 n$ times. First, the prover sends the verifier a random $x \in Z_n^*$ such that x does not have an nth root. Then the verifier chooses a random $r \in Z_n^*$ and a random bit β. The verifier then sends $y \equiv r^n x^\beta \pmod{n}$ to the prover. Next, using the technique due to Benaloh [5] of using cryptographic capsules, the verifier gives a zero-knowledge proof that he knows n and β. Finally, the prover reveals the bit β. The reason this is not perfect zero-knowledge is that the prover must originally produce an nth-nonresidue x, and it's not clear that the simulator can do this. If $q = \gcd(n, \varphi(n))$ is large enough (superpolynomial) though, the simulator could pick $x \in Z_n^*$ at random and it's unlikely that x would be a qth-residue. In this case, the protocol would be statistical zero-knowledge.

2.3 Nongenerators

Suppose p is a prime and g is not a generator of Z_p^*. If the prover knows a $t < p-1$ such that $g^t \pmod{p} \equiv 1$, then she can give a practical statistical zero-knowledge proof that g is not a generator. The proof is statistical zero-knowledge if $s = \frac{p-1}{t}$ is large. The major advantage of the protocol given here over that in [25] is that we do not need to assume that a generator for Z_p^* is publicly available. The set we are concerned with is

$$S = \{(p,g)\mid p \text{ is a prime}, \exists t < p - 1, g^t \equiv 1 \,(\text{mod } p)\}.$$

The values p and g are available to both the prover and the verifier; the value t is initially on the prover's private work tape; and the prover is attempting to convince the verifier that g is not a generator modulo p. Our proof is based on the fact that for every integer r, $g^r \equiv g^{r+tl} \,(\text{mod } p)$ if l is an integer, so the prover can find many discrete logarithms for an element as long as she knows one discrete logarithm. If g was a generator, however, each element would have only one discrete logarithm in the range $[1, p-1]$. The protocol consists of $\log_2 p$ independent repetitions of the following:

1. The prover chooses a random r uniformly from the range $[1, t]$.

2. The prover sends the verifier $h \equiv g^r \,(\text{mod } p)$.

3. The verifier chooses $\beta \in \{0, 1\}$ randomly with equal probabilities and sends β to the prover.

4. If $\beta = 0$, the prover chooses a random z uniformly from $[0, \lceil \frac{s-3}{2} \rceil]$. If $\beta = 1$, the prover chooses a random z uniformly from $[\lfloor \frac{s+1}{2} \rfloor, s-1]$.

5. The prover sends the verifier $r' = r + zt$ who checks that $h \equiv g^{r'} \,(\text{mod } p)$ and that $r' \in [1, \frac{p-1}{2}]$ if $\beta = 0$, or that $r' \in [\frac{p-1}{2} + 1, p-1]$ otherwise.

Notice that in step 5 the prover is revealing a discrete logarithm of h which is less than $\frac{p-1}{2}$ if the verifier's challenge was $\beta = 0$, or greater than $\frac{p-1}{2}$ if $\beta = 1$. If g were a generator, only one discrete logarithm would exist, so for each of the verifier's challenges, the prover would have at most a 50-50 chance of being able to give the correct response. The communication cost of this protocol is $CC_k(N) = O(kN)$.

Let us look at a simulator for this protocol. The simulator would choose a random r uniformly from $[1, p-1]$. The simulator would then run the program for the verifier with the value $g^r \,(\text{mod } p)$ being sent from the prover. The simulator has a 50-50 chance of answering the verifier's question each time simply by revealing r. If it cannot answer, he will backtrack the verifier to the point of choosing r and try another one. so the simulation is expected polynomial time. Both the prover and the simulator choose h to be a random element of the subgroup generated by g, but the distributions of r''s in step 5 are somewhat different depending on whether you have the true prover or the simulator. The true prover never gives r' in the interval $[\frac{p-1}{2} - \frac{t}{2} + 1, \frac{p-1}{2} + \frac{t}{2}]$ if s is odd, but the simulator might. But since s is large, these distributions are statistically close. Let us look at one of the independent repetitions of the above protocol. Let $P(x)$ denote the probability that the true prover reveals x in step 5, and let $S(x)$ denote the probability that the simulator produces x in step 5. For any subset X of $\{1, \ldots, p-1\}$, $|\sum_{x \in X} P(x) - \sum_{x \in X} S(x)| \leq \frac{1}{s}$. Hence for the whole protocol the distributions differ by at most $\frac{\log_2 p}{s}$. Thus this protocol is statistical zero-knowledge on subsets of S of the form

$$S_f = \{(p, g) \mid p \text{ prime}, \exists t < p - 1, g^t \equiv 1 \pmod{p} \text{ and } \frac{p-1}{t} \geq f(\log p)\}$$

where f is superpolynomial.

This restriction to subsets S_f of S is unfortunate. If the prover only proves things from these smaller sets she gives away some information, i.e. that s is large. This does not appear to be much information since if s is small the verifier could himself have found s. But since there is a grey area between large and small, we can't find a uniform simulator that works for all possible magnitudes for s. One solution to this problem is to consider an alternative definition of zero-knowledge. In the GMR-definition we have a simulator which can fool every BPP-distinguisher with probability greater than $1 - \frac{1}{n^c}$ for every c for n sufficiently large. In our definition, we give c to the simulator, which then runs in time polynomial in n^c. Hence the simulator is BPP for fixed c. Otherwise this definition is identical to Oren's [20], and we are using similar notation.

Definition 1 *Let (P, V) be a interactive proof system for L. Then (P, V) is weak zero-knowledge if*

$$\forall V^* : \exists M_{V^*} : \forall x \in L : \forall y : \forall D \in BPP : \forall c :$$

$$|\Pr[D((< P(x), V^*(x, y) >) = 0] - \Pr[D(M_{V^*}(c, x, y)) = 0]| \leq \frac{1}{|x|^c}.$$

It is weak statistical zero-knowledge if, for any subset T of transcripts:

$$\forall V^* : \exists M_{V^*} : \forall x \in L : \forall y : \forall c :$$

$$|\Pr[< P(x), V^*(x, y) > \in T] - \Pr[M_{V^*}(c, x, y)) \in T]| \leq \frac{1}{|x|^c}.$$

We believe that this definition captures the intuition of zero-knowledge.

With this definition we can easily construct a simulator for the nongenerator protocol. It behaves exactly as the old one after testing that $s \geq \log^{c+1} n$. If it finds s and hence t, it proceeds as the real prover would; otherwise it proceeds as the old simulator would.

With this new definition of zero-knowledge we can also remove the assumption, in the protocol in [25] for the same problem, that one generator is publicly known. We can let the prover give the verifier a random generator. This is practical weak zero-knowledge because the simulator can find a generator with probability $1 - \log^{-c} n$ in time polynomial in $\log^c n$. See Appendix B.

The above discussion gives

Theorem 4 *There is a practical interactive proof system for*

$$\{(p,g)|\ p \text{ is a prime}, \exists t < p-1, g^t \equiv 1 \ (\text{mod } p)\}$$

and it is statistical zero-knowledge on

$$\{(p,g)|\ p \text{ is a prime}, \exists t < p-1, g^t \equiv 1 \ (\text{mod } p) \text{ and } \frac{p-1}{t} \geq f(\log_2 p)\},$$

where f is superpolynomial. This protocol has commumication cost $CC_k(N) = O(kN)$. The BPP prover's secret knowledge is t.

Using our new definition we get:

Theorem 5 *There is a practical weak statistical zero-knowledge interactive proof system for*

$$\{(p,g)|p \text{ is a prime}, \exists t < p-1, g^t \equiv 1 \ (\text{mod } p)\}$$

with communication cost $CC_k(N) = O(kN)$. The BPP prover's secret information is t.

The proof system presented above can be extended to work for many other cyclic groups with known order. In particular, when working with the multiplicative group modulo a power of an odd prime, all that is necessary is to substitute $\varphi(q) = p^{n-1}(p-1)$ in place of $p-1$ throughout this exposition.

Furthermore the protocol can be parallelized using techniques similar to those of [4].

In the parallel protocol, Peggy will first choose randomly and uniformly $x \in Z^*_{p-1}$, compute $f \equiv g^x \ (\text{mod } p)$, and send f to Vic. Next, Vic chooses all his challenges $(\beta_1, \beta_2, \ldots, \beta_{\log n})$ and commits to them by choosing a random $s_i \in Z^*_{p-1}$ for each β_i. If $\beta_i = 0$, he lets $t_i \equiv g^{s_i} \ (\text{mod } p)$; otherwise $t_i \equiv f^{s_i} \ (\text{mod } p)$. He sends $(t_1, t_2, \ldots, t_{\log n})$ to Peggy, who now does steps 1 and 2 from the original protocol and sends $(h_1, h_2, \ldots, h_{\log n})$ to Vic. Vic now reveals his challenges by sending $(s_1, s_2, \ldots, s_{\log n})$. Finally Peggy, after checking that Vic did not cheat will send x and $(r'_1, r'_2, \ldots, r'_{\log n})$. Vic checks that $f = g^x$, checks that $x \in Z^*_{p-1}$ and performs the checks corresponding to step 5.

To see that this is still a proof system, observe that the two different commitments come from the same distribution since $x \in Z^*_{p-1}$. Thus receiving these commitments earlier is no help to Peggy.

The simulator is constructed as follows. It does the same as Peggy until Vic reveals all his challenges. Then it backtracks to the point where Vic had just made his commitments. Now the simulator forms its h_i's so that it can answer Vic's questions. If Vic reveals the same old questions, the simulator can answer them. If he reveals another set of questions, the simulator know s, s' such that $g^s = f^{s'}$. This gives

$$g^s = g^{xs'} \Rightarrow g^{s-xs'} = 1.$$

It is easy to see that, since Peggy chose x randomly, $s - xs' \pmod{p-1}$ is a random multiple of t, the order of g. If we run the above procedure twice, either the simulator will be able to successfully perform the simulation or it will get two independent, random multiples of t, at and $a't$. We know from [19] that $\Pr[\gcd(at, a't) = t] = 6/\pi^2$. Thus the simulator will succed in expected polynomial time.

If the modulus has more than one prime factor or is a large power of two, no elements would be generators. One could, however, still ask the question: Does the subgroup generated by the element g have fewer than m elements (for a prime modulus m can be $p-1$)? Then if the prover knows t such that $g^t \equiv 1 \pmod{n}$, and $s = \lfloor m/t \rfloor$ is sufficiently large, one could give a zero-knowledge proof that g only generates a small subgroup.

2.4   Does n have a square factor?

Recall that the protocol given above for proving that an element is a generator of the multiplicative group modulo a prime p only works when $p-1$ is "essentially" square-free. The only problems that arise when $p-1$ has large square factors is that some nongenerators may look like generators. In fact, if g is a generator, the prover can make a nongenerator $g' = g^q$ look like a generator if $p-1 = 2^l p_1^{e_1} p_2^{e_2} \cdots p_k^{e_k}$ and $q = p_1^{f_1} p_2^{f_2} \cdots p_k^{f_k}$ where $0 \le f_i < e_i$ for all i (assuming that $p-1$ and q are such that the verifier cannot determine himself that g' is not a generator). The prover can do this because the prover's algorithm for finding the discrete logarithms given the hint x will still work. Let us call any g' of this form a quasi-generator. Note that if $f_i = e_i$ for any i, then there will be two distinct square roots of x which will work as discrete logarithms, so the protocol is in fact a perfect zero-knowledge proof that g is a quasi-generator, regardless of whether or not $p-1$ is square-free. In the case that $p-1$ is square-free, all quasi-generators are actually generators.

It is possible to use this deficiency in the proof system for primitivity to show that an integer n is not square-free. The set we are concerned with is

$$S = \{n|\ n = q^2 m, q \text{ prime}\}$$

The integer n is available to both the prover and the verifier; the complete factorization of n is initially on the prover's private work tape; and the prover is attempting to convince the verifier that n has a nontrivial square factor. To do this, the prover first finds a prime $p = an + 1$. Assuming the Extended Riemann Hypothesis, one can try random a's which are less than n^2 and expect to find such a prime in time $O(\log n)$.

To see this, consider the following from [12](pp.129, 136). Assuming the Extended Riemann Hypothesis,

$$|\{p\ |\ p \text{ prime}, p \le x, p \equiv 1 \pmod{n}\}| = \frac{\text{li}x}{\varphi(n)} + O(x^{\frac{1}{2}} \log x),$$

where

$$\mathrm{li}x = \int_2^x \frac{1}{\log t}\,dt = \frac{x}{\log x} - \frac{2}{\log} + \int_2^x \frac{1}{\log^2 t}\,dt > \frac{x}{\log x} + O(1).$$

Hence the probability that a random m, chosen so that $m \equiv 1 \bmod n$ and $m \le x$, is prime is

$$\frac{\frac{x}{\varphi(n)\log x} + \frac{O(1)}{\varphi(n)} + O(x^{\frac{1}{2}}\log x)}{\lfloor (x-1)/n \rfloor}$$

We have from [23] that $\varphi(n) \ge C(n/\log\log n)$; hence if $x = n^3$ the above is greater than

$$C'\frac{\log\log n}{\log n} + O(n^{-\frac{1}{2}}\log n).$$

Note that $x = n^{2+\epsilon}$ is sufficient if $\epsilon > 0$.

To find p, one can use Bach's method [3] to produce an appropriate a randomly, along with the complete factorization of a. Since this protocol would be unnecessary if the verifier could find q, we can assume that q does not divide a.

Another way to find an appropriate p is by trying $n+1, 2n+1, 3n+1, \ldots$ until we find a prime. Wagstaff [26] has given an heuristic argument which says that we would usually only have to try up to $O(\log^2 n)$ numbers. Observe that we can factor a since it is so small.

After finding such a prime, the prover will produce an element $h \in Z_p^*$ which is of the form $g^q \pmod p$ for some generator g. The prover can then use the above protocols to show, first, that h is not a generator, and, second, that h is a quasi-generator. Of course, since h is not a generator and it is a quasi-generator, $p-1$ is not square-free. But the verifier can check that $h^{(p-1)/r} \not\equiv 1 \pmod p$ for any prime factor r of a. Thus, the square factor in question must be a factor of n, proving that n does, in fact, have a square factor.

This protocol is obviously not perfect zero-knowledge, or even statistical zero-knowledge unless there is some way for the simulator to produce an h of the required form. Since the simulator does not know q, it seems unlikely that it could produce such an h. We will make the cryptographic assumption that finding the factor q of n is random polynomial time equivalent to distinguishing between random generators and random quasi-generators corresponding to q. This seems reasonable because the known algorithms for testing for primitivity involve factoring $p-1$. In place of the quasi-generators the simulator will produce a random element of Z_p^* which it cannot tell is not a generator (i.e. if r is a factor of a or a small factor of n where small means less than $\log_2^{c+1} p$, then neither h^r nor $h^{(p-1)/r}$ is the identity). With probability $\log_2^{-c} p$ this element is a generator of Z_p^* (see Appendix B). Thus, under the above cryptographic assumption, this protocol is computational zero-knowledge if the verifier can't find q. Now under the usual assumption, that factoring is hard in general, there exists a infinite subset K of S on, which the protocol will be computational weak zero-knowledge. A candidate for a subset of this K is

$$M_\epsilon = \{n| \ \forall \text{primes } p|n \ \exists \text{primes } q_1|p-1, q_2|p+1, q_1, q_2 > n^\epsilon\}$$

since no known factorization algorithm can factor numbers from M_ϵ in expected polynomial time. The protocol does not, however, involve any bit encryption. All previous "natural" zero-knowledge proofs which are neither perfect nor statistical zero-knowledge have used bit encryptions.

The above discussion gives

Theorem 6 *Assuming the Extended Riemann Hypothesis there is a practical interactive proof system for*

$$S = \{n| \ n = q^2 m, q \text{ prime}\},$$

with $CC_k(N) = O(k^2 N)$. The BPP prover secret information is the complete factorization of n.

Let K be a subset of S. For each $n \in K$, we will define the distributions G_n and Q_n as follows. We will choose p randomly and uniformly such that $|p| \leq |n|^3$, p is a prime and $n \mid p-1$. Then choose g at random and uniformly from the set of generators of Z_p^*. Now look at the two distributions

$$G_n = \{(g, p)\} \text{ and } Q_n = \{(g^q, p)\}$$

If

$$\forall D \in BBP \ \forall c \ \exists N \ \forall n \in K : n > N \Rightarrow |\Pr[D(G_n) = 1] - \Pr[D(Q_n) = 1]| \leq \frac{1}{\log^c n}$$

then the protocol is weak zero-knowledge on K.

3 Open Problems

One would like to find practical zero-knowledge proofs for other problems. In particular, we began working on these problems after David Chaum mentioned the problem of finding a practical zero-knowledge proof that an element g generates a large subgroup modulo a composite number n. That problem is still open. We would also like to eliminate the assumption that $p-1$ is "essentially" square-free in the primitivity protocol.

The protocol given here to show that a number is not square-free is zero-knowledge, but not statistical zero-knowledge. A statistical or perfect zero-knowledge protocol for this problem would be interesting.

We would also like to find a practical zero-knowledge proof that a number n is square-free.

4 Acknowledgements

We are very grateful to David Chaum for suggesting the problem mentioned in the last section, to René Peralta for pointing out that proving knowledge of the discrete logarithm is sufficient for step 3 of the primitivity protocol, and to Eric Bach and Kevin McCurley for answering numerous questions on factoring algorithms and the distributions of primes. We would also like to thank Ernie Brickell, Faith Fich, Mark Krentel, Stuart Kurtz, Jeff Shallit, and Janos Simon for helpful discussions.

References

[1] Adleman, L., and M.-D. Huang, *Recognizing primes in random polynomial time*, Proc. 19th ACM Symp. on Theory of Computing, 1987, pp. 462-469.

[2] Adleman, L., K. Manders, and G. Miller, *On taking roots in finite fields*, Proc. 18th IEEE Symp. on Foundations of Computer Science, 1977, pp. 175-178.

[3] Bach, E., *How to generate factored random numbers*, SIAM Journal on Computing, vol. 17, No. 2, April 1988, pp. 179-193.

[4] Bellare, M., S. Micali and R. Ostrovsky, personal communication.

[5] Benaloh, J., *Cryptographic capsules: a disjunctive primitive for interactive protocols*, Advances in Cryptology - Crypto '86 Proceedings, 1987, pp. 213-222.

[6] Berlekamp, E. *Factoring polynomials over large finite fields*, Mathematics of Computations, vol. 24, 1970, pp. 713-735.

[7] Brassard, G., and C. Crépeau, *Non-transitive transfer of confidence: a perfect zero-knowledge interactive protocol for SAT and beyond*, Proc. 27th IEEE Symp. on Foundations of Computer Science, 1986, pp. 188-195.

[8] Brassard, G., C. Crépeau, and J.M. Robert, *All-or-nothing disclosure of secrets*, Advances in Cryptology - Crypto '86 Proceedings, 1987, pp. 234-238.

[9] Chaum, D., *Demonstrating that a public predicate can be satisfied without revealing any information about how*, Advances in Cryptology - Crypto '86 Proceedings, 1987, pp. 195-199.

[10] Chaum, D. J.-H. Evertse, J. van de Graaf, *An improved protocol for demonstrating possession of discrete logarithms and some generalizations*, Advances in Cryptology - EUROCRYPT '87 Proceedings, 1988, pp. 127-141.

[11] Chaum, D., J.-H. Evertse, J. van de Graaf, and R. Peralta, *Demonstrating possession of a discrete logarithm without revealing it*, Advances in Cryptology - Crypto '86 Proceedings, 1987, pp. 200-212.

[12] Davenport, H., *Multiplicative Number Theory*, Markham Publishing Company, 1967.

[13] Feige, U., A. Fiat, and A. Shamir, *Zero-knowledge proofs of identity*, Journal of Cryptology, 1(2), 1988, pp. 77-94.

[14] Goldreich, O., S. Micali, and A,. Wigderson, *Proofs that yield nothing but their validity and a methodology of cryptographic protocol design*, Proc. 27th IEEE Symp. on Foundations of Computer Science, 1986, pp. 174-187.

[15] Goldreich, O., S. Micali, and A,. Wigderson, *Proofs that yield nothing but their validity and a methodology of cryptographic protocol design*, To appear.

[16] Goldwasser, S., and S. Micali, *Probabilistic encryption*, Journal of Computer and System Sciences, vol. 28, 1984, pp. 270-299.

[17] Goldwasser, S., S. Micali, and C. Rackoff, *The knowledge complexity of interactive proof systems*, SIAM Journal on Computing, vol. 18, 1989, pp. 186-208.

[18] Van de Graaf, J., and R. Peralta, *A simple and secure way to show the validity of your public key*, Advances in Cryptology - Crypto '87 Proceedings, 1988, pp. 128-134.

[19] Knuth, D. E. *The Art of Computer Programming* Vol 2, Addison-Wesley, 1969.

[20] Oren, Y. *On the Cunning Power of Cheating Verifiers: some Observations About Zero Knowledge Proofs*, Proc. 28th IEEE Symp. on Foundations of Computer Science, 1987, pp. 462-471.

[21] Rabin, M.O., *Digitalized signatures and public-key functions as intractable as factorization*, Technical Report MIT/LCS/TR-212, M.I.T., January 1979.

[22] Rabin, M.O., *Probabilistic algorithms in finite fields*, SIAM Journal on Computing, vol. 9, 1980, pp. 273-280.

[23] Rosser, J. B., and Schoenfeld, L., *Approximate Formulas for some Functions of Prime Numbers*, Illinois Journal of Math. vol. 6, 1962, pp. 64-94.

[24] Schwarz, W., in American Math. Monthly, vol. 73, 1966, pp. 426-427.

[25] Tompa, M., and H. Woll, *Random self-reducibility and zero knowledge interactive proofs of possession of information*, Proc. 28th IEEE Symp. on Foundations of Computer Science, 1987, pp. 472-482.

[26] Wagstaff, S. S., *Greatest of the Least Primes in Arithmetic Progressions Having a Given Modulus*, Mathematics of Computation, vol. 33 no. 147, July 1979, pp. 1073-1080.

Appendix A

In this appendix we show how the verifier in the primitivity protocol can prove that for $h \in Z_p^*$ and $x \in Z_{p-1}^*$ he "knows" [13] an r such that $g^r \equiv h \pmod{p}$ and $r^2 \equiv x \pmod{p-1}$. Both the verifier and prover can compute $a \equiv g^x \pmod{p} \equiv h^r \pmod{p}$. Hence if the verifier can show that h and a were formed by raising g and h, respectively, to the same power, it will have shown the correct form of h and x. Rather than simply presenting a zero-knowledge proof of knowledge for this problem, we will present a practical zero-knowledge proof of knowledge for something more general. Suppose a prover and verifier (the original verifier will temporarily be acting as a prover) are given the following elements of a commutative group u, v, $a_1, a_2, \ldots, a_n$, and $b_1, b_2, \ldots, b_n$, and that the prover wants to show that she knows integers $e_1, e_2, \ldots, e_n$ such that

$$u = a_1^{e_1} a_2^{e_2} \ldots a_n^{e_n}$$

and

$$v = b_1^{e_1} b_2^{e_2} \ldots b_n^{e_n}.$$

In this zero-knowledge proof, the following will be repeated a number of times equal to the length of the input.

1. The prover chooses random $c_1, c_2, \ldots, c_n$ in the range $[1 \ldots t]$, where t is the order of the group.

2. The prover computes $y = a_1^{c_1} a_2^{c_2} \ldots a_n^{c_n}$ and $z = b_1^{c_1} b_2^{c_2} \ldots b_n^{c_n}$.

3. The prover computes $u' = uy$ and $v' = vz$ and sends these values to the verifier.

4. The verifier chooses $\beta \in \{0, 1\}$ randomly with equal probabilities and sends β to the prover.

5. If $\beta = 0$, the prover reveals $c_1, c_2, \ldots, c_n$, so the verifier can check that u' and v' were formed correctly. If $\beta = 1$, the prover reveals $e_1 + c_1, e_2 + c_2, \ldots, e_n + c_n$, so the verifier can check that the prover knows the same information about u' and v' that she was supposed to know about u and v.

This is clearly a proof of knowledge because an observer seeing the prover respond to both challenges for the same u' and v' could compute $e_1, e_2, \ldots, e_n$ by subtracting. It is also zero-knowledge if the order of the group is known as it would be for the group Z_p^* with which we are mainly concerned. To see this, consider the following simulator which produces transcripts of the proof with exactly the same distribution as would be produced with the legitimate prover. First, the simulator flips a coin. If the result of this coin flip is "heads", the simulator is guessing that the verifier will send $\beta = 0$. Thus, the simulator goes through exactly those steps the prover would in forming u' and v'. Since we are assuming that the simulator knows the order of the group, this is easy. If, when the simulator runs the program for the verifier with this u' and v', the verifier sends $\beta = 0$, the simulator has no problem in revealing the required values. If, however, the verifier sends $\beta = 1$, the simulator will back up the tape over these current u' and v' and will try again with another coin flip. If the result of the coin flip is "tails", the simulator is guessing that the verifier will send $\beta = 1$, so it will produce the random c_i's, the y and the z, but it will send the verifier y and z for u' and v'. When it receives $\beta = 1$ from the verifier, it can simply reveal the c_i's. Of course, if it receives $\beta = 0$, it will have to back up the tape and try again. But it has a 50-50 chance of succeeding each time so the simulation is expected polynomial time, and it is easy to check that it produces transcripts with exactly the same distribution as those produced with the true prover. Thus, this proof system is perfect zero-knowledge.

Appendix B

Let C_n be a cyclic group of order n.

Consider the following procedure.

Construct the set $S = \{p \mid p \text{ prime}, p \leq \log^{c+1} n \text{ and } p|n\}$
repeat
Choose g randomly and uniformly from C_n.
until $\forall p \in S : g^{n/p} \neq 1$
OUTPUT g

Theorem 7 *Prob(g is not a generator)* $< \frac{1}{\log^c n}$

proof :
Suppose $n = p_1 p_2 \ldots p_k q_1 q_2 \ldots q_l$ where each $p_i \leq \log^{c+1} n$ and each $q_i > \log^{c+1} n$.
Let $n' = p_1 p_2 \ldots p_k$ and $T = \{g \mid \text{the procedure can output } g\} = \{g \mid \forall i : g^{n/p_i} \neq 1\}$.
Now for $d|n$ define $A_d = \{x \mid \text{order}(x) = d\}$. $\bigcup_{d|n} A_d = C_n$ and $|A_d| = \varphi(d)$,

$$T = \bigcup_{\substack{d|n \\ n'|d}} A_d.$$

So we get that

$$|T| = \sum_{d|\frac{n}{n'}} \varphi(n'd) = \varphi(n') \sum_{d|\frac{n}{n'}} \varphi(d).$$

Furthermore we have that

$$|G| = |\{g \mid <g> = C_n\}| = \varphi(n) = \varphi(n')\varphi(n/n').$$

Hence we can assume that $k = 0$, i.e. that $q|n$ implies that $q > \log^{c+1} n$.

$$T \setminus G = \bigcup_{i=1}^{l} \{x \mid x^{n/q_i} = 1\},$$

where the cardinality of each term is estimated by

$$|\{x \mid x^{n/q_i} = 1\}| = n/q_i \leq n/\log^{c+1} n.$$

So we get

$$|T \setminus G| \leq l(n/\log^{c+1} n) \leq n/\log^c n.$$

This proves the theorem. $\square$

An alternative to the Fiat-Shamir protocol [*]

Jacques Stern
Équipe de Logique
Université Paris 7
et

Département de mathématiques et informatique
École Normale Supérieure

Abstract

In 1986, Fiat and Shamir [2] exhibited zero-knowledge based identification and digital signature schemes. In these schemes, as well as in
subsequent variants, both the prover and the verifier have to perform
modular multiplications. This paper is an attempt to build identification protocols that use only very basic operations such as multiplication by a fixed matrix over the two-element field. Such a matrix can
be viewed as the parity-check matrix of a linear binary error-correcting
code. The idea of using error-correcting codes in this area is due to
Harari [3] but the method that is described here is both simpler and
more secure than his original design.

1   The signature scheme

The proposed scheme uses a fixed $(n\text{-}k)$-matrix G over the two-element
field. This matrix is common to all users and is originally built randomly.
Thus, considered as a parity-check matrix, it should provide a linear binary
code with a good correcting power. Also common to all users is a family
$w_1, \ldots, w_q$ of words with n bits.

Any user chooses a secret key s which is an n-bit word with a prescribed
number p of 1's. This prescribed number p is also part of the system. Then
he computes his public identification as

$$i = G(s)$$

---

[*]Research supported by the PRC mathématiques et informatique

The sequence w_j will be essentially used as a pseudo-random number generator. Whenever an n-bit word z is given, together with two random integers a, c, with a prime to q, a sequence

$$z_j = z \oplus w_{aj+c} \quad 1 \leq j \leq q$$

is produced.

The identification scheme will rely heavily on the technical notion of a *commitment*. If u is an n-bit word, a *commitment* for u is a sequence

$$G(u_1), \ldots, G(u_l)$$

built from

$$u_1, \ldots, u_l$$

such that

$$\begin{cases} u_i \odot u_j = 0 & 1 \leq i < j \leq l \\ \bigoplus_{i=1}^{l} u_i = u \end{cases}$$

where $\odot$ and $\oplus$ respectively denote bitwise multiplication and bitwise addition modulo 2. Usually, a commitment will be built by choosing randomly a partition of $\{1, \ldots, n\}$ into l pieces. The notion of commitment can be extended to words of length $> n$. By padding randomly, we may restrict ourselves to words whose length is a multiple of n and break these into a sequence of words of length exactly n. Especially, a sequence of integers $n_1, \ldots, n_q$ coded on a fixed number h of bits can be written as a single word of length hq and be given a commitment.

A commitment will be used as a one-way function: in order to disclose it, one announces the sequence

$$u_1, \ldots, u_l$$

from which it was built. Once this is done, anyone can check the correctness of the commitment by applying G to the sequence u_j, recover the original word u and use the information it encodes.

We now describe the interactive protocol that enables any user (which we will call the prover) to identify himself to another one (which we call the verifier). The protocol includes r rounds, each of these being performed as follows.

1. The prover picks a random n-bit word y and sends commitments for y and $y \oplus s$ to the verifier.

2. The verifier computes $x = G(y)$ and $x' = G(y \oplus s)$ and checks that $x = x' \oplus i$. Then, he sends a random n-bit word z to the prover.

3. The prover computes the sequences

$$n_j = |\, y \oplus z_j \,|$$
$$m_j = |\, y \oplus z_j \oplus s \,|$$

where $|\, v \,|$ is the weight of an n-bit word v, that is the number of its ones and where the sequence z_j is produced, as explained above, from integers a and c randomly chosen by the prover. He sends commitments of the sequences $n_j, 1 \leq j \leq q$ and $m_j, 1 \leq j \leq q$, to the verifier.

4. The verifier sends a random element b of $\{0, 1, 2\}$.

5. If b is 0 or 1, the prover announces his commitment for y' where $y' = y \oplus b \cdot s$. He also discloses another of his commitments: the one corresponding to the sequence n_j if b equals 0 and the one for m_j if b is 1. Finally, if b equals 2, the prover reveals both commitments for sequences but no other information.

6. If b equals 0 or 1, the verifier checks that the commitment was correct and that the integers n_j or m_j disclosed from the commitment have been computed honestly. Now, if b is 2, the verifier checks the commitments and computes the average value

$$\mu = \frac{1}{q} \sum_{j=1}^{q} |\, n_j - m_j \,|$$

In the last case, the verifier accepts the round if

$$\mu < 1.07\sqrt{p}$$

The number r of consecutive rounds depends on the required level of security and will be discussed further on as well as the values of the parameters n, k, l, p, q.

2 Soundness of the scheme

We first prove that a fair user will not be rejected. This is not obvious, at least when b sends a 2 and the probabilistic analysis that we need will also be used to establish the security of the scheme. We consider a random n-bit word. Such a random word t can be viewed as a sequence of n independant Bernoüilli trials. We let

$$m = |t| , m' = |t \oplus s|$$

and we consider the distribution of $|m - m'|$. If we let

$$T = \sum_{s(i)=1} t(i)$$

then, it is easily seen that $|m - m'|$ is exactly $|2T - p|$. Thus, we have to study the random variable $|2T - p|$, where T is the sum of p independant Bernoüilli trials. The expectation ν of this variable can be estimated by

$$\nu = \sqrt{\frac{2p}{\pi}} \approx 0.798\sqrt{p}$$

and its standard deviation by

$$\sigma = \sqrt{p - 2p/\pi} \approx 0.603\sqrt{p}$$

Now, the values $|n_j - m_j|$, which are computed in step 6, are precisely values of $|2T - p|$ corresponding to the random choice $t = y \oplus z_j$. We will use the central limit theorem in order to estimate the probability that the computed average value μ does not differ too much from the expectation ν. Of course, this is not quite correct as the w_j's are fixed so that we don't have independant variables. Still, it is heuristically justified as the w_j's have been chosen randomly. Furthermore, no contradiction arises from extensive numerical simulations.

Following these lines, we get

$$\mathbf{P}\{|\mu - \nu| \geq \tau\sqrt{p}\} \leq \int_{\frac{\tau\sqrt{pq}}{\sigma}}^{\infty} e^{-x^2/2} dx$$

$$\leq \frac{2\sigma}{\tau}\sqrt{\frac{2}{\pi pq}} e^{-\tau^2 pq/2\sigma^2}$$

$$\approx \frac{0.962}{\tau\sqrt{q}}(0.253)^{\tau^2 q}$$

Setting $\tau = 0.272$, one finds that the probability of having μ above $1.07\sqrt{p}$ is at most $6.96 \ 10^{-7}$ for $q = 128$. Even if the number of rounds is 60, in which case average values will be computed about 20 times, this makes an overall probability that a fair user is rejected as small as $1.4 \ 10^{-5}$. This can be easily handled, e.g. by giving the prover another chance to identify himself.

3 Security of the scheme

Before we discuss the security of the proposed scheme, let us consider possible collisions between public keys. If s ans s' are secret keys which yield the same public identification, then, $s \oplus s'$ is a codeword of lenth at most $2p$. Now, recall that G was chosen randomly, so that the corresponding code should have a good correcting power. More precisely, it is known, that random codes almost surely satisfy the Gilbert-Varshamov bound ([4]). Granted this fact, we get, for $k = n/2$, that any non-zero codeword has weight at least $0.11\, n$. Thus taking $p < 0.055\, n$ should prevent any collision to happen. Even if we don't fulfill this condition, collisions remain very unliquely. The same argument shows that a commitment essentially bounds the prover to his original choice, provided that the pieces have small weight.

Of course, the security of the scheme relies on the difficulty of inverting the function

$$s \longrightarrow G(s)$$

when its arguments are restricted to valid secret keys. In order to give evidence of this difficulty, let us recall from [1] that it is NP-complete to determine whether a code has a word s of weight $\leq p$ whose image is a given k-bit word i. Let us also observe that, in case no collision of secret keys can happen, finding s is exactly equivalent to finding the codeword w minimizing the weight of $t \oplus w$, when an element t of $G^{-1}(i)$ is chosen. But this is the problem of decoding unstructured codes which is currently believed to be únsolvable.

In order to counterfeit a given signature without knowing the secret key, various strategies can be used.

- Having only y ready for the verifier's query and annoucing something very close to $\mid y \oplus z_j \mid$ in place of m_j. In this case, the false prover hopes that b is 0 or 2 and the probability of success is $(2/3)^{-r}$, where r is the number of rounds. A similar strategy can be defined with $y \oplus s$ in place of y and shifting beetween the two yields the same probability of success.

- Having both y and $y \oplus t$ ready where t is some element such that $G(t) = i$, presumably distinct from s but whose weight is reasonably small. If the cheater realizes he will fail the round that way, he can still go back to the previous strategy for this round.

We will now describe two choices of the parameters such that the corresponding code can resist attacks based on the ability of a cheater to produce elements of $G^{-1}(i)$ of moderate weight. We feel that our analysis gives evidence to the security of our scheme because the cheater's performances that we consider are not met by known algorithms, as far as we are aware.

- We let $n = 51\hat{2}, k = 256, l = 4, p = 30, q = 128$ and we assume that the cheater can produce elements of $G^{-1}(i)$ of weight approximately $0.2n$. With these assumptions, computations similar to the above show that, each time he tries some t, the cheater has a probability of producing an average deviation μ lying below the critic value $1.07\sqrt{30} \approx 5.86$ which is roughly $9 \ 10^{-5}$ and this only increases by a minute fraction the probability of success $2/3$ of the basic cheating strategy, thus yielding a probability $(0.67)^r$ of going through r rounds.

- We let $n = 1024, k = 512, l = 8, p = 40, q = 128$ and we assume that the cheater can produce elements of $G^{-1}(i)$ of weight close to $0.12n$. Assuming that the Gilbert-Varshamov bound holds, this is almost the optimum, except of course if an algorithm can disclose s. With these assumption, any trial has a probability of producing μ below the critic value 6.76 which is roughly $1.03 \ 10^{-3}$ and, once again, it does not increase drastically the probability of success of the basic strategy.

4 Discussion of the scheme

We close the paper by various remarks.

1. We first discuss the amount of information on the secret key s which is disclosed when the scheme is used. Our basic assumption is that it is not possible to break a commitment. Once again, this hypothesis relies on the supposed difficulty of finding a word of small weight whose image by G is given. In our examples, the average length of the words used to build a commitment is 64 and thus, these words are out of reach, given the ability that we have assigned to an opponent in each case. Granted this, the only information that comes out of a round is either a random word, y or $y \oplus s$, or else two distribution of numbers n_j and m_j. Now, if j is fixed, n_j and m_j can be considered as independant random variables following a binomial ditribution. Of course, the family obtained when j varies is not made of independant variables. Still, since the order of appearance is essentially unknown, it seems virtually impossible to undertake any statistical analysis that might reveal s.

2. One of the defects of the proposed scheme is the required amount of memory: in order to store G and the w_j's, one needs 150 kbits in the weaker case considered and 630 kbits in the stronger case. Still, because the operations to perform are very simple, they can be implemented in hardware in a quite efficient way. On the other hand, the communication complexity of the protocol is not much worse than in the Fiat-Shamir scheme: in our smaller example, the n_j's are numbers which almost surely do not differ from 256 by more than 128 and can therefore be coded on 8 bits, so that the sequence of 4 commitments sent by the prover is 4000 bits long.

3. The security of the scheme can be increased by taking q and r larger. Also, the z_j's can be interactively produced instead of being defined in a rather systematic way: all random choices are really independant and this makes the theoretical analysis of the protocol more reliable. On the other hand, the communication complexity becomes much larger and this does not seem to be desirable. Finally, the rounds can be performed in parallel, reducing the number of steps to 6.

4. In order to limit the communication complexity, it is possible to let the prover commit himself to y and $y \oplus s$ by simply announcing $G(y)$ and $G(y \oplus s)$. This actually opens a new way to cheat by trying a lot of members t of $G^{-1}(i)$ instead of one. Since it involves heavy on-line computations, the number of trials can be limited by a timing-out device. Still, it will be necessary to increase q. In order to resist against 10^4 trials, we propose $q = 150$. Accordingly, our statistical requirements can be made a bit more strict (e.g. $1.05\sqrt{p}$ instead of $1.07\sqrt{p}$).

5. It is tempting to lower the value of p. It would be rather dangerous: using the arguments of [5], one can see that secret keys of small weight (e.g. $p = 20$) will presumably be found.

6. **acknowledgement** We wish to thank S. Harari for sending us an early version of his work [3].

References

[1] E.R.Berlekamp, R.J.Mc Eliece and H.C.A. Van Tilborg, On the inherent intractability of certain coding problems, *IEEE Trans. Inform. Theory*, (1978) 384-386.

[2] A. Fiat and A. Shamir, How to prove yourself: Practical solutions to identification and signature problems, *Proceedings of Crypto 86*, Santa-Barbara (1986), 181-187.

[3] S. Harari, Un algorithme d'authentification sans transfert d'information, *Proceedings*, Trois journées sur le codage,Toulon (1988), to appear.

[4] J.N. Pierce, Limit distributions of the minimum distance of random linear codes, *IEEE Trans. Inform. Theory*, (1967) 595-599.

[5] J.Stern, A method for finding codewords of small weight, *Proceedings*, Trois journées sur le codage, Toulon (1988), to appear.

Sorting out zero-knowledge

Gilles BRASSARD †

Département IRO
Université de Montréal

Claude CRÉPEAU ‡

Laboratory for Computer Science
Massachusetts Institute of Technology

1. Introduction

In their 1985 paper, Goldwasser, Micali and Rackoff set forth the notion of zero-knowledge interactive proofs [GMR1]. This seminal paper generated considerable activity around the world. In the span of a few years, a substantial number of results were obtained by different groups of researchers. Among those, the following two theorems make an intriguing pair:

- *Fortnow* [F], together with *Boppana, Hastad and Zachos* [BHZ]: The existence of a perfect zero-knowledge protocol for an **NP**-complete problem would imply that the polynomial hierarchy collapses.

- *Brassard, Chaum and Crépeau* [BCC]: There exists a perfect zero-knowledge protocol for satisfiability.

Nevertheless, the polynomial hierarchy has *not* collapsed! Of course, the resolution of this apparent paradox is that the above two results strongly depend on fundamentally incompatible definitions of what a protocol is.

The purpose of this paper is to explain the various notions involved and to offer a new terminology that emphasizes their differences. There are two *orthogonal* aspects to zero-knowledge interactive proofs. One is the notion of zero-knowledge and the other is the notion of interactive proof. Unfortunately, these two notions are often thought to be inseparable. This confusion is reminiscent of the long lasting confusion among many people between public-key encryption and digital signature. It is clear that interactive proofs make sense independently of zero-knowledge (after all, Babai's Arthur–Merlin games [Ba] were invented independently of [GMR1]), but it is more subtle to see that a protocol could be zero-knowledge without being an interactive proof.

---

† Supported in part by Canada NSERC grant A4107.
‡ Supported in part by an NSERC postgraduate scholarship; part of this research was performed while this author was visiting the Université de Montréal.

d Arguments

rding to [GMR1], an interactive proof is a protocol that takes place between
trarily powerful prover and a computationally limited verifier. The purpose of
protocols is for the prover to convince the verifier of the validity of an assertion.
This notion is very similar to Babai's Arthur–Merlin games [Ba, BM] (not discussed
here). As essential aspect of these protocols is that there is nothing a prover can do in
order to convince the verifier of a false statement, except to blindly hope for an
exponentially small probability of lucky successful cheating. This is the sense in
which it is said of GMR interactive proofs that "A proof is a proof".

In many practical situations, it is nevertheless reasonable to impose computational
limitations on the prover. This setting was investigated independently by Chaum [C2]
(who, on the other hand, allowed unlimited computing power to the verifier — to a
large extent, Chaum's model goes back to [C1]), and by Brassard and Crépeau [BC]
(who restricted both the prover and the verifier to "reasonable" computing power).
In either the Chaum or the Brassard–Crépeau setting, an interactive "proof" may not
be a proof at all. Indeed, a computationally unbounded prover could cheat in convincing the verifier of a false statement. More importantly in practice, many of these
"proofs" rely on an unproved assumption, usually the computational difficulty of computing some specific problem. For instance, even a polynomial-time prover could
succeed at "proving" a false statement with the protocol of [BC], provided that she
has a very efficient factoring algorithm (whose non-existence is still an open problem).

To distinguish interactive proofs in the sense of GMR from the protocols of
[C2, BC], we introduce a new terminology. An interactive protocol is an *argument*
(rather than a proof) if the verifier's faith in the prover's claim must ultimately rest on
an assumption. As we have seen, this assumption could be cryptographic in nature,
such as the assumption that the prover cannot compute a discrete logarithm while the
protocol is in progress [BCC]. It could also be merely that the prover has time-
bounded resources (such as polynomial time), although our current state of knowledge
in computational complexity does not yet allow us to make use of this kind of assumption alone. The assumption behind an argument could also be of a mathematical or
even physical nature, such as the Extended Riemann Hypothesis or the principles of
quantum physics [BB]. Or it could be that there are two provers who initially agree on
their strategy, but with the assumption that they cannot communicate while the protocol
is in progress [BGKW, GKBW]. Other types of assumptions can be dreamt up as well.

The first published use of this new terminology appeared in the title of [BCY].
It should be pointed out that what we now call argument has occasionally been termed
pseudo-proof by others. We wish to make a statement here, as the inventors of the
notion (co-invented by Chaum), that we **very strongly** object to the appellation of
pseudo-proof for the protocols of [C2, BC, BCY].

One may well wonder why settle for arguments when interactive proofs are so much stronger. One reason has to do with the possible zero-knowledge aspect of these protocols (see next section): it is unlikely that perfect zero-knowledge interactive proofs exist for statements concerning **NP**-complete problems [F, BHZ], and even computational zero-knowledge interactive proofs for such statements are not known to exist unless one makes cryptographic assumptions [GMW]. In contrast, perfect zero-knowledge arguments are known for all such statements, and no assumptions are needed to prove that these arguments are perfect zero-knowledge [BCC] — but of course they are not interactive proofs at all in the sense of GMR.

To a large extent, the (probably) computational zero-knowledge interactive proofs of [GMW] and the perfect zero-knowledge arguments of [BCC] are duals of each other. The only aspect of these protocols for which duality fails is that a cheating verifier can work off-line on the transcript of a [GMW] interactive proof and spend as much time as he wishes in attempts to decipher the prover's secret. In contrast, the prover can only cheat a [BCC] argument if she can break the cryptographic assumption on-line, while the protocol is in progress. As a consequence, an algorithm capable of breaking the cryptographic assumption in a few months of CRAY computing time would be bad news for [GMW] proofs but of no real consequence for [BCC] arguments. The reader is referred to section 7 of [BCC] for a more complete discussion of this issue.

Another reason to restrict the computational power of the prover is that it allows one to make sense of the notion of "proof of knowledge" [FFS, TW]. For instance, one such protocol allows a prover to convince a verifier that she (the prover) knows the factors of a given large integer [T]. The point is that it would be absurd for a prover known for her unbounded computing power to convince a verifier that she knows those factors: *of course* she knows them since she can compute them whenever she wishes. These proofs of knowledge play a crucial role in modern identification schemes [FFS] (but read [BBDGQ]). (You may wonder at this point why we call them "proofs of knowledge" rather than "arguments of knowledge" even though the prover's computing power is limited. For one thing, this terminology is well established. More importantly, they *are* GMR-proofs since no assumptions are needed for the proofs to be valid; it is only that they would be of no interest whatsoever if performed by a prover with unbounded computing power or if factoring, in our example, were known to be easy.)

For more discussion on the difference between the settings of [GMR1,2], [C2] and [BC], the reader is referred to [BCC, p. 159].

3. Zero-knowledge

The notion of zero-knowledge was set forward by Goldwasser, Micali and Rackoff [GMR1]. The reader is encouraged to read also the subsequent journal version of their work [GMR2]. Essentially, a zero-knowledge protocol allows a prover to convince a verifier of an assertion without disclosing any information to the verifier beyond the validity of that assertion. In the context of [GMR1,2], all zero-knowledge protocols are interactive proofs, but the notion of zero-knowledge *arguments* makes perfect sense as well. Whenever the prover's computing power is limited, a zero-knowledge protocol will necessarily disclose more than the validity of the assertion: the fact that the prover knows why this assertion is valid is also disclosed. (In the context of interactive proofs, the fact that the prover has this knowledge is implied by her unbounded computing power.) Nevertheless, this additional piece of information revealed when the prover's computing power is limited makes it possible to design protocols that actually reveal *less* than would be possible for any (interesting) interactive proof in which the prover has unbounded computing power: these are the proofs of knowledge discussed at the end of the previous section.

The intent of this paper is not to repeat the formal definitions of zero-knowledge, which can be found in [GMR2]. Rather, we wish to describe briefly several subtly different definitions for the main purpose of contrasting them. The reader not already familiar with the notion of (zero-knowledge) *simulator* is urged to read [GMR2] in order to profit fully from what follows, even though we briefly recall this notion in the next two paragraphs (with apologies to those who *are* familiar with it).

A protocol is perfect zero-knowledge [GMW] if the verifier does not learn anything at all from the interaction beyond the validity of the assertion involved and —if relevant— the fact that the prover knows why it is valid. In order to define this notion more formally, one has to consider the *view* of what the verifier sees during his interaction with the prover. This consists of the outcome of his own random coin tosses as well as of everything that the prover tells him during the interaction. Because of the probabilistic nature of interactive protocols (including random choices made by the prover), a probability distribution is defined on the view of the verifier. A protocol is *perfect* zero-knowledge if, to each polynomial-time verifier, there corresponds a polynomial-time *simulator* capable of producing a view taken from exactly the same probability distribution *without ever talking to the prover*. Intuitively, the existence of this simulator shows that the verifier does not learn anything from the interaction since the prover does not tell him anything that he could not have produced by himself (probabilistically speaking).

If the simulator is only required to produce a view taken from a probability distribution that is polynomially indistinguishable [GM] from the "correct distribution", the protocol is said to be *computational* zero-knowledge [GMR1]. Intuitively, this means that whatever the verifier may obtain from the interaction, he cannot make use of in polynomial time. Finally, the protocol is *statistical* (or *almost perfect*) zero-knowledge [F] if the real and simulated probability distributions are statistically close. When a protocol is simply said to be "zero-knowledge", most authors mean that it is computational zero-knowledge. In the opinion of the current authors, it would be better if the default option were that such protocols are perfect zero-knowledge.

Some researchers have objected to calling *zero* knowledge a protocol that reveals one bit of knowledge, for instance that a given graph is Hamiltonian. For this reason, Galil, Haber and Yung have suggested that such protocols be termed *minimum* knowledge [GHY], meaning that they reveal nothing more than what they absolutely have to by definition of their purpose. By contrast, Feige, Fiat and Shamir wished to keep the term "zero-knowledge", so they introduced the notion of proof of knowledge in order to have a protocol that is truly *zero* knowledge (in their opinion) [FFS] in the sense that it reveals zero bits of information about the "real world". This notion of proof of knowledge was also formalized by Tompa and Woll [TW].

The original definition of zero-knowledge [GMR1] lacked the desirable property that the sequential composition of two zero-knowledge protocols should remain zero-knowledge [GK]. If this property is important, the stronger notion of zero-knowledge *with auxiliary input* [O] should be used. A protocol is perfect zero-knowledge with auxiliary input if, no matter which additional input is given to the verifier before the interaction with the prover, the interaction does not help the verifier learn anything at all that he could not have learnt by himself given the same additional input. The notions of computational and statistical zero-knowledge with auxiliary input are defined similarly.

The formal definition of zero-knowledge [GMR1,2] stipulated that a polynomial-time simulator should exist for each polynomial-time verifier. This begs the following two questions.

1) Should there be a uniform and efficient process by which the simulator can be derived from the verifier?

2) How should the definition be modified if one is also interested in dealing with verifiers that are not limited to polynomial time? (This is an important issue when dealing with arguments rather than interactive proofs.)

We believe that the answer to the first question ought to be "yes". Otherwise, it is hard to support the view that a verifier learns nothing from the interaction if his conversation with the prover could be simulated but only by a simulator that is infeasible to find. We can think of two different definitions that would force the simulator to be efficiently obtainable from the verifier. The first and most obvious definition requires that the verifier be provided using a fixed formalism, such as that of probabilistic interactive Turing machines. In this case, we would insist that the "code" for the simulator (presumably expressed in the same formalism) be efficiently derivable from the code for the verifier.

The second definition, which we prefer for its simplicity even though it is more restrictive, is that of a "black box simulator", first introduced by Oren [O]. This definition requires the simulator to use the verifier as a resettable black box in order to simulate the interaction between that verifier and the prover. In other words, the simulator has no access to the code of the verifier, but has control over its tapes, including its random coin flip tape, and has the ability to bring it to a halt and restart it in its starting state (possibly with different tapes) at any time it wishes. This definition is not as restrictive as it might appear because most simulators in the literature are in fact designed in the black box model. It is known that black box simulation implies zero-knowledge with auxiliary input [O].

This brings us to the second question mentioned above: "How should the definition be modified if one is also interested in dealing with verifiers that are not limited to polynomial time?". One possibility would be to grant the simulator an amount of time comparable (perhaps up to a polynomial factor) to the time allowed to the verifier. Another possibility, which we prefer, involves again the notion of black box simulator: restrict the simulator to polynomial time, but count at unit cost any call on the verifier.

Thus we see that there are two different reasons why there may be protocols that are GMR zero-knowledge but not black box zero-knowledge. Firstly, there may be protocols that are non-uniform zero-knowledge but that cannot be simulated by a black box simulator. Moreover, given the work of [GK], it is not hard to design a protocol that is zero-knowledge against any polynomial-time verifier (although not with auxiliary input), but that would nevertheless allow a super-polynomial-time verifier to get information from an even more powerful prover that he could not compute by himself within his allowed time bound.

All the concepts discussed so far are but minor variations on the original definition of zero-knowledge [GMR1]. Other minor variations have been introduced, such as the notion of zero-knowledge *with respect to a trusted verifier*. We do not discuss them here. Rather, we now discuss a few notions that are more significantly different.

There are practical situations in which it is worthwhile to settle for something *less* than zero-knowledge. This may be the case, for instance, if it allows a substantial increase in efficiency. A nice example of this concept is the version of their identification scheme that Feige, Fiat and Shamir discuss in section 4 of [FFS]. They propose a scheme that is (probably) not zero-knowledge because the interaction reveals information to the verifier that he could (probably) not have computed by himself. Nevertheless, this information is proven in [FFS] not to be enough to help the verifier cheat the system in polynomial time, provided that cheating the protocol in polynomial time is impossible in the first place without the interaction. In other words, partial information is perhaps released on the prover's secret, but this partial information is not sufficient for the verifier to recompute the prover's secret (or any "equivalent" secret that might allow him to defeat the system). Feige, Fiat and Shamir say of such schemes that they "release no useful information".

A similar but stronger notion was introduced by Brassard, Chaum and Crépeau [BCC]. A protocol is *minimum disclosure* if everything that the prover ever says to the verifier is uncorrelated to her secret (or any equivalent secret). Clearly, a minimum disclosure protocol cannot help the verifier compute the prover's secret (or any equivalent secret), hence it "releases no useful information". However, it may not be zero-knowledge because the prover is allowed to reveal irrelevant information even if this information would be hard for the verifier to compute by himself. Contrary to protocols that reveal no useful information, a minimum disclosure protocol is not allowed to reveal partial information, even if such information is not enough to efficiently recompute the secret.

This brings us to a curious notion, which has evolved from discussions between the authors and Silvio Micali. Some protocols are not known to be zero-knowledge nor even minimum disclosure. However, it *is* known that whatever the verifier could learn by deviating arbitrarily from his prescribed behaviour could not be worth more, for instance, than the discrete logarithm of a number of his choosing. (This is the case, in particular, of the parallel version of the main protocol given in [BCC] if the blobs are implemented as suggested in section 6.1.2.) This is *not* to say that the verifier can actually obtain this discrete logarithm from his interaction with the prover, but whatever he can get is worth no more than if he did. Formally, this is so because it would be easy for a simulator to do a perfect job very efficiently were it only given this discrete logarithm. If it can be proven that this discrete logarithm is uncorrelated to the prover's secret (or any equivalent secret), then the corresponding protocol is minimum disclosure. We would like to point out that, because such correlation cannot be ruled out in general, the claim made in [BCC, p. 166] to the effect that "the parallel version of the protocol is minimum disclosure" is not known to hold in all cases.

Given a protocol such as the one we have just discussed, the interesting question is to determine the minimal piece of information that would allow efficient simulation. This information can be thought of as the maximal value of the protocol for any cheating verifier because whatever a verifier can get from the interaction is worth no more than it. An interesting question asked by Silvio Micali is what is the maximal value of the parallel version of the protocol for graph isomorphism [GMW]? Could it be less than the isomorphism itself?

4. Complexity issues

There are several complexity issues pertaining to zero-knowledge protocols. One of the most interesting from a practical point of view is the issue of parallelism. Zero-knowledge protocols usually take place in several rounds of interaction between the prover and the verifier. Each round increases the verifier's confidence in the prover's honesty. Most of these protocols can be reformulated in a natural way in a bounded number of rounds, but they apparently cease to be zero-knowledge when this is done. (As previously mentioned, the identification protocol of [FFS] is an interesting special case: its parallel version may not be zero-knowledge but it releases no useful information, which is just as good for the application that they have in mind.) Nevertheless, it is known that the general protocols of [GMW] and [BCC] *can* be redesigned to be run in parallel and remain zero-knowledge (perfect zero-knowledge in the case of [BCC]). For a full discussion of the fact that everything in **NP** can be argued in perfect zero-knowledge in a bounded number of rounds, read the ICALP version of [BCY]. A summary, including the history of the problem and references to related work, can be found in these EUROCRYPT '89 Proceedings. Following the work of [BCY], Bellare, Micali and Ostrovsky have shown than graph isomorphism can be *proven* in perfect zero-knowledge in a bounded number of rounds (personal communication).

An orthogonal result pertaining to bounded round interactive proofs is due to Fortnow [F] and to Aiello and Hastad [AH]: if L is any language that admits a perfect or statistical zero-knowledge interactive proof system with an arbitrary number of rounds, then both L and its complement also admit a *bounded* round interactive proof system (which is not zero-knowledge in general, however). It is crucial for the Fortnow and Aiello–Hastad results not to limit the prover to polynomial time.

Despite the work mentioned in the first paragraph of this section, it remains interesting to analyse what happens if one runs an arbitrary zero-knowledge protocol in parallel without any additional precaution. In particular, it has not been proven that these protocols cease to be zero-knowledge (it is only the case that no one has been able to design a simulator capable of coping with the situation). If indeed these protocols are not zero-knowledge, what is in general their maximal value (as defined at the end of the previous section)? When are they minimum disclosure? When do they release no useful information?

Other complexity issues pertain to the simulator. One of them is whether it is legitimate to say that a protocol is zero-knowledge if there exists a verifier capable in quadratic time to force the prover to take part in an interaction that no simulator could reproduce in less than cubic time.

Yet another issue pertaining to the simulator is that of *strict* polynomial time versus *expected* polynomial time. Recall that both the verifier and the simulator are probabilistic processes. It is usually the case that the prescribed verifier takes a time that is strictly bounded by a fixed polynomial, regardless of its random choices (perhaps the only exception in the published literature is the prescribed verifier for the zero-knowledge interactive protocol for graph 3-colourability when the number of edges is not a power of 2 [GMW]). Nevertheless, the simulator usually requires polynomial time in the expected sense. Is it possible in general to make do with strict polynomial time for the simulator whenever the verifier is strict polynomial time (or when the black box model is used)? In particular, it is an open question to design a strict polynomial-time simulator that retains the *perfect* zero-knowledge property of the well-known protocol for graph isomorphism [GMW].

ACKNOWLEDGEMENTS

The idea of writing this paper came from several discussions the authors have had with Silvio Micali.

BIBLIOGRAPHY

[AH]   Aiello, W. and Hastad, J., "Perfect zero-knowledge languages can be recognized in two rounds", *Proceedings of the 28th IEEE Symposium on Foundations of Computer Science*, 1987, pp. 439–448.

[Ba]   Babai, L., "Trading group theory for randomness", *Proceedings of the 17th ACM Symposium on Theory of Computing*, 1985, pp. 421–429.

[BM]   Babai, L. and Moran, S., "Arthur–Merlin games: A randomized proof system, and a hierarchy of complexity classes", *Journal of Computer and System Sciences*, vol. 36, 1988, pp. 254–276.

[BBDGQ] Bengio, S., Brassard, G., Desmedt, Y., Goutier, C. and Quisquater, J.–J., "Secure implementation of identification systems", in preparation.

[BB]   Bennett, C. H. and Brassard, G., "Quantum cryptography", in preparation; in the mean time, read chapter 6 in [Br].

[BGKW] Ben Or, M., Goldwasser, S., Kilian, J. and Wigderson, A., "Multi-prover interactive proofs: How to remove intractability assumptions", *Proceedings of the 20th ACM Symposium on Theory of Computing*, 1988, pp. 113–131.

[BHZ]   Boppana, R. B., Hastad, J. and Zachos, S., "Does co–NP have short interactive proofs?", *Information Processing Letters*, vol. 25, 1987, pp. 127–132.

[Br]   Brassard, G., *Modern Cryptology: A Tutorial*, Lecture Notes in Computer Science, vol. 325, Springer-Verlag, 1988.

[BCC]   Brassard, G., Chaum, D. and Crépeau, C., "Minimum disclosure proofs of knowledge", *Journal of Computer and System Sciences*, vol. 37, no. 2, 1988, pp. 156–189.

[BC]   Brassard, G. and Crépeau, C., "Non-transitive transfer of confidence: A *perfect* zero-knowledge interactive protocol for SAT and beyond", *Proceedings of the 27th IEEE Symposium on Foundations of Computer Science*, 1986, pp. 188–195.

[BCY]   Brassard, G., Crépeau, C. and Yung, M., "Everything in **NP** can be argued in *perfect* zero-knowledge in a *bounded* number of rounds", *Proceedings of 16th ICALP Conference*, Stresa, Italy, July 1989, to appear; an extended abstract appears in these *EUROCRYPT '89 Proceedings*.

[C1]   Chaum, D., "Security without identification: Transaction system to make Big Brother obsolete", *Communications of the ACM*, vol. 28, 1985, pp. 1030–1044.

[C2]   Chaum, D., "Demonstrating that a public predicate can be satisfied without revealing any information about how", *Advances in Cryptology – CRYPTO '86 Proceedings*, Springer-Verlag, 1987, pp. 195–199.

[FFS]   Feige, U., Fiat, A. and Shamir, A., "Zero knowledge proofs of identity", *Journal of Cryptology*, vol. 1, no. 2, 1988, pp. 77–94.

[F]   Fortnow, L., "The complexity of perfect zero-knowledge", *Proceedings of the 19th ACM Symposium on Theory of Computing*, 1987, pp. 204–209.

[GHY]   Galil, Z., Haber, S. and Yung, M., "A private interactive test of a Boolean predicate and minimum-knowledge public-key cryptosystems", *Proceedings of the 26th IEEE Symposium on Foundations of Computer Science*, 1985, pp. 360–371.

[GK]   Goldreich, O. and Krawczyk, H., "On sparse pseudo-random distributions", *Advances in Cryptology – CRYPTO '89 Proceedings*, Springer-Verlag, to appear.

[GMW]   Goldreich, O., Micali, S. and Wigderson, A., "Proofs that yield nothing but their validity and a methodology of cryptographic protocol design", *Proceedings of the 27th IEEE Symposium on Foundations of Computer Science*, 1986, pp. 174–187.

[GKBW]   Goldwasser, S., Kilian, J., Ben Or, M. and Wigderson, A., "Efficient identification schemes using two prover interactive proofs", *Advances in Cryptology – CRYPTO '89 Proceedings*, Springer-Verlag, to appear.

[GM]   Goldwasser, S. and Micali, S., "Probabilistic encryption", *Journal of Computer and System Sciences*, vol. 28, 1984, pp. 270–299.

[GMR1]   Goldwasser, S., Micali, S. and Rackoff, C., "The knowledge complexity of interactive proof systems", *Proceedings of the 17th ACM Symposium on Theory of Computing*, 1985, pp. 291–304;

[GMR2]   Goldwasser, S., Micali, S. and Rackoff, C., "The knowledge complexity of interactive proof systems", *SIAM Journal on Computing*, vol. 18, no. 1, 1989, pp. 186–208.

[O]   Oren, Y., "On the cunning power of cheating verifiers: Some observations about zero knowledge proofs", *Proceedings of the 28th IEEE Symposium on Foundations of Computer Science*, 1987, pp. 462–471.

[T]   Tompa, M., "Zero knowledge interactive proofs of knowledge (a digest)", *Second Conference on Theoretical Aspects of Reasoning about Knowledge*, Monterey, CA, 1988; available as Research Report RC 13282 (#59389), IBM Research Division, T. J. Watson Research Center, Yorktown Heights, NY, 1987.

[TW]   Tompa, M. and Woll, H., "Random self-reducibility and zero-knowledge interactive proofs of possession of information", *Proceedings of the 28th IEEE Symposium on Foundations of Computer Science*, 1987, pp. 472–482.

Everything in NP can be argued in *perfect* zero-knowledge in a *bounded* number of rounds

(Extended Abstract)

Gilles BRASSARD †

Département IRO
Université de Montréal

Claude CRÉPEAU ‡

Laboratory for Computer Science
Massachusetts Institute of Technology

Moti YUNG

IBM T. J. Watson Research Center
Yorktown Heights, NY

ABSTRACT

A perfect zero-knowledge interactive protocol allows a prover to convince a verifier of the validity of a statement in a way that does not give the verifier any additional information [GMR, GMW]. Such protocols take place by the exchange of messages back and forth between the prover and the verifier. An important measure of efficiency for these protocols is the number of rounds in the interaction. In previously known perfect zero-knowledge protocols for statements concerning **NP**-complete problems [BCC], at least k rounds were necessary in order to prevent one party from having a probability of undetected cheating greater than 2^{-k}. In the full version of this paper [BCY], we give the first perfect zero-knowledge protocol that offers arbitrarily high security for any statement in **NP** with a constant number of rounds (under a suitable cryptographic assumption). This protocol is a BCC-*argument* rather than a GMR-proof [BC3], as are all the known perfect zero-knowledge protocols for **NP**-complete problems [BCC].

† Supported in part by Canada NSERC grant A4107.

‡ Supported in part by an NSERC postgraduate scholarship; part of this research was performed while this author was visiting the IBM Almaden Research Center.

History, motivation and main result

Much excitement was caused when it was discovered in 1986 by Goldreich, Micali and Wigderson that all statements in **NP** have *computational* zero-knowledge interactive proofs (under the assumption that secure encryption functions exist) [GMW]. See also [BC1]. Such proofs, a notion formalized by Goldwasser, Micali and Rackoff a few years previously, allow an infinitely powerful (but not trusted) prover to convince a probabilistic polynomial-time verifier of the validity of a statement in a way that does not convey any *polynomial-time-usable* knowledge to the verifier, other than the validity of the statement [GMR]. Informally, this means that the verifier should not be able to generate anything in probabilistic polynomial time after having participated in the protocol, that he could not have generated by himself without ever talking to the prover (from mere belief that the statement is true).

A result similar to those of [GMW, BC1] was obtained independently by Chaum, but under a very different model, which emphasizes the unconditional privacy of the prover's secret information, even if the verifier has unlimited computing power [Ch]. Independently, Brassard and Crépeau considered a model (compatible with Chaum's) in which all parties involved are assumed to have reasonable computing power, and they also obtained a protocol unconditionally secure for the prover (meaning that the prover's safety does not depend on unproved cryptographic assumptions) [BC2]. We shall refer to the settings of either [Ch] or [BC2] as the BCC-setting in order to contrast it with the GMR-setting described in the previous paragraph. Protocols in the BCC-setting are called *arguments* rather than proofs because even a polynomial-time prover could cheat them if the cryptographic assumption turns out to be false [BC3]. Joining forces, Brassard, Chaum and Crépeau subsequently showed that everything in **NP** can be argued in perfect zero-knowledge [BCC] (thanks to an idea of Damgaard), which implies that the prover's safety would still be guaranteed even if strong organizations with unknown computing power and algorithmic knowledge were to try to extract her secret and were willing to expend an arbitrary amount of time on this task.

The main motivation behind the work of [GMW, BC1, Ch, BC2, BCC] was a quest for *generality*: how much is it possible to prove in zero-knowledge if little attention is paid to efficiency? Other researchers were willing to sacrifice generality on the altar of efficiency. The best known instance of this approach is Feige, Fiat and Shamir's identification system [FFS], which handles an *ad hoc* problem relevant to the purpose of identification, but could not handle statements about **NP**-complete problems. One reason why the FFS scheme is so attractive in practice is that it requires only a few rounds of interaction between the prover and the verifier. In sharp contrast, the more general protocols of [GMW, BC1, Ch, BC2, BCC] require an unbounded number of rounds in order to achieve an arbitrarily high level of safety. This paper addresses the following question: Is it possible to combine generality and arbitrarily high safety with a small (constant?) number of rounds? (By one "round", we mean two "moves": one message sent by the verifier followed by one message sent by the prover.)

Our answer is that three rounds suffice under the assumption that it is possible to find a prime p with known factorization of $p-1$ such that it is infeasible to compute discrete logarithms modulo p even for someone who knows the factors of $p-1$, or more generally under the assumption that one-way group homomorphisms [IY] exist. Our three-round protocol and a sketch of the proof that it is perfect zero-knowledge can be found in the Proceedings of the 16th ICALP conference [BCY].

It should be pointed out that a similar question has been investigated independently by other researchers. In the GMR-setting, Goldreich and Kahn claim a bounded-round computational zero-knowledge protocol for all statements in **NP** [G]. Feige and Shamir have also developed a bounded-round computational zero-knowledge protocol for all statement in **NP**, but in a setting in which both the prover and the verifier are limited to probabilistic polynomial time [FS]. However, being merely *computational* zero-knowledge, neither of these protocols offer unconditional safety for the prover. Feige and Shamir also claim in [FS] that they have a bounded-round (in fact two rounds) *perfect* zero-knowledge protocol, but they give no detail in the currently available version of their paper (March 1989). Our ICALP paper [BCY] provides the first published bounded-round perfect zero-knowledge protocol for all statements in **NP** (in the BCC-setting).

ACKNOWLEDGEMENTS

It is a pleasure to acknowledge fruitful discussions with Charles H. Bennett, Joan Boyar, David Chaum, Ivan Damgaard, Uri Feige, Oded Goldreich, Joe Kilian, Phillip Rogaway and Adi Shamir.

BIBLIOGRAPHY

[BCC] Brassard, G., Chaum, D. and Crépeau, C., "Minimum disclosure proofs of knowledge", *Journal of Computer and System Sciences*, vol. 37, no. 2, 1988, pp. 156–189.

[BC1] Brassard, G. and Crépeau, C., "Zero-knowledge simulation of Boolean circuits", *Advances in Cryptology – CRYPTO '86 Proceedings*, Springer-Verlag, 1987, pp. 224–233.

[BC2] Brassard, G. and Crépeau, C., "Non-transitive transfer of confidence: A *perfect* zero-knowledge interactive protocol for SAT and beyond", *Proceedings of the 27th IEEE Symposium on Foundations of Computer Science*, 1986, pp. 188–195.

[BC3] Brassard, G. and Crépeau, C., "Sorting out zero-knowledge", *Advances in Cryptology – EUROCRYPT '89 Proceedings*, Springer-Verlag, to appear in this volume.

[BCY] Brassard, G., Crépeau, C. and Yung, M., "Everything in **NP** can be argued in *perfect* zero-knowledge in a *bounded* number of rounds", *Proceedings of 16th ICALP Conference*, Stresa, Italy, July 1989, to appear.

[Ch] Chaum, D., "Demonstrating that a public predicate can be satisfied without revealing any information about how", *Advances in Cryptology – CRYPTO '86 Proceedings*, Springer-Verlag, 1987, pp.195–199.

[FFS] Feige, U., Fiat, A. and Shamir, A., "Zero knowledge proofs of identity", *Journal of Cryptology*, vol.1, no.2, 1988, pp.77–94.

[FS] Feige, U. and Shamir, A., "Zero knowledge proofs of knowledge in two rounds", *Advances in Cryptology – CRYPTO '89 Proceedings*, Springer-Verlag, to appear.

[G] Goldreich, O., personal communication.

[GMW] Goldreich, O., Micali, S. and Wigderson, A., "Proofs that yield nothing but their validity and a methodology of cryptographic protocol design", *Proceedings of the 27th IEEE Symposium on Foundations of Computer Science*, 1986, pp.174–187.

[GMR] Goldwasser, S., Micali, S. and Rackoff, C., "The knowledge complexity of interactive proof systems", *SIAM Journal on Computing*, vol.18, no.1, 1989, pp.186–208.

[IY] Impagliazzo, R. and Yung, M., "Direct minimum-knowledge computations", *Advances in Cryptology – CRYPTO '87 Proceedings*, Springer-Verlag, 1988, pp.40–51.

Zero-Knowledge Proofs of Computational Power

Moti Yung

IBM Research Division

T.J. Watson Research Center

Yorktown Heights, NY 10598

(extended summary)

Abstract

Suppose that the NSA had announced the possession of an efficient factorization algorithm. The cryptology community, after recovering from the initial shock, would demand to see the algorithm and verify it. This request, however, could not be satisfied since the algorithm would probably be classified as top-secret information.

In this note we give a procedure which will satisfy both sides of the above imaginary dispute. This is a way in which one party can prove possession of some "computational power" (e.g., a special-purpose efficient factorization machine) without revealing any algorithmic detail about this computational task (e.g., the factoring algorithm).

1 Introduction

Interactive proofs were originally developed for membership in a language by Goldwasser, Micali and Rackoff [7]. In such a proof one party, the (P)rover, is engaged in an interactive protocol with another party, the (V)erifier. The task of the interaction is to convince the verifier (with overwhelming probability) that indeed the input belongs to some language.

The notion of "zero-knowledge" interactive proof [7] was introduced to capture the fact that an interaction validates the input membership, but does not give any extra advantage to the Verifier. See [7,6,8,3,11] for some of the examples of zero-knowledge interactive proofs. We assume familiarity with the above notions.

Later, interactive proofs "of knowledge" in which a polynomial-time P proves that it "knows" (possesses) a witness for some predicate about the input x was introduced. The formal specifications of such an interactive proof "of knowledge" are difficult to state precisely. Indeed, several such protocols had been proposed in the literature, and used in building cryptographic schemes, without a general definition of their properties. The formal definition and implementation of zero-knowledge interactive proof of knowledge was finally given by Feige, Fiat, and Shamir [5] and Tompa and Woll [13].

Suppose now that a polynomial-time P possesses — not just a witness— but access to some algorithm which gives some computational power. That is, it has some device (which can even be a machine TM to which P has only input-output access). The machine TM as a black box can solve instances of a given problem. We are interested in computations which are hard in the cryptographic sense. Thus, in the context of this paper an easy problem is assumed to be polynomial on the average, and a hard problem is almost always hard for problem size large enough. Here we suggest a way to formalize the notion of proving an access to such a computational device in a zero-knowledge fashion; we call the scheme "a zero-knowledge proof of computational power".

We have the following possible cryptographic applications in mind:

1. The scheme can serve as a method for revealing the discovery of a new cryptanalytic breakthrough in a responsible way. That is, convincing legal users of the broken system that it pays to change their system, but, on the other hand, doing so without publicly revealing the algorithmic details of the discovery. Thus, preventing exploitation of the new information by malicious users.

2. It can also serve consumers as a quality test for a cryptographic device (e.g., a decryption machine), while simultaneously protecting the dealer by preventing the potential customer from employing the device (e.g., using the test to decrypt messages he wants to decipher) during the test procedure.

Our scheme employs interactive proofs of knowledge [5,13,7,6,8], the perfect zero-knowledge interactive argument systems of Brassard, Chaum and Crépeau [3], and generation of random solved instances as defined by Hemachandra, Abadi, Allender, Broder and Feigenbaum [10].

2 From Proving Knowledge to Proving Power

While having a computational power (beyond, say, polynomial-time) is easy to express, the question we ask is how can we model a possession of computational power in the context of interaction between two untrusted parties. A computational power can be checked by revealing an algorithm, and proving its correctness and running time. Doing so in a zero-knowledge fashion was formalized by Blum as proving any Theorem [2]. Such procedure, however requires that the prover has the verified algorithm, and cannot be employed when the prover has only input-output access to the algorithm. Therefore, we change the starting point and try to extend "proof of knowledge" to a proof which demonstrates more than just possession of a witness to some computation, but rather possession of "algorithmic power". We do not attempt to formalize computational power in general, but rather concentrate on the ability to solve problems which are hard in a cryptographic sense (on the average) as power.

2.1 Proof of Knowledge

Next we sketch the formalization of proving knowledge [5,13] which requires, that the polynomial-time P demonstrate the possession of a witness. This is done by forcing it to compute in a way which is tractable only if it has explicit access to the witness.

The two machines P and V share a common input x and the prover has to demonstrate that it possesses a witness w given to it as a private input. The witness is to the fact that $x \in Domain(R)$ for the relation R (i.e., w is such that $(x, w) \in R$). In the factorization example the relation is trivial, x is an integer, and w is its factors.

The interactive proof demonstrates the possession of w, through P's ability to answer questions posed by V, without actually revealing anything about the value of w. Thus, the "proof" is probabilistic in nature, and depends only on the messages sent by the two machines (in turns). The verifier is not able to read the prover's private input tape (which the prover is presumably using in order to compute its answers).

Formally, possession of a witness is modeled by means of a *recovery algorithm*, which is a Turing machine X (for "extractor") that interacts with any possible prover P^* in a special way — typically, while following a modified version of V's program. On input x, the goal of the recovery algorithm is to extract from P a witness. The machine X sends messages to and receives messages from P^*, just as V would do. The only difference is that while performing its computation, X may cause a machine configuration of P^* to be *saved*, and then, perhaps several turns later, X may restore the saved configuration of P^* and send different messages. The fact that X can backtrack the computation models the fact that in the "real" interaction V may send a random message from a set of different ones and P^* should be able to answer them all. At the end of its interaction with P^*, X writes out its guess for w. The idea is that if X is able to extract the knowledge from the interaction when the messages produced by P are available to it then P^* "knows" the witness, and can, (if we modify it,) compute it.

Formally, let P denote the specified machine which has a private input (a witness) and follows the protocol, while P^* denote any machine acting as the legal P. For any input string x, the computation of the interaction denoted $(P, V)[x]$ either accepts or rejects. V's state at the end of the computation confirming the fact that P possesses a witness (accept) or not (reject).

We call a pair of interactive Turing machines (P, V) an *interactive proof-system of knowledge* for the binary relation R with error probability $\delta(k)$ (where k is the security parameter, i.e., the size of the input) if:

1. For any $(x, w) \in R$, if P possesses w then $(P, V)[x]$ accepts with (very high) probability $1 - \delta(k)$.

2. There exists a *recovery algorithm* X (associated with V), as described above, so that for any interactive Turing machine P^*, for any input string x, the probability of the event of both (1) $(P^*, V)[x]$ accepts, and (2) on input x, X interacting with P^* outputs w which does not satisfy $(x, w) \in R$, is vanishingly small (less than $\delta(k)$).

The algorithm $X = X(V)$ works as specified, uniformly for any possible prover P^*, it has a limited access to the communication tapes only, and can backtrack the computation. This extractor works in expected polynomial time, thus the whole interaction of X with P^* is a feasible computation and can be run by P^* itself. This means that, with very high probability, if the interaction with V is an accepting one, then the machine acting as a prover can compute the witness; (if computing the witness is impossible for this machine, we may conclude that it was given to it).

An interaction like this is zero-knowledge if (loosely speaking) given a verifier V^* there is a machine which can compute its view of the interaction (which is a probability space) in expected polynomial time. Thus, its view of the computation does not provide any computational advantage beyond expected random polynomial time computation (in other words, provides "no knowledge").

2.2 Example: proving knowledge of factorization

Tompa and Woll [13] gave a zero-knowledge interactive proof of knowledge of factors of a given number. We start with their example and assume also that n is a composite with two large prime factors $n = pq$. Their proof relies on Rabin's proof of equivalence between extracting square roots and Factorization [12]. We also employ the zero-knowledge proof of residuosity of Goldwasser, Micali and Rackoff [7], formalized as proof of knowledge.

Assume P is given a quadratic residue (square), and is able to give a root. This implies, by Rabin equivalence, that P possesses the factors with probability at least $1/2$. If this is repeated k independent times and P is able to give a root each time then we conclude that P has the factors with probability greater than $1 - (1/2^k)$. In fact, because of the equivalence, an ability of any polynomial-time P^* to give a root means he has or can (with some probability) produce the factors. Namely, this is an interactive proof of knowledge.

The above scheme is not zero-knowledge and actually each time a root is revealed, also the factorization is revealed with probability $1/2$. This is a result of Rabin's equivalence (since P gives a square root, if it is a root not known to the verifier it might enable the factorization). Thus, instead of opening a root the prover will "prove" (in a zero-knowledge fashion) to the verifier that it "knows a root". This is still not a zero-knowledge proof, since the verifier may not choose a square according to the protocol requirements; an act which may give V an advantage. Thus, when it chooses and sends a square it is also required to prove the knowledge of a root— which implies that he indeed chose a square. This makes the protocol zero-knowledge (see [13]).

A high-level description of the protocol may look like this:

Protocol 1: Proving Possession of an Integer Factorization

Given n (which can be checked in random polynomial time not to be a prime power) as input.

For i=1,..,k do

1. The verifier chooses a quadratic residue y_i $(mod\, n)$ and sends it

2. The verifier proves interactively that it knows a square root of y_i $(mod\, n)$

3. The prover proves interactively that it knows a square root of y_i $(mod\, n)$

If all proofs are successful V accepts, otherwise it rejects.

end-of-protocol

2.3 Towards a definition of proof of computational power

Our goal is to extend the "proof of knowledge" to "proof of computational power". In order to do this we must model what is a computational power. Our first observation is that the verifier may be simply modeled as interacting with another Turing machine TM whose "knowledge" of the witness is more complicated than its ability to simply read the bits of w from its private input tape. The value of an answer may be computed by TM in any possible way and we model it by letting the prover interact with a machine which gives answers to queries (say, gives an efficient factorization algorithm, or gives factors, or at least is able to answer queries specified by the verifier). This machine, TM, if restricted to polynomial time is assumed to possess the witness. The definition, however, is oblivious to the way TM computes, and an extractor X (in the definition of proof of knowledge) will produce the witness.

If we could just exhaustively feed the Prover (which has an access to the algorithm in its possession) with all possible inputs and observe that a correct result is given throughout, we could say that all possible witnesses "are possessed" by this prover. For example, asking about the factorizations of all numbers of a given size m, enables us to conclude that this algorithm has the claimed power.

Obviously, however, the number of inputs is prohibitively large for such an interaction which is required to end in polynomial time. Thus we have to relax the interactive proof and turn it into a "statistical test" which samples a random set of inputs and observe how the prover reacts to them.

A statistical test can be based on a law of large numbers. Let $F_k(E)$ be the frequency of the test of k independent trials of event E. For a sample size $k \geq 1$, applying (for example) Bernstein's strong version of the law of large numbers, assuming the actual probability of the parameter is p, then for each $\delta \leq p(1-p)$ the following holds: $prob\{|F_k(E) - p| \geq \delta\} \leq 2e^{-k/(p(1-p))^2}$. This gives a statistical test (by choosing a large enough k) which makes us accept the hypothesis that the probability is indeed p with negligible error. Actually we are interested in the one-sided hypothesis that the actual probability is (almost) one, and we will reject the alternative hypothesis that $p = r$ for some constant r based on the fact that $F_k(E) - r > \delta$.

Next, we start with the factoring example and then the formal definitions.

3 Proving the Power to Factor Integers

Assume that the prover claims that he can factor numbers of a certain size (say, m-bit long numbers). This gives us a probability space which is samplable [1] (i.e., a number with its factorization can be generated at random). We say that a party possesses the ability to factor if he can factor any number (or all but a negligible fraction of the numbers) in the space. While, on the other hand, if he does not have this power we assume that factoring is $(1 - \epsilon) - hard$ for him, that is, he cannot factor more then ϵ fraction of the factors (we call ϵ the tractability fraction).

The intractability assumption of factoring (of composites which are multiplication of two large primes of equal size) is that it is hard on the average. Formally, it says that for any polynomial R, the tractability fraction $\epsilon \leq 1/R(m)$ for integers of all size parameter m large enough. Thus, in particular, we can assume that the tractability fraction ϵ is smaller than $1/2$ (by the intractability assumption, this is true from a certain size m and on).

The nature of a problem being hard on the average then, suggests the following process. Let V sample and check P's ability to factor. V chooses a large enough number of integers uniformly at random from the sample space and P provides factors to all of them. If P possesses TM and can factor, then he should provide factorization to all the sample. On the other hand, for any polynomial-time P*, eventually its probability of factoring is less than $1/2$ for a random choice of an integer. Thus, this constitutes a statistical test.

Let $F_k(E)$ be the frequency of the test of k independent trials of the event that a number of size m (large enough) is being factored by a prover. Assume the probability of factoring is (smaller than or) equal $1/2$. (Recall that by the intractability assumption, this is true for large enough m). For a sample size $k \geq 1$, applying Bernstein's law (we are interested in a one-sided version of it) we get that $prob\{F_k(E) \geq 3/4\} \leq e^{-k/16}$. This gives a polynomial sample size k which with negligible probability will give a successful statistical proof (of the ability to factor all numbers in the given sample space). A polynomial time machine not augmented by TM will fail with overwhelming probability to produce $3/4$ of the requests and the assumption that $p = 1/2$ will be rejected. Thus, a successful test implies that the prover has some computational power and it can factor most of the numbers (which is, by the intractability of factoring, beyond average-P). In addition, the true prover possessing TM, will be able to answer all queries with overwhelming probability, since TM can factor (almost) all numbers in the space.

Again, implementing this suggested statistical test as a protocol is not zero-knowledge, however, the following refinement of it is (by the fact that it uses protocol 1 which is zero-knowledge).

Protocol 2: Proving Power to Factor numbers

On input m size parameter, for i=1,..,k do:

1. V picks a random integer n_i of size m and sends it

2. V uses protocol 1 to prove to P that it knows the factorization of n_i

3. P uses protocol 1 to prove to V that it knows the factorization of n_i

If all proofs are successful V accepts, otherwise it rejects.

end-of-protocol

This is a statistical test, which (as we claimed above) proves that P has the power as claimed. It is also (perfectly) zero-knowledge— given a verifier V^* the space of possible interactions with P can be exactly simulated (or if can factor most of the numbers it can be almost simulated with a negligible error).

4   Interactive Proof of Computational Power

In this section we generalize the above interaction to a more general class of problems which are assumed to be hard on the average; we define the class of problems to which such a proof can be applied.

4.1   Definitions

Let PR be a $(1-\epsilon)-hard$ problem (as defined above) with size parameter m and let it be a samplable problem; denote the sample space $S(m)$. That is, a problem in NP for which it is possible to generate random instances. It can be treated as a collection of relations $\{R(x,w)|x \in S(m), m \in I \subseteq N\}$.

A problem as above is hard on the average and a statistical test can "measure" the difference between a polynomial time machine and a machine with augmented power.

When a witness (a solution) is generated together with the instance, a samplable problem is exactly a problem which has a $(1-\epsilon)-invulnerable$ generation as defined by Hemachandra, Abadi, Allender, Broder and Feigenbaum [10]. This is, generating solved instances which cannot be solved in polynomial-time with some fixed probability $(1 - \epsilon)$.

For our purposes, the requirement of random generation can be weakened. We need not generate in polynomial time a witness, but rather we should be able to generate in polynomial time a random YES-instance with the ability to interactively prove (by a polynomial-time machine, our verifier) that it is indeed an instance which was sampled correctly at random. We call such a problem a *samplable verifiable problem.*

Next we define the interactive proof of computational power. For any input string m, the computation $(P, V)[m]$ either accepts or rejects as before. P is the specified polynomial-time machine with the power to solve instances (it possesses a machine TM with this power), while P^* is any polynomial-time machine.

Let X be a procedure which can interact with the prover P and can backtrack the computation, furthermore, X is given an input from an input machine U which generates random instances; these instances are *unsolved* for X (namely, X does not "know" the witnesses); X tries to use the prover to solve these instances (via the interaction).

We call a pair of interactive Turing machines (P, V) an *interactive proof-system of computational power* for the above problem PR with error probability $\delta(m)$ (where m is the security parameter) if for m large enough:

1. If P is the specified prover which possesses a machine TM with the ability to solve (almost) all instances from $S(m)$, then V accepts (with overwhelming probability $1 - \delta(m)$).

2. For any machine P^* acting as a prover, the probability that $(P^*, V)[x]$ accepts and that an extraction procedure X which samples unsolved instances given by U, using (interacting with) P^*, and does not output the witnesses of all sampled instances is vanishingly small (less than $\delta(m)$).

The algorithm X works as specified, uniformly for any possible prover P*, it has a limited access to the communication tapes only, and can backtrack the computation. This extractor works in expected polynomial time, thus the whole interaction of X with P* is a feasible computation (and can be run by P* itself). This means again that, with very high probability, if the interaction with V is an accepting one, then the machine acting as a prover can compute all the witnesses as required and therefore can pass the statistical test (which means it has some computational power, with respect to the hard problem in question). Note that we use the fact that the instances produced by U are random and the solved instances generated by V are sampled from the same (or almost the same) distribution, therefore they both should fail and succeed with (almost) the same chance.

Now we can state the following

Theorem 1 *Under the intractability assumption of factoring (of large numbers multiplication of two large primes), protocol 2 above is a "perfect zero-knowledge proof system of the computational power to factor".*

4.2   A Protocol Scheme

The following is a general scheme for a samplable (verifiable) problem PR. Recall that an instance of PR is in the domain of a relation R (it has a witness) and the polynomial-time verifier can prove in zero-knowledge the validity of the chosen instance, while the prover has to be able to prove the knowledge of a witness.

Protocol 3: Proving Computational Power for Samplable Verifiable Problem

On input m (size parameter), for i=1,..,k do (k is a function of the required statistical confidence δ):

1. V picks a random solved instance n_i of size m and sends it

2. V uses a perfectly zero-knowledge proof that n_i is in the domain of the relation R

3. P proves in perfect zero-knowledge that it knows a witness w_i such that $R(n_i, w_i)$.

If all proofs are successful V accepts, otherwise he rejects.

end-of-protocol

Theorem 2 *The above protocol scheme is a "(perfect) [computational] zero-knowledge (argument) [proof] system of computational power" for samplable verifiable problems which are $(1 - \epsilon) - hard$.*

The above class of PR problems includes factoring as well as other problems used as candidates for cryptographic applications (Discrete Logarithm, RSA inversion, etc.). These problems have this property with direct verifiability in perfect zero-knowledge, (see [10]). They are assumed to be hard on the average and an overwhelming success in solving random instances of them for large enough security parameter, implies a computational power beyond P. (We remark that proving of computational power can be extended to a larger set of problems, but here we are interested in PR as the set most relevant to applications). The zero-knowledge property of the protocol is derived from the zero-knowledge property of proving knowledge. Note that the exact length and confidence parameters of the proofs in steps 2 and 3 are determined by the confidence parameter δ.

When we allow computational zero-knowledge we can allow proving invulnerable NP-problems in general (an example of such problem of a special size Clique has been recently given by Gurevich and Shelah [9]); this can be done under additional cryptographic assumption.

Next we suggest what to do when the interactive proofs cannot be implemented in perfect zero-knowledge without additional assumption, and an assumption is needed.

The proof of the correct sampling in step 2, is given by the verifier to a prover whose power is not known. In particular, P may hold some advantage in factorization, and our encryption on which the security (zero-knowledge-ness) of the proof is accepted may rely on the same problem (that is, factorization)! In this case encrypting the proof may not be enough, hiding it perfectly is required. Therefore, a natural proof system for an NP-statement to be used in this context is Brassard, Chaum, Crépeau's system of perfect zero-knowledge interactive argument proof [3]. The correctness of the proof relies on V's restriction to polynomial time (which means it can cheat with negligible probability).

In step 3, when P proves to V, if this step can be done perfectly the proof system is perfectly zero-knowledge (otherwise, it is computationally zero-knowledge). Actually, we observe that P can prove using an argument system as well (making the system perfectly zero-knowledge). The proof system is correct based on the underlying cryptographic assumption. If the legal P (possessing the extra power) interacts with V it will follow the protocol, and will not abuse its power. On the other hand, any cheating P^* (which is in polynomial time) will not be able to cheat in the proof under the cryptographic assumption, but with a negligible probability. This completes the proof. The interactive argument system originated in [3] can be implemented under a general assumption that one-way homomorphism exists [11], and it requires only a constant number of iterations [4] which is important for various applications.

References

[1] E. Bach, *Generating Random Numbers with Known Factors*, SIAM Journal on Computing.

[2] M. Blum, *How to Prove a Theorem so No One can Claim It*, International Conf. of. Math., 1987.

[3] G. Brassard, D. Chaum and Crépeau C., *Minimum Disclosure Proofs of Knowledge*, J. Comp. Sys. Sci. 37-2, pp. 156-189.

[4] G. Brassard, Crṕeau C. and M. Yung, *Any NP statement can be proved in perfect zero-knowledge in bounded number of rounds* , ICALP 1989.

[5] U. Feige, A. Fiat and A. Shamir, *Zero-Knowledge Proof of Identity*, Proc. 19th STOC, 1987, pp. 210-217.

[6] Z. Galil,, S. Haber and M. Yung, *A Private Interactive Test of a Boolean Predicate and Minimum-Knowledge Public-Key Cryptosystems*, FOCS, 1985 pp. 360-371.

[7] S. Goldwasser, S. Micali and C. Rackoff, *The Knowledge Complexity of Interactive Proof-Systems*, Proc. 17th STOC, 1985, pp. 291-304.

[8] S. Goldreich, S. Micali and A. Wigderson, *Proofs that Yields Nothing But their Validity, and a Methodology of Cryptographic Protocol Design*, Proc. 27th FOCS, 1986.

[9] U. Gurevich and S. Shelah, , Private Communication.

[10] L. Hemachandra, M. Abadi, E. Allender, A. Broder, and J. Feigenbaum *On Generating Solved Instances of Computational Problems* , Proc. of Crypto 88.

[11] R. Impagliazzo and M. Yung, *Direct Minimum-Knowledge Computations* , Crypto 87.

[12] M. O. Rabin, *Digital Signatures and Public Key Functions as Intractable as Factoring*, Technical Memo TM-212, Lab. for Computer Science, MIT, 1979.

[13] M. Tompa and H. Woll, *Random Self-reducibility and Zero-Knowledge Interactive Proofs of Possession of Information*, FOCS, 1987, pp 472-482.

More Efficient Match-Making and Satisfiability
The Five Card Trick

Bert den Boer

Centrum voor Wiskunde en Informatica

Kruislaan 413, 1098 SJ Amsterdam, The Netherlands.

Abstract. A two-party cryptographic protocol for evaluating any binary gate is presented. It is more efficient than previous two-party computations, and can even perform single-party (i.e. satisfiability) proofs more efficiently than known techniques. As in all earlier multiparty computations and satisfiability protocols, commitments are a fundamental building block. Each party in our approach encodes a single input bit as 2 bit commitments. These are then combined to form 5 bit commitments, which are permuted, and can then be opened to reveal the output of the gate.

A Matchmaking Example

Alice and Bob never met before and wish to find out whether they have some particular mutual interest. But naturally each refuses to show interest first, because of the risk of getting an embarrassing "no" from the other. More formally, Alice has a secret bit a and Bob has a secret bit b and a protocol is needed that reveals exactly the logical "AND" of the two bits. Consequently, if Bob's bit is zero he should learn nothing about Alice's bit; if his bit is one, he cannot fail to learn Alice's bit because in that case her bit has the same value as the AND.

One way to achieve the desired protocol is by physical means—more precisely, five cards. The back of all cards are, as usual, the same. The face side of two of the cards are identical, say the two-of-hearts, and the face of the other three are identical, say the two-of-spades.

Initially, each party is given one card of each type and the remaining spade is put face down on the table and Bob then puts his cards face down on top of the initial spade. His secret choice of ordering for the two cards encodes his bit b: heart on top means *1* and the other way round means *0*.

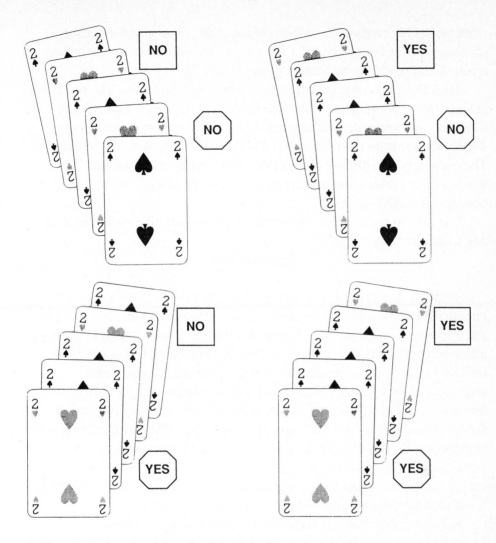

Fig. 1. Match making. If the table had eyes, the table would see five cards: the two of spades in the middle, Bob's two cards on top (the order indicates his choice (in the square)), Alice's two cards at the bottom of the stack (the order indicates her choice (in the octagon)). Three cases are cyclic permutations of each other, so indistinguishable after cutting the cards. The other case is two times yes.

Alice places her cards at the bottom of the stack. The secret order she chooses encodes her bit *a*, in a way that mirrors Bob: heart on the bottom means *1* and spade on the bottom means *0* (see Fig I)

Then Bob and Alice each take turns "cutting" the cards. The result is a random cyclic permutation known to neither. After each is satisfied that the other does not know the cyclic permutation remaining in the cards, they display the cards on the table in a radial pattern like spokes of a wheel. There are only two distinct results apart from cyclic rotations (see Fig I where the first three cases are cyclic permutations of each other), and they correspond to *AND (a, b)*.

It is easy to see that the two hearts are on consecutive spokes exactly when both bits *a* and *b* are *1*.

Introduction

What we just saw is an encoding of two bits into a five bit vector by putting the two bits, their inversion, and an extra zero bit in a certain order and applying a cyclic permutation on this vector. Such a vector is decoded into *1* when it is a cyclic shift of *(1, 0, 0, 0, 1)* or into *0* if it is a cyclic shift of *(0, 1, 0, 1, 0)*. Instead of working with bits we can make sequences of five "blobs". This enables us to make a satisfiability protocol that is more efficient than the similar protocols [BC86]. Our protocol is described in the section: "The main protocol". Satisfiability protocols are as explained in [BC86] ,[BCC] and [GMW86] roughly speaking a way to proof a claim such that the other party cannot prove the same claim by just exploiting the data from the conversation. The above number of five blobs is our gain over the number of twelve blobs used in [BC86].

Before we describe our protocol we exhibit in the section: "Assumptions on Blobs" basicly one extra demand on blobs which we need for efficiency reasons. In order that our idea give rise to a protocol more efficient then the protocol in [BC86] we require that there exist "direct minimal disclosure proofs" for blob equality and blob inequality. In the section: "A Particular Example for a Blob" we suggest a blob which can be made with just one squaring and one computing of a Jacobi-symbol. In the section "Multi party Computations" we exploit our idea for a general multi party computation.

Assumptions on Blobs

In this section we describe requirements for blobs and give a suitable general formula.

For a protocol using blobs to be efficient we require that both creating blobs and opening blobs (showing that a particular blob encrypts a particular bit) are efficient. Bobs are defined, as in [BCC], to satisfy the following: no blob can be opened both as a zero as well as a one (authenticity) and blobs cannot be opened by the opponent (secrecy).

Our blobs should be such that the prover can prove that two blobs hide a different bit (this protocol is called "blob inequality") and also our building block requires that we have a protocol to show that two blobs are encryptions of the same bit without giving information which bit this is (called "blobequality"). It is known that given a protocol for one, a protocol for the other can be constructed.

We also want those protocols to be efficient in the sense that they are non-interactive zero-knowledge with perfect soundness and perfect completeness (see [GMS]). We envisage such a protocol for say blobinequality just by a message of the prover stating that two blobs hide different bits and containing some extra data upon which the verifier can easily verify the statement that the two hidden bits are different. Nonetheless the verifier gets no bias towards which blob hides the zero bit value.

In other words we want a simple protocol where one round gives hundred percent certainty in the involved statement under the assumption that no blob itself can be opened by the prover in two ways (and the verifier has to assume that for any protocol exploiting commitments).

In practice but still in a general setting the way to achieve a bit encoding with a fast protocol for blob equality is the following:

Let G and H be commutative groups,
f a homomorphism from G to H,
$K \in H$, such that the encoder can not find a $c \in G$ such that $f(c) = K$.
The encoding is done in the following way:
take a random element r of G and encode 0 by $f(r)$ and encode 1 by $Kf(r)$.

The reason the encoder cannot find such a c is either because no such value exists or because a suitable value is infeasible to find (because of a cryptographic assumption like the encoder cannot factor or take discrete logs). In the first case the opponent should not be able to distinguish random elements of the image of f with random elements of the coset which includes K because of a cryptographic assumption. In the second case the opponent might know a solution c for the equation $f(c) = K$ but then in his view any blob could be equally likely encode a zero as well as a one. Both cases give different flavors of zero knowledge protocols. In the second case the encoder is unconditionally protected and the opponent can only be fooled if the encoder could crack the underlying problem during the time of the protocol. In the first case the opponent is unconditionally protected but he could learn the satisfying assignment later on by cracking the underlying problem.

Blob inequality is done by giving the fact that two blobs hide different bits and giving some element s to the verifier then the verifier checks that the product of the blobs is equal to $Kf(s)$. Blob equality is done by giving a element u of G such that $f(u)$ is the quotient of the two blobs. It is easy to see that those techniques meets the demands for blob (in) equality.

The Main Protocol

In this section we exhibit our construction for a satisfiability protocol.

The idea is that the prover P and the verifier V have agreed on some circuit consisting of two-input gates realizing the Boolean expression implied by the assignment. All bit values outside and inside the circuit are probabilisticly encoded by P. The output bit of a gate might be used in more than one subsequent gate. The idea is that P encodes the input bits for the whole circuit (the bits corresponding to his secret: the satisfying assignment) and subsequently all bits inside the circuit. The final bit (which is equal to one) at the outside of the circuit has to correspond with the encodings of the last two bits at the input side of the last gate in the circuit. Now it is sufficient if the verifier is convinced that for each gate the input bits and the output bit satisfy the equation implied by this gate. First we deal with those eight gates which are of degree two. Basicly all eight are similarly dealt with. We show how to deal with the so-called NAND-gate. As is known the whole circuit could be built with NAND-gates with at most five times more gates than in a circuit where all ten possible two-input gates are allowed.

Let us denote the input bits of the NAND-gate a and b and the output bit by c. P encodes the output bit c just by a residue C independent of the encodings A and B of a and b. To prove the consistency of the blobs A, B and C the prover will convince the verifier that the bit vectors $(b \oplus 1, a, 0, b, a \oplus 1)$ and $(c, c \oplus 1, 0, c \oplus 1, c)$ are cyclic permutations of each other. To do this P creates another cyclic shift $(d_0, d_1, d_2, d_3, d_4)$ of this vector. Then P encodes those bits d_i by blobs D_i. Those five blobs are given to the verifier and he has the choice out of two questions.

On the first question the prover gives a residue k modulo five and elements s_0, s_1, s_2, s_3, s_4 of the group G such that the following equations hold:

$$D_k = f(s_0) C,$$
$$D_{(k+1) \bmod 5} \, C = K f(s_1),$$
$$D_{(k+2) \bmod 5} = f(s_2),$$
$$D_{(k+3) \bmod 5} \, C = K f(s_3),$$
$$D_{(k+4) \bmod 5} = f(s_4) C.$$

In case the verifier chooses for the second question it means that the prover has to give a residue m mod 5 and five elements $v_0,...,v_4$ of G such that

$$D_m B = K f(v_0),$$
$$D_{(m+1) \bmod 5} = f(v_1) A,$$
$$D_{(m+2) \bmod 5} = f(v_2),$$
$$D_{(m+3) \bmod 5} = f(v_3) B,$$
$$D_{(m+4) \bmod 5} \, A = K f(v_4).$$

The eight gates of degree two can be described by $gate(a,b) = NAND(a \oplus c, b \oplus d) \oplus e$, where c, d and e are bits. For $c = 1$ we just interchange the positions of a and $a \oplus 1$, for $d = 1$ we just interchange the positions of b and $b \oplus 1$. For $e = 1$ we decimate the positions of all elements in the five bit vector with a factor plus or minus two modulo five.

For the exclusive or gate we take $XOR(a, b) = (a \oplus 1, b, 0, b \oplus 1, a)$. If we decimate the position of the elements in this vector with a factor minus two and if we assume that the middle element is at position zero we get $EQ(a, b) = (b \oplus 1, a \oplus 1, 0, a, b)$. This way we have dealt with all two bit input gates

For each gate several sets of five residues like $(D_0,..., D_4)$ are presented to V. Then all first sets are gathered in one class all second sets in another class etcetera. For each class the verifier V request to see either equivalence of each set of five blobs with either the input side of the involved gate or to see equivalence of each set of five blobs with the output side of the involved gate.

<u>Proof of security</u>: Suppose for a class of sets of residues like $(D_0,...,D_4)$ the prover has to show equivalence of each set of five residues with the input side of the involved gate. The prover can only survive this challenge if either he knows a satisfying assignment and made his blobs according to the protocol or in case P does not know P has made all his sets of residues in this class to be consistent with the inputside of involved gates. In the last case he does not know the equivalence for at least one set of five residues with the output side. So his chance of cheating is 2^{-k} by guessing or predicting the question of V and making the sets of five residues like $(D_0,..., D_4)$ accordingly. Otherwise we have to assume that for each set of five residues like $(D_0,..., D_4)$ in at least one class P can proof equivalence with both input side and output side. We may assume the gate is a NAND-gate then this means that the elements

$(b \oplus 1, a, 0, b, a \oplus 1)$ and $(c, c \oplus 1, 0, c \oplus 1, c)$ of $GF(2)^5$ are cyclic shifts of each other.

Now we will have to prove that is precisely the case when $c = NAND(a, b)$. Note that for $a = b = 1$ we do not have two consecutive ones in the cyclic ordering of the bits in the first vector. This can only be the case in the second vector if $c = 0$. For all three other possible values of (a, b) there are two consecutive ones in the first vector and such a vector can be brought in the form $(1, 0, 0, 0, 1)$ by a cyclic shift. This can only happen for $c = 1$.

$\square$

To increase the efficiency more efficient protocols can be made for the special cases (gates of degree one) of the XOR gate and the EQ gate.. We will show it in case the gate is the XOR gate. The first improvement on the protocol is to create blobs E_0 and E_1 and showing equality for the exclusive-or sum of the encoded bits with either the pair A and B or with the pair C and the element $f(1)$(the obvious encoding of zero, where 1 is the unit element of G). The equality of the sum of the bits hidden by E_0 and E_1 and the sum of the bits hidden by A and B is shown by either blobequality of the blobs A and E_0 and blobequality of B and E_1 or by blobinequality of those pairs. The prover decides whether he proves blobequality twice or blobinequality twice and he does not say in advance what his choice is (and of course he has no choice otherwise it means that he can open blobs in two ways which violates our cryptographic assumption).

An even faster protocol is possible in case K^2 can be written as $f(d)$ and d is easy to find by the prover. Then the exclusive-or sum of the bits a an b is encoded by the product $A\,B$ assuming a and b are encoded by the blobs A and B. This is usual the case where G and H are subgroups of the multiplicative group of residues modulo a composite N and f is taking squares.

A final remark need to be made about the complexity to deal with the NAND gate. The equivalence between the five blobs D_i and the single blob C can be checked with five applications of the homomorphism f and six (even with five) multiplications in H and the same holds for the equivalence between the five blobs D_i and the two blobs A and B. Assuming the prover has already created blobs A, B and C the prover can create new blobs D_i and prepare possible answers on queries without having to do divisions in group G or H.

A Particular Example for a Blob

In [BC86] and [BCC] several implementations of blobs are given. We will give a slightly different one. This one is based on multiplications modulo a composite number N whose factorization has special properties (those numbers are often called Blum integers, the first reference for such composites is [Wil80]):

the number N is made by the verifier V.
the group G is the set of residues with positive Jacobi-symbol,
H is the set of squares modulo N,
K is an element of which the prover only knows a square root with negative Jacobi-symbol (and for which the verifier has proved with zero-knowledge techniques that he also knows a square root with positive Jacobi-symbol)

Multiparty Computations

Our idea can be exploited for multi party computations. The setting is as is known a number of participants each which secret bits and some say one bit function they want to compute. The model we have is using one outsider which does not collude with any of the other parties. The outsider makes a product of two primes, publishes this composite number plus one non-quadratic residue K with positive Jacobi-symbol. Then the outsider proves with an interactive zero-knowledge protocol the involved properties. Then everybody encodes their bits using QRA [GM] with the modulus and residue K of the outsider, they pool their blobs, and they all go as one entity in interaction with the outsider. They produce sequences of five blobs, multiply each blob with a random square, apply a random cyclic permutation and a random decimation (in two of the four cases resulting in inverting the encoded bit), and offer the resulting sequence of five blobs to the outsider. The outsider responds with a single blob encoding the same bit as the five blobs. Using those answers the group can compute the value they want by sending the final sequence of five blobs (which encodes the final output or the inverse (the group knows but not the outsider)) and ask to open this blob. If this protocol is well designed the outsider will learn nothing and the participants only the final value. One can put the outsider to the test by asking him to prove that all his reductions from five blobs to one were good and such a protocol is basicly the same as our main protocol. The outsider could be a computer program which data can be erased by the group after the protocol. The group members can only find out secrets of each other when they violate QRA.

In principle our technique can be used to compute with blobs. The blob we get is then a sequence, having a length that is a five power, of single blobs. Two sets of five blobs (containing two commitments for 1 and three commitments for 0) encodes a different bit if after a decimation with a factor two and a cyclic shift of one of the sets the corresponding blobs can shown to encode the same bit. And blobequality is done likewise but without the decimation. To change a blob (of length 5^{L+1})into a blob which encodes the inverted bit (**blob inversion**) we just rearrange all blobs of length 5^L by decimation with a factor two. Only blob inversion for the single blob might not be possible in general for the opponent but in the restricted general case we adopted for blobs it just the quotient of K with the blob. For the full general case the creator of the blob might have to create blobs to encode the inverted input values as well. Furthermore in order to compute with those sequences of blobs for each gate both sequences have to be made of the same length. This is done by blowing up the smallest of them with enough powers of five. To increase the length with a factor of five one replaces all "atom blobs" A by (A, B, O, B, A) where B is a blob encoding a different bit then A does and where O encodes the zero bit. Then any body can compute with those blobs to arrive at a very large blob.

Open Problems and Conclusion

It remains open how the ideas presented above can be used to conduct a general multiparty computation protocol with unconditional secrecy.
It may be concluded that in principle computing with blobs is possible (an open question in [BC86]) and now the question is can it also be done without increasing the length.

References

[BCC]   Brassard, G., Chaum, D. and Crepeau, C. " Minimal disclosure proofs of knowledge" . Journal of computer and system sciences, **37**, 2 , october 1988, 156-189.

[BC86]   Brassard, G and Crepeau, C. "Zero-Knowledge Simulation of Boolean circuits." Advances in Cryptology - Crypto '86, A.M. Odlyzko, Lecture Notes in Computer Science 263, 223-233, Springer-Verlag.

[GM] Goldwasser, S., and Micali, S., "Probabilistic encryption", JCSS, **28**, 2, 1984, 270-299.

[GMW86] Goldreich, O., Micali, S., Wigderson, A. "How to prove all NP statements in Zero-knowledge" Advances in Cryptology - Crypto '86, A.M. Odlyzko, Lecture Notes in Computer Science 263, 171-185, Springer-Verlag.

[GMS] Goldreich, O., Mansour, Y., Sipser, M. "Interactive Proof systems: Provers that never fail and random selection". Symp. on Found. of Comp. Sc., **28**, oct 87, 449-461.

[Wil80]   Williams, H.C., "A modification of the RSA public-key encryption procedure." IEEE Trans. Inform. Theory, **26** (6), 726-729, November 1980.

Section 4

Applications

A SINGLE CHIP 1024 BITS RSA PROCESSOR

André Vandemeulebroecke[1], Etienne Vanzieleghem[2], Tony Denayer[1] and Paul G. A. Jespers[3]

This work has been supported by the Région Wallonne of Belgium and by the FNRS (National Fund for Scientific Research).

[1] Mietec N.V., Raketstraat 62, B-1030 Bruxelles, Belgium - Phone (32) - 2 - 242.50.10.

[2] Alcatel Bell Telephone, Francis Wellesplein n°1, B-2018 Antwerpen, Belgium - Phone (32) - 3 - 240.40.11.

[3] Microelectronics Laboratory, Université Catholique de Louvain, Place du Levant 3, B-1348 Louvain-la-Neuve, Belgium - Phone (32) - 10 - 47.25.40.

ABSTRACT

A new carry-free division algorithm will be described; it is based on the properties of RSD arithmetic to avoid carry propagation and uses the minimum hardware per bit i.e. one full-adder. Its application to a 1024 bits RSA cryptographic chip will be presented. Thanks to the features of this new algorithm, high performance (8 kbits/s for 1024 bits words) was obtained for relatively small area and power consumption (80 mm2 in a 2 μm CMOS process and 500 mW at 25 MHz).

1. INTRODUCTION

With the constant growth of data communications, security is becoming an increasingly important feature. Cryptography provides two ways to solve some aspects of information protection: secret or public key cryptosystems. DES (Data Encryption Standard) is the most popular system of the first category. Integrated versions can attain high processing speeds (20 Mbits/s, [1]) but the key unicity entails the problem of key management. In order to solve this problem, public key cryptosystems use different keys for enciphering and deciphering. The RSA cryptosystem (Rivest, Shamir and Adelman [2]) has emerged among all proposed solutions during the last decade.

The basic RSA operation is a modular exponentiation involving large numbers (typically 200 to 1000 bits). Such a size is mandatory to protect the system against cracking. Modular exponentiation can be split into successive modular multiplications. While powerful multiplication algorithms are well-known, carry-free modular computation (in fact a division which in remainder is kept) algorithms only begin to emerge [3]. As the size of the processed numbers is very large, it is mandatory to overcome the carry propagation problem, in order to keep the addition time as short as possible relative to the clock frequency. Solutions to this problem were already proposed [4] but the required hardware per bit was too big (more than two equivalent full-adders) and consequently not suitable for a 1024 bits RSA chip.

In this paper, we present a new carry-free division scheme based on the Redundant Signed Digit (RSD) representation and using the most reduced hardware (one full-adder by bit). This technique has been applied to the design of a 1024 bits RSA single chip whose performances are best compared to all known solutions published up to now [5, 6, 7, 8, 9]. In sections II to IV, basic concepts of RSD arithmetic are presented and the carry-free division method is described. In sections V to X, the 1024 bits RSA chip is presented; it is compared to other currently available implementations and some specific features of its architecture are detailed.

2. CARRY SAVE AND RSD ARITHMETIC

The conventional addition of two numbers in classical binary form is usually the bottleneck for speed improvement in digital integrated circuits. While other digital operations may be performed in parallel (shift, data storage...), addition requires a carry propagation.

The most significant digit of the sum of two n-digit numbers written in conventional form actually depends on the 2.n data. This case occurs with a full carry propagation from the LSB to the MSB. The classical addition of two binary numbers is sketched in fig 1.

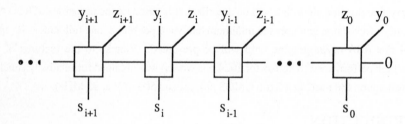

Fig. 1. Classical addition with carry propagation

As the global delay of the adder depends on the carry delay, the basic full adder cell is optimized regarding this criteria. Different solutions proposed to solve the carry propagation problem use redundant representations for the operands and the result. The most well known method is the use of carry save arithmetic.

Carry save adders are widely used in fast arithmetic processors, because of their performance in terms of speed and silicon area. The basic principle of the carry save addition is to reduce the sum of three binary numbers to the sum of two binary numbers without carry propagation. Let us consider the two operands Y (in redundant form) and Z (in classical form). Y is represented as the sum $Y^* + Y^{**}$ and y_i^*, y_i^{**}, $z_i \in \{0, 1\}$. With one level of classical full adders, we can reduce the sum of the three binary numbers: $Y^* + Y^{**} + Z$ to a sum of two binary numbers: $S^* + S^{**}$ (Fig. 2).

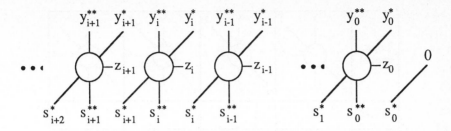

Fig. 2. Carry save addition scheme

The addition process is fully parallel, hence the basic cell has to be optimized for both carry and sum delays. Such a cell is sketched by a ring full adder, to distinguish it from the above square full adder (fig. 1). For each of them, one can write :

$$y_i^* + y_i^{**} + z_i = 2 \cdot s_{i+1}^* + s_i^{**}$$

Defining S as $S^* + S^{**}$, we obtain $Y + Z = S$ with Y and S in redundant carry save form and Z in classical binary form. This scheme is extensively used in parallel multipliers.

The Redundant Signed Digit arithmetic (RSD) has been introduced in the late 50's by Avizienis [10] and is quite similar to the carry save one. In the RSD representation, a number X can be viewed as the difference between two positive binary numbers X^* and X^{**}. We have:

$$X = \sum_{i=0}^{n} x_i \cdot 2^i = \sum_{i=0}^{n} (x_i^* - x_i^{**}) \cdot 2^i \text{ with } x_i^*, x_i^{**} \in \{1, 0\}.$$

Obviously, each digit x_i of X belongs to the set $\{1, 0, \bar{1}\}$ where the upper bar indicates a negative value. With such a definition, a number may have more than one representation. As an example, the number 7 expressed with four bits has the representations: $(0\ 1\ 1\ 1)$, $(1\ 0\ 0\ \bar{1})$, $(1\ 0\ \bar{1}\ 1)$ and $(1\ \bar{1}\ 1\ 1)$.

Clearly, RSD and carry save representations are in essence quite similar. The main difference is the fact that RSD doesn't need use of two's complement form to handle negative numbers. This representation is thus more "natural" if positive and negative numbers have to be processed. Therefore, the multiplication and division will be easier performed with this representation and the sign test will be directly obtained by testing the first nonzero digit.

There has been little use of RSD arithmetic up to now, perhaps due to the inherent complexity of the basic adder cell [3, 4, 10]. In the next section, we will propose a new addition scheme using RSD representation which can be implemented in the same way carry save adders are. Therefore, it allows to combine the advantages of the two solutions.

3. RSD ADDITION AND MULTIPLICATION

The conventional 1-bit full adder assumes positive weights to all of its 3 binary inputs and 2 outputs. Such adders can be generalized to four types of adding cells by imposing positive and negative weights to the binary input/output terminals [11]. Figure 3 lists the names and logic symbols of the four types of generalized full adders.

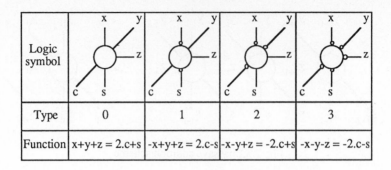

Logic symbol				
Type	0	1	2	3
Function	$x+y+z = 2.c+s$	$-x+y+z = 2.c-s$	$-x-y+z = -2.c+s$	$-x-y-z = -2.c-s$

Fig. 3. Generalized full adders

Each type of full adder is named by the number of its negatively weighted inputs. The implementation of a generalized full adder (GFA) is quite straightforward: it can be shown that a GFA is simply an usual type 0 full adder with inverters replacing the rings. This implies no extra hardware in CMOS compared to a classical full adder as those inverters are always existent.

The addition of two SD (Signed Digit) numbers Y and Z can be performed by cascading two levels of generalized full adders of types 1 and 2 (Fig. 4).

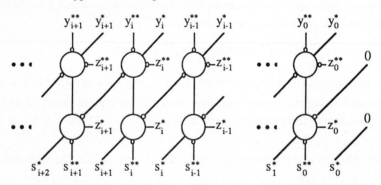

Fig. 4. RSD addition scheme with all data in redundant form

Subtraction may be obtained by permuting the numbers Z* and Z**. The main drawback of this scheme with two RSD numbers is the amount of hardware which is twice than in the carry save case.

Let $Y = Y^* - Y^{**}$ be a number in redundant form, Z a number in two's complement form and $S = S^*$ - S^{**} the result of the operation $Y + Z$ or $Y - Z$, with y_i^*, y_i^{**}, z_i, s_i^*, $s_i^{**} \in \{0, 1\}$. In the following, we treat the sign of Z by using the well known property of 2's complement numbers [11]:

$$\text{if} \quad Z = -z_n.2^n + \sum_{i=0}^{n-1} z_i.2^i = (z_n\ z_{n-1}\ ...\ z_0).$$

$$\text{then} \quad -Z = -(1-z_n).2^n + \sum_{i=0}^{n-1} (1-z_i).2^i + 1 = (1-z_n\ 1-z_{n-1}\ ...\ 1-z_0) + 1.$$

So a subtraction reduces to an addition with an inversion of all bits and adding of 1. The details of these two operations are sketched in fig 5 and 6.

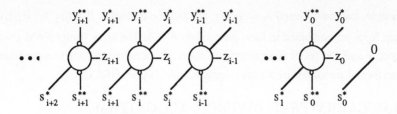

Fig. 5. RSD addition with one operand in 2's complement form

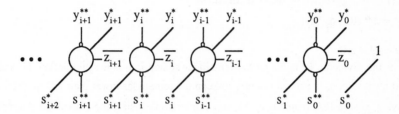

Fig. 6. RSD subtraction with one operand in 2's complement form

In the same way as with the carry save adder, the computation scheme can be implemented with one level of full adders, using in this case generalized full adders of type 1. The addition is fully parallel, thus requiring no carry propagation.

These generalized full adders are used in Pezaris array type multipliers [11] to directly implement the multiplication of two numbers in two's complement form. Pezaris multipliers are the optimal choice for the implementation of classical array multipliers. But as the internal structure is irregular, it is impossible to derive an associated serial-parallel algorithm using one row of adders. This is obviously possible when using rows of RSD adders, as shown in fig. 7.

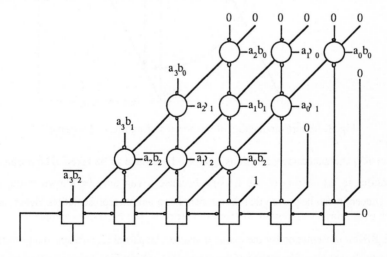

Fig. 7. RSD multiplication scheme

In this example, the multiplicand A = (a_2 a_1 a_0) and the multiplier B = (b_3 b_2 b_1 b_0) in two's complement form are combined to form the product A * B. The array (ring GFA's) contains only GFA of type 1 and the result is in redundant form. If necessary, the back conversion to two's complement form is performed by a carry propagate adder (square GFA's).

4. A NEW CARRY FREE DIVISION ALGORITHM

The classical non restoring division is usually performed using the add/subtract-and-shift-left algorithm [11]. It is governed by the next equations:

$$R^{j+1} = 2 * (R^j - q_j * D)$$

$$-2.D < R^j < 2.D$$

where R^0 is the dividend, R^j is the partial remainder at step j, q_j is the jth quotient bit and D is the divider. At each step, q_j is chosen in order to keep the remainder lower than 2.D in absolute value. Practically, q_j = sign (R^j) hence $q_j \in \{1, \bar{1}\}$. An example using the Robertson diagram [11] is sketched in fig. 8.

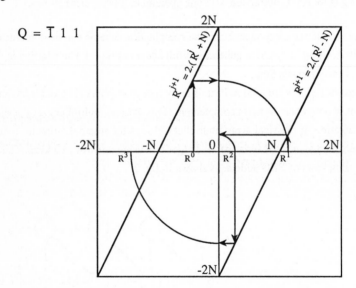

Fig. 8. Robertson diagram of the non restoring division scheme

Starting from R^0, the successive decisions $q_j = \bar{1}$, 1, 1 have to be taken. The problem of this algorithm resides in the testing of the R^j sign because it can only be known through a carry propagation. Indeed, it can be proven that the quotient has a unique representation. Hence, an error in the evaluation of q_j ruins the algorithm. Try for example $q_0 = 1$ in the above example.

If we allow a RSD representation for the quotient and for the partial remainders, the problem can be solved in a quite elegant way. The testing of the three MSD's (Most Significant Digit) of R^j only gives a range which in it resides. For example: R^j is surely negative or R^j is lower than D in absolute value.

Due to the redundant nature of R^j, the two decisions may occur when the ranges overlap. The choice for q_j is given hereunder:

If $R^j < 0$ then $q_j = \overline{1}$.

If $-D < R^j < D$ then $q_j = 0$.

If $R^j > 0$ then $q_j = 1$.

An example using the Robertson diagram is given in fig. 9.

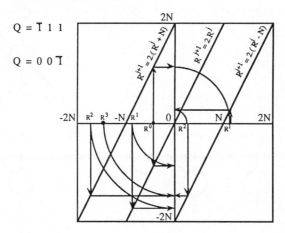

Fig. 9. Robertson diagram of the RSD division scheme

Starting from R^0, we could decide $q_j = 0$ or $\overline{1}$. In fact, this has no effect on the final result: indeed, R^3 is the same because $Q = (\overline{1}\ 1\ 1) = (0\ 0\ \overline{1})$. This technique has been applied with success to a completely different topic: the design of a RSD analog-to-digital converter [12].

More formally, suppose that R^0 is in RSD form and D in two's complement form. Therefore :

$$D = -2^0 .d_0 + \sum_{i=0}^{n} d_i.2^{-i}$$

with $d_i \in \{0, 1\}$ and $d_0 \neq d_1$. Hence: $0.5 \leq |D| \leq 1$.

$$R^0 = \sum_{i=0}^{n+m} r_i^0 .2^{-i} = \sum_{i=0}^{n+m} \left(r_i^{0\,*} - r_i^{0\,**} \right) 2^{-i}$$

with $r_0^0 = 0$, $r_i^0 \in \{1, 0, \overline{1}\}$ and $r_i^{0\,*}, r_i^{0\,**} \in \{1, 0\}$. Hence: $-1 < R^0 < 1$. RSD division steps are sketched at fig. 10.

The physical implementation implies that 3 bits are lost after each division step. The classical division scheme with carry propagation is built in order that they can be forgotten. But due to the redundant nature of the intermediate results, this is not guaranteed with the use of RSD even if the arithmetic conditions of the division algorithm are met. This is known as the representation overflow or consistency problem. It implies a further restriction on the choice of the q_j.

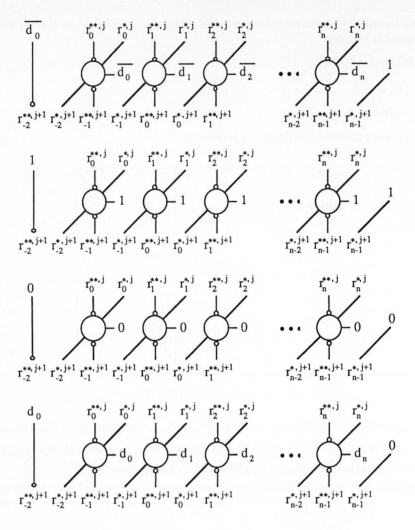

Fig. 10. RSD division steps for $q_j = 1, \ 0, \ \overline{0}, \ \overline{1}$

The division algorithm in a redundant number system is governed by two conditions. At each step :

$$- 2.|D| \ < \ R^{j+1} \ < \ 2.|D| \ . \qquad \text{(Arithmetic condition)}$$

$$R^{j+1} = \sum_{i=0}^{n+m-j} r^{j+1}_i . 2^{-i} = \sum_{i=0}^{n+m-j} \left(r^{j+1}_i {}^* - r^{j+1}_i {}^{**} \right) 2^{-i} \qquad \text{(Consistency condition)}$$

with $r^{j+1}_i {}^*, r^{j+1}_i {}^{**} \in \{1, 0\}$ and $r^{j+1}_i \in \{1, 0, \overline{1}\}$. The consistency condition is met when all r^{j+1}_i

are equal to 0 for $i < 0$.

In the following, we suppose that the divisor D is strictly positive. The generalization to negative divisors is immediate. Table I demonstrates the fact that the examination of the three MSD's (Most

Significant Digit) of R^j is sufficient to insure the respect of the two above conditions. The representation overflow is also avoided. From the values of the three most significant bits of R^j, the correct choice of q_j is given as well as the values of the known most significant digits of R^{j+1} in the two distinct cases $d_2 = 0$ or 1. Notice that $R_0^0 = 0$.

r_0	r_1	r_2	q_j	$d_2 = 0$ r_0^{**}	r_0^{*}	r_1^{**}	$d_2 = 1$ r_0^{**}	r_0^{*}	r_1^{**}
1	0	1	1				0	1	1
1	0	0	1				0	0	0
1	0	$\bar{1}$	1				0	0	1
1	$\bar{1}$	1	1				1	1	1
1	$\bar{1}$	0	1				1	0	0
1	$\bar{1}$	$\bar{1}$	0				0	0	0
0	1	1	1	1	1	0	1	1	1
0	1	0	1	1	1	1	1	0	0
0	1	$\bar{1}$	0	0	0	0	0	0	0
0	0	1	0	1	1	0	1	1	0
0	0	0	0	1	1	1	1	1	1
0	0	0	$\bar{0}$	0	0	0	0	0	0
0	0	$\bar{1}$	$\bar{0}$	0	0	1	0	0	1
0	$\bar{1}$	1	$\bar{0}$	1	1	1	1	1	1
0	$\bar{1}$	0	$\bar{1}$	0	0	0	0	1	1
0	$\bar{1}$	$\bar{1}$	$\bar{0}$	0	0	1	0	0	0
$\bar{1}$	1	1	$\bar{0}$				1	1	1
$\bar{1}$	1	0	$\bar{1}$				0	1	1
$\bar{1}$	1	$\bar{1}$	$\bar{1}$				0	0	0
$\bar{1}$	0	1	$\bar{1}$				1	1	0
$\bar{1}$	0	0	$\bar{1}$				1	1	1
$\bar{1}$	0	$\bar{1}$	$\bar{1}$				1	0	0

Table I. Decision table and partial remainders when examining the 3 MSD's

5. RSA ALGORITHM

The RSA cryptosystem is based on the modular exponentiation of large numbers (typically a few hundreds of bits). Let m be a message to encipher. The enciphered message c is obtained with the key (e, N) by : $c = m^e \pmod N$. The deciphering key (d, N) allows to recover the original message by $m = c^d \pmod N$. Both operations are thus identical. Let:

$$e = \sum_{i=0}^{n-1} e_i.2^i = e_{n-1}.2^{n-1} + ... + e_2.2^2 + e_1.2 + e_0 \text{ with } e_i \in \{0, 1\}.$$

$$m^e \pmod{N} = (m)^{e_0}.(m^2)^{e_1}.(m^{2^2})^{e_2}...(m^{2^{n-1}})^{e_{n-1}} \pmod{N}$$

The algorithm of the modular exponentiation can be derived from the properties of the modular arithmetic. This leads to the right-to-left binary modular exponentiation and can be expressed as a C program.

```
current = m; result = 1;
for (i = 0; i <= n-1; i++)
{  if (e_i == 1) then result = result * current (mod N);

   current = current * current (mod N);

}
```

The number of modular multiplications steps is $n + v(e)$ where $v(e)$ is the number of 1 in the exponent e. For example, with an exponent equal to $65537 = 2^{16} + 1$ (the fourth Fermat number), the number of modular multiplications is 19.

Our implementation is based on the RSD algorithms developed in the previous sections. The module operation (mod) is simply a division in which the remainder is kept as result. The basic modular multiplication step is then a multiplication followed by a division (Fig. 11).

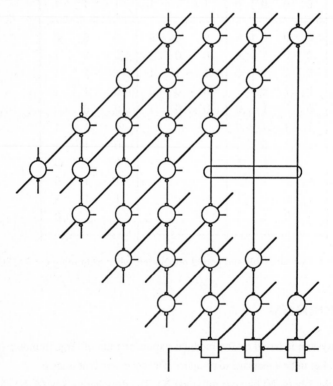

Fig. 11. Modular multiplication step of RSA algorithm

As the size of the operands is large (n = 200 ... 1000 bits), it is impossible to integrate parallel multipliers and dividers on a single chip. Therefore, serial/parallel algorithms are usually used with an adder as basic ALU. Considering a unity add time for the addition (one clock cycle), both multiplication and division are ideally performed in n cycles. In the worst case (all $e_i = 1$), the total number of additions is $4.n^2$ and the associated ideal baudrate is (fclock / 4.n).

The result in redundant form is then converted back to 2's complement form by a carry propagation adder. This conversion is heavily time consuming ($\approx$ 1 μs or 40 cycles with a 25 MHz clock). As this operation seldom occurs, its relative slowness is not a coarse limitation.

6. STATE OF THE ART

The large size of numbers involved in the RSA algorithm leads to large design and processing time. While multiplication algorithms with linear behavior are well known (carry save technique), division usually involves carry propagation because tests are necessary. This considerably degrades the ideal baudrate.

In order to solve this problem, most known realizations use non regular techniques leading to a waste of resources. In two implementations [5, 7], the carry save technique seems to be used for the multiplications while no information is given on the way carry propagation is avoided during division. Anyway, the given throughput does not indicate they use a carry free technique. A recent realization [6] uses a "block" carry save technique while the multiplication and division steps are interleaved. Therefore, the number of divisions steps is higher than N and from the results presented in the paper, the actual baudrate is 2 to 3 times less than the ideal one. On the other hand, the complexity of the basic ALU cell is approximately 1.5 full adder. British Telecom [8, 9] exploits the fact that the maximum carry length is statistically proportional to log(n). By testing each cycle if the carry propagation is completed, the addition time is statistically log(n) cycles. Hence this technique leads to a lower than ideal baudrate (fclock / 4.n.log(n)) while extra circuitry is needed for the test.

7. CHIP ARCHITECTURE

A powerful IC implementation of the RSA algorithm has been carried out, using the method described in the first part of this paper. It is based on a bitslice structure. Each slice contains the minimum hardware to perform such operations i.e. one RSD full-adder, one left/right shift register, 6 registers for keys and partial results storage, one shift register dedicated to I/O and one Manchester carry chain to complete RSD to classical binary form conversion (Fig. 12).

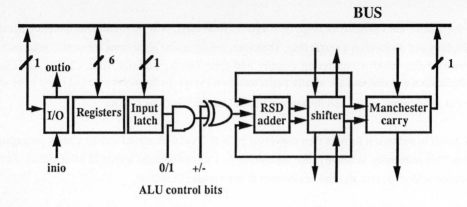

Fig. 12. Bitslice architecture

A combinatorial circuit is dedicated to the control of the ALU (addition or subtraction of 0 or 1) considering the multiplicand bits during the multiplication and the MSB's of the partial remainders during the division. Unlike in other implementations [6, 8], the interleaving between multiplication and division is avoided in order to save extra add cycles. Therefore the LSB's of the partial results must be stored in a RAM (Fig. 13).

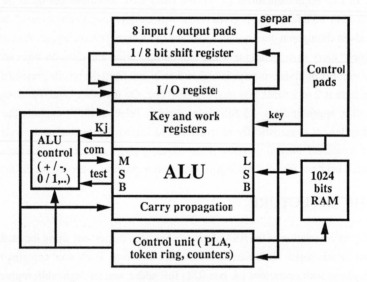

Kj : bit of exponent, multiplier
com : + / -, 1 / 0, shr, shl, ...
test : three MSB digits

Fig. 13. Chip architecture

The RSD technique leads to a very regular and powerful bitslice realization with a reduced hardware. The final baudrate is near the optimum (fclock / 4.n), except for the few percent lost by the RSD to classical binary form conversion (carry propagation). To our best knowledge, RSD arithmetic is used for the first time for RSA computation.

The two next sections are devoted to in depth study of specific design problems that have been encountered during the design of the RSA processors.

8. OPERATIVE PART DESIGN

The two main constraints in the design of this RSA chip were a reduced chip area and a low power consumption. These have been achieved by using static memory points and static devices wherever it was possible. As example, two optimizations will be described in some depth: the use of a static I/O shift register and the design of a bitslice structure with a RAM-like access to the registers.

The design of a good interface between a chip and the outside world is a delicate task. The difficulty lies in the fact that the outside world works in an asynchronous way (clocked by ψ_1 and ψ_2) while the chip works strictly synchronously (with ϕ_1 and ϕ_2). The problem has been solved in a quite elegant way by using a static shift register (Fig. 14).

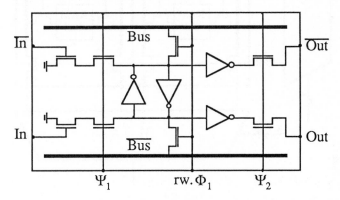

Fig. 14. Static I/O shift register

At the beginning of a RSA computation, a new data is read from the shift register and a result is written to it. During this phase, ψ_1 is low and the accesses are controlled by the command rw.ϕ_1.

Then, this command remains low, disconnecting the shift register from the rest of the chip. The cell acts then as a classical shift register controlled by the two non overlapping phases (ψ_1, ψ_2), asynchronous with the chip's own clock. In parallel with a RSA computation, a new data can be shifted in while the stored result is shifted out. This solution is very flexible while quite economical in terms of silicon area.

In the case of multiplication and especially in case of squaring, the same data register must read twice: the first time in parallel (to be written in the input register of the ALU) and serially for the second time, starting with the LSB (to compute the serial/parallel multiplication). A classical way to perform this function is to use a shift register as presented by fig. 14. This solution has two main drawbacks:

- the area of such a register is quite high; notice that, in the RSA algorithm, at least 3 of these registers are necessary;

- in a 1024 bits datapath, the power consumption of these registers becomes critical; indeed with a state change probability of 0.5, 512 cells per register sink current each cycle; this could lead to a dramatical power consumption.

To avoid these drawbacks, we have implemented a RAM-like hardware sketched in fig. 15.

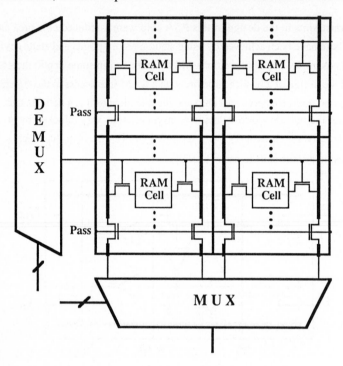

Fig. 15. RAM-like access to the bitslice registers

The registers cell is a 6 T static RAM cell connected to the datapath bus. During the parallel access, the **pass** command is low and each bitslice uses its own bus. Then **pass** is set to high and the entire bitslice acts like a RAM matrix, with wordlines (RW register command) controlled by a demultiplexor and busses as bitlines. The datapath is organized in 8 rows of 128 bits. Each time the serial reading procedure is activated, 128 bits are read from the bitslice and stored at the input of the multiplexor; data are then multiplexed in order to choose the desired bit of the multiplicator. Due to the number of switches on the bitlines, the reading process is rather slow. This is not a real drawback as 128 cycles are available. Compared with shift registers, the power consumption is reduced theoretically by a factor of n (the number of bits); the area inside the datapath is reduced by a factor of 2. At the counterpart, the control of this structure is more complicated than for shift registers but represents only a small silicon area overhead.

9. CONTROL UNIT DESIGN

The RSA program to be implemented had the following features:
- it is rather sequential;
- the number of flags tested at the same step is small (no more than 3);
- loops must be performed.

Aside from the main program, some subprocesses must be executed during multiplication and division loops. As mentioned earlier, reading from bitslice memory is executed 8 times during a 1024 bits multiplication; in the same way, reading/writing to the multiplication tail memory is executed 128 times; these processes are executed in more than one cycle. The control of these processes is devoted to special-purpose units best implemented through "token ring" solution.

The control unit is built around a PLA, a counter datapath and two "token ring" units. The choice of a PLA as heart of the control unit was not so obvious due to the inherent characteristics of the program. A non-classical "token ring" unit seemed to be well-suited. It was in fact rejected for two main reasons:
- it is difficult to adapt the pitch of both command generation plane and "token ring" part;
- this architecture doesn't lead by itself to a regular layout and automatic generation of such units is not easy.

Classical way to design CMOS synchronous PLA's uses multi-phase logic (due to the logical structure of PLA's, it may be shown that at least 3 phases are mandatory) [13]. To fit the PLA in our 2 non-overlapping phases clock scheme, we used the schematic presented at fig. 16; a clocked inverter was inserted between the two planes (AND/OR).

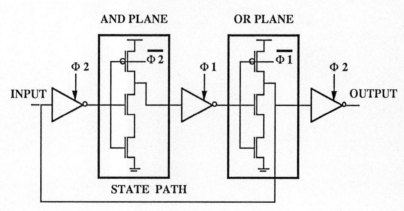

Fig. 16. PLA electrical schematic

These two are both precharged NOR planes which are the fastest logic gates. The area penalty due to this clocked inverter was rather low in our case because of the PLA size (12 flags, 6 state bits, 42 outputs, 82 minterms).

10. MAIN FEATURES

The chip's performances were first evaluated from simulations. All critical cells have been simulated for the worst case at 30 MHz (SPICE) and switch level simulation has proven the correct working of the chip's parts (bitslice, control unit, ...). The chip has been processed and tested (fig. 17-19).

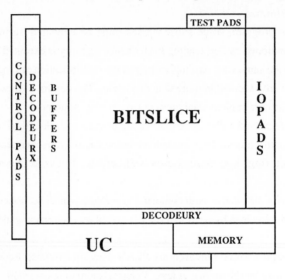

Fig. 17. Chip floorplan

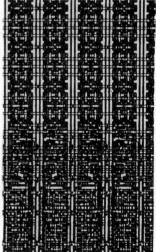

Fig. 18. Chip microphotograph   Fig. 19. Detail of the bitslice

The chip is 95 % functional and tests results are in good concordance with the expected baudrates. Table II lists the main technical characteristics of the chip.

Technology:	CMOS 2 μm with 2 metal layers
Transistor count:	180.000.
Chip size:	80 mm^2 (9.3 x 8.7).
Power dissipation at 25 MHz:	500 mW under 5 V.
Baudrate (all 1024 bits words):	8 kbits/s with a 25 MHz clock.
Key storage:	Static, 2 pairs on the chip.
I/O:	Serial or 8 bits parallel, asynchronous.
Control:	On chip.

Table II. Main features of the RSA chip

Coarse comparisons between different realizations is not always obvious. A typical example is the maximum size of numbers that can be processed by one or more IC's. Our implementation and some others [5, 8, 9] can process numbers up to the size of the ALU. In our case, we feel that 1024 bits is sufficient to meet code security requirements, but 256 is surely not! In the case of British Telecom [8, 9], the choice seems to have been dictated by a British law prohibiting the use of RSA with keys greater than 256 bits. On the other hand, the Cylink chip [7] is cascadable in order to process numbers up to 16 kbits. Finally, the CNET chip allows the processing of 1024 bits words with a 256 bits ALU through the use of a multi-precision arithmetic. Therefore, a refined discussion of respective advantages and drawbacks of all realizations will not be presented in this paper.

Hereafter is listed all available information published on RSA single chips (Table III). As cryptographic products appear to be as secret as information they process, full data usually are not available. The number of bits indicates the size of the ALU physically implemented on chip, thus the size of the datapath in bitslice realizations. Baudrates estimations are for 512 bit words, considering they are inversely proportional to the wordlength.

	Technology	#bits	# Trans.	Die size	Clock	Baudrate
MIT (1980)	NMOS 4μm	512	40.000	44 mm^2	4 MHz	1.2 kbits/s
CNET-CNS (1988)	CMOS 2 μm	256	100.000	88 mm^2	10 MHz	1.6 kbits/s
BR. TEL. (1988)	CMOS 2 μm	256				2.5 kbits/s
CYLINK (1987)		1028				5 kbits/s
UCL (1988)	CMOS 2 μm	1024	180.000	80 mm^2	25 MHz	16 kbits/s

Table III. Comparison between single chip RSA processors

11. CONCLUSIONS

A new carry-free division algorithm has been described. Its features like regularity (its hardware structure is identical to parallel multipliers), minimum hardware per bit (one full-adder), small test unit make it suitable for powerful integration in applications like DSP processors, floating point units,...

A 1024 bits RSA chip has also been presented. Thanks to the division algorithm characteristics, high throughput (8 kbits/s) and integration of a 1024 bits datapath in a standard 2 μm technology were achieved on a single chip. This chip includes also security features (no readable keys, no possible attempt to keys integrity,...) needed in cryptographic applications.

Speed improvement by using Booth's algorithm for multiplication and the interleaving between multiplication and division steps is currently studied. This could lead to a four times faster implementation for a 30 % silicon area increase.

REFERENCES

[1]   I. Verbauwhede, F. Hoornaert, J. Vandewalle and H. de Man, "Security and performance optimization of a new DES data encryption chip," in *Proc. ESSCIRC '87,* pp 31-34, 1987.

[2]   W. Diffie, "The first ten years of public key cryptography", *Proceedings of the IEEE,* Vol. 76, N° 5, pp 560-577, May 1988.

[3]   S. Kuninobu, T. Nishiyama, H. Edamatsu, T. Taniguchi and N. Takagi, "Design of high speed MOS multiplier and divider using redundant binary representation," in *Proc. of the 8th Symp. on Computer Arithmetic,* IEEE, pp 80-86, 1987.

[4]   H. Edamatsu, T. Taniguchi, T. Nishiyama and S. Kuninobu, "A 33 MFlops floating point processor using redundant binary representation", in *Proc. ISSCC 1988,* pp 152-153, 1988.

[5]   R. L. Rivest, "A description of a single chip implementation of the RSA cipher", *Lambda,* Fourth Quarter 1980, pp 14-18.

[6]   Ph. Gallay and E. Depret, "A cryptography processor", in *Proc. ISSCC 1988,* pp 148-149, 1988.

[7]   D. B. Newman, J. K. Omura and R. L. Pickholtz, "Public key management for network security", *IEEE Network Magazine,* Vol. 1, N° 2, pp 11-16, April 1987.

[8]   P. Ivey, A. Cox, J. H. Arbridge and J. Oldfield, "A single chip public key encryption sub-system", in *Proc.ESSCIRC 1988,* pp 241- 242, 1988.

[9]   P. Ivey, A. Cox, J. H. Arbridge and J. Oldfield, "A single chip public key encryption sub-system", IEEE J. Solid-State Circuits, Vol. SC-24, N° 4, pp. 1071-1075, June 1988.

[10] A. Avizienis, "Signed-digit number representations for fast parallel arithmetic", *IRE Trans. Electron. Computers,* Vol. EC-10, pp 389-400, Sept. 1961.

[11] K. Hwang, *Computer Arithmetic Principles, Architecture and Design,* John Wiley & Sons, 1979.

[12] B. Ginetti, A. Vandemeulebroecke and P. Jespers, "RSD cyclic analog-to-digital converter," in *Proc. 1988 Symposium on VLSI Circuits,* Tokyo, Japan, 1988, pp 125-126.

[13] N. Weste and K. Eshraghian, *Principles of CMOS VLSI Design,* Addison-Wesley, 1985.

CRYPTEL - THE PRACTICAL PROTECTION OF AN EXISTING ELECTRONIC MAIL SYSTEM

Hedwig Cnudde

CRYPTECH NV/SA
Project Management
B-1050 Brussels, Belgium

ABSTRACT

This paper describes the practical protection of an existing Electronic Mail System using the RSA-algorithm.

I. INTRODUCTION

BISTEL, which stands for "Belgian Information System by Telephone", is an office automation system set up by the Belgian Government. This system aims at making the access to information and its treatment more efficient and easier.

BISTEL, operational since 1984 combines wordprocessing systems, communication systems and systems for consulting data bases.

The present system consists of a central host computer and terminals spread over the various ministerial departments. Each terminal is connected to the central host computer through the public switched telephone network, the public X.25-network or the telex-network.

As a part of BISTEL, a system of Electronic Mail using ordinary terminals together with PC's as Computer Based Terminals allows the users to communicate with each other, in all possible ways used for day-to-day correspondence; this system can also be applied for sending documents such as the agendas and the documents for the Council of Ministers and for Ministerial Committees.

It was decided to secure the Electronic Mail in the BISTEL-system. This project has been called CRYPTEL (CRYPtography bisTEL). INTERSYS (today CIG-Intersys), the main contractor for the BISTEL-system, assigned this job to CRYPTECH, a company specialized in data security systems.

CRYPTECH is not only the Main Contractor of the CRYPTEL project but also developed most of the required equipment.

II. THE USE OF RSA

A preliminary study ([1]) by the ESAT laboratory of the 'Katholieke Universiteit Leuven' showed that a public key crypto-system (based on the well-know RSA-algorithm of Rivest, Shamir and Adleman) would be the best choice for the protection of the Electronic Mail in the BISTEL-system. This algorithm lends itself very well to guarantee the privacy of a message as well as to verify the integrity of the message and the authenticity of the sender.

The RSA-algorithm (used with a key length of 672 bit) implies the manipulation of blocks of about 80 characters.

III. SEVERAL ENCRYPTION/DECRYPTION RUNS

A distinction must be made between:
   . Communication between 2 correspondents.
   . Communication between the host computer and a correspondent.

The former communication is protected by means of end-to-end encryption, while the latter is protected by link encryption. Starting from this distinction one chooses a combination of encryptions:

. The first operation is a double encryption of a message which will only be decrypted by the correspondent. This allows for both privacy and authenticity protection of the original message.

. The second operation is a double encryption of the communication with the host computer. As such the privacy and the authenticity of this communication are being guaranteed.

Special practical problems that had to be solved, were:

. privacy and authenticity protection of messages by means of RSA;
. verifying the authenticity of people and equipment by means of RSA;
. secure storage of RSA keys using a Smart Card;
. authenticity protection of public keys;
. distribution of public keys;
. local file ecncryption.

IV. THE USE OF PC'S

When using RSA, all characters are manipulated and transmitted in blocks of a certain length. Thus, characters are buffered until a block is filled or until a transmit signal is given.

Use of cryptography has been restricted to Computer Based Terminals (say PC's). Eventually these may be equipped with a Word Processor to allow for the local editing of a message to be transmitted. Anyhow, without any doubt, a PC based solution offers many more advantages:

. PC's are the workstations of the future because of their flexibility;

. the use of PC's offers many more possibilities than just securing Electronic Mail: e.g. files can be stored locally in an encrypted form;

. these PC's can easily be reused in other secure systems.

V. AT THE TERMINAL'S SIDE: THE PC-RSA SECURITY CARD

A software implementation of the RSA-algorithm is considered to be much slower than a hardware version. Since the algorithm must be executed on-line and very quickly, a hardware implementation has been chosen. As a consequence, a special RSA-chip has been developed by CRYPTECH.

The PC-RSA Security Card, a microprocessor based board, contains an RSA chip-set together with control logic and firmware as well as all communication and key loading interfaces.

So, for the crypto-unit in the Computer Based Terminal (i.e. a PC), one opted for a built-in single-board system (occupying 1 slot) connected to an input device for the cryptographical keys: The PC-RSA Security Card.

VI. SECURITY PHILOSOPHY REGARDING THE NETWORK

For each type of transmission, the following rule of thumb can be used: "The encryption of a message (text or control) can be executed at that very place where the information should no longer be present in clear".

Analogously the data must be decrypted at the other side there were the controller text must be readable again. E.g., the headers of the packets of the X.25 protocol can not be encrypted unless one moves the encryption to the network which again means a lowering of the security.

The encryptions and decryptions should preferably be done as close as possible to the user.

VII. AT THE HOST'S SIDE: THE BULK-ENCRYPTOR

The same hardware as in the Computer Based Terminal, in a more complex form, can be used at the level of the host computer:

. This device makes sure that the information intended for the central computer is converted to plain text or that the outgoing information is being encrypted.

. The real contents of a document remains authenticated and encrypted for storage and is only being converted into plain by the end user.

VIII.THE KEY MANAGEMENT: CRYPTO-ADMINISTRATOR AND SMART CARDS

The key management (generation, distribution, storage, transmission, renewal, ...) is a task which should be kept separated from the rest of the network. Some of these operations are executed on a separate computer: a Crypto-Administrator (i.c. a PC).

Only the required keys are being transferred to the Crypto-Units in question. As the capacity of a simple magnetic card is insufficient, the cryptographical keys are stored, for the user, into a Smart Card. A further advantage of the use of Smart Cards is that they can be applied for Access Control.

So, each PC used as a Computer Based Terminal will be equipped with a Smart Card Reader. The CRYPTEL Bulk-Encryptor will be equipped with a Smart Card Reader for the Smart Card of the operator.

The CRYPTEL Crypto-Administrator will be equipped with 2 Smart Card Readers:

. one Reader will be used for the Smart Card of the operator,
. the other one for the Smart Card to be personalized.

REFERENCES

[1] J. Vandewalle, R. Govaerts, W. De Becker and M. Decroos, *Definitief rapport betreffende encryptering systeem voor de elektronische postdienst en de dokumenttransmissie van het BISTEL informatika systeem van de Belgische regering*, April 1985.

TECHNICAL SECURITY: THE STARTING POINT

Jan Van Auseloos

S.W.I.F.T., Avenue E. Solvay 81
B–1310 La Hulpe, Belgium

Abstract. Cryptographic security measures for encryption, authentication, non repudiation are important ... but not sufficient. My intention is to make the reader aware of non-technical security issues.

I. S.W.I.F.T.

I.1 What is S.W.I.F.T. ?

The Society for Worldwide Interbank Financial Telecommunication (S.W.I.F.T.) is a non-profit co-operative society of some 1600 member banks in more than 60 countries, dedicated to provide EFT (transmission, storage, arbitration) in a standardized manner. Between the 3000 connection points, we handle 1.000.000 messages every day: customer transfers, bank transfers, foreign exchange confirmations, credit/debit confirmations, ...

Banks are connected locally to a regional processor, which acts as concentrating point in a country. Every RGP is connected to one of the two Operation centers. An OPC validates, acknowledges, stores and controls the delivery of every input message. SWIFT II, the new generation of systems will provide for more flexibility, functionality and capacity. The EFT (SWIFT I today) will be one of the applications of SWIFT II.

I.2 What is security in S.W.I.F.T. ?

Security is a service (like any other service) which relates to the following objectives:

- Integrity of the network and messages: detect and prevent unauthorized manipulation of system and messages.
- Confidentiality of the network data and messages: restrict all unauthorized access to sensitive operational data and messages (while stored, processed and transmitted).
- Availability of the network and messages: no authorized user will be denied the normal service.
- Accountability: when integrity, confidentiality and availability are compromised the ability to measure the damage and to reconcile.

Since the world (science and reality) is changing all the time, the word security should be seen as an objective, rather than a state.

II. Practical Security in a service company

II.1 Practical Security (top down)

When a user accepts the services presented by a service company, like S.W.I.F.T., a contract appears between the service provider and the service user. This agreement is written down because moral, religion, common sense, loyalty, fair play, ... will change from individual to individual, from company to company.

After signing this contractual agreement, both parties know what they can expect and what they have to provide. It binds every user with S.W.I.F.T. and with all the other users, since a network service depends heavily on the cooperation of all users. As a consequence, R&L (Responsibility & Liability) boundaries are defined and an arbitrator is assigned. Three types of parties are now defined by the R&L boundaries: the service users, the service provider and the arbitrator. Each of them will try to work according to the R&L rules set forward in the contract and will have to spend money on preventive measures or insurance coverage to guarantee its obligations. E.g. to guarantee the above security definitions, S.W.I.F.T. has to implement preventive and corrective controls, they can be technical, procedural, organizational of contractual. Examples of controls are: encryption, authentication, access control, segregation of duties, dual authorization, audit trails, insurance coverage, separate security administration department, ...

When new services are added or when a service changes, the contract has to be changed, the R&L boundaries shift and the costs to guarantee security are re-distributed.

In S.W.I.F.T. this contract is called the User Handbook. In the policy volume you can find:

- description of the services provided,
- where each one's responsibility starts and stops,
- what each one's reaction should be in case of problems,
- what security measures are to be performed by each party,
- ...

The Chief Inspector acts as an arbitrator and handles all claims between users and between users and S.W.I.F.T.

II.2 Practical Security (bottom up)

Risks and threats are a fact of life. In every operational environment, they are predictable or unpredictable. When we have defined the risks and threats, applicable to our operational environment, we can start organizing it to be able to prevent or recover from them. In the bottom up approach we have to select a security related threat, analyse it and find solutions. Let's take the example of message tampering/viewing on the local S.W.I.F.T. connection: every EFT network carries information generated by the senders and should deliver this information to the receiver. The users give their messages to a 'non owned' environment (network of another bank, PTT lines or radiowave or even satellite). This EFT information is ideal for passive/active attacks:

Passive attack: traffic analysis. This involves recording traffic without interfering. From this data, specific figures or names are noted or statistical analysis is done. This is than interpreted. (e.g. traffic analysis, text analysis)

Active attack: traffic manipulation. Now somebody or something intervenes actively in the flow of messages with the intention to:

a. disrupt normal operations or to harm sender, receiver or network.

b. benefit personally (money, knowledge, prestige, ...).

The techniques used are redirect, reorder, delay, insert, remove, change, obstruct, replace parts of traffic. Countermeasures are sequence checking, session reports, standards, authentication, access control, encryption, ...

The quality of the countermeasures against these known risks/threats is sometimes called 'level of security'. Secure means high quality measures and complete coverage of all known risks/threats, today; insecure means: no implemented measures or very poor quality of them.

Even when we imagine that our environment is static and that we documented all threats/risks, technical security methods cannot make complete secure systems for two reasons:

- countermeasures or combinations of them are never 100% effective: they sometimes introduce new risks/threats or endanger other measures (e.g. standardization to avoid disputes can be exploited by a cryptanalyst, reliance on key management, ...).
- for some known risks or threats countermeasures still have to be invented (e.g. virus, terminal identification, ...). Therefore, to achieve a certain level of security, a combination of 3 basic types has to be implemented: prevention, detection with correction and contracts with claims and arbitration.

Prevent: this type of measures is always resident in the system, they are automatically invoked, because they are part of the normal processing and they stop incorrect actions at the moment they occur.

Detect and correct: when prevention is not practical or can/did not work, detective measures detect an occurred risk/threat and try to reconstruct a correct situation (backups, insurance, ...).

Contract, claim and arbitration: when preventive and detective measures are not feasible or to costly, every party will try to shift responsibility to the other party. R&L boundaries will be created. Sometimes one of the parties claims being damaged by another party and asks arbitration.

III. Security administration, audit and development in a service company

Both approaches (bottom up and top down) lead to the same conclusion. They both lead to formulating R&L boundaries and therefore an independent arbitrator. This arbitrator can neither be part of the service company, nor can he be a user. When investigating claims he/she may not be biased when interpreting the contract.

Because of his independent and privileged position, he can do more:

III.1 Security administration

This is best explained with an example. Since he is not a user and since he is not part of the service company structure, he is in an ideal position to generate access codes which only the user will know and a security module in the network. This job cannot be given to a user, knowing the other one's passwords he could be tempted to make fraudulent use of it. Also this job cannot be given to the service company, since then we can expect internal fraud. He is the ideal trusted party in every key-scheme.

III.2 Security development

New services are created within the existing framework, new security inventions are made or new threats/risks pop up. This all requires development work, which should be monitored or initiated by this department.

I would like to make following comment here. Again the discussion has started to use either standardized or non standardized (own) security algorithms. On a standard algorithm, multiple, independent audits have been done, but people will keep on trying to break it. If suddenly broken, this will get worldwide press coverage, forcing the user to change overnight. Even worse, sometimes people claim to have broken it, the press blows this up and the service provider is embarrassed. A private, not widely published algorithm is not a prime candidate for breaking. The cryptanalyst does not have all the information about the algorithm itself, the design principles and the mathematical assumptions and cannot feed himself with research reports from other parties. The chance that a private algorithm is suddenly broken is therefore less. However developing a proprietary algorithm costs money. Why not take the way in between ?

III.3 Security audits

Reviewing security and reporting on the efficiency of it is a typical activity of a security audit department. In this department this role can be larger in scope: the arbitrator can also report to the user community on the security of the service provider or on the security of some service users.

IV. Conclusion

Let me now make a closing remark and perhaps a challenge to all cryptographers.

Let's focus on authentication for example. To get to an acceptable level of security it is necessary that the user community exchanges authenticator keys without communicating them to the service provider. It has always been suggested that a key change must only be known by the sender and receiver and, as a consequence, not auditable by the arbitrator. Key changes are therefore not auditable and completely up to the discretion of all kinds of users (security conscious or not). Would a Boolean function $f[\text{aut}(\text{key1}, \text{txt1}), \text{aut}(\text{key2}, \text{txt2})]$ $= \text{def} = (\text{key1} = \text{key2})$ not increase security indirectly ?

SECURITY IN OPEN DISTRIBUTED PROCESSING

Charles Siuda

Ascom Systems Ltd.
CH-3000 Bern, Switzerland

ABSTRACT

This paper describes two main aspects of security in open distributed processing: Embedment of security capabilities in reference to the OSI Reference Model and mapping of application layer components onto implementations in end-systems. The paper is based on a strategic commitment to the adoption of existing and evolving international standards. Therefore, the work is based on the security architecture (OSI 7498-2) and on the security concept described in ECMA TR/46 - Security in Open Systems - a Security Framework. This framework proposes a set of security facilities which are the building blocks for security services for use in the OSI application layer. One of the main targets of this paper is to show how security services can be embedded in the OSI communication environment. Two new types of common Application Service Element (ASE) are proposed to provide the security services in the application layer: the security information ASE and the security control ASE. The X.400 message handling system has been chosen as an example for securing a productive application.

I.   INTRODUCTION

The convergence in media communications and information technologies requires coordination and cooperation between separate organizations as well as within a single organization. Electronic trading, for example, leads to distributed systems spanning traders, customers and banks. This level of integration is much harder, since no single authority can control the entire activity of the system. Integration through open distributed processing will lead to heterogeneous systems containing a wide variety of computer and networking technologies, supplied by a multiplicity of vendors.

Providing security for computers and communications is quite complex. Security must be incorporated not only into the Open Systems Interconnection (OSI) framework to ensure interoperability and compatibility of equipment and services but also into the Open Distributed Processing (ODP) to ensure secure information processing activities in which discrete components of the overall processing activities may be located in more than one system at more than one location.

Many different security needs can be met by a common set of security services to be provided outside application processes. These services will affect the interactions between users and productive applications, and between productive applications and supportive applications. They will also affect the installation, maintenance and management of applications and of the underlying system. These services, security information, their interactions and their management constitute the scope of security in open distributed processing.

II EMBEDMENT OF SECURITY CAPABILITIES IN REFERENCE TO THE OSI REFERENCE MODEL

2.1 Overview

The basic assumption of this work is that all security functions are located above OSI layer 6, i.e. within the application layer and/or outside the OSI environment. Three approaches have been identified to define the security services in the application layer. The first approach is to define the security services within the scope of specific application service elements. The second approach is to define the security services within other common application service elements, for example within ACSE and ROSE. The third approach defines common application service elements for security services that are accessed by specific application service elements like FTAM, X.400 etc. The third approach seems to have more merits. Therefore, it has been selected as the basic concept for embedment of security services in the application layer.

The message handling applications have been chosen as an example for securing productive applications.

The following supportive applications constitute the high level support environment for the productive applications and for the users of these applications:
- security information provision,
- security control,
- systems management,
- directory.

Only the productive applications are visible to the human user, and are especially used by him.

The following application service elements are required in every set of ASEs:
- Association Control Service Element (ACSE)
- Remote Operations Service Element (ROSE)

Figure 1 shows the application layer components.

2.2   Application Layer Components for Security Systems

The application layer components proposed for security systems are located within the Application Entity (AE). An AE is represented by a set of communication capabilities, called Application Service Elements (ASE). An ASE is typically defined by its own service definition and protocol specification. The Association Control Service Element (ACSE) supports the establishment and release of an application association between a pair of AEs. The Remote Operations Service Element (ROSE) supports the request/reply paradigm of the abstract operations that occur at the ports in the abstract model. The AE contains also two additional functions:

- The Single Association Control Function (SACF) models the coordination of the interactions among the ASEs contained in the Single
  Association Object (SAO) and also models the coordination of their
  use of the presentation layer. The rules concerning these interactions are defined by the application - context of the application
  - association.

- The cryptographic support function provides cryptographic services
  used primarily by security-related ASEs and by other productive ASEs
  which need cryptopgraphic functions for their operations. The
  cryptographic support function provides users with the following
  cryptographic functions:
  - data confidentiality functions,
  - data integrity functions,

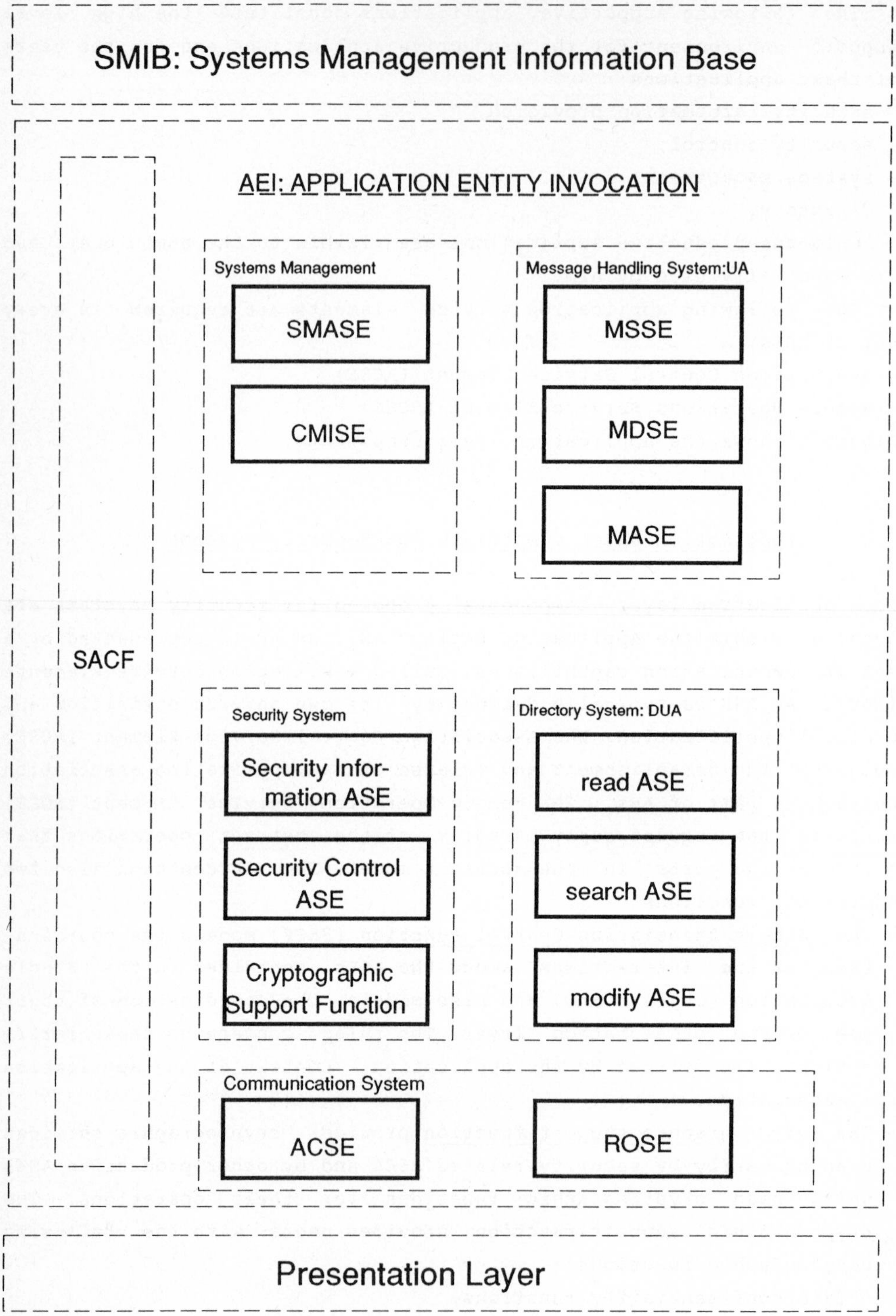

Figure 1- Application Layer Components

- data origin authentication,
- non-repudiation of origin,
- non-repudiation of receipt,
- authentication functions,
- key management functions.

The cryptographic support functions may also be used by systems management processes and directory access processes to support their cryptographic requirements.

2.2.1   Security-related ASEs

Security-related ASEs are a new type of more general ASEs in the application layer to provide the security services based on the ECMA security facilities described in ECMA TR46 and the application layer structure (ISO DP 9545).

There are two classes of security services that are directly and actively involved with implementing security policy:
- the security information providing class of service and
- the security control class of service.

The security information providing class of service provides security information, the security control class of service utilises security information. In accordance with these two classes of security service two security-related ASEs are proposed:
- the security information ASE, and
- the security control ASE
These ASEs may also be used by other productive ASEs.

a) Security Information ASE

The security information ASE provides two security services:
- the peer-entity authentication service and
- the security attribute service.

The peer-entity authentication service provides for the authentication of subjects e.g. human user's as well as active applications. Authentication requires the use of credentials to verify authentication information supplied by a subject. Depending on the trust relationships involved, different authentication mechanisms may be used. In general, authentication relies on an entity proving its identity by showing that it is in possession of some piece of

secret information. Authentication is an end-to-end operation between the two entities concerned. The result of authentication is a
mapping between a communicating entity (the subject) and a verified
identity.

The security attribute service provides attributes for entities
on presentation of the entity's authenticated identity. This service will also map attributes where required.

b) Security Control ASE

Security control has to do with information collection (e.g. reading a human user's credential), with information checking (as in
access authorization) and with information transformation (as in
exchanges between domains). Such information processing provides
the basic operation of security control.

The security control ASE provides the following security control
services:

- Subject sponsor service to support the interaction with human
  users.
- Secure association service to control the security aspects of
  associations between application processes.
- Authorization service to provide access control decisions.
- Interdomain service to provide the functionality for mapping
  policy elements between different domains.
- Cryptographic support service. This service is executed by the
  cryptographic support function.

2.2.2 System Management ASEs

2.2.2.1 Integration of Security Management
into OSI Systems Management

OSI systems management is a set of activities that manage a communications environment conforming to open systems interconnection standards. Security management is used to monitor and control the operation of OSI security services and mechanisms in accordance with its
security policy. Security management is the set of procedures used to

manage the objects and attributes within the Systems Management Information Base (SMIB). These objects are similar to other objects which
are managed by OSI management. What distinguishes security management
from the other Specific Management Functional Areas (SMFAs) is the
class of objects managed and not the operations used to manage them.
For this reason, security management does not define its own directives for the use of Common Management Information Services (CMIS), but
uses those already defined by other SMFAs. Thus, for example the directives used to configurate security objects and security attributes
are exactly those which are defined by the configuration management
SMFA. Therefore, security management is an integral part of OSI
systems management.

2.2.2.2 Security Management Facilities

Security management provides the set of management facilities needed
to operate on security objects so as to allow the implementation of
the security policy. These facilities may be provided through the use
of directives in other SMFAs which in turn use operations of CMIS. The
facilities provided or used by security management are: security-related object management and security-related event and audit trail management.

a) The Security-related object management facility manages security-
realted objects, attributes, states and relationships on a collection of open systems. It includes the facilities defined in configuration management (ISO-9595): object configuration, attribute
management, state management and relationship management.

b) The security event and audit trail management facility is used to
create and communicate security-related events, event logs and
audit trails on a collection of open systems. The facilities required for this function are similar to those defined in fault management (ISO 9595): spontaneous event reporting, cumulative event
gathering, event threshold alarm and management service control.

2.2.2.3 Communication Aspects of Security Management

Management processes perform the management activities in a distributed manner. The interactions which take place between management processes are abstracted in terms of directives issued by one processor to the other. The interactions between managing and agent processes are realized through the exchange of management information. This communication is accomplished using OSI protocols.

The Systems Management Application Service Element (SMASE) defines the semantics and abstract syntaxes of the information transferred as relevant to OSI management in Management Application Protocol Data Units (MAPDUs). The MAPDU is the OSI protocol realization of the abstract notation of directives exchanged between management application processes.

The communication service used by the SMASE may be provided by the Common Management Information Service Element (CMISE) or other ASEs such as Transaction Processing (TP). The use of CMISE also implies the presence of the Remote Operaton Service Element (ROSE) and other ASEs. CMISE specifies the service and procedures for transfer of Common Management Information Protocol Data Units (CMIPDUs). It provides a means for the exchange of information and commands for management purpose in a common manner.

2.2.3 Message Handling ASEs

Access to the Message Transfer System (MTS) abstract service is supported by three application service elements, each supporting a type of port between an MTS-user and the MTS in the abstract model.

The Message Submission Service Element (MSSE) supports the services of the submission-port; and the Message Delivery Service Element (MDSE) supports the services of the delivery-port; and the Message Administration Service Element (MASE) supports the services of the administration-port. Similarly, access to Message Store (MS) abstract service is supported by three application-service-elements: the MSSE supports the indirect-submission-port; the Message Retrieval Service Element (MRSE) supports the services of the retrieval-port; and the MASE supports the services of the administration-port.

These application service elements are in turn supported by other application service elements: The MSSE, MDSE, MRSE and MASE provide the mapping functions of the abstract-syntax notation of an abstract service onto the services provided by the ROSE.

2.2.4   Directory ASEs

The directory system stores information about objects in the real world which is provided, on request, to directory users to facilitate communication between objects. Users of the directory, including human user's and application processes, are represented by Directory User Agents (DUAs); these are application processes that enable users to read or modify information in the Directory Information Base (DIB). This information base consists of object entries which are distributed between several DSAs (Directory System Agents).

The DUA accesses the directory on behalf of the human user or of the application process. All directory services are requested and received by the DUA through access points, each of which corresponds to three different types of port. These ports define several directory operations which are used to read, modify, or search for entries in the directory information base.

The DUA connects to the directory via bind-operations (Remote Operations Service Element, ROSE) and requests directory services from the Directory System Agent (DSA) through the directory application service elements.

When a DUA is in a different open system from a DSA with which it is interacting, these interactions are supported by the Directory Access Protocol (DAP), which is an OSI application layer protocol. Similarly, when a pair of DSAs which are interacting are in different open systems, the interactions are supported by the Directory System Protocol (DSP).

Both the DAP and the DSP are protocols to provide communication between a pair of application processes.

The directory-specific ASEs (read ASE, search ASE, modify ASE) provide the mapping function of the abstract-syntax notation of the directory abstract-service onto the services provided by the ROSE.

The directory bind-operations are mapped onto the ACSE and the directory application service elements are mapped onto remote operation service elements, ROSE, as specified in the directory access protocols.

2.2.5 Mapping Security Facilities onto Security-related ASEs

Security in a distributed system requires the implementation of many different functions. These functions, which are modeled as security facilities, are described in ECMA TR/46 Security in Open Systems – a Security Framework. A short description of the security facilities is given in the table shown in Figure 2. These security facilities are used as building blocks of the proposed security-related ASEs. Figure 3 shows the mapping of the security facilities onto security-related ASEs.

Security Facility	Short Description
Subject Sponsor Facility	Intermediary between the security subject and the other security facilities.
Authentication Facility	Accepts and checks subject authentication information (subject credentials).
Association Management Facility	- Authorization for the two entities to communicate. - Assurance of the identity of both entities.
Security State Facility	Maintaines a view of the current security state of the system.
Security Attribute Management Facility	This facility manipulates the security attributes of subjects and objects.
Authorization Facility	Authorizes or denies requested accesses by subjects to objects.
Inter-domain Facility	Maps one domain's interpretation of security attributes into another domain's interpretation
Security Audit Facility	Receives and analyses event information from other security facilities.
Security Recovering Facility	Acts on information received from the audit facility.
Cryptographic Support Facility	Provides cryptographic services used by other security facilities and by productive services and applications to secure data in storage and transit.

Figure 2: Short description of ECMA security facilities

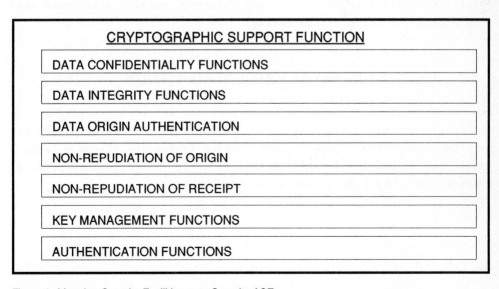

Figure 3- Mapping Security Facilities onto Security ASEs

III.   MAPPING APPLICATION LAYER COMPONENTS
       ONTO IMPLEMENTATIONS IN END-SYSTEMS

3.1   Use of the Directory

The directory system stores information about objects in the real
world (e.g. people, applications) which is provided, on request, to
directory users to facilitate communication between objects. In general, the directory itself has no responsibility for the information
placed in the directory. The directory is simply a repository for information. The directory assures the consistency of that information,
making the appropriate updates in the distributed directory information base and the availability of that information.

3.1.1   The Directory Used as an Attribute Store

Each entry of the directory information base contains at least the
following attributes:
- Common name which identifies the corresponding person, application
  process, application entity, etc.
- Public-key certificate which is used to authenticate two communicating parties.
- Security privilege attributes and security control attributes used
  in access control decision making.

3.1.2   The directory used as a Key Management System

Key management is achieved by using security features provided in the
directory. The key management system is based on the use of public key
cryptosystems. The X.509 authentication framework allows a user's
public key to be stored in its directory entry. The directory provides
a way to reliably link a name with a public key.

3.2   Client-Server Model of a Distributed Application

A distributed application consists of two parts. The part collocated
with the user is referred to as the client entity, the remote part of
the application as the server system. A client entity and a server
system communicate over the network by means of an access protocol. In
order to comply with the OSI Reference Model, the client and the server are considered application processes and are extended with Application Entities (AE). The AEs are parts of the application layer and
contains sets of ASEs. The ASEs provide the communication functions,
in accordance with the service definition, to the client and the server, and implement the access protocol. In doing so, an ASE may use
services provided by other ASEs in the same AE, and by the presentation layer of the OSI Reference Model.

3.3   Mapping Application Layer Components
onto Implementations in End-systems

Security service applications are real, possibly distributed applications, that can be named and registered for use in open systems. These
applications consist of one or more elements that are distributed over
a number of physical systems. These elements are referred to as logical servers. A single physical end-system may support one or more logical servers of different types. The application layer components to
be mapped on implementations in end-systems are shown in Figure 1.

3.3.1   Mapping the Security Information ASE onto Supportive Servers

The security information ASE may be mapped directly onto supportive
servers. Their function is to accept information, process it and return results; which is the classic action of a supportive server,
which is called "security server". The data base of this server is
embedded in the Directory System Agent (DSA).

3.3.2   Mapping the Security Control ASE onto End-system Components

The four kinds of services embedded in the security control ASE are:
authorization services, interdomain services, secure association services and finally subject sponsor services. Secure association is a
function that needs to be performed in each of the two end-systems involved in an association. A subject sponsor service maps to a local
function in each end-system which is directly accessible to human
users or external applications that need a subject sponsor as entry
point into the secure distributed system. Authorization services and
interdomain services may be mapped onto a local function in each end-
system.

3.3.3   Mapping Directory ASEs onto
End-system Components and Logical Servers

The directory user agent maps to a local function in each end-system
which is directly accessible to human users. The directory system
agent maps onto logical servers.

3.3.4   Mapping Message Handling System ASEs
onto End-system Components and Servers

The User Agent (UA) maps to a local function in each end-system which
is directly accessible to human users. The Message Transfer Agent
(MTA) maps onto logical servers.

3.3.5   Mapping Communication ASEs onto End-system Components

The following communication ASEs are required in each set of ASEs:
- The Association Control Service Element (ACSE) supports the estab-
  lishment and release of an application association between a pair of
  AEs.
- The ROSE supports the request/reply paradigm of the abstract opera-
  tions.
Both ASEs are mapped to local functions in all user's end-systems and
server end-systems.

3.4   User's End-system Components

The user's end-system components which are shown in Figure 4 can be grouped in five groups:
- security system,
- management system,
- message handling system,
- directory system,
- communication system.

The security system contains the security facilities which are the building blocks of the security system. The cryptographic support facility is implemented using hardware (the so-called crytographic processor). The subject sponsor facility is the intermediate between the human user and the rest of the security facilities.

The management system, which is located in the user's end-system contains the client part of the client/server management system. Each user's end-system can in principle initiate management operations. However these operations will only be available to application proces-ses with the appropriate privilege attributes. The management system contains the following functional blocks:
- system security management,
- security service management,
- security mechanism management,
- security of OSI management.

The message handling system contains the User Agent (UA) and the Message Store (MS).

The directory system contains the Directory User Agent (DUA).

The communication system contains ACSE and ROSE.

3.5   Components of Servers

All types of server contain the communication system and the security system. The security server, the interdomain and audit server, the security state and recovery server contain in addition to the above described components the directory system agent, which is used as a data base for the servers. The message handling server contains the Message Transfer Agent (MTA). Figure 5 shows the components of a se-cure message handling system.

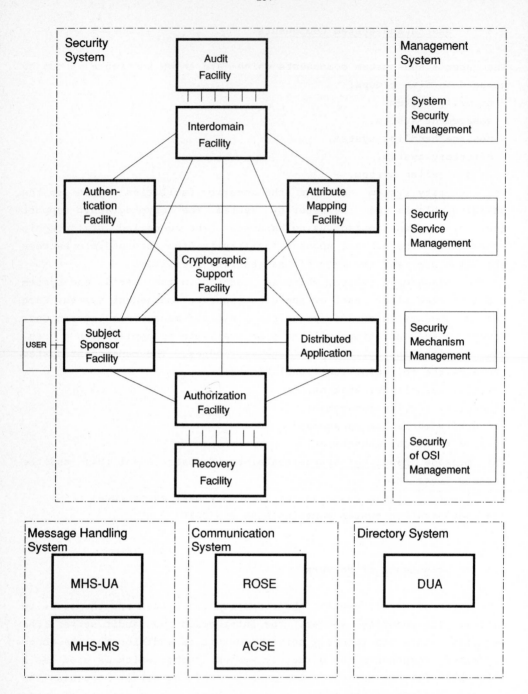

Figure 4- Components of user's end-system

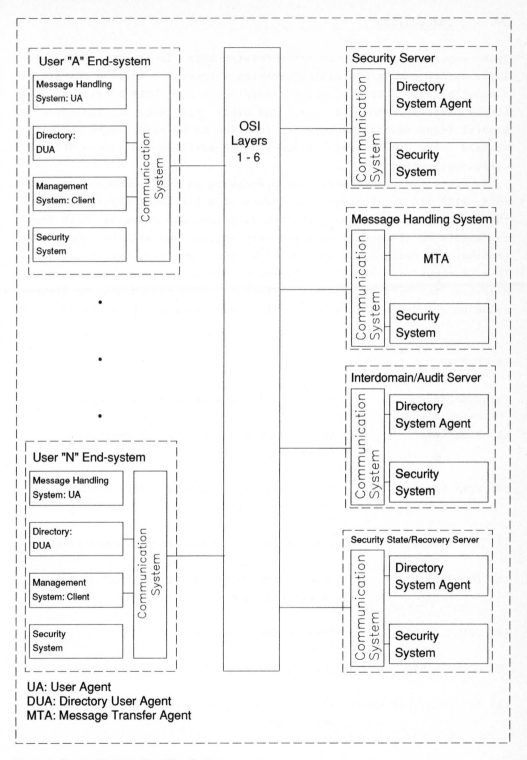

Figure 5- Secure Message Handling System

IV.   CONCLUSIONS

This paper proposes two security-related ASEs in the OSI application layer as new common application-service elements to realize OSI secure communications. These proposed security-related ASEs provide a very simple interface to the user. The user only needs to indicate the general requirements for security function to be used. These security-related ASEs can be used for securing of existing and future OSI-specific applications.

Security in open distributed processing is very complex. The diamond with its great number of facets of brilliance should represent this complexity of security in open distributed processing. Each facet represents an element of this security system, and all the facets work together to reflect the structure of the underlying overall concept. This concept is a vision of the whole security system, implemented in hardware and software in an architecture which should provide maximum value to the users and is adaptable to meet user's future demands.

REFERENCES

(1) ECMA/TR42   Framework for Distributed Office Applications

(2) ECMA/TR46   Security in Open Systems - A Security
                         Framework

(3) ECMA Standard XXX   Security in Open Systems - Data Elements
                         and Service Definitions

(4) ISO 7498/2   Basic Reference Model, Security Architecture

(5) ISO 9545.1   Open System Interconnection, Application
                         Layer Structure

(6) CCITT X.400   Message Handling System

(7) CCITT X.500   The Directory

(8) ISO/JTC1/SC21 N2688 Part 7: Security Management Service
                         Definition

A EUROPEAN CALL FOR CRYPTOGRAPHIC ALGORITHMS: RIPE; RACE INTEGRITY PRIMITIVES EVALUATION

J. Vandewalle[1]   D. Chaum[2]   W. Fumy[3]
C. Jansen[4]   P. Landrock[5]   G. Roelofsen[6]

[1]Katholieke Universiteit Leuven, E.S.A.T.
Kard. Mercierlaan 94, B-3030 Heverlee, Belgium

[2] C.W.I., Kruislaan 413
NL-1098 SJ Amsterdam, The Netherlands

[3] Siemens AG, Gunther-Sharowsky-Str. 2
P.O. Box 3240, D-8520 Erlangen BRD

[4] Philips Usfa B.V., Meerenakkerweg 1
NL-5600 MD Eindhoven, The Netherlands

[5] Dept. of Mathematics, Aarhus University
DK-8000 Aarhus, Denmark

[6] PTT Dr. Neher Laboratories, P.O. Box 421
NL-2260 AK Leidschendam, The Netherlands

ABSTRACT

The first aim of this paper is to situate the call for integrity and authentication algorithms within research on cryptography and within evolution of telecommunication. Motivations for submitting primitives and details on the submission process are also given.

I . BACKGROUND AND DIFFICULTIES IN STANDARDIZING CRYPTOGRAPHIC TECHNIQUES

Last year an interesting collection [1-6] of status reports on cryptography appeared in the proceedings of IEEE. These papers contain more details on a number of important issues like Kerckhoff's assumption (the security of a cipher must entirely reside in the secret key), secret and open research in cryptology, the status of cryptanalysis, standardization efforts, controversy and public acceptance.

In this context it is important to mention that widespread use of cryptography in an open network requires interoperability, and hence some standardization. Such standardization of public algorithms can then be combined with a public scientific evaluation and may lead to wide acceptance, which again stimulates the market. On the other hand, the public nature of the algorithms and their widespread use increases the visibility and the target and hence will attract more attacks. In the international scientific community as present at Crypto and Eurocrypt it is generally agreed that open research on cryptography should produce secure and practical algorithms that can withstand even massive attacks. The DES is such an algorithm that has been analysed extensively and is widely used and standardized [2]. There is a general consensus today that DES is a rather good algorithm with an unfortunately small key [1]. If such algorithms are used internationally for data confidentiality, however, there may be a conflict with national interests [1,7].

II . A BREAKTHROUGH IN EUROPEAN TELECOMMUNICATION

By 1992 the European Community plans to set up a unified European market of about 300 million customers. In view of this market integrated broadband communication (IBC) is planned for commercial use in 1995. This IBC will provide high speed channels (64 kbps, 2 Mbps and more) of image, voice, sound and data communications and will support a broad spectrum of services like telex, telefax, telephony, teletex, videotex, electronic mail, telenewspaper, teleconferencing, videoconferencing, cable TV, telebanking, teleshopping, home banking, EFT, POS, mobile telephony, paging, alarm service, directory services, etc. These services can be home based, office based, (private or public) manufacturing or mobile. They may include dialogue service or messaging or retrieval or a distribution service.

It is clear that the majority of these services offered in future networks are crucially dependent on cryptography for security. Figure 1 indicates the relationship between on the one hand the cryptographic algorithms and their modes of use, and on the other hand the security mechanisms, security services and applications as they are desired in IBC. Data confidentiality is not always required, but integrity is needed for authentication, non repudiation, access control etc.

In order to pave the way towards commercial use of Integrated Broadband Communications (IBC) in Europe by 1995, the commission of European communities has launched the RACE program (Research and development in Advanced Communications technologies in Europe) [8,9]. Under this RACE program pre-competitive and pre-normative work is going on. After a RACE definition phase (1 January 1986 to 31 December 1986) several RACE projects have started at the end of 1987. Within RACE, the RIPE project (RACE Integrity Primitives Evaluation) will put forward an ensemble of techniques to meet the anticipated integrity requirements of IBC. The members of the RIPE project are: Centre for Mathematics and Computer Science, Amsterdam (prime contractor); Siemens AG; Philips Usfa BV; PTT Research, The Netherlands; Katholieke Universiteit Leuven; Aarhus University.

The project's motivation is the unique opportunity to attain consensus on openly available integrity primitives for the future IBC communication network.

III . THE RIPE CALL FOR INTEGRITY PRIMITIVES

In RIPE, it is advocated that the best way to achieve wide acceptance for a collection of algorithms for integrity and authentication is by an open call for such algorithms, similar to the call in the U.S., which has produced DES. These submissions will then be evaluated by RIPE. The project has put significant effort in creating the optimal conditions for standardization of integrity prim-itives. The scope of the project and the evaluation procedure were fixed after having reached consensus with the main parties involved. Also there is a coop-eration between the RIPE project and the two other RACE projects on integrity (working on the functional specification of integrity and on techniques for in-tegrity mechanisms, respectively). Therefore it is the project's firm belief that this work will lead to European standardization.

Submissions can be any digital integrity primitive, from conventional hash functions, one-way functions and message authentication algorithms, through digital signature techniques, all the way to protocols for providing security ser-vices. The scope excludes data confidentiality. Direct benefits for algorithm proposers are expected to include: lead time to develop implementations, reten-tion of intellectual property protection and possible European standardization. Moreover the submitted integrity primitives will be treated confidentially during the evaluation. In exchange, those submissions finally selected must be made public and available for use in IBC on a non-discriminatory, but not necessarily royalty-free basis within the EEC.

In view of the potential use in IBC the submissions will be evaluated with respect to three aspects: functionality, modes of use, and performance. The evaluation will comprise computer simulation, statistical verification, and anal-ysis of mathematical structure, particularly to verify their integrity properties.

All requests for further information and for the mandatory submission kits should be addressed to: Gert Roelofsen, PTT Research; P.O. Box 421; 2260 AK Leidschendam; The Netherlands; Telephone +31(70)332 64 10; Telex 311236 prnl nl; Fax +31(70)332 64 77; email g_roelofsen@pttrnl.nl. Apart from detailing the required form of submission, administrative information and formal procedures, the kit states some general conditions for submitting an integrity primitive and describes the RIPE project's commitments with respect to the confidentiality of submissions. The deadline for submissions is September 15, 1989. The evaluation results will be available by the end of 1990.

In conclusion, it is important to mention that the widest possible encouragement to submit should be given to individuals as well as companies, both within and outside the European Community. At the macro-economic level, this call provides a unique occasion for the international scientific community to see its work used widely, in accordance with an open scientific approach. At the micro-economic level it provides some direct benefits to submitters.

REFERENCES

[1] Massey J., "An introduction to contemporary cryptology", *Proc. IEEE*, Vol. 76, no. 5, May 1988, pp. 533-549.

[2] Smid M.E. and Branstad D.K., "The data encryption standard: past and future", *Proc IEEE*, Vol. 76, no. 5, May 1988, pp. 550-559.

[3] Diffie W., "The first ten years of public-key cryptography", *Proc. IEEE*, Vol. 76, no. 5, May 1988, pp. 560-577.

[4] Brickell E. and Odlyzko A., "Cryptanalysis: a survey of recent results", *Proc. IEEE*, Vol. 76, no. 5, May 1988, pp. 578-593.

[5] Simmons G., "A survey of information authentication", *Proc. IEEE*, Vol. 76, no. 5, May 1988, pp. 603-620.

[6] Abrams M.D. and Powell H.D., *"Tutorial computer and network security"*, IEEE Computer Society Press, Los Angeles 1987.

[7] OTR100, *"Draft RACE Workplan"*, Commission of the European Communities, 1987, Rue de la Loi 200, B-1049, Brussels, Belgium.

[8] OTR200, *"RACE Workplan"*, Commission of the European Communities, 1988, Rue de la Loi 200, B-1049, Brussels, Belgium.

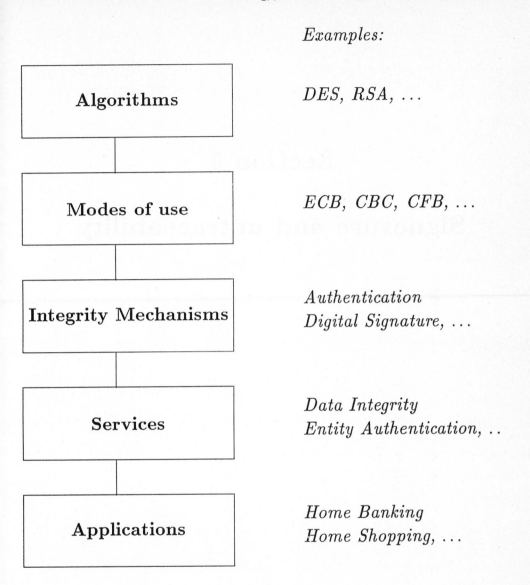

Figure 1. Security architecture and relationship with modes of use and cryptographic algorithms

Section 5

Signature and untraceability

LEGAL REQUIREMENTS FACING NEW SIGNATURE TECHNOLOGY

Mireille Antoine Jean-François Brakeland
Marc Eloy Yves Poullet

Centre de Recherches Informatique et Droit (C.R.I.D.)
Namur, Belgium

Abstract

The recognition of electronic signature and electronic message creates many legal problems in its application in the different law systems (ex: Automatic Teller Machines, Odette, ...).

The electronic signature could be declared inadmissible before the Courts; even if declared admissible, the parties will still have to demonstrate its evidence value before a judge, not aware of the technique and who therefore will be suspicious. In fact, the electronic signature, even if so far ignored by the lawmaker, offers a guarantee of quality and certitude unrivalled by the manuscript signature.

Nevertheless, if legal texts are unsuited to reality, a simple concealment of these texts is not an adequate answer to the problems faced. Therefore some legal modifications must be done.

Jurists are aware of this necessity: International Institutions both public and private are now fighting to encourage the recognition of standard norms in specific areas (customs, banks, ...).

What are the legal requirements related to the definition of the traditional signature ? How can the definition of an electronic signature fulfill these legal requirements ? From the legal point of view, which lessons could be drawn from the comparison between the two types of signature ? What is the technical impact of the norms proposed by International Institutions (UNCID) about ETD (Electronic Transfer of Data) ? These are some of the questions which will be dealt within the context of this paper.

I. INTRODUCTION

It is a common assertion that law has always some delay with respect to reality. In regards to the electronic signature, it is our duty to confess the truth of this assertion.

Nevertheless, the lawyer is able to develop *appropriate* regulations, when he has understood the techniques. Such a regulation will not strain the technical developments but impose the committment to use secure technical means according to the evolving level of technical knowledge.

The paper presented here is a collective work made by a computer scientist and lawyers of the research center for computer and law of Namur.

In a first step, we will make a comparison between traditional and electronic signature. In a second step, we propose to make a short analysis of the recent UNCID-rules which can be considered as an international standard for EDI transactions. And finally, we will pay a particular attention on the way the signature and encryption techniques could be used regarding these UNCID-rules.

II. COMPARISON BETWEEN TRADITIONAL AND ELECTRONIC SIGNATURE

1. Traditional Signature

1.1. Functions

Although there is no universal legal definition of the signature, jurists nevertheless recognize that it fulfills a double function[1]. First, in the identification of the signatory and, second, in the expression of a will to accept the content of the document.

Identification of the signatory: The signature, the very "personal and unique trademark" of a person, when appended to a document, can determine the author of the act, as well as his physical presence.

Expression of will: By signing the document, the signatory Party expresses his will to become an integral part of the legal consequences it implies.

1.2. Form

The signature must achieve the same end in all legal regimes, but the actual physical requirements for it to fulfill these ends (the form), may change from one

---

[1]M. Van Quickenborne,"Quelques réflexions sur la signature des actes sous seing privé", note sous Cass. 28 juin 1982, *R.C.J.B.*, 1985, 68-69.

legal system to the other. For example, in common law countries, the signature must be physically and immediately readable.

2. Electronic Signature

2.1. Functions

The electronic signature is an irreproducible "mark" (as a series of characters) appended to a document. It is the output of a complex algorithm of which the input data can be grouped in three different classes of variables: 1) secret data peculiar to the sender, 2) data known by both the sender and the addressee; and 3) data related to the content of the document itself. In these conditions, the electronic signature will allow:

- to identify the sender of the document;
- to authenticate the content of the document (which means that one could verify if the document had been modified).

2.2. Form

An electronic signature is a short series of characters pasted at the end of a document.

With respect to the requirements of Civil Law systems, it is obvious that the regulatory requests in regards to the form of the signature are not met in all encryption techniques so far. Firstly in certain cases, the electronic signature is readable in a first step by the machine and not directly by the receiver and secondly, in other cases electronic signatures are not readable at all.

In many Civil Law countries, another requirement can be considered as an argument against the recognizance of a legal value of the electronic signature. According to certain courts (the Belgian Supreme Court for example), the signature must express the personality of the sender[2]. This requirement is met by certain techniques of authentication like electronic recognition of physical characteristics (iris, blood, face, ... which present the advantage of identifying the person and not only the holder of the access device) but not by electronic signature.

Certain regulations (like Luxembourg regulation in 1986[3]) have been expressly modified in order to accept the new authentication procedures, like electronic signature. Luxembourg regulation defines the signature as all the means for identifying the sender of a message including secret code. The problem of the regulatory form of the signature can be viewed definitively as the major difficulty for receiving electronic signature as admissible before the courts.

3. Comparing the Traditional Signature With the Electronic One

The introduction of a new identification technique, the electronic signature, will force the business world to adapt to it. So before major changes are forced upon, one has to make sure that the traditional and the electronic signature can fulfill the same functions. A comparison analysis of the different functions will help us to check this.

---

[2]Cass. 7 janvier 1955, *Ann. not.*, 1955, 307.

[3]Loi du 22 décembre 1986 sur la preuve des actes juridiques, *Journal Officiel du Grand-Duché de Luxembourg*, 30 décembre 1986, 2745.

3.1. Comparative schema

functions	traditional signature	electronic signature
Parties Identification – sender – receiver	yes no	yes no (yes)
Content of the document	no	yes
Will to agree to the content of the document	yes	yes

3.2. Explanation of the schema

a. Identification of the parties

Identification of the sender: The "manual signature" is the traditional authentication form of a document. Therefore, it is possible to recognize the person who wanted to agree with the content. However, the security of the signature is not absolute: a manual signature may fluctuate with some circumstances; it may also be imitated etc.

As far as the security factor is involved, the electronic signature is far more reliable, and this for two reasons: On one side, the automated signature is systematically checked and, on the other side, it is intimately related to the content of the document.

Identification of the receiver: Moreover, the electronic signature could be done in a such way that it is also related not only to the information about the sender and the content, but also to some data on the receiver. In that case the signature would also identify the person it is sent to. Something which obviously is impossible to attain with traditional signature.

b. Will to agree with the content of the document

An identification technique should show the approval of the signatory to the content of the document. But this can only be assured of if one meets a specific condition:
The signature must be a distinct operation from the writing of the text itself[4]. The electronic signature must be initiated through a different process; most likely by the person (or legal representative of this person) whose signature it should

---

[4]M. Van Quickenborne, *op. cit.*, 80.

represent. Therefore, the electronic signature, just as the traditional one, would require the physical presence of a human being, but instead of the person having to be there during the actual signature, the electronic signature could be just initiated by the person.

This will transpire in the fact that the electronic signature will be physically separated from the corpus of the text, showing at the end of the page.

c. Content of the document

The signature shows that a particular person has actually agreed with the content of the document. This can only be the case if the document has not been modified, by error or by fraud, after its signature. The electronic signature will allow, in addition to the traditional functions of the manual signature, to assure this authentication function. This is made possible because of the definition of the electronic signature according to the content of the signed document. The coded digital signature is therefore the best feasible guarantee one can get against modifications to the text.

d. Towards New Norms

Even if the electronic signature fulfills the different functions of the manual signature and gives us a much better reliability and security degree than the traditional signature, its recognition requires an adaptation by the legal system, regarding the problem of form requirements.

Up to now, as long as there is no case law on it, it is very difficult to predict whether a court would give probative value in the event of a dispute to the most recent techniques of authentication. However, with regard to predicting the legal value of the most recent means of authentication, it is useful to know the degree of reliability, courts have granted to older techniques of authentication, that is to say, those used for the conclusion of agreement by telex. Two opposite decisions can be quoted thereabout: the first one is Italian, the second one, German.

- In Italy, with regard to the telex contract, the District Court of Ascoli Piceno held that it is possible to show that a telex sent by a teleprinter is, in fact, written by the sender himself to an addressee. But, this is a presumption that could be broken by different means[5].

---

[5]Soc. Socona v. Soc. Sioler-Tronto, District Court of Ascoli Piceno, September 7, 1980, (1982) E.E.C., 317, cited by B. Amory and M. Schauss, "La formation de contrats par des moyens

- In Germany, on the other hand, an expert said before a Court that the identification of the sending and the addressee teleprinter doesn't prove that the person who sent the message is the genuine[6].

It is obvious that these previous case law regarding an older and not so secure technique of authentication allows to envisage that more secure means of authentication like electronic signature will be considered by most of the courts as admissible but saying that, at our opinion the debate is not closed.

Indeed two legal solutions could resolve the problem of admissibility. First, one could abolish all legal evidence requirements, and, therefore give way to free evidence techniques. This, in fact, would be in our opinion a false solution. First, the business world would be in a favourable position compared to the private person, and second, the judge would have to weight the value of the evidence without having any fixed or legal criteria, allowing a dangerous discretion.

A better solution would be to assess and determine a priori, certain characteristics of an electronic signature that would have the same value as the traditional one. This second solution was recommended by the Council of Europe in 1981. Therefore, if a document was electronically signed according to the rule, the judge would have no option but to accept the document as a "prima facie proof". He will take it into account in his overall weighing of the evidence, as the case would be for a traditionally signed document insofar it remains possible for the plaintiff to prove that the signed document is false and criminal sanctions exist in these cases.

No rules have as yet been developed to establish the evidence value of an electronically signed document, but we can rely in part on some standards for the credibility for Electronic Data Interchange (EDI) that were adopted by International Organizations. We will study here the International Chamber of Commerce (ICC) rules known as UNCID rules, Uniform Rules of Conduct for Interchange of Trade Data by Teletransmission.

électroniques", *Dr. de l'informatique*, nr. 4, 1987, 211.

[6]OLG Karlsruhe, 12 juin 1973, *Neue Juristische Wochenschrift*, 1973, Heft 36, 1011. cited by B. Amory and M. Schauss, op. cit., 211.

III. THE UNCID-RULES

1. Background of the UNCID-rules

The UNCID-rules must be understood in the framework of international trade, where security plays such an important role[7].

They were drafted by a special Joint Committee of the International Chamber of Commerce (a non-governmental organisation), composed of members of the United Nations Commission on International Trade Law (UNCITRAL), the Customs Co-operation Council, the OECD, the EEC commission and the International Organisation for Standardisation (ISO), ...

The UNCID-rules don't deal with the problem of legal acceptance, they don't affect the law on admissibility[8]. If a written document is required by law, the UNCID-rules are powerless.

Instead, the rules are to provide a model contract for users of EDI, a foundation on which one can build a "communication agreement". The form and details of these agreements will differ according to the size and type of user groups: for example, UNCID-rules were used for ODETTE (Organisation for Data Exchange through Teletransmission in Europe) and DISH (Data Interchange for Shipping: group of exporters, freight forwarders and shipping lines who took part in an EDI pilot project).

However, UNCID-rules are more than a mere starting point: they also define an accepted level of professional behaviour, a "code of good conduct"[9]. By defining some *technical* requirements for EDI, the UNCID-rules will thus increase the security of this form of communications, and therefore the credibility of an "electronic document" in the eyes of a judge.

2. Principles of the UNCID-rules

According to the introductory note written by the ICC[10], the Joint Committee based its work on five basic principles. The rules should:

---

[7]B. Wheble, "Think data, not documents", *International Financial Law Review*, June 1988, 37.
[8]In opposition to the Recommendation R20 of the Council of Europe.
[9]B. Wheble, *op. cit.*, 38.
[10]ICC Publication Nr. 452.

4.1. SIGNATURE as identification of the signatory

One of the functions of the electronic signature permits the identification of the signatory.

Let's see how the UNCID rules have tackled with this. The following rules deal with the subject:

6.a) *A trade data message may relate to one or more trade transactions and should contain the appropriate identifier for each transaction and means of verifying that the message is complete and correct according to the TDI-AP concerned.*

6.b) *A transfer should identify the sender and the recipient: it should include means of verifying, either through the technique used in the transfer itself or by some other manner provided by the TDI-AP concerned, the formal completeness and authenticity of the transfer.*

The UNCID rules recommend the identification of the participants to the transfer (the sender and the recipient) and also the identification of messages and transactions. The electronic signature and other classical techniques of numeration and sequentialization of messages permit such an identification.

4.2. SIGNATURE as inalterability (integrity) of a document

Another function of the electronic signature allows the control of inalterability of the content of a message. In fact, we must say that the electronic signature only detects the fraudulent or erroneous modifications, but it doesn't prevent from any risks. The following rules deal with the subject:

5.a) *Parties applying a TDI-AP should ensure that their transfers are correct and complete in form, and secure, according to the TDI-AP concerned and should take care to ensure their capability to receive such transfers.*

5.b) *Intermediaries in transfers should be instructed to ensure that there is no unauthorized change in transfers required to be retransmitted and that the data content of such transfers is not disclosed to any unauthorized person.*

6.a) *A trade data message may relate to one or more trade transactions and should contain the appropriate identifier for each transaction and means of verifying that the message is complete and correct according to the TDI-AP concerned.*

6.b) *A transfer should identify the sender and the recipient: it should include means of verifying, either through the technique used in the transfer itself* *or by some other manner provided by the TDI- AP concerned, the formal completeness and authenticity of the transfer.*

7.c) *If a transfer received appears not to be in good order, correct and complete in form, the recipient should inform the sender thereof as soon as possible.*

7.d) *If the recipient of a transfer understands that it is not intended for him, he should take reasonable action as soon as possible to inform the sender and should delete the information contained in such transfer from his system, apart from the trade data log.*

10.a) *Each party should ensure that a complete trade data log is maintained of all transfers as they were sent and received without any modification.*

10.d) *Each party shall be responsible for making such arrangements as may be necessary for the data referred to in paragraph (b) of this article to be prepared as a correct record of the transfers as sent and received by that party in accordance with paragraph (a) of this article.*

10.e) *Each party must see to it that the person responsible for the data processing system of the party concerned, or such third party as may be agreed by the parties or required by law, shall, where so required, certify that the trade data log and any reproduction made from it is correct.*

Question 8 — *Should there be rules on "signature" ?*

Articles (5.a), (6.a) et (6.b) state that the transactions done and the messages exchanged must be correct, complete and secure. Completeness and authenticity could be verified by the transfer protocol used. We have here a typical example of application for the techniques of authentication and, more particularly, the electronic signature.

Articles (7.c) et (7.d) are a particular case of the three preceeding ones. As a transfer (or a message) is received in an incomplete or incorrect form or if it has not been delivered to the right person, the recipient must inform the sender as soon as possible and must also destroy the message received. These characteristics of completeness, correction and security could be verified by means of identification techniques.

Article (5.b) mentions that no modification could be done on the content of a message by the intermediaries of the transfer. How can we be sure that no modification has been done ? By means of identification and signature techniques, of course !

About the problem of conservation of transactions traces in the trade data log (log file), articles (10.a), (10.d) et (10.e) impose that the conservation must be done in a correct way and without any modification on the content of transfers or messages (kept as traces) as well as on the support of these. Once again, we're in the domain of application of authentication and signature techniques.

Finally, we can see that writers of the UNCID rules have been aware of problems about the "signature" as explained in their eighth question and that they are in favour of an agreement about the admissibility of an electronic signature.

4.3 Encryption

We will now analyze how the UNCID rules have considered the problem of encryption of documents.

5.a) *Parties applying a TDI-AP should ensure thar their transfers are correct and complete in form, and secure, according to the TDI- AP concerned and should take care to ensure their capability to receive such transfers.*

5.b) *Intermediaries in transfers should be instructed to ensure that there is no unauthorized change in transfers required to be retransmitted and that the data content of such transfers is not disclosed to any unauthorized person.*

6.a) *A trade data message may relate to one or more trade transactions and should contain the appropriate identifier for each transaction and means of verifying that the message is complete and correct according to the TDI-AP concerned.*

6.b) *A transfer should identify the sender and the recipient: it should include means of verifying, either through the technique used in the transfer itself or by some other manner provided by the TDI-AP concerned, the formal completeness and authenticity of the transfer.*

7.c) *If a transfer received appears not to be in good order, correct and complete in form, the recipient should inform the sender thereof as soon as possible.*

7.d) *If the recipient of a transfer understands that it is not intended for him, he should take reasonable action as soon as possible to inform the sender and should delete the information contained in such transfer from his system, apart from the trade data log.*

9.a) *The parties may agree to apply special protection, where permissible, by encryption or by other means, to some or all data exchanged between them.*

9.b) *The recipient of a transfer so protected should assure that at least the same level of protection is applied for any further transfer.*

10.a) *Each party should ensure that a complete trade data log is maintained of all transfers as they were sent and received without any modification.*

10.d) *Each party shall be responsible far making such arrangements as may be necessary for the data referred to in paragraph (b) of this article to be prepared as a correct record of the transfers as sent and received by that party in accordance with paragraph (a) of this article.*

10.e) *Each party must see to it that the person responsible for the data processing system of the party concerned, or such third party as may be agreed by the parties or required by law, shall, where so required, certify that the trade data log and any reproduction made from it is correct.*

Question 5 - *Should there be rules on secrecy or other rules regarding the substance of the data exchanged ?*

Question 7 - *Should there be rules on encryption or other security measures?*

First, we must remark that the UNCID rules have expressly taken into account the problem of encryption as explained by articles (9.a), (9.b) and questions (5 and 7).

Articles (5.a) and (5.b) are dealing with the problem of security and of non disclosure of a transfer or a message to an unauthorized person. The only way to be protected against such disclosure (or wire tapping) is to assure the confidentiality of the document content and therefore to use encryption techniques.

Articles (6.a), (6.b), (7.c) and (7.d) are dealing with the completeness and the correction of transfers and messages. How is it possible to detect any error or fraud during the transfer ? Beside the use of authentication techniques, the encryption allows that detection. Indeed, if a modification has been made on a previously encrypted message, the recipient of the message will be aware of the problem because it will be impossible for him to decrypt it correctly. This is guaranteed by one particularity of encryption techniques which is: if an error occurs during the transfer of an encrypted message, the decryption of the erroneous message will give as result another message profoundly and completely different from the original one. The end result will be an incomprehensible message.

There is a major problem limiting the use of encryption techniques regarding the UNCID rules (7.c) and (7.d). In fact, if someone receives a message with errors or wrongly addressed, he may not be able to decrypt the message in order to find the correct address of the sender. And therefore, it will be impossible for him to inform the sender about the error as he is required by the two articles.

And finally, the log files, described in articles (10.a), (10.d) and (10.e), could also be protected from an eavesdropper by encryption techniques.

IV. CONCLUSION

We conclude with referring to two opinions of American judges.

1. As asserted by the Supreme Court of Nebraska in 1965, it is time now to "bring the reality of business and professional practice into the courtroom". It is time to consider as *admissible* by the courts an electronic signature like a traditional one.

2. In respect to the opinion of another American judge in 1976, "as one of the many who have received computerized bills and dunning letters for accounts paid since a long time, I am not prepared to accept the product of a computer as the equivalent of the Holy writ". In other terms, the *credibility* of an electronic signature will depend on the security level that the chosen techniques are able to prove.

3. In this respect, the court will take into consideration not only the technical but also the organisational measures taken by the responsible of the computer system, according to the importance and the characteristic of the transactions concluded by this system.

4. Finally, in order to avoid uncertainty as to the legal acceptance of modern means of authentication, it is strongly recommended that parties who intend to use them, agree in advance in writing on the validity of the technology they will use for concluding transactions. From this point of view, the UNCID code can be viewed as an adequate model.

Online Cash Checks

David Chaum

Centre for Mathematics and Computer Science
Kruislaan 413 1098 SJ Amsterdam

INTRODUCTION

Savings of roughly an order of magnitude in space, storage, and bandwidth over previously published online electronic cash protocols are achieved by the techniques introduced here. In addition, these techniques can increase convenience, make more efficient use of funds, and improve privacy.

"Offline" electronic money [CFN 88] is suitable for low value transactions where "accountability after the fact" is sufficient to deter abuse; online payment [C 89], however, remains necessary for transactions that require "prior restraint" against persons spending beyond their available funds.

Three online schemes are presented here. Each relies on the same techniques for encoding denominations in signatures and for "devaluing" signatures to the exact amount chosen at the time of payment. They differ in how the unspent value is returned to the payer. In the first, all change is accumulated by the payer in a single "cookie jar," which might be deposited at the bank during the next withdrawal transaction. The second and third schemes allow change to be distributed among unspent notes, which can themselves later be spent. The second scheme reveals to the shop and bank the maximum amount for which a note can be spent; the third does not disclose this information.

DENOMINATIONS AND DEVALUING

For simplicity and concreteness, but without loss of generality, a particular denomination scheme will be used here. It assigns the value of 1 cent to public exponent 3 in an RSA system, the value of 2 cents to exponent 5, 4 cents to exponent 7, and so on; each successive power-of-two value is represented by the corresponding odd prime public exponent, all with the same modulus. Much as in [C 89], a third root of an image under the one-way function f (together with the pre-image modulo the bank's RSA composite) is worth 1 cent, a 7th root is worth 4 cents, and a 21st root

5 cents. In other words, a distinct public prime exponent is associated with each digit of the binary integer representation of an amount of payment; for a particular amount of payment, the product of all those prime exponents corresponding to 1's in the binary representation of the amount is the public exponent of the signature.

A signature on an image under f is "devalued" by raising it to the public powers corresponding to the coin values that should be removed. For instance, a note having a 21st root could be devalued from its 5 cent value, to 1 cent, simply by raising it to the 7th power.

In earlier online payment systems [C 89], the number of separate signatures needed for a payment was in general the Hamming weight of the binary representation of the amount. Since online systems would be used for higher-value payments (as mentioned above), and extra resolution may be desired to provide interest for unspent funds [C 89], an average of roughly an order of magnitude is saved here.

COOKIE JAR

In this first scheme the payer periodically withdraws a supply of notes from the bank, each with the system-wide maximum value. Consider an example, shown in Figure 1.1, in which two notes are withdrawn. The n_i and r_i are random. The r_i "blind" (from the bank) the images under the public, one-way function f. The bank's signature corresponds to taking the hth root, where $h = 3 \cdot 5 \cdot 7 \cdot 11$. As in all the figures, the payer sends messages from the left and the bank sends from the right.

$$f(n_1)\, r_1^h,\ f(n_2)\, r_2^h$$

$$\longrightarrow$$

$$f(n_1)^{1/h} r_1,\ f(n_2)^{1/h} r_2$$

$$\longleftarrow$$

Fig. 1.1. Cookie-jar withdrawal

In preparing the first payment, the payer divides r_1 out. The signature is then raised to the 55th power to devalue it from 15 cents to 5 cents. Figure 1.2 shows this first payment. Of course the shop is an intermediary between the payer (left) and the bank (right) in every online payment, but this is not indicated explicitly. Also not shown in the figures are messages used to agree on the amounts of payment.

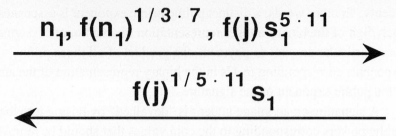

$$n_1, f(n_1)^{1/3 \cdot 7}, f(j) s_1^{5 \cdot 11}$$

$$f(j)^{1/5 \cdot 11} s_1$$

Fig. 1.2. First cookie-jar payment

The first two residues sent in paying, n_1 and its signed image under f, are easily verified by the bank to be worth 5 cents. The third residue is a blinded "cookie jar," a blinded image under f of a randomly chosen value j. This cookie jar is modulo a second RSA composite that is only used for cookie jars. Once the bank verifies the funds received, and that n_1 has not been spent previously, it signs and returns the blinded cookie jar (under the cookie-jar modulus) with public exponents corresponding to the change due.

The second payment, shown in figure 1.3, is essentially the same as the first, except that the amount is 3 cents and the cookie jar now has some roots already on it. If more payments were to be made using the same cookie jar, all resulting signatures for change would accumulate.

$$n_2, f(n_2)^{1/3 \cdot 5}, f(j)^{1/5 \cdot 11} s_2^{7 \cdot 11}$$

$$f(j)^{1/5 \cdot 11 \cdot 7 \cdot 11} s_2$$

Fig. 1.3. Second cookie-jar payment

The cookie jar might conveniently be deposited, as shown in figure 1.4, during the withdrawal of the next batch of notes. It is verified by the bank much as a payment note would be: the roots must be present in the claimed multiplicity and the pre-image under f must not have been deposited before.

$$j, f(j)^{1/5 \cdot 7 \cdot 11 \cdot 11}$$

Fig. 1.4. Cookie-jar deposit

The cookie-jar approach gives the effect of an online form of "offline checks" [C 89], in that notes of a fixed value are withdrawn and the unspent parts later credited to the payer during a refund transaction.

DECLARED NOTE VALUE

Figure 2 depicts a somewhat different scheme, which allows change to be spent without an intervening withdrawal transaction. Withdrawals can be just as in the cookie-jar scheme, but here a single modulus is used for everything in the system. The products of public exponents representing the various amounts are as follows: d is the amount paid, g is the note value, the "change" c is g/d, and h is again the maximal amount, where $d \mid g \mid h$. A payment (still to the bank through a shop) includes first and second components that are the same as in the cookie-jar scheme. The third component is the amount of change c the payer claims should be returned. The fourth is a (blinded) number m, which could be an image under f used in a later payment just as n is used in this one.

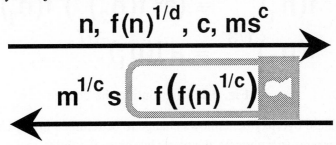

Fig. 2. Declared note value payment

The signature returned contains a "protection" factor (shown inside the padlock). This factor ensures that the payer actually has the cth root of $f(n)$, by requiring that the payer apply f to it before dividing the result out of the signature. Without such protection, a payer could get the system-wide maximum change, regardless of how much change is actually due; with it, the change claimed can only be recovered if the corresponding roots on n are in fact known to the payer.

DISTRIBUTING CHANGE

The change returned in a payment can be divided into parts that fill in missing denominations in notes not yet spent. Suppose, for example, that the last payment is spent with $d = 5 \cdot 11$, $c = 3 \cdot 7$, and that m is formed by the payer as shown in the first line of Figure 3.1. Then unblinding after the payment yields the a shown in the second line.

$$m \equiv f(n_1)^3 f(n_2)^7$$
$$a \equiv m^{1/21}$$

Fig. 3.1. Form of change returned

From a, the two roots shown in the last two lines of Figure 3.2 are readily computed. (This technique is easily extended to include any number of separate roots.) Thus the values unused in the last payment fill in roots missing in notes n_1 and n_2.

$$u = 3^{-1} \bmod 7$$
$$v = 3u \ \mathrm{div} \ 7$$
$$f(n_1)^{1/7} \equiv (a^3 f(n_2)^{-1})^u f(n_1)^{-v}$$
$$f(n_2)^{1/3} \equiv a \, f(n_1)^{-1/7}$$

Fig. 3.2. Distributing the change

Because overpayment allows change to be returned in any chosen denominations (not shown), the payer has extra flexibility and is able to use all funds held. This also increases convenience by reducing the need for withdrawals.

HIDDEN NOTE VALUE

Although the combination of the previous two subsections is quite workable, it may be desirable for the payer not to have to reveal c to the shop or the bank. Figure 4 shows a system allowing this. The payment message is just as in the declared note value protocol above, except that c is not sent. The protection factor (shown again in a lock) is also placed under the signature, but it is missing the extra f and is raised to a random power z chosen by the bank.

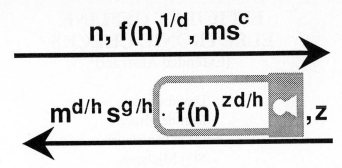

Fig. 4. Hidden note value payment

If z were known to the payer before payment, then the payer could cheat by including $f(n)^{-z}$ in the third component; this would yield the payer the system-wide maximum change, even if none were due. Consider a single change exponent q. If $z \bmod q$ is guessed correctly by a cheating payer, then the payer improperly gets the corresponding coin value. Thus, the chance of successful cheating is $1/q$. If, however, the divisors of h are chosen sufficiently large, quite practical security can be achieved. When the possibilities of distributing change and refunding are included, this scheme's privacy surpasses that of a coin system.

CONCLUSION
Combining online coins improves efficiency, use of funds, convenience, and privacy.

REFERENCES

Chaum, D., "Privacy Protected Payments: Unconditional Payer and/or Payee Anonymity," in Smart Card 2000, North-Holland, 1989, pp. 69–92.

Chaum, D., A. Fiat, & M. Naor, "Offline Electronic Cash," Proceedings of Crypto '88.

EFFICIENT OFFLINE
ELECTRONIC CHECKS
(Extended Abstract)

David Chaum
Bert den Boer
Eugène van Heyst
Stig Mjølsnes
Adri Steenbeek

Centre for Mathematics and Computer Science
Kruislaan 413,   1098 SJ Amsterdam,   The Netherlands

Abstract. Chaum, Fiat, and Naor proposed an offline check system [1], which has the advantage that the withdrawal and (anonymous) payment of a check are unlinkable. Here we present an improved protocol that saves 91% of the signatures, 41% of the other multiplications, 73% of the divisions, and 33% of the bit transmissions.

1. Introduction

In this paper we present a payment system that has the following advantages:
- the bank has only to give one signature per payment.
- the shop does not need online connection with the bank.
- during payment the payer has to do no computations.
- a payer can refund several checks at once, to keep the bank from learning the amount spent for each check.
- the withdrawal and payment of a check are unconditionally unlinkable, but if a check is spent twice, the identity of the cheater will be revealed with high probability.
- payments and refunds are unlinkable, except from that little that can be learned from the total number of each type of unspent denomination.
- refund and withdrawal can be made unlinkable.

In this paper we will only consider the withdrawal of one check. The payer (Alice) creates some candidates in a special way. The bank chooses randomly some of them, and the inner arguments of those selected candidates must be revealed by Alice (the "cut-and-choose" protocol). This is needed to prevent cheating users, or to be sure that there are not to many bad candidates left (i.e. candidates not created in the proper way). If more checks are withdrawn at the same time, less candidates have to be opened (to get the same security).

Because of the anonymous payment, checks must be bought for the maximal amount. Therefore the signature of the bank can be split by the user into two parts: one part is the check, which can be used for a payment; and the other is a refund part, which is used for refunding the unspent value.

2. Setting and Overview

The numbers which are used in this paper (such as the number of candidates in the denomination part, in the challenge part, or to be opened) *are examples.*

The underlaying scheme of this payment system, is an RSA scheme. As in the original paper [1], let f,g be two-argument, collision-free, one-way functions, with g such that if the first argument is fixed, the mapping is l-to-1 from the second argument onto the range. Let h be an injective one-way function with one argument and k an injective one-way function with 40 arguments. Let u_i be Alice's account number concatenated with a counter.

All calculations are modulo N, the factorization of which only the bank knows. In our formulas, we will omit the modular reduction. We use $\oplus$ to denote bitwise exclusive-or and $\parallel$ for concatenation.

The check is a product of 20 terms, which are ordered in value: the first 10 are used for the amount to be spent in a shop (denomination part) and the last 10 to prevent the check from being spent twice (challenge part). These parameters can be varied, but the exponent then also needed to be changed (in order to prevent certain attacks). According to [2], we have the following table:

number of terms in denomination part +1	exponent
1,...,3	5
4,5	7
6,...,15	17
16,17	19
18,...,27	29

Three basic changes over [1] were made to improve efficiency :

(i) Alice initially sends to the bank *not* the candidates for the check, but the value of a one-way function with the candidates as its arguments. Hence she avoids sending half of the candidates.

(ii) The terms in the check are ordered, so each can have almost the same root.

(iii) Alice sends a blinded product of the major and minor terms, so she needs only half as many blinding factors, half as much bandwidth is needed, and the bank makes only one signature. When she receives the signed check, though, she must do some calculations to separate the major and minor terms to separate the product of the signed minor terms for refund. Also an additional one-way function must be introduced for this.

3. Transactions

The payment system consists of three parties (bank, user Alice and shop) and four transactions: withdrawal, payment, deposit, and refund. Each of these transactions is a protocol between two of the three parties. The transactions do not need to be in this order; Alice can first refund a part of the check and later spend the rest of the check in a shop. Also the refund of a check can be done earlier than its deposit.

3.1. Check Withdrawal Transaction

(1) Alice first makes 40 candidates. For each, she chooses at random: $r_i, a_i, b_i, c_i, d_i, e_i$, ($1 \leq i \leq 40$) and computes:
$$x_i = g(a_i \| b_i, c_i),$$
$$y_i = g(a_i \oplus u_i, d_i),$$
$$M_i = f(x_i, y_i), \qquad \text{(called the \textit{major term})}$$
$$m_i = h(g(b_i, e_i)), \qquad \text{(called the \textit{minor term})}$$
$$\alpha_i = M_i^{3^{10}} \cdot m_i^{17} \cdot r_i^{17 \cdot 3^{10}} \quad \text{(called the \textit{candidate})}.$$

All computations can be made before connection with the bank, during which the hash value $k(\alpha_1, \alpha_2, ..., \alpha_{40})$ is sent.

(2) The bank splits the integers $1, ..., 40$ randomly between two unordered partitions S_o and S_c, both of 20 elements. The partitioning is then sent to Alice.

(3) Alice orders the elements in the partition S_c by the M_i value of the corresponding candidate, forming the ordered set T_c. This set, the α_i's of the candidates in T_c, and the $r_i, a_i, b_i, c_i, d_i, e_i$ of the candidates in S_o are sent to the bank. The candidates in S_o are said to be "opened".

(4) The bank verifies that the hash of candidates revealed equals $k(\alpha_1, \alpha_2, ..., \alpha_{40})$ and that every element of the opened partition is correctly formed. (The opened partition can now be discarded by both parties.) The bank takes a random integer R and computes:

$$D = \prod_{i=1}^{10} (\alpha_{t(i)})^{\frac{1}{17 \cdot 3^{11-i}}} \cdot \prod_{i=11}^{20} (\alpha_{t(i)})^{\frac{1}{17}} \cdot R^{\frac{1}{10^{3}}},$$

where $t(i)$ is the ith element of T_c; sends D and R to Alice; and diminishes Alice's account with the value of the check ($=2^{10}-1$). The bank also stores R with Alice's account number.

(5) Alice verifies the validity of the signature on the check by testing whether

$$D^{17 \cdot 3^{10}} \stackrel{?}{=} \left(\left(\left(\prod_{i=11}^{20} (\alpha_{t(i)}) \right)^{3} \cdot \alpha_{t(10)} \right)^{3} \dots \right)^{3} \cdot \alpha_{t(1)} R^{17}.$$

She unblinds the signature on the check and divides it into two parts:

$$C = \prod_{i=1}^{10} (M_{t(i)})^{\frac{3^{i-1}}{17}} \cdot \prod_{i=11}^{20} (M_{t(i)})^{\frac{3^{10}}{17}},$$

$$C' = R^{\frac{17}{10^{3}}} \cdot \prod_{i=1}^{10} (m_{t(i)})^{\frac{3^{11-i}}{3}},$$

where C is the signed check and C' is used to get a refund for any part of C that is unspent. These can be computed any time before payment.

3.2. Payment Transaction

Alice can use a check to pay in a shop. The first 10 terms of a check C (remember that a check is the product of 20 ordered elements) have denominations $2^9, 2^8, .., 2, 1$ respectively. These must be in decreasing order, because the denominations in the refund part C' *must* be in decreasing order; otherwise, Alice can claim more refund with C', because from $e^{7/3^i}$ she can easily compute $e^{7/3^j}$, $(0 \le j \le i)$. Every amount smaller than 1024 can be paid with C; let $(w_1, \dots, w_{10})$ be the binary representation of the amount of payment.

(1) Alice gives C to the shop.
(2) The shop generates a binary challenge-vector $(w_{11}, \dots, w_{20})$ and sends it to Alice.
(3) Alice gives a partial opening of the check C:
        if $w_i=1$, she reveals the corresponding $a_i \| b_i, c_i, y_i$;
        if $w_i=0$, she reveals $x_i, a_i \oplus u_i, d_i$.
(4) The shop verifies the partial opening, the check's signature, and the ordering of the M_i's.

3.3. Deposit Transaction

(1) The shop sends to the bank: C, the vector $\underline{w} = (w_1, .., w_{20})$ (amount and challenge vector), and the partial opening.

(2) The bank verifies the signature and the partial opening, just as the shop did.

(3) The bank stores the check C in its searchable (batch) list of spent checks, the corresponding partial opening a_i or $a_i \oplus u_i$ ($1 \le i \le 20$) in an archive list, and the revealed b_i's in its searchable (batch) list of revealed minor terms. The bank consults the searchable lists to be sure that no b_i is already refunded or that no check is spent twice (possibly by sorting them sometime later). When double spending of a check is found, the u_i can be reconstructed from any difference in the corresponding vectors $\underline{w}$ and $\underline{w}'$, thereby revealing the cheater's account number.

3.4. Refund Transaction

(1) After each payment, but before refunding, the minor terms of the checks are accumulated and $g(b_i, e_i)$ is computed for each $w_i = 1$. Alice sends the bank the product of the C', the R's, and in addition the $g(b_i, e_i)$ for each denomination spent, and the b_i, e_i for the denominations not spent.

(2) The bank verifies the opened minor terms and their signature C' similar to the way checks are verified. The bank also verifies if the R's were stored with Alice's account number.

(3) The bank verifies that the b_i's are not listed and stores them on its list of revealed minor terms together with Alice's account number, to prevent their later use.

Notice that in case of refunding multiple checks, Alice keeps the bank from learning the amount spent for each check; only the total number of each type of unspent denomination is revealed.

Alice can cheat by ordering the set improperly. She can get the maximal profit by spending the 5 highest denominations and telling the bank that she has spent the 5 lowest: she can spend a check of value $2^{10}-1$ for $2^{11}-2^6$. But in [2] it is proven that Alice can not change the ordering of the denomination part, so this kind of cheating is not possible.

4. Storage

For one check Alice stores the following integers in her card computer:
before withdrawal: $r_i, a_i, b_i, c_i, d_i, e_i, [M_i, m_i, \alpha_i]$ for each candidate;
after withdrawal: C, C', R and $a_i, b_i, c_i, d_i, e_i, [x_i, y_i]$ for each unopened candidate;
after payment: C', R and $b_i, e_i, [g(b_i, e_i)]$ for each unopened candidate.

If the card computer has only a very limited amount of memory, it is possible not to store all the integers for each candidate; for instance not the integers between the square brackets. But then before payment or refund, Alice has to compute these integers, in order to do no computations during payment or refund. The memory that comes available after a transaction can be used to store candidates as they are be generated in advance.

5. Demo

Hans Beuze and Peter Sliepenbeek implemented this system on an Apple Macintosh, and Adri Steenbeek wrote the numerical part. Diskettes with this demo are available from the authors. There is also an earlier version available for an IBM pc.

6. Number representation

We suggest the following number representation:

$$
\begin{array}{ll}
u_i, a_i & : \qquad 32 \text{ bits,} \\
b_i & : \qquad 128 \text{ bits,} \\
c_i, d_i, r_i, e_i, N & : \qquad 512 \text{ bits,} \\
f & : 128 \times 128 \rightarrow 512 \text{ bits,} \\
g & : 128 \times 512 \rightarrow 128 \text{ bits,} \\
h & : \qquad 128 \rightarrow 512 \text{ bits.}
\end{array}
$$

7. Improvements

7.1. Anonymous refund

In order to have anonymous refund, the payment system can be changed into the following way: in the withdrawal part, Alice chooses at random: $r_i, a_i, b_i, c_i, d_i, e_i, z_i$, $(1 \leq i \leq 40)$ and computes:

$$
\begin{aligned}
x_i &= g(a_i \| b_i \| z_i, c_i), \\
y_i &= g(a_i \oplus u_i, d_i), \\
M_i &= f(x_i, y_i), \\
m_i &= h(g(b_i \| (z_i \oplus u_i), e_i)), \\
\alpha_i &= M_i^{3^{10}} \cdot m_i^{17} \cdot r_i^{17 \cdot 3^{10}}.
\end{aligned}
$$

If Alice tries to spent a check twice or if she tries to ask refund for an already spent denomination, her identity will be revealed.

7.2. Combining challenge and denomination bits

We combine a challenge and a denomination bit into one term in the following way: in the withdrawal part Alice chooses at random: $r_i,a_i,b_i,c_i,d_i,e_i,z_i$, $(1\le i\le 40)$ and computes:

$$x_i = g(b_i, c_i),$$
$$y_i = g(b_i, d_i),$$
$$v_i = g'(a_i, x_i),$$
$$w_i = g'(a_i \oplus u_i, y_i),$$
$$M_i = f(v_i, w_i),$$
$$m_i = h'(h(b_i, z_i)),$$

$$\alpha_i = M_i^{3^7} \cdot m_i^{17} \cdot r_i^{17 \cdot 3^7}.$$

So the major term M can be expressed by the tree:

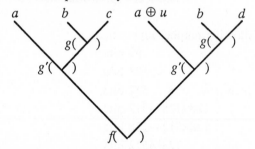

Let during the payment transaction $(k_1,\ldots,k_{10})$ be the binary representation of the amount of payment, and $(l_1,\ldots,l_{10})$ be the binary challenge-vector. Alice gives the following partial opening of the check:

l_i	k_i	revealed
0	0	$a_i \oplus u_i, v_i, y_i$
0	1	$a_i \oplus u_i, v_i, b_i, d_i$
1	0	a_i, w_i, x_i
1	1	a_i, w_i, b_i, c_i

The major term M can be expressed in the following tree, in which the dotted lines indicate the numbers revealed during the payment transaction: in each of the main subtrees, you start at a certain level (which depends on the challenge bit). If the payment bit is a one, then in one of the main subtrees, you go one level deeper.

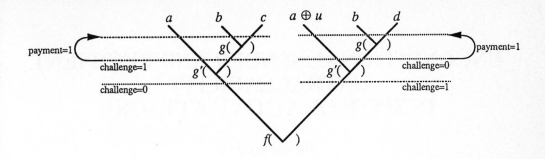

References

[1] David Chaum, Amos Fiat, and Moni Naor, "Untraceable Electronic Cash", to appear in *Advances in Cryptology—Crypto '88*, Lecture Notes in Computer Science, Springer-Verlag.

[2] Jan-Hendrik Evertse and Eugène van Heyst, "Which RSA signatures can be computed from some given RSA signatures?", in preparation, 1989.

UNCONDITIONAL SENDER AND RECIPIENT UNTRACEABILITY IN SPITE OF ACTIVE ATTACKS

Michael Waidner

Institut für Rechnerentwurf und Fehlertoleranz, Universität Karlsruhe
Postfach 6980, D-7500 Karlsruhe 1, F. R. Germany

ABSTRACT

A protocol is described which allows to send and receive messages anonymously using an arbitrary communication network, and it is proved to be unconditionally secure.

This improves a result by CHAUM: The DC-net guarantees the same, but on the assumption of a reliable broadcast network. Since unconditionally secure Byzantine Agreement cannot be achieved, such a reliable broadcast network cannot be realized by algorithmic means.

The solution proposed here, the DC$^+$-net, uses the DC-net, but replaces the reliable broadcast network by a fail-stop one. By choosing the keys necessary for the DC-net dependently on the previously broadcast messages, the fail-stop broadcast can be achieved unconditionally secure and without increasing the complexity of the DC-net significantly, using an arbitrary communication network.

I. OVERVIEW

In [4] CHAUM describes a technique, the DC-net, which should allow to send and receive messages anonymously using an arbitrary communication network.

Section II gives a short and slightly generalized description of the sending mechanism of the DC-net, called *superposed sending* here. Some known efficient and anonymity preserving multi-access protocols for using the multi-access channel superposed sending offers are described.

In [4] the untraceability of senders and recipients of messages is proved to be *unconditional*, but this proof implicitly assumes a *reliable* broadcast network, i.e. each message broadcast by an honest participant is received by each other participant without being changed. Since *unconditional* Byzantine Agreement (i.e. Byzantine Agreement in spite of an attacker with unlimited computational power who may control an arbitrary number of participants) is impossible, such a network cannot be realized by cryptographic means. Thus the assumption may be rather unrealistic.

In Section III it is shown how the sending of a specific participant X can be traced by an active attacker who is able to alter the messages received by X and who controls the current communication partner of X. A number of countermeasures, called *fail-stop broadcast* schemes, are proposed, and it is proved that each one will achieve the desired *unconditional* untraceability in spite of active attacks, independent of the underlying communication network. Superposed sending together with fail-stop broadcast is called a *DC$^+$-net*.

Without any further measures, the *serviceability* of the multi-access channel of superposed sending is not good: each faulty or dishonest participant can untraceably and enduringly disturb the channel.

Unfortunately no measures are known which can guarantee serviceability while preserving *unconditional* untraceability using *arbitrary* communication networks. Therefore this problem is not further considered here.

A scheme which guarantees *nearly unconditional untraceability* and computationally secure serviceability (i.e. untraceability if the attacker cannot prevent the honest participants from communicating and serviceability if additionally the attacker is computationally restricted) is described in [27, 28].

II. UNCONDITIONAL SENDER UNTRACEABILITY

Section II.1 describes the basic mechanisms of the DC-net, superposed sending and broadcast, and defines the notation used throughout this paper. Section II.2 describes anonymity preserving multi-access protocols for superposed sending, and in Section II.3 some general remarks on sender untraceability schemes are given.

II.1. SUPERPOSED SENDING

Assume that a number of participants want to exchange messages over an arbitrary communication network. A computationally unlimited attacker, who is able to eavesdrop communication between any two of the participants (e.g. because he collaborates with the network operator) and who controls an arbitrary subset of the participants, tries to trace the messages exchanged between the participants to their senders and recipients.

If all messages are delivered to each participant, the attacker is not able to trace the *intended* recipient of a message. Therefore unconditionally reliable broadcast guarantees *unconditional recipient untraceability*.

It is important to notice that in this section, as in [4], attackers are assumed to be unable to manipulate the consistency of broadcast.

Sender untraceability is guaranteed by *superposed sending*, which realizes an anonymous multi-access channel:

Let $P = \{P_1, ..., P_n\}$ be the set of all participants and let G be an undirected self-loop free graph with nodes P. Let $(F, \oplus)$ be a finite abelian group. The set F is called the *alphabet*.

To be able to perform a single sending step, called a *round*, each pair of participants P_i, P_j who are directly connected by an edge of G choose a key K_{ij} from F randomly[1]. Let $K_{ji} := K_{ij}$. Participants P_i and P_j keep their common key secret. The graph G is called *key graph*, the matrix K of all keys is called *key combination*.

Each participant P_i chooses a message character M_i from the alphabet F, outputs his *local* sum

$$O_i := M_i \oplus \sum_{\{P_i, P_j\} \in G} \text{sign}(i-j) \cdot K_{ij} \tag{2.1}$$

and receives as input the *global* sum

$$S := \sum_{j=1}^{n} O_j \tag{2.2}$$

---

[1] In the following, the term "X is randomly chosen from a set M" is abbreviated by "$X \in_R M$". This means that X is a uniformly distributed random variable which is independent of "all other variables". What is meant by "all other variables" should always be clear from the context.

As usual, the symbolic operation $\pm 1 \cdot K_{ij}$ is defined by $+1 \cdot x := x$ and $-1 \cdot x := -x$.

Superposed sending realizes an *additive multi-access channel* with (digital) collisions, which is stated in the following lemma.

Lemma 2.1   If the local sums are computed according to (2.1), then the global sum defined in (2.2) is equal to the sum of all message characters:

$$S = \sum_{j=1}^{n} M_j \qquad (2.3)$$

Proof. In (2.2) each key is both added and subtracted exactly once. □

If exactly one character M_i has been chosen unequal to 0, this character is successfully delivered to all participants. Otherwise a (digital) collision occurs which has to be resolved by a multi-access protocol, cf. Section II.2.

Superposed sending guarantees *unconditional sender untraceability*. Let A denote the subset of participants controlled by the attacker. If the graph $G \setminus (P \times A)$ is connected, the attacker gets no additional information about the characters M_i besides their sum.

Lemma 2.2   *Superposed sending.* Let A be the subset of participants controlled by the attacker and assume $G \setminus (P \times A)$ to be connected. Let $(O_1, ..., O_n) \in F^n$ be the output of a single round.

  Then for each vector $(M_1, ..., M_n) \in F^n$ which is consistent with the attacker's a priori knowledge about the M_i and which satisfies

$$\sum_{j=1}^{n} O_j = \sum_{j=1}^{n} M_j \qquad (2.4)$$

the same number of key combinations exist which satisfy Equation (2.1) and which are consistent with the attacker's a priori knowledge about the K_{ij}.

  Hence the conditional probability for $(M_1, ..., M_n)$ given the output $(O_1, ..., O_n)$ (i.e. the a posteriori probability) is equal to the conditional probability for $(M_1, ..., M_n)$ given the sum in (2.4) only (i.e. the a priori probability).

This is stated and proved in [4] for $F = \mathrm{GF}(2)$ by a technique which can easily be applied to any finite field. In [18 Sect. 2.5.3.1] and in the following, Lemma 2.2 is proved for any finite abelian group F. (The general applicability of finite abelian groups was also mentioned in [19].)

Proof. Let $M' := (M'_1, ..., M'_n) \in F^n$ be another vector which satisfies (2.4) and which is consistent with the attacker's a priori knowledge about the M_i.

  To prove Lemma 2.2, a finite sequence $M^0, M^1, ...$ of vectors from F^n is defined, which all satisfy Eq. (2.4) and which differ in only two components. Let $M^k = (M^k_1, ..., M^k_n)$.

  Let $M^0 := (M_1, ..., M_n)$, hence M^0 satisfies Eq. (2.4). If $M^k = M'$ then stop. Now assume $M^k \neq M'$. Since both M^k (by induction hypothesis) and M' satisfy Eq. (2.4) there are at least two different indices i, j with $M^k_i \neq M'_i$ and $M^k_j \neq M'_j$, and since both M^k and M' are consistent with the attacker's a priori knowledge $P_i, P_j \notin A$. Define

$$M^{k+1}_i := M'_i$$
$$M^{k+1}_j := M^k_j \oplus M^k_i \oplus -M'_i \qquad (2.5)$$
$$M^{k+1}_h := M^k_h \text{ for all } h \notin \{i, j\}$$

Obviously M^{k+1} satisfies (2.4). After a maximum of $n - 1$ steps the sequence stops with $M^k = M'$.

Let K^k be the set of all key combinations which satisfy (2.1) for the vector M^k and which are consistent with the attacker's a priori knowledge. Between each pair K^k, K^{k+1} a bijection ϕ^k is defined. Hence $|K^k| = |K^{k+1}|$ for all k and therefore $|K^0| = |K^{n-1}|$ where $M^{n-1} = M'$.

To define ϕ^k consider the equations (2.5). Let $\Delta := M^{k+1}{}_i \oplus -M^k{}_i$. Then $M^{k+1}{}_i = M^k{}_i \oplus \Delta$ and $M^{k+1}{}_j = M^k{}_j \oplus -\Delta$.

Because of the connectivity of $G \setminus (P \times A)$ a path $(P_i = P_{k_1}, ..., P_{k_m} = P_j)$ exists with $P_{k_h} \notin A$ and $(P_{k_h}, P_{k_{h+1}}) \in G \setminus (P \times A)$. Let $K \in K^k$. Then $\phi^k(K)$ is defined by changing the keys on this path appropriately:

$$\forall h = 1, ..., m-1: \quad \phi^k(K)_{k_h k_{h+1}} := K_{k_h k_{h+1}} \oplus -\Delta \cdot \text{sign}(k_h - k_{h+1}),$$

$$\phi^k(K)_{k_{h+1} k_h} := \phi^k(K)_{k_h k_{h+1}}$$

and

$$\forall (f, g) \notin \{ (k_h, k_{h+1}), (k_{h+1}, k_h) \mid h=1, ..., m-1 \}: \quad \phi^k(K)_{fg} := K_{fg}$$

The construction of ϕ^k is depicted in Figure 1.

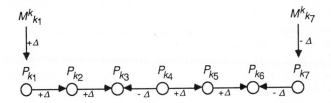

Figure 1   Construction of ϕ^k from a path with $m = 7$. The vertical arrows indicate the change of $M^k{}_{k_h}$, $h=1$, 7, the horizontal arrows the numerical order of the k_h, and the $\pm\Delta$ the change of $K_{k_h k_{h+1}}$:

$\underset{+\Delta}{P_{k_h} \longrightarrow P_{k_{h+1}}}$:$\Leftrightarrow k_h < k_{h+1}$ and therefore $\phi^k(K)_{k_h k_{h+1}} := K_{k_h k_{h+1}} \oplus \Delta$

Obviously, the local outputs of P_{k_h}, $h = 1, ..., 7$, are not changed by ϕ^k.

It can easily be checked that $\phi^k(K)$ satisfies (2.4). Because $\phi^k(K)$ differs from K only in keys which are unknown to the attacker, $\phi^k(K)$ is necessarily consistent with the attacker's a priori knowledge. Since ϕ^k is simply a translation of the group $F^{|G|}$, the bijectivity of ϕ^k is obvious. $\square$

II.2. EFFICIENT AND ANONYMITY PRESERVING MULTI-ACCESS PROTOCOLS

To use the multi-access channel superposed sending offers, it is necessary to regulate the participants' access to the channel by an appropriate, i.e. efficient and anonymity preserving protocol.

For an in depth discussion of possible protocols cf. [18 Sect. 3.1.2]. In the following, some protocols are mentioned, and two protocols are described in detail: a reservation map technique and superposed receiving.

The first step for each multi-access protocol is to combine a fixed number c of characters into a *message*. Each message is transmitted in c consecutive rounds, which are called a *slot*.

In the following, rounds are numbered from 1 to a maximum number t_{max}. Parameter t_{max} is necessary for technical reasons only. Usually $\text{ld}(|F|) \cdot t_{max}$, i.e. the maximum number of transmitted bits, can be assumed to be very large, e.g. $\text{ld}(|F|) \cdot t_{max} = 10^{25}$. Even with a rather unrealistic transmission rate of 10^{15} bps this is sufficient for about 317 years of superposed sending.

The character and output of participant P_i in round t are named M_i^t and O_i^t, respectively, the global sum in round t is named S^t.

The simplest protocol is the well known (slotted) *ALOHA* [4, 24 Sect. 3.2]: If P_i has a message to send, he simply does so in the next slot. If another participant has decided to send a message, too, a collision occurs, which is detected by P_i. After waiting a random number of slots, P_i retransmits his message. Obviously ALOHA preserves anonymity, but wastes the transmission capacity of the network.

II.2.1. RESERVATION MAP TECHNIQUE

To avoid collisions of messages, a simple *reservation map technique* can be used: a slot of r rounds, the *reservation frame*, is used to reserve the following up to r slots [19 Sect. 2.2.2.2].

Let m be an arbitrary integer $\geq n$, and let F be the additive group of integers modulo m. For each message P_i plans to send, he chooses an index k from $\{1, ..., r\}$ at random and outputs 1 as his k-th character for the reservation frame. The resulting reservation message consists of three classes of characters: 0, indicating an unreserved slot, 1, indicating a reserved slot, and $\{2, ..., m\text{-}1\}$, indicating collisions. Since all message slots with corresponding reservation character $\neq 1$ are of no use, they are skipped, i.e. the reservation frame is followed only by as many message slots as there are successful reservations. A slot with reservation character $=1$ is used by the participant who has sent a 1 in the corresponding reservation round.

A similar reservation technique, the *bit-map reservation technique*, is described in [4]: instead of using a relatively large group F to enable the detection of multiple collisions, the superposition is done in $F = GF(2)$ and a value r in the order of the square of s_{max}, the maximum number of reservations, is used to make multiple collisions of an odd number of reservations rather unlikely. Therefore the scheme requires s_{max}^2 additional bits per s_{max} sent messages.

II.2.2. SUPERPOSED RECEIVING

The following two *collision resolution techniques* are based on the observation that
- collisions on this channel, in contrast to an analog one, carry useful information, namely the sum of all collided messages, and
- it is possible to *compute s* collided message characters from s well-defined "collisions", i.e. sums.

Therefore these techniques are subsumed under the name *superposed receiving*.

The first one is an algorithm suggested by PFITZMANN in [18 Sect. 3.1.2] and called *tree-like collision resolution with superposed receiving*. It improves CAPETANAKIS collision resolution algorithm described in [14] using the fact that the set of s collided characters M_i can be computed from each set of s linearly independent sums of these characters. (A performace evaluation of the following can be found in [13].)

Let s_{max} be the maximum number of collided messages, e.g. $s_{max} = n$, and $\{0, 1, 2, ..., M_{max}\} \subset \mathbb{Z}$ the set of all *allowed* message characters. The alphabet F is chosen to be the ring of integers modulo m where m is greater than $s_{max} \cdot M_{max}$. As usual each character $M \in F$ can be interpreted as an integer. A message consists of two characters: For a participant who has to send a message, the first character is always 1 and the second is his message character. For a nonsending participant both are always 0.

Now assume that a new round of the protocol starts and an a priori unknown number of participants have decided to send a message. Let SP denote the set of all sending participants, Σ the sum of their characters M_i modulo m, and $s := |SP|$.

Thus the first slot contains the pair (s, Σ). A number $s \geq 2$ indicates a (digital) collision. To resolve it, each participant computes the average message $M_A := \lfloor \Sigma/s \rfloor$, which is possible, since the modulus m has been chosen so large that Σ is also the sum of the characters in $\mathbb{Z}$.

This average is used to deterministically divide the set SP into two disjoint subsets SP_1 and SP_2: SP_1 consists of all participants $P_i \in SP$ with $M_i \leq M_A$, SP_2 consists of all other sending participants. For $i = 1, 2$ define s_i and Σ_i in analogy to s and Σ.

All participants $P_i \in SP_1$ immediately repeat their messages $(1, M_i)$ in the next slot, hence each user *receives* the pair (s_1, Σ_1) and can *compute* the pair $(s_2, \Sigma_2) = (s \oplus -s_1, \Sigma \oplus -\Sigma_1)$.

Given the rare case $s_2 = 0$, the protocol terminates after the second slot: each participant $P_i \in SP$ has sent the same character $M_i = M_A$. Otherwise, i.e. $s_2 \neq 0$, the sets SP_1 and SP_2 are both nonempty and the collision resolution procedure is recursively applied to (s_i, Σ_i), $i = 1, 2$.

To resolve a collision of s messages, the protocol deterministically needs a maximum of s slots, i.e. $s \cdot (2 \cdot \log(s_{max}) + \log(M_{max}))$ bits.

The second one is suggested by BOS and DEN BOER in [2] and called *root-finding collision resolution with superposed receiving*. It is based on the observation that the collision of s different characters can be resolved using the sums of the first s powers of the characters (where F is a sufficiently large finite field).

The collision is resolved by computing the coefficients of a polynomial whose zeros are exactly the collided characters, and by factoring this polynomial. Since factoring is an expensive task [22] the computational complexity of this technique is much higher than that of tree-like collision resolution. On the other hand it needs only $s \cdot \log(M_{max})$ bits.

To allow long messages to be sent, either the alphabet F could be made large enough to represent a long message by a single character, or superposed receiving could be used as a reservation technique (*reservation by superposed receiving*):

Each participant willing to send a message chooses a reservation message RM_j at random and sends it in the next possible reservation phase. The collision of all s reservation messages is resolved, after which each P_i sorts the received reservation characters RM_j according to their numerical values. The order of the RM_j naturally defines an order of all reserving participants, according to which each P_j sends his real message in the appropriate one of the next s slots. (To increase the fairness of the reservation scheme, this order can be shifted cyclically by an index randomly chosen by all participants together [2].)

The probability of collisions is exponentially small in $\log(|F|)$.

II.3. SOME REMARKS ON SENDER UNTRACEABILITY SCHEMES

Given the very strong assumption of an unlimited attacker (i.e. there may be an arbitrary number of attackers $|A| < |P|$, there are no computational restrictions) the fundamental restrictions of superposed sending as far as performance and reliability are concerned are a consequence of its sender untraceability: In order to make the physical behaviour of a participant meaningless, it is necessary that a participant P_i who is willing to send a character M_i

- does this in an encrypted way,
- each other participant P_j outputs a character, too, and
- the attacker is not able to learn anything about M_i before knowing all the outputs.

Because the attacker is assumed to be an insider it follows from the last fact that the result of such a single sending step cannot contain more information than the last of the participant's output does. Therefore any unconditional sender untraceability scheme realizes a multi-access channel and superposed sending offers the best possible channel capacity as far as only a single round is concerned [17].

To guarantee the unconditional sender untraceability, the global output of the realized multi-access channel has to depend on each participant's output, therefore any unconditional sender untraceability scheme can be untraceably disturbed by each participant.

As far as I know, superposed sending is the only *unconditional* sender untraceability scheme which withstands an unlimited attacker.

There are two other untraceability schemes known from literature, the MIX-net [3] and the concept of physical unobservability [16]. Both can only withstand weaker attackers than superposed sending. The first is based on the use of a public-key cryptosystem and the existence of a number of network stations, called

MIXes, at least one of which has to be trustworthy. The second assumes that the attacker only controls a very small number of participants.

In [1, 7, 5] very general techniques for information theoretically secure fault tolerant distributed computations are described. In general these techniques can be used for implementing a sender and recipient untraceability scheme, but they can only withstand attackers with $3 \cdot |A| < |P|$ and are therefore not further considered here. Also, an untraceability scheme based on such a general technique would be far more expensive than superposed sending with fail-stop key generation described in Section III.2.

To reduce the tremendous number of randomly chosen keys for superposed sending which have to be exchanged by the participants, one can use keys which are generated by pseudorandom bitgenerators (PRBG). If the PRBG used is cryptographically strong, i.e. if distinguishing the PRBG from a true random source in random polynomial time is provably equivalent to solving a (hopefully) hard problem [26], tracing becomes equivalent to this hard problem, too, but the *unconditional* sender untraceability is lost.

Because of the growing importance of public telecommunication networks, it seems necessary to look for efficient implementations of untraceability schemes resulting in networks without user observability. For details about the motivation and the more practical aspects of this task cf. [6, 19, 21, 20, 18].

III. ACTIVE ATTACKS ON UNTRACEABILITY

The power of an *active* attack is based on a very simple observation: for services using two-way communication it is impossible to realize unconditional sender untraceability without unconditional recipient untraceability and vice versa.

To see this, assume that one of the participants controlled by the attacker, say P_a, communicates with some honest participant X, and that X will answer a message M by sending a message M'. If the attacker is able to identify the *sender* of M', he can identify the *recipient* of M and vice versa. If the attacker doesn't control P_a, the same is true for light traffic; then the attacker can identify both communication partners.

In general if sending and receiving is correlated (which is usually the case) the attacker can always learn something about recipients from identifying senders and vice versa.

If active attacks are possible, superposed sending doesn't guarantee recipient untraceability and therefore it doesn't guarantee sender untraceability:

Let I_i (I_i^t) be the input character which participant P_i receives (in round t) and which should always be equal to the global sum S (S^t).

Assume that the attacker is able to deliver an arbitrary character I^*_i to each participant P_i instead of the correct character I_i. This may be possible e.g. if the DC-net is implemented using a star whose centre collaborates with the attacker. Further assume that participant P_a, who is controlled by the attacker, communicates with the honest participant X according to some protocol. P_a knows that X will always answer to a received message M within a given time by sending a message M'.

If the attacker delivers message M consecutively to a single participant only, and a meaningless message to all others, he can always identify X by checking whether he receives M' or not. Instead of delivering M to a single participant only, he can deliver it to a subset of the participants. By successively partitioning the participants he can identify X in $\log(n)$ rounds, provided that the protocol between X and P_a consists of at least $\log(n)$ interactions (on average $\log(n) / 2$ interactions would suffice).

One could argue that this attack can be avoided by using networks for which it is physically more difficult to manipulate the input characters of *all* participants, e.g. trees or rings.

But even if the attacker can only manipulate what a *single* participant P_i receives, at least the unobservability of this participant is essentially decreased, since the attacker can test whether he is communicating with P_i or not.

In case of a network where the participants' stations actively forward other participant's messages (e.g. the ring implementation suggested in [4]), this attack can be performed by the neighbors of P_i without any technical manipulation. Generally, it can be performed by physically disturbing the channel of P_i, or by physically disconnecting P_i from the real network and connecting it to a simulated one with similar physical characteristics.

(If for the case of a ring network one assumes that this attack is not possible, or that observability of single participants is acceptable, i.e. that generally neither participants attack their neighbors nor attackers are able to manipulate cables between participants, then it is more efficient to use the concept of physical unobservability [16] to realize untraceability than to use superposed sending.)

If it were *guaranteed* that in all rounds $t = 1, ..., t_{max}$ each participant not controlled by the attacker receives the same input character, then superposed sending would guarantee unconditional sender and recipient untraceability in the presence of arbitrary active attacks. Such a network is called a DC^+ -net.

For an a priori given number t_{max} of rounds this is the well known problem of *reliable broadcast*. Instead of using a fixed t_{max} one can also try to limit t_{max} adaptively: if in round t two honest participants receive different characters then t_{max} is set to t; this is called *fail-stop broadcast* here.

III.1. RELIABLE BROADCAST

Reliable broadcast is defined by the following two properties [15]: in each round t
  i.   every two honest participants P_i and P_j receive the same character, i.e. $I_i^t = I_j^t$, and
  ii.   if the "sender" X is honest, then each honest participant receives the character sent by X.
If superposition of local sums is done by a central station, e.g. the centre of a star network, which delivers the global sum to all participants, only the centre has the function of a "sender". If each participant receives the local sum of each other and computes the global sum locally for himself, each participant acts as "sender".

Some types of networks, e.g. satellite networks, offer reliable broadcast without any additional protocol, but because of their bandwidth limitations they are not very usual in two-way telecommunication. Also the DC-network is meant to be usable with a variety of underlying communication networks, e.g. rings, therefore a cryptographic solution should be preferred to a physical one.

The problem of achieving reliable broadcast on a network which does not provide it automatically is also known as the Byzantine Generals problem, its solution by protocols as *Byzantine Agreement* [15, 12].

It has been proved that *information-theoretically* secure protocols for reliable broadcast exist iff the number of honest participants is greater than twice the number of dishonest participants, i.e. $|P| > 3 \cdot |A|$, and the attacker is not able to prevent communication between honest participants [12]. All protocols for information-theoretically secure reliable broadcast implicitly make use of perfect authentication codes [10, 23] and therefore require a large number of additional secret keys exchanged by the participants.

Based on the existence of secure signatures there are reliable broadcast protocols for arbitrary numbers $|A| < |P|$ [12]. An adaptive Byzantine Agreement protocol, i.e. one which withstands an attacker with $3 \cdot |A| < |P|$ or $|A| < |P|$ and A is computationally restricted, is described in [27, 28].

Because of its severe limitation $3 \cdot |A| < |P|$ reliable broadcast does not seem to be a useful technique for the desired unconditional recipient untraceability and is therefore not further considered here.

Fail-stop broadcast combines both advantages: it can be implemented in a more efficient way than reliable broadcast and it is unconditionally secure in spite of arbitrary attackers.

III.2. FAIL-STOP BROADCAST

The goal of fail-stop broadcast is to stop message transmission as soon as two honest participants receive different input characters.

If such a difference is detected by an honest participant P_i, the fail-stop can easily be performed: P_i simply disturbs the superposed sending in the subsequent rounds by choosing his outputs randomly from F instead of following Eq. (2.1). Then the global sums of all subsequent rounds are independent of the message characters.

In Section III.2.1 the most obvious, but inefficient, implementation of this idea by a comparison protocol is discussed.

In Section III.2.2 fail-stop key generation schemes are described: they generate keys for superposed sending dependent on the received input characters and ensure that two participants who have received different input characters will use completely independent keys (at least with high probability) and thus will stop message transmission.

It is shown that the most efficient key generation scheme (Sect. III.2.2.2) does not affect the performance and reliability characteristics of pure superposed sending.

III.2.1. COMPARISON OF INPUT CHARACTERS

To detect a difference, the participants can explicitly compare their input characters using an additional protocol: After each round of superposed sending each participant P_i sends his input character I_i to all participants P_j with $j > i$. If an honest participant P_j receives an input character unequal to I_j from another participant P_i, or if he receives nothing from a P_i with $i < j$, he will disturb superposed sending in all subsequent rounds.

Such test phases are well known from Byzantine Agreement protocols.

To make the tests dependable, communication between P_i and P_j should be protected by a perfect authentication scheme [10, 23], i.e. a scheme which allows the attacker to successfully forge a message with probability at the most $1 / \sqrt{|F|}$, if F is used as key space. An additional message and a secret key are therefore necessary for each test.

The number of tests necessary can be determined according to the attacker's assumed power: define G^* to be an undirected graph whose nodes are the participants. Two participants P_i and P_j are directly connected in G^* iff P_i and P_j compare their input characters. In analogy to superposed sending, the following Lemma 3.1 holds:

Lemma 3.1   Let A be the subset of participants controlled by the attacker and assume $G^* \setminus (P \times A)$ to be connected.

       If two honest participants P_i and P_j receive different input characters I_i, I_j, then there exists a pair of honest participants $P_{i'}$ and $P_{j'}$ who are directly connected in G^* and who also receive different input characters.

       Hence either $P_{i'}$ or $P_{j'}$ detects the difference and disturbs superposed sending.

Proof. Because of the connectivity of $G^* \setminus (P \times A)$ there exists a path $(P_i = P_{k_1}, ..., P_{k_m} = P_j)$ with $P_{k_z} \notin A$ and $(P_{k_z}, P_{k_{z+1}}) \in G^* \setminus (P \times A)$. It is assumed that $I_i \neq I_j$, hence there exists an index z such that $I_{k_z} \neq I_{k_{z+1}}$. Choose $(i', j') = (k_z, k_{z+1})$. $\square$

Obviously the connectivity of $G^* \setminus (P \times A)$ is a necessary condition.

The scheme requires $|G^*|$ additional messages in each round, which is usually in the order of $O(n^2)$. If $G = G^*$, and if it is assumed that for each test message the authentication scheme requires a key chosen from F, the number of privately exchanged keys is increased by a factor of two in comparison with pure superposed sending.

In a physical broadcast environment, the number of test messages can be reduced to $O(n)$ broadcast messages by using a digital signature scheme [8, 11] instead of an authentication scheme. But this results in scheme which is computationally secure only.

III.2.2. MESSAGE DEPENDENT KEY GENERATION

III.2.2.1. DETERMINISTIC FAIL-STOP KEY GENERATION

A more efficient realization of fail-stop broadcast is obtained by combining the tasks of detecting differences and stopping the network: if the keys K_{ij} and K_{ji} used for superposed sending depend completely (but not exclusively) on the characters received by P_i and P_j, then a difference between I_i and I_j will automatically disturb superposed sending, thereby stopping message transmission.

Define $\delta_{ij}{}^t := K_{ij}{}^t \oplus -K_{ji}{}^t$ and $\varepsilon_{ij}{}^t := I_i{}^t \oplus -I_j{}^t$ for all i, j, t. A key generation scheme for superposed sending is required which guarantees for all P_i and P_j directly connected in G:

SS   *Superposed sending:* If for all rounds $s = 1, ..., t-1$ the equation $I_i{}^s = I_j{}^s$ holds, then the keys $K_{ij}{}^t$ and $K_{ji}{}^t$ for round t are *equal* and randomly selected from F. More formally:

$$[\forall\, s \in \{1, ..., t-1\}: \varepsilon_{ij}{}^s = 0] \Rightarrow K_{ij}{}^t \in_R F \text{ and } \delta_{ij}{}^t = 0$$

Then superposed sending works as usual.

FS   *Fail-stop:* If there exists an index $s < t$ with $I_i{}^s \neq I_j{}^s$, then the keys $K_{ij}{}^t$ and $K_{ji}{}^t$ for round t are *independently* and randomly selected from F. More formally:

$$[\exists\, s \in \{1, ..., t-1\}: \varepsilon_{ij}{}^s \neq 0] \Rightarrow K_{ij}{}^t \in_R F \text{ and } \delta_{ij}{}^t \in_R F$$

Superposed sending is disturbed by any such pair, i.e. the global sum is independent of the message characters sent. Because of the connectivity of $G \setminus (A \times P)$ this realizes the fail-stop property according to Lemma 3.1 (with $G = G^*$).

In the rest of Section III.2.2 an arbitrary, but fixed, key pair (K_{ij}, K_{ji}) with $P_i \notin A$ and $P_j \notin A$ is considered. Therefore indices i, j are often omitted.

The most powerful attacker is assumed: he is able to observe the values of $K_{ij}{}^t$ and $K_{ji}{}^t$ for each round t directly and he can deliver arbitrary input characters $I_i{}^t$ and $I_j{}^t$ to P_i and P_j. Participants P_i and P_j are assumed to by unsynchronized, hence the attacker can wait for $K_{ij}{}^{t+1}$ before he delivers $I_j{}^t$ to P_j.

Let $(F, +, \bullet)$ be a finite field and let $a^1, a^2, ..., a^{tmax}$ and $b^1, b^2, ..., b^{tmax-1}$ be two sequences whose elements are randomly selected from F and privately exchanged by P_i and P_j. Define for $t = 1, ..., t_{max}$

$$K_{ij}{}^t := a^t + \sum_{k=1}^{t-1} b^{t-k} \bullet I_i{}^k$$

$$K_{ji}{}^t := a^t + \sum_{k=1}^{t-1} b^{t-k} \bullet I_j{}^k$$

(3.1)

Lemma 3.2   The key generation scheme defined by Equation (3.1) satisfies the two conditions SS and FS formulated above.

Proof. Since $a^t \in_R F$, and since $\Sigma := K_{ij}{}^t - a^t$ is independent of a^t, $K_{ij}{}^t \in_R F$.

Assume $\varepsilon_{ij}{}^s = 0$ for all $s < t$. Then obviously $\delta_{ij}{}^t = 0$ and condition SS is satisfied.

Now assume that s is the first round with $\varepsilon_{ij}{}^s \neq 0$. For simplicity let $\varepsilon^u := \varepsilon_{ij}{}^u$ and $\delta^u := \delta_{ij}{}^u$. The differences δ^u are formed according to the following system of linear equations:

$$\delta^u = 0 \text{ for } u = 1, ..., s$$

$$
\begin{pmatrix} \delta^{s+1} \\ \delta^{s+2} \\ \dots \\ \delta^{t-1} \\ \delta^{t} \end{pmatrix} = \begin{pmatrix} \varepsilon^s & 0 & \dots & 0 & 0 \\ \varepsilon^{s+1} & \varepsilon^s & \dots & 0 & 0 \\ \dots & \dots & \dots & \dots & \dots \\ \varepsilon^{t-2} & \varepsilon^{t-3} & \dots & \varepsilon^s & 0 \\ \varepsilon^{t-1} & \varepsilon^{t-2} & \dots & \varepsilon^{s+1} & \varepsilon^s \end{pmatrix} \cdot \begin{pmatrix} b^1 \\ b^2 \\ \dots \\ b^{t-s-1} \\ b^{t-s} \end{pmatrix}
$$

Since $\varepsilon^s \neq 0$, the matrix is regular and defines a bijective mapping. Since all $b^u \in_R F$, all δ^u are uniformly and independently distributed in F. The independence of all $K_{ij}^1, \dots, K_{ij}^t$ and $\delta^{s+1}, \dots, \delta^t$ follows from the independence of all $a^1, \dots, a^t$ and $\delta^{s+1}, \dots, \delta^t$. $\square$

The additional expenditure of this key generation scheme is given by
- the $2 \cdot t_{max} - 1$ privately exchanged keys a^t, b^t for each pair P_i, P_j directly connected in G (instead of only t_{max} for pure superposed sending),
- the storage of all $t_{max} - 1$ received input characters, and
- the $(t-1)$ field additions and multiplications for computing the key for round t.

From the last fact it follows that the scheme requires an average of $t_{max} / 2$ field additions and multiplications per round. Hence the scheme does not seem to be very practical.

Given the assumption that there is no additional communication between P_i and P_j about their current states the scheme is *optimal* with respect to the number of exchanged keys and additional storage requirements.

Lemma 3.3   The key generation scheme defined by Equation (3.1) is optimal with respect to the number of exchanged keys and additional storage requirements, i.e. each key generation scheme which *deterministically* satisfies conditions SS and FS requires at least
- the storage of all $t_{max} - 1$ received input characters and
- $2 \cdot t_{max} - 1$ privately exchanged keys.

Proof. The first limit is obvious: the scheme has to distinguish between all possible sequences of t_{max} input characters, hence all input characters have to be stored.

For proving the second limit, let Z be the secret key shared by P_i and P_j and used for generating the keys $K_{ij}^t, K_{ji}^t, t = 1, \dots, t_{max}$. Let $H(I_i), H(I_j), H(K_{ij}^1), H(K_{ij}^{(2,t_{max})}), H(K_{ji}^{(2,t_{max})}), H(Z)$ be the entropy of the random variables $I_i = (I_i^1, \dots, I_i^{tmax}), I_j = (I_j^1, \dots, I_j^{tmax}), K_{ij}^1 (=K_{ji}^1), K_{ij}^{(2,t_{max})} = (K_{ij}^2, \dots, K_{ij}^{tmax}), K_{ji}^{(2,t_{max})} = (K_{ji}^2, \dots, K_{ji}^{tmax})$, and Z, respectively [9, Chapter 2].

By applying standard rules of information theory

$$
H(K_{ij}^1 K_{ij}^{(2,t_{max})} K_{ji}^{(2,t_{max})} \mid I_i I_j) \leq H(Z K_{ij}^1 K_{ij}^{(2,t_{max})} K_{ji}^{(2,t_{max})} \mid I_i I_j)
$$

$$
= H(Z \mid I_i I_j) + H(K_{ij}^1 K_{ij}^{(2,t_{max})} K_{ji}^{(2,t_{max})} \mid Z I_i I_j)
$$

Since Z is chosen independently of the attacker's input characters $H(Z \mid I_i I_j) = H(Z)$, and since the keys are completely determined by Z and $I_i, I_j, H(K_{ij}^1 K_{ij}^{(2,t_{max})} K_{ji}^{(2,t_{max})} \mid Z I_i I_j) = 0$.
Hence it follows

$$
H(Z) \geq H(K_{ij}^1 K_{ij}^{(2,t_{max})} K_{ji}^{(2,t_{max})} \mid I_i I_j)
$$

Since only a lower bound is proved, it can be assumed that the attacker chooses I_i^1 and I_j^1 differently. Then the keys K_{ij}^1, and K_{ij}^t, K_{ji}^s for $t, s = 2, \dots, t_{max}$ are independently chosen, i.e.

$$
H(K_{ij}^1 K_{ij}^{(2,t_{max})} K_{ji}^{(2,t_{max})} \mid I_i I_j) = H(K_{ij}^1 K_{ij}^{(2,t_{max})} K_{ji}^{(2,t_{max})})
$$

$$
= H(K_{ij}^1) + H(K_{ij}^{(2,t_{max})}) + H(K_{ji}^{(2,t_{max})})
$$

Hence

$$
H(Z) \geq H(K_{ij}^1) + H(K_{ij}^{(2,t_{max})}) + H(K_{ji}^{(2,t_{max})})
$$

i.e. Z must consist of at least $1 + (t_{max}-1) + (t_{max}-1) = 2 \cdot t_{max} - 1$ keys. $\square$

III.2.2.2. PROBABILISTIC FAIL-STOP KEY GENERATION

To get a more efficient key generation scheme, it seems necessary to switch to a probabilistic version of FS: For a given fail-stop mechanism, let $Prob_A$ be the attacker's *probability of success*. The attacker is successful if, in spite of choosing $I_i^s \neq I_j^s$ for a $s < t_{max}$, there exists an index t, $s < t \leq t_{max}$, such that the global sum S^t and the message characters M_i^t, $i = 1, ..., n$, are *not* independent.

For each $d \in \mathbb{N}$ define

    FS_d If two honest participants receive two different input characters in round t, they will disturb superposed sending for the following d rounds.

The maximum number d for which FS_d is satisfied is a random variable with probability distribution $Prob(d)$.

Let $a^1, a^2, ..., a^{tmax}, b^3, b^4, ..., b^{tmax}, e$ be randomly and privately selected elements of the finite field F. Let $b^1 = b^2 = 0$ and let $K_{ij}^0 = K_{ji}^0 = 0$ and $I_i^0 = I_j^0 = 0$. Then define for $t = 1, ..., t_{max}$

$$K_{ij}^t := a^t + b^t \cdot K_{ij}^{t-1} + e \cdot I_i^{t-1}$$

$$K_{ji}^t := a^t + b^t \cdot K_{ji}^{t-1} + e \cdot I_j^{t-1}$$

(3.2)

Lemma 3.4   The key generation scheme defined by Equation (3.2) satisfies condition SS. The maximum number d for which FS_d is satisfied is a geometrically distributed random variable:

$$Prob(d) = \frac{1}{|F|} \cdot (1 - \frac{1}{|F|})^{d-1}$$

The attacker's probability of success is

$$Prob_A \leq 1 - (1 - \frac{1}{|F|})^{tmax}$$

Proof. Since $a^t \in_R F$, and since $\Sigma := K_{ij}^t - a^t$ is independent of a^t, $K_{ij}^t \in_R F$.

Assume $\varepsilon_{ij}^s = 0$ for all $s < t$. Then obviously $\delta_{ij}^t = 0$ and condition SS is satisfied.

Now assume that s is the first round with $\varepsilon_{ij}^s \neq 0$. For simplicity let $\varepsilon^v := \varepsilon_{ij}^v$ and $\delta^v := \delta_{ij}^v$.

    In the next round $\delta^{s+1} = e \cdot \varepsilon^s$. Since $\delta^v = 0$ for all $v \leq s$ the attacker has no information about the actual value of e before round $s+1$. By assumption $\varepsilon^s \neq 0$, hence δ^{s+1} is uniformly distributed in F.

    Now consider the rounds $s + u + 1$ with $u \geq 1$. If $\delta^{s+u} = 0$, then $\delta^{s+u+1} = e \cdot \varepsilon^{s+u}$. From round $s+1$ the attacker knows the value of e, hence δ^{s+u+1} is *not* independently distributed in F. If $\delta^{s+u} \neq 0$, then $\delta^{s+u+1} = b^{s+u+1} \cdot \delta^{s+u} + e \cdot \varepsilon^{s+u}$. Since b^{s+u+1} is uniformly distributed in F, δ^{s+u+1} is uniformly distributed, too, and since b^{s+u+1} is only used in that round, δ^{s+u+1} is independent of all other δs.

    Therefore the actual value of d is given by the lowest value $d \geq 1$ for which $\delta^{s+d} = 0$. Since δ^{s+1} is uniformly distributed,

$$Prob(\delta^{s+1} \neq 0) = 1 - \frac{1}{|F|}$$

and since for $\delta^{s+d} \neq 0$, δ^{s+d+1} is uniformly distributed,

$$Prob(\delta^{s+d+1} \neq 0 \mid \delta^{s+d} \neq 0) = 1 - \frac{1}{|F|}$$

From this it follows

$$Prob(d) = \frac{1}{|F|} \cdot (1 - \frac{1}{|F|})^{d-1}$$

The independence of all $K_{ij}^1, ..., K_{ij}^t$ and $\delta^{s+1}, ..., \delta^d$ follows from the independence of all $a^1, ..., a^t$ and $\delta^{s+1}, ..., \delta^d$.

The probability of success is simply the probability that $s + d \leq t_{max}$:

$$Prob_A = Prob(d \leq t_{max} - s)$$

Since $s \geq 0$,

$$Prob_A \leq Prob(d \leq t_{max}) = 1 - Prob(d > t_{max}) = 1 - (1 - \frac{1}{|F|})^{t_{max}}$$

$\square$

Since d is geometrically distributed, the average value of d is $|F|$ [25 p. 579]. Hence $|F|$ must be chosen considerably larger than t_{max}.

Corollary. Assume the key generation scheme of Eq. (3.2). Then

$$Prob_A \leq 1 - (\frac{1}{4})^{t_{max} / |F|}$$

Proof. From Lemma 3.4 it follows

$$Prob_A \leq 1 - (1-\frac{1}{|F|})^{t_{max}} = 1 - (1 - \frac{1}{|F|})^{|F| \cdot t_{max} / |F|}$$

The sequence $(1 - \frac{1}{x})^x$ increases monotonously. Since $|F| \geq 2$

$$Prob_A \leq 1 - (\frac{1}{4})^{t_{max} / |F|}$$

$\square$

Obviously with a decreasing value of $t_{max} / |F|$ the probability $Prob_A$ vanishes. From the corollary it follows for each $0 \leq L < 1$

$$\frac{t_{max}}{|F|} \leq \frac{1}{2} \cdot ld(\frac{1}{1-L}) \quad \Rightarrow \quad Prob_A \leq L$$

E.g. for $L = 10^{-9}$

$$\frac{t_{max}}{|F|} \leq 7 \cdot 10^{-10}$$

is sufficient, which is satisfied e.g. by $|F| = 2^{108}$ and $t_{max} = 10^{23}$. These values allow the transmission of

$$t_{max} \cdot ld(|F|) = 10^{23} \cdot 108 \text{ bits} \approx 10^{25} \text{ bits}$$

For a transmission speed of 10^{15} bits/s (which is far beyond today's technology) this would be sufficient for about 317 years.

The key generation of Eq. (3.2) requires as many privately exchanged keys as the scheme defined by Eq. (3.1), i.e. $2 \cdot t_{max} - 1$.

To evaluate Eq. (3.2) for round t, it is only necessary to store the last key, K_{ij}^{t-1} (in contrast to the last $t-1$ keys for Eq. (3.1)) and to perform 2 field additions and multiplications. In contrast to the scheme of Eq. (3.1), only large fields are suitable.

III.2.2.3. COMBINATION OF KEY GENERATION AND EXPLICIT TESTS

If the multi-access protocol guarantees that for some slots only one participant is allowed to choose a nonzero message, this participant can test the network:

Assume that superposed sending is stopped after a broadcast inconsistency by one of the key generation schemes described above, i.e. the global sums are randomly distributed. Then each participant P_i who is allowed to use a slot exclusively, and sends a message randomly selected from F^c, will receive a wrong message with probability $1 - |F|^{-c}$. Thus he detects the disturbance with the same probability and can explicitly stop superposed sending by choosing his following output characters randomly from F instead of according to Eq. (2.1).

If it is guaranteed that each participant sends a test message within a fixed number s of slots, and if there are at least two honest participants, this makes it unnecessary to consider more than the last $(s-1) \cdot c$ input characters for key generation: after $s-1$ slots, superposed sending will be explicitly disturbed with high probability by some honest participant who received a disturbed test message instead of the one he sent.

The required fairness of the multi-access protocol can deterministically be satisfied by superposed receiving and in a probabilistic sense by each reservation technique (Sect. II.2). If e.g. each participant reserves exactly one test message and at the most one real message in each reservation phase, each participant tests the network within $s = 4 \cdot n$ slots.

Obviously the fairness of this can only be guaranteed if all participants behave fairly, i.e. each unfair (and therefore dishonest) participant can prevent some honest participants from successfully doing their required reservation. Therefore each honest participant who cannot send a message within s slots should disturb superposed sending.

The additional rules do not help the attacker: Assume that an honest participant P_i detects a disturbance, i.e. $I_i^t \neq M_i^t$, and stops sending. Nevertheless the attacker is not able to observe the sending of P_i.

If the disturbance detected by P_i was a consequence of a previous broadcast inconsistency, the sending was stopped anyway, hence there is nothing to show. Otherwise, and if all honest participants receive the same input character, the unobservability of P_i follows from Lemma 2.2, and if the attacker manipulates the broadcast property for round t, sending is stopped by the key generation scheme anyway, independent of P_i's test.

Proper modifications to the key generation schemes will be discussed in the following two sections.

The advantages and disadvantages of the combination are the same in both schemes:

- For key generation, the parameter t_{max} is replaced by $(s-1) \cdot c$, which decreases the number of additional secret keys from t_{max} to $(s-1) \cdot c$, and for deterministic key generation the computation complexity from $O(t_{max}^2)$ to $O(s^2 \cdot c^2)$ operations and from $O(t_{max})$ to $O(s \cdot c)$ required storage.
- Some honest participants may be forced to send meaningless test messages, thus the throughput of the DC-net is decreased. The number of additional test messages depends on the participants' sending rates.

III.2.2.3.1. COMBINATION OF DETERMINISTIC KEY GENERATION AND EXPLICIT TESTS

Assume that the deterministic scheme of Eq. (3.1) is used in combination with explicit tests.

If round u is the first disturbed round, the attacker has no information about the privately exchanged keys b^v, $v = 1, \ldots, u$. After round $u + (s-1) \cdot c$, it is highly probable that the DC^+-net will be disturbed by at least one honest participant who has detected the disturbance. Hence instead of $t_{max} - 1$ additional keys, a maximum of $(s-1) \cdot c$ are really necessary:

$$K_{ij}{}^t := a^t + \sum_{k=t-(s-1)\cdot c}^{t-1} b^{t-k} \cdot I_i{}^k$$

$$(3.3)$$

$$K_{ji}{}^t := a^t + \sum_{k=t-(s-1)\cdot c}^{t-1} b^{t-k} \cdot I_j{}^k$$

Lemma 3.5   The key generation scheme defined by Equation (3.3) satisfies condition SS. Together with the additional rules for testing and disturbing it ensures the fail-stop property in a probabilistic sense: Let h be the number of honest participants, $h \geq 2$. Then

$$Prob_A \leq \frac{1}{|F|^{c\cdot(h-1)}}$$

Proof. Since $a^t \in_R F$, and since $\Sigma := K_{ij}{}^t - a^t$ is independent of a^t, $K_{ij}{}^t$ is uniformly distributed in F. Assume $\varepsilon_{ij}{}^\mu = 0$ for all $u < t$. Then obviously $\delta_{ij}{}^t = 0$ and condition SS is satisfied.

Now assume that u is the first round with $\varepsilon_{ij}{}^\mu \neq 0$. According to Lemma 3.3 (with $t_{max} = (s-1)\cdot c-1$) the global sums of the following $(s-1)\cdot c-1$ rounds are all randomly chosen from F. Since it is assumed that during the s slots each participant tests the network, the attacker's only chance is that during the first $s-1$ slots none of the at least $h-1$ honest participants detects the disturbance. The probability that a single test doesn't detect a disturbance is $|F|^{-c}$, hence the attacker's probability is less than $|F|^{-c\cdot(h-1)}$. $\square$

The scheme requires only $(s-1) \cdot c$ additional keys instead of the $t_{max}-1$ of the key generation scheme of Section III.2.2.1.

The number of field operations per round is in the order of $(s-1) \cdot c - 1$. To avoid unnecessarily expensive field computations, $F = GF(2)$ should be chosen. Here, with $h \geq 2$, $Prob_A \leq 1 / 2^c$.

Since each of the n participants should send a message within s slots, s should be in the order of n. In this case, the scheme requires $O(n \cdot c)$ operations. For $F = GF(2)$ and therefore $c \approx -\log(Prob_A)$ this is equal to $O(n \cdot -\log(Prob_A))$.

III.2.2.3.2. COMBINATION OF PROBABILISTIC KEY GENERATION AND EXPLICIT TESTS

Assume that the probabilistic scheme of Eq. (3.2) is used in combination with explicit tests.

By the same argumentation as above it follows that instead of $t_{max} - 1$ additional keys, a maximum of $(s-1) \cdot c$ are really necessary, i.e. it is possible to use the $(s-1) \cdot c$ keys $b^0, ..., b^{(s-1)\cdot c-1}$ cyclically:

Let $a^1, ..., a^{t_{max}}, e, b^0, ..., b^{(s-1)\cdot c-1}$ be randomly chosen keys. Then

$$K_{ij}{}^t := a^t + b^{t \bmod (s-1)\cdot c} \cdot K_{ij}{}^{t-1} + e \cdot I_i{}^{t-1}$$

$$(3.4)$$

$$K_{ji}{}^t := a^t + b^{t \bmod (s-1)\cdot c} \cdot K_{ji}{}^{t-1} + e \cdot I_j{}^{t-1}$$

Lemma 3.6   The key generation scheme defined by Equation (3.4) satisfies condition SS. Together with the additional rules for testing and disturbing it ensures the fail-stop property in a probabilistic sense:

$$Prob_A \leq 1 - (1 - \frac{1}{|F|})^{(s-1)\cdot c}$$

Proof. The first part is proved as in Lemma 3.4. The worst case for the second part, i.e. the best case for an attacker, is that out of all testing participants only the last two are honest. Then the attacker is unsuccessful iff the actual value of d (defined as for Eq. (3.2)) is greater than $(s\text{-}2) \cdot c$, and the test detects the disturbance. Hence

$$Prob_A \le 1 - \sum_{j=1}^{c-1} \left(Prob(d=(s\text{-}2)\cdot c+j) \cdot (1 - \frac{1}{|F|^j}) \right) - Prob(d \ge (s\text{-}1)\cdot c) \cdot (1 - \frac{1}{|F|^c})$$

$$\le 1 - Prob(d \ge (s\text{-}1)\cdot c) \cdot (1 - \frac{1}{|F|^c})$$

$$= 1 - (1 - \frac{1}{|F|})^{(s\text{-}1)\cdot c\text{-}1} \cdot (1 - \frac{1}{|F|^c})$$

$$\le 1 - (1 - \frac{1}{|F|})^{(s\text{-}1)\cdot c}$$

☐

Again only large fields F (e.g. $ld(|F|) \approx 150$) are suitable.

IV. FINAL REMARKS

Superposed sending together with one of the discussed fail-stop key generation schemes (Sect. III.2.2) guarantees the desired unconditional sender and recipient untraceability.

If one tries to transform this nice theoretical result into a real communication network, a lot of practical problems must be solved, but none of them becomes really harder if fail-stop broadcast is used in addition to normal superposed sending.

For this, consider the performance of superposed sending measured by
- the number of exchanged keys per message transmitted,
- its communication complexity,
- its computational complexity,
- and the reliability of the scheme.

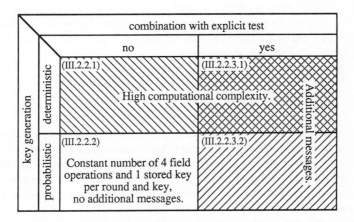

Figure 2   Comparison of fail-stop key generation schemes

The *number of additional keys* is increased by a factor of two at the most. This was shown to be the optimal value for deterministic key generation schemes without explicit tests. In theory this seems to be acceptable, and in practice pseudorandomly generated keys one will mostly be chosen anyway (and as a result of this, unconditional untraceability will be lost).

Communication complexity (Fig. 2). None of the pure key generation schemes (Sect. III.2.2.1, III.2.2.2) requires additional messages to be sent.

If combinations of key generation and explicit tests (Sect. III.2.2.3) are used, some honest participants may-be forced to send meaningless test messages. The number of additional test messages depends on the participants' sending and testing rates. If real messages are encrypted end-to-end, they appear to be randomly selected from F^c, i.e. they can be used instead of explicit test messages.

Computational complexity (Fig. 2). The key generation requires some additional time and memory for each exchanged key. For that reason the schemes with deterministic key generation (Sect. III.2.2.1, III.2.2.3.1) seem to be less practical, but if one uses one of the schemes with probabilistic key generation (Sect. III.2.2.2, III.2.2.3.2), the computation requires only the storage of the last key and two field additions and multiplications per round and exchanged key.

All schemes except that of Sect. III.2.2.1 realize *probabilistic* untraceability only, i.e. there is a small probability that an attacker will successfully transmit different messages to different participants. But all four schemes don't rely on any unproved assumptions.

For probabilistic key generation (Sect. III.2.2.2, III.2.2.3.2) only large fields F are suitable, but this is no hard restriction:

- Usually the cardinality $|F|^c$ of the set of all transmission units "message" will be relatively large. It doesn't matter whether one uses a small field and a large c or a large field and a small c.
- The reservation map technique and (reservation by) superposed receiving (Sect. II.2) require a large cyclic group $(F,\oplus)$, anyway. It is important to notice that the group $(F,\oplus)$ used for superposed sending need not be the additive (or multiplicative) group of the finite field $(F,+,\cdot)$ used for key generation. E.g. one can use the field $F = GF(2^m)$ for key generation and, by interpreting the elements of $GF(2^m)$ as binary encoded integers, the additive group of integers modulo 2^m for superposed sending.

The *transmission delay* introduced by key generation could be decreased by parallelizing the key generation for different rounds. This can be done in two ways.

One can use $k > 1$ DC$^+$-nets, say DC$^+_0$, ..., DC$^+_{k-1}$, in a time division technique, i.e. in round t the DC$^+$-net DC$^+_{t \bmod k}$ is used. To preserve untraceability, each interaction between participants should be completely performed using only a single DC$^+$-net, i.e. each participant should answer a message only by that DC$^+$-net by which he has received the message.

The other possibility is to use just one DC$^+$-net, but to make the keys for round t dependent not on the directly preceding rounds $t-i$, $i = 1, 2, ..., t-1$, but on the rounds $t-i$, $i = k, k+1, ..., t-1$ for a $k > 1$. To preserve untraceability, each participant has to wait at least $k-1$ rounds before he answers to a received character.

Naturally the fail-stop property decreases the *reliability* of the network, since every inconsistent broadcast will immediately stop the network independent of whether it was caused by an attacker or a physical fault. But most transient faults in a network can be tolerated by usual data link protocols [24 Chapter 4], and if a permanent fault occurs (e.g. if a participant's station is damaged or all links between two participants are cut), superposed sending is disturbed and the network is stopped anyway. Therefore reliability is not essentially reduced by the discussed fail-stop schemes.

The problem of combining untraceability and *serviceability* in spite of active attacks is discussed in [27, 28].

Hence, the pure probabilistic key generation scheme (Sect. III.2.2.2) with an appropriately large field F seems to be the most practical choice.

ACKNOWLEDGEMENTS

I'm pleased to thank Birgit Baum-Waidner and Andreas Pfitzmann for lots of stimulating discussions and Birgit Pfitzmann for her efficacious support, and I'm grateful to Manfred Böttger, Klaus Echtle, and Tina Johnson for many valuable suggestions, and the German Science Foundation (DFG) for financial support.

The results presented here are also contained in [27], which was written in cooperation with Birgit Pfitzmann.

REFERENCES

[1]   M. Ben-Or, S. Goldwasser, A. Wigderson: Completeness theorems for non-cryptographic fault-tolerant distributed computation; 20th STOC, ACM, New York 1988, 1-10.

[2]   J. Bos, B. den Boer: Detection of Disrupters in the DC Protocol; Abstracts of Eurocrypt '89.

[3]   D. Chaum: Untraceable Electronic Mail, Return Addresses, and Digital Pseudonyms; CACM 24/2 (1981) 84-88.

[4]   D. Chaum: The Dining Cryptographers Problem: Unconditional Sender and Recipient Untraceability; J. of Cryptology 1/1 (1988) 65-75 (draft received May 13, 1985).

[5]   D. Chaum: The Spymasters Double-Agent Problem – Multiparty Computations Secure Unconditionally from Minorities and Cryptographically from Majorities; rec. July 21, 1989, announced for Crypto '89.

[6]   D. Chaum: Security without Identification: Transaction Systems to make Big Brother Obsolete; CACM 28/10 (1985) 1030-1044.

[7]   D. Chaum, C. Crépeau, I. Damgård: Multiparty unconditional secure protocols; 20th STOC, ACM, NY 1988, 11-19.

[8]   W. Diffie, M. E. Hellman: New Directions in Cryptography; IEEE Trans. on Information Theory 22/6 (1976) 644-654.

[9]   R. G. Gallager: Information Theory and Reliable Communication; John Wiley & Sons, New York 1968.

[10] E. N. Gilbert, F. J. Mac Williams, N. J. A. Sloane: Codes which detect deception; BSTJ 53/3 (1974) 405-424.

[11] S. Goldwasser, S. Micali, R. L. Rivest: A Digital Signature Scheme Secure Against Adaptive Chosen-Message Attacks; SIAM J. Comput. 17/2 (1988) 281-308.

[12] L. Lamport, R. Shostak, M. Pease: The Byzantine Generals Problem; ACM TOPLAS 4/3 (1982) 382-401.

[13] E. Marchel: Leistungsbewertung von überlagerndem Empfangen bei Mehrfachzugriffsverfahren mittels Kollisions-auflösung; Diplomarbeit am Institut für Rechnerentwurf und Fehlertoleranz, Universität Karlsruhe 1988.

[14] J. L. Massey: Collision-Resolution Algorithms and Random-Access Communications; Multi-User Communication Systems; Edited by G. Longo; CISM Courses and Lectures No. 265, International Centre for Mechanical Sciences; Springer-Verlag, Wien 1981, 73-137.

[15] M. Pease, R. Shostak, L. Lamport: Reaching Agreement in the Presence of Faults; JACM, 27/2 (1980) 228-234.

[16] A. Pfitzmann: A switched/broadcast ISDN to decrease user observability; 1984 International Zürich Seminar on Digital Communications, IEEE, 1984, 183-190.

[17] B. Pfitzmann: private communication, 1987.

[18] A. Pfitzmann: Diensteintegrierende Kommunikationsnetze mit Teilnehmer-überprüfbarem Datenschutz; Dissertation Fakultät für Informatik, Universität Karlsruhe 1989 (will be published in: IFB, Springer-Verlag, Berlin 1989).

[19] A. Pfitzmann: How to implement ISDNs without user observability - Some remarks; Fakultät für Informatik, Interner Bericht 14/85, Universität Karlsruhe 1986.

[20] A. Pfitzmann, B. Pfitzmann, M. Waidner: Datenschutz garantierende offene Kommunikationsnetze; Informatik-Spektrum 11/3 (1988) 118-142.

[21] A. Pfitzmann, M. Waidner: Networks without user observability -- design options; Eurocrypt '85, LNCS 219, Springer-Verlag, Berlin 1986, 245-253; revised version: Computers & Security 6/2 (1987) 158-166.

[22] M. O. Rabin: Probabilistic Algorithms in Finite Fields; SIAM J. Comput. 9/2 (1980) 273-280.

[23] G.J. Simmons: A Survey of Information Authentication; Proc. IEEE 76/5 (1988) 603-620.

[24] A. S. Tanenbaum: Computer Networks; 2nd ed., Prentice-Hall, Englewood Cliffs 1988.

[25] K. S. Trivedi: Probability and Statistics with Reliability, Queuing, and Computer Science Applications; Prentice-Hall, Englewood Cliffs 1982.

[26] U. V. Vazirani, V. V. Vazirani: Efficient and Secure Pseudo-Random Number Generation (extended abstract); Crypto '84, LNCS 196, Springer-Verlag, Berlin 1985, 193-202.

[27] M. Waidner, B. Pfitzmann: Unconditional Sender and Recipient Untraceability in spite of Active Attacks – Some Remarks; Fakultät für Informatik, Interner Bericht 5/89, Universität Karlsruhe 1989.

[28] M. Waidner, B. Pfitzmann: The Dining Cryptographers in the Disco: Unconditional Sender and Recipient Untraceability with Computationally Secure Serviceability; presented at Eurocrypt '89; final version: these proceedings.

Detection of Disrupters in the DC Protocol

Jurjen Bos

Bert den Boer

Centrum voor Wiskunde en Informatica, Amsterdam

Introduction

The Dining Cryptographers algorithm [Chaum 87] allows participants to send messages in a way ensuring that they cannot be traced.
Such sender anonymity, however, gives a new problem: participants can send so many messages that others may not send. We call this *disruption*. The interesting part of the problem is finding disrupters without reducing the untraceability of the other senders. A theoretical solution appears in [Chaum 87], but it is quadratic in the number of participants, and thus infeasible.
Here, a efficient solution is presented that is practical.

Overview

We will first briefly introduce the known protocol. We then discuss several different kinds of disruption and efficient ways to prevent them. Finally, we introduce a new way to solve the collision problem without retries.

The protocol

The protocol is "unconditionally untraceable". This means that there is no mathematical way to determine the sender of a message, given the outputs of all participants. (There are ways to find the sender, of course, like asking everybody whether the message was theirs.)

Keys

We assume that the messages sent are, instead of field elements [Chaum 87], elements of an abelian group (written additively). This idea was proposed by

[Pfitzmann 87]. The transmissions are divided into *rounds*. In each round a group element will be sent..

The protocol relies on the use of *keys*. A key is a uniformly distributed random group element that is known by two participants. A pre-arranged one of the participants will use the key, the other one will use the negative of the key. In the picture, we depict this with an arrow: the arrow points to the user who uses the positive value. This guarantees that the sum of all key values used is zero. Every participant then sends (effectively, broadcasts) the sum of her message with all the keys she shares. This is depicted with a gray oval. If she wants to send nothing, she should send zeroes (she broadcasts the sum of her keys only). The sum of the outputs of all participants—which can be computed from public information—is the sum of the messages of the participants.

We do not need to assume that every pair of participants shares keys. For details about the key sharing, see [Chaum 88].

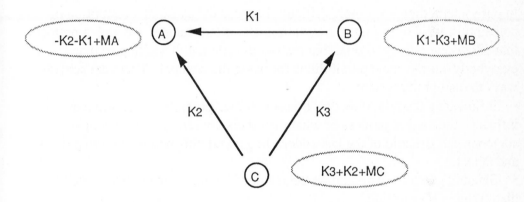

Bus system

The protocol can be viewed as a bus system. In this system, the output is accessible directly, while the input has to be coded. (This coding is the addition of the keys.)

Using this analogy, we can apply the current literature on bus systems.

There are two differences between our bus system and those currently used:

• If two participants send simultaneously (this is a *collision*), their outputs will be added. In current systems the result is just garbage. We will use this fact for making more advanced collision resolve schemes.

• It is impossible to trace a sender (this was a design goal).

Bus systems need a protocol to assign the bus fairly to senders. We will call such a system a *transmission rule*. Current bus systems use the following properties for transmission rules:
- every user must be able to send a fair amount of data;
- the bus should be used as efficiently as possible.

In our problem, the rule needs one extra feature:
- the protocol may not limit sender anonymity.

Disruption and Collisions

Disruption is the sending of messages that disturb other messages. A collision is a number of messages that disturb each other, but are sent following to the transmission rule. A lot of current systems use collisions (for example, the ALOHA protocol).

In our system, things are more difficult, because the cause of a disruption or collision is difficult to find. To get a fair use of the channel, we seek a protocol that allows everybody to send, but restrict the amount of sending so that everybody has the same possibilities for using the channel. There are several ways to disrupt the system:
- Influencing the total directly. It is sometimes possible for a participant to influence the total regardless of what the others are sending, for example if someone can get hold of the last adder. In general, this will be easy to prevent and detect.
- Sending garbage. This can be the result of a faulty sender. This kind of disruption is easy to find.
- Sending too much. There does not exist a transmission rule that allows any participant to send as many messages as she likes. The amount that can be sent by a participant must be restricted in some way. Sending more messages than is allowed by the rule is also considered disruption.

Opening

Our only weapon in the fight against disrupters is *opening*. Opening a round of the protocol means that all information passed in this round must be made public. It is enough to only open the keys, because the other information is already public. It is only possible to open a round which does not contain any private information, so we need some trick to guarantee that there are rounds to open that do not compromise privacy, but that will catch a disrupter sending.

The use of opening can be demonstrated by an example. We assume time is divided into periods, and that every sender is forbidden to send more than once in a period. We then have the following method that will sometimes catch a disrupter:
• If the number of messages in a period is greater than the number of participants, there is certainly a disrupter. He now can be caught.
• Open an anonymous voting protocol in which every participant claims a block. This reserved block is assumed to be her message and will not be opened for privacy reasons. Participants that did not send a block choose a random block to be reserved.
• At this point, there must be some unclaimed blocks left over. These can be opened without harming the privacy of non-disrupters. This will certainly yield a disrupter.

This system has the disadvantage that there can be collisions. The collisions are caused by the fact that senders send at a random moment (not to reveal their identity).

Order of computation

The total of the outputs from the participants can be computed in several ways. One of the first ways that comes to mind is to let the participants add up the total themselves, in a chain fashion:
• The first participant sends out her total.
• The later participants send out the sum of their total with the previous total.
• The output of the last participant is then the total to be broadcasted.
This method has the important drawback that the last participant completely determines the value of the total. She can use this fact to disrupt the protocol without being noticed.
Another way, which does not have this drawback, is a tree structure: The messages from two branches are added up using a stream adder to each new branch.
A second advantage of this system is that the turnaround time (the time it takes to compute the total from the bits of the participants) is only logarithmic in the number of participants.

Transmission rules

We now give an overview of several transmission rules.

Slot reservation

This is the protocol originally proposed in [Chaum 88]. The idea is to fix in advance the time somebody is going to use. It works like this:

- There is a "reservation phase" that consists of a long stream of group elements. Everybody is obliged to send exactly one 1 value in this phase.
- If this phase does not contain as many 1's as there are participants, there was a collision and the whole process is repeated again.
- The order of participants is the order in which they sent 1's.

This system has the disadvantage that participants can influence their time of sending, and that the overhead for the reservation phase is huge. The reservation phase has to have $200n^2$ bits to get a 99% probability of a collision-free reservation phase.

Collision detect

It is a good idea to try to adjust the well-known ALOHA [Davies 79] system to our case: it works very well in practice in the case of a low average load. The idea is that a participant sends whenever she has something to send. If the average load is low, this will not cause a collision in most cases, so there is no delay. If there is a collision, she must try again after some time. If we limit the amount of sending, we can catch disrupters with opening, as we showed before. In this case, collisions cause a very long delay, because of the limitation on the sending.

The collision resolve algorithm

The additive property of the system gives us the opportunity to use the result of a collision. By prepending each message with a *header*, it is possible to find out the order in which the participants have to send without any further messages.

Like in the slot reservation protocol, the size of the field must be sufficient to give a reasonable probability that the header messages are different. (For example, to have a probability of 99% that ten different headers are different, we need at least 4481 field elements).

We let participant i send as header the message $R_i, A_i, A_i^2, A_i^3, ..., A_i^k$.
Assume now that n participants collide, so that the added-up result of those n

simultaneous headers is $\sum_{i=1}^{n} R_i$, $\sum_{i=1}^{n} A_i$, $\sum_{i=1}^{n} A_i^2$, $\sum_{i=1}^{n} A_i^3$, ..., $\sum_{i=1}^{n} A_i^k$. Here we assume (without loss of generality) that the participants are numbered 1 to n.

We will write the sums of the A_i^j as $S_1, S_2, ..., S_k$.

We now need an algorithm that assigns each of the n colliding participants a unique number in the range $1...n$. We will construct the algorithm from a standard technique [Lidl 71] to find the A_i.

Observe the polynomial $P(x) = \prod_{i=1}^{n}(1-xA_i)$, which has as roots the reciprocals of the A_i. We can write this out as $\sum_{j=0}^{n} \sigma_j x^j$, with

$\sigma_0 = 1$,

$\sigma_1 = -\sum_{i=1}^{n} A_i$,

$\sigma_2 = \sum_{i=1}^{n-1} \sum_{i'=i+1}^{n} A_i A_{i'}$, and so on.

We can rewrite these equations as (*)

$S_1 - \sigma_1 = 0$,

$S_2 - \sigma_1 S_1 + 2\sigma_2 = 0$, up to

$S_n - \sigma_1 S_{n-1} + ... + (-1)^{n-1} \sigma_{n-1} S_1 + (-1)^n n\sigma_n = 0$.

In this case, the equations are easy to solve, because it is a linear triangular system. We will see later on that in the binary case it is different. Having the coefficients of the polynomial P, we can find its zeroes. Because P completely factors in monomials, and $P(0) \neq 0$, we can use that $P(x) \mid x^{p-1} - 1$. Because $x^{p-1} - 1 = (x^{\frac{p-1}{2}} + 1)(x^{\frac{p-1}{2}} - 1)$, we have $P(x) = \gcd(P, x^{\frac{p-1}{2}} + 1) \cdot \gcd(P, x^{\frac{p-1}{2}} - 1)$. We will call $P_1'(x) = \gcd(P, x^{\frac{p-1}{2}} + 1)$ and $P_2'(x) = \gcd(P, x^{\frac{p-1}{2}} - 1)$.

To now define an order on the factors of P, we define one of the two factors as the "earlier" factor, and the other as the "last" factor. Later on we will define a way to do this. If a participant wants to know the place of her value A_i, she can quickly find out in which of the two parts her factor $x - A_i$ occurs by finding out whether $P_1'(A_i)$ or $P_2'(A_i)$ is zero. If her factor occurs in the factor which is defined as the "last" factor, then the order number is the degree of the "first" factor plus the order number in the "last" factor. This means that further investingation of the "first" factor is not necessary. Repeating the process on the factor that contains $x - A_i$ yields the order number of the A_i in expected $^2\log n$ steps. We can repeat the process using the factorizations $x^p - x = ((x + it)^{\frac{p-1}{2}} + 1)((x + it)^{\frac{p-1}{2}} - 1)(x + it)$ for a prearranged t, and i increasing from 1 upwards.

We now have an algorithm that allows a participant to find the place of her A_i , but is is yet *predictable*: a sender can (approximately) influence the resulting order by choosing another value. We can solve this by letting each participant send another random value R in the header (so that the header looks like $R, A_i, A_i^2, A_i^3, ..., A_i^k$), and using the sum of those R's determine the resulting order. For example, one can use the bits of the sum of the R to determine whether $P_1'(A_i)$ or $P_2'(A_i)$ is defined to be the "earlier" factor.and also take t to be the sum of the R (assuming this sum is not zero).

The polynomials we use do not guarantee that the process will end quickly; if this is a problem, we can define an "escape rule" for those A_i that do not get separated after a prearranged number of steps. This means that participant who factor out their monomial in this number of steps will not notice the problem. The participants who need extra steps can use random choices to factor the remaing polynomial, and order the subset of the A_i just in normal increasing order. If necessary, we can then use another rule for making the order unpredictable.

The binary case

We now assume we have a field of 2^r elements. In this case, the values of S_j for even j are useless (because $S_{2j} = S_j^2$). This means that we have to send $A_i, A_i^3, A_i^5, ..., A_i^{2k-1}$ to have enough equations in (*). The resulting system for the σ_i is now harder to solve; a nice solution occurs in [Burton 71]. Factoring the polynomial is also a little bit different.

We factor the polynomial $x^{2^r} - x$ into $G(x)(G(x) + 1)$ where G is the absolute trace function [Lidl, page 170]:

$$G(x) = \sum_{i=0}^{r-1} x^{2^i} = x + x^2 + x^4 + ... + x^{2^{r-1}} . \text{ Using this factorization and}$$

succesively the factorizations $G(y^i x)(G(y^i x) + 1)$ of the polynomial $x^{2^r} - x$ for i in $1, \ldots, r\text{-}1$ (where $1, y, y^2, \ldots, y^{r-1}$ is a basis for $GF(2^r)$), we can apply the same method as above to find the position of the A_i . An advantage is that the factorization will end in a relatively short time, so that an escape rule is not necessary.

References

[Burton 71]   Herbert O. Burton: "Inversionless Decoding of Binary BCH Codes", IEEE transactions on Information Theory, Vol. IT-17, No. 4, July 1971.

[Chaum 87]   D. Chaum: "The Dining Cryptographers Problem: Unconditional sender and Recipient Untraceability". Journal of Cryptology (1988) 1:65-75.

[Davies 79]   D. W. Davies, D. L. A. Barber, W. L. Price, C. M. Solomides: "Computer Networks and their Protocols", John Wiley and Sons, 1979.

[Lidl 83]   Rudold Lidl, Harald Niederreiter: "Finite fields","Encyclopedia of mathematics and its applications", Vol. 20, section Algebra.

[Pfitzmann 87] A. Pfitzmann: "Dienseintegrierende Kommunaktionsnetze mit Teilnehmer-überprüfbarem Datenschutz", Ph. D. thesis, 1987.

Section 6

Cryptanalysis

RANDOM MAPPING STATISTICS

Philippe Flajolet
INRIA Rocquencourt
F–78150 Le Chesnay (France)

Andrew M. Odlyzko
AT&T Bell Laboratories
Murray Hill, NJ 07974 (USA)

Abstract. Random mappings from a finite set into itself are either a heuristic or an exact model for a variety of applications in random number generation, computational number theory, cryptography, and the analysis of algorithms at large. This paper introduces a general framework in which the analysis of about twenty characteristic parameters of random mappings is carried out: These parameters are studied systematically through the use of generating functions and singularity analysis. In particular, an open problem of Knuth is solved, namely that of finding the expected diameter of a random mapping. The same approach is applicable to a larger class of discrete combinatorial models and possibilities of automated analysis using symbolic manipulation systems ("computer algebra") are also briefly discussed.

1 Introduction

Random maps occur in many problems of discrete probability. Consider for instance the following assertions:

1. Throw n balls into m urns at random. Then, a proportion of about $e^{-n/m}$ of the urns will usually be empty. [Hashing].

2. A room contains 23 persons. It is a good idea (the odds are 50.7% in your favour!) to bet that two persons in the room have the same birthdate. [Birthday paradox].

3. You buy chocolate bars that contain coupons and there are n different possible coupons. Expect to buy (and possibly eat!) about $n \log n$ chocolate bars in order to obtain a full collection. [Coupon collector problem].

4. When using a middle–square random number generator (or some other "randomly" designed random number generator) operating with ℓ digits, the generator is likely to cycle after about $2^{\ell/2}$ steps. ["Random" random number generators].

5. Pollard's integer factorization algorithm is likely to find a factor of a composite integer n within $\approx n^{1/4}$ steps. [Pollard's rho–method].

6. There are n spies that attend a cryptography conference and leave their hats at the cloakroom. When the lecture is over, each spy picks up a hat at random. Then, there is a probability close to e^{-1} that nobody has his hat on his head. [Derangement problem].

These assertions are all classical. A moment's reflection shows that they convey some information on (random) functions from a finite set to a finite set. We thus let $\mathcal{F}_n^{<m>}$ denote the collection of all functions from a finite n–set domain to a finite m–set range, and use $\mathcal{F}_n \equiv \mathcal{F}_n^{<n>}$ to denote the special case where $m = n$, in which situation we merely consider an arbitrary function of a finite set into itself.

Situations where we deal with $\mathcal{F}_n^{<m>}$ are commonly known as *occupancy problems* in discrete probability theory. Models where we consider random elements of $\mathcal{F}_n$ are known as *random mappings* models.

Assertions 1 and 2 are typical of statistical properties of random elements of $\mathcal{F}_n^{<m>}$, i.e., occupancy problems. Assertion 1 is typical of a whole range of problems that present themselves when analyzing the expected performance of hashing algorithms [22]. Assertion 2 is the classical "birthday paradox" and it owes its celebrity to the rather counterintuitive low value of 23. Assertion 3 constitutes the classical "coupon collector problem". It is slightly more complicated than the earlier ones, since now m is itself a random variable in the process. However, if we look at the probability that a fixed number m of bars suffice for a full collection, it reduces to a standard statistical problem over $\mathcal{F}_n^{<m>}$.

Assertion 4 brings us closer to the subject of this paper, since it deals with the iteration structure of a finite set into itself. It is an assertion concerning $\mathcal{F}_n^{<m>}$ with $m = n = 2^\ell$. What it says in essence is that a random *mapping* $f \in \mathcal{F}_n$ will tend to "cycle" after about $\sqrt{n}$ steps. As is quite well known, this fact, combined with an idea of Floyd for testing random number generators, gave rise to Pollard's rho–method [33] for integer factoring (cf. Assertion 5). This eventually led to the factorization of the eighth Fermat number $F_8 = 2^{2^8} + 1$, see [3].

The last assertion, number 6, is related to random permutations which form a special subset of $\mathcal{F}_n$.

The model of random functions —where every function from $\mathcal{F}_n^{<m>}$ or $\mathcal{F}_n$ is taken equally likely— may be either "exact" (# 1,2,3,6) or "heuristic" in which case (# 4,5) we postulate, on the basis of simulations, that properties of a special class of functions (e.g. quadratic function models) should be asymptotically the same as properties of the class of *all* functions[1].

Our purpose here is to describe a unified framework for analyzing a number of statistical[2] properties of random mappings. A probabilistic problem to be analyzed is first specified symbolically in terms of a collection of suitable combinatorial constructions. If this specification succeeds, then combinatorial theory guarantees that *generating functions* for parameters of interest can be found. We then recover *asymptotic* information from these generating functions using *complex analysis*, and more precisely, using the local behaviour of generating functions around their singularities.

This approach is effective in analyzing a large number of "decomposable" parameters of random mappings. With it, we are able to derive in a uniform manner a number of results otherwise obtained by a variety of probabilistic or combinatorial arguments. We also demonstrate the effectiveness of our approach by solving an open problem of Knuth [23], namely that of estimating the expected diameter of random mappings.

Note. We refer to Knuth's book [23] for background information on random number generators. Random mappings are the subject of a vast collection of works; Mutafčiev's survey [26] cites 113 references! For general presentations, we direct the reader to the classic paper of Harris [19], the papers by Arney and Bender [1], and Stepanov [44]. In this area, the contribution of the "Russian school" which uses essentially probabilistic methods, as shown by Kolchin's book *Random Mappings* [24], is notable.

For completeness, we mention several recent papers not referenced in [24], namely [4, 7, 11, 20, 21, 30, 32]. In addition, there is now a growing literature on random mapping patterns, and we refer to [27] for a comprehensive list of references on this subject.

This paper is an (extended!) abstract. In particular, statements of Section 4 regarding extremal statistics should be taken as preliminary announcements of results: Several of the proofs there (Theorems 7,8) are extremely delicate and, at the time of this writing, have not appeared in full detail. The reader interested in quantitative estimates on random mappings rather than methodology can proceed directly to the self contained statements of Theorems 2–8.

---

[1] In the case of Pollard's algorithms and iteration of quadratic functions modulo integers, a notable advance is due to Bach [2] who proved recently that—in *initial* stages— quadratic functions behave asymptotically like random functions. Bach's result ultimately relies on the Weil–Deligne theorem establishing the truth of the "Riemann hypothesis" for zeta functions of algebraic curves!

[2] The term "statistics" is to be understood in the sense of discrete probability.

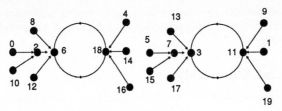

Figure 1. The functional graph associated to the map $\varphi(x) = x^2 + 2 \pmod{20}$. The functional graph comprises two connected components each containing a cycle of length 2. The function $x^2 + 2 \pmod{n}$ is one of a restricted set of polynomial functions whose iteration structure can be precisely described. For general polynomials, essentially, the only known approach is heuristic where one postulates that a polynomial behaves like a random mapping. (See however [2] for one of the very few rigorous results in this domain.)

2   Methods

Any element of $\mathcal{F}_n^{<m>}$ can be viewed as a word over an m–ary alphabet of length m. Thus, there are m^n mappings from an n–set into an m–set. Specializing this observation, we find that the cardinality of $\mathcal{F}_n \equiv \mathcal{F}_n^{<n>}$ is n^n. We are going to rederive this trivial result by means of generating functions. If $\{f_n\}_{n \geq 0}$ is a sequence of numbers, then its (exponential) *generating function* (GF) is defined to be

$$f(z) = \sum_{n \geq 0} f_n \frac{z^n}{n!}. \tag{1}$$

Proceeding in such a simple case as the enumeration of $\mathcal{F}_n$ via generating functions may seem a complicated detour. However, it has the advantage of illustrating, without unnecessary complications, a complete chain in the approach we propose to follow for appreciably harder problems. In this way, we shall be able to give a unified presentation of a number of problems otherwise treated by a variety of *ad hoc* methods.

As is well known, there are two components in the use of generating functions.

A. First, it is classical that a number of *combinatorial constructions* translate directly into generating function equations. Thus, by properly specifying a counting problem by means of these constructions, we are able to derive mechanically a collection of generating function equations that —in principle, at least— solve our problem exactly.

B. Second, the *singularities* of generating functions (now treated as analytic objects) condense most of the *asymptotic* information needed to recover their coefficients.

We refer to [18, 43] for background knowledge related to combinatorial analysis (Part A). General references for asymptotic methods can be found in [6, 29] and our approach follows closely our paper [13].

Our treatment of random mappings is based not on the direct representation of mappings by sequences of choices but instead on their decomposition as *functional graphs*.

Let φ be an element of $\mathcal{F}_n$. Consider the directed graph whose nodes are the elements $[1..n]$ and whose edges are the ordered pairs $\langle x, \varphi(x) \rangle$, for all $x \in [1..n]$. If we start from any u_0 and keep iterating φ, i.e., we consider the sequence $u_1 = \varphi(u_0), u_2 = \varphi(u_1) \ldots$, we are going to find, before n iterations, a value u_j equal to one of $u_0, u_1, \ldots, u_{j-1}$. In graphical terms, starting from any u_0, the iteration structure of φ is described by a simple path that connects to a cycle. The length of the path (measured by the number of edges) is called the *tail length* of u_0 and is denoted by $\lambda(u_0)$. The length of the cycle (measured by the number of edges or nodes) is called the *cycle length* of u_0 and is denoted by $\mu(u_0)$. We also call *rho-length* of u_0 the quantity $\rho(u_0) = \lambda(u_0) + \mu(u_0)$ which is the length of the non repeating trajectory of the point u_0.

If we now consider all possible starting points u_0, paths exhibit confluence and form into trees; these trees, grafted on cycles, form components; finally, a collection of (connected) components forms a functional graph (see Fig. 1).

2.1 Combinatorial Enumerations

Looking at Figure 1, a computer scientist could be tempted to give a description of functional graphs of the following form

```
type    FunGraph   = set(Component);
        Component  = cycle(Tree);
        Tree       = Node * set(Tree);
        Node       = Latom(1).  %Comment: atom of size 1 (labelled)
```

In other words a functional graph is a *set* of connected components; a component is a *cycle* of trees; a tree is recursively defined by appending a node to a *set* of trees; a node is a basic atomic object (of size 1), and labelled by an integer.

Let us adopt here the convention that if C is a class of combinatorial structures, then C_n (or c_n) denotes the number of elements in the class which have size n —i.e., n nodes. As seen already, we let

$$C(z) = \sum_{n \geq 0} C_n \frac{z^n}{n!}$$

denote the corresponding (exponential) generating function. Thus, we use the same letters or groups of letters to denote structures (C), counting sequences (C_n or c_n) and generating functions ($C(z)$ or $c(z)$).

Recent formalization of the process of combinatorial counting (see e.g., [18]) offers the possibility of translating directly specifications of the type above into generating function equations. Here, they provide for the collection of equations:

$$\begin{aligned}
FunGraph(z) &= \exp(Component(z)); \\
Component &= \log(1 - Tree(z))^{-1}; \\
Tree(z) &= Node(z) \times \exp(Tree(z)); \\
Node(z) &= z.
\end{aligned} \tag{2}$$

Comparison between the formal specification and the collection of equations reveals that we have used the translation mechanism

$$\begin{aligned}
\texttt{set} &\mapsto \exp(.) \\
\texttt{cycle} &\mapsto \log(1 - (.))^{-1} \\
* &\mapsto \times \text{ (ordinary product)}
\end{aligned} \tag{3}$$

This mechanism (3) is quite powerful and of course completely general [17]. We will not attempt here to redo the whole theory that underlies such derivations. Let us just indicate that if $\mathcal{F}$, $\mathcal{G}$ and $\mathcal{H}$ are three classes of labelled structures related by $\mathcal{F} = \mathcal{G} * \mathcal{H}$, then the corresponding counting sequences satisfy

$$f_n = \sum_{k=0}^{n} \binom{n}{k} g_k h_{n-k}.$$

In the equation above, index k selects the size of the $\mathcal{G}$ component (there are g_k possibilities for this component and h_{n-k} possibilities for the $\mathcal{H}$ component), and the binomial coefficient represents the number of ways of distributing labels $[1..n]$ between the the two components. At the GF level, this relation on coefficients gives $f(z) = g(z) \cdot h(z)$. The rule for sets, for instance, follows from similarly interpreting the expansion

$$\exp(g(z)) = 1 + \frac{1}{1!}(g(z))^1 + \frac{1}{2!}(g(z))^2 + \frac{1}{3!}(g(z))^3 + \cdots$$

as meaning that a set over $\mathcal{G}$ has either 0 or 1 or 2 or 3, etc. elements.

If we abbreviate our generating functions for FunGraph, Component and Tree by $f(z)$, $c(z)$ and $t(z)$, we obtain a more readable form of our basic set of equations (3):

$$\begin{cases} f(z) &=& e^{c(z)} \\ c(z) &=& \log \dfrac{1}{1 - t(z)} \\ t(z) &=& ze^{t(z)} \end{cases} \qquad (4)$$

which expresses generating functions of interest in terms of the implicitly defined tree function $t(z)$.

We briefly digress here to indicate how exact counting results are hidden behind such equations. Function $t(z)$ was considered by Eisenstein and Cayley (amongst others). The Lagrange inversion theorem furnishes the number of trees of size n in the form $t_n = n^{n-1}$ (Cayley's theorem); the same theorem gives the explicit expansion of $f(z) = (1 - t(z))^{-1}$, and one gets as expected $f_n = n^n$.

2.2 Asymptotic Analysis

Probabilistic problems on random mappings are usually more complicated than the plain enumeration results that we have just discussed. Fortunately fairly synthetic methods exist that also permit one to extract directly the asymptotic form of coefficients of a complicated generating function from its singularities.

These methods take their roots in the work of Darboux in the last century [29] and we shall make use here of the approach called *singularity analysis* which originates in [28] and [12], and which is exposed by us in [13].

If we first observe the asymptotic form of coefficients[3] of standard functions

$$[z^n]\frac{1}{1 - 3z} = 3^n, \qquad (5)$$

$$[z^n]\frac{1}{1 - 4z} = 4^n, \qquad (6)$$

$$[z^n]\frac{1}{\sqrt{1 - 4z}} \sim \frac{4^n}{\sqrt{\pi n}}, \qquad (7)$$

we notice from (5,6) that the location of a singularity of the function (at $\frac{1}{3}$ or $\frac{1}{4}$) determines the dominant exponential behaviour of its coefficients (as 3^n or 4^n). Comparison of (6) and (7) reveals further that a singularity of a square–root type yields a subexponential factor also of a square root type, namely $1/\sqrt{\pi n}$.

Our previous observations were based on functions with Taylor coefficients of a simple explicit form. What is of interest in our context, is that it is sufficient to determine local *asymptotic* expansions near a singularity, and such expansions can be "transferred" to coefficients in the same way as before. This is the heart of the method called singularity analysis in [13]. The precise formulation of one of the results of that paper that we shall need is as follows:

Theorem 1 (Singularity Analysis) *Let $f(z)$ be a function analytic in a domain*

$$\mathcal{D} = \{z \mid |z| \leq s_1, \ |\mathrm{Arg}(z - s)| > \frac{\pi}{2} - \eta\},$$

---

[3]We let as usual $[z^n]f(z)$ denote the coefficient of z^n in the expansion of $f(z)$

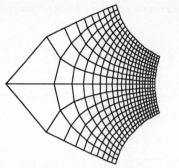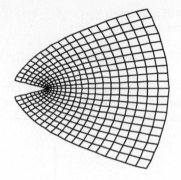

Figure 2. Two conformal representations by

$$f(z) = (1-z)^{1/2} \quad \text{and} \quad g(z) = (1-z)^{9/5}$$

of the unit square $Q = \{z = x + iy \mid -1 \le x \le +1, \ -1 \le y \le +1\}$. The two different types of singular behaviours at $z = 1$ (left "angles" on the diagrams) are reflected by different growths of coefficients, namely

$$f_n \equiv [z^n] f(z) \approx n^{-1-1/2} = n^{-3/2} \quad \text{and} \quad g_n \equiv [z^n] g(z) \approx n^{-1-9/5} = n^{-14/5}.$$

---

where s, $s_1 > s$, and η are three positive real numbers. Assume that, with $\sigma(u) = u^\alpha \log^\beta u$ and $\alpha \notin \{0, -1, -2, \ldots\}$, we have

$$f(z) \sim \sigma\left(\frac{1}{1 - z/s}\right) \qquad \text{as } z \to s \text{ in } \mathcal{D}.$$

Then, the Taylor coefficients of $f(z)$ satisfy

$$[z^n] f(z) \sim s^{-n} \frac{\sigma(n)}{n\Gamma(\alpha)}.$$

For instance, using Theorem 1, we find:

$$[z^n] \frac{e^z}{\sqrt{1 - 4z}} \sim e^{1/4} \frac{4^n}{\sqrt{\pi n}}, \tag{8}$$

$$[z^n] \sqrt{\frac{1}{4z} \log(1 - 4z)^{-1}} \frac{1}{\sqrt{1 - 4z}} \sim \sqrt{\log n} \frac{4^n}{\sqrt{\pi n}}. \tag{9}$$

To obtain the first relation (8), observe that the only singularity of $h(z) = e^z/\sqrt{1 - 4z}$ is at $z = \frac{1}{4}$, and there $h(z) \sim e^{1/4}/\sqrt{1 - 4z}$, the asymptotic form of the coefficients being then given by (7). The second relation (9) illustrates the variety of singular behaviours that can be treated by singularity analysis, and here a $\sqrt{\log}$ on the function transfers into a $\sqrt{\log}$ on the coefficients.

Random Mappings. Let us apply this technology to functions involved in the analysis of random mappings. We are then required to determine the singularities of the function $t(z)$ which determines all other functions in (4). We have seen that

$$t(z) = z \exp(t(z)), \tag{10}$$

which defines $t(z)$ implicitly.

Proposition 1 *The tree function $t(z)$ defined by (10) is analytic in the domain $\mathcal{D}$ formed by the complex plane slit along $(e^{-1}, +\infty)$. For z tending to e^{-1} in $\mathcal{D}$, $t(z)$ admits the singular expansion,*

$$t(z) = 1 - 2^{1/2}\sqrt{1 - ez} - \frac{1}{3}(1 - ez) + O((1 - ez)^{3/2}). \tag{11}$$

Proof. In fact, implicitly defined functions normally have square root type singularities. Equation (10) is a particular case of the general scheme

$$F(z, y(z)) = 0, \tag{12}$$

which determines $y(z)$ as a function of y. It is known —by the *implicit function theorem*— that, if we have a solution (z_0, y_0) of (12), then we can "continue" it in a neighbourhood of (z_0, y_0) provided that

$$\left. \frac{\partial}{\partial y} F(z, y) \right|_{\substack{z=z_0 \\ y=y_0}} \neq 0. \tag{13}$$

In other words, if $F(z_0, y_0) = 0$ and $F_y(z_0, y_0) \neq 0$, then a branch of $y(z)$ satisfies $y(z_0) = y_0$, and that branch is regular at z_0. Observe also that locally, the dependency between z and y is expressed by

$$(z - z_0) F_z(z_0, y_0) + (y - y_0) F_y(z_0, y_0) \sim 0 \tag{14}$$

corresponding, as expected, to a locally linear dependence between z and y.

In contrast, if condition (13) ceases to be valid, then the dependence between z and y assumes the form

$$(z - z_0) F_z(z_0, y_0) + \frac{1}{2}(y - y_0)^2 F_{yy}(z_0, y_0) + \text{smaller order terms} = 0. \tag{15}$$

Solving (15) for y, we thus find between z and y a square–root dependency:

$$y \sim y_0 \pm \left(2 \frac{F_z(z_0, y_0)}{F_{yy}(z_0, y_0)} \right)^{1/2} \sqrt{z_0 - z}. \tag{16}$$

The brief discussion above shows the paradigm of a singularity analysis of implicitly defined functions [10, Chap V]. The fundamental ideas, in the realm of asymptotic counting, seem to go back to Pólya, and Meir and Moon derived in this way a number of statistical properties of random trees (see e.g., [25]).

In the case of the tree function $t(z)$, we can apply our previous discussion with $F(z, y) = y - ze^y$. The singularities of $t(z)$ are thus amongst numbers z_0 which satisfy the system of two equations in two unknowns,

$$y_0 - z_0 e^{y_0} = 0 \quad \text{and} \quad 1 - z_0 e^{y_0} = 0$$

which provides $y_0 = 1$ and $z_0 = e^{-1}$. The singularity of $t(z)$ that we need to consider is thus $z = e^{-1}$; around this point, the singular expansion (11) is easily derived from the model (15,16). ∎

We can now apply singularity analysis to $t(z)$ and the functions that depend on it. By Theorem 1, considering functions $t(z), 1/(1 - t(z))$, etc., we find

$$\begin{aligned}
\frac{t_n}{n!} &= [z^n] t(z) \sim \frac{e^n}{\sqrt{2\pi n^3}} \\
\frac{c_n}{n!} &= [z^n] c(z) \sim \frac{e^n}{2n} \\
\frac{f_n}{n!} &= [z^n] f(z) \sim \frac{e^n}{\sqrt{2\pi n}}.
\end{aligned} \tag{17}$$

The result concerning f_n is expected, and by a complicated detour, we have rediscovered Stirling's formula! In view of Cayley's result that $t_n = n^{n-1}$, the first line is also equivalent to Stirling's formula. However, the asymptotic form of c_n already represents a non obvious asymptotic result.

We shall see in the next section that, once this basis has been established, many asymptotic estimates follow very easily.

3 Additive Parameters

We now follow the approach of Section 2, in order to derive expected values of several parameters of interest in the study of random mappings: First, set up generating function equations; second, analyze locally the singularities of these generating functions. We consider here *additive parameters* whose values can be determined by simple (essentially additive) rules from the structural decomposition of random mappings into functional graphs. It proves convenient to subdivide additive parameters into two classes:

direct parameters (e.g., the number of connected components) represent the number of certain distinguished configurations in mappings;

cumulative parameters (e.g., expected distance to cycle, λ) represent characteristics of mappings as seen from a random point.

Note. Estimates given in this section are essentially classical. The first results on random mappings appear to have been found in the 1950's by a variety of methods including exact enumerations, discrete probability or generating functions. The paper by Harris [19] provides a first extensive approach to problems discussed in this section. Further results are given by Stepanov [44] or Arney and Bender [1], and our presentation follows similar lines.

3.1 Direct Parameters

Let $\xi[\varphi]$ be a parameter of functional graph (or equivalently, mapping) φ, such as the number of connected components. We introduce the quantities

$$\xi_n = \sum_{\varphi \in \mathcal{F}_n} \xi[\varphi] \qquad \text{and} \qquad \Xi(z) = \sum_{n \geq 0} \xi_n \frac{z^n}{n!}, \tag{18}$$

called respectively the total value (over $\mathcal{F}_n$) and the (exponential) generating function associated to parameter ξ. Observe that, with $\mathcal{F} = \cup_n \mathcal{F}_n$, the generating function $\Xi(z)$ has the alternative form

$$\Xi(z) = \sum_{\varphi \in \mathcal{F}} \xi[\varphi] \frac{z^{|\varphi|}}{|\varphi|!}, \tag{19}$$

and the expected value of ξ taken over $\mathcal{F}_n$ is nothing but

$$\mathbf{E}\{\xi \mid \mathcal{F}_n\} = \frac{\xi_n}{n^n} = \frac{n!}{n^n} [z^n] \Xi(z). \tag{20}$$

Thus, once $\Xi(z)$ is known, the expectation analysis of ξ becomes similar to counting problems encountered earlier.

Theorem 2 (Direct Parameters) *The expectations of parameters number of components, number of cyclic points, number of terminal points, number of image points, and number of k-th iterate image points[4] in a random mapping of size n have the asymptotic forms, as $n \to \infty$,*

(i)	*# Components*	$\frac{1}{2} \log n$
(ii)	*# Cyclic nodes*	$\sqrt{\pi n / 2}$
(iii)	*# Terminal nodes*	$e^{-1} n$
(iv)	*# Image points*	$(1 - e^{-1}) n$
(v)	*# k-th iterate image points*	$(1 - \tau_k) n$,

where the τ_k satisfy the recurrence $\tau_0 = 0$, $\tau_{k+1} = e^{-1+\tau_k}$.

---

[4]In the sequel, we use the term "point" as synonym for "node". Parameter *number of components* refers to the number of connected components; a point is *cyclic* if it belongs to a cycle; x is *terminal* if it has no preimage ($\varphi^{(-1)}(x) = \emptyset$), and it is an *image point* otherwise. A k-th iterate image point of φ is an image point of the k-th iterate $\varphi^{(k)}$ of φ. Clearly, (*iii*) and (*iv*) are equivalent results, and (*iv*) is a particular case of (*v*).

Proof. *The algebra of generating functions.* Introduce temporarily bivariate generating functions which for a given parameter ξ are defined as

$$\xi(u, z) = \sum_{\varphi \in \mathcal{F}} u^{\xi[\varphi]} \frac{z^{|\varphi|}}{|\varphi|!}. \tag{21}$$

We can view variable u as "*marking*" parameter ξ. The generating function associated to (mean value of) ξ is nothing but

$$\Xi(z) = \frac{\partial}{\partial u} \xi(u, z) \Big|_{u=1}. \tag{22}$$

We shall illustrate the method of proof in cases (i), (ii) and (iii). For the number of components and number of cyclic points, we find respectively as values of $\xi(u, z)$:

$$\begin{array}{rcl}
\xi_1(u, z) &=& \exp\left(u \log \frac{1}{1-t(z)}\right) \\
\xi_2(u, z) &=& \exp\left(\log \frac{1}{1-u\,t(z)}\right).
\end{array} \tag{23}$$

For the number of terminal nodes (points without preimages), we have instead a two level scheme,

$$\left\{ \begin{array}{rcl}
\xi_3(u, z) &=& \exp\left(\log \frac{1}{1-t(u,z)}\right) \\
t(u, z) &=& z e^{t(u,z)} + (u-1)z,
\end{array} \right. \tag{24}$$

where $t(u, z)$ is the GF for trees with u marking leaves.

Eqs. (23,24) derive from a simple extension of the translation schemes of Section 2.1, introducing only an auxiliary variable u which "marks" configurations of interest.

Applying principle (22) to bivariate GF's (23,24), we find for the corresponding GF's of total values the forms

$$\begin{array}{rcl}
\Xi_1(z) &=& \dfrac{1}{1-t(z)} \cdot \log\left(\dfrac{1}{1-t(z)}\right) \\[2mm]
\Xi_2(z) &=& \dfrac{t(z)}{(1-t(z))^2} \\[2mm]
\Xi_3(z) &=& \dfrac{z}{(1-t(z))^3}.
\end{array} \tag{25}$$

The Analysis of Generating Functions. All GF's above are expressible in terms of the tree function $t(z)$. Singularity analysis as $z \to e^{-1}$ is now immediate from the discussion of Section 2.2. Consider for instance case (i) dealing with the number of components. From Eq. (11), we find *directly* for $\Xi_1(z)$ the singular expansion

$$\Xi_1(z) \sim \frac{1}{2\sqrt{2}} \frac{1}{\sqrt{1-ez}} \log \frac{1}{1-ez}. \tag{26}$$

Analytic continuation beyond the circle of convergence is guaranteed by continuation properties of $t(z)$ (cf. Section 2.2). Thus, we are justified in applying the singularity analysis theorem, and we get

$$[z^n]\, \Xi_1[z] \sim \frac{1}{2\sqrt{2}} \frac{e^n \log n}{\sqrt{\pi n}}, \tag{27}$$

from which part (i) of the theorem follows after normalization by $n!/n^n$.

Finally case (iv) is a direct variant of case (iii). Case (v) follows simply by adapting the argument used for counting terminal nodes, with the help of the GF's of trees of bounded height which we discuss in Section 4. ∎

3.2 Cumulative Parameters

We now turn to the study of random mappings in $\mathcal{F}_n$ as seen from a random point (any of the n nodes in the associated functional graph is taken equally likely). Let now $\xi[\varphi, \nu]$ be a parameter of point ν in mapping $\varphi \in \mathcal{F}$. An example of such a parameter is the distance of point ν to its cycle in φ. We introduce the quantities

$$\xi_n = \sum_{\substack{\nu \in \varphi \\ \varphi \in \mathcal{F}_n}} \xi[\varphi, \nu] \quad \text{and} \quad \Xi(z) = \sum_{n \geq 0} \xi_n \frac{z^n}{n!}, \tag{28}$$

called again total value of ξ and generating function associated with ξ. The expected value of ξ is now to be taken over the set $[1..n] \times \mathcal{F}_n$ (which has cardinality n^{n+1}) and is

$$\mathbf{E}\{\xi \mid \mathcal{F}_n\} = \frac{\xi_n}{n^{n+1}} = \frac{n!}{n^{n+1}} [z^n] \Xi(z). \tag{29}$$

Theorem 3 (Cumulative Parameter Estimates) *Seen from a random point in a random mapping of $\mathcal{F}_n$, the expectations of parameters[5] tail length, cycle length, rho–length, tree size, component size, and predecessors size have the following asymptotic forms:*

(i)	*Tail length* (λ)	$\sqrt{\pi n/8}$
(ii)	*Cycle length* (μ)	$\sqrt{\pi n/8}$
(iii)	*Rho length* $(\rho = \lambda + \mu)$	$\sqrt{\pi n/2}$
(iv)	*Tree size*	$n/3$
(v)	*Component size*	$2n/3.$
(vi)	*Predecessors size*	$\sqrt{\pi n/8}.$

Proof. We shall just give the main steps in the proof in the case of the cycle length parameter (ii).

The algebra of generating functions. The bivariate GF

$$\log \frac{1}{1 - ut(z)}$$

is a GF of connected components, where variable u marks the number of cyclic elements. If we consider

$$z \frac{\partial^2}{\partial z \partial u} \log \frac{1}{1 - ut(z)} \bigg|_{u=1}, \tag{30}$$

then we have a generating function for weighted single–component mappings where a component of size n with k cyclic points has weight $n \cdot k$.

The expression in (30) is equal to $zt'/(1-t)^2$. We then cumulate these weights over all components of random mappings; we can prove generally that this operation corresponds to multiplication of the single–component generating function by $1/(1 - t)$. Thus, the GF associated to cycle length is

$$\xi(z) = \frac{zt'(z)}{(1 - t(z))^3}.$$

The analysis of generating functions. From our basic expansion (Proposition 1 and (11)) of $t(z)$ around the singularity $z = e^{-1}$, we find that $t'(z) \sim 2^{-1/2} e(1 - ez)^{-1/2}$. Thus, we have

$$\xi(z) \sim \frac{1}{4}(1 - ez)^{-2} \quad \text{as } z \to e^{-1},$$

and the result for cycle length follows from Theorem 1.

Analogous methods can be employed to cope with the other five cases. ∎

---

[5]Tail length, cycle length and rho–length are defined at the beginning of Section 2. The *tree size* parameter of node ν means the size of the maximal tree (rooted on a cycle) containing ν; *component size* means the size of the connected component that contains ν. The *predecessors size* of ν is the size of the tree rooted at ν or equivalently the number of iterated preimages of ν.

3.3 Probability Distributions

Though this is not our main purpose here, it is also of interest to consider various characteristics of probability distributions of random mapping parameters. Variance and higher moments can be determined by the same methods as have been employed earlier in this section, though often at a higher computational cost.

Exact probability distributions in random mappings usually have (asymptotic) limit forms, a number of them, like in the case of simpler parameters of Section 3.1 being either Gaussian (with density $e^{-x^2/2}$) or Rayleigh (with density $xe^{-x^2/2}$) in the limit. Let us examine for instance the parameter number of components from Section 3.1; asymptotic normality was first derived by Stepanov [44]. First, the variance estimate is easily derived by differentiation of the function $\xi_1(u, z)$ given in (23) and we find that the standard deviation is $\sim \frac{1}{2}\log n$. In [16], the authors derive Stepanov's result as a particular case of a general law for coefficients of bivariate generating functions of the form $e^{uf(z)}$: The idea, which is applicable to several other parameters, is to extract the coefficient of z^n in the bivariate GF using singularity analysis, and taking u complex in the vicinity of 1, we estimate in this way the characteristic function of the discrete distribution of interest.

The methods we have already introduced can also be used to derive refined counting results like the number of cycles of size r (for a fixed integer r) in a random mapping.

Theorem 4 (r–configurations) *For any fixed integer r, the parameters[6] number of r-nodes, number of predecessor trees of size r, number of cycle trees of size r and number of components of size r, have the following asymptotic mean values:*

$$
\begin{array}{lll}
(i) & r\text{-nodes:} & ne^{-1}/r! \\
(ii) & r\text{-predecessor trees:} & nt_r e^{-r}/r! \\
(iii) & r\text{-cycle trees:} & (\sqrt{\pi n/2}) \cdot t_r e^{-r}/r! \\
(iv) & r\text{-cycles:} & 1/r \\
(v) & r\text{-components:} & c_r e^{-r}/r!,
\end{array}
$$

where t_r is the number of trees having r nodes, $t_r = r^{r-1}$, and $c_r = r![z^r]c(z)$ is the number of connected mappings of size r.

Proof. Generating functions result from the marking techniques of Section 3.1. For instance, the GF of functional graphs with u marking r–cycles (case (iv)) is

$$
f(u, z) = \exp\left(\log\frac{1}{1 - t(z)} + (u - 1)\frac{t(z)^r}{r}\right).
$$

Computation of the coefficients of

$$
f_u(1, z) = \frac{1}{(1 - t(z))^2}t_r\frac{z^r}{r}
$$

by singularity analysis yields the result. ∎

We thus see that node degrees in a random mapping are approximately Poisson distributed with parameter 1, a result consistent with our earlier estimate of the number of terminal nodes. The expected number of r–cycles decreases as $1/r$, a property similar to that of random permutations: For instance, a random mapping has on the average 1 fixed point. (Notice however that the implied error terms are not uniform; a random permutation has an average of $\log n$ cycles, while a random mapping has only $\frac{1}{2}\log n$.) Contour integration techniques will usually provide useful estimates when one needs to let r vary as a function of n.

---

[6]An r–node is a node of indegree r; a *cycle tree* is a tree rooted on a cycle; a *predecessor tree* is an arbitrary tree in the functional graph.

4 Extremal Statistics

The purpose of this section is to examine extremal statistics on random mappings. We consider questions which, in the perspective of random number generators are like: *"Are there good seed values that lead to long periods?"*. In particular for ξ one of the parameters discussed in Section 3.2 —λ, μ, $\rho = \lambda + \mu$, tree size or component size— we consider $\xi^{\max}$ defined by

$$\xi^{\max}[\varphi] = \max_{\nu \in \varphi} \xi[\varphi, \nu]. \tag{31}$$

The generating function approach works fairly well for these parameters. As in Section 3.1, we introduce the generating function associated with an extremal parameter $\xi^{\max}$,

$$\Xi(z) = \sum_{n \geq 0} \xi_n \frac{z^n}{n!}, \qquad \text{where} \quad \xi_n = \sum_{\varphi \in \mathcal{F}_n} \xi^{\max}[\varphi]. \tag{32}$$

Thus $n! \, n^{-n} \, [z^n] \, \Xi(z)$ represents the expectation $\mathbf{E}\{\xi^{\max} | \mathcal{F}_n\}$.

The approach to the determination of Ξ goes through a class of generating functions $f^{[k]}(z)$ where $f^{[k]}(z)$ is a "subseries" of the generating function of all functional graphs defined by

$$f^{[k]}(z) = \sum_{\varphi \in \mathcal{F}_n^{[k]}} \frac{z^{|\varphi|}}{|\varphi|!}, \qquad \text{with} \quad \mathcal{F}_n^{[k]} = \{\varphi \in \mathcal{F}_n \mid \xi^{\max}[\varphi] \leq k\}. \tag{33}$$

By a classical formula[7], Ξ is expressed in terms of the $f^{[k]}$ by

$$\Xi(z) = \sum_{k \geq 0} [f(z) - f^{[k]}(z)]. \tag{34}$$

However, the analytic treatment of the $f^{[k]}$ and of the associated sum in (34) becomes appreciably more difficult than in our earlier examples. Corresponding generating functions lead to two sorts of analytic problems:

Truncated series. We need to find uniform estimates for truncated Taylor series near their dominant singularity [μ, tree size, component size].

Singular iteration. We need to estimate uniformly the convergence of iteration schemes near a singularity of the fixed point [λ and ρ].

We distinguish two categories of parameters, longest paths (λ, μ, ρ) and largest components (trees and connected components).

4.1 Longest Paths

The case of the longest cycle in a random functional graph will serve to introduce the subject. The expectation was first determined by Purdom and Williams [35]. These authors use a result of Shepp and Lloyd [41] which is based on deep Tauberian methods and which describes the distribution of the longest cycle in a random permutation. Our derivation proceeds instead directly from generating functions using singularity analysis.

---

[7]The argument is a generating function version of the following well known formula for the mean value of a discrete random variable X:

$$\mathbf{E}\{X\} = \sum_{k \geq 1} k \Pr\{X = k\} = \sum_{k \geq 1} \Pr\{X \geq k\} = \sum_{k \geq 0} [1 - \Pr\{X \leq k\}].$$

Theorem 5 *The expectation of the maximum cycle length in a random mapping of $\mathcal{F}_n$ satisfies*

$$\mathbf{E}\{\mu^{\max}|\mathcal{F}_n\} \sim c_1\sqrt{n},$$

where $c_1 \approx 0.78248$ is given by

$$c_1 = \sqrt{\frac{\pi}{2}} \int_0^\infty [1 - e^{-E_1(v)}]\,dv,$$

and $E_1(v)$ denotes the exponential integral

$$E_1(v) = \int_v^\infty e^{-u}\frac{du}{u}.$$

Proof. (Sketch) Generating functions in this problem involve the *truncated logarithm*,

$$\ell_k(u) = \sum_{j=1}^k \frac{u^j}{j}. \tag{35}$$

Let $f^{[k]}(z)$ denote the GF of functional graphs, all of whose cycles have length at most k. Then, we have

$$f^{[k]}(z) = \exp\left(\ell_k(t(z))\right), \tag{36}$$

with $t(z)$ again the tree function.

Introduce the generating function $\Xi(z)$ associated to parameter $\mu^{\max}$ in the sense of (32). This GF is readily determined from (36):

$$\Xi(z) = \sum_{k\geq 0}[\frac{1}{1-t(z)} - f^{[k]}(z)] = \frac{1}{1-t(z)}\sum_{k\geq 1}[1 - e^{-r_k(t(z))}], \tag{37}$$

where $r_k(u)$ is the complement of the truncated logarithm,

$$r_k(u) = \log\frac{1}{1-u} - \ell_{k-1}(u) = \sum_{j\geq k}\frac{u^j}{j}.$$

The problem rests now on the determination of the asymptotic behaviour of $\Xi(z)$ as $z \to e^{-1}$. Set $t(z) = e^{-x}$. By conformal mapping properties of $t(z)$ related to its square–root singularity, when z lies in a suitable indented domain that includes the disk $|z| < e^{-1}$ (a $\mathcal{D}$ domain in the sense of Theorem 1), we have $|t(z)| < 1$ so that x lies in the half plane $\Re(x) > 0$.

The main steps for the estimation of $\Xi(z)$ are:

$$\Xi(z) = \frac{1}{1-t(z)}\sum_{k\geq 1}[1 - \exp(-\sum_{j\geq k}\frac{e^{-jx}}{j})] \tag{38}$$

$$\sim \frac{1}{1-t(z)}\sum_{k\geq 1}[1 - \exp(-\int_{kx}^\infty e^{-v}\frac{dv}{v})] \tag{39}$$

$$\sim \frac{1}{(1-t(z))^2}\int_0^\infty [1 - \exp(-\int_u^\infty e^{-v}\frac{dv}{v})]\,du. \tag{40}$$

Once Eq. (40) is established, the theorem follows immediately by singularity analysis. Now the transition from (38) to (39) results from approximating a sum by an integral, i.e. by Euler–Maclaurin summation. (It is important that we should have convergence of the integral, but this is granted since $\Re(x) > 0$, which also allows us to change the upper limit of integration from $x\infty$ to $+\infty$ using Cauchy's theorem.) The transition from (39) to (40) follows similarly by Euler summation, noting that the step x in the discrete sum (39) is $\sim 1 - t(z)$ as $z \to e^{-1}$. The only details that are omitted from this proof are the derivation of uniform error bounds. ∎

In passing, observe that the same method permits one to estimate the expected length of the longest cycle in a random permutation, thereby avoiding the delicate Tauberian arguments of [41]. Related distribution results are discussed by Stepanov in [44].

The next theorem concerns the expected value of $\lambda^{\max}$. Results concerning distribution estimates were first derived by Sachkov [39] and Proskurin [34] using multivariate probabilistic methods. The derivation that follows brings a "singular iteration problem", and the corresponding methods are also useful for $\rho^{\max}$ estimates.

Theorem 6 *The expectation of the maximum tail length $(\lambda^{\max})$ in a random mapping of $\mathcal{F}_n$ satisfies*

$$\mathbf{E}\{\lambda^{\max}|\mathcal{F}_n\} \sim c_2\sqrt{n},$$

where $c_2 \approx 1.73746$ is given by

$$c_2 = \sqrt{2\pi}\log 2.$$

Proof. (Sketch) Let $t^{[h]}(z)$ denote the GF of trees with height at most h. (Height is measured by the number of edges along a longest branch, so that a one node tree has height 0.) These generating functions are given by the recurrence

$$t^{[0]}(z) = z, \qquad t^{[h+1]}(z) = z\exp(t^{[h]}(z)). \tag{41}$$

We note that, as $h \to \infty$, we have $t^{[h]}(z) \to t(z)$, at least in the sense of convergence in the ring of formal power series. The GF for mappings with $\lambda^{\max} \leq h$ follows again from the techniques of Section 2, and it is

$$f^{[h]}(z) = \exp(\log\frac{1}{1-t^{[h]}(z)}) = \frac{1}{1-t^{[h]}(z)}. \tag{42}$$

The GF associated to $\xi^{\max} = \lambda^{\max}$ is then found by Eq. (34) to be

$$\Xi(z) = \sum_{h\geq 0}[\frac{1}{1-t(z)} - \frac{1}{1-t^{[h]}(z)}]. \tag{43}$$

The analytic problem now lies with determining the nature of the approximation of $t(z)$ by the $t^{[h]}(z)$ when z is in the vicinity of e^{-1}. This is a *singular iteration* problem. For instance, for z real, $0 < z < e^{-1}$, the convergence is geometric. On the opposite, for $z > e^{-1}$, we have a case of strong hyperexponential divergence. At exactly $z = e^{-1}$, convergence is extremely slow being of order $1/h$. In other terms, we approximate a function, $t(z)$ with an algebraic singularity (branch point), by a collection of entire functions $t^{[h]}(z)$, and we need to find uniform estimates in z and h in a neighbourhood of the singularity of the limit[8].

Approximations similar to those needed for the proof of Theorem 6 are provided by us in [12], where we analyze the expected height of random trees of various sorts in this manner (see also [38] for closely related results). Imitating the method of proof of [12], we define

$$\epsilon = \epsilon(z) = 2^{1/2}\sqrt{1-ez} \qquad \text{and} \qquad e_h(z) = t(z) - t^{[h]}(z).$$

We also introduce the "indented" domain

$$D = \{z \mid |z| \leq e^{-1}, |\text{Arg}(e^{-1} - z)| \leq \pi/2 + \delta\} \tag{44}$$

for some $\delta > 0$. The first step, whose rather involved proof we omit, is to show that in the region

$$\{z \mid |z| \leq e^{-1} + \delta, z \notin D\}$$

---

[8]It turns out that the $t^{[h]}(z)$ converge to $t(z)$ for all z in $\{z \mid z = \zeta e^{-\zeta}, |\zeta| \leq 1\}$. This follows from recent results in iteration of entire functions due to Devaney [9, 8]; however, these results do not seem to provide the necessary quantitative information we need.

we have $e_h(z)$ small and $t^{[h]}(z)$ bounded away from 1, so that $\Xi(z)$ is analytic there. Therefore, in order to apply Theorem 1, we only need to study $\Xi(z)$ for $z \in D$.

The main step in the proof of the theorem is to establish that for $z \in D$, the $e_h(z)$ are approximated by an explicit function of h and ϵ,

$$e_h(z) \approx 2\epsilon \frac{(1-\epsilon)^h}{1-(1-\epsilon)^h}. \tag{45}$$

This approximation shows both the slow convergence of $t^{[h]}(e^{-1})$ to $t(e^{-1})$ (in fact, it shows that $e_h(e^{-1}) \sim 2/h$), and the geometric convergence for $\epsilon \neq 0$. The proof of a precise form of (45) proceeds along lines similar to those of [12], although there are some additional technical complications.

We define

$$w(z) = \frac{1}{1-e^{-z}} - \frac{1}{z} - \frac{1}{2}, \tag{46}$$

so that $w(z)$ is analytic in $|z| < 2\pi$. The basic recurrence for the $t^{[h]}(z)$ shows that

$$e_{h+1}(z) = t(z)(1 - e^{-e_h(z)}), \tag{47}$$

and therefore, by normalizing and the trick of "taking inverses"[9],

$$\frac{t(z)^j}{e_j(z)} = \frac{t(z)^{j-1}}{e_{j-1}(z)} + \frac{1}{2}t(z)^{j-1} + t(z)^{j-1}w(e_{j-1}(z)), \tag{48}$$

from which it follows that

$$\frac{t(z)^h}{e_h(z)} = \frac{1}{2}\frac{1-t(z)^h}{1-t(z)} + \frac{1}{t(z)-z} + \sum_{j=0}^{h-1} t(z)^j w(e_j(z)). \tag{49}$$

Equation (49) is the basic tool used to estimate $e_h(z)$, with the first term in (49) corresponding to approximation (45). Another argument shows that if the δ in the definition (44) of D is taken to be small enough, then $|e_h(z)| < 5$, say for all $h \geq 0$ and all $z \in D$. This means that the expansion (49) holds for all $z \in D$, without singularities arising from the w function. Sharp bounds for the error in the approximation (45) for $e_h(z)$ are obtained by iterated use of (49): First the crude bound for $e_j(z)$, when inserted into (49), gives a more refined estimate which is then used to obtain an improved estimate for the sum on the right hand side of (49), yielding the final approximation result.

As an illustration, we develop the case where $z = e^{-1}$. The reduced form of (49), with $t(e^{-1}) = 1$ and $e_h \equiv e_h(e^{-1})$ reads

$$\frac{1}{e_h} = \frac{h}{2} + \frac{1}{1-e^{-1}} + \sum_{j=0}^{h-1} w(e_j). \tag{50}$$

We start "bootstrapping" with the information that $0 < e_h < 1$. Eq. (50) provides

$$\frac{h}{2} + O(1) \leq \frac{1}{e_h} \leq \frac{h}{2} + \frac{h}{10} + O(1),$$

which guarantees that $1/e_h$ is upper and lower bounded by terms that are of order h. Thus $e_h = \Theta(\frac{1}{h})$. Reinserting this information inside (50), and using the fact that $w(x) = x/12 + O(x^3)$ for small x, we get the improved estimate

$$\frac{h}{2} + O(1) \leq \frac{1}{e_h} \leq \frac{h}{2} + O(\log h),$$

---

[9]This technique amounts to comparing a non linear slowly converging iteration to a homographic recurrence, $u_{n+1} = (au_n + b)/(cu_n + d)$. A good illustration is de Bruijn's treatment of the iterates of the function $\sin(x)$ in [6, Ch. 8].

so that $e_h \sim 2/h$. Continuing in this fashion, we obtain

$$\frac{1}{e_h} \equiv \frac{1}{e_h(e^{-1})} = \frac{h}{2} + \frac{1}{12}\log h + O(1),$$

and an expansion to an arbitrary order can be generated in this way.

Returning now to $\Xi(z)$, we have

$$
\begin{aligned}
\Xi(z) &= \frac{1}{1-t(z)}\sum_{h\geq 0}\frac{e_h(z)}{1-t^{[h]}(z)} \\
&= \frac{1}{(1-t(z))^2}\sum_{h\geq 0}\frac{e_h(z)}{1+\frac{e_h(z)}{1-t(z)}}.
\end{aligned}
\tag{51}
$$

Setting, like in the preceding theorem, $t(z) = e^{-z} \sim 1 - \epsilon$, the computation develops as follows:

$$\Xi(z) \sim \frac{2}{\epsilon}\sum_{h\geq 0}\frac{(1-\epsilon)^h}{1+(1-\epsilon)^h} \tag{52}$$

$$\sim \frac{2}{\epsilon}\sum_{h\geq 0}\frac{e^{-hx}}{1-e^{-hx}} \tag{53}$$

$$\sim \frac{2}{\epsilon^2}\int_0^\infty \frac{e^{-y}}{1-e^{-y}}\,dy \tag{54}$$

$$\sim \frac{2\log 2}{\epsilon^2}. \tag{55}$$

A crucial step there consists of justifying the use of the approximation (45) inside the exact form (51) resulting in Eq. (52) or its equivalent form (53). Once (52) and (53) are established, (54) follows by Euler–Maclaurin summation. The final result (55), when subjected to singularity analysis yields the statement of the theorem. ∎

The last result in this subsection concerns the parameter $\rho^{\max}$ also called sometimes, in accordance with graph theoretic terminology, the *diameter*. It provides an answer to an open problem of Knuth ([23], Ex. 3.1.14, p. 519). As could be expected from the nature of the parameter $\rho^{\max}$, the proof combines the tools developed for Theorems 5 ($\mu^{\max}$) and 6 ($\lambda^{\max}$), and in particular it strongly relies on the estimates of $t^{[h]}(z)$ and $e_h(z)$.

Theorem 7 *The expectation of the maximum rho length ($\rho^{\max}$) in a random mapping of $\mathcal{F}_n$ satisfies*

$$\mathbf{E}\{\rho^{\max}|\mathcal{F}_n\} \sim c_3\sqrt{n},$$

where $c_3 \approx 2.4149$ is given by

$$c_3 = \sqrt{\frac{\pi}{2}}\int_0^\infty \left[1 - e^{-E_1(v)-I(v)}\right]dv,$$

with $E_1(v)$ denoting the exponential integral and

$$I(v) = \int_0^v e^{-u}[1 - \exp(\frac{-2u}{e^{v-u}-1})]\frac{du}{u}.$$

Results from the previous sections indicate that, in a random mapping, most of the points tend to be grouped together in a single giant component. This component might therefore be expected to have very tall trees and a large cycle. Thus, the inequality

$$c_3 = 2.4149... < c_1 + c_2 = 2.5199...$$

is rather interesting as it says that, with non zero asymptotic probability, the tallest tree in a functional graph is not rooted on the longest cycle.

Proof. (Sketch) Due to the intrinsically technical proof, we shall content ourselves here with a brief description of the major points of the analysis.

The generating function of functional graphs with rho–length at most k is, in accordance with (32),

$$f^{[k]}(z) = e^{v_k(z)} \qquad \text{where} \quad v_k(z) = t^{[k-1]}(z) + \frac{1}{2}(t^{[k-2]}(z))^2 + \cdots + \frac{1}{k}(t^{[0]}(z))^k, \qquad (56)$$

with $v_0(z) = 0$. This form is easily justified, since in order to build a connected component with rho–length $\leq k$, we either graft a tree of height $\leq k - 1$ on a 1 node cycle, or two trees of height at most $k - 2$ on a 2 node cycle etc. Thus the GF of $\rho^{\max}$ is

$$\Xi(z) = \sum_{k \geq 0} \left[\frac{1}{1 - t(z)} - e^{v_k(z)}\right].$$

Let now $\Xi_0(z)$ be the GF associated with the longest cycle parameter defined in (37). Several routes are conceivable. A convenient one starts by considering the difference

$$\Delta(z) = \Xi(z) - \Xi_0(z) = \sum_{k \geq 0} [e^{\ell_k(t(z))} - e^{v_k(z)}]$$

which is associated to $\rho^{\max} - \mu^{\max}$. Factoring out the quantity $e^{\ell_k(t(z))}$ in the general term, we find:

$$\Delta(z) = \frac{1}{1 - t(z)} \sum_{k \geq 0} e^{-\sum_{j>k} t(z)^j/j}[1 - e^{-w_k(z)}], \qquad (57)$$

where the w's are given by

$$
\begin{aligned}
w_k(z) &= \sum_{l=1}^{k} \frac{1}{l}[t(z)^l - (t^{[k-l]}(z))^l] \\
&= \sum_{l=1}^{k} \frac{t(z)^l}{l}\left[1 - \left(\frac{t^{[k-l]}(z)}{t(z)}\right)^l\right] \\
&= \sum_{l=1}^{k} \frac{t(z)^l}{l}\left[1 - \exp\left(-l(t(z) - t^{[k-l-1]}(z))\right)\right] \\
&= \sum_{l=1}^{k} \frac{t(z)^l}{l}\left[1 - \exp\left(-le_{k-l-1}(z)\right)\right].
\end{aligned}
\qquad (58)
$$

Taken together, the last form in (58) and Eq. (57) summarize the algebraic forms of generating functions needed for asymptotic analysis.

From this exact form, the analysis proceeds, setting again $t(z) = e^{-x}$, so that $x \sim 2^{1/2}\sqrt{1 - ez}$. We use the $\approx$ symbol to emphasize the fact that error terms are not made explicit (and may be dominant in some eventually unessential regions).

First it can be proved that the dominant terms in the sum (57) of $\Delta(z)$ are for those values of k such that $kx = \Theta(1)$.

A crucial step is to approximate $w_k(z)$. We have from (58)

$$w_k(z) \approx x \sum_{l=1}^{k} \frac{e^{-lx}}{lx}\left[1 - \exp\left(-le_{k-l-1}(z)\right)\right],$$

where, by the general approximation of (45),

$$le_{k-l-1}(z) \approx 2lx\frac{e^{-(k-l)x}}{1 - e^{-(k-l)x}}. \qquad (59)$$

We now appeal to a continuous model for these sums based on Euler–Maclaurin summation. Setting $kx = v$, $lx = u$, we derive for $w_k(z)$ the approximation

$$w_k(z) \approx \int_0^v e^{-u}\left[1 - \exp\left(-2u\frac{e^{-(v-u)}}{1 - e^{-(v-u)}}\right)\right]\frac{du}{u}$$
$$\approx I(kx).$$

(60)

Injecting this form inside the main formula (57) for $\Delta(z)$ leads us to

$$\Delta(z) \approx \frac{1}{1 - t(z)}\sum_{k\geq 0} e^{-E_1(kx)}[1 - e^{-I(kx)}],$$

which yields to a final assault of Euler Maclaurin:

$$\Delta(z) \sim \frac{1}{x^2}\int_0^\infty e^{-E_1(v)}[1 - e^{-I(v)}]\,dv.$$

(61)

There are of course considerable technical difficulties in actually organizing the proper approximations with their error terms. The form (61) combined with the information gathered in (40) regarding the GF of $\mu^{\max}$ shows that

$$\Xi(z) \sim \frac{1}{x^2}\int_0^\infty [1 - e^{-E_1(v)-I(v)}]\,dv.$$

(62)

At this stage, the result falls as a ripe fruit by singularity analysis. ∎

4.2 Largest Configurations

We consider here the analysis of the largest tree and of the largest component in a random mapping. The analysis given here will be only partial since we shall appeal to a *smoothness hypothesis* (which is intuitively clear, but harder to establish rigorously).

Generating function equations here involve series truncation operators that we have already used implicitly when dealing with longest cycle. Let $a(z) = \sum_n a_n z^n$ be a power series. We introduce two operators called *truncation* $\mathbf{T}_m$ and *remainder* $\mathbf{R}_m$ that are defined by

$$\mathbf{T}_m[a(z)] = \sum_{n\leq m} a_n z^n, \qquad \mathbf{R}_m[a(z)] = \sum_{n > m} a_n z^n.$$

(63)

Let ξ^{max} be one of the parameters of random mappings, largest tree size or largest component size. We shall say that the parameter is *smooth* if the following condition is satisfied:

$$\textit{There exists } \delta \textit{ such that } \quad \delta = \lim_{n\to\infty}\frac{1}{n}\mathbf{E}\{\xi^{\max}|\mathcal{F}_n\}.$$

(64)

If δ exits, then by standard Abelian theorems [45, Chap. 7], $\Xi(z)$ satisfies $\Xi(z) \sim \delta_1(1 - ez)^{-3/2}$ when z tends to e^{-1} along the real axis $z < e^{-1}$, for some δ_1 directly related to δ (actually $\delta_1 = 2\sqrt{2}\delta$). Thus if we find that, limited to the real line inside its circle of convergence, $\Xi(z)$ has the proper behaviour, then we are able to deduce the value of δ:

$$\delta = \frac{1}{2\sqrt{2}}\lim_{z\to e^{-1}}\Xi(z)(1 - ez)^{3/2}.$$

The smoothness assumption thus dispenses with finding local expansions in a *complex neighbourhood* of e^{-1}. The reason why we introduce it here is to bypass some intrinsic difficulty in the singular behaviour of truncated Taylor series. Indeed, Jentzsch's theorem [45, p. 238] states that, *for every power series, every point of the circle of convergence is a limit–point of zeros of partial sums.* For largest components, the generating functions $f^{[k]}$ of (33) involve truncated Taylor series and thus exhibit a very irregular behaviour on the circle $|z| = e^{-1}$. The validity of our singular expansions is then restricted to the interior of the disk of convergence $|z| < e^{-1}$. It is probable that a more refined analysis (e.g. using different integration contours for different terms in the GF $\Xi(z)$) would enable us to dispense with the smoothness condition, but this is presently not obvious.

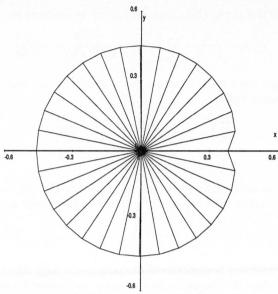

Figure 3. The star diagram of zeros of the polynomial $U_{32}(z)$, where

$$U_m(z) = 1 - \sum_{n=1}^{m} n^{n-1} \frac{z^n}{n!}.$$

This polynomial is a "truncation" of $1 - t(z)$, with $t(z)$ being the tree function, and its zeros appear in the analysis of largest tree size. In accordance with Jentzsch's theorem, the zeros tend to accumulate around the circle $|z| = e^{-1}$.

Theorem 8 *Assuming the smoothness condition, the expected value of the size of the largest tree*[10] *and the size of the largest connected component in a random mapping of $\mathcal{F}_n$ are asymptotically*

$$\begin{array}{lll} (i) & \text{Largest tree:} & d_1 n \\ (ii) & \text{Largest component:} & d_2 n, \end{array} \tag{65}$$

where $d_1 \approx 0.48$ and $d_2 \approx 0.75782$ are given by

$$d_1 = 2 \int_0^\infty \left[1 - \frac{1}{1 + \frac{1}{2\sqrt{\pi}} \int_x^\infty e^{-v} v^{-3/2}\, dv} \right] dx$$

$$d_2 = 2 \int_0^\infty \left[1 - \exp\left(\frac{1}{2} \int_x^\infty e^{-v} v^{-1}\, dv\right) \right] dx.$$

Proof. (Sketch) The generating functions associated to the two cases under discussion are respectively

$$\Xi_1(z) = \sum_{m \geq 0} \left[\frac{1}{1 - t(z)} - \frac{1}{1 - \mathbf{T}_m[t(z)]} \right]$$

$$\Xi_2(z) = \sum_{m \geq 0} \left[\frac{1}{1 - t(z)} - e^{-\mathbf{T}_m[c(z)]} \right].$$

To approximate them, we set $z = e^{-1-y}$.

---

[10]Interesting distribution properties of the size of the largest tree are discussed in [24, p. 164] and [31].

Consider the case of largest tree (Ξ_1). Then, the GF can be rewritten as

$$\Xi_1(z) = \frac{1}{1 - t(z)} \sum_{m \geq 0} \left[1 - \frac{1}{1 + \mathbf{R}_m[t(z)]/(1 - t(z))} \right]. \tag{66}$$

When m is large enough, and y small, using $1 - t(z) \sim 2^{1/2} y^{1/2}$, we get by Stirling's approximation and Euler Maclaurin summation:

$$\frac{\mathbf{R}_m[t(z)]}{1 - t(z)} \approx 2^{-1/2} y^{-1/2} \sum_{n > m} \frac{e^{-ny}}{\sqrt{2\pi n^3}}$$

$$\approx \frac{1}{2\sqrt{\pi}} \int_{my}^{\infty} e^{-v} v^{-3/2} \, dv. \tag{67}$$

The final step consists in transporting approximation (67) inside Eq. (66), and using a further step of Euler–Maclaurin summation. The derivation for maximum component size is similar. $\blacksquare$

5 Extensions

The methodology discussed here is applicable to the analysis of a large class of combinatorial structures, roughly speaking those that can be specified using the combinatorial constructions of Section 2. It is also systematic enough that some of these analyses can be automated using computer algebra systems.

5.1 Alternative Models

Harris [19] already discusses mappings without fixed points. In the context of Section 2.1, this means that the specification of functional graphs (FunGraph) has to be altered by prohibiting cycles of length equal to 1 inside components:

```
type    FunGraph   = set(Component);
        Component  = cycle(Tree,card>1);
        Tree       = Node*set(Tree);
        Node       = Latom(1);
```

It is a simple exercise to derive the modified form of Eqns. (2) in this case:

$$
\begin{aligned}
FunGraph(z) &= \exp(Component(z)); \\
Component &= \log(1 - Tree(z))^{-1} - Tree(z); \\
Tree(z) &= Node(z) \cdot \exp(Tree(z)); \\
Node(z) &= z,
\end{aligned}
\tag{68}
$$

and in the equation for $Component(z)$, we have taken out the possibility of an isolated tree on a (size 1) cycle. In other words, the equation for modified functional graphs is

$$f_1^*(z) = \frac{e^{-t(z)}}{1 - t(z)}. \tag{69}$$

Following Meir and Moon [25], Arney and Bender [1] discuss random mappings with constraints on the degrees of nodes. In fact, if we consider the functional graph attached to a quadratic transformation $\phi(x) = x^2 + c \bmod n$ for n prime, we see that, with a single exception $x = c$, all nodes have degree 0 or 2. This justifies interest in binary functional graphs, where the only (in)degrees of nodes allowed are 0 or 2. In that case, the specification only needs editing:

```
type    FunGraph    = set(Component);
        Component   = cycle(Node*BinTree);
        BinTree     = Node + Node * set(BinTree, card = 2);
        Node        = Latom(1),
```

The equation determining the GF $f_2^*(z)$ of these modified mappings becomes thus

$$f_2^*(z) = \frac{1}{1 - zb(z)}, \qquad \text{with} \quad b(z) = z + \frac{1}{2}zb^2(z). \tag{70}$$

Solving the quadratic equation for $b(z)$, we then find that

$$f_2^*(z) = \frac{1}{\sqrt{1 - 2z^2}},$$

and in particular, there are $2^{-n}(2n)!\binom{2n}{n}$ binary functional graphs of size $2n$. Algebraically, the case of general degree restrictions can be treated with comparable ease, and the corresponding analytic treatment involves the general discussion on singularity analysis of implicitly defined functions given in Section 2.2.

It is then a simple task to adjust the approach taken in earlier sections (especially Sect. 3) to such modified models. Analysis reveals that, in this case, though multiplicative constants are quite sensitive to such changes, the basic orders of growth of parameters remain essentially unaffected. An example in sharp contrast with this situation is treated in the next section as an illustration of the capabilities of an automatic analysis system.

5.2   Automatic Analysis

The methodology that we have followed in order to analyze additive parameters of random mappings is general enough, so as to make it amenable to some form of automatization. Together with B. Salvy and P. Zimmermann, the first author has developed a system named $\Lambda\Upsilon\Omega$ (Lambda–Upsilon–Omega), which takes as inputs specifications of combinatorial structures and characteristic parameters, and produces (in a number of cases) automatically the expected values of the parameters. The system makes extensive use of resources of the computer algebra system MAPLE [5].

Such an approach proves useful when analyzing complex models. A description of the current state of the system is given in [15] and it will only be illustrated by treating a *"sensitivity analysis"* problem due to Michèle Soria who discusses systematically such phenomena in her thesis [42].

The analysis below is produced automatically by the $\Lambda\Upsilon\Omega$ system. The session presents the analysis of a variant of the model of random mappings: We modify the classical definition of functional graphs by forcing all nodes on cycles to have indegree 2 exactly. In other words, we consider special functional graphs (the Sfungraph type) made of sets of cycles of special *planted* trees (Stree). The problem consists in determining to what extent essential parameters are sensitive to such a change in the model. Standard functional graphs have on average $\frac{1}{2}\log n + O(1)$ components (cycles) and $\sim \sqrt{\pi n}$ nodes on cycles. The small change in the specifications somewhat unexpectedly results in a rather drastic change of stochastic properties of these graphs.

The $\Lambda\Upsilon\Omega$ system accepts as inputs structural descriptions of "decomposable" structures in the style of Section 2 and of our earlier formal specifications. Thus, our class of special functional graphs will be specified quite naturally by:

```
type    Sfungraph   = set(Scomponent);
        Scomponent  = cycle(Stree);
        Stree       = product(Node,Tree);
        Tree        = product(Node,set(Tree));
        Node        = Latom(1);
```

The $\Lambda\Upsilon\Omega$ system is primarily designed to estimate the average case complexity of algorithms. In order to analyze parameters like the number of components, we therefore write a procedure whose *complexity* is precisely equal to the parameter to be analyzed. The second part of the input thus reads:

```
procedure number_of_components(f : Sfungraph);
begin
        forall c in f do
                count;
end;

procedure number_of_cyclic_points(f : Sfungraph);
begin
        forall c in f do
                forall d in c do
                        count;
end;

measure count:1;
```

where the last line specifies that count is a counter with constant complexity equal to 1.

Systematic translation mechanisms allow us to compile such specifications into equations over generating functions. This task is achieved by the ALGEBRAIC ANALYZER of ΛϒΩ written in the Caml language [46].

```
Counting generating functions:
    Tree(z)=Node(z)*exp(Tree(z))
    Stree(z)=Node(z)*Tree(z)
    Scomponent(z)=L(Stree(z))
    Sfungraph(z)=exp(Scomponent(z))
    Node(z)=z

Complexity descriptors:
    tau_number_of_components(z)=(exp(Scomponent(z))*1*Scomponent(z))+0
    tau_number_of_cyclic_points(z)=(exp(Scomponent(z))*1*Stree(z)/(1-Stree(z)))
```

The second batch (labelled "complexity descriptors") represents generating functions of procedures' costs. These equations are then solved by a SOLVER programme written in Maple. The solution is here expressed in terms of $L(y) \equiv \log(1-y)^{-1}$ and of Maple's W–function which is defined (implicitly) by $W(z)e^{W(z)} = z$.

```
    tau_number_of_components(z) = exp(L(- z W(- z))) L(- z W(- z))


                                    exp(L(- z W(- z))) z W(- z)
    tau_number_of_cyclic_points(z) = - ------------------------------
                                          1 + z W(- z)
```

At this stage, an ANALYTIC ANALYZER with extensive asymptotic capabilities, takes control of the asymptotic analysis [40]. It is built on a large library of Maple programmes (currently about 7000 lines), and on this problem, it selects a strategy based on singularity analysis. The number of special functional graphs (Sfungraph) of size n then appears in its raw form produced by the system as:

```
... n! times:
            3/2 1/2           2
      exp(-1)   2    exp(1/2) exp(-1/2) exp(n)          exp(n)
(1/2 ---------------------------------------------) + (O(--------))
                   2    1/2 3/2                            2
            (1 - exp(-1))  Pi   n                          n
```

a formula which, after going through Maple's simplifier and LaTeX interface yields *verbatim*

$$(n!) \times \left[\left(\frac{\sqrt{2}e^{n-1}}{2(e^{-1}-1)^2 \sqrt{\pi} n^{3/2}} \right) + \left(O(\frac{e^n}{n^2}) \right) \right].$$

The system then computes total costs (i.e., total values of parameters over all structures of size n) via their generating functions. From there, mean value estimates folllow. For instance, in the case of the *average* number of components, we get the following message

```
Floating point evaluation:
                                    1
        (1.458675144) + (O(------))
                                  1/2
                                 n
```

BS21G182.OUTPS main cycle 0 ——— : 20 millions values

Figure 4. A rendering due to Quisquater and Delescaille of the "giant component" in a functional graph representing an iteration structure of the DES cryptosystem. The DES is used here as an iterator on a set of cardinality 2^{56} by letting its "output" loop on its "key" entry (keeping the "message" fixed). The drawing represents a skeleton graph where approximately 1 in every 10^6 points is sampled. Such graphs are discussed in [36, 37].

where the symbolic form of the constant 1.4586 was also determined in passing by the system:

$$1 - \log(1 - e^{-1}).$$

Similarly, for the number of cyclic points, we obtain

```
Floating point evaluation:
                                 1
            (2.163953412) + (O(------))
                                1/2
                               n
```

with the symbolic form of the constant being

$$-\frac{e^{-1} + 1}{e^{-1} - 1}.$$

In total, within a few minutes of symbolic computations, the system, starting from formal specifications, has determined first symbolically, then numerically, that: (i) the expected number of components is ~ 1.45; (ii) the expected number of points lying on cycles is ~ 2.16. This example demonstrates an unusual case of model sensitivity (compare with the corresponding values of $O(\log n)$ and $O(\sqrt{n})$ for unconstrained random mappings). The precise capabilities of the system are described in [14, 15, 40, 46].

6 Conclusions

We have seen a systematic approach to the analysis of a large number of parameters of random mappings (or functional graphs) using a coherent generating function framework.

In a random mapping of size n, cycles presents themselves after about $\sqrt{n}$ iteration steps (Section 2), and this phenomenon is fairly unavoidable since the expected diameter is also $O(\sqrt{n})$ (Section 3). Also, random functional graphs tend to have one giant component and a few large trees.

These facts are well illustrated by extensive computations performed by J-J. Quisquater with the DES cryptographic system (see Fig. 4 and [36, 37]). Simulations with shift register sequences [1] or with Pollard's algorithm [33] (i.e., quadratic functions), as well as Bach's theoretical results [2] also confirm the frequent validity of predictions based on the heuristic random mapping model for various applications in cryptography, random number generation, computational number theory, or the analysis of algorithms.

Acknowledgements. This research was supported in part by the ESPRIT II Basic Research Actions Program of the EC under contract No. 3075 (project ALCOM).

References

[1] J. Arney and E. D. Bender. Random mappings with constraints on coalescence and number of origins. *Pacific J. Math.*, 103:269–294, 1982.

[2] E. Bach. Toward a theory of Pollard's rho–method. *Information and Computation*, to appear, 1989.

[3] R. P. Brent and J. M. Pollard. Factorization of the eighth Fermat number. *Mathematics of Computation*, 36:627–630, 1981.

[4] A. Z. Broder. Weighted random mappings; properties and applications. Technical Report STAN-CS-85-1054, Computer Science Dept., Stanford University, 1985. (Author's PhD Thesis).

[5] B.W. Char, K.O. Geddes, G.H. Gonnet, M.B. Monagan, and S.M. Watt. *MAPLE: Reference Manual*. University of Waterloo, 1988. 5th edition.

[6] N. G. de Bruijn. *Asymptotic Methods in Analysis*. North Holland, third edition, 1958. Reprinted by Dover, 1981.

[7] J. M. Delaurentis. Components and cycles of a random function. In C. Pomerance, editor, *Advances in Cryptology*, volume 293 of *Lecture Notes in Computer Science*, pages 231–242, 1988. (Proceedings of CRYPTO'87, Santa–Barbara.).

[8] R. L. Devaney. Dynamics of entire maps. in *Proc. International Conference on Dynamics*, Stefan Banach Center, Warsaw (to appear).

[9] R. L. Devaney. Julia sets and bifurcation diagrams for exponential maps. *Bulletin of the American Mathematical Society*, 11:167–171, 1984.

[10] M. A. Evgrafov. *Analytic Functions*. Dover, New York, 1966.

[11] P. Flajolet, D. E. Knuth, and B. Pittel. The first cycles in an evolving graph. *Discrete Mathematics*, 75:167–215, 1989.

[12] P. Flajolet and A. Odlyzko. The average height of binary trees and other simple trees. *Journal of Computer and System Sciences*, 25:171–213, 1982.

[13] P. Flajolet and A. M. Odlyzko. Singularity analysis of generating functions. *SIAM Journal on Discrete Mathematics*, 3(1), February 1990. To appear. Also available as INRIA Research Report 826, 1987, 25p.

[14] P. Flajolet, B. Salvy, and P. Zimmermann. Lambda–Upsilon–Omega: An assistant algorithms analyzer. In T. Mora, editor, *Applied Algebra, Algebraic Algorithms and Error–Correcting Codes*, volume 357 of *Lecture Notes in Computer Science*, pages 201–212, 1989. (Proceedings AAECC'6, Rome, July 1988).

[15] P. Flajolet, B. Salvy, and P. Zimmermann. Lambda–Upsilon–Omega: The 1989 Cookbook. Research Report 1073, Institut National de Recherche en Informatique et en Automatique, August 1989.

[16] P. Flajolet and M. Soria. Gaussian limiting distributions for the number of components in combinatorial structures. *J. Combinatorial Theory*, 1989. To appear. Available as INRIA Research Report 809, March 1988.

[17] D. Foata. *La série génératrice exponentielle dans les problèmes d'énumération.* S.M.S. Montreal University Press, 1974.

[18] I. P. Goulden and D. M. Jackson. *Combinatorial Enumeration.* John Wiley, New York, 1983.

[19] B. Harris. Probability distributions related to random mappings. *Annals of Mathematical Statistics*, 31(2):1045–1062, 1960.

[20] J. Jaworski. *Random Mappings.* PhD thesis, A. Mickiewicz University, 1985. (In Polish).

[21] I. B. Kalugin. A class of random mappings. *Proceedings of the Steklov Institute of Mathematics*, 177(4):79–110, 1988. (Issue on *Probabilistic Problems of Discrete Mathematics*).

[22] D. E. Knuth. *The Art of Computer Programming*, volume 3: Sorting and Searching. Addison-Wesley, 1973.

[23] D. E. Knuth. *The Art of Computer Programming*, volume 2: Seminumerical Algorithms. Addison-Wesley, 2nd edition, 1981.

[24] V. F. Kolchin. *Random Mappings.* Optimization Software Inc., New York, 1986. Translated from *Slučajnye Otobraženija*, Nauka, Moscow, 1984.

[25] A. Meir and J. W. Moon. On the altitude of nodes in random trees. *Canadian Journal of Mathematics*, 30:997–1015, 1978.

[26] L. R. Mutafčiev. On some stochastic problems of discrete mathematics. In *Mathematics and Education in Mathematics* (Sunny Beach), pages 57–80, Bulgarian Academy of Sciences, Sophia, Bulgaria, 1984.

[27] L. R. Mutafčiev. Limit theorems concerning random mapping patterns. *Combinatorica*, 8:345–356, 1988.

[28] A. M. Odlyzko. Periodic oscillations of coefficients of power series that satisfy functional equations. *Advances in Mathematics*, 44:180–205, 1982.

[29] F. W. J. Olver. *Asymptotics and Special Functions.* Academic Press, 1974.

[30] A. I. Pavlov. On an equation in a symmetric semigroup. *Proceedings of the Steklov Institute of Mathematics*, 177(4):121–128, 1988. (Issue on *Probabilistic Problems of Discrete Mathematics*).

[31] Yu. L. Pavlov. The asymptotic distribution of maximum tree size in a random forest. *Theory of Probability and Applications*, 22:509–520, 1977.

[32] Yu. L. Pavlov. On random mappings with constraints on the number of cycles. *Proceedings of the Steklov Institute of Mathematics*, 177(4):131–143, 1988. (Issue on *Probabilistic Problems of Discrete Mathematics*).

[33] J. M. Pollard. A Monte Carlo method for factorization. *BIT*, 15(3):331–334, 1975.

[34] G. V. Proskurin. On the distribution of the number of vertices in strata of a random mapping. *Theory of Probability and Applications*, 18:803–808, 1973.

[35] P. Purdom and J. Williams. Cycle length in a random function. *Transactions of the American Mathematical Society*, 133:547–551, 1968.

[36] J.-J. Quisquater and J.-P. Delescaille. Other cycling tests for DES. In C. Pomerance, editor, *Advances in Cryptology*, volume 293 of *Lecture Notes in Computer Science*, pages 255–256. Springer-Verlag, 1988. (Proceedings of CRYPTO'87, Santa–Barbara.).

[37] J.-J. Quisquater and J.-P. Delescaille. How easy is collision search? New results and applications to DES. In *Proceedings of CRYPTO'89*, Lecture Notes in Computer Science. Springer-Verlag, 1989. To appear.

[38] A. Rényi and G. Szekeres. On the height of trees. *Australian Journal of Mathematics*, 7:497–507, 1967.

[39] V. N. Sachkov. Random mappings with bounded height. *Theory of Probability and Applications*, 18:120–130, 1973.

[40] B. Salvy. Fonctions génératrices et asymptotique automatique. Research Report 967, Institut National de Recherche en Informatique et en Automatique, 1989.

[41] L. A. Shepp and S. P. Lloyd. Ordered cycle lengths in a random permutation. *Transactions of the American Mathematical Society*, 121:340–357, 1966.

[42] M. Soria. *Méthodes d'analyse pour les constructions combinatoires et les algorithmes*. Doctorat ès sciences, Université de Paris–Sud, Orsay, 1989.

[43] R. P. Stanley. *Enumerative Combinatorics*, volume I. Wadsworth & Brooks/Cole, 1986.

[44] V. E. Stepanov. Limit distributions of certain characteristics of random mappings. *Theory of Probability and Applications*, 14:612–626, 1969.

[45] E. C. Titchmarsh. *The Theory of Functions*. Oxford University Press, 2nd edition, 1939.

[46] P. Zimmermann. Alas: un système d'analyse algébrique. Research Report 968, Institut National de Recherche en Informatique et en Automatique, 1989.

Factoring by electronic mail

Arjen K. Lenstra
Department of Computer Science
The University of Chicago
1100 E 58th Street
Chicago, IL 60637
arjen@gargoyle.uchicago.edu

Mark S. Manasse
DEC Systems Research Center
130 Lytton Avenue
Palo Alto, CA 94301
msm@src.dec.com

Abstract. In this paper we describe our distributed implementation of two factoring algorithms, the elliptic curve method (ecm) and the multiple polynomial quadratic sieve algorithm (mpqs).

Since the summer of 1987, our ecm-implementation on a network of MicroVAX processors at DEC's Systems Research Center has factored several most and more wanted numbers from the Cunningham project. In the summer of 1988, we implemented the multiple polynomial quadratic sieve algorithm on the same network. On this network alone, we are now able to factor any 100 digit integer, or to find 35 digit factors of numbers up to 150 digits long within one month.

To allow an even wider distribution of our programs we made use of electronic mail networks for the distribution of the programs and for inter-processor communication. Even during the initial stage of this experiment, machines all over the United States and at various places in Europe and Australia contributed 15 percent of the total factorization effort.

At all the sites where our program is running we only use cycles that would otherwise have been idle. This shows that the enormous computational task of factoring 100 digit integers with the current algorithms can be completed almost for free. Since we use a negligible fraction of the idle cycles of all the machines on the worldwide electronic mail networks, we could factor 100 digit integers within a few days with a little more help.

1. Introduction

It is common practice to begin a paper on integer factoring algorithms with a paragraph emphasizing the importance of the subject because of its connection to public-key cryptosystems; so do we. We refer to [3] for more information on this point.

This paper deals with the practical question

how big are the integers we can factor with our present algorithms?

This question is rather vague, because we did not specify how much time and/or money we are willing to spend per factorization. Before making our question more precise, let us illustrate its vagueness with four examples which, in the summer of 1988, represented the state of the art in factoring.

(i) In [7, 24] Bob Silverman *et al.* describe their implementation of the multiple polynomial quadratic sieve algorithm (mpqs) on a network of 24 SUN-3 workstations. Using the idle cycles on these workstations, 90 digit integers have been factored in about six weeks (elapsed time).

(ii) In [21] Herman te Riele *et al.* describe their implementation of the same algorithm on two different supercomputers. They factored a 92 digit integer using 95 hours of CPU time on a NEC SX-2.

(iii) 'Red' Alford and Carl Pomerance implemented mpqs on 100 IBM PC's; it took them about four months to factor a 95 digit integer.

(iv) In [20] Carl Pomerance *et al.* propose to build a special purpose mpqs machine 'which should cost about $20,000 in parts to build and which should be able to factor 100 digit integers in a month.'

In all these examples mpqs was being, or will be used as factoring algorithm, as it is the fastest general purpose factoring algorithm that is currently known [19]. In order to compare (i) through (iv), we remark that for numbers around 100 digits adding three digits to the number to be factored roughly doubles the computing time needed for mpqs; furthermore mpqs has the nice feature that the work can be evenly distributed over any number of machines. It follows that 100 digit integers could be factored in about one month, using

- the idle time of a network of 300 SUN-3 workstations, or

- one NEC SX-2, or

- 1200 IBM PC's, or

- one $20,000 special purpose processor.

Returning to our above question let us, in view of these figures, fix one of the resources by stipulating that we want to spend at most one month of elapsed time per factorization. Apparently, the answer then depends on the amount of money we are willing to spend. Clearly, if we use mpqs, and if we start from scratch, then the last alternative is definitely the cheapest of the four possibilities listed above. However, there is no reason at all to start from scratch.

If we really want to find out where our factorization limits lie nowadays, we should take into account that the natural habitat of the average computer scientist has changed considerably over the last few years: many people have access to some small number of small machines, and many of those small machines can communicate with each other by means of electronic mail. What would happen if someone exploited the full possibilities of his environment? The current factorization algorithms, when parallelized, do not require much inter-processor communication. Therefore, electronic mail could easily take care of the distribution of programs and data and the collection of results. So, if someone writes a factorization program, mails it to his friends along with instructions how to run it on the background, and convinces his friends to do the same with their friends up to an appropriate level of recursion, then the originator of the message could end up with a pretty powerful factorization machine. It is not unlikely that he will be able to factor 100 digit integers in much less than one month, and *without spending one single penny*.

Thus we rephrase our question as follows:

> *how big are the integers we can factor within one month of elapsed time, if we only want to use computing time that we can get for free?*

So far, we have only used the multiple polynomial quadratic sieve algorithm (mpqs) in our running time estimates. As we noted above, the reason for this is that mpqs is the fastest *general purpose* factoring algorithm that we currently know of, *i.e.*, it is the fastest algorithm whose running time is, roughly speaking, completely determined by the size of n, and not by any other properties that n might have. Thus, for mpqs we can fairly accurately predict the precise moment at which a factorization will be found, once the computation has been set up. This implies that, if we decide to use mpqs, the answer to the above question depends solely on the amount of computational power we are able to get. Furthermore, if we are able to factor some integer of some given size within a month, then we can factor *any* integer of about that size in about the same amount of time. This holds irrespective of how 'difficult' the number might be considered to factor, like RSA keys which are usually chosen as products of two primes of about the same size [22].

This does not imply that, given an arbitrary integer n to be factored which is not 'too big' for us, we immediately apply mpqs. That only makes sense if one knows that the number in question is 'difficult.' Ordinarily, one should first try methods that are good in finding factors with special properties; examples of such methods are trial division (small factors), Pollard's 'rho'-method (bigger small factors) [15, 18], the $p-1$ and $p+1$-methods and their variants (factors having various smoothness properties) [2, 16], and the elliptic curve method (factors of up to about 35 digits) [13]. The more time one invests in one of these methods, the higher its probability of success. Unfortunately it is difficult to predict how much time one should spend, because the properties of the factors are in general unknown. And, the time invested in an unsuccessful factorization attempt using one of these methods is completely wasted; it only gives a weak conviction that n does not have a factor of the desired property, but no proof (with the exception of trial division, which yields a proof that no factor less than or equal to the trial division bound exists if nothing has been found).

To return again to our question, let us assume that we will use mpqs as the method of last resort. As remarked above, the answer then depends on the free computational power we can organize, since mpqs cannot be lucky (up to a certain minor point fow which we refer to [24]). So, we should concentrate on methods to get as much free computing time as we can, which we do as described above: try to get volunteers on the electronic mail networks to run our program at times that their machines would otherwise be idle. To attract the attention of the maximum number of possible contributors, and to make them enthusiastic about the project, we adopted the following strategy:

- Write an mpqs-program that is as portable as we can make it, and ask some friends and colleagues working in the same field to experiment with it. In that way we should get a good impression of what is possible, and what should be avoided, and we get some experience in running an 'electronic mail multiprocessor' on a small scale.

- Use this program, the forces that we can organize at DEC's Systems Research Center (SRC), and the (still) relatively small external power to achieve some moderately impressive factorization results.

- Publish those results in the sci.math newsgroup to attract attention from possible contributors, and ask for their help.

- Use electronic mail to distribute the program to people who express interest.

- Get more impressive factorization results.

- Repeat the last three steps as long as we are still interested in the project.

Once such a distributed mpqs implementation has been set up it is a minor effort to include some other useful factorization features. For instance, we plan to include the elliptic curve method (ecm) in the package we distribute; we already have considerable experience with a distributed ecm program at SRC, but at the time of writing this paper we had not included it in the program we have distributed worldwide.

Here we should remark that ecm can easily be distributed over any number of machines, as it consists of a number of independent factorization trials. Any ecm trial can be lucky, and find a factorization, independent of any other trial. The probability of success per trial depends on the size of the factor to be found, and is therefore difficult to predict; see Section 2 for details. The mpqs algorithm works completely differently. There the machines compute so-called relations, which are sent to one central location. Once sufficiently many relations have been received, the factorization can be derived at the central location; see Section 3 for details.

Given such an extended package, a typical factorization effort would proceed as follows. Upon receipt of a new number, all machines on the factorization network do some specified number of ecm trials. A successful trial is immediately reported to us, and we broadcast a message to stop the current process. If all ecm trials have failed and the number in question is not too big, the machines move to mpqs and start sending us relations. We then wait until we have sufficiently many relations to be able to derive the factorization.

At the time of writing this paper we have factored two 93, one 96, one 100, one 102, and one 106 digit number using mpqs, and we are working on a 103 digit number; for all these numbers extensive ecm attempts had failed. The 100 digit number took 26 days (of elapsed time). About 85% of the work of this factorization was carried out at SRC. For the 106 digit number the external machines contributed substantially more than 15% of the total factorization effort, namely 30%. The 103 digit number will be done exclusively by external machines, while the machines at SRC are working on other factorizations. As the factorization network is growing constantly, it is difficult to predict how many machines we will eventually be able to get. Consequently, at the present moment we are still unable to give a reliable answer to our question, but we have the impression that the present approach should enable us to factor 110 digit numbers.

It is a natural question to ask what consequences this could have for the security of the RSA cryptosystem. The well-known RSA challenge is to factor a certain 129 digit number concocted by Rivest, Shamir and Adleman. At the time they presented this challenge, and generously offered to pay $100 for the factorization, it appeared to be far out of reach. The analysis of mpqs however, shows that factoring 129 digit numbers is 'only' about 400 times as hard as factoring 100 digit numbers. Now that 100 digit numbers can be factored, this does not seem like a very secure safety margin. The $100 prize will not be much of an incentive, however: postage costs and the fees for currency conversion will make dividing the spoils counterproductive.

RSA cryptosystems are being used nowadays with keys of 512 bits, which amounts to 155 digits. Factoring 155 digit numbers is about 40000 times harder than factoring 100 digit numbers, which sounds pretty safe. But even given the present factoring algorithms, it is unclear how long this will remain safe. There are zillions of idle cycles around that can be used, as we have seen, by anyone who has access to the electronic mail network; this network is growing rapidly, and its future computational power is difficult to predict. More worrisome is that determined and powerful adversaries could in principle organize some gigantic factoring effort by linking together all their machines. Factoring 155 digit numbers is then not as impossible as many people would like it to be.

These estimates do not even take into account that the average processor will get much faster than it is now. Currently an average workstation operates at 1 to 3 million instructions per second (mips); a fast workstation operates in the range from 6 to 10 mips. Soon new workstations will be released running at 20 to 25 mips, and it is to be expected that within five years a moderately priced workstation will run at 100 mips. The reader can easily figure out what consequences this will have for the safety of 512 bit RSA keys.

Of course, we are not factoring actual RSA keys. For one thing, when we report our previous successes to try to gather helpers to work on the project, we might inadvertently attract the attention of the owner of the key, who might be unhappy at discovering this. Secondly, the only reason to factor RSA keys is to impersonate the owner. The people helping us factor the number might demand their share and we do not know anyone with a small enough key and a large enough bank balance to make this worthwhile. To avoid any suspicion, we factor only numbers from the Cunningham project [6], or other numbers with short algebraic descriptions and of some mathematical significance [5].

Another concern of our helpers might be that they have accepted a Trojan horse [25]. For this reason our implementations are relatively straightforward; helpers might not be able to verify the correctness of the algorithm, but they can certainly see that nothing too strange is going on. We could have written something that is very clever, which can find helpers by itself. Such a program would be a virus [1, 10], and while it might help us in conquering a single large number, it would probably hurt our standing in the community in the long run. Someone who is not concerned about propriety might be less scrupulous.

With all these negative possibilities, why should anyone take the risks of helping us? So far, the people who helped us are our friends and are not inclined to ascribe base motives to our work. Additionally, we have managed to help them get their names in the newspapers. In the future we will have to make sure that the inconveniences involved in running the program are outweighed by the good feelings generated by whatever share of the limelight we can place our helpers in.

We do not claim that in the long run our approach is the most cost-effective way of factoring large integers on a regular basis. For that purpose we believe that Pomerance's machine mentioned in (iv) above, or a couple of them, is the most promising attack. For the security of RSA cryptosystems, however, we think that our approach of building an ad hoc network is more threatening; there it counts what we *can* do, not what we can do on a regular basis. As we mentioned above, the power of such a network is difficult to estimate, and our setup has the advantage that it can almost immediately profit from any advances in technology (faster machines) or theory (faster algorithms).

The rest of this paper is organized as follows. In Section 2 we give a rough outline of the elliptic curve method and its expected behavior, and we present some of the results we obtained using this method. The same will be done for the multiple polynomial quadratic sieve algorithm in Section 3. Some details of the distribution techniques that we have used and that we are planning to use are given in Section 4.

2. The elliptic curve method

The elliptic curve method (ecm) consists of a number of independent factorization trials. Any trial can be lucky and find a factorization, independent of any other trial. The larger the smallest factor of the number to be factored, the smaller the probability of success per trial. The elliptic curve method is a *special purpose* factoring algorithm in the sense that it can only be expected to work if the number to be factored has a reasonably small factor.

Indeed, the elliptic curve method is a very useful method to find small factors of large numbers. For instance, we will see that if a 100 digit number has a 38 digit factor, it is probably faster to find this factor using ecm than using the multiple polynomial quadratic sieve algorithm. A problem, however, is that one does not know beforehand whether the number to be factored indeed has a small factor. If it has a small factor, then applying ecm has a reasonable probability of success; if there is no small factor, then ecm will have virtually no chance. This naturally leads to the question how much time one should spend on an attempt to factor a number using ecm.

This question would probably be easier to answer if the results from a failed ecm attempt would be useful for other purposes. Unfortunately however, a failed ecm trial does not contribute anything that is useful for the rest of the computation, or that might be helpful for other factorization attempts. Consequently, the time invested in an unsuccessful ecm factorization attempt is completely wasted. It is even the case that, if ecm fails to detect a small factor after some number of trials, this does not guarantee that there *is* no small factor, although the existence of a small factor becomes less 'likely'. Also, the expected remaining computing time grows with the time that has been spent already.

Those less desirable properties are not exactly fitted to make the ecm a very popular method. Who is, after all, willing to spend his valuable cycles on a lengthy computation that will probably not produce anything that is useful. This would not be much of a problem if those cycles are *not* valuable. On big mainframes such cycles might be difficult to find, but on the average workstation they are abundant, as workstations are usually idling at least half of the time. A disadvantage is that one workstation is probably slower than a big mainframe. Many workstations together, however, should give at least the same computational power as one big machine, when applied to a method that can easily be parallelized like ecm.

Therefore, in order to find out what can still be done using ecm, and what is out of reach, and without getting complaints about wasted computing time, we should run ecm in the background on a large number of workstations. Because we cannot predict the sizes of the factors of the numbers we attempt to factor, our attempts will quite often be fruitless; there is not much of a difference between wasting cycles on unsuccessful ecm attempts or idling, so nobody will complain too seriously. Every now and then we will hit upon a lucky number and find a factorization, thus making ourselves, the contributors of the cycles (we hope), and the authors of [6] (in any case) happy.

In the rest of this section we will discuss some aspects of our MicroVAX implementation which is written in Pascal with VAX assembly language for the multiprecision integer arithmetic (for a wider distribution of the ecm program we have a C-version as well). The ecm is described in detail in [13]; descriptions that focus more on practical aspects and implementations of ecm can be found in [4, 15]. For the purposes of this paper the following rough description of the elliptic curve method suffices.

Elliptic curve method. Given an integer n to be factored, randomly select an elliptic curve modulo n and a point x in the group of points of this elliptic curve.

First stage: Select an integer m_1, and raise x to the power k, where k is the product of all prime powers $\leq m_1$. If this computation fails because a non-trivial factor has been found, then terminate. Otherwise, continue with the second stage.

Second stage: Select an integer $m_2 > m_1$, and try to compute x^{kq} for the primes q between m_1 and m_2 in succession. If this computation fails because a non-trivial factor has been found, then terminate. Otherwise, start all over again.

Given the sizes of the m_i and the smallest factor p of n, the probability that one iteration is successful in factoring n can be derived. Given a choice for the m_i, the expected number of iterations to find a factor of a certain size with a certain probability then easily follows. Asymptotically the expected running time of ecm to find a factor p of n is $O((\log n)^2 e^{(1+o(1))\sqrt{2\log p \log\log p}})$, summed over all trials. Every iteration, or trial, of ecm is completely independent of every other trial. This means that we can expect to achieve an s fold speed-up by running the ecm program on s independent identical machines, as long as we make sure that those machines make different random choices.

The second stage as formulated here has the disadvantage that it needs a table containing the differences between the consecutive primes up to m_2. For huge values of m_2 this might become problematic, especially in a set-up where memory efficiency is an important aspect. Therefore we use the so-called *birthday paradox* version for the second stage as described in [4, section 6]. Our implementation also incorporates other ideas mentioned in the same paper which make it even more time and space efficient. We refer to [4, sections 7, 9.1, 9.3, and 9.4] for a description of these improvements.

Given the program and given the relative speeds of the two stages, it remains to analyze how the parameters should be chosen (and modified) during execution. For this purpose it is useful first to optimize those choices, *given* the size of the smallest prime factor p of n. For a fixed size of p the optimum parameter choice can be found by slightly changing the analysis given in [4] to take the various improvements into account. The resulting analysis is somewhat different from the one in [4], but it is sufficiently similar that we do not give any details of our computation and instead refer to [4].

In this paper we are only interested in the resulting running time estimates on a MicroVAX II processor. Our unit of work is one multiplication in $\mathbf{Z}/n\mathbf{Z}$, i.e., one multiplication of two numbers in $\{0, 1, ..., n-1\}$ followed by a reduction modulo n. This operation can be performed at the cost of roughly two multiplications of integers of about the same size as n by representing the numbers as suggested in [14]. This representation of integers resulted in a 20 percent speed-up, as compared to the ordinary representation. On a MicroVAX II processor, which operates at about 1 mips, one unit of work for a 100 digit number takes about 0.0045 seconds.

The amount of work needed to find a factor of a certain size with probability at least 60 percent is given in Table 1. We have also indicated how many millions of seconds this would take on a 1 mips computer for various sizes of n; notice that 30 million seconds is about one year. Furthermore, the table lists the optimal values for m_1 in the first stage, and the number of iterations to achieve a 60 percent probability of success; it appeared that the effective value for m_2 for our implementation is about $30 \cdot m_1$.

Table 1

Amount of work, optimal parameter choice,
and millions of seconds on a 1 mips computer to find
smallest factor p of n with a success probability of 60 percent

$\log_{10}p$	$\log_{10}\text{work}$	m_1	trials	millions of seconds for $\log_{10}n =$			
				80	90	100	110
25	8.6	65000	300	1	2	2	2
26	8.9	85000	400	2	3	3	4
27	9.1	115000	500	3	4	5	6
28	9.3	155000	650	6	7	9	11
29	9.5	205000	750	9	12	15	18
30	9.7	275000	950	15	20	24	28
31	9.9	360000	1100	25	30	37	45
32	10.1	480000	1350	40	50	60	72
33	10.3	625000	1600	60	75	95	115
34	10.5	825000	1950	95	120	150	180
35	10.7	1100000	2300	150	190	230	280
36	10.9	1400000	2800	230	290	360	430
37	11.1	1800000	3300	350	450	550	670
38	11.3	2350000	3900	540	680	840	1020
39	11.5	3000000	4700	830	1040	1290	1560
40	11.6	3850000	5600	1300	1600	1960	2400

What does this mean for a network of approximately 100 workstations, each consisting of five MicroVAX processors? Assuming that the average workstation is idle from 5.00 PM to 9.00 AM, we should be able to get about 28 million seconds per weekday. This should enable us to find 30 digit factors of 100 digit numbers in about one day. During the weekend we expect to work 1.5 times faster.

The numbers we attempted to factor all came from the appendix of unfactored numbers from the Cunningham Tables [6, 8]. In Table 2 we list some of the most and more wanted numbers from the Cunningham Tables we factored with ecm. As the 'expected time' in Table 2 is based on a 60 percent probability of success, it is only a very rough indication. Furthermore, the term *expected time* is misleading, because we can only expect that time *given* the resulting factorization.

Table 2

Some factorizations obtained with ecm

(time in millions of seconds on a 1 mips computer)

$\log_{10}n$	$\log_{10}p$	time expected	time observed	elapsed days	n is factor of
99.2	26.1	3	47	2.4	$10^{137}-1$
96.3	26.3	3	5	0.3	$11^{97}+1$
103.4	27.6	7	3	0.3	$3^{229}+1$
100.0	27.7	8	6	0.5	$10^{101}-1$
100.8	28.8	14	15	0.5	$7^{137}-1$
98.0	30.7	32	25	1.4	$6^{137}+1$
89.7	31.6	40	15	0.8	$2^{361}-1$
105.2	31.7	60	21	0.9	$10^{143}-1$
100.3	33.5	120	21	1.2	$2^{353}-1$
88.6	33.9	110	130	5.7	$5^{157}-1$
93.7	34.2	160	31	1.8	$10^{116}+1$
87.8	35.6	240	89	4.1	$2^{353}+2^{127}+1$

As might be clear from this table, the elliptic curve method is indeed very useful to find small factors, and it does so in a reasonable amount of time. Except for the first two entries, we have been quite lucky several times.

Of course, there were numerous failures as well. Some of the failures for which the factorization was later found by others (or by ourselves) are listed in Table 3. The meaning of the 'expected time' column in Table 3 is: 'the time ecm should have spent for a 60 percent probability of success'. As usual, p denotes the smallest prime factor of n.

Table 3

Some failed ecm attempts

(time in millions of seconds on a 1 mips computer)

$\log_{10}n$	$\log_{10}p$	time expected	time spent	elapsed days	n is factor of	factor found by
79.4	29.4	11	23	1.4	$2^{483}+1$	Silverman, mpqs
91.4	33.1	83	130	7	$6^{131}-1$	te Riele, mpqs
92.1	33.1	84	140	7.4	$2^{368}+1$	Ruby, ecm
105.8	36.6	530	130	7	$2^{353}+1$	see below, mpqs
92.4	37.1	570	130	7	$10^{109}+1$	see below, mpqs
89.5	40.7	>1500	160	9	$5^{160}+1$	Silverman, mpqs
100.0	40.9	>2000	145	8	$11^{104}+1$	see below, mpqs
94.6	43.2	>10000	130	7	$2^{332}+1$	Alford, mpqs
101.1	46.4	>10000	130	7	$2^{391}-1$	see below, mpqs
95.3	47.5	>10000	270	14	$11^{107}-1$	see below, mpqs

For the first three entries of Table 3 we clearly have had some bad luck; failure of ecm does not at all imply that there is no small factor. The last five entries were not actual failures, in the sense that the smallest factor was too big for ecm to find.

In practice it appears that ecm is relatively insensitive to the choice of the parameters. If m_1 is chosen too large for the (unknown) smallest prime factor of n, the probability of success per curve increases, and consequently the number of trials decreases. And vice versa, if m_1 is too small, the probability per curve decreases, and the number of trials increases. In both cases the product of m_1 and the number of trials, which is up to a constant a good indication for the work done, will be close to optimal. This practical observation is supported by the theoretical analysis. For instance, if one uses the parameters that are optimal for $\log_{10}p = 30$ for an n that happens to have a 32 digit smallest factor, the optimal $\log_{10}$work of 10.12 increases to only 10.13 (i.e., 2400 trials with $m_1 = 275000$ instead of 1350 trials with $m_1 = 480000$).

Consequently, the parameter choices do not matter too much, at least within certain limits. In our experience $m_1 = 1.001^{i-1} \cdot 300000$ at the ith trial worked satisfactorily, as did slightly larger (and smaller) choices for the growth rate and m_1.

3. The multiple polynomial quadratic sieve algorithm

The quadratic sieve algorithm is described in [19]. Descriptions of the multiple polynomial variation of the quadratic sieve algorithm (mpqs) can be found in [20, 24].

Like the elliptic curve method, the quadratic sieve algorithms are probabilistic algorithms. But unlike ecm, the quadratic sieve algorithms do not depend on certain properties of the factors to be found. Furthermore, their success does not depend on a *single* lucky choice, which, as we have seen in Section 2, is sometimes unlikely ever to occur. Instead, they work by combining many small instances of luck, each of which is much more likely to take place.

Once the computation has been set up, one can easily see how 'lucky' the method is on the average, the progress can easily be monitored, and the moment it will be completed can fairly accurately be predicted. This is completely independent of how 'difficult' the number to be factored is considered to be.

The quadratic sieve algorithms are called *general purpose* factoring algorithms because, up to a minor detail (cf. [24]), their run time is solely determined by the size of the number to be factored. Their expected running time to factor a number n is $e^{(1+o(1))\sqrt{\log n \log \log n}}$; this is independent of the sizes of the factors to be found. Notice that this is the same as the asymptotic running time of the elliptic curve method if the smallest factor p is close to $\sqrt{n}$.

The quadratic sieve algorithms consist of two stages, a time consuming first stage to collect so-called *relations*, and a relatively easy second stage where the relations are combined to find the factorization. In the multiple polynomial variation of the quadratic sieve algorithm the first stage can easily be distributed over almost any number of processors, in such a way that running the algorithm on s identical processors at the same time results in an s fold speed-up. In the second stage the combination is found using Gaussian elimination, which is usually done on one machine.

The following rough description might be helpful to understand the factorization process. Let n be the number to be factored. First one chooses an integer $B > 0$ and a factor base $\{p_1, p_2, ..., p_B\}$, consisting of $p_1 = -1$ and the first $B-1$ primes p for which n is a square modulo p. Next one looks for relations, which are expressions of the form

$$v^2 \equiv q^t \cdot \prod_{j=1}^{B} p_j^{e_j} \bmod n,$$

for $t \in \{0,1\}$, $e_j \in \mathbb{Z}_{\geq 0}$, and q a prime not in the factor base. Below we will explain how these relations can be found. We will call a relation with $t = 0$ a *small* relation, and a relation with $t = 1$ a *partial* relation. Two partial relations with the same q can be combined to yield a relation of the form

$$w^2 \equiv q^2 \cdot \prod_{j=1}^{B} p_j^{e_j} \bmod n,$$

(where w is the product of the v's of the two partial relations, and the e_j are the sums of the exponents of the two partial relations); such a relation will be called a *big* relation. For n

having 90 to 110 digits B will approximately range from 17000 to 100000.

Given a total of more than B small and big relations, we will be able to find a dependency among the exponent vectors $(e_j)_{j=1}^B$, and therefore certainly a dependency modulo 2 among the exponent vectors modulo 2; this can be done by means of Gaussian elimination on a bit matrix having B columns and $> B$ rows. Such a dependency modulo 2 then leads to a solution x, y to $x^2 \equiv y^2 \bmod n$. Given this solution, one has a reasonable probability of factoring n by computing $\gcd(x-y,n)$. More relations lead to more dependencies, which will lead to more pairs x, y. Therefore we can virtually guarantee to factor n, if the relations matrix is slightly over-square.

There are various ways to generate relations. We will show how small relations can be generated; how partial relations can be generated easily follows from this. In the original description of the quadratic sieve algorithm, Pomerance proposes to use the quadratic polynomial $f(X) = ([\sqrt{n}]+X)^2-n$, and to look for at least $B+1$ integers m such that $f(m)$ is smooth; here we say that an integer is smooth if it can completely be factored over the factor base $p_1, p_2, ..., p_B$. Because $f(m) \equiv ([\sqrt{n}]+m)^2 \bmod n$, a smooth $f(m)$ produces a relation. From $f(m) \approx 2m\sqrt{n}$ (for small m), and the heuristic assumption that the $f(m)$'s behave approximately as random numbers with respect to smoothness properties, we can derive an integer m_B such that we expect that there are at least $B+1$ distinct m's with $|m| \leq m_B$ for which the corresponding $f(m)$'s are smooth.

Given a list of values of $f(m)$ for $|m| \leq m_B$, those that are smooth can be found as follows. Let $p > 2$ be a prime in the factor base that does not divide n, then the equation $f(X) \equiv 0 \bmod p$ has two solutions $m_1(p)$ and $m_2(p)$ (because n is a square modulo p), which can easily be found. But $f(m_i(p)) \equiv 0 \bmod p$ implies that $f(m_i(p)+kp) \equiv 0 \bmod p$ for any integer k. To find the $f(m)$ on our list that are divisible by p it suffices therefore to consider the locations $m_i(p)+kp$ for all integers k such that $|m_i(p)+kp| \leq m_B$ and $i = 1, 2$. Finding the $f(m)$ that are smooth can therefore be done by performing this so-called *sieving* for all p in the factor base.

Here we should remark that in practice one does not set up a list of $f(m)$'s for $m = -m_B, ..., -1, 0, 1, ..., m_B$, to divide the values at appropriate locations by the primes. Instead a list of zero's is set up, to which approximations to the logarithms of the primes are added. For locations m which contain sufficiently big numbers after sieving with all primes, one computes and attempts to factor $f(m)$; how big this value should be also depends on the maximal size of the big prime (the q) one allows in a partial factorization (see above), as $f(m)$'s that do not factor completely possibly lead to a partial relation. Of course, the whole interval will not be processed at the same time, but it will be broken up into pieces that fit in memory.

A heuristic analysis of this algorithm leads to the expected asymptotic running time mentioned above. The algorithm could in principle be parallelized by giving each machine an interval to work on. Several authors (Davis in [9] and Montgomery in [24]) suggested a variation of the original quadratic sieve algorithm which is not only better suited for parallelization, but which also runs faster (although the asymptotic running time remains $e^{(1+o(1))\sqrt{\log n \log\log n}}$). Instead of using one polynomial to generate the smooth residues, they suggest using many different suitably-chosen polynomials. Evidently, this makes parallelization easier, as each machine could work on its own polynomial(s). Another advantage of using many polynomials is that one can avoid the performance degradation that mars the original quadratic sieve algorithm ($|f(m)|$ grows linearly with $|m|$, so that the probability of smoothness decreases as $|m|$ grows), by considering only a fixed interval per polynomial before moving to the next polynomial. This more than outweighs the disadvantage of having to initialize many polynomials. We used the multiple polynomial variation as suggested by Montgomery; for details we refer to [11] and [24].

To determine the optimal value of B, the size of the factor base, we have to balance the advantages of large B's (many smooth residues) against the disadvantages (we need at least $B+1$ smooth residues, they are quite costly to find, and the final reduction gets more expensive). The crossover point is best determined experimentally; in practice anything that is reasonably close to the optimal value works fine.

This does not imply that we will always use a B that is close to optimal. For n around 103 digits, the optimal B gets so large that it would cause various memory problems. In the first place the memory required by the program to generate relations grows linearly with B. In our setup we should also be able to run the program on small workstations, which implies that B cannot get too big. In the second place, storing all relations that we receive for large B (in particular the partial relations) gets problematic. And finally, storing the matrix during the Gaussian elimination becomes a problem (apart from the fact that the elimination gets quite slow, see below). As a consequence we will be using suboptimal values of B for n having more than 103 digits, at least until we have solved these problems. See below for the values of B that we have used for various n.

During the first stage of the algorithm, one not only collects small relations, residues that factor completely over the factor base, but one looks for partial relations as well, i.e., residues that factor over the factor base, except for a factor $q > p_B$. The reason for this is, as we have seen above, that two partial relations with the same q can be combined into one so-called big relation, which is just as useful as one small relation during the final combination stage. The more partials we can get, the more bigs we will find, thus speeding up the first stage of the algorithm. But evidently, one needs quite some number of partial relations to have a reasonable probability to find two with the same q (cf. birthday paradox).

In principle we could easily keep all partials with q prime and $q < p_B^2$, and collect those during the first stage. Doing that would however not be very practical, as we would get an enormous number of partial relations. Furthermore, most matches will be found among the smaller q's. For that reason, we only keep partial relations for which q is not too big. An upper bound of 10^8 on q works quite satisfactorily. With this bound we found about fifteen times as many partials as smalls for $B = 50000$, which at the end of the first stage resulted in about three big relations for every two smalls. For $B = 65500$ and the same bound on q we got about thirteen partials per small relation, and at the end we had five bigs for every three smalls.

The smalls and partials we receive (and generate) at SRC are processed in the following straightforward way. About once every few days the new relations are verified for correctness. This verification is done for obvious reasons: who knows what people send us, and who knows what mailers do with the messages they are supposed to send; we never received faulty relations that were in the right format, but we got quite some totally incomprehensible junk, badly mutilated by some mailer. After verification, the smalls are sorted by their weight, and merged into the file of 'old' small relations, thereby deleting the double relations (sometimes we get one message several times). The partials are sorted by their q value, and merged into the sorted file of 'old' partials. If we find two relations with the same q, one is kept in or inserted in the sorted file of partial relations, and the combination of the two is computed and merged into the file of big relations. Again, doubles are deleted during the merging, and a combination of two identical partial relations is rejected. Of course, during these processes, we check to see if we are so lucky as to have found a small or big relation with even exponents, as such a relation could immediately lead to a factorization; we have not been so lucky yet, as was to be expected.

Figure 1 illustrates the progress of the first stage for the factorization of a 100 digit number with $B = 50000$ and $q \leq 10^8$. The first stage started at noon, September 15, 1988, and it was completed on October 9, when we reached a total of more than 50000 small and big relations.

It can be seen that the number of big relations grows faster and faster: the more partial relations we get, the higher probability we have to find a double q, and therefore a big relation. The number of small relations grows more or less linearly, as was to be expected. Among the about 320000 relations we had found on October 9 there were about 20500 smalls, and the remaining partials produced about 29500 big relations. For the other numbers we factored the graphs of the numbers of small and big relations behaved similarly to Figure 1.

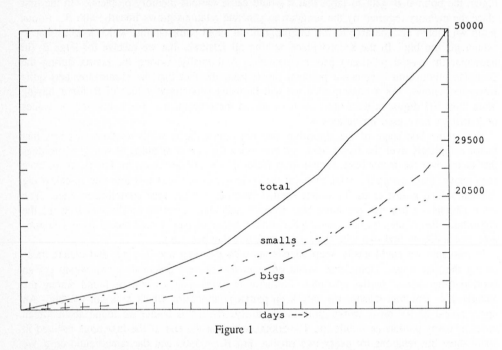

Figure 1

Given these figures, it will not come as a surprise that, once the first stage of the mpqs program has been set up on some number of machines (and keeps running on those machines!), the moment that the total number of small and big relations exceeds B can fairly accurately be predicted. As mentioned above, it then remains to find a dependency modulo 2 among the exponent vectors of the small and big relations. For our initial experiments we wrote a straightforward Gaussian elimination algorithm [12], which was perfectly able to handle the sparse matrices resulting from factor bases with B up to about 35000 that we had chosen for the smaller numbers we factored. For larger B our algorithm definitely gets very slow; for $B = 50000$ the Gaussian elimination took for instance about 1.5 days on a VAX 8800. We are still working on the implementation of more advanced methods for the Gaussian elimination, like Wiedemann's method [26] and extensions of Odlyzko's intelligent Gaussian elimination techniques [17]. We should note that we never needed $B+1$ rows in the matrix to produce a factorization; it appeared that about $0.99 \cdot B$ is enough in practice.

We conclude this section with some of the results we obtained with the multiple polynomial quadratic sieve algorithm. Because we wanted to be able to run the program at as many sites as possible, we decided to write our programs completely in C. Of course this causes some inefficiencies, because we did not use any assembly language code. We could have provided such code for various popular types of machines, but we did not do that yet.

While carrying out our experiments, we attempted bigger and bigger numbers. Bigger numbers have two important advantages: collecting relations takes at least a few weeks, so that we do not have to change our (worldwide) inputs often, and the resulting factorizations attract more attention, and therefore more contributors. In this way we collected relations for the factorizations of the numbers mentioned in Table 4; as usual, p denotes the smallest prime factor of n. All numbers in Table 4 were most or more wanted in the Cunningham Tables [6]. At the time of writing this paper we are collecting relations for a 103 digit number, the last entry in Table 4.

Table 4
Some factorizations obtained with mpqs

$\log_{10}n$	$\log_{10}p$	B	elapsed days	external help	n is factor of
92.1	35.3	25000	13	$\leq 1\%$	$7^{139}+1$
92.4	37.1	25000	11.5	5 %	$10^{109}+1$
95.3	47.5	37900	24	5 %	$11^{107}-1$
100.0	40.9	50000	26	15 %	$11^{104}+1$
101.1	46.4	65500	45	21 %	$2^{391}-1$
105.8	36.6	65500	120	30 %	$2^{353}+1$
102.88	?	65500	?	100 %	$10^{122}+1$

4. Distributing the factorization process

In the summer of 1987 we implemented elliptic curve factorization, distributed over the network of approximately 100 Firefly workstations in use at SRC. The overall structure of the distribution was similar to Silverman's system for quadratic sieve: one central system coordinates the factorization, searching out idle systems and supplying them with tasks. The internal structure is quite different, however.

To support research in distributed computation and recompilation, Ellis [23] built 'mi' and 'dp' — the machine information and distant process servers. 'mi' keeps track of the utilization of every workstation, the period of time since the last keystroke or button press, and an availability predicate provided by the owner of the workstation. Most machines have the default predicate, which allows distant processing only if the machine has been unused for at least half an hour on weekends or in the evening, or two hours during the workday. The distant process mechanism starts a program on a selected machine, transparently connecting open file descriptors for standard input and output back to the originating workstation.

In addition, the standard SRC environment includes a stub compiler for remote procedure calls. One constructs the interface in our standard programming language and supplies an implementation of the service. 'flume', the stub compiler, and the RPC runtime system then allow client programs to make calls to the server just by calling procedures in the interface. The RPC runtime includes provisions for reasonably prompt notification in the event that the client address space or workstation crashes.

Using these facilities, our factoring program was easy to write. We constructed a standalone executable which, when supplied with a random seed and an exponent bound, runs a trial of the elliptic curve algorithm. If it discovered a factorization, it printed it and exited. If it received a Unix interrupt signal, it printed a snapshot of the current state of the computation. The program accepted a description of a partially completed curve as an input to resume factorization. We then needed to invoke this using the distant process machinery.

However, Fireflies are multiprocessors; each has five processors. Thus, we wanted our program to run multiple times on each Firefly.

To facilitate this, we wrote a driver program which spawns one factoring task per processor.

The driver program keeps the workstation busy by using RPC to request a factorization task whenever some processor is idle. It also reports factorizations and snapshots to the central server. Additionally, this program watches for the workstation to come back into use. Should this happen, we take one of two actions. If the owner of the workstation has returned and pressed a button on his mouse or keyboard, we signal the workers to exit, collect their final reports, and exit the driver. If another distant process appears on the workstation, we defer to it by suspending execution of the workers until it completes.

The main program services requests for tasks and records reports of factorizations and snapshots. Periodically, it queries the 'mi' server for the list of all machines which are available, and some statistics about the machines. For each machine which is not currently running a factoring driver, and which has enough free virtual memory that we won't cause the machine to abort user programs, 'dp' is used to invoke a driver.

Using this setup, we accumulated an average computing power roughly equivalent (for our purposes) to a dedicated Cray-1. Following a processor upgrade to the Fireflies, we find ourselves running at a continuous rate of roughly 500 mips. This is sufficient, as reported above, for the discovery of 30+ digit factors in a day.

Subsequently, we came to the realization that our needs could easily be satisfied by a much looser communications system. Our bandwidth requirements are small for elliptic curve -- we just need to send out the number to be factored, and collect the factorization if it ever appears. In particular, we realized that the current electronic mail networks should easily handle the traffic needed to run an elliptic curve factorization.

Political and technical realities made it preferable to attempt this with quadratic sieve, instead. As described above, elliptic curve is much more sensitive to the quality of implementation, and needs to be carefully tuned to run well on a given machine. Moreover, although the rate of progress is slower with quadratic sieve, it is certain, and the factorizations are impressive.

In the summer of 1988, we implemented a new worker program that runs quadratic sieve, printing the relations it finds. The driver and central server were modified to handle both elliptic curve and quadratic sieve. We used this setup to factor a 93 digit number and gain experience with the complete quadratic sieve process.

The worker was also capable of being run by hand and sending results by electronic mail. We distributed the program to a few of our friends, and asked them to tell us how it failed to be sufficiently portable.

We then refined the program, and factored a 96 digit number with more help from outside SRC. We then factored a 100 digit number, with the assistance of most of the prominent members of the factoring community. At this point, everything seemed sufficiently stable that we started to develop an analogue of the driver program that was simple and portable, which we announced the availability of on netnews, along with our most recent results.

The current implementation consists of the quadratic sieving program, some shell scripts, and the utilities we need to check the relations we receive for correctness and to find linear combinations that lead to factorizations.

At SRC, we have a special user account, factor@src.dec.com, that receives all the mail containing relations, requests for copies of the program, and requests for lists of tasks. A shell script examines the mail as it arrives, processing it as appropriate. This takes advantage of the feature of the Berkeley mail delivery program, which allows mail to be delivered as input to a user program.

At helper sites, we run a script that is a simple infinite loop, waking up once each minute. It checks that the factoring process is still running. If not, it waits a little while (in case it failed due to some persistent hardware problem) and starts a new one. It fetches the task to perform out of a list which can be shared by many computers, if the list is on a file system which can be read and written by those computers. The script also checks that the load average on the machine isn't too high, and that there's no recent typing activity by users. If either of these fails, the worker is suspended. On Unix systems, the script periodically enqueues itself in the 'at' queue, so that the program will survive a crash of the machine.

While these scripts are not particularly elegant, they are sufficiently portable that we have workers running on almost every known type of Unix system. A simple variant of these scripts also runs on VMS. As of this writing, we find that we have access to roughly 1000 mips of sustained computing power for factoring. We have not yet run up against the limits of what the mail systems will handle; we estimate that we could easily handle a factor of 10 more mail without imposing a burdensome load on our mail transport system. This would allow us to factor a 103 digit number in about a week. Since the number of messages is largely independent of the number being factored, we can factor larger numbers in roughly the same amount of time, if we can get our hands on enough computers; we might need to stretch the computation out to a month to factor 130 digit numbers if we didn't want to add load to the mail system.

Acknowledgments. Many thanks to John Ellis for building 'dp' and 'mi'. Without 'dp' and 'mi' we would not even have started this factorization project. Also many thanks to the people at SRC who made their machines available or shut them off for distant processes; they provided us with lots of cycles, and feedback about things we should **not** do. The first author would like to thank SRC for its great hospitality and support during the summers of 1987 and 1988.

Acknowledgments are due to Sam Wagstaff who tirelessly provided us with his endless lists of 'most wanted', 'more wanted', and 'most wanted unwanted' numbers.

And then it is our pleasure to acknowledge the support of our colleagues on the worldwide factorization network. At the time of writing this paper, the following people had donated their cycles: Dean Alvis, Indiana University at South Bend, South Bend, Indiana; Bean Anderson, Peter Broadwell, Dave Ciemiewicz, Tom Green, Kipp Hickman, Philip Karlton, Andrew Myers, Thant Tessman, Michael Toy, Silicon Graphics, Inc., Mountain View, California; Per Andersson, Christopher Arnold, Lennart Augustsson, Chalmers University of Technology, Gothenburg, Sweden; Butch Anton, Hewlett-Packard, Cupertino, California; Greg Aumann, Telecom Australia Research Laboratories, Melbourne, Australia; Gregory Bachelis, Robert Bruner, Wayne State University, Detroit, Michigan; Bob Backstrom, Australian Nuclear Science & Technology Organization, Lucas Heights, Australia; Richard Beigel, The Johns Hopkins University, Baltimore, Maryland; Anita Borg, Sam Hsu, Richard Hyde, Richard Johnsson, Bob Mayo, Joel McCormack, Louis Monier, Donald Mullis, Brian Reid, Michael Sclafani, Richard Swan, Terry Weissman, DEC, Palo Alto, California; Richard Brent, Australian National University, Canberra, Australia; Brian Bulkowski, Brown University, Providence, Rhode Island; John Carr, Jonathan Young, MIT, Cambridge, Massachusetts; Enrico Chiarucci, Shape Technical Centre, The Hague, The Netherlands; Ed Clarke, United Kingdom; Derek Clegg, Island Graphics Corporation, San Rafael, California; Robert Clements, BBN Advanced Computers Inc., Cambridge Massachusetts; Chris Cole, Peregrine Systems, Inc., Irvine, California; Leisa Condie, Australia; Bob Devine, DEC, Colorado Springs, Colorado; John ffitch, Dave Hutchinson, University of Bath, Bath, United Kingdom; Seth Finkelstein, MathSoft, Inc., Cambridge, Massachusetts; Mark Giesbrecht, University of Toronto, Toronto, Canada; Peter Graham, University of Manitoba, Winnipeg, Canada; Ruud Harmsen, ?, The Netherlands; Paw Hermansen, Odense University, Odense, Denmark; Phil Hochstetler, ?; Robert Horn, Memotec Datacom, North Andover, Massachusetts; Scott Huddleston, Tektronix, Beaverton, Oregon; Bob Johnson, Logic Automation Incorporated, Beaverton, Oregon; Arun Kandappan, University of Texas, Austin, Texas; Jarkko Kari, Valtteri Niemi, University of Turku, Turku, Finland; Gerard Kindervater, Erasmus Universiteit, Rotterdam, The Netherlands; Albert Koelmans, University of Newcastle upon Tyne, Newcastle, United Kingdom; Emile LeBlanc, Hendrik Lenstra, Kenneth Ribet, Jim Ruppert, University of California, Berkeley, California; Andries Lenstra, Amsterdam, The Netherlands; Paul Leyland, Oxford University, Oxford, United Kingdom; Walter Lioen, Herman te Riele, Dik Winter, Centrum voor Wiskunde en Informatica, Amsterdam, The Netherlands; David Lilja, Soren Lundsgaard, Kelly Sheehan, University of Illinois, Urbana-Champaign, Illinois; Barry Lustig, Advanced Decision Systems, Mountain View,

California; Ken Mandelberg, Emory University, Atlanta, Georgia; Peter Montgomery, University of California, Los Angeles, California; Francois Morain, Institut National de Recherche en Informatique et en Automatique, Le Chesnay, France; Marianne Mueller, Sun, Mountain View, California; Duy-Minh Nhieu, John Rogers, University of Waterloo, Waterloo, Canada; Gary Oberbrunner, Thinking Machines Corp., Cambridge, Massachusetts; Andrew Odlyzko, AT&T Bell Laboratories, Murray Hill, New Jersey; Michael Portz, Rheinisch-Westfaelische Technische Hochschule Aachen, Aachen, West-Germany; Jim Prescott, Rochester?; Ulf Rehmann, Universität Bielefeld, Bielefeld, West Germany; Mike Rimkus, The University of Chicago, Chicago, Illinois; Mark Riordan, Michigan State University, East Lansing, Michigan; Michael Rutenberg, Reed College, Portland, Oregon; Rick Sayre, Pixar, San Rafael, California; Charles Severance, ?; Jeffrey Shallit, Dartmouth College, Hanover, New Hampshire; Hiroki Shizuya, Tohoku University, Sendai, Japan; Robert Silverman, the MITRE Corporation, Bedford, Massachusetts; Malcolm Slaney, Apple, Cupertino, California; Ron Sommeling, Victor Eijkhout, Ben Polman, Katholieke Universiteit Nijmegen, Nijmegen, The Netherlands; Bill Sommerfeld, Apollo Division of Hewlett Packard, Chelmsford, Massachusetts; Mitchell Spector, Seattle University, Seattle, Oregon; Pat Stephenson, Cornell University, Ithaca, New York; Nathaniel Stitt, ?, Berkeley-area?, California; Marc-Paul van der Hulst, Universiteit van Amsterdam, Amsterdam, The Netherlands; Jos van der Meer, Sun Nederland, Amsterdam, The Netherlands; Frank van Harmelen, Edinburgh University, Edinburgh, Scotland; Brick Verser, Kansas State University, Manhattan, Kansas; Charles Vollum, Cogent Research, Inc., Beaverton, Oregon; Samuel Wagstaff, Purdue University, West-Lafayette, Indiana;

Our sincere apologies to everybody who contributed but whose name does not appear on the above list. Please let us know! Keeping track of the names of all our helpers proved to be the hardest part of this project.

References

1. L. Adleman, "The theory of computer viruses," Proceedings Crypto 88, 1988.

2. E. Bach. J. Shallit, "Factoring with cyclotomic polynomials," *Proceedings 26th FOCS*, 1985, pp 443-450.

3. G. Brassard, *Modern Cryptology*, Lecture Notes in Computer Science, vol. 325, 1988, Springer Verlag.

4. R.P. Brent, "Some integer factorization algorithms using elliptic curves," Australian Computer Science Communications v. 8, 1986, pp 149-163.

5. R.P. Brent, G.L. Cohen, "A new lower bound for odd perfect numbers," Math. Comp., *to appear*.

6. J. Brillhart, D.H. Lehmer, J.L. Selfridge, B. Tuckerman, S.S. Wagstaff, Jr., *Factorizations of $b^n\pm1$, $b = 2, 3, 5, 6, 7, 10, 11, 12$ up to high powers, second edition*, Contemporary Mathematics, vol. 22, Providence: A.M.S., 1988.

7. T.R. Caron, R.D. Silverman, "Parallel implementation of the quadratic sieve," J. Supercomputing, v. 1, 1988, pp 273-290.

8. A.J.C. Cunningham, H.J. Woodall, *Factorisation of $(y^n\mp1)$. $y = 2, 3, 5, 6, 7, 10, 11, 12$ up to high powers (n)*, London: Hodgson (1925).

9. J.A. Davis, D.B. Holdridge. "Factorization using the quadratic sieve algorithm," Sandia National Laboratories Tech Rpt. SAND 83-1346, December 1983.

10. P.J. Denning, "The Science of Computing: Computer Viruses," *American Scientist*, v. 76, May-June 1988.

11. A.K. Lenstra, H.W. Lenstra, Jr, "Algorithms in number theory," in: J. van Leeuwen, A. Meyer, M. Nivat, M. Paterson, D. Perrin (eds.), *Handbook of theoretical computer science*, to appear; report 87-8, The University of Chicago, Department of Computer Science, May 1987.

12. A.K. Lenstra, M.S. Manasse, "Compact incremental Gaussian elimination over $Z/2Z$," report 88-16, The University of Chicago, Department of Computer Science, October 1988.

13. H.W. Lenstra, Jr., "Factoring integers with elliptic curves," *Ann. of Math.*, v. 126, 1987, pp. 649-673.

14. P.L. Montgomery, "Modular multiplication without trial division," *Math. Comp.*, v. 44, 1985, pp 519-521.

15. P.L. Montgomery, "Speeding the Pollard and elliptic curve methods of factorization," *Math. Comp.*, v. 48, 1987, pp 243-264.

16. P.L. Montgomery, R.D. Silverman, "An FFT extension to the p-1 factoring algorithm," manuscript, 1988.

17. A.M. Odlyzko, "Discrete logarithms and their cryptographic significance," pp. 224-314; in: T. Beth, N. Cot, I. Ingemarsson (eds), *Advances in cryptology*, Springer Lecture Notes in Computer Science, vol. 209, 1985.

18. J.M. Pollard, "A Monte Carlo method for factorization," *BIT*, v. 15, 1975, pp 331-334.

19. C. Pomerance, "Analysis and comparison of some integer factoring algorithms," pp. 89-139; in: H.W. Lenstra, Jr., R. Tijdeman (eds), *Computational methods in number theory*, Mathematical Centre Tracts 154, 155, Mathematisch Centrum, Amsterdam, 1982.

20. C. Pomerance, J.W. Smith, R. Tuler, "A pipeline architecture for factoring large integers with the quadratic sieve algorithm," *SIAM J. Comput.*, v. 17, 1988, pp. 387-403.

21. H.J.J. te Riele, W.M. Lioen, D.T. Winter, "Factoring with the quadratic sieve on large vector computers," report NM-R8805, 1988, Centrum voor Wiskunde en Informatica, Amsterdam.

22. R.L. Rivest, A. Shamir, L. Adleman, "A method for obtaining digital signatures and public-key cryptosystems," *Commun. ACM.*, v. 21, 1978, pp. 120-126.

23. E. Roberts, J. Ellis, "parmake and dp: Experience with a distributed, parallel implementation of make," Proceedings from the Second Workshop on Large-Grained Parallelism, Software Engineering Institute, Carnegie-Mellon University, Report CMU/SEI-87-SR-5, November 1987.

24. R.D. Silverman, "The multiple polynomial quadratic sieve," *Math. Comp.*, v. 48, 1987, pp. 329-339.

25. K. Thompson, "Reflections on Trusting Trust," *Commun. ACM*, v. 27, 1984, pp. 172-80.

26. D.H. Wiedemann, "Solving sparse linear equations over finite fields," *IEEE Transactions on Information Theory*, v. 32, 1986, pp. 54-62.

CRYPTANALYSIS OF SHORT RSA SECRET EXPONENTS

Michael J. Wiener

Bell-Northern Research Ltd.

P.O. Box 3511 Station C
Ottawa, Ontario, Canada K1Y 4H7

Abstract[1]

A cryptanalytic attack on the use of short RSA secret exponents is described. This attack makes use of an algorithm based on continued fractions that finds the numerator and the denominator of a fraction in polynomial time when a close enough estimate of the fraction is known. The public exponent e and the modulus pq can be used to create an estimate of a fraction that involves the secret exponent d. The algorithm based on continued fractions uses this estimate to discover sufficiently short secret exponents. For a typical case where $e < pq$, $\gcd(p-1, q-1)$ is small, and p and q have approximately the same number of bits, this attack will discover secret exponents with up to approximately one-quarter as many bits as the modulus. Ways to combat this attack, ways to improve it, and two open problems are described. This attack poses no threat to normal case RSA where the secret exponent is approximately the same size as the modulus. This is because this attack uses information provided by the public exponent and, in the normal case, the public exponent can be chosen almost independently of the modulus.

---

[1]For the full paper, the reader is referred to the IEEE Transactions on Information Theory, Vol. 36, No. 3, May 1990, p. 553-558.

HOW TO BREAK THE DIRECT RSA-IMPLEMENTATION OF MIXES

Birgit Pfitzmann Andreas Pfitzmann

Institut für Rechnerentwurf und Fehlertoleranz, Universität Karlsruhe
Postfach 6980, D-7500 Karlsruhe 1, F. R. Germany

ABSTRACT

MIXes are a means of untraceable communication based on a public key cryptosystem, as published by David Chaum in 1981 (CACM 24/2, 84-88) (=[6]).

In the case where RSA is used as this cryptosystem directly, i.e. without composition with other functions (e.g. destroying the multiplicative structure), we show how the resulting MIXes can be broken by an active attack which is perfectly feasible in a typical MIX-environment.

The attack does not affect the idea of MIXes as a whole: if the security requirements of [6] are concretized suitably and if a cryptosystem fulfils them, one can implement secure MIXes directly. However, it shows that present security notions for public key cryptosystems, which do not allow active attacks, do not suffice for a cryptosystem which is used to implement MIXes directly.

We also warn of the same attack and others on further possible implementations of MIXes, and we mention several implementations which are not broken by any attack we know.

I. INTRODUCTION: MIXES

Basically, a MIX-network [6] is a means of sender anonymity, which can at the most be computationally secure.

Meanwhile, other sender anonymity schemes have been published, which are information-theoretically secure, namely superposed sending (DC-net) [7, 8] and, against more limited attackers, RING-networks [18]. Nevertheless, MIXes are still a matter of interest, since their communication overhead is much smaller. More precisely, in the other schemes, each participant has to send about as much in the physical sense as all the participants together want to send in the logical sense. With MIXes, the overhead is at most about the product of what the participant himself wants to send and the number of MIXes he uses; for long messages in some MIX-schemes there is nearly no overhead. Therefore, MIXes seem to be the only way to provide sender anonymity for telephony using the cables of conventional telephone networks, i.e., the only way complete privacy can be introduced in public communication networks in the near future [20].

The idea behind MIX-networks is that a, hopefully trustworthy, station called MIX collects a number of messages from their senders, performs a cryptographic operation on each of them to change their outlooks, and outputs them to their addressees in a different order. Thus an attacker should not be able to find out which outgoing message corresponds to which incoming one (except possibly his own).

Of course, the recipients should be able to read their messages in spite of the change in outlook. Therefore, in the basic scheme (which is part of nearly all more sophisticated schemes), the MIX achieves this change by deciphering using a public key cryptosystem, and the senders must prepare their messages by enciphering them with the public key of the MIX. (If every message passed only one MIX, the scheme could be changed so that the MIX would use a private key with each sender. But usually, several MIXes in a row are used for every message, because in this case it suffices that one of them is trustworthy, and those in the middle may know neither sender nor recipient.)

So far, since the encryption function was assumed to be deterministic in [6], like that of RSA, everybody could take an output message, encrypt it again and check which input message they obtain. To avoid this reencryption attack, nondeterminism was introduced by attaching a random part to each message before encryption. After decryption, the MIX deletes this part. Another simple attack, replay, was avoided by having the MIX discard repeated input messages. Otherwise, an attacker could find out what becomes of an input message by repeating it and observing which output message is repeated.

To avoid confusion about which kinds of MIXes we break and which not, we distinguish (top-down) the basic idea of a MIX-network, MIX-schemes (e.g. the described basic scheme, or the return address scheme [6]), implementations of MIX-schemes (e.g. additional use of redundancy), and the choice of the cryptosystem to be used. (The "unsealing" operation in [6] is just the cryptographic operation the MIX performs in the basic MIX-scheme, "sealing" the corresponding operation the sender performs.)

In [6], the random part is always implemented as a large random string in front of the message. (This does not imply that no other operations can be performed, but that they would have to be considered as a part of the cryptosystem; this is no loss in generality if the requirement "$K(K^{-1}(X))=X$", which implies bijectivity, is dropped. For our purposes it is better to discuss such operations seperately as implementation.)

As an example of a public key cryptosystem, RSA is mentioned, and at that time there seems to have been little choice.

Without knowing about further attacks, one might have considered it natural to use RSA directly as the cryptosystem in the implementation with the random string in front. We call this the *direct RSA-implementation of MIXes*. For this case, the operation of a MIX with the modulus $m = p \cdot q$, the public exponent c for enciphering, and the private exponent d for deciphering is shown in Figure 1. N_1 and N_2 are two messages, R_1 and R_2 the attached random strings, and the commas denote concatenation.

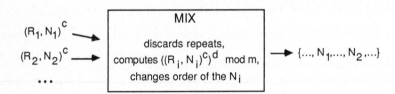

Figure 1 Operation of a MIX of the basic scheme in its direct RSA- implementation

II. THE ATTACK

The attack as described in this section is specific for the direct RSA-implementation of the basic MIX-scheme (cf. Section I). In Section III we discuss which other possible schemes, implementations, and cryptosystems are vulnerable.

II.1. HISTORY

The attack is based upon the well-known attack on RSA, which exploits the fact that RSA is a multiplicative homomorphism, by Davida [9] in the version by Judy Moore (according to [11]).

It has been adapted to other situations before (no guarantee on completeness): to cryptosystems with some abstract properties in [10, 17], to signatures with some redundancy in [16], or to yield a factoring algorithm [12]. It was put to positive use for blind signatures and, thereby, for untraceable credentials and payments (e.g. [7]). It also forms a small substep in proofs of the security of single bits of RSA, from [15] to [1]).

II.2. THE IDEA

In our case, the difficulty with the well-known RSA-attack lies in the fact that the random string in a decrypted message is not output, and that an attacker who forms his message according to the attack, instead of according to the MIX-scheme, does not know which output corresponds to his own input. (This last fact is also the reason why the system is not trivially broken by the active attacks of [15], for which an oracle outputting the last bit of a message suffices.)

All following congruences are modulo m. $L := \lceil \log_2(m) \rceil$ is the block length of the cipher. Let the first b bits be reserved for the random strings and the remaining B bits for the messages, i.e. a message looks like this:

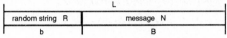

Thus the encrypted form M of a message N with attached random string R is

$$M \equiv (R \cdot 2^B + N)^c \bmod m.$$

Consider an attacker who wants to trace such a message M which was input to the MIX (see Figure 2). He chooses a "small" (cf. II.4.) factor f, forms $M^* := f^c \cdot M$ and inputs M^* to the MIX. On the one hand, the MIX decrypts M^* and interprets it as a message N^* with an attached random string R^*, i.e. as

$$M^{*d} \equiv R^* \cdot 2^B + N^*,$$

of which it outputs N^*.

BATCH 1:

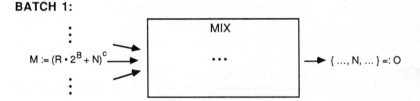

BATCH 2:

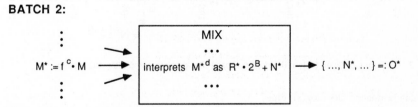

Figure 2   Scenario of the attack; the two batches are not necessarily distinct

On the other hand, the attacker knows that

$$M^{*d} \equiv f^{c \cdot d} \cdot M^d \equiv f \cdot (R \cdot 2^B + N).$$

Together this implies $N^* - f \cdot N \equiv f \cdot R \cdot 2^B - R^* \cdot 2^B$. Since $\gcd(2^B, m) = 1$ and 2^B therefore has an inverse modulo m, this becomes

$$(N^* - f \cdot N) \cdot 2^{-B} \equiv f \cdot R - R^*. \tag{$*$}$$

On the one hand, the attacker knows f and 2^{-B}, and that N and N* are in the sets O and O* of output messages of the respective batches. Hence if the batch size is reasonably small (this is always true in practice, cf. II.3.), he can compute all possible left sides of Equation (✳), i.e. all $(N_2 - f{\cdot}N_1) \cdot 2^{-B}$ for $N_1 \in O, N_2 \in O^*$.

On the other hand, R, R* $\in \{0,..., 2^b-1\}$, thus for the right side of (✳)

$$f \cdot R - R^* \in \{-2^b+1,..., f{\cdot}(2^b-1)\}.$$

The attacker now tries to find out which pair (N_1, N_2) of output messages is (N, N^*) by computing the left side of Equation (✳) for each of them and testing for this condition. If f has been chosen to be suitably small (cf. II.4.), the condition will not hold for most other values $(N_2 - f{\cdot}N_1) \cdot 2^{-B}$.

II.3. OVERHEAD

One can ask whether the batch size s can be made so large that the attacker cannot consider all pairs of output messages. For some time-critical services this seems impossible anyway, because there would not be enough messages to collect within the permitted time [19]. Anyway, the attacker can also speed up his attack by computing the sets of values which occur in the left sides of Equation (✳), i.e.

$$V_1 := \{f \cdot N_1 \cdot 2^{-B} \mid N_1 \in O\} \quad \text{(including } f \cdot N \cdot 2^{-B})$$

and

$$V_2 := \{N_2 \cdot 2^{-B} \mid N_2 \in O^*\} \quad \text{(including } N^* \cdot 2^{-B})$$

(all numbers reduced modulo m) and sorting them. Now he has to check for which $v_1 \in V_1$ there is a $v_2 \in V_2$ such that

$$v_2 - v_1 \in \{-2^b+1,..., f{\cdot}(2^b-1)\} \bmod m. \qquad (\diamond)$$

This might look like Figure 3.

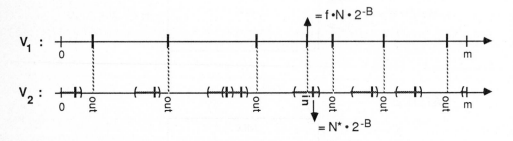

Figure 3 Test for Condition (✧); in the depicted case the attack is completely successful

All this can be done in $O(s \cdot \log(s))$ time.

So the attacker mainly has to carry out 1 modular exponentiation and division, 3·s modular multiplications, and to sort s numbers modulo m twice. The MIX itself has to carry out s modular exponentiations and to sort s numbers modulo m once (and many more, if the key is not changed with every new batch). Thus the complexity of the attack is nearly the same as that of legal mixing. Therefore the attack is always feasible.

II.4. CHOOSING THE FACTOR f AND THE PROBABILITY OF SUCCESS

The attacker is completely successful in identifying the message N, if $f \cdot N \cdot 2^{-B}$ is the only v_1 to fulfil Condition ($\diamondsuit$). (It does not matter whether $N^* \cdot 2^{-B}$ is also the only v_2, because the attacker need not identify N^*.)

Assume that the values $v_1 \in V_1$ and $v_2 \in V_2$, except $N^* \cdot 2^{-B}$, were chosen from $\{0,..., m-1\}$ randomly and independently. The probability that a specific $v_1 \neq f \cdot N \cdot 2^{-B}$ is near any of the v_2's in the sense of ($\diamondsuit$), i.e. in the union of the intervals around them (cf. Figure 3), is then bounded by

$$\mathrm{Prob}(v_1) < s \cdot |\{-2^b+1,..., f \cdot(2^b-1)\}| \ / \ m \ = \ s \cdot ((f+1) \cdot 2^b - f) \ / \ m$$

$$\approx s \cdot f \cdot 2^b \ / \ 2^{b+B} \ = \ s \cdot f \ / \ 2^B.$$

The messages N_1, N_2 can be assumed to be end-to-end encrypted and thus to be chosen randomly and independently. However, they are only chosen from $\{0,..., 2^B-1\}$. Therefore the assumption is not really justified. Nevertheless, not knowing better, we will maintain it and hope that it is not too unreasonable, since the messages are multiplied by the rather large number 2^{-B} (it is at least $\approx 2^b$). The probability of complete success, i.e. that none of the s-1 other v_1's fulfils the condition, is then bounded by

$$\mathrm{Prob}(\mathrm{success}) > 1 - (s-1) \cdot \mathrm{Prob}(v_1) \approx 1 - s^2 \cdot f \ / \ 2^B.$$

Note that this probability is independent of the length b of the random string. Hence if B is large, say $B > 200$, the attacker can choose any factor $f < 2^{100}$, and success will be quite certain since the batch size s can be assumed to be smaller than 10^{10}.

Thus those implementations that seemed most efficient when only the previously known attacks were considered, are insecure: One would have chosen $L > 600$ to ensure the security of the RSA modulus against factoring, and would have considered $b \approx 100$ sufficient to prevent an attacker from trying all random strings in a reencryption attack. The remaining $B > 500$ bits would have been used for the messages.

The attack can still succeed if, in most cases, more than one v_1 fulfils Condition ($\diamondsuit$): The attacker can choose several factors f to attack the same message. For each f, messages $N_1 \in O$ which cannot be N according to ($\diamondsuit$) can be excluded from further consideration. In this paragraph, we denote everything which corresponds to a given factor f by an additional subscript f.

Assume that the attacker inputs each M^*_f in a different batch. Then the corresponding sets $V_{2,f}$ are independent except for the values $N^*_f \cdot 2^{-B}$. Thus the conditions ($\diamondsuit$) for the different $v_{1,f}$'s arising from a fixed $N_1 \neq N$ are nearly independent. The probability that the attacker cannot exclude N_1 from the possible values of N is therefore the product of the probabilities $\mathrm{Prob}(v_{1,f})$ over all chosen values f. He can minimize it by choosing as many small f's as possible, i.e. $f := 2, 3,....$ If 2^B is not much larger than s, he must stop before $f > 2^B/s$.

The attacker can do even better. If ($\diamondsuit$) holds for a given $v_{1,f}$, this is usually the case for just one $v_{2,f}$. He can use this $v_{2,f}$ to compute a tighter bound on R, which in turn makes the interval in ($\diamondsuit$) smaller for the following factors f: If $v_{1,f}$ and $v_{2,f}$ really correspond to N and N^*, resp., then $v_{2,f} - v_{1,f} \equiv f \cdot R - R^*_f$. This gives $f \cdot R \equiv v_{2,f} - v_{1,f} + R^*_f$. Let x denote the right side of this congruence reduced mod m, i.e. $0 \leq x < m$. Since $0 \leq f \cdot R < m$, too (otherwise ($\diamondsuit$) would always be true), the equation $f \cdot R = x$ holds in $\mathbb{Z}$. This means $R = x / f$. Since only R^*_f is unknown in x, x lies in an interval of size 2^b. Thus the attacker obtains an interval of size $2^b/f$ in which R lies. This knowledge can be used in ($\ast$) to improve ($\diamondsuit$).

To be safe from this attack, $\mathrm{Prob}(v_1)$ must be about 1 even for small f. This means approximately $2^B < s$. Choosing B so small in the basic MIX-scheme would not only reduce efficiency, the main advantage of MIXes, but is rather senseless, because it means that in every batch all possible messages occur, or the same message occurs several times. Thus the recipients would receive no information (in other words: mostly, not even addressing would be possible, to say nothing of any message content). Other MIX-schemes where this does not hold are mentioned in Section III.

II.5. FEASIBILITY OF THE ACTIVE PART OF THE ATTACK

In principle, MIXes are better candidates for active attacks than most other users of a public key system for two reasons: First, they are forced to output much larger parts of messages they decrypted. Secondly, these messages are meaningless to them, because they are usually end-to-end-encrypted. Hence a MIX cannot detect an active attack by means of natural redundancy in the messages, as other users sometimes might. Nevertheless, there could be two problems for the attacker.

First, the attacker must see the message M he wants to trace before he can form a suitable M*. In analogy to the terminology of [14], one could call this a *directed* chosen ciphertext attack, directed against a particular input message.

If the MIX uses the same key for many batches, this is no problem. If a new key is used for each new batch, M* must be submitted to the MIX in the same batch as M. If the input messages are really collected, i.e. if they can arrive at any time and are stored by the MIX, the attacker still has a lot of time for his attack, at least against the early messages. (This would not hold for an *adaptive* chosen ciphertext attack, where the attacker would need an output from the MIX to form another input, e.g. if one could transform one of the attacks in [15] into a MIX-attack).

The attack can be prevented (without changing the scheme, implementation or cryptosystem of the MIX), if the MIX-network is changed so that the participants themselves store their messages and send them to the MIX at a prearranged time. But care must still be taken because the attacker only needs to perform one modular multiplication between seeing M and submitting M*. Hence if the messages arrive via the same time-division network, or from different local networks using different local clocks, the time might still suffice. So all the participants must be forced to send (or to commit to, cf. [5]) considerable parts of their messages before the first one has completed his. Also, this measure can no longer be applied if messages pass several MIXes: Here, each MIX must be considered as a potential attacker against its successor (this is why several MIXes are used), and of course it receives the messages early enough to perform an attack on them.

Secondly, if the MIXes are arranged as cascades in the sense that several MIXes must be passed in fixed order, only the first MIX is accessible to normal users directly. But this does not prevent an active attack on one of the others. Firstly, the previous MIX is a potential attacker (see above). Secondly, an attacker can prepare a message to attack the i-th MIX so that it passes through the previous i-1 MIXes correctly.

III. VARIATIONS

III.1. OTHER MIX-SCHEMES

The same attack can be used against the basic return address scheme [6, p.85] (if the same implementation and cryptosystem are used), because there, the address part of a message is just like a message in the basic scheme.

Many other schemes are based upon the idea in [6, p. 87] to use the basic scheme only for the first block of a message input to a MIX, and to include a private key in this block. The MIX uses this key to change the outlook of the rest of the message. Such schemes can enable better untraceability, increased performance, and fault tolerance [6, 19].

If, as in the original version in [6, p. 87], the address of the recipient or the following MIX is included in the first block, this address has the same effect as the message in the basic scheme. Thus the attack can be carried out. It will succeed if the batch size is usually smaller than the address space (cf. II.4.). This is quite likely e.g. if every station can act as a MIX. In this case, the attacker can even find out the enclosed secret key: Once he is able to trace messages, he can use the MIX as an oracle outputting the last bits of decrypted messages in Algorithm 1 of [15]. Of course, this only provides additional valuable information if the same key is to be used for future messages which have no obvious connection with the one already traced.

If the address is included in one of the other blocks, or if MIX-cascades must be passed in fixed order and therefore all but the last one output no address, this kind of attack is impossible.

Even if the implementation of the basic scheme used for the first block is secure, the danger of another, quite trivial, attack on such a scheme should not be overlooked: The secret key should not belong to a block cipher in electronic code book mode. Otherwise, patterns of equal blocks within a message would be repeated in the output. This is not necessarily checked by the MIX, because only repeats of the first blocks need to be discarded to prevent the replay attack (cf. Section I), and even if the senders took care to avoid this situation, an active attacker could replace some blocks in a message by repeats of others.

III.2. OTHER IMPLEMENTATIONS OF THE BASIC SCHEME

The same kind of attack is applicable (even easier), if the random string is attached at the end of the message instead of in front: A message is prepared as $M \equiv (N \cdot 2^b + R)^c$, and instead of Equation ($\ast$), one has $(N^* - f \cdot N) \cdot 2^b \equiv f \cdot R - R^*$.

A weaker form of the attack is still possible if the random part appears as bits inserted at predefined positions between the message bits, and either the random part is shorter than the message or there exists a large uninterrupted block of message bits:

Let "ins(R, N)" denote the operation of inserting the bits of R into N. An attacker who chooses M^* as in II.2. knows that, similar to ($\ast$), $\text{ins}(R^*, N^*) \equiv f \cdot \text{ins}(R, N) \mod m$. He can transform this into equations in $\mathbb{Z}$, namely

$$f \cdot \text{ins}(R, N) = \text{ins}(R^*, N^*) + k \cdot m \quad \text{for some } k \in \{-1, \ldots, f-1\}.$$

If f is chosen to be a power of 2, $f \cdot \text{ins}(R, N)$ is just the bit pattern of ins(R, N), shifted to the left. If, moreover, f is very small, then for each $N_1 \in O$, $N_2 \in O^*$, and each possible k, the attacker can test whether it is possible that this equation holds: For fixed k he has a sum in $\mathbb{Z}$, where some bits are missing in two of the three numbers. He must test if the remaining patterns match.

Assume, e.g., that there is a block of 10 adjacent bits of N in ins(R, N), and that f=2. Then there are 9 adjacent bit positions where the bits of all the three numbers are known. Since there are two possible carries from the right, the probability that the equation holds on these bits is only 2^{-8} for randomly chosen N_1, N_2. In this way, each block of at least three adjacent bits can be tested.

Even if these tests don't rule out any message pairs and k's at once, the attacker can be successful. If, e.g., exactly every other bit is a message bit and f=2, then with each pair (N_1, N_2), the equation yields unique solutions R and R^*. Using these, the attacker can reencrypt ins(R, N_1) and check if it is M.

Nevertheless there are easy countermeasures to prevent this kind of attack. One possibility is to merge the random string and the message part completely before encryption, e.g. by encryption in another, unrelated cryptosystem (where the key need not be secret).

The other is to make the active attack infeasible by adding redundancy to the messages, as if active attacks on RSA itself are to be prevented (e.g. the image of N under a one-way function, again merged with the message somehow), so that the MIX will usually not accept $f^c \cdot M$ as properly formed.

Especially with MIXes, one could try to discard not only repeats, but also multiples of previous messages by small factors. "Is $M^* = f^c \cdot M$ for some small f?" can be tested as "Is $f := M^{*d} \cdot M^{-d}$ small?", thus with an expense of mainly one multiplication per message pair. If n is the number of messages mixed using the same key, this increases the complexity of mixing from $O(n \cdot \log(n))$ to $O(n^2)$ (ignoring $\log(m)$-factors). For some services this is still feasible, at least if the key is changed with every new batch. For the basic scheme, "f is small" should be defined as "the check ($\diamondsuit$) (cf. II.3.) provides information". Thus f is considered large enough if ($\diamondsuit$) holds for all possible pairs (v_1, v_2). This means $(f+1) \cdot 2^b - f \geq m$, i.e. about $f > 2^B$. The probability that no legal messages collide in this sense is at least about $1 - n^2 \cdot 2^B / m \approx 1 - n^2 / 2^b$, if the M^d's are chosen randomly (cf. II.4.). This is tolerable for some parameters. Nevertheless the more common redundancy schemes seem more convenient for both users and MIXes.

III.3. OTHER CRYPTOSYSTEMS

Also the direct use of another cryptosystem in the otherwise unchanged implementation would not necessarily be helpful (or in an implementation without the random string, if the cryptosystem is already probabilistic). Some public key cryptosystems which are provably secure against passive attacks are definitely not suitable, because they are known to be vulnerable to active attacks, and versions of these attacks can still be applied directly in the MIX-environment:

With the quadratic residuosity system of [13], the attack of Example 6 in [15], where the attacker inserts an encrypted bit of someone else's message into his own message in a disguised form, is possible.

With the system of [3] (the version as secure as factorization) one can apply one of the attacks which the authors themselves probably mean when stating that the system is insecure against active attacks: (Remember that encryption means that a pseudorandom number generator is run, which repeatedly squares a number and outputs the last bits each time. This output is added to the message from right to left. At the end, the following square is appended to the encrypted message, so that the recipient can recover the seed.)

To obtain the last bit of a certain square root of a given number z, the attacker uses z like the seed for the pseudorandom number generator, but squares z fewer times than prescribed. Now he adds the resulting (too short) bit string to the left part of an arbitrary message N (with or without an attached random string R, resp.). The MIX subtracts the same string again. Thus the attacker can recognize N by the unchanged left part, if he has chosen a sufficiently long string. The changes in the following $\log(\log(m))$ bits of N are the last bits of the square root of z. The last one of these was desired. Hence the attacker has a suitable oracle to factor using Algorithm 5 in [15]. (For an attack against the cryptosystem alone, all the squaring to recognize N is not needed.)

The second attack, being adaptive, can be avoided if the key is changed with every new batch. Redundancy in the messages alone would not help, because in each step the attacker can determine the message the MIX sees nearly completely. For the remaining $\log(\log(m))$ bits he can try every combination. This, in its turn, can be avoided by interdicting repeats of seeds for the pseudorandom number generator and of the intermediate squares.

There does not seem to be much use in discussing these special improvements to the implementation of the basic MIX-scheme further, because many seem secure against this special kind of attack and, at the moment, none of them seems provably secure.

Of course, the final consequence of discussions about both random strings and redundancy is that a public key cryptosystem used in the basic MIX-scheme should be probabilistic and secure against active attacks (cf. [2], but we have heard doubts about it; an interactive system, on the other hand, is quite unwieldy if several MIXes in a row are used [4]).

We are happy to thank Manfred Böttger and Michael Waidner for helpful comments, and Michael for lots of other support, too.

REFERENCES

[1]   W. Alexi, B. Chor, O. Goldreich, C. P. Schnorr: RSA and Rabin functions: Certain parts are as hard as the whole; SIAM J. Comput. 17/2 (1988) 194-209.

[2]   M. Blum, P. Feldman, S. Micali: Non-interactive zero-knowledge and its applications; 20th STOC, ACM, New York 1988, 103-112.

[3]   M. Blum, S. Goldwasser: An Efficient Probabilistic Public-Key Encryption Scheme Which Hides All Partial Information; Crypto '84, LNCS 196, Springer-Verlag, Heidelberg 1985, 289-299.

[4]   M. Böttger: Untersuchung der Sicherheit von asymmetrischen Kryptosystemen und MIX-Implementierungen gegen aktive Angriffe; Studienarbeit am Institut für Rechnerentwurf und Fehlertoleranz, Universität Karlsruhe 1989.

[5]   G. Brassard: Modern Cryptology - A Tutorial; LNCS 325, Springer-Verlag, Berlin 1988.

[6]   D. L. Chaum: Untraceable Electronic Mail, Return Addresses, and Digital Pseudonyms; CACM 24/2 (1981) 84-88.

[7] D. Chaum: Security without Identification: Transaction Systems to make Big Brother Obsolete; CACM 28/10 (1985) 1030-1044.

[8] D. Chaum: The Dining Cryptographers Problem: Unconditional Sender and Recipient Untraceability; J. of Cryptology 1/1 (1988) 65-75.

[9] G. Davida: Chosen Signature Cryptanalysis of the RSA (MIT) Public Key Cryptosystem; TR-CS-82-2, University of Wisconsin, Milwaukee (October 1982) (quoted in [11]).

[10] R. A. DeMillo, M. Merritt: Chosen Signature Cryptanalysis of Public Key Cryptosystems; Technical Memorandum, School of Information and Computer Science, Georgia Institute of Technology, Atlanta 1982. (quoted in [17]).

[11] D. E. Denning: Digital Signatures with RSA and Other Public-Key Cryptosystems; CACM 27/4 (1984) 388-392.

[12] Y. Desmedt, A. M. Odlyzko: A chosen text attack on the RSA cryptosystem and some discrete logarithm schemes; Crypto '85, LNCS 218, Springer-Verlag, Heidelberg 1986, 516-522.

[13] S. Goldwasser, S. Micali: Probabilistic Encryption; J. of Computer and System Sciences 28 (1984) 270-299.

[14] S. Goldwasser, S. Micali, R. L. Rivest: A Digital Signature Scheme Secure Against Adaptive Chosen-Message Attacks; SIAM J. Comput. 17/2 (1988) 281-308.

[15] S. Goldwasser, S. Micali, P. Tong: Why and How to establish a Private Code On a Public Network; 23rd FOCS, IEEE Computer Society, 1982, 134-144.

[16] W. de Jonge, D. Chaum: Attacks on Some RSA Signatures; Crypto '85, LNCS 218, Springer-Verlag, Berlin 1986, 18-27.

[17] M. John Merritt: Cryptographic Protocols; Ph. D. Dissertation, School of Information and Computer Science, Georgia Institute of Technology, February 1983.

[18] A. Pfitzmann: A switched/broadcast ISDN to decrease user observability; 1984 International Zurich Seminar on Digital Communications, IEEE, 1984, 183-190.

[19] A. Pfitzmann: How to implement ISDNs without user observability - Some remarks; Fakultät für Informatik, Universität Karlsruhe, Interner Bericht 14/85.

[20] A. Pfitzmann, B. Pfitzmann, M. Waidner: Datenschutz garantierende offene Kommunikationsnetze; Informatik-Spektrum 11/3 (1988) 118-142.

AN INFORMATION-THEORETIC TREATMENT OF HOMOPHONIC SUBSTITUTION

Hakon N. Jendal, Yves J.B. Kuhn & James L.Massey

Institute for Signal and Information Processing
Swiss Federal Institute of Technology
CH-8092 Zürich, Switzerland

1. INTRODUCTION

The history of cryptology shows that most secret-key cipher systems that have been broken were broken by exploiting the departure of the plaintext statistics from those of a completely random sequence. The technique of "homophonic substitution" is an old technique for converting an actual plaintext sequence into a (more) random sequence. At EUROCRYPT '88, Günther [1] introduced an important generalization of homophonic substitution, which we will call "variable-length homophonic substitution". The purpose of this paper is to give an information-theoretic treatment of Günther's type of homophonic substitution.

In Section 2, we give a rather careful discussion of Shannon's concept of a "strongly-ideal" cipher system, as this provides the motivation for any type of homophonic substitution. Section 3 gives the precise definition of variable-length homophonic substitution together with the necessary and sufficient condition for such substitution to be perfect, i.e., to create a completely-random sequence. Section 4 shows that perfect homophonic substitution can be achieved by the introduction of less than 2 bits of entropy into each source letter that is coded, and Section 5 shows that such perfect homophonic substitution can be realized using less than 4 random bits per letter coded. Section 6 indicates certain obvious generalizations of the previous results and mentions their implications for source coding (or "data compression").

The information-theoretic results used in this paper are quite basic and may be found in any good textbook on information theory, e.g., the book by Gallager [2].

2. STRONGLY-IDEAL AND UNBREAKABLE CIPHER SYSTEMS

The purpose of "homophonic substitution" can be explained by considering a secret-key cipher system as diagrammed in Fig. 1. For ease of notation, let X^n and Y^n denote the plaintext and ciphertext sequences $[X_1, X_2,..., X_n]$ and $[Y_1, Y_2, ..., Y_n]$, respectively. As customary and as Fig. 1 suggests, we assume always that the secret key Z is statistically independent of the plaintext sequence X^n for all n. We shall call the cipher <u>non-expanding</u> if the plaintext digits and ciphertext digits take values in the same D-ary alphabet and there is an increasing infinite sequence of positive integers $n_1, n_2, n_3, ...$ such that, when Z is known, X^n and Y^n uniquely determine one another for all $n \in S = \{n_1, n_2, n_3, ...\}$. We shall also call a sequence of D-ary random variables <u>completely random</u> if each of its digits is statistically independent of the preceding digits and is equally likely to take on any of the D possible values. The following proposition is proved in the Appendix by elementary information-theoretic arguments.

<u>Proposition 1</u>: If the plaintext sequence encrypted by a non-expanding secret-key cipher is completely random, then the ciphertext sequence is also completely random and is also statistically independent of the secret-key.

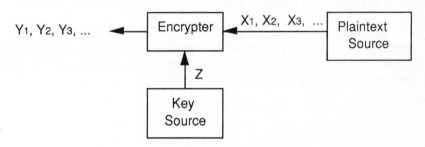

Fig. 1: A secret-key cipher system

Shannon [3] has defined the <u>key-equivocation</u> function f(n) of a secret-key cipher system to be the conditional entropy of the key given the first n digits of ciphertext, i.e., $f(n) = H(Z \mid Y^n)$. The key-equivocation function f(n) is thus a measure of the number of values of the secret key Z that are consistent with the first n digits of ciphertext. Because f(n) can only decrease as n increases, Shannon called a cipher system <u>ideal</u> if f(n) approaches a non-zero value as n tends toward infinity, and <u>strongly ideal</u> if f(n) is constant, i.e., if $H(Z \mid Y^n) = H(Z)$ for all n, which is equivalent to the statement that the ciphertext sequence is statistically independent of the secret key.

Corollary 1 to Proposition 1: If the plaintext sequence encrypted by a non-expanding secret-key cipher is completely random, then the cipher system is strongly ideal (regardless of the probability distribution for the secret key).

Virtually all useful non-expanding ciphers have the property, which we call "non-degeneracy", that changing the value of the secret key Z, without changing the value of the plaintext sequence X^n, will change the value of the ciphertext sequence for all n sufficiently large, except for a negligibly small fraction (often 0) of possible key values for any given value of X^n. Equivalently, a non-expanding cipher is non-degenerate if

$$H(Y^n \mid X^n) \approx H(Z)$$

holds for all sufficiently large n and all probability distributions for X^n when all possible values of the secret key Z are equally likely. But, as shown in the Appendix,

$$H(Y^n \mid X^n) = H(X^n \mid Y^n)$$

holds for all n in a non-expanding cipher when the plaintext sequence X_1, X_2, ... is completely random. The following conclusion is immediate.

Corollary 2 to Proposition 1: If the plaintext sequence encrypted by a non-expanding secret-key cipher is completely random and all possible key values are equally likely, then the conditional entropy of the plaintext sequence given the ciphertext sequence satisfies

$$H(X^n \mid Y^n) \approx H(Z)$$

for all n sufficiently large.

This corollary implies in particular that, in a ciphertext-only attack, the cryptanalyst can do no better to find X^n than by guessing at random from among as many possibilities as there are possible values of the secret key Z. In other words, the cipher system is unbreakable in a ciphertext-only attack when the number of possible key values is large.

The foregoing has shown that virtually any non-expanding secret key cipher can be used as the cipher in an unbreakable cipher system, provided that the plaintext source emits a completely random sequence. But it is precisely the goal of "homoponic substitution" to convert a source not of this type into such a source. When the homophonic coding is "perfect", it is then a trivial task to build unbreakable secret-key cipher systems in the form shown in Fig. 2.

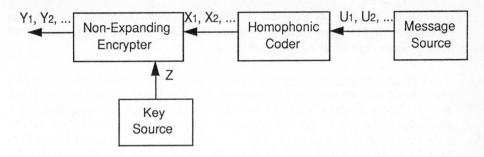

Fig. 2: Use of homophonic substitution within a secret-key cipher system

3. VARIABLE-LENGTH HOMPHONIC SUBSTITUTION

Here and hereafter, we will look upon the plaintext source of the previous section as the result of coding the actual <u>message source</u>, whose output sequence we denote by $U_1, U_2, U_3, ...$, into the D-ary sequence $X_1, X_2, X_3, ...$. We assume that the random variables U_i take values in an alphabet of L letters where $2 \le L < \infty$. Until further notice, we assume that the source is <u>memoryless and stationary</u> or, equivalently, that $U_1, U_2, U_3, ...$ is a sequence of independent and identically-distributed (i.i.d.) L-ary random variables. The coding problem for the actual message source then reduces to the coding problem for the single random variable $U = U_1$. To avoid uninteresting complications, we assume hereafter that all L values of D have non-zero probability.

Note that, when $L = D^w$ for some positive integer w and when all L possible values of U are equally likely, the simple coding scheme of assigning a different one of the D^w D-ary sequences of length w to each value of U makes the codeword $X_1, X_2, ..., X_w$ completely random. <u>Conventional homophonic substitution</u> attempts to achieve this same result when the values of U are not equally likely by choosing (if possible) an appropriate w with $D^w > L$, partitioning the D^w D-ary sequences of length w into L subsets, placing these subsets in correspondence with the values of U in such a manner that the number of sequences in each subset is proportional to the probability of the corresponding value of U, and then choosing the codeword for a particular value u of U by an equally-likely choice from the subset of sequences corresponding to u. (Successive letters from the message source are independently coded in this manner.) When such a partitioning of the D-ary sequences of length w is possible, the codeword $X_1, X_2, ..., X_w$ is equally likely to be any of the D-ary sequences of length w so that the sequence $X_1, X_2, ..., X_w$ is completely random. The different codewords that represent the same value u of U are traditionally called the "homophones" for U, but we shall soon use this terminology in a slightly different and more fundamental sense. It is easy to see that conventional homophonic substitution for which $X_1, X_2, ..., X_w$ is completely random is possible if and only if each value u_i of U has probability n_i/D^W for some integer n_i, in which case n_i is the number of homophones that must be assigned to u_i.

Variable-length homophonic substitution, introduced by Günther [1], generalizes the conventional scheme in that the D-ary sequences used can have different lengths, and the seqences in the subset corresponding to a given value u of U can be selected with unequal probabilities as the codeword for u. The length W of the codeword $X_1, X_2, ..., X_W$ for U can thus be a random variable. For an arbitrary probability distribution for U, Günther [1] gave an algorithm for such variable-length homophonic substitution with D = 2 that makes the resulting binary codeword $X_1, X_2, ...,$ X_W completely random. He also noted that, when L = 2^n so that the "natural coding" of a value of U would be a binary sequence of length n, his algorithm sometimes gave an expected codeword length E[W] less than n so that his algorithm also performed "data compression".

Fig. 3 diagrams a coding scheme of sufficient generality to include conventional homophonic substitution and variable-length homophonic substitution, as well as conventional source coding (or "data compression"). By the homophonic channel of Fig. 3, we mean a memoryless channel whose input alphabet $\{u_1, u_2, ..., u_L\}$ coincides with the set of possible values of U, whose output alphabet $\{v_1, v_2, v_3, ...\}$ is either finite or countably infinite, and whose transition probabilities $P(V=v_j | U=u_i)$ have the property that for each j there is exactly one i such that $P(V=v_j | U=u_i) \neq 0$. We shall consider those v_j for which $P(V = v_j | U = u_i) > 0$ to be the homophones for u_i, rather than considering the codewords into which these v_j are encoded to be the "homophones." By the D-ary prefix-free encoder of Fig. 3, we mean a device that assigns a D-ary sequence to each v_j under the constraint that this codeword is neither the same as another codeword nor forms the first part (or "prefix") of a longer codeword. This provision, which is satisfied by Günther's coding scheme [1], ensures that, when $X_1, X_2, ...$ is a sequence of codewords, the end of each codeword can be recognized without examining any following symbols in the sequence. It is well-known in information theory (cf. [2, p.49]) that such coding is general in the sense that for any D-ary uniquely-decodable code there is a D-ary prefix-free code with exactly the same codeword lengths.

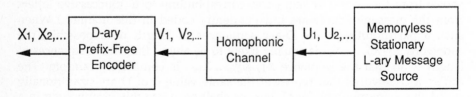

Fig. 3: A general scheme for homophonic substitution

387

When the homophonic channel of Fig. 2 is <u>deterministic</u> in the sense that all non-zero transition probabilities are 1 (so that we might as well say V = U), then Fig. 3 depicts the usual source coding (or "data compression") situation considered in information theory. When the homophonic channel is non-trivial but the binary encoding is <u>trivially prefix-free</u> because all codewords have the same length m (i.e., the code is a "block code"), then Fig. 3 depicts conventional homophonic substitution. In the case where both the homophonic channel is deterministic and the binary encoding is non-trivially prefix-free, then Fig. 3 depicts variable-length homophonic substitution as introduced by Günther [1].

Fig. 4 gives two examples of the general homphonic-substitution scheme illustrated in Fig. 3, both for the same binary (i.e., L=2) message source. We will soon see that both schemes in Fig. 4 are perfect. The upper system exemplifies conventional homophonic substitution into binary sequences of length w = 2. The lower system illustrates Günther's variable-length homophonic substitution. Note that the variable-length scheme has an expected codeword length of E[W] = 3/2 digits compared to E[W] = 2 for the conventional scheme. This reduction of coded symbols is an advantage offered by perfect variable-length homophonic substitution even when perfect conventional homophonic substitution is possible.

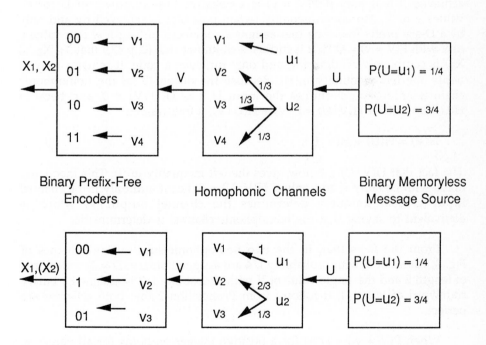

Fig. 4: Two examples of perfect homophonic substitution for the same binary memoryless message source

We will call a homophonic-substitution scheme <u>perfect</u> if the encoded D-ary sequence $X_1, X_2, ...$ is completely random. For the memoryless (source and channel) case considered in Fig. 3, this is equivalent to the condition that the codeword $X_1, X_2, ...,X_W$ for $V = V_1$ be completely random. Hereafter, all entropies are assumed to be in bits and all logarithms are understood to be to the base 2.

<u>Proposition 2:</u> For the homphonic-substitution scheme of Fig. 3,

$$H(U) \leq H(V) \leq E[W] \log D \tag{1}$$

with equality on the left if and only if the homophonic channel is deterministic, and with equality on the right if and only if the homophonic-substitution scheme is perfect. Moreover, there exists a D-ary prefix-free coding of V such that the scheme is perfect if and only if $P(V = v)$ is a negative integer power of D for all possible values v of V. When this condition is satisfied, the scheme is perfect if and only if $P(V = v_i) = D^{-w_i}$ holds for all values v_i of V where w_i is the length of the D-ary codeword assigned to v_i.

<u>Proof:</u> It is well-known in information theory that $H(V) \leq E[W] \log D$ holds for every D-ary prefix-free coding of U (cf. [2, p.50]) and that equality can be achieved if and only if $P(V = v)$ is a negative integer power of D for all values v of V. Moreover, equality (when possible) is achieved by and only by a D-ary prefix-free code that assigns a codeword of length w to a value v of V with $P(V = v) = D^{-w}$. It is further well-known, (cf. [2, p.47]) that $X_1, X_2, ..., X_W$ is completely random for, and only for, such a code. It remains only to verify the left inequality in (1). Because the output V of the homophonic channel uniquely determines the input U, i.e., $H(U|V) = 0$, and because $H(U,V) = H(U) + H(V|U) = H(V) + H(U|V)$, it follows that

$$H(V) = H(U) + H(V|U). \tag{2}$$

The fact that $H(V|U) \geq 0$ now gives the left inequality in (1). This inequality holds with equality if and only if $H(V|U) = 0$, i.e., if and only if the channel input U also uniquely determines the channel output V, which is equivalent to saying that the homophonic channel is deterministic.

From the facts that, in the two homophonic-substitution schemes of Fig. 3, values of V with probability 1/4 are assigned binary $(D = 2)$ codewords of length 2 and the single value of V with probability 1/2 is assigned a binary codeword of length 1, it follows from Proposition 2 that both schemes are perfect.

When $P(V = v_i) = D^{-w_i}$ for a positive integer w_i holds for all values v_i of V, it is well-known (cf. [2, p.48]) that a D-ary prefix-free code in which the codeword for v_i has length w_i may be simply constructed as follows: Choose any distinct D-ary sequences of length 1 to be codewords for those v_i (if any) with $P(V = v_i) = D^{-1}$, choose any distinct D-ary sequences of length 2 not

h is bijective on A and thus h is a permutation on A. Because of this $h(P_i)$ and $P_{i+1}(h^{-1})$ are permutations too.

There are $\phi(M)$ different isomorphisms and $n - 1$ possible positions where an isomorphism can be inserted. Thus the key space is reduced by the factor $\phi(M)^{n-1}$.

The remaining keys are truly different. This is proved by the fact that the key can be deduced within a chosen ciphertext attack (discussed below) with the exception of the isomorphisms mentioned above. The number of usable keys is again reduced because the machine is self synchronizing when operating in deciphering mode. It is useful to consider permutations as not significantly different if they can be reached by cyclic shifts. Thus the number of keys is reduced by the factor M_n.

The total number is: $\frac{M!^n}{(M^n * \phi(M)^{n-1})}$ (where ϕ is the Euler phi-function)

To characterize the deciphering mechanism we define a set of recursive functions as follows:

$$f_n(i) := c[i]$$
$$\vdots$$
$$f_r(i) := P_{r+1}^{-1}(f_{r+1}(i)) - P_{r+1}^{-1}(f_{r+1}(i - 1))$$
$$\vdots$$
$$f_1(i) := P_2^{-1}(f_2(i)) - P_2^{-1}(f_2(i - 1))$$
$$f_0(i) := P_1^{-1}(f_1(i)) - P_1^{-1}(f_1(i - 1))$$

$f_r(i)$ deciphers the last $n - r$ rotors and calculates the value $a_r[i]$. There is one thing that should be mentioned:

$f_r(i)$ depends on the last $n + 1 - r$ ciphertext bytes: $c[i], c[i - 1], \ldots, c[i - n + r]$.

The flowchart for the deciphering mechanism looks somewhat different because the feedback loops are substituted by delay elements:

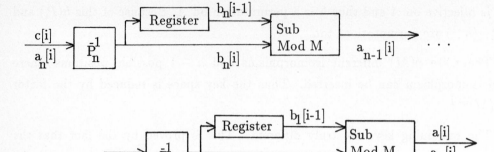

It is easy to see, that the machine is self-synchronizing when operating in deciphering mode. In the deciphering scheme one can also see what seems to be the main problem of the cipher:

Constant ciphertext effect

If the machine is fed with a constant ciphertext we have after n characters $b_j[i] = P_j^{-1}(0)$ $(j = 1, 2, \ldots, n - 1)$. Thus the machine can be forced into a well defined state only by a given ciphertext and without knowing the key (the permutations). Later we will show that this fact makes it possible to easily break the cipher with a chosen-ciphertext-attack.

Proof of constant ciphertext effect:

Because of the constant ciphertext for each $j = 2, \ldots, n$ the input to $f_j(i)$ is equal to the input of $f_j(i - 1)$. So we have:

$$f_j(i) = f_j(i - 1) \qquad \Leftrightarrow$$
$$P_j^{-1}(f_j(i)) = P_j^{-1}(f_j(i - 1)) \qquad \Leftrightarrow$$
$$P_j^{-1}(f_j(i)) - P_j^{-1}(f_j(i - 1)) = 0 = f_{j-1}(i) = a_{j-1}[i] = P_j^{-1}(b_{j-1}[i]) \quad \Rightarrow$$
$$f_j(i) = 0 \qquad j = 1, 2, \ldots, n - 1$$
$$b_j[i] = P_j^{-1}(0) \quad j = 1, 2, \ldots, n - 1.$$

2 Attacks

2.1 Ciphertext-only-attack

We suppose it is very difficult to break the cipher under this conditions. The easiest way we found is to make a statistical analysis over M^{n+1} different elements. Generally M^{n+1} is very large and thus this attack will usually be impossible.

2.2 Given-plaintext-attack

A solution using backtracking may be possible but we found that even for small M's and n's the searching tree is much too deep and too wide. Another possibility is a solution using equations. With a given plaintext all $b_1[i]$ are given too. We have:

$$b_1[i] = P_1^{-1}(f_1(i)) \tag{1}$$

With $b_1[i] = b_1[j]$ we get:

$$P_1^{-1}(f_1(i)) = P_1^{-1}(f_1(j)) \Leftrightarrow (P_1^{-1} \text{is bijective}) \tag{2}$$

$$f_1(i) = f_1(j) \tag{3}$$

with a little luck it is possible to find a position j with $c[j] = c[j+1] = \ldots = c[j+n-1]$ (n constant ciphertext chars). Because of the constant ciphertext effect we have $f_1(j) = 0$ and get:

$$
\begin{aligned}
P_2^{-1}(f_2(i)) - P_2^{-1}(f_2(i-1)) &= 0 & &\Leftrightarrow \\
P_2^{-1}(f_2(i)) &= P_2^{-1}(f_2(i-1)) & &\Leftrightarrow (P_2^{-1} \text{ is bijective}) \\
f_2(i) &= f_2(i-1) & &\Leftrightarrow \\
P_3^{-1}(f_3(i)) - P_3^{-1}(f_3(i-1)) &= P_3^{-1}(f_3(i-1)) - P_3^{-1}(f_3(i-2))
\end{aligned}
$$

It can be easily shown that the above trick does not work a second time. We found no further way to simplify the equations. Thus (treating $P_3^{-1}(f_3(i))$ as an unknown) a linear system with M^{n-2} unknowns has to be solved. It will usually be impossible to do this within an appropriate amount of time. This attack does not use the the internal structure of these 'unkowns'. Thus it may be possible to solve the equations much faster and much easier.

2.3   Chosen-plaintext-attack

The idea is to feed the machine with a constant (zero) plaintext and to wait until the output becomes cyclic. To analyze the output we first consider the output of the first $n - 1$ rotors ($a_{n-1}[i]$). Let m be the length of a cycle of $a_{n-1}[i]$. Let s be the sum over a cycle of $a_{n-1}[i] : s := a_{n-1}[1] + \ldots + a_{n-1}[m]$. Suppose s and M are coprime. Then the length of a complete output cycle is $s * M$. We have $c[i + j * m] = P_n(b_{n-1}[i] + j * s)$. Because s and M are coprime $j * s$ covers whole A. Because of this $c[i + j * m] = P_n(b_{n-1}[i] + j * s)$ covers A too. Now $c[i + j * m]$ run through the same cycle for each i (except for the starting point). If we use s as the '1-element' in A we get:

$$P_n(0) \ = \ c[1]$$
$$P_n(1) \ = \ c[1 + m]$$
$$\vdots$$
$$P_n(M - 1) \ = \ c[1 + m * (M - 1)]$$

Now the last rotor is completely known (except for rotation identical rotors).

This is only if s and M are coprime. If they are not coprime it is usually impossible to split the output into cycles covering A. Seeing a complete output cycle we can tell if it is possible to extract the last rotor. If it is impossible we have to restart the attack with a short random sequence before the zero plaintext to put the machine in an undefined state. The restart of the attack does not require the machine to be reset, because we simply want to put the machine into an undefined state. If the last rotor is known we can eliminate him from the equations and restart the whole attack to get the next rotor. After all rotors are known we need to compute the initial position $b_i[0]$ of each rotor. To do this we encipher a random sequence $a[1], \ldots, a[n]$ and get the ciphertext $c[1], \ldots, c[n]$. Using the defining equations we now can compute the values in the following order:

$$a_n[1], a_n[2], \ldots, a_n[n], \qquad (a_n[i] = c[i])$$
$$b_n[1], b_n[2], \ldots, b_n[n], \qquad (b_n[i] = P_n^{-1}(a_n[i]))$$
$$a_{n-1}[2], a_{n-1}[3], \ldots, a_{n-1}[n], \qquad (a_{n-1}[i] = b_n[i-1] - b_n[i])$$
$$\vdots$$
$$b_k[n+1-k], b_k[n+2-k], \ldots, b_k[n], \qquad (b_k[i] = P_k^{-1}(a_k[i]))$$
$$a_{k-1}[n+2-k], a_{k-1}[n+3-k], \ldots, a_{k-1}[n], \quad (a_{k-1}[i] = b_k[i-1] - b_k[i])$$
$$\vdots$$
$$b_1[n], \qquad (b_1[i] = P_1^{-1}(a_1[i]))$$
$$b_1[n-1], b_1[n-2], \ldots, b_1[0], \qquad (b_1[i] = b_1[i+1] - a[i+1])$$
$$a_1[n-1], a_1[n-2], \ldots, a_1[0], \qquad (a_1[i] = P_1(b_1[i]))$$
$$\vdots$$
$$b_k[n-k], b_k[n-k-1], \ldots, b_k[0], \qquad (b_k[i] = b_k[i+1] - a_{k-1}[i+1])$$
$$a_k[n-k], a_k[n-k-1], \ldots, a_k[1], \qquad (a_k[i] = P_k(b_k[i]))$$
$$\vdots$$
$$b_n[0] \qquad (b_n[i] = b_n[i+1] - a_{n-1}[i+1])$$

By applying cyclic shifts to the Permutations we can get to $b_k[0] = 0$ to conform with the above definition. Unfortunately the period of the output will be about M^{n-1}. Hence, this attack will often not be realizable.

There is still another possibility for a chosen plaintext attack: All previously discussed attacks tried to guess the last rotor. Now we try to attack the first rotor at the beginning. We found that this attack has almost no chance to break the machine and so this will only be a short sketch of how the attack works. With a short random sequence we put the machine into an undefined state. The we keep $b_1[i] = x$ constant and look for the length of an output cycle. After doing this several time we can try to classify $P_1(x)$ and get $gcd(M, P_1(x))$. Having classified all numbers we can try to classify sums of two or more numbers. Continuing with these classifications we can compute the first rotor. The main problem of this attack is that we found no way to make any classifications if more than 4 rotors are used. Another problem is that the length of the output cycles is very big (about M^{n-1}). So in most cases this attack will be useless.

2.4 Chosen-ciphertext-attack

The basic idea for this attack is to turn the last rotor step by step. It is easy to see that it is sufficient to find a short plaintext sequence that does nothing but turn the last rotor. We will use the 'constant ciphertext effect' mentioned before to find such a sequence. It is sufficient to turn the last rotor for an amount of steps that is coprime to M, because every primitive element can be chosen as the '1' element.

Method: We first decipher the ciphertext $0^n 1^n$ and get a plaintext $d[1], \ldots, d[2n]$. The only difference in the state of the machine after deciphering 0^n and $0^n 1^n$ is the position of the last rotor. Now the newly constructed plaintext is: $d[1], \ldots, d[n], (d[n+1], \ldots, d[2n])^{M-1}$ After each n-character block the machine is in the same state with the exception that the last rotor has turned for a constant number of steps. The number of steps is $s := P_n^{-1}(0) - P_n^{-1}(1)$. The output will be a sequence of the form: $(P_n(x))^n (P_n(x+s))^n (P_n(x+2s))^n \ldots (P_n(x+(M-1)s))^n$, where $x := P_n^{-1}(0)$ If s is a unit in $(Z_M, +)$ the output will cover whole Z_M and give the last permutation (except for some isomorphisms as mentioned above). If s is not a unit in $(Z_M, +)$ the output will cover only a part of Z_M. In this case we try an other ciphertext e.g. $0^n 2^n$. After the last rotor is known we can attack the next rotor. Finally we compute the initial position of the rotors as shown above. This attack is very efficient and depends only polynomially on the values of n and M.

3 Conclusion

The implementation found in some operating systems uses $n = 3$ and $M = 256$; i.e. byte encryption with 3 rotors. It can be easily broken with a known plaintext using equations (256 unknowns). It seems that, if 5 ore more rotors are used, the only way to break the cipher is a choosen ciphertext attack. So this cipher is, in many environments, a tool to provide good (not optimal) security. A further advantage of this cipher is that it can be implemented very efficiently in both hardware and software.

CRYPTANALYSIS OF VIDEO ENCRYPTION
BASED ON SPACE-FILLING CURVES

Michael Bertilsson[1] Ernest F. Brickell[2]
Ingemar Ingemarsson[1]

[1] Dept of Electrical Engineering
Linköping University
581 83 LINKÖPING, Sweden

[2] Sandia National Laboratories
Albuquerque
NM 87185, USA

ABSTRACT

In this paper we cryptanalyze a method for enciphering images constructed by Adi
Shamir and Yossi Matias. The enciphering is done using space-filling curves. We present an algorithm doing the cryptanalysis. We also present results achieved using
the algorithm on a part of an image.

I. INTRODUCTION

In CRYPTO'87 Adi Shamir and Yossi Matias presented a video scrambling technique
[1]. This technique uses space-filling curves (SFC) to encipher images and sequences of images. Shamir and Matias state that standard cryptographic techniques
are inadequate for enciphering of images. There are three basic reasons for this
claim.

1 The transmitted signal is analog
2 The transmission rate is very high
3 The allowable bandwidth is limited

The main idea presented in [1] is to store an image in a frame buffer and then
scan the image along a pseudo-random SFC. Several algorithms generating SFC:s are
presented in [1]. Independently of the algorithm used they get SFC:s looking something like the one in Figure 1.

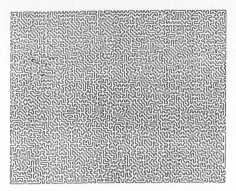

Figure 1

We show in this paper that it is not a good idea to use SFC:s for enciphering sequences of images. This is true independently of the algorithm used to generate the SFC. Our algorithm only use the fact that the enciphering is done using SFC:s and is not dependent on the algorithm used to generate the SFC. We assume that the encipherer chooses a new SFC for every frame.

This is not an attack on the use of SFC:s for enciphering still images. Our method is not possible to use on still images.

II.PROPERTIES OF IMAGES

In this section we give some properties of images and sequences of images that is used in the cryptanalysis.

Transmission of sequences of images is usually done at a rate of approximately 25 frames per second. This rate is needed when there is a lot of motion in the sequence. But sometimes there is no motion in the sequence. When there is no motion the same image is transmitted in a number of consecutive frames. We assume that the encipherer chooses a new SFC for every new frame. This means that the same image will be scanned by different SFC:s if there is no motion in the sequence.

Pixel values are non-uniformly distributed in a quantized image. This is also valid for adjacent pixels. In particular there are some pairs of adjacent pixel that will occur only once in the image. The pairs occurring only once will be called unique pairs.

We define the distance between two points p1 and p2 in a square lattice as follows

$$d(p1,p2) = p_v + p_h$$

where p_v is the number of points between p1 and p2 vertically and p_h the number of points between them horizontally. This distance is similar to the Lee distance used in coding theory.

Using this distance we can calculate a maximum possible distance between unique pairs even if we only have a SFC scan of the image. Suppose that we have two unique pairs, up1 and up2 in a SFC scan. We count the number of pixels, n(up1,up2), between the two unique pairs. This is the same as the maximum possible distance between the two unique pairs in the original image. If we have several SFC scans of the same image we can get a better result. We then calculate the maximum possible distance, d_{max}, between the two unique pairs as

$$d_{max}(up1,up2) = min \ n(up1,up2)$$

where the min is over all the SFC scans containing both up1 and up2. We define n(up1,up2)=0 if the two unique pairs have one pixel in common.

III.CRYPTANALYSIS

The cryptanalysis is divided into several steps which can be performed separately.

First we have to find the unique pairs in the different scans of the image. We are faced with two problems when we do this. Both problems come from the fact that we do not get all the pairs of pixels in the image when we are scanning it using a SFC. We only get approximately half of the pairs in the original image in our SFC scan.

There is a risk that we miss a unique pair in a SFC scan. We can solve this problem if we have many scans of the same image. If the same image is transmitted several times and the encipherer selects a new SFC for every frame we have many scans of the same image. Since it is probable that a unique pair occurs in at least one of the scans we will find most of the unique pairs.

Some pairs of pixels will occur only a few times in an image. If we miss all of these pairs but one we get a pair occurring only once in this specific SFC scan. This means that in this scan the pair will look like a unique pair. This problem is also solved using the different scans we have of the same image. We can detect these false unique pairs in two different ways. First it is likely that this false unique pair will occur two times or more in at least one of the other scans and if so it is removed from the unique pairs. Secondly a pixel belonging to a unique pair can have only four values on the adjacent pixels, one of which is the value of the other pixel in the unique pair. So if a pixel belonging to a possible unique pair has five or more different values on its adjacent pixels the pair can not be unique.

The second step in the cryptanalysis is to synchronize the different scans locally around the unique pairs. Different SFC:s scan the surrounding of the unique pairs in different ways. Eight of these are interesting in this step in the cryptanalysis. These are shown in Figure 2a and 2b.

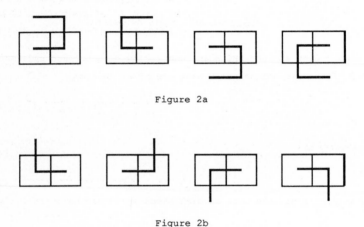

Figure 2a

Figure 2b

If we have the leftmost scan in Figure 2a in one SFC and the leftmost scan in Figure 2b in another we get the values of the two pixels above the unique pair. If we have the rightmost scan in 2a and 2b in two different SFC:s we get the values of the pixels below the unique pair. Similarly for the two scans in the middle. If we can get the value of both the pair above and the pair below the unique pair we are usually able to get the values of all the six pixels surrounding the unique pair, see Figure 3. This is possible if we have four different pixel values on pixels adjacent to each of the two pixels in the unique pair.

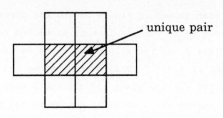

Figure 3

When we have constructed the small surroundings around the unique pairs we try to put these surroundings together. We search the marked pairs in Figure 4 looking for unique pairs. This is done for all the surroundings.

If we among these pairs find a unique pair we join these two surroundings. This is repeated until the surroundings around all the unique pairs have been searched. This gives us a set of pieces of the image.

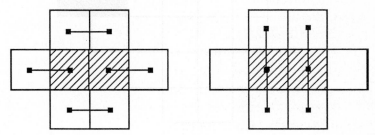

Figure 4

We would like to know how far apart these different pieces are. The exact distance is not possible to calculate. But we can calculate the maximum possible distance between unique pairs, see section 2. Using this and the fact that we know to which piece the unique pairs belong we calculate the maximum possible distance, D_{max}, between two pieces p1 and p2. This is done using

$$D_{max}(p1,p2) = min\ D_{max}(upi,upj)$$

where min is over all distances between a unique pair belonging to p1, upi, and a unique pair belonging to p2, upj. We also know between which unique pairs in the two pieces the distance holds.

Before we use the distances we try to make each piece created above larger . We use the different scans of the same image. The method is explained using Figure 5.

We have the scan showed in Figure 5a and the piece shown in Figure 5b. Pixels 1 and 2 form a unique pair. The pixels both above and below pixel 1 in Figure 5b are different from pixel 3. So the only place where we can put pixel 3 is in the empty place to the right of pixel 2. This means that we can expand the piece a little bit. This is repeated for all the scans and all the pieces. After this step the pieces will be a little larger which makes it easier to fit them together.

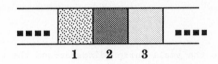

Figure 5a

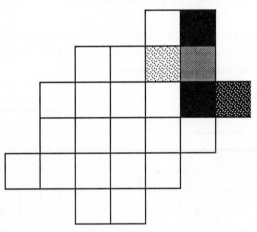

Figure 5b

Now we use the distances between pieces to join the different pieces. A typical situation when the distance between pieces is zero is shown in Figure 6.

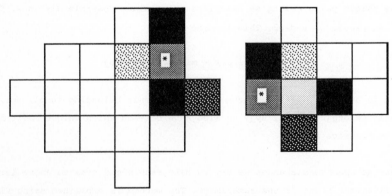

Figure 6

In Figure 6 we have two different pieces. The pieces have $D_{max}(p1,p2)=0$. The pixel marked with a * in the two pieces is common to the two unique pairs which are at distance zero. Knowing this we have eight different ways to fit the pieces together, rotation 90 degrees four times and mirroring. Hopefully only one of these eight is possible. If we have more than one possible way to join two pieces we do not join them. In this case only one of them is possible so we can join the two pieces into one. This way we try to join all the different pieces with mutual distance zero . We also try to do this for larger distances. Since we only know the maximum possible distance between the pieces we have to try to fit the pieces together for all distances shorter than the one calculated. Because of this we do not try to fit pieces together with mutual distance larger than five.

The last two steps can be repeated a desired number of times alternating between the two. This can be done both for the same distance and for increasing distances in the last step.

IV.LOCATION OF UNIQUE PAIRS

One important question is where the unique pairs occur in an image. Suppose we are able to create one large piece consisting of all the unique pairs. Are we then able to see what the image looks like ? If not it is not interesting to do the cryptanalysis described in section 3. In this section we claim that you actually get a good idea of what the original image looked like if you can create a piece consisting of all unique pairs.

In a part of the image the pixels may have the same or almost the same values, i.e. low spatial frequency. In this part of the image there will be few or no unique pairs. On the other hand in a part of the image with edges (high spatial frequencies) we have many unique pairs. This is illustrated in Figure 7. In this Figure we show the original image "Lenna" and an image showing the unique pairs in "Lenna". We see that we find the unique pairs were there is a lot of "information" in the image.

 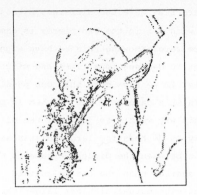

Figure 7

V.RESULTS

Using the algorithm described in section 3 we tried to cryptanalyze a part of the
image "Lenna". The size was 64x64 pixels. We used 25 scans of the same image. The
result from the cryptanalysis was the four large pieces shown in Figure 8 and some
smaller pieces. From the calculation of distances we knew that piece 1 was close to
piece 2 and piece 2 was close to piece 3. We also knew that piece 4 was distant from
the other three pieces. The small pieces was not used since most of them consisted
of only one unique pair.

Figure 8

We showed a person the four pieces in Figure 8. We also told her which pieces should be close to each other. Piece number four was not used. Knowing only this she was able to put piece number 1, 2 and 3 together into the piece shown in Figure 9. The size of this piece is 64 pixels from top to bottom and approximately 45 pixels from left to write.

Figure 9

VI.CONCLUSIONS

We have shown that space-filling curves are unfit for secure enciphering of sequences of images. In the experiment on a part of the image "Lenna" the result was so good that you could guess what that part of the original image looked like. This was done using only 25 scans of the same image. We have also indicated what the result would be if we were to perform the analysis on the whole image.

REFERENCE

[1]   Yossi Matias, Adi Shamir, *A Video Scrambling Technique Based On Space Filling Curves*, Proc. of CRYPTO'87 pp 398-417

Impossibility and Optimality Results on Constructing Pseudorandom Permutations

(Extended Abstract)

Yuliang Zheng, Tsutomu Matsumoto and Hideki Imai

Division of Electrical and Computer Engineering

Yokohama National University

156 Tokiwadai, Hodogaya, Yokohama, 240 Japan

Summary   Let $I_n = \{0,1\}^n$, and H_n be the set of all functions from I_n to I_n. For $f \in H_n$, define the *DES-like transformation* associated with f by $F_{2n,f}(L,R) = (R \oplus f(L), L)$, where $L, R \in I_n$. For $f_1, f_2, \ldots, f_s \in H_n$, define $\psi(f_s, \ldots, f_2, f_1) = F_{2n,f_s} \circ \cdots \circ F_{2n,f_2} \circ F_{2n,f_1}$. Our main result is that $\psi(f^k, f^j, f^i)$ is *not* pseudorandom for any positive integers i, j, k, where f^i denotes the i-fold composition of f. Thus, as immediate consequences, we have that (1) none of $\psi(f, f, f)$, $\psi(f, f, f^2)$ and $\psi(f^2, f, f)$ are pseudorandom and, (2) Ohnishi's constructions $\psi(g, g, f)$ and $\psi(g, f, f)$ are optimal. Generalizations of the main result are also considered.

1. Introduction

Random generation is of supreme importance for cryptography, and has recently received extensive investigation by many computer scientists [GGM] [S] [Y]. As mentioned in [LR], if polynomial-time computable pseudorandom invertible permutations are available, then we can design ideal secret-key block ciphers that are provably secure against the chosen plaintext attack. This paper also deals with the construction of pseudorandom (invertible) permutations.

The set of positive integers is denoted by $\mathcal{N}$. For each $n \in \mathcal{N}$, let $I_n = \{0,1\}^n$. Denote by $s_1 \oplus s_2$ the bit-wise XOR of two strings $s_1, s_2 \in I_n$, and by H_n the set of all

2^{n2^n} functions from I_n to I_n. The *composition* of two functions f and g in H_n, denoted by $f \circ g$, is defined by $f \circ g(x) = f(g(x))$ where $x \in I_n$. And in particular, $f \circ f$ is denoted by f^2, $f \circ f \circ f$ by f^3, and so on.

Associate with $f \in H_n$ a function $F_{2n,f} \in H_{2n}$ defined by $F_{2n,f}(L,R) = (R \oplus f(L), L)$ for all $L, R \in I_n$. (Note that our definition for $F_{2n,f}$ is *notationally* different from that given in [LR] and [S]. However, the difference is *not* essential, and does *not* affect the results to be proved below.) $F_{2n,f}$ is a permutation in H_{2n}, and called the *DES-like transformation* associated with f [NBS] [FNS]. Furthermore, for $f_1, f_2, \ldots, f_s \in H_n$, define $\psi(f_s, \ldots, f_2, f_1) = F_{2n,f_s} \circ \cdots \circ F_{2n,f_2} \circ F_{2n,f_1}$. We say that $\psi(f_s, \ldots, f_2, f_1)$ consists of s rounds of DES-like transformations.

In their wonderful paper [LR], Luby and Rackoff showed that permutations $\psi(h, g, f)$, where $f, g, h \in_R H_n$, cannot be efficiently distinguished from an $r \in_R H_{2n}$, here by $x \in_R X$ we mean that x is drawn randomly and uniformly from a finite multiset X. In other words, from *three* independent random functions $f, g, h \in H_n$, one can construct, by three applications of DES-like transformations, a permutation in H_{2n} which cannot be efficiently distinguished from a truly random function in H_{2n}.

Ohnishi [O] observed that *two* independent random functions are sufficient in Luby and Rackoff's construction. In particular, he showed that both $\psi(g, f, f)$ and $\psi(g, g, f)$, where $f, g \in_R H_n$, cannot be efficiently distinguished from an $r \in_R H_{2n}$. See Appendix for more information on the proof of it.

In the thesis, Ohnishi also showed that neither $\psi(f, f, f)$ nor $\psi(f, g, f)$ are pseudorandom. This result was independently obtained by Rueppel in [R].[1] However, it still remains open whether or not permutations like $\psi(f, f, f^2)$ and $\psi(f^2, f, f)$ are pseudorandom. The technique used in [O] and [R], which is described in the final section of this paper, is not applicable to these cases.

In the remaining part of this paper, we first introduce the notion of pseudorandomness, then show that for any f and for any $i, j, k \in \mathcal{N}$, there is a circuit that distinguishes $\psi(f^k, f^j, f^i)$ from an $r \in_R H_{2n}$. Thus, as immediate consequences, we have that (1) none of $\psi(f, f, f)$, $\psi(f, f, f^2)$ and $\psi(f^2, f, f)$ are pseudorandom and, (2) Ohnishi's constructions $\psi(g, f, f)$ and $\psi(g, g, f)$ are optimal among the pseudorandom

---

[1] At Eurocrypt'88, Schnorr [S] *erroneously* claimed that $\psi(f, f, f)$, where $f \in_R H_n$, cannot be efficiently distinguished from an $r \in_R H_{2n}$.

permutations $\psi(f_3^k, f_2^j, f_1^i)$ where $i, j, k \in \mathcal{N}$ and $f_1, f_2, f_3 \in H_n$ such that for any $1 \leq s, t \leq 3$, either $f_s = f_t$ or f_s is independent of f_t. We also investigate generalizations of our main result.

2. Notion of Pseudorandomness

Let $n \in \mathcal{N}$. An *oracle circuit* T_n is an acyclic circuit which contains, in addition to ordinary AND, OR, NOT and constant gates, also a particular kind of gates — *oracle gates*. Each oracle gate has an n-bit input and an n-bit output, and it is evaluated using some function from H_n. The output of T_n, a single bit, is denoted by $T_n[f]$ when a function $f \in H_n$ is used to evaluate the oracle gates. The size of T_n is the total number of connections in it. Note that one can view an oracle circuit as a circuit with no inputs or as a circuit with inputs to which constants are assigned.

A family of circuits $T = \{T_n \mid n \in \mathcal{N}\}$ is called a *statistical test for functions* if each T_n is an oracle circuit whose size is bounded by some polynomial in n.

Assume that S_n is a multi-set consisting of functions from H_n. Let $S = \{S_n \mid n \in \mathcal{N}\}$ and $H = \{H_n \mid n \in \mathcal{N}\}$. We say that T is a *distinguisher* for S if for some polynomial P and for infinitely many n, we have $|Pr\{T_n[s] = 1\} - Pr\{T_n[h] = 1\}| \geq 1/P(n)$, where $s \in_R S_n$ and $h \in_R H_n$. We say that S is *pseudorandom* if there is no distinguisher for it. (See also [GGM], [LR] and [Y].)

In this paper we are only concerned with pseudorandom permutations, i.e., pseudorandom functions $S = \{S_n \mid n \in \mathcal{N}\}$ where each S_n consists of permutations from H_n. It is convenient to say that an $s \in_R S_n$ is pseudorandom whenever S is pseudorandom, and not pseudorandom (or can be distinguished from an $r \in_R H_n$) otherwise.

3. Main Result

This section proves our main result on permutations $\psi(f^k, f^j, f^i)$ where $f \in H_n$ and $i, j, k \in \mathcal{N}$. For $i, j, k \in \mathcal{N}$, let $\Psi_{2n}(i, j, k)$ be the multi-set consisting of all functions $\psi(f^k, f^j, f^i) \in H_{2n}$ where $f \in H_n$, and let $\Psi(i, j, k) = \{\Psi_{2n}(i, j, k) \mid n \in \mathcal{N}\}$.

[**Theorem 1**] *For any* $i, j, k \in \mathcal{N}$, *there is a distinguisher* $T = \{T_{2n} \mid n \in \mathcal{N}\}$ *for* $\Psi(i, j, k)$, *i.e.,* $\Psi(i, j, k)$ *is not pseudorandom. Each* T_{2n} *has* $(m_1 + m_2 + 1)$ *oracle gates, where* $m_1 = (i + j)/d$, $m_2 = (j + k)/d$ *and* $d = \gcd(i + j, j + k)$.

Proof: Denote by $O_0, O_1, \ldots, O_{m_1+m_2}$ the $(m_1 + m_2 + 1)$ oracle gates, by (X_{s1}, X_{s2}) and (Y_{s1}, Y_{s2}) the input to and output of O_s respectively, and by 0^n the all-0 string in I_n. The structure of T_{2n} is as follows. (See also Figure 1.)

DESCRIPTION OF T_{2n} :

(1) The input to O_0 is $(X_{01}, X_{02}) = (0^n, 0^n)$.

(2) The input to O_1 is $(X_{11}, X_{12}) = (0^n, Y_{01})$. And if $m_1 > 1$, then for each $1 < p \leq m_1$, the input to O_p is $(X_{p1}, X_{p2}) = (0^n, X_{(p-1)2} \oplus Y_{(p-1)1})$.

(3) The input to O_{m_1+1} is $(X_{(m_1+1)1}, X_{(m_1+1)2}) = (Y_{02}, 0^n)$. And if $m_2 > 1$, then for each $m_1 + 1 < t \leq m_1 + m_2$, the input to O_t is $(X_{t1}, X_{t2}) = (X_{(t-1)1} \oplus Y_{(t-1)2}, 0^n)$.

(4) Finally, T_{2n} outputs a bit 1 iff $Y_{m_12} = X_{(m_1+m_2)1} \oplus Y_{(m_1+m_2)2}$.

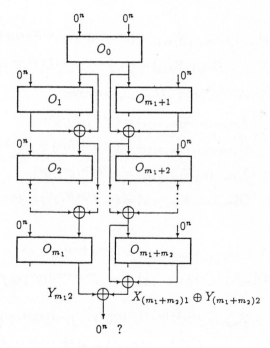

Figure 1: Structure of Oracle Circuit T_{2n}

Obviously, the size of T_{2n} is of polynomial in n. Now we analyze the behavior of T_{2n} in the following two cases: CASE-1, where a function $\psi(f^k, f^j, f^i) \in \Psi_{2n}(i, j, k)$ is used to evaluate the oracle gates, and CASE-2, where a function drawn randomly and uniformly from H_{2n} is used to evaluate the oracle gates. We show that in the former case, the probability that T_{2n} outputs a bit 1 is 1 and, in the latter case, the probability is less than $1/2^{n-1}$. Thus $T = \{T_{2n} \mid n \in \mathcal{N}\}$ is a distinguisher for $\Psi(i, j, k)$.

CASE-1: Notice that $\psi(f^k, f^j, f^i)(L, R) = (R \oplus f^i(L) \oplus f^k(L \oplus f^j(R \oplus f^i(L))), L \oplus f^j(R \oplus f^i(L)))$. Thus the output of O_0 is $(Y_{01}, Y_{02}) = (f^i(0^n) \oplus f^{k+j+i}(0^n), f^{j+i}(0^n))$.

Denote by $\sim$ a string which we do not care. The inputs to and outputs of $O_1, O_2, \ldots, O_{m_1}$ are as follows:

$$O_1: (X_{11}, X_{12}) = (0^n, f^i(0^n) \oplus f^{k+j+i}(0^n)),$$
$$(Y_{11}, Y_{12}) = (f^{k+j+i}(0^n) \oplus f^{2k+2j+i}(0^n), \sim);$$
$$O_2: (X_{21}, X_{22}) = (0^n, f^i(0^n) \oplus f^{2k+2j+i}(0^n)),$$
$$(Y_{21}, Y_{22}) = (f^{2k+2j+i}(0^n) \oplus f^{3k+3j+i}(0^n), \sim);$$

$$\ldots\ldots\ldots\ldots$$

$$O_{m_1}: (X_{m_11}, X_{m_12}) = (0^n, f^i(0^n) \oplus f^{m_1 k + m_1 j + i}(0^n)),$$
$$(Y_{m_11}, Y_{m_12}) = (\sim, f^{m_1 k + (m_1+1)j + i}(0^n)).$$

Similarly, for $O_{m_1+1}, O_{m_1+2}, \ldots, O_{m_1+m_2}$, we have:

$$O_{m_1+1}: (X_{(m_1+1)1}, X_{(m_1+1)2}) = (f^{j+i}(0^n), 0^n),$$
$$(Y_{(m_1+1)1}, Y_{(m_1+1)2}) = (\sim, f^{j+i}(0^n) \oplus f^{2j+2i}(0^n));$$
$$O_{m_1+2}: (X_{(m_1+2)1}, X_{(m_1+2)2}) = (f^{2j+2i}(0^n), 0^n),$$
$$(Y_{(m_1+2)1}, Y_{(m_1+2)2}) = (\sim, f^{2j+2i}(0^n) \oplus f^{3j+3i}(0^n));$$

$$\ldots\ldots\ldots\ldots$$

$$O_{m_1+m_2}: (X_{(m_1+m_2)1}, X_{(m_1+m_2)2}) = (f^{m_2 j + m_2 i}(0^n), 0^n),$$
$$(Y_{(m_1+m_2)1}, Y_{(m_1+m_2)2}) = (\sim, f^{m_2 j + m_2 i}(0^n) \oplus f^{(m_2+1)j + (m_2+1)i}(0^n)).$$

Thus

$$Y_{m_12} = f^{m_1 k + (m_1+1)j + i}(0^n) = f^{m_1(k+j)+j+i}(0^n)$$
$$= f^{\frac{i+i}{d}(k+j)+j+i}(0^n) = f^{j\frac{k+j}{d} + i\frac{k+j}{d} + j + i}(0^n)$$
$$= f^{j m_2 + i m_2 + j + i}(0^n) = f^{(m_2+1)j + (m_2+1)i}(0^n)$$
$$= X_{(m_1+m_2)1} \oplus Y_{(m_1+m_2)2},$$

and the probability that T_{2n} outputs a bit 1 is 1.

CASE-2: There are two sub-cases to be analyzed: $(Y_{01}, Y_{02}) = (0^n, 0^n)$ and $(Y_{01}, Y_{02}) \neq (0^n, 0^n)$.

1) When $(Y_{01}, Y_{02}) = (0^n, 0^n)$, we have $Y_{m_1 2} = X_{(m_1+m_2)1} \oplus Y_{(m_1+m_2)2}$. But $Pr\{(Y_{01}, Y_{02}) = (0^n, 0^n)\} = 1/2^{2n}$.

2) When $(Y_{01}, Y_{02}) \neq (0^n, 0^n)$, we have $(X_{11}, X_{12}) \neq (X_{(m_1+1)1}, X_{(m_1+1)2})$, and hence $(X_{11}, X_{12}) \neq (0^n, 0^n)$ or $(X_{(m_1+1)1}, X_{(m_1+1)2}) \neq (0^n, 0^n)$. Suppose that $(X_{11}, X_{12}) \neq (0^n, 0^n)$. (The other case is similar.) Then (Y_{11}, Y_{12}), and hence (Y_{21}, Y_{22}), (Y_{31}, Y_{32}), $\ldots$, $(Y_{m_1 1}, Y_{m_1 2})$ are all random strings in I_{2n}. These strings are independent of $(X_{(m_1+1)1}, X_{(m_1+1)2})$, and hence of $(Y_{(m_1+1)1}, Y_{(m_1+1)2}), (Y_{(m_1+2)1}, Y_{(m_1+2)2}), \ldots, (Y_{(m_1+m_2)1}, Y_{(m_1+m_2)2})$. So when $(Y_{01}, Y_{02}) \neq (0^n, 0^n)$, the probability that $Y_{m_1 2} = X_{(m_1+m_2)1} \oplus Y_{(m_1+m_2)2}$, i.e., T_{2n} outputs a bit 1, is $1/2^n$.

Thus, for CASE-2, we have

$$Pr\{T_{2n}[f] = 1\}$$
$$= Pr\{T_{2n}[f] = 1 \mid (Y_{01}, Y_{02}) = (0^n, 0^n)\} \cdot Pr\{(Y_{01}, Y_{02}) = (0^n, 0^n)\}$$
$$+ Pr\{T_{2n}[f] = 1 \mid (Y_{01}, Y_{02}) \neq (0^n, 0^n)\} \cdot Pr\{(Y_{01}, Y_{02}) \neq (0^n, 0^n)\}$$
$$= 1 \cdot 1/2^{2n} + 1/2^n \cdot (1 - 1/2^{2n})$$
$$< 1/2^{n-1}.$$

This completes the proof. ∎

As a consequence of Theorem 1, we know that none of $\psi(f, f, f)$, $\psi(f, f, f^2)$ and $\psi(f^2, f, f)$, where $f \in_R H_n$, are pseudorandom.

Next we discuss the optimality of $\psi(g, g, f)$ and $\psi(g, f, f)$ where $f, g \in_R H_n$. Apparently, $F_{2n,f}$ can be distinguished from an $r \in_R H_{2n}$. It was proved in [LR] that *two* applications of DES-like transformations cannot obtain a pseudorandom permutation. In particular, Luby and Rackoff showed that $\psi(g, f)$, where $f, g \in H_n$, can be easily distinguished from an $r \in_R H_{2n}$.

Thus, by putting together Theorem 1 and Ohnishi's observations mentioned above, we see that to get a pseudorandom permutation in H_{2n}, *two* independent random functions from H_n and three applications of DES-like transformations are not only *sufficient* but also *necessary*, as far as our construction is restricted to the permutations

$\psi(f_3^k, f_2^j, f_1^i)$, where $i, j, k \in \mathcal{N}$ and $f_1, f_2, f_3 \in H_n$ such that for any $1 \le s, t \le 3$, either $f_s = f_t$ or f_s is independent of f_t. In other words, under the above condition, pseudorandom permutations $\psi(g, f, f)$ and $\psi(g, g, f)$ proposed by Ohnishi [O], where $f, g \in_R H_n$, are *optimal* in the sense that they consist of the minimal rounds of DES-like transformations, and "consume" the minimal number of independent random functions from H_n.

4. Generalizations

This section extends in two directions Theorem 1 to the case of *generalized DES-like transformations*.

Let $\ell \in \mathcal{N}$ with $\ell \ge 2$. Following [FNS, pp.1547-1549] and [S], we associate with an $f \in H_n$ a function $F_{\ell n, f} \in H_{\ell n}$ defined by $F_{\ell n, f}(B_1, B_2, \ldots, B_\ell) = (B_2 \oplus f(B_1), B_3, \ldots, B_\ell, B_1)$, where $B_i \in I_n$. Call $F_{\ell n, f}$ the *generalized DES-like transformation* associated with f.

For $f_1, f_2, \ldots, f_s \in H_n$, define $\theta(f_s, \ldots, f_2, f_1) = F_{\ell n, f_s} \circ \cdots \circ F_{\ell n, f_2} \circ F_{\ell n, f_1}$. It is easy to show that when $s < 2\ell - 1$, $\theta(f_s, \ldots, f_2, f_1)$ can be distinguished from an $r \in_R H_{\ell n}$. By modifying the proof for the Main Lemma of [LR], it can be shown that when $s = 2\ell - 1$, $\theta(f_s, \ldots, f_2, f_1)$ is pseudorandom where $f_1, f_2, \ldots, f_s \in_R H_n$.

Now we prove an impossibility result on $\theta(f_{2\ell-1}, \ldots, f_2, f_1)$. For $(2\ell - 1)$ integers $i_1, i_2, \ldots, i_{2\ell-1} \in \mathcal{N}$, let $\Theta_{\ell n}(i_1, i_2, \ldots, i_{2\ell-1})$ be the multi-set consisting of all functions $\theta(f^{i_{2\ell-1}}, \ldots, f^{i_2}, f^{i_1}) \in H_{\ell n}$ where $f \in H_n$, and let $\Theta(i_1, i_2, \ldots, i_{2\ell-1}) = \{\Theta_{\ell n}(i_1, i_2, \ldots, i_{2\ell-1}) \mid n \in \mathcal{N}\}$.

[**Theorem 2**] *For any* $i_1, i_2, \ldots, i_{2\ell-1} \in \mathcal{N}$, *there is a distinguisher* $T = \{T_{\ell n} \mid n \in \mathcal{N}\}$ *for* $\Theta(i_1, i_2, \ldots, i_{2\ell-1})$. *Each* $T_{\ell n}$ *has* $(m_1 + m_2 + 1)$ *oracle gates, where* $m_1 = (i_1 + i_2 + \cdots + i_\ell)/d$, $m_2 = (i_2 + i_3 + \cdots + i_{\ell+1})/d$ *and* $d = \gcd(i_1 + i_2 + \cdots + i_\ell, i_2 + i_3 + \cdots + i_{\ell+1})$.

Proof: There are two cases to be treated: $\ell = 2$ and $\ell > 2$. The former has been proved in Theorem 1. The proof for the latter is similar to that for the former.

As in the proof of Theorem 1, denote by $O_0, O_1, O_2, \ldots, O_{m_1+m_2}$ the $(m_1 + m_2 + 1)$ oracle gates, by $(X_{s1}, X_{s2}, \ldots, X_{s\ell})$ and $(Y_{s1}, Y_{s2}, \ldots, Y_{s\ell})$ the input to and output of O_s respectively, and by 0^n the all-0 string in I_n.

DESCRIPTION OF $T_{\ell n}$:

(1) The input to O_0 is $(X_{01}, X_{02}, \ldots, X_{0\ell}) = (0^n, 0^n, \ldots, 0^n)$.

(2) The input to O_1 is $(X_{11}, X_{12}, \ldots, X_{1\ell}) = (0^n, Y_{03}, 0^n, \ldots, 0^n)$.
And if $m_1 > 1$, then for each $1 < p \le m_1$, the input to O_p is
$(X_{p1}, X_{p2}, \ldots, X_{p\ell}) = (0^n, X_{(p-1)2} \oplus Y_{(p-1)3}, 0^n, \ldots, 0^n)$.

(3) The input to O_{m_1+1} is $(X_{(m_1+1)1}, X_{(m_1+1)2}, \ldots, X_{(m_1+1)\ell}) = (Y_{02}, 0^n, \ldots, 0^n)$. And if $m_2 > 1$, then for each $m_1 + 1 < t \le m_1 + m_2$, the input to O_t is $(X_{t1}, X_{t2}, \ldots, X_{t\ell}) = (X_{(t-1)1} \oplus Y_{(t-1)2}, 0^n, \ldots, 0^n)$.

(4) $T_{\ell n}$ outputs a bit 1 iff $Y_{m_1 2} = X_{(m_1+m_2)1} \oplus Y_{(m_1+m_2)2}$.

See also Figure 2 for the structure of $T_{\ell n}$. Analysis necessary is also similar to Theorem 1, and omitted here. ∎

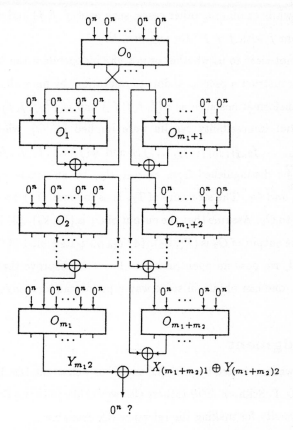

Figure 2: Structure of Oracle Circuit $T_{\ell n}$

Further analysis of the proof for Theorem 2 reveals that even given $(\ell - 1)$ independent random functions from H_n, it is not guaranteed that one can always obtain pseudorandom permutations in $H_{\ell n}$, by $(2\ell - 1)$ applications of generalized DES-like transformations. This is formally stated below.

Let $i_1, i_2, \ldots, i_{\ell+1} \in \mathcal{N}$, and let $\widetilde{\Theta}_{\ell n}(i_1, i_2, \ldots, i_{\ell+1})$ be the multi-set consisting of all functions $\theta(f_{\ell-1}, \ldots, f_3, f_2, f_1^{i_{\ell+1}}, \ldots, f_1^{i_2}, f_1^{i_1}) \in H_{\ell n}$ where $f_1, f_2, \ldots, f_{\ell-1} \in H_n$, and let $\widetilde{\Theta}(i_1, i_2, \ldots, i_{\ell+1}) = \{\widetilde{\Theta}_{\ell n}(i_1, i_2, \ldots, i_{\ell+1}) \mid n \in \mathcal{N}\}$.

[Theorem 3] *For any $i_1, i_2, \ldots, i_{\ell+1} \in \mathcal{N}$, $\widetilde{\Theta}(i_1, i_2, \ldots, i_{\ell+1})$ is not pseudorandom.*

5. Concluding Remarks

Our consideration has been restricted to the case of $\psi(f_3^k, f_2^j, f_1^i)$ where $i, j, k \in \mathcal{N}$ and $f_1, f_2, f_3 \in H_n$ such that for any $1 \leq s, t \leq 3$, either $f_s = f_t$ or f_s is independent of f_t. It is worth while examining other cases, such as $\psi(\hat{f}, f, f)$ and $\psi(f, f, \hat{f})$ where $\hat{f}$ is constructed from f with $\hat{f} \neq f^m$ for any $m \in \mathcal{N}$.

Also, it is not clear to us whether or not *one* independent random function $f \in H_n$ can be used to construct a pseudorandom permutation, by more than three applications of DES-like transformations such as $\psi(f, f, f, f^2)$ and $\psi(f^2, f, f, f)$.

Some partial impossibility results were implied in [O], where Ohnishi showed that both $\psi(f_s, \ldots, f_2, f_1, f_0, f_1, f_2, \ldots, f_s)$ and $\psi(f_s, \ldots, f_2, f_1, f_1, f_2, \ldots, f_s)$, where $f_i \in H_n$, can be distinguished from an $r \in_R H_{2n}$ by an oracle circuit $\bar{T}_{2n}$ with two oracle gates O_1 and O_2. The structure of $\bar{T}_{2n}$ is as follows: (1) Choose $X_1, X_2 \in I_n$. (2) Input (X_1, X_2) to O_1. Assume that the output of O_1 is (Y_1, Y_2). (3) Input (Y_2, Y_1) to O_2. Assume that the output of O_2 is (Z_1, Z_2). (4) $\bar{T}_{2n}$ outputs a bit 1 iff $(X_1, X_2) = (Z_2, Z_1)$.

To the end, we pose an open problem: Prove or disprove that from *one* random function in H_n, one can obtain in some way a pseudorandom (invertible) permutation in H_{2n}.[2]

Acknowledgment

This work was motivated by [LR], [O] and [S] as well as the Dec.1988 communication with Professor C. P. Schnorr. The authors also would like to thank Professor A. Maruoka of Tohoku University for making the reference [O] available.

---

[2] Schnorr believed that the answer to the problem would be affirmative. (private conversation at Eurocrypt'89, April 1989.)

References

[FNS] H. Feistel, W. A. Notz and J. L. Smith: "Some cryptographic techniques for machine-to-machine data communications," *Proceedings of IEEE* , Vol. 63, No. 11, (1975), pp.1545-1554.

[GGM] O. Goldreich, S. Goldwasser and S. Micali: "How to construct random functions," *Journal of ACM*, Vol. 33, No. 4, (1986), pp.792-807.

[LR] M. Luby and C. Rackoff: "How to construct pseudorandom permutations from pseudorandom functions," *SIAM Journal on Computing*, Vol. 17, No. 2, (1988), pp.373-386. (A preliminary version including other results appeared in *Proceedings of the 18th ACM Symposium on Theory of Computing*, (1986), pp.356-363.)

[NBS] *Data Encryption Standard*, Federal Information Processing Standards (FIPS) Publication 46, National Bureau of Standards, U.S. Department of Commerce, (1977).

[O] Y. Ohnishi: "A study on data security," *Master Thesis* (in Japanese), Tohoku University, Japan, (March, 1988).

[R] R. A. Rueppel: "On the security of Schnorr's pseudorandom generator," *Abstracts of EUROCRYPT'89*, Houthalen, (April 10-13, 1989).

[S] C. P. Schnorr: "On the construction of random number generators and random function generators," *Advances in Cryptology — EUROCRYPT'88*, LNCS Vol. 330, Springer-Verlag, (1988), pp.225-232.

[Y] A.C. Yao: "Theory and applications of trapdoor functions," *Proceedings of the 23rd IEEE Symposium on Foundations of Computer Science*, (1982), pp.80-91.

Appendix

In [O], Ohnishi showed that both $\psi(g, f, f)$ and $\psi(g, g, f)$, where $f, g \in_R H_n$, cannot be efficiently distinguished from an $r \in_R H_{2n}$. He obtained the result by carefully modifying the proof for the Main Lemma of [LR]. The major modification begins with the definition of B-gate$_i$ [LR,p.382]. Now we describe the definition for the case of $\psi(g, g, f)$. The case of $\psi(g, f, f)$ is similar.

Let $\Omega = \{0, 1\}^{3nm}$, and $\omega = \omega_1, \cdots, \omega_{3nm} \in \Omega$. For $1 \leq i \leq m$, define $X_i(\omega), Y_{2i-1}(\omega)$ and $Y_{2i}(\omega)$ as follows:

$$X_i(\omega) = \omega_{(i-1)n+1} \bullet \cdots \bullet \omega_{(i-1)n+n},$$

$$Y_{2i-1}(\omega) = \omega_{mn+(2i-2)n+1} \bullet \cdots \bullet \omega_{mn+(2i-2)n+n},$$

$$Y_{2i}(\omega) = \omega_{mn+(2i-1)n+1} \bullet \cdots \bullet \omega_{mn+(2i-1)n+n}.$$

Also let

$$X(\omega) = < X_1(\omega), \ldots, X_m(\omega) >,$$

$$Y(\omega) = < Y_1(\omega), \ldots, Y_{2m}(\omega) > .$$

The ith oracle gate is computed as follows:

B-gate$_i$:

The input is $L_i(\omega) \bullet R_i(\omega)$,

$\ell \leftarrow \min\{j : 1 \leq j \leq i, R_i(\omega) = R_j(\omega)\}$,

$\alpha_i'(\omega) \leftarrow L_i(\omega) \oplus X_\ell(\omega)$,

$\ell \leftarrow \min\{\{2j - 1 : 1 \leq j \leq i, \alpha_i'(\omega) = \alpha_j'(\omega)\} \bigcup$

$\quad \{2j : 1 \leq j \leq i - 1, \alpha_i'(\omega) = \beta_j'(\omega)\}\}$,

$\beta_i'(\omega) \leftarrow R_i(\omega) \oplus Y_\ell(\omega)$,

$\ell \leftarrow \min\{\{2j - 1 : 1 \leq j \leq i, \beta_i'(\omega) = \alpha_j'(\omega)\} \bigcup$

$\quad \{2j : 1 \leq j \leq i, \beta_i'(\omega) = \beta_j'(\omega)\}\}$,

$\gamma_i'(\omega) \leftarrow \alpha_i'(\omega) \oplus Y_\ell(\omega)$,

The output is $\beta_i'(\omega) \bullet \gamma_i'(\omega)$.

Note that the same function g is applied in both the second and the third rounds of DES-like transformations of $\psi(g, g, f)$. So the key point is that each input to g should be compared with *all* previous inputs to it, no matter which round they appear in.

The remaining portion of the proof proceeds in the same way as [LR], with some obvious modifications introduced by the above defined B-gate$_i$.

On the Security of Schnorr's Pseudo Random Generator

Rainer A. Rueppel

Crypto AG, P.O. Box 474, 6301 Zug

Switzerland

Abstract

At Eurocrypt 88 Schnorr [8] proposed a pseudo random generator for which he claimed that it could not be distinguished from a truly random source with less than $2^{o(n)}$ output bits, even when unlimited computing power was available. We show that this generator can, in fact, be distinguished with only $4n$ bits of output. Moreover, we present an efficient (linear-time) algorithm which recovers the key from a substring only slightly larger than the generator's keysize. Consequently, the generator is insecure.

1 Introduction

There are two ways in which one can (conceptually) limit the opponent's capabilities (we assume that the description of the generator is public, and that the opponent is allowed to obtain pure keystream).

1. The opponent has limited computing resources. For instance, one may assume that only polynomial-time attacks are feasible.

2. The opponent has limited access to the keystream bits. For instance, one may assume that it is unfeasible to collect more than l consecutive keystream bits, or, that it is unfeasible to collect a total of more than m keystream bits.

The basic notion of security for a pseudo random generator is that of indistinguishability [2, 3, 6, 10]. Ideally, one would like to design keystream generators that, below the unicity distance, cannot be distinguished from a random source, even with unlimited computing power; and that, beyond the unicity distance, cannot be distinguished from a random source under the assumption of reasonably bounded resources.

A complexity-theoretic framework has emerged over the past few years [2, 3, 6, 9, 10] that defines a pseudo random generator to be *perfect* or *cryptographically secure* if it passes all polynomial-time statistical tests. So far, it is not clear whether perfect generators do exist. Nevertheless, if one introduces some reasonable complexity hypothesis, one can prove "perfectness" of a generator. For instance, "perfect" generators have been postulated based on the discrete log [3], based on quadratic residuosity [2], based on one-way functions [10], based on RSA [1, 6]. At Eurocrypt 88 Schnorr proposed a pseudo random generator and a different notion of security such that the security of the generator is not based on any such unproven assumption [8]. His construction makes use of the permutation function generator proposed in [4]. This permutation generator consists of a m-round DES-like structure, where in each round i a different (pseudo) random function f_i is applied. It is shown that 3 rounds suffice to prove perfectness (polynomial-time indistinguishability) of the resulting permutation generator, provided the functions f_i are indistinguishable.

Schnorr's pseudo random generator $G = \{G_k\}$ is defined as follows:

Input x: the key (seed) is a random function $f : I_n \to I_n$; size of description $n2^n$.

1. set $y_i^0 = i$ for $i = 0, 1, ..., 2^{2n} - 1$.

2. For $j = 0, 1, 2$ do
$$y_i^{j+1} = (R(y_i^j), L(y_i^j) \oplus f(R(y_i^j))))$$

{L and R mean left and right half of argument}.

Output $G_k(x)$: the sequence of $y_i^3, i = 0, 1, 2, ..., 2^{2n} - 1$. Note that the key x has size $k = n2^n$ (the function description) and that $G_k(x)$ has length $2n2^{2n}$. Thus, the generator stretches a seed of length k into roughly k^2 pseudo random bits.

The main claim in [8] is that this generator passes all statistical number tests (even those with unlimited time bound) that depend on at most $2^{n/3 - (\log n)^2}$ bits of $G_k(x)$. This result refers to the situation where the opponent has limited access to the keystream bits, but otherwise has unlimited computing power. In Section 2 it is shown that Schnorr's claim is erroneous; more precisely, a statistical test is exhibited which distinguishes G with as little as $4n$ bits of $G_k(x)$. This fact was independently discovered by Ohnishi [7], Zheng, Matsumoto, and Imai [11], who were studying the construction of pseudo-random permutations, as proposed by Luby and Rackoff [4]. Maurer and Massey [5] identified and took up the constructive problem hidden in Schnorr's approach, namely to design what they called *perfect local randomizers*; these are algorithms that stretch a short string of random bits into a long pseudo-random sequence of bits with the property that every subset of e output bits is completely random (in the information-theoretic sense). They showed that such perfect local randomizers can elegantly be constructed using coding-theoretic tools. However, in a practical application the assumption that the opponent cannot observe more than e bits may be too restrictive. Therefore, if a local randomizer is proposed for direct use as keystream generator [8], it is mandatory to analyze its computational security. For Schnorr's Generator we present an attack that recovers the key f from a small subset of $n2^n + O(n)$ bits of the output sequence of $G_k(x)$. Moreover, this algorithm runs in time $O(n2^n)$. Note, that this attack reaches the performance limits of any attack, since simply reading the key from memory requires linear time and linear space in the size of the key. It is also demonstrated that some generalizations of Schnorr's generator are prone to the same kind of efficient attack.

2 Results

Let some $y_0 = (l, r)$, and let F_f denote the permutation of I_{2n} as indexed by the key f. Then
$$Ly^3 = LF_f(l, r) = r \oplus f(l \oplus f(r))$$
$$Ry^3 = RF_f(l, r) = l \oplus f(r) \oplus f(r \oplus f(l \oplus f(r)))$$

We will drop the subscript f whenever the same key f is used in every round of the generator. Let O_F be the oracle that evaluates $F(y)$ at specific arguments y (in one computational step). We are interested in distinguishing the 2 cases, (1) F is a random function, (2) $F = F_f$, as used in G.

Theorem 1 *Schnorr's generator G can, for any number of rounds, m, be distinguished from a truly random source by 2 oracle calls, i.e., by observing $4n$ output bits of $G_k(x)$.*

Proof: let $F_m(l,r)$ denote the m-round permutation function on I_{2n}. Then, for any $m > 0$, F_m and its inverse F_m^{-1} are related as follows:

$$F_m^{-1}(l,r) = P_H(F_m(P_H(l,r)))\tag{1}$$

P_H denotes the permutation function that switches left and right half of its argument. The proof is by induction; assume (1) holds for m, and let $F_m^{-1}(l,r) = (a,b)$, then

$$F_{m+1}^{-1}(l,r) = (b \oplus f(a), a)$$

$$F_{m+1}(P_H(l,r)) = (a, b \oplus f(a))$$

Thus (1) holds also for $(m+1)$. For $m = 1$

$$F_1^{-1}(l,r) = (r \oplus f(l), l)$$

$$F_1(l,r) = (r, l \oplus f(r))$$

Thus 1 holds for any $m > 0$. It follows

$$(l,r) = F_m \circ F_m^{-1}(l,r) = F_m \circ P_H \circ F_m \circ P_H(l,r)$$

To distinguish $G_k(x)$ from a purely random string, we apply P_H to an arbitrary argument (l,r), call the oracle O_F, apply P_H a second time, and call the oracle a second time. The result will be (l,r) with probability 1 if F_m was used by the oracle. For a random function the probability that (l,r) will be the result is 2^{-2n}. $\Box$

Corollary 1 *If two functions f,g are used alternately, then G can still be distinguished by observing $4n$ output bits of $G_k(x)$, provided the number of rounds is odd.*

Proof: Let g be the function used in odd rounds 1,3,5,.. and let f be the function used in even rounds 2,4,6,..; then

$$F_{gf,1}(l,r) = (r \oplus g(l), l)$$

$$F_{gf,1}^{-1}(l,r) = (r, l \oplus g(r))$$

$$F_{gf,3}(l,r) = [r \oplus f(l \oplus g(r)), l \oplus g(r) \oplus g(r \oplus f(l \oplus g(r)))]$$

$$F_{gf,3}^{-1}(l,r) = [r \oplus g(l) \oplus g(l \oplus f(r \oplus g(l))), l \oplus f(r \oplus g(l))]$$

Hence, relation (1) still holds. $\Box$ Ohnishi [7] has independently discovered that this result holds for any palindromic arrangement of functions, that is, for any sequence of functions where reversing the order of the functions does not change the sequence of the functions. In the sequel, the computational security of G shall be analyzed for the case where the opponent has acquired a keystream-segment whose length is comparable to the keylength. This seems also interesting in the light of the generator's close structural ties to DES. The following algorithm will recover the key from a substring of size about $n2^n$ of $G_k(x)$. It is based on the observation that one can always find an $i \in [0,1,..,2^n - 1]$ such that

$$LF(i,r) = i\tag{2}$$

For any r there must exist a unique i such that $f(r) = i \oplus r$ which implies $f(i \oplus f(r)) = i \oplus r$ which in turn implies (2).The converse is not true in general, since f need not be invertible, that is, for an image $i \oplus r$ there may be more than one preimage. The case $LF(l,r) = l$ but $f(r) \neq l \oplus r$ is called a "false alarm". Such a "false alarm" can be easily eliminated by noting that

$$LF(LF(j,r), l \oplus r \oplus j) = j, \qquad j = 0, 1, 2, ..., 2^n - 1$$

has to hold with probability 1 if $f(r) = l \oplus r$. But if $f(r) \neq l \oplus r$ the check equation will only hold with probability 2^{-n} for any j. Thus, one additional test will usually discover a "false alarm". After verification that $f(r_0) = l_0 \oplus r_0$ key and keystream are related through the simple equation:

$$f(j) = LF(j \oplus f(r_0), r_0) \oplus r_0 \qquad \forall j \neq r_0$$

Algorithm: *Recover key f*

1. fix some r_0; set $i = 0$

2. test if $LF(i, r_0) = i$ {lock-in condition}
   if false set $i = i + 1$ and test again.

3. set $i_0 = i$
   {define $u_j = j \oplus i_0 \oplus r_0$ and $v_j = LF(j, r_0)$; these variables serve to check for correct lock-in}
   test if $LF(v_j, u_j) = j$ for $j = 0, 1, ..$
   {usually one test suffices to rule out a false lock-in}
   if false set $i = i_o + 1$ and go back to (2).

4. $f(r_o) = i_o \oplus r_o$
   for all $j \neq r_o$ set $f(j) = LF(j \oplus f(r_o), r_o) \oplus r_o$
   {the key f is recovered}

The number of bits required from the output sequence $G_k(x)$, in order to recover the complete key, is $n2^n + O(n)$, that is, just a little more than the keysize. The running time is $O(2^n)$, if n-bit operations are counted. The algorithm cannot be faster, since it has to look at each entry of the key f. From [4] it is known that the use of 3 independent random functions g, f, h results in a pseudo random permutation $F_{g,f,h}$. According to [11], it was proved in [7] that two independent random functions g, f used either in order g, f, f or g, g, f suffice to guarantee pseudo randomness of $F_{g,f}$. Consider the corresponding generalization of Schnorr's generator where F_f is replaced by $F_{g,f}$. Then the local randomness property of G' directly follows from the indistinguishability of $F_{g,f}$. Now the computational security of the generalized generator G' shall be addressed. For G' it holds

$$Ly^3 = LF_{g,f}(l,r) = r \oplus f(l \oplus g(r))$$

$$Ry^3 = RF_{g,f}(l,r) = l \oplus g(r) \oplus f(LF_{g,f}(l,r))$$

Algorithm (Sketch): *Recover key g,f*

1. fix some r_0;
   get $F_{g,f}(l, r_0)$ for $l = 0, 1, ..., 2^n - 1$.

2. define $f_0(l) = LF(l, r_0) \oplus r_0$ and $f_i(l) = f_0(l \oplus i)$;
   $\{$if $g(r_0) = i$ then f must equal f_i and $RF_{g,f} = l \oplus i \oplus f_i(LF_{g,f})\}$
   for $i = 0, 1, ..$ assume $g(r_0) = i$;
   if $RF_{g,f} = l \oplus i \oplus f_i(LF_{g,f})$ for $l = 0, 1, ..$ (only few checks are necessary) then
   $g(r_0) = i$ and $f = f_i$.

3. fix some l_0
   get $F_{g,f}(l_0, r)$ for $r = 0, 1, ..., 2^n - 1$
   compute $g(r) = RF_{g,f}(l_0, r) \oplus l \oplus f(LF_{g,f}(l_0, r))$ for all r.

The number of bits required from the output sequence $G'_k(x)$, in order to recover the complete key f, g, is $4n2^n$, that is, about double the keysize. The running time is $O(n2^n)$, that is, it is linear in the keysize. Again, the algorithm cannot be faster, since it has to look at each entry of the key f.

3   Concluding Remarks

Schnorr's generator offers neither a provable local randomization nor a reasonable resistance against cryptanalytic attacks. In fact, it can be predicted in linear time. Hence, it should not be used for encryption purposes. But it raises some interesting questions:

1. The local randomization problem, identified and taken up by Maurer and Massey [5]:
   Can one construct perfect local randomizers $G_k(x)$ which stretch a key x of length k into a pseudo random sequence of length $l > k$ in such a way that any $e < k$ bits of the pseudo random sequence are truly random? They showed that this problem can elegantly be solved using coding-theoretic tools. In fact, they obtained much better results than what Schnorr hoped for using the complexity-theoretic approach. The following example compares Schnorr's generator with a linear local randomizer of identical parameters:
   Example [5]: take an $[N, K]$ extended Reed-Solomon code over $GF(2^{2n})$ where $N = 2^{2n}$ and $K = 2^{n-1}$; if this code is mapped into $GF(2)$ a binary code with parameters $N' = 2n2^{2n}$ and $K' = 2n2^{n-1} = n2^n$ is obtained. Such a code has the property that every subset of $e \geq K = n2^{n-1}$ columns of the encoding matrix G is linearly independent. Consequently, if the $n2^n$ information bits are chosen independently and uniformly, the corresponding codeword of length $2n2^{2n}$ has the property that every subset of $e \geq 2^{n-1}$ bits is completely random. Note that this is about the third power of Schnorr's bound.

2. The pseudo random permutation construction problem [11]: is it possible to construct, with only one random function f, a permutation function F_f which is provably pseudo random? Then the local randomness property would be inherited by a bit generator construction such as Schnorr's.

But, what one would really like is a generator with both a proof of local randomization and a proof of unpredictability (under the assumption of reasonably bounded resources)

Acknowledgment

I wish to thank Jim Massey and Ueli Maurer for helpful discussions on the topic of the local randomizer. I would also like to thank Prof. C.P. Schnorr and Prof. T. Matsumoto for their comments.

References

[1] W. Alexi, B. Chor, O. Goldreich, and C.P. Schnorr, "RSA and Rabin Functions: Certain Parts are as Hard as the Whole", SIAM Journal on Comput. 17 (1988), pp.194-209.

[2] L. Blum, M. Blum, and M. Shub, "A simple unpredictable pseudo-random number generator", SIAM J. Comput. 15 (1986), pp. 364-383.

[3] M. Blum, S. Micali, "How to generate cryptographically strong sequences of pseudo-random bits", SIAM J. Comput. 13 (1984), pp. 850-864.

[4] M.Luby, C. Rackoff, "How to construct pseudorandom permutations from pseudorandom functions", SIAM J.Comput. 17 (1988), pp. 373-386.

[5] U.Maurer, J.L.Massey, "Perfect Local Randomness in Pseudo-random Sequences", submitted to Crypto 89.

[6] S. Micali, C.P. Schnorr, "Efficient, perfect random number generators" Preprint MIT, University of Frankfurt, 1988.

[7] Y. Ohnishi, "A study on data security", Master Thesis (in Japanese), Tohuku University, Japan, 1988.

[8] C.P. Schnorr, "On the construction of random number generators and random function generators", Proc. Of Eurocrypt 88, Lecture Notes in Computer Science 330, Springer Verlag, 1988.

[9] A. Shamir, "On the generation of cryptographically strong pseudo-random sequences", 8th International Colloquium on Automata, Languages, and Programming, Lecture Notes in Computer Science 62, Springer Verlag, 1981.

[10] A.C.Yao, "Theory and applications of trapdoor functions", Proc. of the 25th IEEE Symp. on Foundations of Computer Science, New York, 1982.

[11] Y. Zheng, T. Matsumoto, H. Imai, "Impossibility and Optimality Results on Constructing Pseudorandom Permutations", Proceedings of Eurocrypt 89, this Volume, Lecture Notes in Computer Science, Springer Verlag, 1989.

How easy is collision search? Application to DES

Jean-Jacques Quisquater   Jean-Paul Delescaille

Philips Research Laboratory Belgium

Avenue Albert Einstein, 4

B–1348 Louvain-la-Neuve, Belgium

jjq@prlb.philips.be — jpdesca@prlb.philips.be

(Extended summary)

1   About collisions

Given a cryptographic algorithm f (depending upon a fixed message m and a key k), a pair of keys with collision k_1 and k_2 (in short, a *collision*) are keys such that

$$f(m, k_1) = f(m, k_2).$$

The existence of collisions for a given cryptographic algorithm means that this algorithm is not *faithful* in a very precise technical sense (see [3]). It is important to know if it is *easy* to find collisions for a given cryptographic algorithm. Indeed, the existence of such easy-to-find collisions means that this algorithm (or, maybe, its mode of use) is not secure for many applications related to hashing functions used in the context of digital signatures.

While there is a large probability that DES, in its basic mode, has collisions, nobody has found a collision for DES until now. It is thus a challenging problem to find only one. We found 21 collisions with the same plaintext ($=$ identical m).

The used algorithm is based on the so-called theory of *distinguished points* (see the abstract in the proceedings of CRYPTO '87 by the same authors). The result was obtained thanks to efficient implementations of DES on VAXes and SUN's and intensive use of the idle time from 35 workstations at our laboratory.

2 Algorithms

2.1 A naive algorithm

Here we will use DES(m, k) for denoting DES in its basic mode, with m as the input message (64 bits), k as the key (56 bits); the obtained result has 64 bits.

If we suppose that DES can be modelled as a *random mapping*, the following algorithm works. Given a fixed message m, compute about 2^{32} values DES(m, k_i), where the k_i's are all different. Sort the obtained values. With a high probability, we will obtain one collision. The problem with such a method is the need of a very large memory (disk and RAM) for storing the values. The associated problem of sorting a lot of data is not so simple. This method is not feasible as an annex task in a network of workstations.

2.2 Algorithm without memory

There exist very efficient algorithms to find cycles in periodic functions mapping some finite domain D into D (see [1]). If we take a random element x from the finite set D and generate the infinite sequence $f^0(x) = x, f^1(x) = f(x), f^2(x) = f(f(x)), ...$, then we know that the sequence becomes cyclic. That is, there exists some value l that $f^{l+c}(x) = f^l(x)$ (the *point of contact* common to the leader and the cycle) and $f^{l+c-1}(x) \neq f^{l-1}(x)$ (one value on the leader and one on the cycle). That is, by definition, we found a collision for such a f.

It is simple to modify such algorithms to find cycles when the domain D has less elements than the codomain. That is, the input k has less elements than the output. We need some projection function g for mapping the output onto the next input. However, we have a new problem. The common point is not necessary the result of a collision. In fact, for DES, the probability of having found a collision is one out of 256. We now call a pair of antecedent points of such a common point a *pseudo-collision*.

So the algorithm becomes the following one. Given m and an initial value x_0, find a first pseudo-collision. Verify if it is a *true* collision. If no, try again with a new initial value x_1, aso. A new problem with such an algorithm is that we compute many times the same values due to the fact that we are computing values on the same cycle with high probability (see the paper by Flajolet and Odlyzko, in these proceedings). An effective technique to overcome this problem is the use of distinguished points.

2.3 Effective algorithm in use

Figure 1 describes the algorithm we are using. The two variables *pseudo_collision* and *pseudo_cycle* are first set to false. The variable i is a counter used for the number of the current initial value. At each call of *new_init*, a new and different value for y is chosen. The counter k is used for computing the number of computed values since the last call of *add*, that is, since the last time we found a distinguished point. The procedure *distinguished_point* is true if the input y is a distinguished point, that is a value with some attribute fast to compute (for instance, we used the attribute that the value y had 20 bits set to 0 at the left). The procedure *add* puts the value y into TABLE by checking if there is another entry already there with the same value y. It is a fast operation (comparisons with elements in a small table). If yes, then *add* puts the variable *pseudo_collision* to true. We have then detected a pseudo-collision but we do not know its exact value. We will find that in a next phase. The variable *limit_k* is used for avoiding the problem of looping due to a cycle without any distinguished point. After some time, TABLE contains a large number of values indicating pseudo-collisions. We are now in position to find out if there are some collisions in this set. For that we compute the effective values of the pseudo-collisions. Sometimes it is a collision.

3 Results

The first collision has been found January 13, 1989, the birthday of the first author (another application of the birthday paradox!) after 3 weeks of computation.

Here is this first one:

PLAIN = 0404040404040404 (in hexadecimal)
k_1 = 4A5AA8D0BA30585A (idem)
k_2 = 46B2C8B62818F884 (idem)
RESULT = F02D67223CEAF91C (for the two keys)

```
pseudo_collision ← false;
pseudo_cycle ← false;
i ← 0;
repeat
    y ← new_init(i);
    k ← 0;
    repeat
        y ← f(y);  k ← k + 1;
        if (distinguished_point(y)) then
            begin
                add(y, pseudo_collision);  k ← 0 ;
            end
        if k > limit_k then pseudo_collision ← true ;
    until pseudo_collision or pseudo_cycle;
until i > limit_i;
```

Figure 1: The pseudo-collision detecting algorithm

The workstations worked in parallel on the same problem (= same m) with distinct initial points and distinct projection functions g. In this context, the number of found collisions is proportional to the square of the used time if we consider that the studied cryptographic function acts as a random mapping. The projection functions were simply different sets of 56 bits out of the outputs of 64 bits.

We have found 21 collisions for DES (March 13, 1989). The table at the end of this paper gives the complete values.

An algorithm was implemented to draw the mappings resulting from these DES computations (see an example in the paper by Flajolet and Odlyzko, in these proceedings).

References

[1] Robert Sedgewick, Thomas G. Szymanski and Andrew C. Yao, *The complexity of finding cycles in periodic functions*, SIAM J. Comput., vol. 11, 2, pp. 376–390, 1982.

[2] Jean-Jacques Quisquater and Jean-Paul Delescaille, *Other cycling tests for DES*, Springer Verlag, Lecture notes in computer science **293**, Advances in cryptology, Proceedings of CRYPTO '87, pp. 255–256.

[3] Burton Kaliski, Ronald Rivest and Alan Sherman, *Is the Data Encryption Standard a group? (Results of cycling experiments on DES)?*, J. Cryptology, vol. 1, 198, pp. 3–36.

PLAIN	KEY 1	KEY 2	CIPHER
0404040404040404	4a5aa8d0ba30585a	46b2c8b62818f884	f02d67223ceaf91c
	d296c2ca66be3c60	1680b00c1c22c6b4	e20332821871eb8f
	6edaa03254d2a298	22a64edc20e07032	7237f9e44466059f
	cc3adc3616cc1c32	620e08e886aa8c1c	345d8975676ffde0
	a2aa9adc56a60ad6	b41ebe7a88c4a8c8	301c9a64b903048d
	5888c640ee3016d4	8654a2b862a82486	8f4a67da0852722d
	1e620c46682e325c	0ed86014328cf2da	96f0faf4f80b6b29
	780a76586c7c0ca4	92f69c5aa2c84ee8	1d901196097a93f4
	46f422a832ac0c18	1680f2049484b4b2	85795a73b4af5d78
	3eb8406c969c9c84	e4f06aaea2022e02	46184d44b739a147
	28e8161878343ea0	36a0f03afe48c226	c5ed963b29a48bf6
	060c0e048614bc42	5c4afa4ae0c62a84	c931dab489f515a1
	d0e4aa90baba681c	d8fc6cba3c0a946c	a3c7d6d33eb1400d
	36da7e6010d6a07e	2c2c5a243cd882fa	6a5d431ed4863421
	7aac9c602e9854b6	ac78ca74c6a0ea6e	2edeaaa86e5141af
	ce806eee7cfcd2ec	ae8838904874c606	150e0b6ff35b4f0e
	366cf4baa8cc6c80	76f6527c54447ade	77964b1e86be688e
	6ece1e20bef2b0f8	be827240c8bc3e6a	f29fdbc8dc6c174a
	5e301c2452d88476	5406c60cb4d6f0c8	c6120f53b62eed0d
	0e5ebe562c961274	b45e08326ea40e10	ef5293f14f84fc4f
	624e36aa48926a2e	a862d2aef0c06c54	7dd3c3d34ea30c2f

Table of known collisions for DES.

Section 7

Sharing and authentication schemes

Prepositioned Shared Secret and/or Shared Control Schemes[1]

Gustavus J. Simmons
Sandia National Laboratories
Albuquerque, New Mexico 87185 USA

Abstract

Secret sharing is simply a special form of key distribution.

Introduction

This is the third in a series of papers devoted to the analysis and realization of much extended capabilities for shared secret and/or shared control schemes over and above what can be achieved by simple (k,ℓ)-threshold schemes [42,43]. Since an essential first step has been to understand as clearly as possible the underlying principles on which shared control is based, all of the papers -- this one included -- have also been concerned with the formulation of a simple unifying model of sufficient generality to encompass all of the extensions.

The need for additional capabilities has thus far occurred in three main areas: in the enforcement of command objectives, either of the national command authority or of the subordinate military commands, in the implementation of multinational controls of treaty controlled actions and as described in other sources, in meeting the needs of the financial and banking community. In both military command and in banking applications, there is frequently a requirement for the control scheme to accommodate more than one class of participants with differing capability to initiate the controlled action. While it is certainly true that no two members of the Joint Chiefs of Staff equal the President (in terms of their command authority), one can easily conceive of circumstances in which the President might wish to prearrange for them to be able to act in his stead. For example, the President anticipating circumstances under which he would be unable to act might wish to give an order of the form: "If two of you agree that this circumstance has occurred, then this is what you should do..." -- the gravity of the action being so great that he wants to be certain that at least two of them concur before the action can be initiated. Similarly, in military command schemes, there is a frequent need to make the command structure be less vulnerable to a decapitation type attack, which can be achieved by making it possible for lower levels of command under some circumstances to autonomously initiate actions that normally could only be initiated by higher echelons of command. In most instances, the only acceptable way this can be done is to compen-

1. This work performed at Sandia National Laboratories supported by the U. S. Department of Energy under contract no. DE-AC04-76DP00789.

sate for the lower level of authority (and hence of responsibility) by requiring a higher level of concurrence before the action can be initiated. This could be done by implementing separate and independent simple k_i-out-of-ℓ_i threshold schemes for each level of command (control). However, in almost all such multilevel control schemes it is required that capability be hereditary, meaning that any member of a higher level of capability class should be able to function as a member of a lower (less capable) class, with the diminished capability of a member of that class. In other words, if the concurrence of any two colonels, or of any three majors, suffices to reconstitute the launch enable code for a tactical missile, then it is inconceivable that one colonel and two majors acting together should not also be able to launch the missiles. Precisely the same situation arises in a banking setting in which any two vice presidents or any three senior tellers can authenticate an electronic funds transfer. In such a scheme, any vice president along with any pair of senior tellers should be able to do so also.

In the case of treaty controlled actions, irrespective of whether the controlled action is the direct object of the treaty or only associated with verifying compliance with the conditions of the treaty, there are at least two parties with differing, and oftentimes competing, objectives. In the simplest situation of this type, two parties have agreed that it should be impossible to initiate an action unless they both agree. At the place (or time) at which the control is effected, each party has its own control team to insure that their national input to the control scheme will only be made under the agreed upon circumstances. The type of concurrence each nation requires of its control team members in order for their national input to be made can vary widely. In analogy to the U.S. practice of the two-man control of nuclear weapons (and associated enabling information) or of the launch control of strategic missiles, a national control team could consist of two members, neither of whom has the capability to act alone, but who can jointly make their nation's input to the control scheme. It could equally well be a multilevel control scheme of the sort discussed in the preceding paragraphs. The important point is that in these control schemes there are two or more parties, none of whom can substitute for any of the others.

The first paper in this trilogy, "How to Really Share a Secret" [42] was devoted to these two extensions to threshold schemes: multilevel and multipart(y) control schemes. Only later did we realize that these were simply special cases of a single category of extended capability; namely, general concurrence schemes. In the most general formulation of concurrence schemes, there is a set of participants (the insiders) each of whom has a piece of private information related to a secret piece of information, not all of which is known to any other participant nor to any outsider. A concurrence scheme specifies those subsets of the participants that are supposed to be able to recover the secret or to initiate the controlled action by pooling their private pieces of information. Simple threshold schemes or even multilevel and/or multiparty schemes are distinguished only in that in these cases there is a concise way of describing the authorized subsets of participants. Ideally, any subset of the participants that doesn't include one of the authorized groupings should have no better chance of recovering the secret than does an outsider. Schemes that satisfy this condition have been called perfect by Schellenberg and Stinson [40], a terminology we will adopt here also. Clearly in a perfect shared control scheme the information content of the part of each participant's private piece of information that he must keep secret is at least as great as the information content of the secret information itself. If there were any participant for which this wasn't true, then an unauthorized subset of the participants which could be made into an authorized subset by having the participant in question join with them would have less uncertainty about the secret than an outsider: contradicting the assumption that the scheme is perfect. This quantity of information is commonly referred to as a share. All of the extended capability schemes described in [42], i.e., multilevel and multiparty schemes and simple combinations of these, are characterized by each participant only having to keep secret a single share of information. Shared secret schemes of this sort are said to be ideal. At Crypto'88, Benaloh showed that there were concurrence schemes which could not be realized by any ideal scheme [3]. The paper by Brickell "Some Ideal Secret Sharing Schemes" [19] addresses the question of characterizing ideal shared secret schemes in general. Although generalized concurrence is not the subject of this paper, a brief Appendix to this paper concludes our treatment of multilevel and multipart shared secret schemes begun in [42]. The point to the remarks of the last several pages is that one direction of generalization for shared secret and/or shared capability schemes is to be able to realize much more general concurrences than unanimity or simple k-out-of-ℓ threshold schemes.

In most applications for shared control schemes, the consequences of an authorized concurrence are immediate. For example, if two vice presidents of a bank must enter their private pieces of information into the locking mechanism of the vault in order for the secret combination to be reconstituted and for the door to open, they know immediately whether their action has been successful or not since the vault door either opens or else remains shut and locked. Similarly, the missile launch control officers also have an immediate confirmation of the correctness of their inputs if the missile is launched. The second paper of this trilogy, "Robust Shared Secret Schemes or 'How to be Sure You Have the Right Answer Even Though You Don't Know the Question'" [43], was devoted to the problems that arise if the consequences of a shared control scheme are distant in either time or place from where the private pieces of information are input. If the shared secret scheme controls the enabling of a missile warhead, it is clearly desirable to know that the warhead has not been enabled prior to launch as opposed to learning that it wasn't after its arrival over the target. In this application, it would be easy enough to provide a simple go/no-go indication of status, but there are applications in which this isn't feasible. Perhaps the most convincing example is provided by a scheme to accomplish the remote activation/deactivation of preemplaced smart mines. Conceivably it would be possible to have the mines report their status on request, but this could give away their location to an opponent passively eavesdropping on the communication. On the other hand, the information needed to deactivate the mines is at least as important (militarily) as the effort required to locate and destroy them, which justifies its shared control. In this case the action (the mine's function) is distant in both time and place from the point from which they are controlled so it is vitally important for the host nation to know that the mines have indeed been turned off before he runs a convoy through the field, and equally important that he know that they have been reactivated after the convoy's passage.

Basically, this class of problems is analogous to the collection of communications problems addressed by error detecting and correcting codes, the most obvious difference being that the signal is tested for correctness on receipt in the one case and prior to transmission in the other. In the application to shared control schemes, we want to be able to verify (in probability) that the correct secret value has been reconstructed (error detection) and, if it hasn't, to be able to recover from the error (error correction). Since there are many applications for shared control schemes in which the controlled action is distant in either time or place or both from the place at which the control is effected -- such as the enabling or turn-off of space-based defense systems, self-destruct commands to aberrant missiles and/or space shots, the activation and deactivation control of smart mines as described above, etc.; the second area of extended capability for shared control schemes has to do with providing means to verify whether the correct value for the secret has been reconstructed and, if not, to recover from the error. In "Robust Shared Secret Schemes," a general technique was described that makes it possible for the participants to verify (in probability) whether they have reconstituted the correct value for the secret information, even though they have no way of knowing either before or after the fact what the secret value is.

In this paper, we consider a third class of extended capabilities to shared control schemes that are difficult to describe separated from a detailed description of the implementation of the schemes. Roughly speaking, in the applications of interest here we wish to separate the private pieces of information, whose function it is to reveal a secret piece of information under appropriate circumstances, from the secret piece of information they conceal. This almost sounds paradoxical; however, the need for such a capability arises in several real-world applications for shared control schemes.

We mentioned earlier in connection with a description of multilevel concurrence schemes the problem of a decapitation attack on command. Although this terminology is suggestive, it may not be self-explanatory. If there is information which is held by a higher level of command that must be communicated to lower levels of command in order for some action to be initiated -- such as arming warheads, launching missiles, etc. -- an adversary may attempt to prevent the action from occurring by destroying the higher command in a surprise attack before the information can be disseminated to the lower levels of command. This is called a decapitation attack. There is another aspect to decapitation attacks in that if the higher levels of command are destroyed, the lower levels may either not know what to do or else be ineffectual in their response. We are not concerned with these consequences to a decapitation attack, though, but only with the situation in which the lower levels of command are rendered incapable of carrying out an action, that they are otherwise able to do, because one or more pieces of information needed to initiate the action were prevented from reaching them by the attack on the superior command. Multilevel shared control schemes were devised specifically to solve this problem in situations in which the lower level of authority (and hence responsibility) at a lower level of command could be satisfactorily compensated for by requiring an increased level of concurrence for the action to be initiated. There are actions, however, in which either the law doesn't permit a delegation of authority (and hence capability) irrespective of the concurrence that might be required for it to be exercised, or else in which the party(ies) holding the capability are not willing to delegate the capability under normal circumstances. The crucial words in this description are "under normal circumstances" implying that there are circumstances that would either permit such a delegation to be made or in which such a delegation would be acceptable. The problem, from the standpoint of the shared secret or shared control scheme, is the same in either case.

Consider a missile battery at which there are a dozen officers. The consequences of a missile being launched without proper authority would be so great that in normal times (peacetime or in lower levels of alert) the capability to initiate such an action is to be held at a higher level of command: in other words, the policy is that even if all of the officers at the battery believe that a missile should be launched, they should not be able to do so without requesting authorization from the superior commander (and more importantly, could not do so without being given the launch enable codes). In the absence of a shared control scheme, the only way that the superior commander could protect against a decapitation attack on his headquarters (and him) would be to preemptively enable the missiles as a part of going to an advanced state of alert. But these are precisely the circumstances in which there is the greatest concern that something might go wrong and a missile be launched when it shouldn't have been. Requiring the concurrence of k of the battery officers, say two of them, for a launch is a way of increasing confidence in the proper execution of the plan of battle. What is needed is a scheme in which, under normal circumstances, only the superior commander has the capability to enable a missile launch, but which would allow him, when intelligence inputs or other early warnings indicate, to delegate a 2-out-of-12 shared control of the launch to the battery to prevent a decapitation attack from succeeding. The problem is: How does he establish the 2-out-of-12 shared control scheme at the time the battery goes to an advanced state of alert? If no advance arrangements have been made, the twelve pieces of private information would have to be communicated to the battery officers in a secure and authenticated manner, at a time (advanced stage of alert) when communications are apt to be both congested and disrupted. Even if the information could be communicated to the battery at that time, the risk of human error in dealing with unfamiliar codes of a size at the limits of mnemonic aids to memorization would be high. Ideally, it should be possible to distribute the private pieces of information in advance of a need to use the shared control scheme, but with the constraint that in normal circumstances even if all of the participants were to violate their trust and pool their private pieces of information they would still have no better chance of recovering the secret than an outsider would have of simply guessing it. In such a scheme, at the time the battery is put in an advanced state of alert, a single piece of information (one share in the terminology introduced earlier) would need to be communicated by the superior commander to activate the prepositioned 2-out-of-12 shared control scheme so that any two of the officers would thereafter be able to launch the missiles. The important point to this discussion is that almost all of the information needed to implement the shared control (the private pieces of information) could be communicated in advance of the need during a time of low tension and reliable communications. Since there is no special urgency during this set-up phase, the communication could even be handled by courier or by having the officers in the subordinate command come to headquarters to be given their private pieces of information. At a time when the battery is going to a state of advanced alert, i.e., a period of high tension when communications will be at a premium, only a single share (the minimum amount) of information needs to be communicated to activate the scheme.

This same example (a missile battery) can also be used to illustrate the other extended capability to shared control schemes which is also the subject of this paper. There are two ways this need can arise. First, consider the case in which not all of the missiles have the same launch enable code. The problem in this case is to devise a scheme which can be prepositioned that will allow any one, or any selected subset, of the launch enable codes to be activated in a shared control scheme without affecting the quality of control of the unreleased missiles. Clearly, this could be accomplished by prepositioning a shared control scheme for each launch enable code, almost equally clearly this would be a completely unacceptable solution since each participant would be required to remember several private pieces of information, each of which is near the limit of even mnemonically aided recall. What is needed is a way that the same pieces of private information can be used to recover different pieces of secret information.

Another equally important problem has to do with how the battery can stand down from an advanced state of alert, where standing down means reverting to the kind of control that existed prior to the alert. In order for this to be possible, the scheme must provide both a capability for the superior commander to activate the shared control scheme (delegate authority) and to deactivate it, i.e., to rescind his delegation of authority, if the circumstances change so that an advanced state of alert is no longer warranted. If the system is to truly revert to the same type and quality of control after a recall that it had prior to the alert without changing the private pieces of information involved in the shared control scheme, then both the activating information which is to be sent by the superior commander and the enabling code that the missile will respond to must change with each delegation of authority, irrespective of whether the delegated capability was exercised or not. Although it is inappropriate to the purpose of this paper to say much about the practical problems of implementing shared control schemes, it is perhaps worthwhile remarking that there are two ways (at least) to achieve this. The simplest scheme would be for the enable codes to change automatically as a function of time; say once each day. Another approach would be for the mechanism in the missile control-ler that carries out the calculation of the secret information from the private pieces of information to have a stored list of enabling values, only one of which would be operational at a time. In a scheme of this type, the activating piece of information that is sent by the superior commander would have to correspond to the current value of the secret if the shared control is to be operable. If a recall is received (from the superior command), its entry would advance the store to the next stored value for the secret and output a piece of information that could only be obtained by executing this protocol. This would return the missile to a condition wherein only the superior commander could enable it for a launch, or delegate its release if the battery was later put in an advanced state of alert again. The old value of the secret information would become invalid so that stale values of the activating information would not be operable. The unique output which could only be obtained by properly carrying out the recall protocol could be returned to the supe-rior command to verify that the control system had been returned to its prealert status.

The point of this lengthy discussion of the simple, but plausible, example of a missile battery was to illustrate as clearly as possible the two essential features to the schemes that are the subject of this paper:

a) It should be possible to preposition all of the private information needed for the shared control subject to the condition that even if all of the participants were to violate the trust of their position and collaborate with each other, they would have no better chance of recovering the secret information than an outsider has of guessing it.

b) It should be possible to activate the shared control scheme once it is in place by communicating a single share of information, and for many applications, it should also be possible to reveal different secrets (using the same prepositioned private pieces of information) by communicating different activating shares of information.

There are numerous applications that require these capabilities of shared control schemes, however, we will not discuss the details of these applications further but concentrate instead on a discussion of how such extended capabilities can be achieved. Since this depends critically on the implementation of the shared control schemes, we must first digress to describe the general model for shared control schemes.

General Model for Shared Secret and/or Shared Control Schemes

There are two essentially different ways in which pieces of information related to another, secret, piece of information can be constructed and distributed among a group of participants so that designated subsets of the participants can recover the secret piece of information, while no collection of participants that doesn't include one of these subsets can. One of these classes of shared secret schemes can be adapted to provide the extended capabilities that are the subject of this paper, while the other cannot. We begin our development of the general model with a discussion designed to clarify the essential difference between the two classes of schemes.

In some schemes, the set of possible values for the secret, consistent with all of the private pieces of information that have been exposed, remains unchanged until the last required piece of private information becomes available, at which point the unique value for the secret suddenly becomes the only possibility. In others, as each successive piece of the private information is exposed, the range of possible values that the secret could assume narrows, until finally when the last required piece of private information becomes available, the secret will have been isolated and identified. There are numerous examples of each type of system. It is easiest to illustrate this behavior using a pair of small examples.

First, consider the simplest possible example of a shared secret scheme, a 2-out-of-ℓ scheme:

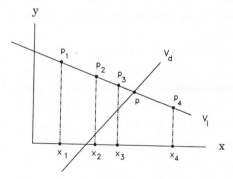

Figure 1.

The private pieces of information are points on a line V_i, whose intersection with a line V_d is the index point, p, at which the shared secret information is defined: $p = (x_p, y_p)$. As was pointed out in "How to (Really) Share a Secret" [42], the information needed to specify one of the points, p_i, is not all of the same type insofar as its security requirements are concerned. If we use the obvious specification of the point p_i in the affine plane AG(2,q) by its coordinates (x_i, y_i), then it is sufficient (for the security of the shared secret scheme in this example) that each participant keep secret one of the coordinate values (say y_i), and that he merely insure the integrity of the other coordinate value against substitution, modifications, etc. With this convention for partitioning the private pieces of information into secret and nonsecret parts, V_i cannot be parallel to the y axis since in that case $x_i = x_p$ for all i and V_i could be deduced from the exposed (non-secret) parts of any pair of the private pieces of information.

The point we wish to make using this simple scheme has to do with the probability of some improper (i.e., unauthorized) collection of persons recovering the secret. An outsider who knows only the public parts of the private pieces of information, but none of the secret parts and the geometrical nature of the scheme, i.e., the line V_d and that there is a line V_i whose intersection with V_d determines the unknown point p, cannot restrict the possible values for p beyond the fact that it is on V_d. Since each of the q points of V_d has the same number of lines on it that are not parallel to the y axis and hence which could be the unknown line V_i, it should be obvious that the opponent can be held to an uncertainty about the secret of

$$H(p) = \log(q) \quad , \tag{1}$$

i.e., his "guessing probability" of choosing p in a random drawing using a uniform probability distribution on the points of V_d.

Now consider the uncertainty faced by one of the participants; an insider. He knows his private piece of information, the point p_i on V_i, the public abscissas x_j, $j \neq i$, for the other participant's private pieces of information and the line V_d. Each point, p', on V_d determines a unique line lying on both p' and p_i which could be the unknown (to the participant) line V_i. Clearly, his uncertainty about the secret is the same as that of an outsider who has no access to any privileged information

$$H(p) = \log(q) \quad . \tag{2}$$

These are the only two meaningful improper groupings of persons in this example since no combination of outsiders with an insider is more capable (in improperly recovering the secret) that is the insider alone. Consequently, this is a perfect 2-out-of-ℓ scheme.

We next consider another 2-out-of-ℓ shared secret scheme:

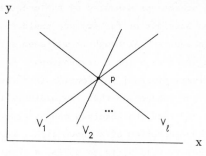

Figure 2.

In this case, the private pieces of information are lines all of which are concurrent on, the secret point p. To an outsider, every point in the plane is equally likely to be the point p, hence his uncertainty about p is

$$H(p) = \log(q^2) = 2\log(q) \quad . \tag{3}$$

An insider, on the other hand, knows that p must be a point on the line which is his private piece of information. Hence, his uncertainty is only

$$H(p) = \log(q) \quad . \tag{4}$$

Consequently, this scheme is not perfect, since the insiders have an advantage (in cheating) over an outsider. Both schemes, however, provide the same minimum level of security against unauthorized recovery of the secret information. From that standpoint alone, they would appear to be equally good. There are other factors that need to be considered, such as the amount of secret information each participant must be responsible for -- or even the information content of the part whose integrity must be insured -- and the information content of the secret itself. Since the plane is 2-dimensional in both points and lines, it is easy to see that in either example the participant need only keep the equivalent of one coordinate value, i.e., $\log_2(q)$ bits, secret about his private piece of information and to insure the integrity of a like amount of information. Thus, the two schemes are equivalent with respect to these parameters. It is at least plausible to define the information content of the secret to be the least uncertainty faced by any unauthorized person or grouping of persons about it; in other words, $\log_2(q)$ bits in both examples as well.

The bottom line to this discussion of the two small examples is that while they are certainly different (not just superficially in the geometrical implementations) they are also alike in important respects. It is the differences that we wish to understand in order to better understand shared secret schemes.

In the first example, where the secret was defined at the point p on the line V_d, the security of the scheme was measured by the uncertainty about p, which, as we saw, was the same for an outsider or for any one insider. We now examine this example from a different standpoint. Although it may seem strange at first to do so, the line V_i can be thought of as being a cryptographic key which encrypts the plaintext (ordinate) of a private piece of information into the ciphertext (abscissa). This is consistent with our convention that the ordinate is the information being protected (kept secret) and that the abscissa can be exposed. In this case, the secret information is the decryption of a known cipher text (the ordinate, x_p, of the point p) using the key V_i. If V_i is to define a cryptotransformation, it must not be parallel to either the y axis or the x axis in order for it to define a one-to-one mapping (i.e., a nonsingular linear transformation) of the y axis onto the x axis.

While it is essential for the simple shared secret scheme described here, that V_i be restricted to not be parallel to the y axis -- since as we explained earlier the exposed parts of the private pieces of information would reveal V_i in that case -- it isn't necessary that V_i be resticted to not be parallel to the x axis as well. If V_i is parallel to the x axis, then the secret coordinate of the point p would satisfy $y_p = y_i$ for all i, but this could only be discovered if two or more of the participants compared (exposed) their secret pieces of information. But in this example, any two participants have been assumed to have the capability to reveal the secret, not just to recover V_i. Therefore, there is no necessity to exclude lines parallel to the y axis in this example since we are only using the one-way nature of the encryption operation without any requirement that it also be invertible -- which will be satisfied so long as to each choice of a plaintext y_p, every ciphertext is an equally likely preimage.

Looked at in this way, we can calculate the uncertainty about the (secret) key to the various combinations of individuals. An outsider knows only that V_i is a line in the plane not parallel to the y axis, i.e., one of the q^2+q lines in the plane less the q lines parallel to the y axis, or q^2 lines in all. Hence his uncertainty about the key is

$$H(V_i) = \log(q^2) = 2\log(q) \qquad (5)$$

which is twice his uncertainty about y_p, the encrypted value of the ordinate of p. Note that in this interpretation V_d has effectively been restricted to be the line parallel to the y axis lying on x_p. It should be noted that the outsiders' uncertainty about p, $H(p)$, in the first example is the same as his uncertainty about V_i, $H(V_i)$ in this example.

There are q+1 lines through each point on the line V_d, one of which is V_d itself, and hence not a candidate to be the key V_i. Therefore the q^2 potential keys (lines in AG(2,q) not parallel to the y axis) are uniformly distributed q at a time on each of the q points of V_d; hence

$$H(p) = \log(q) \quad . \tag{6}$$

In other words, since the set of q^2 lines that could be the unknown key, V_i, are uniformly distributed on the points on V_d (q on each point) and are all equally likely to be the key, an opponent's chance of determining the secret by "guessing" at the value of the key is exactly the same as his chance of "guessing" the value of the secret in the first place: $\log(q)$ in either case.

Next consider the situation with an insider. He knows a point on the unknown key, V_i. There are q+1 lines through this point, q of which are potential keys. Consequently, there is a one-to-one association between the potential keys (given his insider information) and the possible values for the secret cipher. Thus for the insider

$$H(V_i) = H(p) = \log(q) \quad . \tag{7}$$

The point is that in the first example it was the uncertainty about the key that was eroded with the exposure of successive pieces of the private information, i.e., of plaintext/ciphertext pairs in the present setting, (only one such pair is possible in this small example; we are anticipating the general case in this remark), however the uncertainty about the secret index point, $H(p)$ or more precisely $H(y_p)$, remains the same for any grouping other than one able to uniquely identify the key. So long as the surviving candidate keys uniformly map each plaintext into all possible ciphers, the uncertainty about the secret plaintext remains the same, even though the uncertainty about the key decreases with each successive piece of private information that becomes available. The second example has no intermediate key, so it is the uncertainty about the secret point, p, that is directly eroded by the exposure of successive pieces of private information. When viewed in this way, a very close relationship exists between cryptanalysis in depth (with the key as the depth component) and shared secret schemes.

The entire purpose of this discussion was to support the following conclusion: the information contained in each of the private pieces of information constrains the values that some other variable can take. If this variable is the secret, then the shared secret system cannot be perfect, since in that case, unauthorized groupings of insiders would necessarily have an advantage over outsiders in guessing at the value of the secret. If, however, the variable is an intermediate function, out of a family of functions, satisfying suitable constraints such as being entropy preserving over the space in which the secret is located, then the scheme can be perfect. Although we won't make direct use of the principle here, we are in fact faced with the problem of devising cryptosystems with the unusual property that they are immune to cryptanalysis in depth (against the key as the depth component) for all "improper" groupings of plaintext/ciphertext pairs, but cryptanalyzable with certainty of success in recovering the key given any set of plaintext/ciphertext pairs that includes at least one of the prescribed concurrences.

The example shown in Figure 1 contains all of the essential features for the
general model we will use for shared control schemes, irrespective of how complex
the required concurrence may be. Essentially there is one geometric object (an
algebraic variety -- generally a linear subspace in some higher dimensional space),
which can be determined given any subset of the points in it that includes at least
one of the specified concurrence groupings, which intersects another object in a
single point, p, at which the secret is defined. While p is a point in both of the
sets, the lines V_i and V_d in the example, the first set is always secret (until it
is reconstructed by an authorized concurrence among the participants) while in many
applications the other is publicly known, a priori. We therefore refer to the geometric object (set of points) whose determination isn't shared amongst the participants as the domain variety, V_d, since the secret (argument) can be thought of as
being a point concealed in its domain. The object determined by the shared information can be thought of as indicating (in the sense of pointing to) the secret
point p in V_d. We therefore call this object the indicator (variety), V_i. A single
shared secret scheme may have either several indicators, or several domains, depending on the nature of the concurrence that is being realized. In the example of
Figure 1, the indicator was the line V_i, determined by any pair of the points on it.
V_i could have been equally well replaced by a quadratic curve (determined by any
three of its points) or a cubic (determined by any four of its points not lying on a
quadratic curve, i.e., of rank four), etc. This, in fact, is the implementation of
the shared secret scheme originally proposed by Shamir [41]. In this paper, we will
only consider cases in which both V_i and V_d are linear subspaces of some higher
dimensional containing space, S. For most applications, S would be a projective
space PG(n,q) over some field GF(q), however, all of our examples in this paper will
be constructed in affine spaces AG(n,q), mainly because of the closer analogy to the
more familiar Euclidean spaces. Our constructions make essential use of a simple
result in projective geometry known as the rank formula:

$$r(U) + r(V) = r(U \cap V) + r(U \cup V) \qquad (8)$$

true for all subspaces U and V of the containing space $S = PG(n,q)$. r(x) denotes
the rank of the subspace x. Note that r(x) = dim(x) + 1, and that the empty subspace has rank 0, and consequently dimension -1. It is easy to see that (8) does
not hold in affine spaces. In AG(3,q) there are pairs of parallel lines, i.e.,
pairs of lines which do not intersect, but whose union is only a plane:
$r(U \cap V) = 0$ and $r(U \cap V) = 3$ so that

$$r(U) + r(V) = 2 + 2 \neq 0 + 3 = r(U \cap V) + r(U \cup V) \quad .$$

From the standpoint of geometric intuition (8) is more accessible if rank is
replaced by dimension:

$$dim(U) + dim(V) = dim(U \cap V) + dim(U \cup V) \quad . \tag{8'}$$

We are only interested in cases in which $V_i \cap V_d = p$, p a point, i.e., in which $dim(V_i \cap V_d) = 0$. The following example, which we will make extensive use of, indicates the usefulness of (8'). In a 4-dimensional space, **S**, any pair of planes that do not lie in a common 3-dimensional subspace, intersect in a single point. Since they do not lie in a common 3-dimensional subspace, the $dim(U \cup V) = 4$, so that we have

$$dim(U \cap V) = dim(U \cup V) - dim(U) - dim(U) = 4 - 2 - 2 = 0 \quad ,$$

hence, $U \cap V = p$, p a point. We will represent this 4-dimensional construction with the figure:

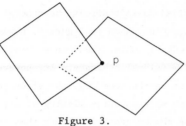

Figure 3.

Similarly, we will represent the intersection of a 3-dimensional subspace (of a 4-dimensional space **S**) with a line by the figure:

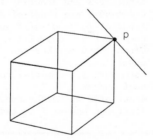

Figure 4.

Although the method of construction is completely general and independent of the dimensionality of the containing space **S**, all of the examples in this paper will be constructed using only the four lowest dimensional configurations:

S	U	V
plane	line	line
3-space	line	plane
4-space	line	3-space
4-space	plane	plane

In the Appendix to this paper, we analyze several shared control schemes that can be realized even within these small examples. The detailed discussion has been relegated to the Appendix so as to not interfere with the main line of development in the paper: devising prepositioned schemes that must await an activating piece of information before they become operational. There is one point in connection with those examples, though, that must be made here; namely, the concept of dual configurations. In Figure 1, both V_i and V_d were lines, and hence in some sense, indistinguishable. In other words, given two lines that intersect at the desired point, p, in the plane, it is immaterial which line is chosen to provide the concealment for p, V_d, and which is chosen to be shared, V_i. Similarly, in Figure 3 it is immaterial which of the planes is identified as V_d and which as V_i. In a sense, both of these constructions are self-dual in the sense that interchanging the role of the two varieties involved doesn't affect the control or security characteristics of the scheme. The construction shown in Figure 4, however, is different. If V_i is the line, then the only control scheme that can be realized is a 2-out-of-ℓ one to indicate a point in a 3-dimensional subspace. If, however, V_i is chosen to be the 3-space pointing to a point on the line, V_d, the situation is dramatically different. We won't even attempt an exhaustive listing of all of the possibilities in this case, but some of the shared control schemes that can be realized are:

Table 1.

1.	Simple threshold:		4-out-of-ℓ
2.	Two levels:		4-out-of-ℓ and 3-out-of-ℓ
		or	4-out-of-ℓ and 2-out-of-ℓ
3.	Three levels:		4-out-of-ℓ, 3-out-of-ℓ and 2-out-of-ℓ
4.	Two parts:		each part a 2-out-of-ℓ scheme. Both parts must concur, but neither can act in the stead of the other.
5.	≥ 2 parts:		each part a 2-out-of-ℓ scheme. Two of the parts must concur.
6.	Three parts:		4-out-of-ℓ where the concurrence must include at least one member from each part.
	Etc.		

Clearly there is an enormous difference, in terms of the kinds of control that can be realized, depending on whether V_i is chosen to be the line or the 3-space in the configuration shown in Figure 4. If the dimension of the containing space, S, is even, there will always be a self-dual geometric configuration as we've already seen for the cases n = 2 and 4 in Figures 1 and 4, respectively. All other configurations have dual interpretations depending on which subspace is chosen to be the indicator: i.e., the line or the plane when dim(S) = 3 or the line or the 3-space when dim(S) = 4 as indicated in Figure 4. These small configurations suffice to permit us to illustrate (in the Appendix) all of the essential details of multilevel and multipart shared control.

Returning now to the discussion of shared control schemes given in the first part of this paper where we argued that in a perfect scheme it had to be the "key" that was revealed when an authorized concurrence of participants occurred. In the example used there, the dimension of the space, S, was only two and the configuration was self-dual both of which obscured several important points. We remarked that the plane was 2-dimensional in either points or lines, so that the uncertainty about the key was 2-dimensional to an outsider (2 log(q)) and 1-dimensional to an insider, but that the uncertainty about the secret was only 1-dimensional to either of them. These notions need to be generalized to arbitrary varieties. Stripped of the inessential (to the present discussion) details of how sets of points can be chosen in a variety V_i so that any of the designated subsets of them will suffice to reconstruct all of V_i, the configurations of interest are of the form:

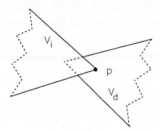

Figure 5.

where $\dim(V_i) + \dim(V_d) = \dim(S)$ and $\dim(V_i \cap V_d) = 0$.

Consider now the simple case in which $\dim(S) = 3$ and $\dim(V_i) = 1$, implying that $\dim(V_d) = 2$.

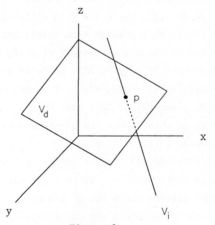

Figure 6.

In $AG(n,q)$, there are q^n points, so there are q^2 possible values for p -- given that V_d is known a priori. On the other hand, the total number of lines in $AG(3,q)$ is $q^2(q^2+q+1)$ of which q^4 have a single point in common with V_d, and hence are candidates to be the unknown key. q^2 of these lines lie on each of the q^2 points in V_d. Similarly, considering the case in which one of the insiders is attempting to cheat, or in which one of the insiders' private piece of information has been exposed, there are q^2+q+1 lines through the point, q^2 of which have a single point in common with V_d, and hence are candidates to be the unknown key. One of these lines lies on each point in V_d, so that the uncertainty about the key is the same as the uncertainty about the secret, $H(V_i) = H(p) = \log(q)$, in this case.

In the first example, Figure 1, we saw the uncertainty about the key go from $O(q^2)$ to $O(q)$ to 1 as the concurrence progressed from an outsider to a single insider and finally to an authorized concurrence. Similarly, in the example just analyzed, we saw that the uncertainty about the key went from $O(q^4)$ to $O(q^2)$ to 1 for the same progression of concurrences. Since both examples were of 2-out-of-ℓ threshold schemes, the only difference is in the dimension of the subspace in which the secret point p is concealed: one in the first example and two in the second. As we shall see, this is no coincidence.

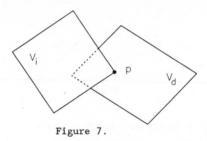

Figure 7.

Consider the self-dual construction shown in Figure 7 in which $\dim(S) = 4$ and $\dim(V_i) = \dim(V_2) = 2$. In $AG(4,q)$, there are $q^2(q^2+1)(q^2+q+1)$ planes. $(q^2+1)(q^2+q+1)$ of these planes pass through each point, but if we assume that the point is known to be in one of the planes (either V_i or V_d by duality) then the number of planes that intersect it only at that point is only q^4. This is easy to see since there are $(q^2+1)(q^2+q+1) = q^4+q^3+2q^2+q+1$ planes on the point in all, one of which is the containing plane (V_i or V_d): call this plane π. There are $q(q+1)$ lines in π itself, on each of which there are q^2+q+1 planes, one of which is π again. There are, therefore, $(q+1)(q^2+q)$ planes on the point that have only a line in common with π. The difference, q^4, is the number of planes in $AG(4,q)$ which have only the single point p in common with π. Since there are q^2 points in π, at each of which there are q^4 such planes, there are q^6 planes in $AG(4,q)$ that have only a single point in common with V_d and hence are candidates to be the unknown key, V_i. By hypothesis, each plane counted at one of the points of π has no other point in common with π, so no plane has been counted twice.

By a similar line of argument, given a point, x, not on π, and a point, p, in π, there are q^2+q+1 planes lying on both x and p, i.e., containing the line $<p,x>$.

There are q+1 lines in π lying on p, each of which defines a plane containing the point x. Therefore there are q^4 planes lying on the point x (i.e., consistent with one of the private pieces of information) that intersect π in only a single point, and hence candidate keys.

Finally, there is a single plane lying on any two of the independent points (meaning that they are on a line skew to π) and a point in π. Since there are q^2 points in π, there are q^2 planes on any pair of the private points which intersect π in a single point, and hence which could be a key. Just as in the previous two examples, we see that the uncertainty about the key has gone from $O(q^6)$ to $O(q^4)$ to $O(q^2)$ and finally to 0, as the concurrence has progressed from an outsider, to one insider to two insiders and, finally, to an authorized concurrence of three insiders.

What we have observed in the preceding examples can be stated precisely. Let dim(S) = n, dim(U) = k, dim(V) = n-k and dim(U $\cap$ V) = 0; i.e., a configuration of the type depicted in Figure 5, then we have the following.

Theorem:

The dimension of the space of (n-k)-flats in S lying on δ independent points, $0 \leq \delta \leq$ n-k is

$$k(n-k-\delta+1) \quad .$$

Although the statement of the theorem is altered for our purposes, a proof can be found in many sources; Somnerville [A3], for example.

Prepositioned Schemes

It should be obvious by this point that one way to preposition a shared control scheme in such a way that the participants will be powerless to recover the secret until they are later enabled to do so, is to go ahead and field the private pieces of information, but to withhold the identification of the domain V_d until such time as the scheme is to be activated. That way even if all of the insiders should conspire to pool their private pieces of information in an attempt to recover the secret before the domain is revealed, the most that they can do is to reconstruct the indicator V_i and hence to learn that p is a point in the subspace V_i instead of possibly being any point in S, which is all that an outsider knows. There is a problem, though, with this simple approach which is best illustrated using two small examples of shared secret schemes analyzed earlier.

In the example of a 2-out-of-ℓ shared secret scheme shown in Figure 1, both V_i and V_d were lines in the plane S. Since a plane is 2-dimensional in both points and lines, i.e., it requires two shares of information (two coordinate values) to specify either one, the same amount of information would have to be communicated to identify p as a general point in **S** as would have to be communicated to identify V_d. This might seem to indicate that two shares of information would be needed to activate the scheme, instead of the obvious minimum of a single share. Given V_i (or else V_d) p is no longer an arbitrary point in the plane but rather an unknown point on a line, whose specification requires only one share of information. Similarly, given that p is constrained to be a point on V_i, V_d no longer need be free to be an arbitrary one of the q^2 lines in the plane not parallel to the y-axis, but can instead be restricted to be one out of a set of q lines in which one line lies on each point of V_d. One easy way to do this would be to preposition an x-coordinate, x_d, different than the point at which V_i intersects the x-axis at the time the scheme is set up. Later, when the scheme is to be activated, the y intercept of the line V_d through the points x_d and p is all that would need to be communicated to permit V_d to be determined. Thereafter, any two of the participants could recover V_i using their private points, and hence recover the secret p.

If we consider the 3-out-of-ℓ scheme shown in Figure 7 in which **S** is 4-dimensional and V_d is a plane, the problem is much more difficult to deal with since 4-space is 6-dimensional in planes while the secret is only 2-dimensional in information content. While a similar, but more complex, resolution is possible in which four out of the six needed shares of information would be prepositioned along with the private pieces of information required to set up the shared secret scheme, and the remaining two shares communicated at the time the scheme is to be activated, there is a more efficient (and general) way to implement such schemes. Table 2, tabulating the dimension of the space of m-flats in an n-dimensional space, suggests how difficult this problem can become. If V_i and V_d were both three dimensional, which only permits a 4-out-of-ℓ shared control, the space of 3-flats is already 12-dimensional, meaning that twelve shares of information are required to identify V_d.

Table 2.

Dimension of the space of m-flats in an
n-dimensional space.

n	m 0	1	2	3	4	5	6	7
1	1							
2	2	2						
3	3	4	3					
4	4	6	6	4				
5	5	8	9	8	5			
6	6	10	12	12	10	6		
7	7	12	15	16	15	12	7	
8	8	14	18	20	18	14	8	

The problem, arising in our earlier identification of V_i with a cryptographic key, is that the revelation of the secret was equated with the identification of the point p at which the secret is determined. p is not itself the secret, but rather some entropy preserving function evaluated with p as an argument reveals the actual secret. In several examples this function was taken to be either the projection of p onto one or more of the natural coordinate axes or else the value of the variables parameterizing a surface at p. Instead, consider p to be a normal cryptographic key, say a 56-bit key for the DES, and let the information that is to be communicated to enable the system be a cipher, which when decrypted with the key will reveal the secret plaintext. Clearly, this implementation solves both of the objectives of a prepositioned shared secret scheme. If the participants cheat and misuse their private pieces of information, all that they can do is recover the shared cryptographic key. Since the cipher hasn't yet been communicated, they have no information whatsoever about the secret plaintext. On the other hand, any plaintext whatsoever can be revealed without having to change the private pieces of information, simply by communicating the cipher that will decrypt with the fixed (shared) key into the desired text.

To illustrate this implementation, consider again the simple 2-out-of-ℓ scheme shown in Figure 1 when used with the DES encryption algorithm. The plane in this case would be $AG(2,2^{56})$. Each private piece of information would consist of 112 bits, 56 of which would have to be kept secret by the participant, and 56 of which need only be protected against substitution, alteration or destruction or loss. V_d, or rather two shares of information (112 bits) adequate to determine V_d, would be prepositioned at the time the scheme was set up. Once this has been done, any two of the participants, using their private pieces of information, could determine V_i and hence recover p which in this case would be a 2-tuple in $AG(2,2^{56})$. There is no reason to not use the simplest entropy preserving function available; namely, let the secret DES key be the y coordinate of p, since by the constraints on the construction in this example all 2^{56} possible values are equally likely. Thus an authorized concurrence could recover the DES key at any time after the scheme was set up, however they could not recover the secret(s) until such time as the cipher was communicated. In those applications where it was either tolerable or acceptable that a proper concurrence be able to recover the secret at any time after the scheme was fielded, the cipher(s) could be prepositioned along with the private pieces of information. In situations such as are addressed in this paper, the cipher(s) could be withheld until it is desired that the scheme be enabled, at which time the minimum of one share of information would have to be communicated for each secret that is to be revealed.

Conclusion

The conclusion was the abstract for this paper: Secret sharing is simply a special form of key distribution.

Appendix

The purpose of this appendix is to bring together the simple geometrical results on which the construction of shared secret schemes depends and to exhibit in detail several small examples illustrating such things as duality, multilevel and multipart schemes, and which also show the way in which the key is gradually revealed as a function of the concurrence involved. We will work only in the finite affine spaces AG(n,q). Theorems will be stated without proof, since the proofs are easily found in many sources; Sommerville [A3] or Hirschfeld [A1,A2], for example.

Define a quantity

$$\varphi(n,m;q) = \begin{cases} \prod_{i=0}^{m-1} \dfrac{(q^n - q^i)}{(q^m - q^i)} & 1 \le m \le n \\ \text{or} \\ 1 & m = 0 \end{cases}$$

then we have the following enumerative theorems.

Theorem 1:

The number of distinct m-flats in AG(n,q) is

$$q^{n-m} \, \varphi(n,m;q) \quad .$$

Theorem 2:

The number of distinct m-flats in AG(n,q) passing through k+1 linearly independent points, $k \ge 0$, is

$$\varphi(n-k,m-k;q) \quad .$$

Theorem 2 is normally stated in terms of the number of m-flats passing through a given k-flat, but since in shared control schemes the private pieces of information are normally points in the space, it is illuminating to state the result in terms of linearly independent sets of k+1 points, i.e., of a collusion of k+1 insiders.

Example 1, which is the same as the example shown in Figure 1 of the text, shows the format we will use for all of the examples. The concurrences are self-explanatory. The second column will, for a single level scheme, show the number of flats of the same dimension as the indicator lying on the points exposed by the concurrence. Theorem 1 gives this value for an the outsider and Theorem 2 for all other concurrences. If V_d is unknown, then this is the equivocation faced by the collusion (concurrence) in guessing the value of the key. If V_d is known, then only those potential keys that intersect V_d in a single point are candidates to be the actual key. In Example 1 as we have already pointed out in the text, the outsider knows that the indicator must be a line, i.e., one out of the $q^2 + q$ total lines in AG(2,q), if he also knows V_d, then he need not consider any of the q lines in

Example 1.

S is 2-dimensional.

V_d is 1-dimensional

2-out-of-ℓ scheme.

(V_i is 1-dimensional).

concurrence	lines lying on the exposed points	candidate keys	possible values for the secret
outsider	$q^2 + q$	q^2	q
1 insider	$q + 1$	q	q
≥ 2 insiders	1	1	1

AG(2,q) that do not intersect V_d. In other words, only the q^2 lines that do intersect V_d in a single point are then candidates to be the key. This same format is
used in all of the other examples as well. The essential point in all of the examples is that the candidate keys are (by design) uniformly distributed on the set of
possible values the secret could take for all improper concurrences of the participants.

Examples 2 and 3 demonstrate the principle of duality in the smallest configuration in which it occurs. The reader should note the difference in the uncertainty
as to the identity of the key from among the candidate keys in the two cases. Example 4, based on the same geometric configuration as examples 1 and 2 is again the
smallest configuration in which a multilevel shared control scheme can be realized;
in this case a two-level scheme in which level 1 is a 2-out-of-ℓ scheme and level 2
is a 3-out-of-ℓ scheme. The private pieces of information for the level 1 participants are points on the line V_{i_1} lying in the plane V_{i_2} and intersecting V_d in the
point p, i.e., indicating p. Similarly, the private pieces of information for the

Example 2.

S is 3-dimensional

V_d is 2-dimensional

2-out-of-ℓ scheme.

(V_i is 1-dimensional).

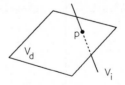

concurrence	lines lying on the exposed points	potential keys	possible values for the secret
outsider	$q^2(q^2+q+1)$	q^4	q^2
1 insider	q^2+q+1	q^2	q^2
≥ 2 insiders	1	1	1

Example 3.

S is 3-dimensional.

V_d is 1-dimensional

3-out-of-ℓ scheme.
(V_i is 2-dimensional).

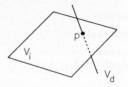

concurrence	planes lying on the exposed points	candidate keys	possible values for the secret
outsider	q^3+q^2+q	q^3	q
1 insider	q^2+q+1	q^2	q
2 insiders	$q+1$	q	q
$\geq$ 3 insiders	1	1	1

Example 4.

S is 3-dimensional

V_d is 1-dimensional

Two Levels:
    Level One, 2-out-of-ℓ scheme.
        (V_{i_1} is 1-dimensional).

    Level Two, 3-out-of-ℓ scheme.
        (V_{i_2} is 2-dimensional).

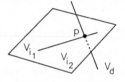

concurrence	lines lying on exposed points	planes lying on exposed points	candidate key lines lines	candidate key planes	pv for the secret
outsider	$q^2(q^2+q+1)$	$q(q^2+q+1)$	$q(q^2+q)$	q^3	q
1 level one insider	q^2+q+1	q^2+q+1	q	$q(q+1)$	q
1 level two insiders	q^2+q+1	q^2+q+1	q	$q(q+1)$	q
1 level one, and 1 level two insider		$q+1$		q	q
2 level two insiders		$q+1$		q	q
$\geq$ 2 level one insiders	1		1		q
$\geq$ 3 level two insiders		1		1	

Example 5.

S is 4-dimensional

V_d is 3-dimensional

2-out-of-ℓ scheme.
(V_i is 1-dimensional).

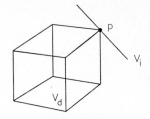

concurrence	lines lying on the exposed points	potential keys	possible values for the secret
outsider	$q^3(q^3+q^2+q+1)$	q^6	q^3
1 insider	q^3+q^2+q+1	q^3	q^3
$\geq$ 2 insiders	1	1	1

level two participants are points in general position (no three collinear) in the plane V_{i_2}. Although it is not germane to this paper, we remark that no pair of the level 2 points can be collinear with a level 1 point, otherwise the three participants who held those points would only be capable of reconstructing a line in V_{i_2} skew to V_d, and hence they would be unable to recover the secret, contrary to the requirement that any person from level 1 in cooperation with any two from level 2 should be able to do so. In a separate paper [44], the author has solved the problem of optimally choosing the sets of private points in a two-level scheme of the type depicted in Example 4. In the table, the blank entries in the array for potential or candidate keys for V_{i_1} when a collusion involves two persons merely indicates that the additional parties to the concurrence in this case do not further restrict the possibilities over what a single participant did.

Examples 5 and 6 are included primarily because this dual configuration is the smallest that can be used to illustrate multipart shared control schemes, as shown in Example 7. Lines V_1 and V_2 in the 3-dimensional indicator V_i are skew, both with respect to each other and, of course, by construction with respect to V_d. The participants in each part would be given points on their parties' line as their private pieces of information to form two independent 2-out-of-ℓ shared control schemes. In such a control scheme it should require the concurrence of at least two persons from each part to recover the secret. The construction obviously satisfies the sufficiency of such a concurrence since the pairs of participants suffice to reconstruct the skew lines V_1 and V_2 which span the 3-flat V_i which in turn indicates the point p on V_d. Necessity however is somewhat more delicate. It is conceivable that the line defined by the point held by a member of part 1 and the point held by a member of part 2 could intersect V_d, in which case the concurrence of only two persons would know with certainty that they had recovered the secret (this presupposes that V_d is known a priori). Similarly, it is conceivable that a plane defined by one of

Example 6.

S is 4-dimensional.

V_d is 1-dimensional

4-out-of-ℓ scheme.

(V_i is 3-dimensional).

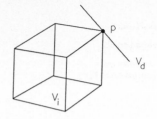

concurrence	planes lying on the exposed points	candidate keys	possible values for the secret
outsider	$q^4+q^3+q^2+q$	q^4	q
1 insider	q^3+q^2+q+1	q^3	q
2 insiders	q^2+q+1	q^2	q
3 insiders	$q+1$	q	q
$\geq$ 4 insiders	1	1	1

the lines (V_1 or V_2) and a point on the other line (a concurrence of only three persons) might intersect V_d, which would have to be the point p since the plane is in the 3-flat V_i which itself has only a single point of intersection with V_d. It might appear that it would be a difficult problem to avoid all of these unacceptable configurations, that we might even be compelled to test for all possible unacceptable concurrences and eliminate from consideration for use as private pieces of information, points on the lines V_1 and V_2 that result in such configurations. In fact, we do this, but without any necessity for testing. We use a special case of a much more general result from classical geometry [A3].

Theorem:

Given a pair of skew lines in a 3-space and a point not on either of the lines, there is a unique line lying on all three.

In this restricted form, the result is easy to see; the line V_1 and the point p determine a plane in V_i, say π_1. Similarly, V_2 and p determine a plane π_2. π_1 and π_2 intersect in a line lying on p and intersecting V_1 in a point p_1 and V_2 in a point p_2. p_1 cannot be used as one of the private pieces of information for the participants in part 1, and p_2 cannot be used in part 2. With these points eliminated, clearly no collusion of two or three participants can recover the secret since the flats their points determine will all be skew to V_d.

This is a particularly interesting example since it was devised to make it possible to share control of a treaty controlled action between two national control teams in such a way that at least two participants from each control team must con-

Example 7.

S is 4-dimensional.

V_d is 1-dimensional

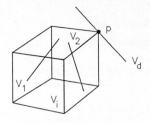

Two Parts:
    Part 1, 2-out-of-ℓ scheme.
          (V_1 is 1-dimensional).

    Part 2, 2-out-of-ℓ scheme.
          (V_{2i} is 2-dimensional).

$$V_1 \cap V_2 = \phi \qquad \dim(V_1 \cup V_2) = \dim(V_i) = 3$$

concurrence	3-spaces lying on the exposed points	candidate keys	possible values for the secret
outsider	$q(q^3+q^2+q+1)$	q^4	q
1 insider from either part	q^3+q^2+q+1	q^3+q^2	q
1 insider from each part	q^2+q+1	q^2+q	q
$\geq$ 2 insiders from either part	1	1	1

cur before the controlled action can be initiated; say between the U.S. and the
U.S.S.R.

It is an easy matter to extend this example to include other parties in such a
way that so long as at least two participants from at least two of the parties concur the controlled action could be initiated. For example, the UN could be a third
party and the action should be possible to initiate so long as at least two out of
the UN, U.S. or U.S.S.R. concur. To see how this can be achieved, let ℓ be the line
lying on p, p_1 and p_2. Let V_3 be any line in V_i skew to V_1 and V_2 which intersects
ℓ in a single point p_3. All of the other points on V_3 are available for use as private pieces of information for the UN team. By the same sort of argument given
above, it is easy to see that no improper concurrence of participants from among the
three parties can define (with their private points) a flat that intersects V_d, hence
no unauthorized concurrence can recover the secret. On the other hand, any collection of participants that includes at least two members from each of two of the
control teams will define a pair of skew lines in V_i, and hence, V_i, and thence p.

There are a very large number of other control schemes which can be realized
using this geometrical configuration, some of which were described in Table 1. We
mention only one, since it is indicative of a different type of partitioning of
capability. In V_i, choose a plane V_1 skew to V_d. Points in general position in
this plane can be used to realize a 3-out-of-ℓ scheme, not to recover p but to
recover the plane V_1. Another participant of higher, but not absolute, authority is

Example 8.

S is 4-dimensional.

V_d is 1-dimensional

Two Parts:

    Part 1, 2-out-of-ℓ scheme.
        (V_1 is 1-dimensional).

    Part 2, 2-out-of-ℓ scheme.
        (V_2 is 1-dimensional).

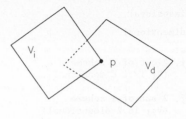

$$V_1 \cap V_2 = \phi \qquad \dim(V_1 \cup V_2) = \dim(V_i) = 3$$

concurrence	3-spaces lying on the exposed points	candidate keys	possible values for the secret
outsider	$q(q^3+q^2+q+1)$	q^4	q
1 insider from either part	q^3+q^2+q+1	q^3+q^2	q
1 insider from each part	q^2+q+1	q^2+q	q
≥ 2 insiders from either part	1	1	1

given a point in V_i not in V_1 and not collinear with p and any point used in the plane V_1, nor coplanar with p and any pair of points used in V_1. These conditions are easily satisfied using the more general version of the theorem used to construct the usable point sets in Example 7. It now requires the concurrence of any three members of the one class and of the unique participant to initiate the controlled action. Clearly, no unauthorized concurrence, which could include the worst case of the unique participant and any two of the other class, could define a flat that intersected V_d, and hence they could not recover the secret. It should be obvious that for even such a simple configuration as shown in Example 7 that the range of possible control schemes is enormous.

Example 8 is included for a double purpose -- besides illustrating a perfect shared control scheme; it also illustrates a commonly used imperfect scheme, which we describe as a convincing argument that these abstract geometrical configurations are already "real world" solutions to problems of shared control. In the body of the paper, we mentioned the U.S. policy of the two-man control of nuclear weapons, and of the information needed to enable and hence to use these weapons. For some classes of weapons this controlling information takes the form of four symbols from an appropriate alphabet, i.e., 1-out-of-q symbols. In order to satisfy the two-man rule, this information is partitioned so that one team knows only two of the symbols, i.e., their private piece of information is a plane in a 4-space consisting of all of the 4-tuples of symbols in which the two that they know are fixed. Similarly, the other team knows another plane in the 4-space. Since by the rank formula, any two planes in a 4-space that do not lie in a common 3-dimensional subspace intersect in a single point, the two planes intersect in a single point: the combination. This scheme is imperfect since insiders have a better chance of guessing the secret than outsiders, but it is a shared control scheme (two party) based on the geometrical configuration shown in Example 8.

Example 9.

S is 4-dimensional

V_d is 2-dimensional

Two Levels:
    Level One, 2-out-of-ℓ scheme
        (V_{i_1} is 1-dimensional)

    Level Two, 3-out-of-ℓ scheme
        (V_{i_2} is 2-dimensional)

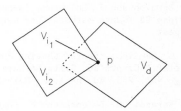

concurrence	lines lying on exposed points	planes lying on exposed points	candidate key lines	candidate key planes	pv for the secret
outsider	$q^3(q^3+q^2+q+1)$	$q^2(q^2+1)(q^2+q+1)$	$q^2(q^3+q^2)$	q^6	q^2
1 level one insider	q^3+q^2+q+1	$(q^2+1)(q^2+q+1)$	q^2	q^4	q^2
1 level two insiders	q^3+q^2+q+1	$(q^2+1)(q^2+q+1)$	q^2	q^4	q^2
1 level one and 1 level two insider		q^2+q++1		q^2	q^2
2 level two insiders		q^2+q+1		q^2	q^2
≥ 2 level one insiders	1			1	1
≥ 3 level two insiders		1		1	

References

A1. J.W.P. Hirschfeld, _Finite Projective Spaces of Three Dimensions_, Oxford Mathematical Monographs, Oxford University Press, New York, 1985.

A2. J.W.P. Hirschfeld, _Projective Geometries Over Finite Fields_, Oxford Mathematical Monographs, Oxford University Press, New York, 1979.

A3. D.M.Y. Sommerville, _An Introduction to the Geometry of N Dimensions_, Dover Publications, Inc., New York, 1958.

Note: This bibliography includes all of the papers on shared secret or threshold schemes which the author is aware of (April 1989). Although only a few of the references appearing here are cited in this paper, it has been included for its own value to other researchers.

1. C. A. Asmuth and G. R. Blakley, "Pooling, Splitting and Reconstituting Information to Overcome Total Failure of Some Channels of Communication," Proc. IEEE Computer Soc. 1982 Symp. on Security and Privacy, Oakland, CA, Apr. 26-28, 1982, pp. 156-169.

2. C. Asmuth and J. Bloom, "A Modular Approach to Key Safeguarding," IEEE Trans. Info. Theory, Vol. IT-29, No. 2, Mar. 1983, pp. 208-210.

3. J. Benaloh and J. Leichter, "Generalized Secret Sharing and Monotone Functions," Crypto'88, Santa Barbara, CA, Aug. 21-25, 1988, Advances in Cryptology, to appear.

4. J. C. Benaloh, "Secret Sharing Homomorphisms: Keeping Shares of a Secret Secret," Crypto'86, Santa Barbara, CA, Aug. 11-15, 1986, Advances in Cryptology, Vol. 263, Ed. by A. M. Odlyzko, Springer-Verlag, Berlin, 1986, pp. 251-260.

5. L. Berardi, M. DeFonso and F. Eugeni, "Thereshold Schemes Based on Criss-Cross Block Designs," Private Communication.

6. L. Berardi and F. Eugeni, "Geometric Structures, Cryptography and Security Systems Requiring a Quorum," Proc. 1987 ATTI del Primo Simposio Nazionale su Stato e Prospettive della Ricerca Crittografica in Italia, Rome, Italy, Oct. 30-31, 1987, pp. 127-133.

7. A. Beutelspacher and K. Vedder, "Geometric Structures as Threshold Schemes," Proc. 1987 IMA Conference on Cryptography and Coding Theory, Cirencester, England, Oxford University Press, to appear.

8. A. Beutelspacher, "Enciphered Geometry: Some Applications of Geometry to Cryptography," Proceedings of Combinatorics'86, Annals of Discrete Mathematics, 37, North-Holland, 1988, pp. 59-68.

9. A. Beutelspacher, "How to say 'No'," Eurocrypt'89, Apr. 11-13, 1989, Houthalen, Belgium, Advances in Cryptology, Springer-Verlag, Berlin, to appear.

10. G. R. Blakley and R. D. Dixon, "Smallest Possible Message Expansion in Threshold Schemes," Crypto'86, Santa Barbara, CA, Aug. 11-15, 1988, Advances in Cryptology, Vol. 263, Ed. by A. M. Odlyzko, Springer-Verlag, Berlin, 1986, pp. 266-274.

11. G. R. Blakley and C. Meadows, "Security of Ramp Schemes," Crypto'84, Santa Barbara, CA, Aug. 19-22, 1984, Advances in Cryptology, Vol. 196, Ed. by G. R. Blakley and D. Chaum, Springer-Verlag, Berlin, 1985, pp. 411-431.

12. G. R. Blakley and L. Swanson, "Security Proofs for Information Protection Systems," Proc. IEEE Computer Soc. 1981 Symp. on Security and Privacy, Oakland, CA, Apr. 27-29, 1981, pp. 75-88.

13. G. R. Blakley, "One-time Pads are Key Safeguarding Schemes, Not Cryptosystems: Fast Key Safeguarding Schemes (Threshold Schemes) Exist," Proc. IEEE Computer Soc. 1980 Symp. on Security and Privacy, Oakland, CA, Apr. 14-16, 1980, pp. 108-113.

14. G. R. Blakley, "Safeguarding Cryptographic Keys," Proc. AFIPS 1979 Nat. Computer Conf., Vol. 48, New York, NY, June 1979, pp. 313-317.

15. J. R. Bloom, "A Note on Superfast Threshold Schemes," preprint, Texas A&M Univ., Dept. of Mathematics, 1981.

16. J. R. Bloom, "Threshold Schemes and Error Correcting Codes," Am. Math. Soc., Vol. 2, 1981, pp. 230.

17. J. Box, D. Chaum and G. Purdy, "A Voting Scheme," Crypto'88, Santa Barbara, CA, Aug. 21-25, 1988, Advances in Cryptology, to appear.

18. E. F. Brickell and D. R. Stinson, "The Detection of Cheaters in Threshold Schemes," 18th Annual Conference on Numerical Mathematics and Computing, Sept. 29-Oct. 1, 1988, Winnipeg, Manitoba, Canada, Congressus Numerantium, Vol. 68-69, to appear 1989.

19. E. G. Brickell, "Some Ideal Secret Sharing Schemes," 3rd Carbondale Combinatorics Conference, Oct. 31, 1988, Carbondale, IL, J. Combinatorial Mathematics and Combinatorial Computing, to appear.

20. D. Chaum, Claude Crepeau and I. Damgard, "Multiparty Unconditionally Secure Protocols," 4th SIAM Conference on Discrete Mathematics, San Francisco, CA, June 13-16, 1988, abstract appearing in SIAM Final Program Abstracts: Minisymposia, #M-28/3:20pm, pp. A8.

21. D. Chaum, "How to Keep a Secret Alive: Extensible Partial Key, Key Safeguarding, and Threshold Systems," Crypto'84, Santa Barbara, CA, Aug. 19-22, 1984, Advances in Cryptology, Vol. 196, Ed. by G. R. Blakley and D. Chaum, Springer-Verlag, Berlin, 1984, pp. 481-485.

22. D. Chaum, "Computer Systems Established, Maintained, and Trusted by Mutually Suspicious Groups," Memo. No. UCB/ERL/M79/10, Univ. of Calif, Berkeley, ERL 1979; also, Ph.D. dissertation in Computer Science, Univ. of Calif., Berkeley, 1982.

23. B. Chor, S. Goldwasser, S. Micali and B. Awerbuch, "Verifiable Secret Sharing and Achieving Simultaneity in the Presence of Faults," Proc. 26th IEEE Symp. Found. Comp. Sci., Portland, OR, Oct. 1985, pp. 383-395.

24. G. I. Davida, R. A. DeMillo and R. J. Lipton, "Protecting Shared Cryptographic Keys," Proc. IEEE Computer Soc. 1980 Symp. on Security and Privacy, Oakland, CA, Apr. 14-16, 1980, pp. 100-102.

25. M. De Soete and K. Vedder, "Some New Classes of Geometric Threshold Schemes," Eurocrypt'88, May 25-27, 1988, Davos, Switzerland, Advances in Cryptology, Vol. 330, Ed. by C. G. Günther, Springer-Verlag, Berlin, pp. 57-76.

26. A. Ecker, "Tactical Configurations and Threshold Schemes," preprint (available from author).

27. Paul Feldman, "A Practical Scheme for Non-iteractive Verifiable Secret Sharing," Proc. 28th Annual Symp. on Foundations of Comp. Sci., Los Angeles, CA, Oct. 12-14, 1987, IEEE Computing Soc. Press, Washington, D.C., 1987, pp. 427-437.

28. S. Harari, "Secret Sharing Systems," Secure Digital Communications, Ed. by G. Longo, Springer-Verlag, Wien, 1983, pp. 105-110.

29. M. Ito, A. Saito and T. Nishizeki, "Secret Sharing Scheme Realizing General Access Structure," (in English) Proc. IEEE Global Telecommunications Conf., Globecom'87, Tokyo, Japan, 1987, IEEE Communications Soc. Press, Washington, D.C., 1987, pp. 99-102. Also to appear in Trans. IEICE Japan, Vol. J71-A, No. 8, 1988 (in Japanese).

30. M. Ito, A. Saito and T. Nishizeki, "Multiple Assignment Scheme for Sharing Secret," preprint (available from T. Nishizeki).

31. E. D. Karnin, J. W. Greene and M. E. Hellman, "On Secret Sharing Systems," IEEE International Symposium on Information Theory, Session B3 (Cryptography), Santa Monica, CA, Feb. 9-12, 1981, IEEE Trans. Info. Theory, Vol. IT-29, No. 1, Jan. 1983, pp. 35-41.

32. S. C. Kothari, "Generalized Linear Threshold Scheme," Crypto'84, Santa Barbara, CA, Aug. 19-22, 1984, Advances in Cryptology, Vol. 196, Ed. by G. R. Blakley and D. Chaum, Springer-Verlag, Berlin, 1985, pp. 231-241.

33. K. Koyama, "Cryptographic Key Sharing Methods for Multi-groups and Security Analysis," Trans. IECE Japan, Vol. E66, No. 1, 1983, pp. 13-20.

34. C. Matsui, K. Tokowa, M. Kasahara and T. Namekawa, "Notes on (K,N) Threshold Scheme," Proc. Joho Riron To Sondo Ooyo Kenkyukai, VII-th Symposium, Kinugawa, Japan, Nov. 5-7, 1984, pp. 158-163 (in Japanese); The VII-th Symposium on Information Theory and Its Applications, English translation available from G. J. Simmons.

35. R. J. McEliece and D. V. Sarwate, "On Sharing Secrets and Reed-Solomon Codes," Com. ACM, Vol. 24, No. 9, Sept. 1981, pp. 583-584.

36. C. Meadows, "Some Threshold Schemes Without Central Key Distributors," Crypto'88, Santa Barbara, CA, Aug. 21-25, 1988, Advances in Cryptology, to appear.

37. M. Merritt, "Key Reconstruction," Crypto'82, Santa Barbara, CA, Aug. 23-25, 1982, Advances in Cryptology, Ed. by D. Chaum, R. L. Rivest and A. T. Sherman, Plenum Press, New York, 1983, pp. 321-322.

38. M. Mignotte, "How to Share a Secret," Workshop on Cryptography, Burg Feuerstein, Germany, Mar. 29-Apr. 2, 1982, Cryptography, Vol. 149, Ed. by T. Beth, Springer-Verlag, Berlin, 1983, pp. 371-375.

39. R. von Randow, "The Bank Safe Problem," Discrete Applied Mathematics, 4, 1982, pp. 335-337.

40. P. J. Schellenberg and D. R. Stinson, "Threshold Schemes from Combinatorial Designs," submitted to the Journal of Combinatorial Mathematics and Combinatorial Computing.

41. A. Shamir, "How to Share a Secret," Massachusetts Inst. of Tech. Tech. Rpt. MIT/LCS/TM-134, May 1979. (See also Comm. ACM, Vol. 22, No. 11, Nov. 1979, pp. 612-613.)

42. G. J. Simmons, "How to (Really) Share a Secret," Crypto'88, Santa Barbara, CA, Aug. 21-25, 1988, Advances in Cryptology, to appear.

43. G. J. Simmons, "Robust Shared Secret Schemes or 'How to be Sure You Have the Right Answer Even Though You Don't Know the Question'," 18th Annual Conference on Numerical Mathematics and Computing, Sept. 29-Oct. 1, 1988, Winnipeg, Manitoba, Canada, Congressus Numerantium, Vol. 68-69, to appear 1989.

44. G. J. Simmons, "Sharply Focused Sets of Lines on a Conic in PG(2,q)," 20th Southeastern International Confeence on Combinatorics, Graph Theory & Computing, Feb. 20-24, 1989, Boca Raton, FL, Congressus Numerantium, to appear 1989.

45. D. R. Stinson and S. A. Vanstone, "A Combinatorial Approach to Threshold Schemes," Crypto'87, Santa Barbara, CA, Aug. 16-20, 1987, <u>Advances in Cryptology</u>, Ed. By Carl Pomerance, Springer-Verlag, Berlin, 1988, pp. 330-339.

46. D. R. Stinson and S. A. Vanstone, "A Combinatorial Approach to Threshold Schemes," <u>SIAM J. Disc. Math</u>, Vol. 1, No. 2, May 1988, pp. 230-236. (This is an expanded version of the paper appearing in <u>Advances in Cryptology: Proceedings of Crypto'87</u>, Vol. 293, Ed. By Carl Pomerance, Springer-Verlag, Berlin, 1988.)

47. M. Tompa and H. Woll, "How to Share a Secret with Cheaters," Crypto'86, Santa Barbara, CA, Aug. 19-21, 1986, <u>Advances in Cryptology</u>, Vol. 263, Ed. by A. M. Odlyzko, Springer-Verlag, Berlin, 1986, pp. 261-265; also <u>Journal of Cryptology</u>, Vol. 1, No. 2, 1988, pp. 133-138.

48. H. Unterwalcher, "A Department Threshold Scheme Based on Algebraic Equations," <u>Contributions to General Algebra</u>, 6, Dedicated to the memory of Wilfried Nöbauer, Verlag B. G. Teubner, Stuttgart (GFR), to appear Dec. 1988.

49. H. Unterwalcher, "Threshold Schemes Based on Systems of Equations," <u>Österr. Akad. d. Wiss</u>, Math.-Natur. Kl, Sitzungsber. II, Vol. 197, 1988, to appear.

50. H. Yamamoto, "On Secret Sharing Schemes Using (k,L,n) Threshold Scheme," <u>Trans. IECE Japan</u>, Vol. J68-A, No. 9, 1985, pp. 945-952, (in Japanese); also published as "Secret Sharing System Using (k,L,n) Threshold Scheme," <u>Electronics and Communications in Japan</u>, Part 1, Vol. 69, No. 9, 1986, pp. 46-54.

51. H. Yamamoto, "On Secret Sharing Communication Systems with Two or Three Channels," <u>IEEE Trans. Info. Theory</u>, Vol. IT-32, No. 3, May 1986.

52. H. Yamamoto, "Coding Theorem for Secret Sharing Communication Systems with Two Noisy Channels," <u>IEEE Trans. Info. Theory</u>, to appear.

53. T. Uehara, T. Nishizeki, E. Okamoto and K. Nakamura, "Secret Sharing Systems with Matroidal Schemes," <u>Trans. IECE Japan</u>, Vol. J69-A, No. 9, 1986, pp. 1124-1132, (in Japanese; English translation available from G. J. Simmons) presented at the 1st China-USA International Conference on Graph Theory and its Applications, Jinan, China, June 1986. English summary by Takao Nishizeki available as Tech. Rept. TRECIS8601, Dept. of Elect. Communs., Tohoku University, 1986.

Some Ideal Secret Sharing Schemes

Ernest F. Brickell*

Sandia National Laboratories

Albuquerque, NM 87185

Abstract

In a secret sharing scheme, a dealer has a secret. The dealer gives each participant in the scheme a share of the secret. There is a set Γ of subsets of the participants with the property that any subset of participants that is in Γ can determine the secret. In a perfect secret sharing scheme, any subset of participants that is not in Γ cannot obtain any information about the secret. We will say that a perfect secret sharing scheme is ideal if all of the shares are from the same domain as the secret. Shamir and Blakley constructed ideal threshold schemes, and Benaloh has constructed other ideal secret sharing schemes. In this paper, we construct ideal secret sharing schemes for more general access structures which include the multilevel and compartmented access structures proposed by Simmons.

1 Introduction

Given a set of n participants and a set Γ of subsets of the participants, a *secret sharing scheme* for Γ is a method of distributing shares to each of the participants such that any subset of the participants in Γ can determine the secret, but any subset of participants that is not in Γ cannot determine the secret. The *share* of a participant refers specifically to the information that the dealer sends in private to the participant. If any subset of participants that is not in Γ cannot determine any information about the secret, then the secret sharing scheme is said to be perfect. Given a secret sharing scheme in which S is the set of possible secrets and T is the set of possible shares, we define the *information rate*, ρ, of the scheme as $\rho = \log|T|/\log|S|$. For example, if the secret is a random element of $\mathrm{GF}(q)$, and all shares are elements of $\mathrm{GF}(q)$, then the information rate is 1. Simmons [5] defined a related notion. He called a secret sharing scheme *extrinsic* if the set T of possible shares is the same for all participants. We will say that a secret sharing scheme is *ideal* if it is perfect and has information rate 1.

---

*This work performed at Sandia National Laboratories and supported by the U.S. Department of Energy under contract No. DE–AC04–76DP00789.

The first constructions of secret sharing schemes were due to Blakley [2] and Shamir [4]. Their schemes are called threshold schemes because they have the property that for some t, only the subsets of participants of cardinality at least t can determine the secret. Both the Blakley and the Shamir schemes are perfect and can be ideal as we will demonstrate later in this section.

Given a secret sharing scheme, the access structure, Γ, is defined as the set of subsets of participants that can determine the secret. In this paper, we will restrict our attention to secret sharing schemes in which Γ is monotone, that is if $B \in \Gamma$, and if B is contained in C, then $C \in \Gamma$.

Ito, Saito, and Nishizeki [3] have shown that for any monotone set of subsets, Γ, there exists a perfect secret sharing scheme for which Γ is the access structure. Benaloh [1] has proven this result using a construction that has a lower information rate than the construction of Ito, et.al. although his construction is far from ideal for arbitrary Γ. Benaloh has also shown that there exist monotone sets Γ, which cannot be the access structure for an ideal secret sharing scheme. We will say that a monotone set of subsets, Γ, is an *ideal access structure* if there is some ideal secret sharing scheme for which Γ is the access structure.

The motivation for the current paper is to find ideal secret sharing schemes with access structures that are more general than threshold access structures.

Simmons [5] has described an access structure that arises in a practical application of secret sharing. A *multilevel access structure* is one in which each participant is assigned a level which is a positive integer and the access structure consists of those subsets which contain at least r participants all of level at most r. In other words, 2 participants of level 2 can determine the secret, as can 3 of level 3. But also 1 participant of level 2 and 2 participants of level 3 can determine the secret. Simmons asked whether all multilevel access structures are ideal access structures.

In this paper, we answer Simmons' question in the affirmative. Specifically, in Theorem 1, we show that given any multilevel access structure, there exists Q such that for any q a prime power with $q > Q$, there is an ideal secret sharing scheme realizing this access structure over $\mathrm{GF}(q)$.

One drawback to the construction given in Theorem 1 is that it requires the dealer to check many (possibly exponentially many) matrices to see that they are nonsingular. In Theorem 2, we give a different construction for realizing multilevel access structures that removes this undesirable property.

Simmons also pointed out that there were potential applications for compartmented access structures. In a *compartmented access structure*, there are different compartments, say $C_1, \ldots, C_u$, and positive integers $t_1, \ldots, t_u$ and t. The access structure consists of all subsets containing at least t_i participants from C_i for $1 \le i \le u$, and a total of at least t participants. Simmons' original notion of compartmented schemes had $t = \sum_{i=1}^{u} t_i$, but we have generalized his notion slightly since we have been able to construct more general ideal secret sharing schemes. In section 3, we show that for any compartmented access structure, there exists a Q, such that for $q > Q$, there exists an ideal secret sharing scheme for Γ over $\mathrm{GF}(q)$.

We conclude this section with a brief description of the threshold schemes of Shamir and Blakley.

The scheme of Shamir [4] is based on polynomials over GF(q). Let $f(x) = \sum_{i=0}^{t-1} a_i x^i$. The secret is $f(0) = a_0$. Participant P_j will receive an ordered pair $(x_j, f(x_j))$. It is easy to show that this is a threshold scheme, since for any t participants, there is only one polynomial of degree $t - 1$ passing through their t points. Also it is a perfect threshold scheme since for any $t - 1$ participants and any point $(0, a)$, there is a polynomial of degree t passing through their $t - 1$ points and $(0, a)$. This scheme will be ideal if the value of x_j is publicly revealed so that the share of participant P_j is just the value of $f(x_j)$.

The scheme of Blakley [2] is based on geometries over finite fields. Let V be a t-dimensional vector space over GF(q) and let e_1 be the t-dimensional vector $(1, 0, \ldots, 0)$. The dealer picks a 1-dimensional flat, g, that is not perpendicular to e_1 and a $(t - 1)$-dimensional flat, H, such that g and H intersect in a single point, P. The secret will be the first coordinate of P. g will be made public but H will be kept secret. The dealer will pick n points $p_i, i = 1, \ldots, n$ such that these points together with P are in general position, that is any t of the points generate a $(t - 1)$-dimensional flat. Participant P_i will receive the point p_i. This is a perfect secret sharing scheme since any t of the participants can use their points to determine the hyperplane H, but for any $t - 1$ of the participants, there is a hyperplane passing through their points and any given point on g. The Blakley scheme can be modified slightly so that it is ideal. Let g be the first coordinate axis. When the dealer gives point p_i to participant P_i, he can make public all the coordinates except the first coordinate, and give only the first coordinate to P_i in secret. So P_i's share is only the first coordinate.

2   A Basic Secret Sharing Scheme

In this section, we give a slight generalization of the Shamir and Blakley schemes which is guaranteed to have information rate 1, and in Proposition 1, give sufficient conditions for it to be perfect and thus an ideal secret sharing scheme.

The Basic Secret Sharing Scheme: The secret is an element in some finite field GF(q). The dealer chooses a vector $\mathbf{a} = (a_0, \ldots, a_t)$ for some t, where each $a_j \in$ GF(q), and a_0 is the secret. Denote the participants by P_i for $1 \leq i \leq n$. For each P_i, the dealer will pick a t-dimensional vector $\mathbf{v}_i$ over GF(q). All of the vectors $\mathbf{v}_i$ for $1 \leq i \leq n$ will be made public. The share that the dealer gives to P_i will be $s_i = \mathbf{v}_i \cdot \mathbf{a}$. Let $\mathbf{e}_i$ denote the i'th t-dimensional unit coordinate vector (i.e. $\mathbf{e}_1 = (1, 0, \ldots, 0)$).

Proposition 1 Let $\gamma = \{P_{i_1}, \ldots, P_{i_k}\}$ be a set of participants.

(1)   The participants in γ can determine the secret if the subspace $\langle \mathbf{v}_{i_1}, \ldots, \mathbf{v}_{i_k} \rangle$ contains $\mathbf{e}_1$.

(2)   The participants in γ receive no information about the secret if the subspace $\langle \mathbf{v}_{i_1}, \ldots, \mathbf{v}_{i_k} \rangle$ does not contain $\mathbf{e}_1$.

Proof   Let M be the matrix with rows $\mathbf{v}_{i_1}, \ldots, \mathbf{v}_{i_k}$. Let $\mathbf{s} = (s_{i_1}, \ldots, s_{i_k})$. To prove (1), let $\mathbf{w}$ be the vector such that $\mathbf{w}M = \mathbf{e}_1$. Then $\mathbf{w}M\mathbf{a} = a_0$. Hence $\mathbf{w} \cdot \mathbf{s} = a_0$.

To prove (2), let $\mathbf{w}_0, \ldots, \mathbf{w}_t$ be the column vectors of M. If $\mathbf{w}_0 \notin \langle \mathbf{w}_1, \ldots, \mathbf{w}_t \rangle$, then there exists $\mathbf{d}$ such that $\mathbf{d} \cdot \mathbf{w_i} = 0$ for $1 \le i \le t$ and $\mathbf{d} \cdot \mathbf{w}_0 = 1$. So $\mathbf{d}M = \mathbf{e}_1$, but this contradicts the assumption that $\mathbf{e}_1 \notin \langle \mathbf{v}_{i_1}, \ldots, \mathbf{v}_{i_k} \rangle$. Hence, $\mathbf{w}_0 \in \langle \mathbf{w}_1, \ldots, \mathbf{w}_t \rangle$. So there exists $\mathbf{b}$ such that $M\mathbf{b} = \mathbf{0}$ and $b_0 \ne 0$. The only information the participants in γ have about a_0 is that $M\mathbf{a} = \mathbf{s}$. But $\mathbf{s} = M\mathbf{a} = M(\mathbf{a} + \alpha\mathbf{b})$ for all $\alpha \in \mathrm{GF}(q)$. Consequently, given any $c_0 \in \mathrm{GF}(q)$, there exists $\mathbf{c} = (c_0, \ldots, c_t)$ with $c_i \in \mathrm{GF}(q)$ for $1 \le i \le t$ such that $M\mathbf{c} = \mathbf{s}$. Therefore, the participants in γ cannot rule out any element of $\mathrm{GF}(q)$ as a possibility for a_0. $\square$

3　Multilevel Schemes

In this section, we give an existence proof that any multilevel access structure can be achieved in an ideal secret sharing scheme. Then we give a different construction that requires less computation on the part of the dealer.

The Basic Multilevel Scheme:　Let Γ be a multilevel access structure with levels $l_1 < l_2 < \ldots < l_R$. Let N_r be the number of participants of level l_r. Denote the participants by P_i for $1 \le i \le n$, and let L_i be the level of P_i. We will use the basic secret sharing scheme. So we need only specify how the dealer will choose the vectors $\mathbf{v}_i$. For each P_i, the dealer will pick an $x_i \in \mathrm{GF}(q)$. Let $\mathbf{v}_i$ be the l_R-dimensional vector $(1, x_i, x_i^2, \ldots, x_i^{L_i-1}, 0, \ldots, 0)$. Note that if $l_1 = 1$ and P_i is a participant with $L_i = 1$, then $\mathbf{v}_i = \mathbf{e}_1$. Define polynomials $f_j(x) = \sum_{i=0}^{j-1} a_i x^i$. The share s_i that the dealer gives to P_i will satisfy $s_i = f_{L_i}(x_i)$.

To complete the proof that there exists an ideal secret sharing scheme for any multilevel access structure, we need only to show that for any multilevel access structure, there is a method for the dealer to choose the x_i so that $\langle \mathbf{v}_{i_1}, \ldots, \mathbf{v}_{i_k} \rangle$ contains $\mathbf{e}_1$ iff $\{P_{i_1}, \ldots, P_{i_k}\} \in \Gamma$. In the remainder of this section, we give three different methods for the dealer to choose the x_i.

Theorem 1　*Let Γ be a multilevel access structure with levels $l_1 < l_2 < \ldots < l_R$. Let N_r be the number of participants of level l_r. Let n be the total number of participants. If $q > (l_R - 1) \begin{pmatrix} n \\ l_R - 1 \end{pmatrix}$, then there is an ideal secret sharing scheme for Γ over $\mathrm{GF}(q)$.*

Proof　We will use the basic multilevel scheme construction. We only need to show how the dealer will choose the x_i. Let $\mathbf{v}_0 = \mathbf{e}_1$ (although there is no participant P_0). Suppose the dealer has chosen x_i for all i, $0 \le i < h$. Let Ω be the set of subspaces spanned by some subset of size $L_h - 1$ of the vectors $\{\mathbf{v}_i \mid 0 \le i < h\}$. $|\Omega| < \begin{pmatrix} h \\ L_h - 1 \end{pmatrix}$. The dealer then picks x_h so that the L_R-dimensional vector $\mathbf{v}_h = (1, x_h, x_h^2, \ldots, x_h^{L_h-1}, 0, \ldots, 0)$ is not in any of the subspaces in Ω. To see that this is possible, let $H \in \Omega$, and let $\mathbf{w} = (w_0, w_1, \ldots, w_{L_h-1}, 0, \ldots, 0)$ be a normal vector to H. Then $\sum_{i=0}^{L_h-1} w_i x^i = 0$ has at most $L_h - 1$ solutions over $\mathrm{GF}(q)$.

Suppose that k participants, $P_{i_1}, \ldots, P_{i_k}$, of level at most k try to recover the secret and suppose that there is no subset of this set which contains l participants of level at most l for any $l < k$. The vectors $\mathbf{v}_{i_1}, \ldots, \mathbf{v}_{i_k}$ are independent and are contained in the k-dimensional space spanned by $\mathbf{e}_1, \ldots, \mathbf{e}_k$. Hence, $\mathbf{e}_1 \in \langle \mathbf{v}_{i_1}, \ldots, \mathbf{v}_{i_k} \rangle$ and so by Proposition 1, these participants can determine the secret.

Suppose now that a set $\gamma \notin \Gamma$ of participants try to recover the secret. Let $\gamma = \{P_{i_1}, \ldots, P_{i_k}\}$. Since the vectors $\mathbf{e}_1, \mathbf{v}_{i_1}, \ldots, \mathbf{v}_{i_k}$ are independent, by Proposition 1, these participants cannot obtain any information about a_0. $\square$

The Blakley scheme can also be modified to implement a multilevel access structure. The dealer again picks g to be the first coordinate axis and a sequence of flats F_i satisfying: $F_1 \subset F_2 \subset \ldots \subset F_R$, $F_1 \cap g$ is nonempty, and $g \not\subset F_R$. The secret is $P = F_1 \cap g$. A person of level r will be given a point on F_{r-1}. The points should be selected so that any r participants of rank at most r can determine the point P, and also so that for the flat, F, generated by a group of participants in which for any r there is no subset of r participants who all have rank at most r, $F \cap g$ must be empty. This construction was also discovered by Simmons [6].

One other issue to consider is the amount of computation needed for the dealer to construct a system. For the original Blakley system, the dealer must do a check to make sure that the points are in general position. An obvious way to do this requires $\binom{n}{k}$ time, although if the points are carefully selected, no such check is necessary. Also, no such check is needed for the Shamir scheme. Unfortunately, this nice property does not hold in the above construction for multilevel schemes. The obvious way to implement the scheme presented in Theorem 1 would require many checks to be sure that the points are in general position. We have however found some constructions which do not require checking.

The first construction we will mention is only feasible if there are not too many levels involved. We will use the basic multilevel scheme and so we simply need to describe how the dealer will pick the x_i. For illustration, suppose that we want to allow levels 2 or 3. Pick $q = p^2$. Let α be algebraic of degree 2 over $GF(p)$ (i.e. α satisfies an irreducible polynomial of degree 2 over $GF(p)$). The dealer picks an element y_i in $GF(p)$ for each participant P_i so that if $i \neq j$ and $L_i = L_j$, then $y_i \neq y_j$. For a participant of level 3, the dealer sets $x_i = y_i$. For a participant of level 2, he uses $x_i = \alpha y_i$. This system will have the desired properties. To see that three participants $P_{i_1}, P_{i_2}, P_{i_3}$ with $L_{i_1} = 2$, and $L_{i_2} = L_{i_3} = 3$ can determine the secret, consider the matrix M formed by $\mathbf{v}_{i_1}, \mathbf{v}_{i_2}, \mathbf{v}_{i_3}$. The determinant of this matrix is a polynomial in α of degree at most 1. It can be shown that the constant term in this polynomial is nonzero. Since α is algebraic of degree 2, the value of the polynomial must be nonzero.

In the more general setting, with levels $l_1 < \ldots < l_R$, the dealer picks $\alpha_1, \ldots, \alpha_{R-1}$, where α_r satisfies an irreducible of degree $\left\lfloor \frac{l_r^2}{2} \right\rfloor + 1$ over

$$GF\left(p^{\displaystyle\prod_{j=r+1}^{R-1} \left(\left\lfloor \frac{l_j^2}{2} \right\rfloor + 1 \right)}\right).$$

The dealer then sets $x_i = \alpha_{L_i} y_i$. The proof that this system has the desired properties is an extension of the above arguement. We will not include the arguement here because the following theorem constructs ideal multilevel schemes in a more efficient manner.

Theorem 2 *Let Γ be a multilevel access structure with levels $1 = l_0 < l_1 < \ldots < l_R$. Let q be a prime satisfying $q > N_r + 1$ for $1 \leq r \leq R$. Let $\beta = Rl_R^2$. Then there is an ideal secret sharing scheme for Γ over $GF(q^\beta)$ which can be constructed in time polynomial in $(N_1, \ldots, N_R, q\ )$.*

Proof Once again, we just need to show how the dealer will pick the x_i to use in the basic multilevel scheme. If there is no participant of level 1, add a participant P_0 with $L_0 = 1$. The dealer selects a y_i for each P_i so that $y_i \neq y_j$ if $L_i = L_j$ and $i \neq j$. Define $\rho(i)$ to be the integer j such that $L_i = l_j$. The dealer also picks an α that satisfies an irreducible of degree Rl_R^2 over $GF(p)$. Let $x_i = y_i \alpha^{R-\rho(i)}$.

Let $\gamma = \{P_{i_1}, \ldots, P_{i_k}\}$, be a set of k participants each of whom has level at most k and suppose that there is no subset of γ which contains more than l participants of rank at most l for any $l < k$. Let n_j be the number of these participants of rank l_j. Let $M'(\gamma)$ be the matrix whose rows are the vectors $\mathbf{v}_{i_1}, \ldots, \mathbf{v}_{i_k}$. Let $M(\gamma)$ be the matrix consisting of only the first k columns of $M'(\gamma)$. $M(\gamma)$ is essentially the same matrix as $M'(\gamma)$ since all of the columns removed consisted of all zeros.

To show that $M = M(\gamma)$ is nonsingular, we will show that the determinant of M can be written as a polynomial in α of degree less than Rl_R^2. We will show that the polynomial is not identically zero by showing that the constant term is nonzero.

Consider the determinant of M as a polynomial in α. Let $M = (m_{i,j})$. Recall that the determinant is the sum of all elementary signed products of M, where an elementary signed product is the product of the terms $m_{1,c_1}, \ldots, m_{k,c_k}$ with the appropriate sign, where $c_1, \ldots, c_k$ is a permutation of $1, \ldots, k$. Any nonzero elementary signed product will satisfy $c_i \leq L_i$ for $1 \leq i \leq k$. The maximum exponent of α in a row i of M is $(R - \rho(i))(L_i - 1)$. Therefore the maximum exponent of α in an elementary signed product is $\leq \sum_{r=1}^{R-1}(R - r)(l_r - 1)n_r < Rl_R \sum_{r=1}^{R-1} n_r \leq Rl_R^2$.

Let $T_{-1} = 0$, and let $T_j = \sum_{i=0}^{j} n_i$ for $0 \leq j \leq R$. The exponent of α in a nonzero elementary signed product will be $\sum_{i=1}^{k}(c_i - 1)(R - \rho(i))$. This sum achieves its minimum exactly when $\{c_{T_{r-1}+1}, \ldots, c_{T_r}\} = \{T_{r-1} + 1, \ldots, T_r\}$ for $0 \leq r \leq R$. Let D_r be the n_r by n_r submatrix of M generated by the rows and columns $T_{r-1} + 1, \ldots, T_r$. Let z be the minimum exponent of α in the determinant of M. Then the term $\theta \alpha^z$ for $\theta \in GF(q)$ in the determinant of M satisfies $\theta \alpha^z = \prod_{r=1}^{R} |D_r|$. Since each D_r is a multiple of a Van der Monde matrix, $|D_r| \neq 0$. Therefore, the coefficient of α^z is nonzero. Thus, since $M(\gamma)$ is nonsingular, the participants in γ can determine a_0.

Suppose now that γ is a set of $k - 1$ participants each of level at most k and suppose that there is no subset of γ which contains l participants of level at most l for any $l < k$. Let $\gamma' = \gamma \cup \{P_0\}$. Now γ' is a set of k participants each of level at most k and there is no subset of γ' which contains more than l participants of level at most l for any $l < k$. The matrix $M(\gamma')$ will thus be nonsingular. Therefore, $\mathbf{e}_1 \notin \langle \mathbf{v}_i \mid P_i \in \gamma \rangle$. From Proposition 1, the participants in γ receive no information about the value of a_0. $\square$

4   Compartmented schemes

In a compartmented scheme, there are disjoint sets of participants $C_1,\ldots,C_u$. The access structure consists of subsets of participants containing at least t_i from C_i for $i = 1,\ldots,u$, and a total of at least t participants. Let n be the total number of participants.

Theorem 3 *Let Γ be a compartmented access structure. If $q > \dbinom{n}{t}$, then there is an ideal secret sharing scheme for Γ over $GF(q)$.*

Proof WLOG, we may assume that $T = t - \sum_{i=1}^{u} t_i \geq 0$. The dealer chooses a vector $\mathbf{a} = (a_0,\ldots,a_{t-1})$ where a_0 is the secret. Let $T_0 = T$, and let $T_i = T + \sum_{j=1}^{i} t_j$ for $1 \leq i \leq u$. Denote the participants by $P_{r,i}$ where $P_{r,i}$ is in compartment C_r. For participant $P_{r,i}$, the dealer will pick a t-dimensional vector over $GF(q)$ of the form

$$\mathbf{v}_{r,i} = (1, x_{r,i}, x_{r,i}^2, \ldots, x_{r,i}^{T-1}, 1, \ldots, 1, \underbrace{x_{r,i}^T, \ldots, x_{r,i}^{T+t_r-1}}_{\text{coordinates } T_{r-1}+1,\ldots,T_r}, 1, \ldots, 1)$$

for some $x_{r,i} \in GF(q)$. As in Theorem 1, the dealer must be careful in choosing the $x_{r,i}$. Let $\prec$ denote lexicographic ordering on ordered pairs. I.e. $(r,i) \prec (s,j)$ iff $r < s$ or ($r = s$ and $i < j$). Let $\mathbf{v}_{0,0} = \mathbf{e}_1$. Suppose that the dealer has chosen $x_{r,i}$ for all $(r,i) \prec (s,j)$. Then the dealer must choose $x_{s,j} \neq 1$ so that the vector $\mathbf{v}_{s,j}$ is not in any subspace spanned by a set of vectors consisting of at least t_r of the $\mathbf{v}_{r,i}$ for each $r < s$ and at least $t_s^* = \min(t_s - 1, j - 1)$ of the $\mathbf{v}_{s,i}$ for $i < j$ and a total of at most $T + t_s^* + \sum_{r=1}^{s-1} t_r$ of the $\mathbf{v}_{r,i}$ for $(0,0) \preceq (r,i) \prec (s,j)$. Since $q > \dbinom{N}{t}$, it is easy to see that this is possible by using similar arguments to those used in Theorem 1.

A set of participants in Γ can determine the secret since the vectors $\mathbf{v}_{r,i}$ are independent. Conversely, suppose that a set $\gamma = \{P_{r,i} \mid (r,i) \in I\}$ of participants is not in Γ. Suppose there is a C_s such that γ does not contain at least t_s of the participants in C_s. Let M be the matrix with rows $\mathbf{v}_{r,i}$ for $(r,i) \in I$. Let M' be the matrix consisting of columns $1, T_s+1, \ldots, T_s+t_s$ of M. There are only t_s distinct rows in M', namely the rows corresponding to the vectors $\mathbf{v}_{r,i}$ with $r = s$ and $(r,i) \in I$, and the vector $(1,1,\ldots,1)$. Let $\{i_1,\ldots,i_{t_s-1}\} = \{i \mid (s,i) \in I\}$. Let M'' be the matrix consisting of the rows $\mathbf{e}_1, \mathbf{v}_{s,i_1}, \ldots, \mathbf{v}_{s,i_{t_s-1}}$. Then $|M''| = |M_{11}''|$, where M_{11}'' is the matrix M'' with the first row and column removed. But M_{11}'' is just a Van der Monde matrix with row i_j multiplied by x_{s,i_j}^T for $1 \leq j \leq t_s - 1$. So $|M_{11}''| \neq 0$. Therefore $\mathbf{e}_1$ is not in $\langle \mathbf{v}_{r,i} \mid (r,i) \in I \rangle$. If γ contains at least t_r participants from C_r for $1 \leq r \leq u$, but does not contain a total of at least t participants, then the participants in γ receive no information about a_0 since $\mathbf{e}_1$ and the vectors $\mathbf{v}_{r,i}$ for $(r,i) \in I$ are independent. $\square$

The construction presented in Theorem 3 requires that the dealer check exponentially many subspaces. It is easy to give an efficient implementation in the case that $t = \sum_{i=0}^{u} t_i$. The dealer can simply choose a_0 as the secret, and then randomly pick

$b_1, \ldots, b_u$ such that $a_0 = \sum_{i=0}^{u} b_i$. He then uses a threshold scheme with threshold t_i and secret b_i to distribute shares to the participants in C_i. However, we have found no efficient construction for the more general compartmented access structures.

5 Remarks

Benaloh [1] has shown that any set of subsets which can be recognized by a monotone circuit in which all gates and all inputs have fanout 1 can be realized as the access structure of an ideal secret sharing scheme. He also pointed out that since threshold schemes were ideal secret sharing schemes, threshold gates could be added to the circuits as well. Since we have now shown that multilevel schemes and compartmented schemes are ideal, gates realizing these access structures can be added as well.

6 Acknowledgments

I would like to thank Gus Simmons for introducing me to multilevel and compartmented secret sharing schemes and to Dan Davenport for useful conversations concerning this paper.

References

[1] Josh C. Benaloh and Jerry Leichter. Generalized secret sharing and monotone functions. to appear in Advances in Cryptology - CRYPTO88.

[2] G. R. Blakley. Safeguarding cryptographic keys. In *Proceedings AFIPS 1979 National Computer Conference*, pages 313–317, 1979.

[3] M. Ito, A. Saito, and T Nishizeki. Secret sharing scheme realizing general access structure. In *Proceedings IEEE Globcom'87*, pages 99–102, Tokyo, Japan, 1987.

[4] Adi Shamir. How to share a secret. *Communications of the ACM*, 22(11):612–613, Nov 1979.

[5] Gustavus J. Simmons. How to (really) share a secret. to appear in Advances in Cryptology - CRYPTO88.

[6] Gustavus J. Simmons. Robust shared secret schemes. to appear in Congressus Numerantium, Vol. 68-69.

Cartesian Authentication Schemes

M. De Soete[1]   K. Vedder[2]   M. Walker[3]

[1]*MBLE–I.S.G.*, Rue des Deux Gares 82, 1070 Brussel, Belgium
[2]*GAO*, Euckenstraße 12, 8000 München 70, Federal Republic of Germany
[3]*Racal Research Ltd.*, Worton Drive, Reading Berkshire RG2 OSB, England

Abstract. This paper gives a characterisation of perfect Cartersian authentication schemes. It is shown that their existence is equivalent to the existence of nets. Furthermore the paper presents constructions of new authentication schemes derived from generalised n-gons which take on the lowest combinatorial bound for the impersonation attack. They include, as special cases, those based on projective planes and generalised quadrangles which are described in [5] and [3] respectively. It investigates the properties of the encoding rules and contains a brief discussion of questions in connection with key management.

1   A Mathematical Authentication Model

There are three participants in the authentication model introduced by Simmons [7]: a *transmitter*, a *receiver* and an *opponent*. The transmitter wants to communicate certain information to the receiver, whereas the opponent tries to deceive the receiver.

More formally, we have a set of *source states S*, a set of *authenticators* or *authenticated messages M* and a set of *keys K*. A source state $s \in S$ is the information which the transmitter wishes to communicate to the receiver. This is done under a common secret key $l \in K$ which defines the encoding rule e_l used to determine the authenticated message $m = e_l(s)$ sent to the receiver (this means that e_l is a mapping from S to M and hence we investigate codes without splitting). In order for the receiver to be able to uniquely determine the source state from the obtained message, there can be at most one source state which is encoded by any given authenticated message $m \in M$ (i.e. $e_l(s) \neq e_l(s')$ if $s \neq s'$). This means that the encoding rules e_l, $l \in K$, are one-to-one mappings from S to M.

In the authentication codes we are going to construct every message uniquely determines the source state, independently of the key used. Such codes which offer no secrecy are called *Cartesian*. Another way of expressing the property of a scheme which is Cartesian is to say that there exists a map from M to S the restriction of which to $e_l(S) \subseteq M$ is the inverse of the map e_l for every key l.

We shall henceforth assume that every authenticated message is the image of at least one source state under at least one encoding rule. This is no restriction, since the deceiver knows the encoding rules and can thus rule out all other elements in M.

The receiver verifies the validity of a received authenticated message m^* by checking that m^* is contained in $e_l(S)$. If this is the case, then the unique source state s^* with $e_l(s^*) = m^*$ will be accepted. A message m^* not in $e_l(S)$ is rejected as fraudulent. Our definition of the acceptance rule implies that the probability of every source state is non-zero and that the receiver accepts a verified source state independent of its probability.

2 Perfect Authentication Codes

We consider two situations in which an opponent can launch an attack. In the first one he / she tries to make the receiver accept an authenticated message without having intercepted one. This is called *impersonation*. The other one is *substitution*. Here the opponent replaces an intercepted authenticated message by a different one.

We assume that there is a given probability distribution on the set of source states. The transmitter and receiver will determine a probability distribution on K, called an *encoding strategy*. We will denote by P_i the probabilities that the opponent can deceive the transmitter / receiver with an impersonation ($i = 0$) or a substitution ($i = 1$) attack.

Simmons [7] defined a code to be *perfect* if

$$\max(P_0, P_1) = \frac{1}{\sqrt{|E|}},$$

with E ($\subset K$) the set of "distinct" encoding rules or transformations (since different keys can define the same encoding transformation).

Perfect authentication codes were investigated by several authors [5], [7]. In [5] it was shown that for uniform source distribution $P_1 \geq 1/\sqrt{|E|}$. So the best such a scheme can achieve for the transmitter and receiver is $1/\sqrt{|E|}$. We prove the following theorem in this paper.

Theorem 1 *There exists a perfect Cartesian authentication code on r source states, $r \cdot k$ authenticated messages and k^2 encoding rules if and only if there exists a net of degree r and order k (see [1], [2]).*

We remark that this equivalence is implicitly contained in the paper by Gilbert, MacWilliams and Sloane [5] but the proof given here is the first direct approach for arbitrary source distribution.

3 Properties of the encoding rules

When substituting a fraudulent message for an intercepted one, the opponent may have one of the following three aims in mind

(i) to "disturb" the system,

(ii) to have any fraudulent message accepted,

(iii) to have a particular fraudulent message accepted.

In the schemes we consider he can always achieve (i), while (ii) and (iii) have the same probabilities attached.

The probability that the opponent can successfully deceive the receiver depends on the way the sets $e_l(S)$ are related to each other for the various encoding rules. We have to assume that the opponent has complete knowledge not only of the source states and authenticated messages but also of all the encoding rules. That is to say that the security of the scheme depends on the particular choice of e_l (or l) being kept secret. If, for instance, $e_l(S) \cap e_{l'}(S) = \phi$ for any two distinct encoding rules, then the opponent can derive the encoding rule e_l, say, from every intercepted authenticated message m. He could thus replace m by $m' \neq m$, $m' \in e_l(S)$, and he can deceive the receiver with a probability of 1. The opponent's probability to impersonate successfully the transmitter, that is to plant a message without having observed one, is just $|e_l(S)|/|M| = 1/|E|$. In the other extreme we have $|M| = |e_l(S)| = |S|$ for one and hence for all keys. In this situation the opponent can deceive the receiver in either case with a probability of 1.

An intercepted message m provides the opponent with some information about the encoding rule. It has to lie in the set E^m of all those encoding rules which take on m as an image of a source state, that is $E^m = \{e_l \mid e_l \in E \text{ and } m \in e_l(S)\}$. Assuming that all encoding rules are equally likely, the opponent has a probability of $1/|E^m|$ of guessing the correct rule. He can deceive the receiver with a probability of 1 if $\cap e_l(S)$, over all e_l in E^m, contains an authenticated message $m' \neq m$. For a replacement of m by m' would not be noticed by the receiver. This yields the following requirements

(i) $|E^m| \neq 1$ for all $m \in M$, and

(ii) $\cap_{E^m} e_l(S) = \{m\}$ for all $m \in M$.

Let us illustrate this by an example. We take $S = \{A, B, C\}$, $E = \{e_1, \ldots, e_4\}$ and M the entries of the matrix below which define the encoding rules. The only case where the opponent cannot deceive the receiver with a probability of 1 is when he intercepts b_1.

	A	B	C
e_1	a_1	b_1	c_1
e_2	a_2	b_1	c_2
e_3	a_1	b_2	c_1
e_4	a_2	b_3	c_2

Example 1

Let $e_K(s)$ denote the set of images of the source state s under all encoding rules, that is $e_K(s) = \{e_l(s) \mid l \in K\} \subseteq M$. Since, for Cartesian authentication schemes, distinct encoding rules never map distinct source states to the same authenticated message, distinct source states give rise to disjoint sets. It follows that the sets $e_K(s)$, $s \in S$, partition the set of authenticated messages and that there is a natural 1-1 correspondence between S and $\{e_K(s) \mid s \in S\}$. The knowledge of this property gives an opponent a sure way of disturbing the system. Say he intercepts the message $m = e_l(s)$. He then replaces m by a message $m' \in e_K(s)$, $m' \neq m$. Though m' corresponds to the same source state s as the correct message m the receiver cannot decide whether this type of substitution was played or whether m' was substituted for a message $m_1 = e_l(s_1)$ authenticating a source state $s_1 \neq s$.

If the opponent wants to deceive the receiver, then he has to choose a message $m' = e_l(s')$ in the set $e_l(S)$ with $s' \neq s$. Which set he chooses and which message in this set depends primarily on the way the encoding rules "distribute" the source states among the authenticated messages. We will from now on assume that there is a uniform probability distribution on the set of encoding rules, that is all encoding rules are equally likely.

For a message m let n_m denote the number of encoding rules which take on m as an image, that is $n_m = |E^m|$. If the opponent runs an impersonation attack, then he is not restricted in his choice by any message sent by the transmitter. Thus his probability to succeed depends solely on the distribution of the values n_m in M. If he does not mind which fraudulent message gets accepted he picks the message m with $n_m = \max\{n_m \mid m \in M\}$. If he wants to run a *chosen* attack, that is he wants a specific source state s to be accepted, then he chooses the message m with the largest value n_m among $e_K(s)$. Since all encoding rules are equally likely his chances to succeed is $P_0 = n_m/|E|$. This yields the following inequality $n_m/|E| \geq |S|/|M|$. We also note that, if n_m is a constant n, both the chosen and the non-specific attack have the same probability. Counting pairs (e_l, m) with $e_l(s) = m$ for some $s \in S$ we obtain $n \cdot |M| = |E| \cdot |S|$. Hence $n/|E| = |S|/|M|$, which is the lowest possible bound. The situation is different in the case of substitution. Firstly, the opponent, who has intercepted a message m, $m \in e_K(s)$, " cannot" choose a message in the "intercepted" class $e_K(s)$. Secondly, he has obtained some knowledge about the set E^m of encoding rules which might have been used. Let us illustrate this again by giving an example.

	A	B	C	A'	D
e_1	a_1	b_1	c_1	a_1	d_1
e_2	a_1	b_2	c_2	a_2	d_1
e_3	a_1	b_3	c_3	a_3	d_1
e_4	a_2	b_1	c_2	a_3	d_2
e_5	a_2	b_2	c_3	a_1	d_2
e_6	a_2	b_3	c_1	a_2	d_2

Example 2

Suppose that we have a probability distribution on the set of source states which reads as follows:

$$\text{prob}(A) = 0.6, \ \text{prob}(B) = \text{prob}(C) = 0.2.$$

Given this distribution the opponent has a success rate of $1/2$ in 40% of all transmissions, that is when B or C is authenticated. In the remaining 60% of all transmissions his success rate is just $1/3$.

The probability distribution on the source sates can be counteracted by using column A' instead of A which requires one extra authenticated message. In this scheme all transmissions supply the opponent with a success rate of $1/2 > 1/\sqrt{6} = 1/\sqrt{|E|}$. Furthermore, $P_0 = 1/3$ and $P_1 = 1/2$ since $n_m = 2$ for all messages m. The reader should convince himself that this is the best possible value one can achieve on three source states with six encoding rules.

The transmission of a message rules out not only certain encoding rules but also certain messages. Say message $m = b_1$ was sent by the transmitter, then an opponent knows that the receiver will not accept a_2 or c_3 as they are not in $E^m(S)$. So the condition which should be imposed on the encoding rules is that $E^m(S) = M$ and that E^m gives a uniform distribution on all messages not in $e_K(s)$. It is easily seen that this cannot be achieved in a scheme with just three source states. An example of such a scheme can be constructed from columns B, C, A' and D extending them in the obvious way. We note that this is an affine plane of order 3.

Now let all messages in $M \setminus e_K(s)$ lie on the same number n' of encoding rules in E^m. Then by counting pairs (m', e_l) with $m' \notin e_K(s)$ and $e_l(s') = m'$ for some $s' \in S$ and $e_l \in E^m$ we obtain

$$n' \cdot (|M| - |e_K(s)|) = |E^m| \cdot (|S| - 1).$$

We note that

$$\frac{n'}{|E^m|} = \frac{|S|}{|M|} = \frac{n}{|E|}$$

and $P_1 = P_0$ in such a system if and only if

$$|M| = |S| \cdot |e_K(s)|.$$

The last equation holds, by the way, for the perfect schemes described in [5]. They are constructed from projective planes which are the projective closure of affine planes. These are nets of maximal degree.

4   Perfect Schemes and Nets

4.1   Nets

A *finite incidence strucuture* $\mathcal{P}$ is a triple (P, B, I) which consists of two finite, non-empty and disjoint sets P and B and a subset $I \subseteq P \times B$. The elements of I are called *flags* while those of P and B are referred to as *points* and *lines* respectively. I is called the *incidence relation*. We say that a point x and a line L are incident with each other and write $x \in L$ if and only if (x, L) is a flag.

Throughout this paper we shall only be concerned with *linear* incidence structures, which are characterised by the property that distinct points are incident with at most one line. As furthermore every line is incident with at least two points we can identify each line with the set of points it is incident with.

An incidence structure is *tactical* if it has the property that every point is incident with the same number r of lines and every line is incident with the same number k of points. Using $|P| = v$ and $|B| = b$ it is easily seen by counting flags in two different ways that

$$v \cdot r = b \cdot k = |I|.$$

A *net* of degree r and order k (see Bose [1]) is an incidence structure $\mathcal{P}=(P, B, I)$ which satisfies the following axioms

(i) the lines can be partitioned into exactly r disjoint, non-empty "parallel classes" such that

        (a) each point is on exactly one line of each class

        (b) two lines of distinct classes intersect in exactly one point;

(ii) each line is incident with k points.

Then it can be proved that the incidence structure is tactical, so each line of the net contains exactly k distinct points and each point of the net lies on exactly r distinct lines. Furthermore the net has exactly $r \cdot k$ distinct lines and k^2 distinct points.

It is easy to see that $r \leq k + 1$ and $r = k + 1$ if and only if $\mathcal{P}$ is an affine plane. An affine plane is known to exist for every prime power k. Nets with (at least) $r = 4$ parallel classes exist for all values of k.

4.2 Proof of the Theorem

We will first prove that an authentication scheme constructed from a net is perfect. Let (P, B, I) be a net of degree r and order k, and let $\mathcal{P}$ be its set of parallel classes. The construction closely resembles the one given in [5] for projective planes. Let

$$S = \mathcal{P}$$

be the set of source states,

$$M = B$$

the set of authenticated messages and

$$K = P$$

the set of keys. The encoding rule e_l defined by the key l maps a source s to the authenticated message m which is the unique line in the parallel class s which is incident with l. This is well defined since every point of the net is incident with exactly one line of each parallel class. Observe also that, since each line of a net is contained in exactly one parallel class, this authentication scheme is Cartesian.

We now prove that the schemes constructed above are perfect. We show that $P_0 = P_1 = 1/\sqrt{|E|}$ under the assumption that there is a uniform distribution on the set of keys, so $p(l) = 1/|K| = 1/|P|$. Let $q_0(m)$ be the opponents optimal impersonation strategy. Then

$$P_0 = \sum_{m \in M} q_0(m) \cdot \sum_{l\,I\,m} p(l).$$

Setting $p(l) = 1/|K|$ and noting that $\sum_{x\,I\,m} 1 = k$, gives

$$P_0 = \frac{k}{|P|} = \frac{1}{\sqrt{|K|}},$$

where the last equality follows from $|P| = k^2$ for a net.

Suppose now that the opponent has observed the sender's authenticated message m, and let $q_1(m^*|m)$ be his optimal strategy. Clearly, we have $q_1(x|m) = 0$, whenever $x||m$, because parallel messages always correspond to the same source state, and the opponent usually is not interested in substituting m by a message corresponding to the same source state. Then

$$P_1 = \sum_{m \in M} p(m) \cdot \sum_{m^* \in M, m^* \not|| m} q_1(m^*|m) \cdot \sum_{l\, I\, m,\, l\, I\, m^*} p(l|m).$$

Since m and $m*$ are not parallel they intersect in exactly one point. Furthermore $p(l|m) = 1/k$, since there are k points per line and $p(l)$ is uniform. This gives

$$P_1 = \frac{1}{k} = \frac{1}{\sqrt{|K|}}.$$

Now we will prove that every perfect authentication system defines a net. Suppose we are given a Cartesian authentication scheme with a set of source states S, a set of authenticated messages M, a set of keys K and a set of encoding rules $\{e_l : S \to M \mid l \in K\}$. We associate with our authentication scheme an incidence structure $N = (K, M, I)$, with points set K, line set M and an incidence relation defined in the following way:

if $l \in K$ and $m \in M$, then $l\, I\, m \iff$ there exists an $s \in S$ such that $e_l(s) = m$.

Note that the term line above is an abuse of terminology since two "lines" may have more than one point in common.

For each $m \in M$ we denote by $\sigma(m)$ the unique source state which encodes to m. This is well defined for our authentication scheme is Cartesian. The function $\sigma : M \to S$ gives rise to an equivalence relation "$||$" on M, defined by

$m||m'$ if and only if $\sigma(m) = \sigma(m')$.

For each $s \in S$ let $\mathcal{P}_s = \{m \in M \mid \sigma(m) = s\}$ denote the equivalence class of "parallel" lines corresponding to s. Two distinct parallel lines m and m^* in $\mathcal{P}_s$ have no point in common. For otherwise there exists a point (key) l such that $e_l(s') = m$ and $e_l(s^*) = m^*$. As $m||m^*$ it follows that $s = s' = s^*$ and, since e_l is a mapping, $m = m^*$ contradicting $m \neq m^*$. It also follows that for each point l of our incidence structure N there is a unique line $e_l(s)$ in $\mathcal{P}_s$ which is incident with l. So the lines of $\mathcal{P}_s$ partition the points of N and all the lines form a "parallelism".

The above discussion indicates that N has a net–like structure. It remains to show that each line is incident with the same number of points and that two distinct non–parallel lines intersect in a unique point. We denote by $[m]$ the number of points on a line m and by $[m, m^*]$ the number of points the lines m and m^* have in common.

We assume that the key is chosen according to the uniform distribution $p(l) = 1/|K|$. Let $q_0'(m)$ be any impersonation strategy selected by the opponent. Then, since the authentication scheme is perfect, we have

$$P_0' = \sum_{m \in M} q_0'(m) \cdot \sum_{\{l \mid e_l(\sigma(m))=m\}} p(l) \leq \frac{1}{\sqrt{|K|}}$$

where P_0' denotes the probability of a succesful impersonation. Substituting $p(l) = 1/|K|$, and $|\{l \mid e_l(\sigma((m)) = m\}| = [m]$, gives

$$P_0' = \frac{1}{|K|} \sum_{m \in M} q_0'(m) \cdot [m] \leq \frac{1}{\sqrt{|K|}}.$$

Since this inequality holds for every strategy $q_0'(m)$, it follows that

$$[m] \leq \sqrt{|K|},$$

for all $m \in M$. As a consequence we obtain

$$|S| \cdot |K| = \sum_{m \in M} [m] \leq |M|\sqrt{|K|},$$

with equality if and only if $[m] = \sqrt{|K|}$ for each m. In this expression $|S|$ is the number of parallel classes (i.e., the number of source states). The equality in the expression follows because both sides of the equation count the number of flags of the incidence structure.

Suppose now that $q_1'(m)$ is an opponent's substitution strategy, and let $q_1'(m^*|m)$ be the strategy given that m has been intercepted. We shall assume that $q_1'(x|m) = 0$ for all messages x which are parallel to m. Since the authentication scheme is perfect we have

$$P_1' = \sum_{m \in M} p(m) \cdot \sum_{m^* \in M, m \parallel m^*} q_1'(m^*|m) \cdot \sum_{\{l \mid e_l(\sigma(m))=m, \, e_l(\sigma(m^*))=m^*\}} p(l|m) \leq \frac{1}{\sqrt{|K|}},$$

where P_1' is the expected probability of a succesful substitution.
However

$$p(m) = p(\sigma(m)) \sum_{l \, I \, m} p(l)$$

so that substituting for $p(l) = 1/|K|$, $p(l|m) = 1/[m]$ and noting that $|\{l \mid e_l(\sigma(m)) = m, \, e_l(\sigma(m^*)) = m^*\}| = [m, m^*]$, we obtain

$$P_1' = \sum_{m \in M} p(\sigma(m)) \cdot \frac{[m]}{|K|} \cdot \sum_{m^* \in M, m \parallel m^*} q_1'(m^*|m) \cdot \frac{[m, m^*]}{[m]} \leq \frac{1}{\sqrt{|K|}}.$$

This holds for all choices of $q_1'(m^*|m)$. For each m, choose an m' with

$$[m, m'] \geq [m, m^*], \text{ all } m^* \not\parallel m$$

and define $q'_1(x|m) = 1$ if $x = m'$ and 0 otherwise (i.e. the strategy is to substitute a message which has the maximum number of points in common with m). Then we obtain

$$\sum_{m \in M} p(\sigma(m)) \cdot [m, m'] \leq \sqrt{|K|}.$$

But

$$\sum_{m \in M} p(\sigma(m)) \leq \sum_{m \in M} p(\sigma(m)) \cdot [m, m'],$$

since $[m, m'] \geq 1$.

On the other hand

$$\sum_{m \in M} p(\sigma(m)) \geq \frac{|M|}{|S|}.$$

Hence there results

$$\frac{|M|}{|S|} \leq \sum_{m \in M} p(\sigma(m)) \leq \sum_{m \in M} p(\sigma(m)) \cdot [m, m'] \leq \frac{|M|}{|S|}$$

which gives $[m, m'] = 1$ for all m and $|M| = \sqrt{|K|} \cdot |S|$. The first of these equalities means that $[m, m^*] \leq 1$ for all m, m^* whilst the second equality tells us that $[m] = \sqrt{|K|}$ for all m.

It remains to argue that if $m \nmid m^*$, then $[m, m^*] = 1$. This is trivial. Each line $\sigma(m^*)$ which meets m does so in precisely one point. There are $[m] = \sqrt{|K|}$ such lines, and each is incident with $\sqrt{|K|}$ distinct points. Thus each of the $|K|$ points of the incidence structure lies on precisely one of these lines.

5 Authentication Schemes Constructed from Generalised Polygons

5.1 Generalised Polygons

Given an incidence structure P, in the sense of [2] (see also Section 4.1), we denote by $\Delta(P)$ the flag graph of P. Thus $\Delta(P)$ is the bipartite graph having vertex set the collection of points and blocks of P and edges the totality of flags (unordered incident point–block) pairs of P.

A *(thick) generalised polygon* ($n \in N$, $n \geq 2$) is an incidence structure P with the property that $\Delta(P)$ satisfies the following three conditions:

(i) each vertex has valency at least 3;

(ii) each pair of edges is contained in a circuit of length $2n$;

(iii) there is no circuit of length less than $2n$.

In this paper we deal with $n \geq 3$ and use the term "line" instead of block.
Examples of generalised 3- and 4-gons are projective planes and generalised quadrangles which were used in [5] and [3], respectively to construct authentication schemes. The finite generalised n-gons, with which we shall be dealing exclusively, were studied by Feit and Higman [4]. They proved that $n = 3, 4, 6$ or 8. These structures are tactical configurations (which means that every point is incident with the same number of lines) and we follow the convention of denoting the number of points on a line by $s + 1, s > 1$, and the number of lines through a point by $t + 1, t > 1$.

In discussing generalised n-gons the following notation and observations prove to be useful.
Condition (ii) implies that $\Delta = \Delta(P)$ is connected and that the distance $\delta(X, Y)$ between two vertices X and Y is at most n. If $\delta(X, Y) = n$, then X and Y are called *opposite*. If vertices X and Y are such that $\delta(X, Y) < n$, then there is, by condition (iii), a unique path of length $\delta(X, Y)$ which joins X and Y. This path is denoted by $< X, Y >$.
Now, let X be a vertex of Δ, and let $\Delta_d(X) = \{Y \in \Delta \mid \delta(X, Y) = d\}$, the set of all vertices which are at distance d from X. We can interprete Δ as a graph with root X and $n + 1$ levels $0, \ldots, n$, where the vertices at level d are the elements of $\Delta_d(X)$. It is easily seen that a vertex Y at level $d \neq 0, n$ is joined to exactly one vertex at level $d - 1$ and s or t vertices at level $d + 1$, depending on it being a line or a point. We denote those at level $d + i$ from X and level i from Y by $\Delta_{+i}(Y)$. The vertices at level n are joined only to vertices at level $n - 1$. This means that the subgraph with vertex set $\cup_{d \neq n} \Delta_d(X)$ forms a tree (see [6]).
So the number of vertices in $\Delta_d(X)$, $d < n$, is equal to the number of distinct paths of length d which start in X. Hence, if X is a point, then

$$|\Delta_d(X)| = \begin{cases} (t+1)(st)^{(d-1)/2} & d \text{ odd} \\ (t+1)s^{d/2}t^{d/2-1} & d \text{ even.} \end{cases}$$

To obtain $\Delta_n(X)$ we have to divide $\Delta_{n-1}(X)$ by $t + 1$ or $s + 1$, whichever is the degree of the vertices opposite to X. Here it is important to observe that $s = t$ for n odd. For, if X is a point, then an opposite vertex Z is a line and hence there are precisely $s + 1$ distinct paths from Z to X. Thinking of Z as the root there are $t + 1$ paths from X to Z. But the number of paths has to be the same. These observations yield the expressions (2) and (3) given below.

5.2   The Schemes

For every finite generalised n-gon of order $n \geq 3$ with parameters s and t we can define, making use of an arbitrary point X, a Cartesian authentication scheme with the set of source states

$$|S| = |\Delta_1(X)| = t + 1, \tag{1}$$

the set of authenticated messages

$$|M| = |\Delta_d(X)| = \begin{cases} (t+1)t^{d-1} & d \text{ odd} \\ (t+1)s^{d/2}t^{d/2-1} & d \text{ even} \end{cases} \tag{2}$$

and the set of keys

$$|K| = |\Delta_n(X)| = \begin{cases} t^2 & n = 3 \\ s^{n/2}t^{n/2-1} & n \text{ even.} \end{cases} \tag{3}$$

The encoding rule e_Z determined by the key Z maps a source Y to

$$e_Z(Y) = Z_d$$

where Z_d is the vertex at distance d from X in the unique path $< Y, Z >$. A receiver with knowledge of the key Z authenticates Z_d by checking that it is at distance $n - d$ from Z. The corresponding source state Y is simply the vertex adjacent to X in the path $< X, Z_d >$.

Starting with a line instead of a point results in an interchange of the parameters s and t in the above expressions. Note that $s = t$ for $n = 3$. We remark that such a scheme can be constructed from any tree by introducing an extra level $n+1$ and joining vertices of level n with new ones at level $n + 1$.

5.3   Implementation and Security

When implementing a scheme it is important to strike the right balance between security and the problem of handling. Keeping in mind that security is a relative term one may say that the more secure a system becomes the less managable it will be. Furthermore, one should take into account that the security gained by keeping certain additional information secret could be short lived. In our case, a potentional attacker can deduce at least part of this information from intercepted authenticated messages.

In the extreme one could keep everything but the underlying scheme secret, that is not only the key Z but also the level d and the root X could be part of the keying information. This is certainly interesting from a theoretical point of view. It would, however, make the system extremely difficult to handle since the receiver needs to know how the source states are labelled in order to pick the correct authenticated message. Thus all the information about the source states becomes part of the key and the system becomes unmanagable.

As d can be derived from the knowledge of X and an intercepted authenticated message, keeping d secret has no effect on the probability of a successful substitution by an interceptor. It does, however, increase the likelihood of detecting an impersonation attack. For any vertex not equal to X or at distance n from X could be an authenticated message and not only $|S|$ messages at distance d. So the security gained by making d part of the key might well justify the additional overhead.

We will now assume that both X and d are known and that all keys are equally likely. The probability to guess the correct key is $1/|K|$ if no message is intercepted, and $1/|\Delta_{n-d}(m) \cap \Delta_n(X)|$ if one message m is intercepted.

For $d > n/2$ any two authenticated messages uniquely determine the key since there is no circuit of length less than $2n$. This has two important consequences. Firstly, the key has to be changed after each transmission. Secondly, distinct keys define distinct encoding rules since their "corresponding" set of authenticated messages have at most one element in common.

For $d \leq n/2$ the situation depends on the structure of the underlying geometry. We will highlight this in Section 5.4.

5.4 Impersonation and substitution

Obviously, for $d = 1$ every source state is encoded by itself and this encoding is independent of the key. This implies that $P_0 = P_1 = 1$. We will henceforth assume that $d > 1$ and that X is a point. If X is a line, then the values of s and t have to be interchanged in all the following formulas.

We have the following probabilities for the impersonation attack.

$$P_0 = \begin{cases} 1/(st)^{(d-1)/2} & d \text{ odd} \\ 1/s^{d/2}t^{d/2-1} & d \text{ even.} \end{cases}$$

The sets $\Delta_{+(d-1)}(s)$, $s \in S$, partition the set of vertices at level d. For a given key k each of them contains exactly one authenticated message.
We note that $P_0 = |S|/|M|$ which is the lowest possible combinatorial bound for an impersonation attack (see [7]).

For the substitution attack we obtain

$$P_1 = P_0 \text{ for } d \leq n/2$$

whilst for $d > n/2$, we have

$$P_1 = \begin{cases} 1/(st)^{(n-d)/2} & n \text{ even}, d \text{ even} \\ 1/s(st)^{(n-d-1)/2} & n \text{ even}, d \text{ odd} \\ 1/t & n = 3, d = 2. \end{cases}$$

For $d > n/2$ any two authenticated messages uniquely determine the key k. This means that a successful substitution attack is equivalent to determining the key. Let m be the intercepted message. Then choose any vertex in $\Delta_{n-d}(m) \cap \Delta_n(X)$, say k'.

If $k' = k$, then every vertex in $\Delta_{n-d}(k')$ is an authenticated message. Pick your favourite one.

If $k' \neq k$, then except for m no vertex in $\Delta_{n-d}(k')$ is an authenticator.

For $d < n/2$ it follows from Axioms (ii) and (iii) that for every source state s exactly one vertex in $\Delta_{+(d-1)}(s)$ is an authenticated message. We note that in all cases the probability to have one's favourite source state accepted is the same as having any odd source state accepted and that the probabilities are independent of the distribution on the source states (assuming that all of them have non-zero probability).

We conclude this section with a complete list of the schemes for all possible values of n. An entry * in the tables below means that the value depends, as already mentioned, upon the structure of the generalised n-gon.

$n = 3$: projective planes
$|S| = t + 1, |K| = t^2$

d	2
$|M|$	$t(t+1)$
P_0	$1/t$
P_1	$1/t$
$|E|$	t^2

$n = 4$: generalised quadrangles
$|S| = t + 1, |K| = s^2 t$

d	2	3
$|M|$	$s(t+1)$	$st(t+1)$
P_0	$1/s$	$1/st$
P_1	$1/s$	$1/s$
$|E|$	*	$s^2 t$

$n = 6$: generalised hexagon

$|S| = t + 1$, $|K| = s^3 t^2$

d	2	3	4	5
$|M|$	$s(t+1)$	$st(t+1)$	$s^2 t(t+1)$	$(st)^2(t+1)$
P_0	$1/s$	$1/st$	$1/s^2 t$	$1/(st)^2$
P_1	$1/s$	$1/st$	$1/st$	$1/s$
$|E|$	*	*	$s^3 t^2$	$s^3 t^2$

$n = 8$: generalised octagon

$|S| = t + 1$, $|K| = s^4 t^3$

d	2	3	4	5	6	7
$|M|$	$s(t+1)$	$st(t+1)$	$s^2 t(t+1)$	$(st)^2(t+1)$	$s^3 t^2(t+1)$	$(st)^3(t+1)$
P_0	$1/s$	$1/st$	$1/s^2 t$	$1/(st)^2$	$1/s^3 t^2$	$1/(st)^3$
P_1	$1/s$	$1/st$	$1/s^2 t$	$1/s^2 t$	$1/st$	$1/s$
$|E|$	*	*	*	$s^4 t^3$	$s^4 t^3$	$s^4 t^3$

Remark For n even, i.e. $n = 2m$, it is possible to construct perfect schemes when using as root a regular vertex X with $d = m$.

References

[1] R. C. Bose, *Graphs and designs*, in: Finite geometric structures and their applications, ed. A. Barlotti, Ed. Cremonese Roma (1973), 1–104.

[2] P. Dembowski, *Finite Geometries*, Springer Verlag, 1968.

[3] M. De Soete, *Some Constructions for Authentication / Secrecy Codes*, Advances in Cryptology–Proceedings of Eurocrypt '88, Lect. Notes Comp. Science 330, Springer 1988, 57–75.

[4] W. Feit and G. Higman, *The non–existence of certain generalised polygons*, J. Algebra 1 (1964), 114–131.

[5] E. N. Gilbert, F. J. MacWilliams and N. J. Sloane, *Codes which detect deception*, Bell System Technical Journal, Vol. 53–3 (1974), 405–424.

[6] D. Jungnickel, *Graphen, Netzwerke und Algorithmen*, Wissenschaftsverlag Bib. Inst. Zürich, 1987.

[7] G. J. Simmons, *Authentication Theory / Coding Theory*, Advances in Cryptology–Proceedings of Crypto'84, Lect. Notes Comp. Science 196, Springer 1985, 411–432.

HOW TO SAY "NO"

Albrecht Beutelspacher
Mathematisches Institut
Justus-Liebig-Universität Gießen
Arndtstr. 2, D-6300 Gießen
Federal Republic of Germany

ABSTRACT

We pose the question whether there exist threshold schemes with positive and negative votes (shadows), that is threshold schemes in which any qualified minority can prohibit the intended action. Using classical projective geometry, the existence of such systems is proved. Finally, possible attacks on such systems are discussed.

1. INTRODUCTION

Threshold schemes have been introduced in order to control the access to secret data: The secret is divided into n "shadows" such that from any t shadows the secret can be reconstructed, but it is not feasable to retrieve the secret by knowing only t−1 shadows. So, only if a certain numbers t of participants say "yes", the secret will be disclosed. (Threshold schemes have been introduced by Shamir [4]; for an excellent survey on threshold schemes which emphasizes in particular the connections to geometry see [5].)

In many practical situations it is desirable that a qualified minority should also be able to "close" the secret. Think for example of the serious (usually not solved) problem of deactivating a master key after a certain time. Here one would like to have a system which allows a qualified "no". The aim of this note is to introduce such systems. These will be "ordinary threshold schemes" with an additional negative feature. Typically, every user will get a "positive" and a "negative" vote.

First we will introduce some notation and then study several examples. Since we look for constructions using geometry we shall call the shadows *points*.

A (t;s)-*threshold scheme* consists of a collection $P \cup N$ of points such that the following conditions are satisfied.

- Any t points of P together with at most s−1 points of N determine a secret X uniquely;

- if less than t points of P are "active", then the secret X cannot be retrieved (independent how many points of N are active).

- if at least s points of N are active, then it is impossible to determine the secret.

One can think of the points of P as *positive* votes, whereas the points of N represent *negative* votes.

2. EXAMPLES

2.1 Basic Example

Consider a 3-dimensional geometry. We shall restrict ourselves to the 3-dimensional projective space $P = PG(3,q)$ of order q. (For the geometric background see [2] and [3].) Fix a point X of P which will be identified with the secret. Choose a line ℓ_0 of P through X. Now we define

P to be a set of points on a line ℓ intersecting ℓ_0 in X and

N as a set of points distinct from $\ell' \cap <\ell,\ell_0>$ on a line ℓ' skew to ℓ.

Protocol: If a set U of (positive and negative) points is *active*, then the system computes $<U>$ and intersects it with ℓ_0. If $<U> \cap \ell_0$ is a point, the system takes it as the secret.

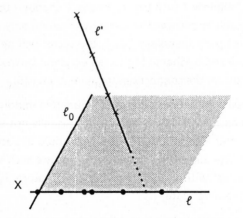

Analysis of the Basic Example:

- If at least two points of P are active, but no point of N, then the system computes ℓ and $\ell \cap \ell_0 = X$.

- If at least two points of P and exactly one point Y of N is active, then the system computes the plane $E = <\ell,Y>$; since $Y \notin <\ell,\ell_0>$, we have $E \neq <\ell,\ell_0>$; so, E intersects ℓ_0 only in X.

- If at least two points of P and at least two points of N are active, then the system computes $<\ell,\ell'> = P$, intersects it with ℓ_0 and gets ℓ_0. The probability of choosing the correct secret point is only $1/(q+1)$. So, the secret cannot be retrieved.

- If at most one point of P and at least two points of N are active, then the system gets a point different from X of ℓ_0 – if any.

Conclusion: The Basic Example provides a (2;2)-threshold scheme.

2.2 A general construction

The above construction can be generalized in the following way.

Consider $P = PG(d,q)$ and sets P and N of subspaces of P. If a subset $U \subseteq P \cup N$ is active, the the system computes $R = <U>$ and intersects R with a prefixed subspace M containing the secret X. If $R \cap M = X$, then the system says "yes".

The trick is the following: If enough elements of P are active, then R will contain X; but if in addition enough elements of N are active, then R will contain all points of M. In this case, for an attacker, all point of M is equally likely; so, he can reconstruct the secret only with the (a priori !) probability $1/(q+1)$.

In the following *example* we will deal only with the situation in which all elements of $P \cup N$ are points.

Let $d = t+s-1$. Fix a $(t-1)$-dimensional subspace T, an $(s-1)$-dimensional subspace S which is skew to T and a line M which intersects T in a unique point X and is skew to S. Let M be spanned by the points X and Y. Now choose a set P of points in T, a set N of points of S in such a way that the set $P \cup N \cup \{X,Y\}$ is an *arc*, which means that any $d+1$ of them generate the whole space. This implies:

- Let U be a subset of $P \cup N$. If U contains at most $t-1$ points of P, then $X \notin <U>$, so the secret cannot be reconstructed.

- If in a subset U of $P \cup N$ there are at most $s-1$ points of N, then $Y \notin <U>$, so the secret may be retrieved - if there are enough points of P in U.

- If U contains at least s points of N and at least t points of P, then $<X,Y> \subseteq U$, so the secret cannot be retrieved.

Hence we get a (t;s)-threshold scheme.

3. GENERALIZATION

Let G be a *geometry* consisting of *subspaces* (the empty set, points, lines, planes,..., i-dimensional subspaces,..., hyperplanes, the whole space) satisfying the following condition: For any set of subspaces $U_1,...,U_a$, there is a unique subspace $<U_1,...,U_a>$ of smallest dimension containing them. Such a geometry is often called a *matroid* [6] and most of the considered geometries are matroids (for instance projective and affine spaces, vector spaces,...)

Fix a point X in the geometry **G** and a line ℓ through it. Furthermore, fix two non-negative integers t and s. Now consider sets P and N of subspaces of **G** satisfying the following properties. For any subspaces $V_1,...,V_a$ of P and $W_1,...,W_b$ of N it holds

- If $\dim V_1 + ... + \dim V_a < t$, then $X \ell < V_1,...,V_a,W_1,...,W_b >$.

- If $\dim V_1 + ... + \dim V_a \geq t$ and $\dim W_1 + ... + \dim W_b < s$, then

  $< V_1,...,V_a,W_1,...,W_b > \cap \ell = \{X\}$.

- If $\dim V_1 + ... + \dim V_a \geq t$ and $\dim W_1 + ... + \dim W_b \geq s$, then

  $\ell \subseteq < V_1,...,V_a,W_1,...,W_b >$.

Clearly, such an arrangement fulfills our initial requirements. *Namely:* The first condition says that if $V_1,...,V_a$ are too small alltogether, then the secret cannot be reached. The second condition guarantees that if $< V_1,...,V_a >$ is big enough, but $< W_1,...,W_b >$ is not too spacious, then the secret will be retrieved automatically. Finally, the last condition says that a great portion $< W_1,...,W_b >$ of subspaces of N can obstruct the system.

We note that this system is very flexible in as much as the elements of P and N may have different dimensions. This corresponds to real-world requirements, since in reality not all men are equal: User i gets subspaces $V_i \in P$ and $W_i \in N$ whose dimensions correspond to the (positive, resp. negative) power (importance, level in the hierarchy,...) of i. In other words, this is a (t;s)-threshold scheme only if the subspaces in $P \cup N$ are just points.

On the other hand, there is an abundance of geometries with the above properties. In section 2 we have indicated that one can use flats for the construction of such systems. But many other objects do the job, for instance curves of a certain degree, surfaces, and so on. In such a way one can generalize also Shamir's original construction [4].

We remark that for ordinary threshold schemes a similar approach has been undertaken in [1].

4. ATTACKS

There is an obvious attack against our models of (t;s)-threshold schemes: Sombody who knows a set $P' \cup N'$ of active positive and negative points (with $|P'| \geq t$) has a good chance to forge the system. If he knows P', he can directly determine the secret. If, on the other hand, he knows 'only' $P' \cup N'$, then he can form all t-element subsets of this set and try each possibility; since the number of all t-element subsets of $P' \cup N'$ is much smaller than the number of points per line, also this is a reasonable attack.

These attacks imply strong restrictions on the machine which evaluates the incoming votes (points). Firstly, the machine must keep all inputs secret. Furthermore, the machine must operate trustworthy: It must not be possible to misuse it in order to obtain the positive or negative votes.

Therefore, there seem to be strong restrictions in using such a system. This is true. But it is only fair to mention that similar restrictions apply to all kinds of threshold schemes.

1. In any threshold scheme, the machine evaluating the shadows has to keep these shadows secret. (Otherwise, everybody who has access to the machine could obtain a sufficient number of shadows.) So, although the machine is not supposed to keep secrets for a long time, it must be able to keep secrets for a short time and then destroying them reliably.

2. Let us look at another example, namely Simmon's [5] realization of *robust shared secret systems*. In its simplest form, the shadows are points on a line ℓ and any two shadows are sufficient to determine the secret $X = \ell \cap \ell_0$. As a guarantee that the points are correct, the system wants a third point; if all three points lie on the same line, the system is convinced that all points are correct.

Of course, such a procedure makes only sense if the machine works trustworthy: Otherwise, given any two shadows, the machine could easily compute a third point on the line given by the two shadows and could so 'convince' itself that the points are correct.

To sum up, the restrictions on the machine, namely secrecy for short-term secrets and trustworthy operation are shared by many, if not all threshold schemes. But of course, there might exist systems (with or without negative votes) which have these properties only to a certain extend or not at all.

5. CONCLUSION

We present an extension of threshold schemes which allows negative votes. The philosophy behind the construction of these new schemes is the following. In an (ordinary) threshold scheme, the secret cannot be reconstructed, if there is too little information. With negative votes, the secret may also not be retrieved, if there is too much information. Using geometric structures we construct several infinite families of such threshold schemes.

REFERENCES

[1] A. Beutelspacher and K. Vedder, *Geometric structures as threshold schemes*. The Institute of Mathematics and its Applications, Conference Series **20**, Cryptography and Coding (ed. by H.J. Beker and F.C. Piper), 1989, 255-268.

[2] A. Beutelspacher, *Einführung in die endliche Geometrie I. Blockpläne*. B.I.-Wissenschaftsverlag, Mannheim - Wien - Zürich, 1982.

[3] P. Dembowski, *Finite Geometries*. Springer-Verlag, 1968.

[4] A. Shamir, *How to share a secret*. Comm. ACM Vol. **22** (1), 612-613 (1979).

[5] G.J. Simmons, *How to (really) share a secret*. To appear in Proc. of Crypto '88.

[6] D.J.A. Welsh, *Matroid theory*. Academic Press, London, New York, San Francisco, 1976.

KEY MINIMAL AUTHENTICATION SYSTEMS
FOR UNCONDITIONAL SECRECY

Philippe Godlewski[*]   Chris Mitchell[**]

Abstract

This paper is concerned with cryptosystems offering unconditional secrecy. For those perfect secrecy systems which involve using key just once, the theory is well established since Shannon's works ; however, this is not the case for those systems which involve using a key several times. This paper intends to take a rigorous approach to the definition of such systems. We use the basic model for a security code developped by Simmons, initially for unconditional authentication. We consider the definition of perfect L-fold secrecy given by Stinson and used by De Soete and others. We consider other definitions : Ordered Perfect L-fold secrety and Massey's Perfect L-fold secrecy, and attempt to classify them. Lower bounds are given for the number of keys in such perfect systems, and characterisation of systems meeting these lower bounds are obtained. The last part of the paper is concerned with discussing examples of key minimal systems providing unconditional secrecy.

1. SCOPE AND PURPOSE

Two of the main applications of cryptography are the provision of secrecy and/or authentication for messages. In 1949 Shannon, [Shan1], showed how to construct systems offering unconditional secrecy, i.e. theoretically perfect secrecy systems, at the expense of the use of very large key spaces. Following this work on secrecy, Simmons, [Simm1], and others, [Bric1], [Gilb1], have considered systems which offer unconditional authentication, again at the expense of requiring very large numbers of keys.

In fact, most practical security systems are not theoretically secure, and could be broken given unlimited computational resources. Such practical security systems are based on reasonable assumptions about the difficulty of certain computational problems, and have the advantage of using manageable numbers of keys.

Nevertheless, unconditionally secure systems do find a use in certain special applications, e.g. the Washington-Moscow 'Hot Line', [Mass1]. It is also interesting to note that, although such 'perfectly secure' systems have been studied for nearly 40 years, the theory is not fully developed, at least in the public domain. It is the purpose of this paper to contribute to the development of this theory.

In particular, it attempts to classify a number of different definitions of perfect secrecy. Developing from this discussion of definitions, lower bounds are given for the number of keys in such perfect systems, and theorems characterising systems meeting these lower bounds are obtained ('key-minimal systems') are obtained. The last part of the paper is concerned with discussing examples of key-minimal systems providing unconditional secrecy.

2. NOTATION

We use the basic model for a security code developed by Simmons, [Simm1], and used by Brickell, [Bric1], De Soete, [DeSo1], [DeSo2] and Stinson, [Stin1], [Stin2].

[*]   Département réseaux, Télécom Paris, 46 rue Barrault
75634 Paris cedex 13 France
[**] Hewlett-Packard Ltd. Filton Road, Stoke Gifford,
Bristol BS12 6QZ, England.

In this model there are three parties: the transmitter, T, the receiver, R, and an opponent, O. The transmitteT wishes to send R one or more pieces of information $s \in S$ in such a way that they cannot be read (secrecy) and/or modified/impersonated (authentication) by O. Users T and R achieve this by using a secret, pre-agreed encoding rule $e \in E$; this encoding rule e may be regarded as the cryptographic transformation corresponding to a secret key. It is always assumed that O knows the system (i.e. the Security Code) completely, the only secret is the encoding rule (i.e. the key) in use. Then T emits $m = e(s)$ which is actually transmitted and, perhaps, intercepted by O. The objective is to design a scheme which protects T and R from O.

More formally, a Security Code consists of three sets: a set S of *Source States,* a set M of encoded messages and a set E of encoding rules. Each encoding rule e is an injective function from S into M (we do not allow *splitting* here). We write k for $|S|$, v for $|M|$ and b for $|E|$ throughout, and, following De Soete, [DeSo1], [DeSo2], write $SC(k,v,b)$ for a Security Code with k source states, v encoded messages and b encoding rules.

We consider various probabilities. We write $p_S(s)$, $p_E(e)$, and $p_M(m)$ for the a priori probabilities of occurrence of source state, encoding rule and message. We suppose $p_S(s)>0$ and $p_E(e)>0$ for every $s \in S$, and $e \in E$.. We write $p_{SIM}(s|m)$, $p_{MIS}(m|s)$, $p_{SIE}(s|e)$, ... for the conditional probabilities. We also abuse this notation slightly by writing $p_S(\underline{s})$ for L-tuple (or ordered set) $\underline{s} = (s_1, s_2, ...s_L)$ or $p_S(S')$ for L-set (unordered) $S' = \{s_1, s_2, ...s_L\}$ in S. We assume that encoding rule e and the different source states $s_1, s_2, ...s_L$ are chosen independently.

A set M' of messages is **allowable** if $p_M(M')>0$. In other words, M' is allowable iff M' could correspond to a set of messages encoded under a single encoding rule.

3. DEFINITIONS OF 'PERFECT' SECRECY
The initial problem that needs to be overcome in a formal study of cryptosystems providing unconditional or 'perfect' secrecy is the fact that existing definitions vary. Therefore, before attempting to study such systems we review the existing definitions, and indicate the relationships between them.

The first definition we give is a slightly modified version of a definition due to Stinson, [Stin1], [Stin2].

Given $L \geq 1$, an SC(k,v,b) is said to provide **Unordered Perfect L-fold secrecy (U(L)-secrecy)** if for every allowable L-subset M' of M and for every L-subset S' of S:
$p_{SIM}(S'|M') = p_S(S')$.

The second definition we give is the unmodified form of Stinson's definition, [Stin1], [Stin2].

Given $L \geq 1$, an SC(k,v,b) is said to provide **Stinson Perfect L-fold secrecy (S(L)-secrecy)** if for every allowable L'-subset M' of M and for every L''-subset S' of S $(L'' \leq L' \leq L)$:
$p_{SIM}(S'|M') = p_S(S')$.

The following lemma is immediate from the definitions:

<u>Lemma 3.1</u> An SC(k,v,b) provides S(L)-secrecy if and only if it provides U(L')-secrecy for every L' satisfying $1 \leq L' \leq L$.

However U(L)-secrecy by itself is not sufficient to guarantee S(L)-secrecy. For example, any SC(k,v,b) provides U(k)-secrecy, but will not necessarily provide S(k)-secrecy. Note that both these definitions are concerned with unordered sets of messages. A scheme providing

S(L)-secrecy protects its users against the opponent O gaining any information about the content of a set of L intercepted messages. However, such a scheme will not necessarily prevent O gaining information about the possible orderings of source states corresponding to observed messages. To provide this stronger notion of secrecy requires the use of a scheme satisfying our third definition, as follows:

Given L$\geq$1, an SC(k,v,b) is said to provide **Ordered Perfect L-fold secrecy (O(L)-secrecy)** if for every allowable L-tuple $\underline{m}$ of distinct messages from M and for every L-tuple $\underline{s}$ of distinct source states from S : $p_{S|M}(\underline{s}|\underline{m}) = p_S(\underline{s})$.

It is then straightforward to establish:

Lemma 3.2 If an SC(k,v,b) provides O(L)-secrecy then it also provides O(L')-secrecy for every L' satisfying 1$\leq$L'$\leq$ L.
We also have:

Lemma 3.3 If an SC(k,v,b) provides O(L)-secrecy then it also provides S(L)-secrecy.

Before proceeding to our fourth (and final) definition it is important to note that all the above definitions relate to 'ciphertext-only' attacks. Essentially, they are all concerned with the situation where the opponent O has intercepted L encoded messages and wishes to deduce information about the corresponding set of L source states. We now consider a definition of perfect security (due to Massey, [Mass1]) based on the concept of a 'known plaintext' attack.

Massey defines a known plaintext attack of order i to be an attack where the opponent O has intercepted i valid and distinct plaintext/ciphertext pairs (i.e. source state/encoded message pairs) all encrypted using the same encoding rule, e say. O is also assumed to have a further encoded message, produced using e and distinct from the messages in the i pairs, for which he wishes to obtain information about the corresponding source state. Then the attack will be said to 'succeed' if, for any source state s distinct from the states in the i pairs, the probability that s corresponds to m given the knowledge of the i pairs is different from the a priori probability of s (given that it is known that it differs from the source states contained in the i pairs).

An SC(k,v,b) is said to provide **Massey Perfect L-fold secrecy (M(L)-secrecy)** if, for any $i < L$, the scheme is secure against an order i known plaintext attack.

Note that the above definition is intended to be precisely the same as Massey's except that what we call M(L)-secrecy is what Massey calls Perfect (L-1)-fold secrecy. We have modified the definition so that it corresponds more closely with the other definitions given here. An equivalent definition of M(L)-secrecy, and one that fits more naturally with the other definitions is as follows:

Consider any SC(k,v,b). Let $\underline{s}$ be any i-tuple of distinct source states and let $\underline{s}'$ be the unique (i-1)-tuple derived from $\underline{s}$ by deleting its last entry. Let $\underline{m}$ be any allowable i-tuple compatible with $\underline{s}'$ for some encoding rule $e \in E$. Then, the SC(k,v,b) provides M(L)-secrecy if and only if for every $i\leq$L, and for every $\underline{s}, \underline{s}', \underline{m}$ as above :
$p(\underline{s}|\underline{m},\underline{s}') = p(\underline{s}|\underline{s}')$.

It is perhaps surprising to discover that Massey's definition is no stronger than the previous one. In fact we have:

Theorem 3.4 If an SC(k,v,b) provides O(L)-secrecy then it also provides M(L)-secrecy.

From now on, although it may be a little more powerful, we use the definition of O(L)-secrecy rather than that of M(L)-secrecy, since it appears to be easier to handle. Before proceeding note also that, for L = 1, all the above definitions coincide and in fact equate to Shannon's notion of perfect secrecy, [Shan1].

4. BOUNDS FOR L-SECURE SYSTEMS

We now consider a variety of bounds which can be established for L-secure systems of various types. We start by considering the weakest form of L-secrecy, namely U(L)-secrecy.

Lemma 4.1 If an $SC(k,v,b)$ provides U(L)-secrecy, then for every allowable L-set of messages M' and for every L-set of source states S' there exists an encoding rule e such that $e(S') = M'$.

It is also straightforward to show:

Lemma 4.2 If an $SC(k,v,b)$ provides U(L)-secrecy, then
$$b \geq |A_L|,$$
where A_L is the set of allowable L-subsets of M.

Using these Lemmas we can now establish the following theorem. Note the bound in this theorem is a special of theorem 5.3 in [DeSo1] for systems providing S(L)-secrecy.

Theorem 4.3 If an $SC(k,v,b)$ provides U(L)-secrecy, then
$$b \geq (v/k).\binom{k}{L},$$
Moreover, if $b = (v/k).\binom{k}{L}$, then:

    (i) For any pair of encoding rules e_1, e_2 either
        $e_1(S) = e_2(S)$
        or $e_1(S)$ and $e_2(S)$ are disjoint.
    (ii) If e_1 and e_2 are encoding rules satisfying
        $e_1(S) = e_2(S)$
        then
        $p_E(e_1) = p_E(e_2) = p_M(M^*)$
        for every M^* in A_L which is also a subset of $e_1(S)$.

We consider examples of schemes possessing U(L)-secrecy in section 5 below. Note that, because S(L)-secrecy implies U(L)-secrecy, the results of Theorem 4.3 also apply to S(L)-secure systems.

If we now consider O(L)-secrecy, then we get a similar set of results as follows:

Lemma 4.4 If an $SC(k,v,b)$ provides O(L)-secrecy then for every allowable L-tuple of distinct messages $\underline{m}$ and for every L-tuple of distinct source states $\underline{s}$ there exists an encoding rule e such that
$$e(\underline{s}) = \underline{m}.$$

It is also straightforward to show:

Lemma 4.5 If an $SC(k,v,b)$ provides O(L)-secrecy, then
$$b \geq |O_L|,$$
where O_L is the set of allowable L-tuples of distinct elements of M.

Using these Lemmas we can now establish the following result. Note that the bound in this theorem was previously established by Massey (equation (5), [Mass1]).

Theorem 4.6 If an $SC(k,v,b)$ provides O(L)-secrecy, then
$b \geq v.(k-1)!/(k-L)!$.
Moreover, if $b = v.(k-1)!/(k-L)!$ then:

(i) For any pair of encoding rules e_1, e_2 , either

$e_1(S) = e_2(S)$

or $e_1(S)$ and e_2 (S) are disjoint.

(ii) If e_1, and e_2 are encoding rules satisfying

$e_1(S) = e_2$ (S)

then

$p_E(e_1) = p_E(e_2) = p_M(\underline{m})$

for every $\underline{m}$ in O_L for which all elements in $\underline{m}$ are in $e_1(S)$.

5. EXAMPLES

We will consider some examples of L-secure systems for which the numbers of encoding rules meet the lower bounds established in section 4 above. It is of interest to construct such systems since, for any security system, it is always desirable to minimise the number of encoding rules and hence the key size. We shall divide our examples into two categories; namely those satisfying the bounds of Theorems 4.3 and 4.6 respectively.

We consider (t,w) homogeneous or transitive set of permutations, Latin Square, Perpandicular Array (using results in [Mull1]), Orthogonal Arrays of Type II ([Rao1]), to characterize and to construct schemes achieving the bounds.

REFERENCES

[Bric1] E.F. Brickell, "A few results in message authentication", *Congressus Numerantium* **43** (1984) 141-154.

[DeSo1] M. De Soete, "Some constructions for authentication-secrecy codes", paper given at *Eurocrypt 88*.

[DeSo2] M. De Soete, "Bounds and constructions for authentication-secrecy codes", paper given at *Crypto 88*.

[Gilb1] E.N. Gilbert, F.J. MacWilliams and N.J.A. Sloane, "Codes which detect deception", *Bell System Technical Journal* **53** (1974) 405-424.

[Mass1] J.L. Massey, "Cryptography - a selective survey", in: *Digital Communications, editors: C. Biglieri and C. Prati*, Elsevier (North-Holland), 1986, pp. 3-21.

[Mull1] R.C. Mullin, P.J. Schellenberg, G.H.J. van Rees and S.A. Vanstone, "On the construction of perpendicular arrays", *Utilitas Mathematica* **18** (1980) 141-160.

[Rao1] C.R. Rao, "Combinatorial arrangements analogous to orthogonal arrays", *Sankhya Series A* **23** (1961) 283-286.

[Shan1] C.E. Shannon, "Communication theory of secrecy systems", *Bell System Technical Journal* **28** (1949) 656-715.

[Simm1] G. Simmons, "Authentication theory/coding theory", in: *Advances in Cryptology: Proceedings of Crypto 84*, Springer-Verlag (Berlin), 1985, pp. 411-431.

[Stin1] D.R. Stinson, "A construction for authentication/secrecy codes from certain combinatorial designs", in: *Advances in Cryptology: Proceedings of Crypto 87*, Springer-Verlag (Berlin), 1988, pp. 355-366.

[Stin2] D.R. Stinson, "Some constructions and bounds for authentication codes", *Journal of Cryptology* **1** (1988) 37-51.

Section 8

Sequences

PARALLEL GENERATION
OF RECURRING SEQUENCES

Christoph G. Günther
Asea Brown Boveri
Corporate Research
CH-5405 Baden, Switzerland

ABSTRACT

In applications, such as radar ranging or test pattern generation, linear recurring sequences are needed at rates that require a parallel generation of the sequences. Two parallelisation methods for the generation of these sequences are discussed and previous results are made applicable to arbitrary degrees of parallelisation and arbitrary sequences. In particular, a previously known technique (sometimes called windmill technique) is shown to be explainable in a very simple way and to be equally appropriate for the parallelisation of non-linear recursions. The method is, furthermore, shown to be suitable for VLSI-realisations and software implementations.

I. INTRODUCTION

Linear recurring sequences for high rate applications can often not be generated directly by the associated linear feedback shiftregister (LFSR), at least not at reasonable costs. Examples of such applications are radar ranging [1], direct sequence spread spectrum (DSSS) communication [2], test pattern generation [3], stream cipher cryptography [4] and decoding by error trapping [5]. In some of these examples, such as radar ranging or the encryption of TV-pictures, the problem is a technological one, which can in principle be solved by faster, low price and low consumption semiconductors. For many applications, however, the difficulty is of a more fundamental nature, since the linear recurring sequences are not needed at a higher rate on an absolute scale but relative to other rates. Examples are the hashing of the data bits by the chipping sequence in DSSS communication [2] or the clocking of the driven register in a binary rate multiplier [6].

Consequently, methods have been developed for the parallel generation of several phases of a decimated sequence. These methods can be subdivided into two classes: those that generate one phase with an appropriate LFSR and construct the other phases by the use of a linear feedforward network [7]-[11] (Fig. 1.a) and those that generate at once all phases in parallel by the use of a network of interconnected shift registers [12]-[17] (Fig. 1.b). Warlick and Hershey have based their approach to the latter class of generators on a windmill-like arrangement of LFSR's [15]. For this reason such generators are sometimes called windmill generators. As we wish to emphasize the characteristics of the two types of generators introduced above, we will, however, prefer to call the generators, from Fig. 1.a and Fig. 1.b, *parallel feedforward (PFF)* and *parallel feedback generators (PFB)*, respectively.

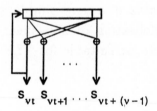

Fig. 1.a. A parallel feedforward generator

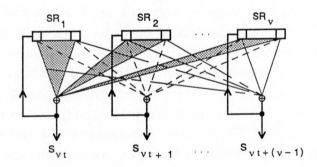

Fig. 1.b. A parallel feedback generator

The main purpose of the paper is to present a simple and intuitive explanation of the PFB-scheme and to give construction rules for such generators which are valid for arbitrary recursions and arbitrary degrees of parallelism (Sec. III).

The importance of these generators is due to their structure: They show a high degree of periodicity. This makes them very suitable for VLSI realisations and for software implementations.

The PFF-generators do not seem to have a corresponding periodicity in their structure and there is also no obvious generalisation to arbitrary recursions. In the linear case, however, we will see in Sec. II that they are as universal as the PFB-generators.

II. PARALLEL FEEDFORWARD GENERATORS

The decimation by ν of a sequence s is a sequence $\tilde{s}$ defined by $\tilde{s}_t :=$ $s_{\nu t}$. In the present context it is natural to also consider $\tilde{s}_t^{(i)} := s_{\nu t+i}$ with $i \in \{0, 1, \ldots, \nu - 1\}$. The objective is to generate the sequences $\tilde{s}^{(0)}, \tilde{s}^{(1)}, \ldots, \tilde{s}^{(\nu-1)}$ in parallel.

We note, that since the period of any sequence decimated by ν divides $T/\gcd(\nu, T)$, there are at most $\gcd(\nu, T)$ different decimated sequences.

Consider sequences s with an irreducible minimal polynomial of period T and degree l. For the decimations $\tilde{s}$ of s by ν, Zierler [18] has proved that the minimal polynominal of $\tilde{s}$ is irreducible, of period $\tilde{T} = T/\gcd(T, \nu)$ and of degree $\tilde{l} = \min\{l : \tilde{T} \mid 2^l - 1\}$ if $\tilde{s} \neq 0$.

Based on this observation, Lempel and Eastman [7] have proposed to generate $\tilde{s}^{(0)}, \tilde{s}^{(1)}, \ldots, \tilde{s}^{(\nu-1)}$ in parallel by the use of ν LFSR's. This was the first step towards PFF-generators.

The first PFF-generator was then introduced by Möhrmann [8]. He has noted that the decimation of an m-sequence s by $\nu = 2, 4, \ldots, 2^{l-1}$ does not change the minimal polynomial and has observed that the sequences $\tilde{s}^{(i)}$ can be obtained by linearly combining $\tilde{s}_{t-l+1}, \tilde{s}_{t-l+2}, \ldots, \tilde{s}_t$. (An m-sequence is a linear recurring sequence of maximal period.) Eier and Malleck [9] have further formalised these results and have indicated a general construction scheme for the feedforward network.

Surböck and Weinrichter[10] have also extended the previous results. They have shown how sequences with irreducible polynomials and non-prime periods can be generated in parallel by PFF-generators of lengths shorter than the original LFSR's if $\nu \mid T$.

For arbitrary rates $\nu \in \mathbf{N}$, we prove in the following that $\tilde{s}_t^{(0)}$, $\tilde{s}_t^{(1)}, \ldots, \tilde{s}_t^{(\nu-1)}$ is a linear function of $\tilde{s}_{t-\tilde{l}+1}, \tilde{s}_{t-\tilde{l}+2}, \ldots, \tilde{s}_t$, with $\tilde{l}$ bounded from above by the linear complexity l of s. (The linear complexity of s is the length of the shortest linear recursion that can generate s.) This statement is obtained in three steps: i) the sequences $\tilde{s}^{(0)}, \tilde{s}^{(1)}, \ldots, \tilde{s}^{(\nu-1)}$ all satisfy the same linear recursion (theorem 1), ii) this recursion has a length $\tilde{l}$ which is bounded from above by l (corollary 2) and iii) there exists an initial condition $\tilde{s}_0, \tilde{s}_1, \ldots, \tilde{s}_{\tilde{l}-1}$ for $\tilde{s}$ such that $\tilde{s}_t^{(0)}, \tilde{s}_t^{(1)}, \ldots, \tilde{s}_t^{(\nu-1)}$ can be obtained by linearly combining $\tilde{s}_{t-\tilde{l}+1}, \tilde{s}_{t-\tilde{l}+2}, \ldots, \tilde{s}_t$ (lemma 3).

Theorem 1 extends Zierler's result [18] to arbitrary sequences and describes the generating polynomial $\tilde{p}(z)$ of the decimated sequence explicitly. For the formulation and the later proof of theorem 1, we define α_k and α_i to be μ-conjugate $(k, i \in J)$ if their images under $z \rightarrow z^\mu$ are conjugate, i.e., if there exists j such that $\alpha_i^\mu = \alpha_k^{\mu 2^j}$. With

$$\tilde{J}_k := \{i \in J : \alpha_i \text{ is } \mu - \text{conjugate to } \alpha_k\}$$

and

$$\tilde{J} := \{k \in J : k = \text{smallest element in } J_k\},$$

this defines a unique partition of J:

$$J = \cup_{k \in \tilde{J}} \tilde{J}_k.$$

Theorem 1. Let $p(z) = \prod_{i \in J} p_i(z)^{m_i}$ be the decomposition of $p(z)$ into distinct irreducible factors and let $p_i(z) = \prod_{j=0}^{d_i-1}(z - \alpha_i^{2^j})$ be the canonical representation of the irreducible factors. For $\nu = 2^\kappa \mu \in \mathbf{N}$, with $2 \nmid \mu$, let

$$\tilde{d}_k = \min\{d : \frac{(2^{d_k} - 1)}{\gcd(\mu, 2^{d_k} - 1)} |2^d - 1\}(|d_k), \tag{1}$$

and let $\tilde{m}_k = \max_{i \in \tilde{J}_k} \lceil \frac{m_i}{2^\kappa} \rceil$. Then the following three assertions hold:

a. The polynominal $\tilde{p}(z)$ *of smallest degree, that generates all decimations* $\tilde{s}^{(i)}$, $i \in \{0, \ldots, \nu - 1\}$ *of all sequences s* which have minimal polynomial $p(z)$ is given by

$$\tilde{p}(z) = \prod_{k \in \tilde{J}} \tilde{p}_k(z)^{\tilde{m}_k}, \tag{2}$$

$$\tilde{p}_k(z) = \prod_{j=0}^{\tilde{d}_k - 1} (z - \alpha_k{}^{\mu 2^j}). \tag{3}$$

b. If ν is relatively prime to the period T of s, we have $\tilde{J} = J$, $\tilde{d}_k = d_k$ and $\tilde{m}_k = m_k$.

c. In this case, the polynomial $\tilde{p}(z)$ *is minimal for all decimations* $\tilde{s}^{(i)}, i \in \{0, \ldots, \nu - 1\}$ *of all sequences* with minimal polynomial $p(z)$.

As far as we could trace it back, Duvall and Mortick [19] were the first to prove that $\prod_{k \in J} \tilde{p}_k(z)^{\tilde{m}_k}$ can generate $\tilde{s}^{(0)}$. Recently, a very elegant and compact proof of the assertion that $\prod_{k \in J} \tilde{p}_k(z)^{m_k}$ generates $\tilde{s}^{(i)}, \forall i$ was given by Niederreiter [20]. Smeets [16] has proved the strongest result, so far, *i.e.*, $\tilde{p}(z)$ generates $\tilde{s}^{(i)}, \forall i$. Based on Niederreiter's result, we give a compact proof of this assertion and a proof of the theorem in the appendix.

The following corollary is an immediate consequence of theorem 1.

Corollary 2. Let $\tilde{s}$ be a sequence s decimated by ν. Then its linear complexity $\tilde{l}$ is related to the linear complexity l of s by

$$\tilde{l} \leq l. \tag{4}$$

If ν is relatively prime to T, we have

$$\tilde{l} = l. \tag{5}$$

By theorem 1, the sequences $\tilde{s}^{(0)}, \tilde{s}^{(1)}, \ldots, \tilde{s}^{(\nu-1)}$ can thus be generated by ν identical LFSR's, each of maximal length $\tilde{l}$. The following lemma shows that in fact only one such LFSR is needed:

Lemma 3. Let $\tilde{p}(z)$ be an arbitrary polynomial of degree $\tilde{l}$, then there is a sequence $\tilde{r}$ generated by $\tilde{p}(z)$ such that any sequence $\tilde{s}$ generated by $\tilde{p}(z)$ can be expressed as

$$\tilde{s}_t = \sum_{i=0}^{\tilde{l}-1} \tilde{\sigma}_i \tilde{r}_{t-i}. \tag{6}$$

The proof of this lemma is easily obtained from observing that the vector of initial conditions $(\tilde{r}_{\tilde{l}-1}, \tilde{r}_{\tilde{l}-2}, \ldots, \tilde{r}_0) = (1, 0, \ldots, 0)$ is transformed into $\tilde{l}$ linearly independent vectors by applying the linear recursion.

The feedforward network is now derived from the initial conditions $s_0, s_\nu, \ldots, s_{\nu(\tilde{l}-1)}; \ldots; s_{\nu-1}, s_{2\nu-1}, \ldots, s_{\nu\tilde{l}-1}$ for $\tilde{s}^{(0)}; \tilde{s}^{(1)}; \ldots; \tilde{s}^{(\nu-1)}$. These are obtained by the use of the original linear recursion and s_0, $s_1, \ldots, s_{l-1}$. The maximal depth of the XOR-tree needed to implement the network is $\log_2 \tilde{l} (\leq \log_2 l)$ and the maximum number of XOR's is $\nu \log_2 \tilde{l} (\leq \nu \log_2 l)$.

In this context, we also observe that the number of memory cells is equal to $\tilde{l} (\leq l)$. Thus, if we compare the parallel implementation by a PFF-generator with the serial implementation by an LFSR, we note that only the number of logical gates is increased by a factor ν.

III. PARALLEL FEEDBACK GENERATORS

Two different forms of the generator shown in Fig. 1.b have been introduced, one by Hsiao [12] and the other by Hurd [13], by Maritsas, Arvillias and Bounas [14], and later by Warlick and Hershey [15]. Hsiao has obtained his generator by considering iterates T^ν of the transfer matrix T, associated with the linear recursion $s_t = \sum_{i=1}^{l} \pi_i s_{t-i}$:

$$T = \begin{pmatrix} 0 & 1 & 0 & \ldots & 0 \\ \vdots & \ddots & \ddots & \ddots & \vdots \\ 0 & \ldots & 0 & 1 & 0 \\ 0 & \ldots & 0 & 0 & 1 \\ \pi_l & \ldots & \pi_3 & \pi_2 & \pi_1 \end{pmatrix}. \tag{7}$$

This simple construction method was, however, not adopted by many authors, probably for two reasons: first the electronic components in the late 60's were not suitable for practical implementations of this construction and secondly the method did not necessarily provide structured generators. In their papers, Hurd, and Maritsas, Arvillias and Bounas, correspondingly used a completely different approach. They started with a particular, well structured generator and searched for conditions under which the given generator yields different phases of a decimated m-sequence. More precisely, they considered a network of cyclically and linearly interconnected shiftregisters as represented in Fig. 2.

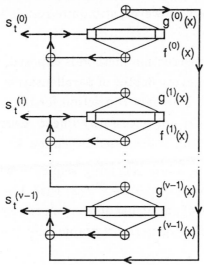

Fig. 2.   Generator with a cyclic arrangement of shiftregisters as considered in [13]-[15]

The sequences obtained from such a network are solutions of the following system of linear recurrence equations

$$f^{(i)}(D) \cdot s_t^{(i)} = g^{(i+1)\mathrm{mod}\nu}(D) \cdot s_t^{(i+1)\mathrm{mod}\nu}, \ i \in \{0, 1, \ldots, \nu - 1\}. \quad (8)$$

Due to the cyclic nature of this system, they are also solutions of

$$p(D) \cdot s_t^{(i)} := \left(\prod_{j=0}^{\nu-1} f^{(j)}(D) + \prod_{j=0}^{\nu-1} g^{(j)}(D) \right) \cdot s_t^{(i)} = 0. \quad (9)$$

The polynomial $p(z)$ thus generates the sequences $s^{(0)}, s^{(1)}, \ldots, s^{(\nu-1)}$. If it is primitive, all sequences are correspondingly phases of one single m-sequence. Warlick and Hershey [15], who started from an even more geometric conception, have obtained similar results for the special case $f_i(z) = 1 + z^t, \ i \in \{0, 1, \ldots, \nu - 1\}; \ g_i(z) = z^l, \ i \in \{0, 1, \ldots, \nu - 2\}$ and $g_{\nu-1}(z) = z^{l+1}$.

In the work of Hurd, of Maritsas, Arvillias and Bounas and of Warlick and Hershey no assertions are made about the relative phases between the sequences $s^{(i)}$. They will usually be non consecutive. This severly limits the applicability of the scheme. Smeets [16] and Smeets and Chambers [17] have indicated sufficient conditions to insure that the sequences $s^{(i)}$ have appropiate phase relations. The assumptions they need in order to prove their assertions are, however, still quite restrictive and their proofs rather involved.

In the following a construction method is exposed, which associates to every sequence s and to every degree of parallelisation ν a PFB-generator for $\tilde{s}^{(0)}, \tilde{s}^{(1)}, \ldots, \tilde{s}^{(\nu-1)}$. The only restriction for the sequence s is that it can be implemented by a not necessarily linear recursion. The simplest case of such a construction is discussed in lemma 4.

Lemma 4. If s is a linear recurring sequence with a characteristic polynomial of the form $p(z) = 1 + z^\lambda f(z)$ with $\lambda > 1$ and $\deg p(z) = l$. Then, for any degree of parallelisation $\nu \leq \lambda$, the generation of s can be parallelised with a PFB-generator, graphically constructed according to algorithm A.

Algorithm A: (see also Fig. 3)

a) draw the indices $0, 1, \ldots, l+\nu-1$ on a line and their values modulo ν underneath,

b) draw the recursions for $s_l, \ldots, s_{l+\nu-1}$,

c) distribute the memory cells of the shiftregister on ν lines according to the partition induced by the indices modulo ν (the cells on line i will contain successive values of $\tilde{s}^{(i)}$),

d) interconnect the successive cells in each line and determine the new content of the first cell of line i by use of the linear recursion for s_{l+i}, drawn in step b).

The idea behind this lemma is the following: due to the particular form of the feedback polynomial the elements s_{t+i}, $i \in \{0, 1, \ldots, \nu - 1\}$ do not depend on the elements s_{t+j}, $j \in \{0, 1, \ldots, i - 1\}$ and, therefore, the linear recursion for these new elements can be evaluated in parallel. The construction of these recursions is straightforward and leads, e.g., to the description given by algorithm A. The completion of these remarks to a proof is trivial.

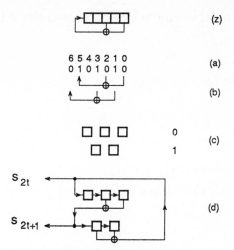

Fig. 3. Algorithm A applied to the linear recursion $p(z) = 1 + z^3 + z^5$ with rate $\nu = 2$. Step (z) shows the LFSR for a serial generation and step (d) the PFB-generator for a generation of two elements in parallel.

We note that in lemma 4 the linearity of the recursion was not used. Furthermore, the independence of the s_{t+i} from the s_{t+j}, with $0 \leq j \leq i - 1$, was not essential either, as the elements s_{t+j}, $0 \leq j \leq i - 1$ can be expressed in terms of the elements s_{t+k}, with $0 \leq k \leq j - 1 \leq i - 2$ and so on:

$$
\begin{aligned}
s_{t+i} &= F(s_{t+i-1}, \ldots, s_{t+i-l}) \\
&= F(F(s_{t+i-2}, \ldots, s_{t+i-l-1}), s_{t+i-2}, \ldots, s_{t+i-l}) \\
&= \ldots \\
&= G_i(s_{t-1}, \ldots, s_{t-l}) , \quad i \in \{0, 1, \ldots, \nu - 1\},
\end{aligned}
\tag{10}
$$

where F is the recursion that generates s, and where G_i is the recursion obtained by iteratively applying F to eliminate all s_{t+j}, $0 \leq j < i$. These remarks lead us immediately to:

Theorem 5. Let F be an arbitrary recursion of length l. Then the generation of any sequence s, associated with F, can be parallelised with a PFB-generator up to an arbitrary degree of parallelisation. The corresponding generator can be constructed using either algorithm B, which requires l memory cells, or algorithm B'. In the latter case the generator has a total number of memory cells d bounded from above by $d \leq l + \nu$.

Algorithm B: (see also Fig. 4.a)

a) unchanged,

b) draw the recursion G_i for s_{l+i}, $i \in \{0, \ldots, \nu - 1\}$, i.e., draw the recursion with the values of s_{l+j}, $0 \leq j < i$ eliminated by the use of the recursion F,

c) draw ν outputs on ν lines and distribute the l memory cells of the shiftregister on the $\min\{l, \nu\}$ upper lines, according to the partition induced by the indices modulo ν,

d) unchanged.

Algorithm B': (see also Fig. 4.b)

a) draw the indices $0, 1, \ldots, l+2\nu-2$ on a line and their values modulo ν underneath,

b) draw the recursion $G_{\nu-1}$ for $s_{l+\nu-1+i}$, $i \in \{0, 1, \ldots, \nu - 1\}$, i.e., draw the recursion with the values of $s_{l+j-1+i}$, $j \in \{0, \ldots, \nu - 1\}$ eliminated by the use of the recursion F,

c) draw ν outputs on ν lines and distribute the memory cells ($\leq l + \nu - 1$) of the shiftregister associated with $G_{\nu-1}$, according to the partition induced by the indices modulo ν,

d) unchanged.

After completion of the contruction, the generators B and B' are loaded with the initial conditions, $s_0, \ldots, s_{l-1}$ and $s_{l+\nu-d}, \cdots s_{l+\nu-1}$, respectively, where $s_l, \ldots, s_{l+\nu-1}$ is obtained by repeated application of the recursion F.

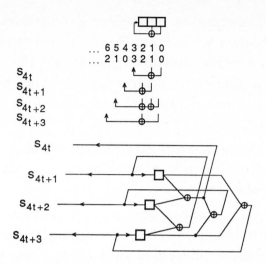

Fig. 4.a. Generator obtained by applying algorithm B to $p(z) = 1 + z^2 + z^3$ and $\nu = 4$.

In the linear case, algorithm B' is equivalent to the contruction of a polynomial $q(z) = 1 + z^\nu f(z)$ which is a multiple of $p(z)$ and has no tabs in the ν leading positions. It is obvious that such a polynomial generates the sequence s if applied to the correct initial conditions. By Lemma 4 this polynomial can, furthermore, be used to generate $\tilde{s}^{(0)}, ..., \tilde{s}^{(\nu-1)}$ in parallel. A complete proof of the theorem is an easy consequence of remarks made prior to its formulation.

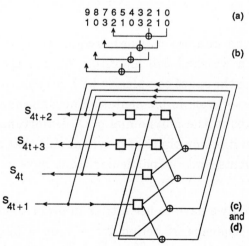

Fig. 4.b. Generators obtained with algorithm B' are always well structured. Algorithm B' was applied to the same example as in Fig. 4.a.

Hsiao's result, described at the beginning of the paragraph, is a particular case of theorem 5/algorithm B, which correspondingly does not necessarily lead to highly structured generators. The structure of the generators is much improved with algorithm B', which, however, requires up to ν additional memory cells and a corresponding extension of the initial conditions.

It is obvious that generators constructed according to algorithm B' can easily be implemented in VLSI-technique and are also well suited for software implementations.

In software implementations, when $\nu \nmid l$, the structure obtained from algorithm B' is further improved by matching the length of the recursion with a multiple of the wordlength w. In the typical case $l \geq w$, it is done by changing the recursion $s_{t+\nu-1} = G_{\nu-1}(s_{t-1}, \ldots, s_{t-i}, \ldots, s_{t-l})$ into $s_{t+\nu-1} = G_{\nu-1}(s_{t-1}, \ldots, F(s_{t-i-1}, \ldots s_{t-i-l}), \ldots, s_{t-l})$ with $i + l + \nu - 1 = nw$, $n \in \mathbf{N}$, $i \in \{1, \ldots, l\}$. In the case $l < w$ this transformation is applied repeatedly. If F is linear, the tranformation corresponds to a change of the polynomial $q(z) = \gamma(z)p(z) = 1 + q_\nu z^\nu + \cdots + q_{l+\nu-1}z^{l+\nu-1}$ into $q'(z) = (\gamma(z) + z^{nw-l})p(z) = q(z) + z^{nw-l}p(z)$ with $nw - l \geq \nu$. Clearly, these recursions have the required form, and the smallest n which fulfills the above inequalities is

$$n = \lceil \frac{l + \nu}{w} \rceil. \tag{11}$$

This is exactly the number of words required anyway. Therefore the adaptation, which we shall call algorithm C, can always be performed without increasing the space complexity with respect to algorithm B'. The complexity of the software implementation becomes mainly dependent on the number of tabs in the polynomial $q(z)$. Unfortunatly, no results are known for constructing multiples of polynominals that have only few non-vanishing coefficients and, in particular, that have ν vanishing leading coefficients.

In this situation we have to use a trick in order to generate in parallel 32 bits of an m-sequence of period $2^{32} - 1$ on a 32-bit machine. We choose a polynomial $p(z) = 1 + z^{25} + z^{29} + z^{30} + z^{32}$, which has the fewest possible number of tabs and no tabs in the 16 leading positions. (We note that when $l \equiv 0 \pmod 8$, no irreducible trinomials exits [21].) Then by squaring $p(z)$ we obtain a polynomial $q(z) = (p(z))^2 = 1 + z^{50} + z^{58} +$

$z^{60} + z^{64}$, which has the same small number of tabs and no tabs in the 31 leading positions. This polynomial $q(z)$ can be used to write a very simple assembler program which generates the sequence wordwise.

IV. SUMMARY AND CONCLUSION

Recurring sequences are often needed at rates that cannot easily be achieved by sequential generation of the elements. Two methods for the parallel generation were correspondingly described: the parallel feedforward (PFF) and the parallel feedback method (PFB).

The parallel feedforward method, which was previously known as the more universal method, could be shown to be applicable to all linear recurring sequences. The parallel feedback method is even more performant: it is not only applicable to linear recurring sequences but to any sequence with a minimal recursion of reasonable length, and is highly suitable for VLSI- and for software-implementations.

In hardware implementations, both methods have a space complexity that is increased by a factor ν with respect to the number of logical gates and that remains essentially unchanged with respect to the number of memory cells. The effective gain in time complexity, $i.e.$, the effective rate, is $\nu\tau_{sr}/\max\{\tau_{sr}, \tau_{gate}\log_2 l\}$, where τ_{sr} and τ_{gate} are the time delays introduced by the shift register cells and the gates, respectively. In software implementations the parallel generation of whole words by the PFB-method is possible at practically no costs in space complexity and no costs in time complexity (rate= ν). This is due to the wordwise processing capability of arithmetic logic units (ALU's).

Finally, we note that parallel feedback generators are easily constructed by either of the algorithms, depending on the implementation intended.

ACKNOWLEDGMENT

I would like to thank Professor James L. Massey for his interest and encouragement.

APPENDIX: Proof of Theorem 1

In order to prove the minimality of $\bar{p}(z)$, we need two preliminary results: theorem A.1 and lemma A.2. With

$$\binom{\xi}{\eta}_2 = \begin{cases} \binom{\xi}{\eta} & \text{if } \xi \geq \eta \\ 0 & \text{else,} \end{cases} \qquad (A.1)$$

and

$$\text{Tr}(\xi) := \sum_{j=0}^{d-1} \xi^{2^j}, \quad \text{if } \xi \in \text{GF}(2^d) \qquad (A.2)$$

theorem A.1 reads:

Theorem A.1. (Milne-Thomson [22] and Ward [23], see also Key [24], Herlestam [25])

Let $p(z)$ be as in theorem 1 then all sequences s generated by the polynomial $p(z)$ have the form

$$s_t = \sum_{i \in J} \sum_{n=1}^{m_i} \binom{2^l - n}{t}_2 \text{Tr}(\gamma_{i,n} \alpha_i{}^t) \qquad (A.3)$$

and the period

$$T = 2^{\max_{i \in J} \lceil \log_2 m_i \rceil} \text{lcm}\{T_i\}_{i \in J}, \qquad (A.4)$$

with $\gamma_{i,n} \in \text{GF}(2^{d_i})$, α_i a root of $p_i(z)$ and T_i the period of α_i. Furthermore, $p(z)$ is the minimal polynomial of s if and only if $\gamma_{i,m_i} \neq 0$, $\forall i \in J$.

For the formulation of lemma A.2, we need the Hasse-derivative. It is defined as follows [26]:

$$D_\zeta^{(j)} \sum_{k=0}^{\infty} a_k \zeta^k := \sum_{k=j}^{\infty} a_k \binom{k}{j} \zeta^{k-j}, \qquad (A.5)$$

and is formaly equivalent to $\frac{1}{j!}(\frac{d}{d\zeta})^j$.

Lemma A.2. Let $\nu = 2^\kappa \mu$ with $2 \nmid \mu$, let $\tilde{l} := l - \kappa$, $\tilde{n} := \lceil \frac{n}{2^\kappa} \rceil$ and let ω be such that $\mu\omega \equiv 1 \mod 2^{\tilde{l}}$. Then

$$\binom{2^l - n}{\nu t}_2 = \sum_{i=0}^{\tilde{n}} c_i \binom{2^{\tilde{l}} - i}{t}_2 \qquad (A.6)$$

where

$$c_i := D_\zeta^{(2^{\tilde{l}}-i)}(1 + (\zeta - 1)^\omega)^{2^{\tilde{l}}-\tilde{n}}|_{\zeta=0}. \qquad (A.7)$$

and $c_1 = 1$.

Proof: Let $x = \sum_i x_i 2^i$ and $y = \sum_i y_i 2^i$ then Lukas' theorem states $\binom{x}{y} = \prod_i \binom{x_i}{y_i}$. This implies

$$\binom{2^l - n}{2^\kappa \mu t}_2 = \binom{2^{\tilde{l}} - \tilde{n}}{\mu t}_2. \qquad (A.8)$$

We thus have only to prove

$$\binom{2^{\tilde{l}} - \tilde{n}}{\mu t}_2 = \sum_{i=0}^{2^{\tilde{l}}-1} c_i \binom{2^{\tilde{l}} - i}{t}_2 \qquad (A.9)$$

and $c_i = 0$, $\forall i > \tilde{n}$. We consider the generating functions of these expressions:

$$\sum_{t=0}^{2^{\tilde{l}}-1} \binom{2^{\tilde{l}} - \tilde{n}}{\mu t}_2 z^t = (1 + z^\omega)^{2^{\tilde{l}}-\tilde{n}} \qquad (A.10)$$

and

$$\sum_{t=0}^{2^{\tilde{l}}-1} \sum_{i=0}^{2^{\tilde{l}}-1} c_i \binom{2^{\tilde{l}} - i}{t}_2 z^t = \sum_{i=0}^{2^{\tilde{l}}-1} c_i (1 + z)^{2^{\tilde{l}}-i}. \qquad (A.11)$$

The substitution $\zeta = 1 + z$ leads to

$$\sum_{i=0}^{2^{\tilde{l}}-1} c_i \zeta^{2^{\tilde{l}}-i} = (1 + (\zeta - 1)^\omega)^{2^{\tilde{l}}-\tilde{n}} \qquad (A.12)$$

and by taking the $(2^{\tilde{l}} - i)$-th Hasse derivative in $\zeta = 0$:

$$c_i = D_\zeta^{(2^{\tilde{l}}-i)}(1 + (\zeta - 1)^\omega)^{2^{\tilde{l}}-\tilde{n}}|_{\zeta=0} \qquad (A.13)$$

In modulo 2 arithmetic this derivative vanishes $\forall i > \tilde{n}$. For $i = \tilde{n}$ we have

$$c_{\tilde{n}} \equiv \omega^{2^{\tilde{l}}-\tilde{n}} \equiv 1 \bmod 2 \qquad (A.14)$$

since $2 \nmid \omega$. This concludes the proof of lemma A.2.

Theorem A.1 and lemma A.2 lead to an explicit representation of the decimated sequence $\tilde{s}^{(0)}$ and with some simple additions to an explicit representation of all decimated sequences. This could have been used to prove theorem 1. We shall, however, extract most results directly from the polynominal, which turns out to be much simpler.

Proof of theorem 1. a. We first prove that $\tilde{p}(z)$ can generate all decimations $\tilde{s}^{(i)}$. Through a very elegant approach, Niederreiter [20] has shown that

$$\tilde{p}(z) \ \bigg| \ \prod_{i \in J} \prod_{j=0}^{d_i-1} (z - \alpha_i^{\nu 2^j})^{m_i}. \tag{A.15}$$

By regrouping terms, $\tilde{p}(z)$ can thus be written in the following form

$$\tilde{p}(z) = \prod_{k \in \tilde{J}} \tilde{p}_k(z)^{m'_k}. \tag{A.16}$$

with an m'_k to be determined.

Let D and $\tilde{D}$ denote the time shift on the undecimated and decimated sequence, respectively. Then

$$\tilde{p}(\tilde{D})\tilde{s}^{(i)}|_t = \tilde{p}(D^\nu)s|_{\nu t+i}. \tag{A.17}$$

The left hand side (lhs) of this equation is zero by definition. Since s has minimal polynomial $p(z)$, a necessary and sufficient condition for this to hold on the right hand side (rhs) is

$$p(z) \mid \tilde{p}(z^\nu)$$

$$= \prod_{k \in \tilde{J}} \prod_{j=0}^{\tilde{d}_i-1} (z^\mu - \alpha_k^{\mu 2^j})^{2^\kappa m'_k} \tag{A.18}$$

Now $\prod_{i \in \tilde{J}_k}(z - \alpha_i^{2^j})|z^\mu - \alpha_k^{\mu 2^j}$ holds since every factor of the lhs divides the rhs and since these factors are relatively prime. The divisibility condition (A.18) is thus satisfied whenever $m_i \leq 2^\kappa m'_i$ and $\tilde{m}_i = \lceil \frac{m_i}{2^\kappa} \rceil$ is the smallest such m'_i.

Next we prove the existence of a sequence s such that $\tilde{s}^{(0)}$ has minimal polynomial $\tilde{p}(z)$. Since $\gcd(\tilde{p}_k(z), \tilde{p}_{k'}(z)) = 1$ if $k \neq k'$, we can restrict ourselves to $p(z) = \prod_{i \in \tilde{J}_k} p_i(z)^{m_i}$ and $\tilde{p}(z) = \tilde{p}_k(z)^{\tilde{m}_k}$. We consider

$$s_t = \sum_{i \in \tilde{J}_k} \binom{2^l - m_i}{t}_2 \text{Tr}(\gamma_i \alpha_i^t) \tag{A.19}$$

with $\gamma_i \neq 0$, which by theorem A.1, has minimal polynomial $p(z)$. By using lemma A.2 the decimation $\tilde{s}_i^{(0)}$ becomes

$$\tilde{s}_i^{(0)} = s_{\nu t} = \sum_{i \in \tilde{J}_k} \sum_{j=0}^{\tilde{m}_i} c_j \binom{2^{\tilde{l}} - j}{t}_2 \mathrm{Tr}(\gamma_i \alpha_k{}^{\mu t}). \qquad (A.20)$$

Define the set of factors with maximal multiplicity $\tilde{I}_k := \{i \in \tilde{J}_k : \tilde{m}_i \geq \tilde{m}_j, \ \forall\, j \in \tilde{J}_k\}$ and the sequence $\tilde{u}$ associated with these factors

$$\tilde{u}_t := \binom{2^{\tilde{l}} - \tilde{m}_k}{t}_2 \sum_{i \in \tilde{I}_k} \mathrm{Tr}(\gamma_i \alpha_k{}^{\mu t}), \qquad (A.21)$$

then by theorem A.1, $\tilde{s}^{(0)} - \tilde{u}$ is generated by $\tilde{p}_k(z)^{\tilde{m}_k - 1}$. If

$$\sum_{i \in \tilde{I}_k} \mathrm{Tr}_{\mathrm{GF}(2^{d_i})|\mathrm{GF}(2^{\tilde{d}_k})}(\gamma_i) := \sum_{i \in \tilde{I}_k} \sum_{j=0}^{\frac{d_i}{d_k} - 1} \gamma_i{}^{2^{\tilde{d}_k j}} \neq 0, \qquad (A.22)$$

the same theorem implies that $\tilde{p}_k(z)^{\tilde{m}_k}$ is minimal for $\tilde{u}$ and thus also for $\tilde{s}^{(0)} = \tilde{u} + (\tilde{s}^{(0)} - \tilde{u})$. The condition can always be fullfilled, $e.g.$, by choosing a set $\{\gamma_i\}_{i \in \tilde{J}_k}$ with $\mathrm{Tr}(\gamma_i) = 1$ for one index i and $\mathrm{Tr}(\gamma_{i'}) = 0$, $\forall i' \in \tilde{J}_k \setminus \{i\}$.

b. We prove the assertions for ν relatively prime to T. $\tilde{J} = J$ and $\tilde{d}_k = d_k$ are trivial. For the proof of $\tilde{m}_k = m_k$ we note that

$$T = 2^{\max_{i \in J} \lceil \log_2 m_i \rceil} \cdot \mathrm{lcm}\{T_i\}_{i \in J}, \qquad (A.23)$$

with T_i the period of α_i (theorem A.1). Thus $\gcd(\nu, T) = 1$, $i.e.$,

$$2^{\min\{\kappa, \max_{i \in J} \lceil \log_2 m_i \rceil\}} \gcd(\mu, \mathrm{lcm}\{T_i\}_{i \in J}) = 1 \qquad (A.24)$$

implies either $\kappa = 0$ $i.e.$, $\tilde{m}_k = m_k$ or $\max_i \lceil \log_2 m_i \rceil = 0$, $i.e.$, $m_i = 1$ and $\tilde{m}_i = \lceil \frac{1}{2^\kappa} \rceil = 1 = m_i$.

c. Finally, we have to prove that $\tilde{p}(D)$ is minimal for all decimations $\tilde{s}^{(i)}$ of all sequences s generated by $p(z)$, whenever $\gcd(\nu, T) = 1$. By the primality condition, there is an e such that $\nu^e \equiv 1 \bmod T$, $i.e.$, an e-fold decimation by ν leads back to the original sequence. Now, degree $\tilde{p}(z) \leq$ degree $p(z)$, for every one of the decimations

$$deg\, p(z) \geq deg\, \tilde{p}(z) \geq deg\, \tilde{\tilde{p}}(z) \geq \ldots deg\, p(z) \qquad (A.25)$$

which is only possible if all degrees are equal.

REFERENCES

[1] W.M. Siebert, "A Radar Detection Philosophy," *IRE Trans. Inform. Theory,* vol. IT-2, pp. 204-221, (1956).

[2] R.L. Pickholtz, D.L. Schilling, L.B. Milstein, "Theory of Spread-Spectrum Communications - A Tutorial," *IEEE Trans. Commun.,* vol. COM-30, pp. 855-884, May 1982.

[3] P.H. Bardell, W.H. McAnney, "Pseudorandom Arrays for Built-In Tests," *IEEE Trans. Comput.,* vol. C-35, pp. 653-658, July 1986.

[4] R.A. Rueppel, *Analysis and Design of Stream Ciphers,* Springer-Verlag, Berlin, Heidelberg 1986.

[5] T. Kasami, "A Decoding Procedure for Multiple-Error-Correcting Cyclic Codes," *IEEE Trans. Inform. Theory,* vol. IT-10, pp. 134-138, April 1964.

[6] W.G. Chambers, S.M. Jennings, "Linear Equivalence of Certain BRM Shift-Register Sequences," *Electron. Lett.,* vol. 20, pp. 1018-1019, Nov. 1984.

[7] A. Lempel, W.L. Eastman, "High Speed Generation of Maximal Length Sequences," *IEEE Trans. Comput.,* vol. C-20, pp. 227-229, Feb. 1971.

[8] K.H. Möhrmann, "Erzeugung von binären Quasi-Zufallsfolgen hoher Taktfrequenz durch Multiplexen," *Siemens Forsch.- u. Entwickl.-Ber.,* vol. 3, pp. 218-224 (1974).

[9] R. Eier, H. Malleck, "Anwendung von Multiplextechniken bei der Erzeugung von schnellen Pseudozufallsfolgen," *NTZ,* vol. 28, pp. 227-231 (1975).

[10] F. Surböck, H. Weinrichter, "Interlacing Properties of Shift-Register Sequences with Generator Polynomials Irreducible over $GF(p)$,"
IEEE Trans. Inform. Theory, vol. IT-24, pp. 386-389, May 1978.

[11] C.F. Woodcock, P.A. Davies, "The Generation of High-Speed Binary Sequences from Interleaved Low-Speed Sequences," *IEEE Trans. Commun.,* vol. COM-35, pp. 115-117, Jan. 1987.

[12] M.Y. Hsiao, "Generating PN Sequences in Parallel," *Proc., 3rd. Ann. Princeton Conf. Inform. Sci. Syst.,* Princeton NJ., pp. 397-401, March 1969.

[13] W.J. Hurd, "Efficient Generation of Statistically Good Pseudonoise by Linear Interconnected Shift Registers," *IEEE Trans. Comput.,* vol. C-23, pp. 146-152, Feb. 1974.

[14] D.G. Maritsas, A.C. Arvillias, A.C. Bounas, "Phase-Shift Analysis of Linear Feedback Shift Register Structures Generating Pseudorandom Sequences," *IEEE Trans. Comput.,* vol. C-27, pp. 660-668, July 1978.

[15] W.W. Warlick Jr., J.E. Hershey, "High-Speed M-Sequence Generators," *IEEE Trans. Comput.,* vol. C-29, pp. 398-400, May 1980.

[16] B.J.M. Smeets, *Some Results on Linear Recurring Sequences,* Ph.D. Thesis, Technical University of Lund, Sweden (1987).

[17] B.J.M. Smeets, W.G. Chambers, "Windmill Generators: A Generalization and an Observation of How Many There Are," *Advances in Cryptology - EUROCRYPT'88,* Lect. Notes in Comp. Science, vol. 330, pp. 325-330, Springer-Verlag (1988).

[18] N. Zierler, "Linear Recurring Sequences," *J. Soc. Indust. Appl. Math.,* vol. 7, pp. 31-48, March 1959.

[19] P.F. Duvall, J.C. Mortick, "Decimation of Periodic Sequences," *SIAM J. Appl. Math.,* vol. 21, pp. 367-372, Nov. 1971.

[20] H. Niederreiter, "A Simple and General Approach to the Decimation of Feedback Shift-Register Sequences," *Probl. of Control and Inform. Theory,* vol. 17, pp. 327-331 (1988).

[21] R.G. Swan, "Factorization of Polynomials over Finite Fields," *Pac. J. Math.,* vol. 12, pp. 1099-1106 (1962).

[22] L.M. Milne-Thomson, *The Calculus of Finite Differences,* Macmillan and Co., London 1951.

[23] M. Ward, "The Arithmetical Theory of Linear Recurring Series," *Trans. Am. Math. Soc.,* vol. 35, pp. 600-628, Jan. 1933.

[24] E.L. Key, "An Analysis of the Structure and Complexity of Nonlinear Binary Sequence Generators," *IEEE Trans. Inform. Theory,* vol. IT-22, pp. 732-736, Nov. 1976.

[25] T. Herlestam, "On Functions of Linear Shift Register Sequences," *Advances in Cryptology - EUROCRYPT'85*, Lect. Notes in Comp. Science, vol. 219, pp. 119-129, Springer-Verlag (1986).

[26] H. Hasse, "Theorie der höheren Differentiale in einem algebraischen Funktionenkörper mit vollkommenem Konstantenkörper bei beliebiger Characteristik," *J. reine angew. Math.*, vol. 175, pp. 50-54 (1936).

KEYSTREAM SEQUENCES WITH A GOOD LINEAR COMPLEXITY
PROFILE FOR EVERY STARTING POINT

Harald Niederreiter

Mathematical Institute, Austrian Academy of Sciences
Dr.-Ignaz-Seipel-Platz 2
A-1010 Vienna, Austria

1. INTRODUCTION AND PROBABILISTIC RESULTS

The linear complexity profile was introduced by Rueppel [7], [8, Ch. 4] as a tool for
the assessment of keystream sequences with respect to randomness and unpredictability
properties. In the following let F be an arbitrary field. We recall that a sequence of elements of F is called a kth-order <u>linear feedback shift register</u> (LFSR)
<u>sequence</u> if it satisfies a kth-order linear recursion with constant coefficients from
F. The zero sequence $0,0,\ldots$ is viewed as an LFSR sequence of order 0. Now let S
be an arbitrary sequence $s_1,s_2,\ldots$ of elements of F. For a positive integer n
the (local) <u>linear complexity</u> $L_n(S)$ is defined as the least k such that
$s_1,s_2,\ldots,s_n$ form the first n terms of a kth-order LFSR sequence. The sequence
$L_1(S),L_2(S),\ldots$ of integers is called the <u>linear complexity profile</u> (LCP) of S.
For basic facts about the LCP see [4], [7], [8, Ch. 4].

The study of the LCP leads to the requirement that the LCP of a keystream sequence should simulate the LCP of a random sequence. Typical features of the LCP of
a random sequence were investigated in [5], [7], [8, Ch. 4]. An additional requirement was pointed out by Piper [6], namely that a keystream sequence should have an
acceptable LCP for every starting point. In other words, if S_m is the shifted sequence $s_{m+1},s_{m+2},\ldots,$ then S_m should have an acceptable LCP for every $m = 0,1,\ldots$.
In the present paper we are mainly concerned with this requirement.

A basic technical device in this work is the identification of the sequence S
with its <u>generating function</u> $\sum_{i=1}^{\infty} s_i x^{-i}$, viewed as an element of the field
$G = F((x^{-1}))$ of formal Laurent series in x^{-1} over F (compare also with [4]).
For the probabilistic theory we take F to be the finite field F_q with q elements,
where q is an arbitrary prime power. Note that in practical stream cipher applications we have the binary case $q = 2$. The uniform probability measure on F_q assigns

the measure $1/q$ to each element of F_q. This probability measure induces the complete product probability measure h on the set F_q^∞ of all sequences of elements of F_q by a standard procedure of probability theory (see [2, Sec. 4]). It is easily seen that the measure h is the same as the Haar measure used in [5] when the latter measure is transferred from the set of all generating functions to the set F_q^∞. We say that a stated property of elements of F_q^∞ holds <u>h-almost everywhere</u> (h-a.e.) if the property holds for all elements of a subset of F_q^∞ of h-measure 1. A property that holds h-a.e. can be viewed as a typical property of a random sequence of elements of F_q.

<u>Theorem 1.</u> Let Π be a property of elements of F_q^∞ that holds h-a.e. Then the set of all $S \in F_q^\infty$ for which Π holds simultaneously for all shifted sequences S_m, $m = 0,1,\ldots,$ has h-measure 1.

<u>Proof.</u> Let P be the set of all elements of F_q^∞ which have the property Π and let V be the set of all $S \in F_q^\infty$ for which $S_m \in P$ for all $m = 0,1,\ldots$. Let τ be the unilateral shift operator on F_q^∞ defined by

$$\tau (s_1,s_2,\ldots) = (s_2,s_3,\ldots) \qquad \text{for} \quad (s_1,s_2,\ldots) \in F_q^\infty .$$

Then $S_m = \tau^m S$, thus $S_m \in P$ if and only if S lies in the mth iterated inverse image $\tau^{-m}P$. Therefore $V = \bigcap_{m=0}^{\infty} \tau^{-m}P$. Now τ is measure-preserving with respect to h by [1, Ch. 1], and so $h(\tau^{-m}P) = h(P) = 1$ for all m. Thus we get $h(V) = 1$. $\square$

<u>Theorem 2.</u> If $F = F_q$, then h-a.e. we have

$$\lim_{n \to \infty} \frac{L_n(S_m)}{n} = \frac{1}{2} \qquad \text{for} \quad m = 0,1,\ldots,$$

$$\limsup_{n \to \infty} \inf \frac{L_n(S_m) - (n/2)}{\log n} = \pm \frac{1}{2 \log q} \qquad \text{for} \quad m = 0,1,\ldots .$$

<u>Proof.</u> This follows from Theorem 1 and [5, Theorems 7 and 10]. $\square$

From Theorem 1 one can deduce various other probabilistic results. For example, all the probabilistic results in [5] hold simultaneously for all shifted sequences S_m, with Theorem 2 just covering two instances that are easy to state. In particular, the results about the distribution of partial quotients in the continued fraction expansion of a random generating function S hold simultaneously for all S_m. In a nutshell, Theorem 1 provides the basis for saying that Piper's requirement is met by random sequences of elements of F_q.

2. SEQUENCES WITH A GOOD LINEAR COMPLEXITY PROFILE

We introduce a class of sequences for which the LCP is close to that of a random sequence and for which this property is retained under shifts. Put $\text{Log } t = \max(1, \log t)$ for $t \geq 1$.

Definition 1. A sequence S of elements of the field F has a good LCP if there exists a constant C (which may depend on S) such that

$$|L_n(S) - \frac{n}{2}| \leq C \text{ Log } n \qquad \text{for} \quad n = 1,2,\ldots .$$

If $F = F_q$, then it follows from [5, Theorem 10] that the property of having a good LCP holds h-a.e. For further analysis we need the connection between the LCP and continued fractions as developed in [4]. Every $S \in G$ has a unique continued fraction expansion

$$S = A_0 + 1/(A_1 + 1/(A_2 + \ldots)) = : [A_0, A_1, A_2, \ldots]$$

with $A_j \in F[x]$ for $j \geq 0$ and $\deg(A_j) \geq 1$ for $j \geq 1$. This expansion is finite for rational S and infinite for irrational S. The polynomials $A_j, j \geq 1$, are called the partial quotients and $A_0 = : \text{Pol}(S)$ is the polynomial part of S. The polynomials P_j and Q_j are defined as in [4] and P_j/Q_j is called a convergent. We note that

$$\deg(Q_j) = \sum_{i=1}^{j} \deg(A_i) \qquad \text{for} \quad j \geq 1. \tag{1}$$

Then we have the following formula [4, Theorem 1].

Lemma 1. For any sequence S of elements of F and any $n \geq 1$ we have $L_n(S) = \deg(Q_j)$, where $j \geq 0$ is uniquely determined by the condition

$$\deg(Q_{j-1}) + \deg(Q_j) \leq n < \deg(Q_j) + \deg(Q_{j+1}). \tag{2}$$

Proposition 1. If the generating function of a sequence S is irrational and $\deg(A_j) \leq C \text{ Log } j$ for all $j \geq 1$, where C is a constant, then

$$|L_n(S) - \frac{n}{2}| \leq \frac{C}{2} \text{ Log } n \qquad \text{for all} \quad n \geq 1.$$

Proof. Consider an n satisfying (2) and first assume $j \geq 2$. Then $L_n(S) = \deg(Q_j)$ by Lemma 1, hence using (1) we get

$$|L_n(S) - \tfrac{n}{2}| = |\deg(Q_j) - \tfrac{n}{2}| \leq \tfrac{1}{2} \max(\deg(A_j), \deg(A_{j+1}))$$

$$\leq \tfrac{C}{2} \text{Log}(j+1) \leq \tfrac{C}{2} \text{Log}(\deg(Q_j) + 1) \leq \tfrac{C}{2} \text{Log } n.$$

The proposition is easily checked for $j = 0$ and $j = 1$. $\square$

Proposition 2. If a sequence S satisfies $|L_n(S) - (n/2)| \leq C \text{ Log } n$ for some constant $C \geq 1$ and all $n \geq 1$, then its generating function is irrational and

$$\deg(A_j) < (4C \text{ Log } C + 8C) \text{ Log } j \qquad \text{for all } j \geq 1. \tag{3}$$

Proof. The given condition on $L_n(S)$ implies $\lim_{n \to \infty} L_n(S) = \infty$, and so the generating function of S is irrational. To prove (3) we proceed by contradiction, and we let j be the least index such that the inequality in (3) does not hold. First assume $j = 1$, so that

$$\deg(A_1) \geq 4C \text{ Log } C + 8C.$$

For $n = \deg(A_1) - 1$ we have $n \geq 4C \text{ Log } C + 7C$ and also $L_n(S) = 0$ by Lemma 1. Thus

$$|L_n(S) - \tfrac{n}{2}| = \tfrac{n}{2} \geq \frac{4C \text{ Log } C + 7C}{2 \log(4C \text{ Log } C + 7C)} \log n$$

since the function $t/\log t$ is increasing for $t \geq e$. By distinguishing between the cases $1 \leq C \leq e$ and $C > e$ one shows that

$$4 \text{ Log } C + 7 > 2 \log(4C \text{ Log } C + 7C) \qquad \text{for all } C \geq 1,$$

and so $|L_n(S) - (n/2)| > C \text{ Log } n$, a contradiction. Now let $j \geq 2$. Then with $C_1 = 4C \text{ Log } C + 8C$ we have $\deg(A_i) < C_1 \text{ Log } i$ for $1 \leq i < j$. Together with (1) we get

$$\log \deg(Q_j) < \log(\deg(A_j) + C_1 \sum_{i=1}^{j-1} \text{Log } i) \leq \log(\deg(A_j) + C_1(j-1) \text{ Log } j).$$

Since the function $t^{-1} \log(t + C_1(j-1) \text{ Log } j)$ is decreasing for $t \geq e$ and since $\deg(A_j) \geq C_1 \text{ Log } j$, it follows that

$$\log \deg(Q_j) < \frac{\log(C_1 j \text{ Log } j)}{C_1 \text{ Log } j} \deg(A_j).$$

For $n = \deg(Q_{j-1}) + \deg(Q_j)$ we have by Lemma 1 and (1),

$$|L_n(S) - \frac{n}{2}| = |deg(Q_j) - \frac{1}{2}(deg(Q_{j-1}) + deg(Q_j))| = \frac{1}{2} deg(A_j)$$

$$> \frac{C_1 \, Log \, j}{2 \, log(C_1 j \, Log \, j)} \, log \, deg(Q_j) > \frac{C_1 \, Log \, j}{2 \, log(C_1 j \, Log \, j)} \, log \, \frac{n}{2}.$$

Now $n \geq deg(A_j) + 2 \, deg(A_{j-1}) \geq C_1 \, Log \, j + 2$, and so

$$|L_n(S) - \frac{n}{2}| > \frac{C_1 \, Log \, j}{2 \, log(C_1 j \, Log \, j)} (1 - \frac{log \, 2}{log(C_1 \, Log \, j + 2)}) \, log \, n.$$

Therefore, to arrive at the contradiction $|L_n(S) - (n/2)| > C \, Log \, n$, it suffices to show that

$$(2 \, Log \, C + 4)(1 - \frac{log \, 2}{log(C_1 \, Log \, j + 2)}) \geq \frac{log(C_1 j \, Log \, j)}{Log \, j}. \tag{4}$$

To prove (4), we first consider $j = 2$. Then (4) attains the simpler form

$$(2 \, Log \, C + 4)(1 - \frac{log \, 2}{log(C_1 + 2)}) \geq log(2C_1).$$

We have

$$(2 \, Log \, C + 4)(1 - \frac{log \, 2}{log(C_1 + 2)}) \geq (2 \, Log \, C + 4)(1 - \frac{log \, 2}{log \, 14}) > (1.47)Log \, C + 2.94.$$

For $1 \leq C \leq e$ it follows that

$$(2 \, Log \, C + 4)(1 - \frac{log \, 2}{log(C_1 + 2)}) > 4.41 > log(24e) \geq log(24C) = log(2C_1).$$

Since the function $(0.47)t - log(t+2)$ is increasing for $t \geq 1$, we obtain $(0.47)t + 0.86 > log(t+2)$ for $t \geq 1$, and so

$$(0.47)log \, C + 0.86 > log(log \, C + 2) \quad for \quad C \geq e.$$

It follows that

$$(1.47)log \, C + 2.94 > log \, C + log \, 8 + log(log \, C + 2) = log(2C_1) \quad for \quad C \geq e,$$

hence (4) is shown for $j = 2$. For $j \geq 3$ we have

$$1 - \frac{log \, 2}{log(C_1 \, Log \, j + 2)} \geq 1 - \frac{log \, 2}{log(12 \, log \, 3 + 2)} > 0.74.$$

Since the function $log(C_1 t \, log \, t)/log \, t$ is decreasing for $t \geq e$, we obtain (4) for $j \geq 3$ if we can show that

$$(1.48)\text{Log } C + 2.96 \geq \frac{\log(3C_1 \log 3)}{\log 3}. \tag{5}$$

For $1 \leq C \leq e$ we have

$$(1.48)\text{Log } C + 2.96 = 4.44 > \frac{\log(36e \log 3)}{\log 3} \geq \frac{\log(3C_1 \log 3)}{\log 3},$$

hence (5) holds. The function $(0.62)t - \log(t+2)$ is increasing for $t \geq 1$, thus $(0.62)t + 0.67 > \log(t+2)$ for $t \geq 1$, and so

$$(0.62)\log C + 0.67 > \log(\log C + 2) \quad \text{for } C \geq e.$$

Adding $\log C + \log(12 \log 3)$ on both sides and then dividing by $\log 3$, we obtain (5). $\square$

Theorem 3. A sequence S has a good LCP if and only if its generating function is irrational and there exists a constant C (which may depend on S) such that $\deg(A_j) \leq C \text{ Log } j$ for all $j \geq 1$.

Proof. This follows from Propositions 1 and 2. $\square$

Note that the generating function of S is irrational if and only if S is not an LFSR sequence (see [4, Sec. 2]). We now use the valuation v on G introduced in [4, Sec. 3]. We also write $\text{Fr}(S) = S - \text{Pol}(S)$ for $S \in G$.

Lemma 2. If $S \in G$ and $f, g \in F[x]$ with $f \neq 0$ and $v(fS - g) < -v(f)$, then $f = DQ_j$ and $g = DP_j$ for some $j \geq 0$, where $D \in F[x]$ and $D \neq 0$.

Proof. We have $v(fS - g) < 0$, so [4, Lemma 3] can be applied and yields $f = \sum_{k=0}^{j} D_k Q_k$, $g = \sum_{k=0}^{j} D_k P_k$, where $D_k \in F[x]$, $v(D_k) < v(A_{k+1})$ for $0 \leq k \leq j$, and $D_j \neq 0$; moreover, if i is the least index with $D_i \neq 0$, then $v(fS - g) = v(D_i) - v(Q_{i+1})$. If we had $i < j$, then

$$v(fS - g) \geq v(D_i) - v(Q_j) \geq -v(Q_j) \geq -v(f),$$

a contradiction. Thus $i = j$, hence $f = DQ_j$ and $g = DP_j$ with $D = D_j$. $\square$

Lemma 3. Let $S \in G$ and $f, g \in F[x]$ with $fg \neq 0$. Let $\frac{f}{g}S = [A_0', A_1', A_2', \ldots]$ and let P_j'/Q_j' be the corresponding convergents. If $v(A_j') > v(f) + v(g)$ for some $j \geq 1$, then there exists an $i \geq 1$ such that $v(A_j') \leq v(A_i) + v(f) + v(g)$ and $v(Q_{i-1}) \leq v(Q_{j-1}') + v(f)$.

Proof. By [4, eq. (6)] we have $v(Q'_{j-1} \frac{f}{g}S - P'_{j-1}) = - v(Q'_{j-1}) - v(A'_j)$, hence $v(A'_j) > v(f) + v(g)$ implies $\bar{v}(fQ'_{j-1} S - gP'_{j-1}) < - v(fQ'_{j-1})$. Then Lemma 2 yields $fQ'_{j-1} = DQ_{i-1}$ and $gP'_{j-1} = DP_{i-1}$ for some $i \geq 1$, where $D \in F[x]$ and $D \neq 0$. It follows that $v(Q_{i-1}) \leq v(DQ_{i-1}) = v(Q'_{j-1}) + v(f)$. Moreover,

$$v(A'_j) \leq v(A'_j) - v(Q_{i-1}) + v(Q'_{j-1}) + v(f)$$

$$= - v(Q_{i-1}) - v(fQ'_{j-1} S - g P'_{j-1}) + v(f) + v(g)$$

$$= - v(Q_{i-1}) - v(DQ_{i-1} S - DP_{i-1}) + v(f) + v(g)$$

$$\leq - v(Q_{i-1}) - v(Q_{i-1} S - P_{i-1}) + v(f) + v(g) = v(A_i) + v(f) + v(g),$$

where we used again [4, eq. (6)] in the last step. $\square$

Theorem 4. If S has a good LCP and $f, g \in F[x]$ with $fg \neq 0$, then the sequence with generating function $Fr(\frac{f}{g}S)$ has a good LCP.

Proof. We use the notation in Lemma 3 and recall that $v(f) = \deg(f)$ for any $f \in F[x]$. We have $Fr(\frac{f}{g}S) = [0, A'_1, A'_2, \ldots]$. The hypothesis and Theorem 3 yield $v(A_i) \leq C \log i$ for all $i \geq 1$. Now let $j \geq 1$ be such that $v(A'_j) > v(f) + v(g)$. Then Lemma 3 implies $v(A'_j) \leq C \log i + v(f) + v(g)$ and $i \leq v(Q_{i-1}) + 1 \leq v(Q'_{j-1}) + v(f) + 1$, thus

$$v(A'_j) \leq C \log(v(Q'_{j-1}) + v(f) + 1) + v(f) + v(g). \tag{6}$$

This holds trivially if $v(A'_j) \leq v(f) + v(g)$, and so (6) holds for all $j \geq 1$. Now let $n \geq 1$ be arbitrary, then with a suitable $j \geq 0$ we get by Lemma 1 and (6),

$$\left| L_n(Fr(\frac{f}{g}S)) - \frac{n}{2} \right| \leq \frac{1}{2} \max(v(A'_j), v(A'_{j+1}))$$

$$\leq \frac{C}{2} \log(v(Q'_j) + v(f) + 1) + \frac{1}{2}v(f) + \frac{1}{2}v(g)$$

$$\leq \frac{C}{2} \log(n + v(f) + 1) + \frac{1}{2}v(f) + \frac{1}{2}v(g) \leq C_1 \log n$$

with a suitable constant C_1. $\square$

Corollary 1. If S has a good LCP, then every shifted sequence S_m, $m = 1, 2, \ldots$, has a good LCP.

Proof. The generating function of S_m is given by $Fr(x^m S)$, and so the desired result follows from Theorem 4. $\square$

For $S \in G$ we put $K(S) = \sup_{j \geq 1} \deg(A_j)$. If the sequence S has an irrational generating function and $K(S) < \infty$, then Theorem 3 shows that S has a good
LCP. Thus by Corollary 1, every shifted sequence S_m has a good LCP. More precisely
we have the following.

Theorem 5. Let $S \in G$ with $K(S) < \infty$ and $f, g \in F[x]$ with $fg \neq 0$. Then

$$K(\frac{f}{g}S) \leq K(S) + \deg(f) + \deg(g).$$

Proof. If in the notation of Lemma 3 we have $v(A_j') > v(f) + v(g)$ for some $j \geq 1$,
then by this lemma we get $v(A_j') \leq K(S) + v(f) + v(g)$, and the latter inequality
holds trivially if $v(A_j') \leq v(f) + v(g)$. Recall also that $v(f) = \deg(f)$ for any
$f \in F[x]$. □

Corollary 2. If $K(S) < \infty$, then $K(S_m) \leq K(S) + m$ for $m = 1, 2, \ldots$.

Proof. This follows from Theorem 5 since the generating function of S_m is $Fr(x^m S)$. □

In particular, Corollary 2 shows that if $K(S) < \infty$, then $K(S_m) < \infty$ for all m.
The result of Corollary 2 is in general best possible, as is proved by considering the
following generalized Rueppel sequence S constructed in [3]. Let $F = F_2$ and let
$s_i = 1$ if $i = 2^j - 1$ for some $j \geq 1$ and $s_i = 0$ otherwise. Then $K(S) = 1$ by
[3, p. 232]. Now let $m = 2^j - 1$ for some $j \geq 1$. Then the generating function of
S_m has the form x^{-2^j} + smaller powers, and so the first partial quotient of S_m
has degree 2^j. Thus $K(S_m) \geq 2^j$. On the other hand, Corollary 2 yields $K(S_m) \leq$
$K(S) + m = 2^j$, hence $K(S_m) = 2^j$. Thus for this sequence S we have $K(S_m) = K(S) + m$
for infinitely many m.

3. SEQUENCES WITH A UNIFORMLY GOOD LINEAR COMPLEXITY PROFILE

It follows from Corollary 1 that if S has a good LCP, then for every $m \geq 0$ there
exists a constant C_m (which may depend on S and m) such that

$$|L_n(S_m) - \frac{n}{2}| \leq C_m \log n \quad \text{for all } n \geq 1.$$

In this context it is of interest to consider the following notion.

Definition 2. A sequence S of elements of the field F has a **uniformly good LCP** if
there exists a constant C (which may depend on S but not on m) such that

$$|L_n(S_m) - \frac{n}{2}| \leq C \text{ Log } n \qquad \text{for all} \quad m \geq 0 \quad \text{and} \quad n \geq 1.$$

Theorem 6. If $F = F_q$, then the set of all sequences with a uniformly good LCP has h-measure 0.

Proof. Let W be the set in question. Let the sequence $S \in F_q^\infty$ be such that it contains somewhere a string of zeros of length $r \geq 1$, and let s_{k+1} be the first term of this string. Then $L_r(S_k) = 0$, hence if also $S \in W$, then it follows from Definition 2 that

$$|L_r(S_k) - \frac{r}{2}| = \frac{r}{2} \leq C \text{ Log } r.$$

This implies that r is bounded from above by a constant which depends only on C. Thus $W \subseteq B$, where B is the set of all $S \in F_q^\infty$ for which all strings of zeros have bounded length. We have $B = \bigcup_{r=1}^\infty B_r$, where B_r is the set of all $S \in F_q^\infty$ for which all strings of zeros have length $\leq r$. Fix $r \geq 1$, and for any integer $t \geq 0$ let $B_r^{(t)}$ be the set of all $S \in F_q^\infty$ with the following property: for each $j = 0,1,\dots,t$ at least one of the terms $s_{j(r+1)+1}, s_{j(r+1)+2},\dots,s_{(j+1)(r+1)}$ is $\neq 0$. Then $B_r \subseteq \bigcap_{t=0}^\infty B_r^{(t)}$. The probability that at least one of any $r + 1$ consecutive s_i is $\neq 0$ is given by $1 - q^{-r-1}$, and so $h(B_r^{(t)}) = (1 - q^{-r-1})^{t+1}$. It follows that

$$h(B_r) \leq h(\bigcap_{t=0}^\infty B_r^{(t)}) \leq (1 - q^{-r-1})^{t+1} \qquad \text{for all} \quad t \geq 0,$$

and letting $t \to \infty$ on the right we get $h(B_r) = 0$. This yields $h(B) = 0$, and so $W \subseteq B$ implies $h(W) = 0$. $\square$

Theorem 6 shows that having a uniformly good LCP is <u>not</u> a typical property of a random sequence of elements of F_q. Sequences with a uniformly good LCP can also be characterized in terms of continued fraction expansions. Let $[0,A_1^{(m)},A_2^{(m)},\dots]$ be the continued fraction expansion of the generating function of the shifted sequence S_m, $m = 0,1,\dots$.

Theorem 7. A sequence S has a uniformly good LCP if and only if its generating function is irrational and there exists a constant C (which may depend on S but not on m) such that $\deg(A_j^{(m)}) \leq C \text{ Log } j$ for all $j \geq 1$ and $m \geq 0$.

Proof. This follows from Propositions 1 and 2. $\square$

It is an open question whether there actually exists a sequence with a uniformly good LCP. Another open question is to decide whether there exists a sequence S for which $\sup\limits_{m=0,1,\ldots} K(S_m) < \infty$. A positive answer to the second question will of course imply a positive answer to the first question.

REFERENCES

[1] U. Krengel: Ergodic Theorems, de Gruyter, Berlin, 1985.

[2] M. Loève: Probability Theory, 3rd ed., Van Nostrand, New York, 1963.

[3] H. Niederreiter: Continued fractions for formal power series, pseudorandom numbers, and linear complexity of sequences, Contributions to General Algebra 5 (Proc. Salzburg Conf., 1986), pp. 221–233, Teubner, Stuttgart, 1987.

[4] H. Niederreiter: Sequences with almost perfect linear complexity profile, Advances in Cryptology – EUROCRYPT '87 (D. Chaum and W. L. Price, eds.), Lecture Notes in Computer Science, Vol. 304, pp. 37–51, Springer, Berlin, 1988.

[5] H. Niederreiter: The probabilistic theory of linear complexity, Advances in Cryptology – EUROCRYPT '88 (C. G. Günther, ed.), Lecture Notes in Computer Science, Vol. 330, pp. 191–209, Springer, Berlin, 1988.

[6] F. Piper: Stream ciphers, Elektrotechnik und Maschinenbau 104, 564–568 (1987).

[7] R. A. Rueppel: Linear complexity and random sequences, Advances in Cryptology – EUROCRYPT '85 (F. Pichler, ed.), Lecture Notes in Computer Science, Vol. 219, pp. 167–188, Springer, Berlin, 1986.

[8] R. A. Rueppel: Analysis and Design of Stream Ciphers, Springer, Berlin, 1986.

On the Complexity of Pseudo-Random Sequences - or: If You Can Describe a Sequence It Can't be Random

Thomas Beth   Zong-Duo Dai[1]

Europäisches Institut für Systemsicherheit (EISS)
Universität Karlsruhe (T.H.)
D – 7500 Karlsruhe, F.R. of Germany

"Any one who considers automatical methods of producing random digits is, of course in a state of sin"

$(J.v.Neumann)_{[15]}$

Abstract

We shall prove in this note that the Turing-Kolmogorov-Chaitin complexity and the Linear Complexity are the same for practically all 0-1-sequences of length n, already for moderately large n.

1   Introduction

The availability of reproducible random - looking sequences is a precondition for many applications of modern cryptology, cf.[4,1,13].

Such sequences have to be generated at the sender's and receiver's side by deterministic automata. For this reason the generated sequences are called *Pseudo-Random Sequences* rather than random sequences. A central topic of research is the problem of determining the complexity of a given sequence S of length n, $n \in N$, over a finite alphabet[13,14]. The view

---

[1]On leave from Academia Sinica,Beijing, China

generally generated in approaching this problem is that of a cryptanalyst who, after a plain-text attack [4,1] having obtained a long random-looking string, has to determine a generating automaton A of size α as small as possible, reproducing S from a so-called seed key k as short as possible. The complexity of the sequence S is then the size α plus the length of the seed key k, which is widely considered as a measure of security of the sequence. A common practice is to choose the generating automaton A from the class $\mathcal{L}$ of linear feedback shift registers (LFSR)[1,13,8]. The so-called Berlekamp-Massey-Algorithm [2,3,5,10] gives a fast procedure to construct a generating LFSR of minimal so-called linear complexity for a given sequence of length n.

Amongst those, who have been working in the area of sequence complexity it has been known that the linear complexity measure λ is only one way of bounding the security from above, which sometimes seems to be far out of the practical relevance, as some well chosen examples show, cf.[7]. In order to overcome such difficulties other complexity measures have been considered, to describe the complexity of sequences and sequence generators, cf.[14].

To overcome the difficulties inherent to approaching the complexity problem within a fixed class of automata an alternative way is to consider the wide class of Turing machines (TM), as a universal class of generators instead. As any Turing machine (TM) can be obtained from programming a universal Turing machine (UTM), the length $\mathcal{X}$ of a shortest program for a UTM making it produce a given sequence S is called the Turing-Kolmogorov-Chaitin-complexity of this sequence. In the remainder of this paper we will show that for the complexity measures the relation $\lambda \sim \mathcal{X}$ holds for all but a set of arbitrary low probability of (0-1)-sequences of length n with growing n. This result can be generalized to arbitrary alphabets.

2   Basic Notations and Facts

In the following technical sections we will make use of the standard concepts of algorithms and complexity, cf. [9], which will lead to the notion of automata.

Let n be a positive integer and $2^n = \{0,1\}^n$ denote the set of all binary strings of length n, whereas $2^* = \{0,1\}^* = \bigcup_{n=0}^{\infty} 2^n$ denotes the set of all sequences over the alphabet $\{0,1\}$.

A deterministic automaton M with binary input/output is given as usually as $M = (S, \Sigma, \Omega, \delta, \alpha, \lambda)$, cf. [9] where

S is the set of states with a special initial state α. The input alphabet I and the output alphabet O coincide with $\Sigma = \{0, 1\}$. $\delta : S \times I \to S$ is the next-state-function, which is naturally extended to a mapping $\delta^{(n)} : S \times I^n \to S$ by setting

$$
\begin{aligned}
\delta(s; \underline{\varepsilon}) &:= \delta^n(s; \varepsilon_{n-1}, ..., \varepsilon_0) \\
&= \delta^{n-1}(\delta(s; \varepsilon_0); \varepsilon_{n-1}, ..., \varepsilon_1) \\
&\qquad \text{for } s \in S \text{ and } \underline{\varepsilon} = (\varepsilon_{n-1}, \cdots, \varepsilon_0) \in I^n.
\end{aligned}
$$

$\lambda : \Omega \to O$ denotes the output function, which generates an output, whenever the state has reached a state in the set of final states $\Omega \subset S$.

A state $s \in S$ can be reached from a state $t \in S$, if there exists a sequence of inputs $\underline{\varepsilon} \in \sigma^*$ such that

$$
t = \delta(s, \underline{\varepsilon}).
$$

An output sequence $\underline{x} \in 2^n$ is said to be *generated by* M iff there exists a sequence $(\omega_1, \omega_2, \cdots, \omega_n) \in \Omega^n$ in which ω_{i+1} can be reached from ω_i for $i = 1, 2, \cdots, n - 1$ and where

$$
\lambda(\omega_j) = x_j, \qquad \text{for } j \in [1, 2, \cdots, n].
$$

The set of all sequences $\underline{x} \in O^*$ generated by M will be denoted by $A(M)$, and $A_n(M) = A(M) \cap \Sigma^n$.

The machine is then called a *sequence generator* for all $n \in \mathbf{N}$.

Let $\mu(M)$ denote the complexity of describing the finite state machine M, i.e. $\mu(M) = \text{size}(\delta) + \log_2(|S|)$ where the size (δ) is the number of bits needed to completely specify the next-state-function δ. Let $\mathcal{M}$ be a class of sequence generators (of a certain type). Let

$$
\mu_{\mathcal{M}} : O^n \longrightarrow \mathbf{N} \cup \{\infty\}
$$

be given by

$$\mu_{\mathcal{M}}(\underline{x}) = \min\{\mu(M) | M \in \mathcal{M}, \underline{x} \in A_n(M)\}.$$

If $\mu_{\mathcal{M}}(\underline{x}) \in \mathbf{N}$ we say that the output sequence $\underline{x}$ with respect to the machine type $\mathcal{M}$ affords a description of length $\mu_{\mathcal{M}}(\underline{x})$.

3 The Well-Known Case : Linear Sequence Generators

Let $\mathcal{L}$ be the class of linear feedback shift registers (LFSR), i.e. the class of linear recursions over $GF(2)$. Each linear recursion L is uniquely described by the feedback polynomial $q(z) \in GF(2)[z]$ and an initial state $(x_0, \dots, x_{l-1})$, where $1 = \deg q(x)$. The sequence of length n generated by the linear automaton L is $(x_0, \dots, x_{l-1}, x_l, \dots, x_{n-1})$, where

$$x_{t+l} = \sum_{i=0}^{l-1} c_i \cdot x_{t+i}, 0 \le i < n - l$$

and $q(x) = x^l + \sum_0^{l-1} c_1 x^i$.

The complexity $\mu(L)$ of describing L within the class $\mathcal{L}$ of LFSR is $\mu(L) = 2 \cdot l$.

For a given sequence $\underline{x} \in 2^n$ we denote by

$$\lambda(\underline{x}) := \mu_{\mathcal{L}}(\underline{x})$$

the *linear complexity* of the sequence $\underline{x}$, which in other words is twice the length of the shortest linear feedback shift register which could have "produced" $\underline{x}$.

Theorem 3.1 *For any given $\underline{x} \in 2^n$ the Berlekamp-Massey-algorithm will compute $\lambda(\underline{x}), p(z)$, and $q(z)$ in $O(n^2 \log n)$ bit operations.*

Proof. This is the well-known result from shift-register analysis, cf. [5,10,13].

□

Theorem 3.2 *Let $k \in [0 : n]$. Let $N_n(k) := |\{\underline{s} \in 2^n | \lambda(\underline{s}) = 2k\}|$. Then*

$$N_n(k) = \begin{cases} 2^{\min(2n-2k, 2k-1)} & \text{if } n \ge k > 0 \\ 1 & \text{if } k = 0. \end{cases}$$

Proof. [13].

4 The Most General Case: Turing Machine As Sequence Generator

Let $\mathcal{T}$ be the class of Turing machines (TM).

A Turing Machine TM can be considered as a finite state machine with an additional infinite tape as work space of tape symbols $Z = \{0, 1, \#\}$, which can be read from or printed to via a tape head that itself is controlled by the finite state control through move-commands +1(left), −1(right) or 0(stop). The formal description extends that of a finite state machine

$$T = (S, I, Z; \delta, \alpha, \Omega).$$

The next-state function δ is extended to a next-move function

$$\delta : S \times I \times Z \to S \times (Z \times \{\mathbf{Z}_3\})$$

which from a given state $\sigma \in S$, an input symbol $i \in I$ and a read symbol $z \in Z$ computes

$$\delta(\sigma, i, z) = (\sigma', z', m)$$

i.e. it changes the state from σ into σ', write a new tape symbol and moves the tape head by $m \in \mathbf{Z}_3$ (left, right or not at all). The work tape is upon start of the TM empty, i.e. filled with blanks $\#$.

The concept of a Turing Machine can now be used to construct the so-called Universal Turing Machine (UTM), which in essence is a general purpose Turing Machine capable of simulating any other TM as given above.

The Universal Turing Machine is a Turing Machine U whose finite state control is designed such that it generates output strings $\underline{x} \in \{0, 1\}^*$ of the form $\underline{y} = (\underline{x}, \underline{p}(T))$ where $\underline{p}(T)$ is an input equal to the binary program which makes U behave like the Turing machine T generating x.

Theorem 4.1 *Any Turing machine T can be simulated in this form by a universal Turing machine U.*

Proof. [9]

Definition 4.2 *Let $\underline{x} \in 2^n$ be a binary sequence. Let T be a given Turing machine generating x. Let $\underline{p}(T)$ be the program , which makes this UTM simulate T.*

Then the number

$$\mu(T) := |\underline{p}(T)|$$

i.e. the length $|\underline{p}|$ of the program $\underline{p}$ is called the size of T.

The number

$$\mathcal{X}(\underline{x}) := \mu_T(\underline{x}) = \min\{\mu(T)|T \text{ generates } \underline{x}, T \in \mathcal{T}\}$$

is called the Turing- Kolmogorov-Chaitin complexity *of the sequence $\underline{x}$.*

Theorem 4.3 *The function $\mathcal{X}$ generally is not computable (by a Turing machine)*

Proof. Suppose $\mathcal{X}$ is computable. Construct a Turing Machine T_k which generates and inspects binary sequences in say lexicographical order until a sequence $\underline{x}$ of complexity $\mathcal{X}(\underline{x}) > K$ larger than the constant K is found and then accepts this sequence $\underline{x}$.

The size of this machine is given by

$$\mu(T) = O(\log K) \,,$$

so that the fact $\mu(T) < K$ for large K creates a contradiction.

$\square$

In spite of this result we derive the following result on the average behaviour of $\mathcal{X}$:

Theorem 4.4 *Let $k \in [1 : n]$. Then*

$$|\{\underline{s} \in 2^n|\mathcal{X} \le k\}| \le 2^{k+1} \,.$$

Proof. Amongst the 2^{k+1} Turing machines T with $\mu(T) \le k$ there are at most 2^{k+1} sequence generators.

$\square$

Corollary 4.5 *On the space 2^n of all 0-1-sequences with equidistributed probability the following holds for all ε, $0 < \varepsilon < 1$:*

$$\text{Prob}\{\underline{s} \in 2^n | \mathcal{X}(\underline{s}) > (1 - \varepsilon)n\} > 1 - 2^{-\varepsilon n + 1} .$$

This corollary shows that practically all sequences of moderately large length have a Turing-Kolmogorov-Chaitin complexity close to the length of the sequence.

In other words this is intuitively clear:

A "true" random sequence of length n has no shorter description than just the sequence itself.

In the next section we shall investigate the relation between the TKC-complexity measure $\mathcal{X}$ and the linear complexity λ.

5 The Main Result

Theorem 5.1 *For any real number ε, $0 < \varepsilon < 1$, arbitrary integer n, and an arbitrary string $\underline{s} \in 2^n$, we have*

1. $\text{Prob}\{(1 - \varepsilon)\lambda(\underline{s}) \leq \mathcal{X}(\underline{s})\} \geq 1 - \frac{8}{3} 2^{-\frac{\varepsilon}{2-\varepsilon}n}$,

2. $\text{Prob}\{(1 - \varepsilon)\mathcal{X}(\underline{s}) \leq \lambda(\underline{s})\} \geq 1 - \frac{1}{3} 2^{-\varepsilon n + (1-\varepsilon)(1+\log n)+1} - \frac{1}{3} 2^{-n}$

Lemma 5.2 *For any positive real numbers t and u, such that*

$$n + t < 2n, \qquad u < n,$$

we have

1. $\#\{s \in 2^n | n + t < \lambda(s)\} \leq \frac{1}{3} 2^{n-t+1} - \frac{1}{3}$,

2. $\#\{s \in 2^n | \mathcal{X}(s) < n - u\} \leq 2^{n-u+1}$,

3. $\#\{s \in 2^n | \lambda(s) \leq n + t, n - u \leq \mathcal{X}(s)\} \geq 2^n - \frac{1}{3} 2^{n-t+1} - 2^{n-u+1}$,

4. $\text{Prob}\{\lambda(s) \leq n + t, n - u \leq \mathcal{X}(s)\} \geq 1 - \frac{1}{3} 2^{-t+1} - 2^{-u+1}$.

Proof.

The assertion "2." and the proposition "1. and 2. $\Rightarrow$ 3. $\Rightarrow$ 4." are trivial. Thus we need only to prove 1. By Theorem 3.2 we have

$$
\begin{aligned}
\#\{s \in 2^n | n + t < \lambda(s)\} &= \sum_{n+t<2i\leq 2n} 2^{2n-2i} \\
&= \sum_{\substack{\lambda=\lfloor\frac{n+t}{2}\rfloor \\ 2\lambda+2\leq 2i\leq 2n}} 2^{2n-2i} \\
&= 2^{2n-2\lambda-2} \sum_{0\geq j\geq n-\lambda-1} 2^{-2j} \\
&= 2^{2n-2\lambda-2} \cdot \frac{1 - 2^{-2(n-\lambda)}}{3/4} \\
&= \frac{1}{3} \cdot 2^{2n-2\lambda} - \frac{1}{3} \\
&\leq \frac{1}{3} \cdot 2^{n-t+1} - \frac{1}{3}.
\end{aligned}
$$

$\square$

Lemma 5.3 *For any positive real number t, we have*

1. $\#\{s \in 2^n | 0 \leq \lambda(s) < n - t\} \leq \frac{1}{3} \cdot 2^{n-t+1} + \frac{1}{3}$,

2. $Prob\{n - t \leq \lambda(s)\} \geq 1 - \frac{1}{3}2^{-t+1} - \frac{1}{3} \cdot 2^{-n}$.

Proof. 2. is the consequence of 1. We need only to prove 1.

$$
\begin{aligned}
\#\{s \in 2^n | 0 \leq \lambda(s) < n - t\} &= 1 + \sum_{1\leq 2i<n-t} 2^{2i-1} \\
&= 1 + \sum_{\substack{\lambda=\lceil\frac{n-t}{2}\rceil \\ 2\leq 2i\leq 2\lambda-2}} 2^{2i-1} \\
&= 1 + 2 \sum_{0\leq i\leq \lambda-2} 2^{2j} \\
&= 1 + 2 \cdot \frac{2^{2(\lambda-1)} - 1}{3} \\
&\leq \frac{1}{3} \cdot 2^{n-t+1} + \frac{1}{3}.
\end{aligned}
$$

$\square$

Proof (of the Theorem 5.1)

1. Take $t := \frac{\varepsilon}{2-\varepsilon}n$, then $(1-\varepsilon)(n+t) = n - t$. Now we have

$$\text{Prob}\{(1-\varepsilon)\lambda(s) \le \mathcal{X}(s)\}$$
$$\ge \text{Prob}\{\lambda(s) \le n+t, (1-\varepsilon)(n+t) \le \mathcal{X}(s)\}$$
$$= \text{Prob}\{\lambda(s) \le n+t, n-t \le \mathcal{X}(s)\}$$
$$\ge 1 - \frac{1}{3} \cdot 2^{-t+1} - 2^{-t+1}$$
$$= 1 - \frac{8}{3} \cdot 2^{\frac{-\varepsilon}{2-\varepsilon}n} \qquad \text{by Lemma 5.2.}$$

2. Take $t = \varepsilon n - (1-\varepsilon)\lceil \log n\rceil$, then $(1-\varepsilon)(n+\lceil\log n\rceil) = n - t$.

$$\text{Prob}\{(1-\varepsilon)\mathcal{X}(s) \le \lambda(s)\}$$
$$\ge \text{Prob}\{\mathcal{X}(s) \le n + \lceil\log n\rceil, (1-\varepsilon)(n+\lceil\log n\rceil) \le \lambda(s)\}$$
$$= \text{Prob}\{n - t \le \lambda(s)\}$$
$$\ge 1 - \frac{1}{3}2^{-\varepsilon n+(1-\varepsilon)(1+\log n)+1} - \frac{1}{3}2^{-n} \qquad \text{(by lemma 5.3)}$$

$\square$

The effect of theorem 5.1 can be written in simplified form as

Corollary 5.4 *For all* ε $(0 < \varepsilon < 1)$

$$P_{\varepsilon,n} := \text{Prob}\{\underline{s} \in 2^n | (1-\varepsilon) \cdot 2\lambda(\underline{s}) \le \mathcal{X}(\underline{s}) \le (1+\varepsilon) \cdot 2\lambda(\underline{s})\} \to 1$$

when $n \to \infty$.

Numerical example 5.5
For $\varepsilon = 0.01$ we have for sequence lengths n.

n	2000	4000
$P_{\varepsilon,n}$	0.998	0.999997.

6 Infinite Sequences

Following Niederreiter [12] we denote by $(\Omega, \mathcal{F}, m, \tau)$ the dynamical system on $\Omega = \{0,1\}^\infty$ with the one–sided Bernoulli–shift τ on the product space with the unique Haar–measure m (identical to the product measure).

For $\underline{s} \in \Omega$ and $m, n \in \mathbb{N}$ with $m \le n$ we denote $\underline{s}_m^n = (s_m, \dots, s_n)$.

Especially let $\underline{s}^n := \underline{s}_0^n$.

Then we formulate the above results in terms of probability as follows.

Theorem 6.1 *For $\underline{s} \in \Omega$*

$$\lim_{n \to \infty} \frac{\mathcal{X}(\underline{s}^n)}{\lambda(\underline{s}^n)} = 2$$

m–almost everywhere.

Proof. Apply the Borel–Cantelli lemma (cf. [6]) to the independant cylinder sets

$$A_{k,\varepsilon} := \{\underline{s} \in \Omega | (1-\varepsilon) \cdot 2\lambda(\underline{s}_{2^{k}-1}^{2^{k}-1}) \leq \mathcal{X}(\underline{s}_{2^{k}-1}^{2^{k}-1}) \leq (1+\varepsilon) \cdot 2\lambda(\underline{s}_{2^{k}-1}^{2^{k}-1})\}$$

for $k \in \mathbf{N_0}$ and $\varepsilon > 0$.

From corollary 5.4. we conclude that

$$\sum_{k=1}^{\infty} m(A_{k,\varepsilon}) = \infty$$

for all $k \in \mathbf{N_0}$ and $\varepsilon > 0$.

Thus the assertion.

$\square$

We are indebted to Prof. Harald Niederreiter for suggesting this formulation of our results.

7 Conclusions

The results of section 6 show that for almost all sequences, which have a short Turing Machine description, cannot be considered as random sequences.

The introduction of the Turing machine concept shows as a consequence that there is no way by which a large subset of truly random looking sequences in 2^n can be generated by a comparatively small finite state machine. In other words, the results show that the use of so called non-linear generators is not gaining much in terms of the complexity of sequences, if the complexity of the generators is evaluated properly.

The implications to the design of new, highly complex secure encipherment algorithms cf. [11] will have to be investigated further, especially in view of circuit or VLSI complexity.

References

[1] Beker, H., Piper, F.: *Cipher Systems*, Northwood, 1982.

[2] Berlekamp, E.R.: *Coding Theory*, McGraw–Hill, 1968.

[3] Beth, T., Gollmann, D.: *Vorlesungsmanuskript: Signale, Codes und Chiffren*, 1988

[4] Beth, Heß, Wirl: *Kryptographie*, Teubner 1983.

[5] Blahut, R.E.: *Theory and Practice of Error Control Codes*, Addison-Wesley, 1983.

[6] Breimàn, L.: *Probability*, Addison–Wesley, 1968.

[7] Fumy, W., Rieß, H.P.: *Kryptographie*, Oldenburg, 1988.

[8] Gollmann, D.: *Kaskadenschaltungen taktgesteuerter Schieberegister als Pseudozufallsgeneratoren*, Dissertation Universität Linz, 1983.

[9] Hopcroft, F.E.; Ullman, F.D.: *Introduction to Automata Theory, Languages and Computation*, Addison-Wesley, 1979.

[10] Massey, J.L.: *Shift Register Synthesis and BCH Decoding*, IEEE Trans. on Information Theory, **15** (1969), pp 122–127.

[11] Micali, S.; Schnorr, C.P.: *Efficient, Perfect Random Number Generators*, June 1988.

[12] Niederreiter, H.: *The Probabilistic Theory of Linear Complexity*, Proc. Eurocrypt 88, LNCS 330, p. 191 – 209.

[13] Rueppel, R.: *Analysis and Design of Stream Ciphers*, Springer, 1986.

[14] Schnorr,C.P.: *On the Construction of Random Number Generators and Random Function Generators*, Eurocrypt 1988, Davos.

[15] Schnorr,C.P., Stimin,H.: *Endliche Automaten und Zufallsfolgen*, Acta Informatica **1**, 345-359, (1972).

[16] Chaitin,G.J.: *Information, Randomness and Incompleteness*, World Scientific Publishing, Singapore 1987.

FEEDFORWARD FUNCTIONS
DEFINED BY DE BRUIJN SEQUENCES

Z.D.Dai[1)]   K.C.Zeng[2)]

[1)] Universität Karlsruhe, Institut für Algorithmen und
Kognitive Systeme
on leave from the Graduate School of USTC, Academia Sinica
[2)] Data and Communications Security Research Center
Graduate School of USTC, Academia Sinica

Abstract

In this paper, we show that feedforward functions defined by de
Bruijn sequences, called de Bruijn functions, satisfy some basic cryptographic requirements. It is shown how the family of de Bruijn feedforward functions could be parametrized by a key space. De Bruijn
feedforward functions are balanced and complete. A lower bound of the
degree of de Bruijn functions is given. A certain correlational weakness
of a class of de Bruijn functions is analyzed and an algebraic method to
meliorate the weakness is also given and it will not cause any substantial
drawbacks withregard to other requirements. The lower bound given in
this paper is by no means discouraging, yet there is hope for substantial
improvements. So, improving the given lower bound is proposed as an
open problem at the end of this paper.

In cryptosystems based on nonlinear feedforward functions of linear feedback shift-register sequences, one applies non-linear functions of the form

$$(a_i, a_{i+1}, \ldots, a_{i+n-1}) \longmapsto c_i = f_k(a_i, a_{i+1}, \ldots, a_{i+n-1})$$

to an LSR-sequence

$$\alpha_k = (a_0, a_1, \ldots, a_i, \ldots,), \quad a_i \in GF(2)$$

to produce the corresponding output sequences

$$\gamma_k = (c_0, c_1, \ldots, c_i, \ldots), \quad c_i \in GF(2).$$

The function f_k, as is well known, can be expressed in an unique way as a polynomial in the indeterminates x_i, $0 \le i \le n-1$, linear in each of them seperately. f_k should satisfy certain cryptographic requirements. The following is a list of the simplest ones among them.

1. The function f_k with value range $\{0,1\}$ should be balanced, i.e. it will assume the value zero exactly 2^{n-1} times over the space $V_n(GF(2))$ of binary sequences of a certain length n, and complete, i.e. it will contain each of the n indeterminates explicitly.

2. It should have a reasonably high total degree, so as to guarantee a high enough linear complexity for the output sequence.

3. It should be free from certain correlational weaknesses which may influence negatively the cryptograghic strength of the output sequence.

4. The family of feedforward functions used should be parametrized by a space $V_l(GF(2))$, in the sense that to every vector k in the space there will correspond a function f_k in the family, and different k's correspond to different functions. The sequences in the space $V_l(GF(2))$ are called the keys and l the key size.

The feedforward function f_k can be regarded as a one time data base installed in the cryptosystem in accordance with the key k we select. Such a data base can be easily realized by storing the value table of the function f_k in a $2^n \times 1$ RAM.

Since de Bruijn sequences rf. [3] are binary sequences, which have periods equal to powers of 2 and can be produced in large numbers, rf. [1,2,4], it is natural to think of defining the value table of the feedforward function f by means of a de Bruijn sequence

$$\beta = (b_0, b_1, \ldots, b_i, \ldots, b_{2^n-1}; \ldots),$$

by putting

$$f(i_0, i_1, \ldots, i_{n-1}) = b_i, \quad i = \sum_{j=0}^{n-1} i_j 2^j, \quad i_j \in \{0,1\}$$

and this function f can be called a de Bruijn feedforward function (or de Bruijn function).

First we give an approach for realizing the key-conditioned installation and alteration of the de Bruijn feedforward functions as described in requirement 4. In doing this, we make use of the terminology and notations introduced in [1], and start with the factor Σ_f of a suitably chosen non-singular n-stage shift-regester $SR(x_n = f)$, together with a fixed reference vertex a belonging to the de Bruijn Good graph G_n. Determine for each cycle σ (not containing the reference vertex a, i.e., $a \notin \sigma$) in Σ_f the edge set $E_a(\sigma)$, write

$$r(\sigma) = \lfloor \log_2 |E_a(\sigma)| \rfloor,$$

and fix in $E_a(\sigma)$ an arbitrary subset $E'_a(\sigma)$, consisting of $2^{r(\sigma)}$ elements. After that it will be easy to set up an one to one correspondence between binary sequences s of length

$$N = \sum_{\sigma \in \Sigma_f, a \notin \sigma} r(\sigma)$$

and spanning trees $T(s)$ of Γ_f, composed by edges belonging to the sets $E'_a(\sigma)$, exclusively.

Now the key-conditioned installation of feedforward functions can be realized in the following way. Choose the key size to be

$$l = N + n,$$

and decompose every key k of length l into a juxtaposition $k = k_1 \# k_2$ of two segments of lengths N and n respectively. Use the former to determine the feedback logic and the latter as the initial state of the de Bruijn sequence to be constructed, and store the sequence thus obtained in a RAM in the way described above, to serve as the value table of the feedforward function f_k. It is easy to see that the mapping $k \longmapsto f_k$ is one to one. It is obvious that de Bruijn functions are balanced. It is interesting that they also turn out to be complete.

Theorem 1 *If $n \geq 2$, then the function f defined by a de Bruijn sequence of degree n is complete.*

Being a balanced function in n indeterminates, the total degree of the function f defined by a de Bruijn sequence cannot exceed $n - 1$. What worries us is how low the degree of f may happen to be. The lower bound obtained in Theorem 2 below shows that the situation here is by no means discouraging.

Theorem 2 *If the function f is defined by a de Bruijn sequence of degree n, then we have*

$$\deg(f) \geq \lfloor \log_2 n \rfloor.$$

But one must keep clearly in mind that functions defined by certain classes of de Bruijn sequences may suffer from systematic correlational weaknesses, which may influence the hardness of the output sequences. We illustrate this point by analyzing the case, where the de Bruijn sequences are obtained by starting with the shift-register $SR(x_n = x_0)$. It is well known that the number of cycles in the factor Σ_{x_0} is

$$Z(n) = \sum_{d \mid n} \phi(n/d) 2^d / n.$$

We write

$$\epsilon = 1/2 - (Z(n) - 1)/2^{n-1}$$

and state the mentioned correlational weakness in the form of the following

Theorem 3 *If the de Bruijn sequence $\beta = (b_0, b_1, \ldots, b_i, \ldots)$ is defined by a spanning tree T of the skeleton Γ_{x_0}, rf. [1], then for any integer $t > 0$ we have*

$$\mathrm{Prob}(b_{i+tn} = b_i) = 1/2 + (2\epsilon)^{t-1}\epsilon.$$

This situation, however, is by no means hopeless. Besides choosing more appropriate de Bruijn sequences β_k, one may, for example, make use of sequences

$$\gamma_k = d(L)\beta_k$$

derived from the sequences β_k mentioned in Theorem 3 by applying a fixed operator of the form

$$d(L) = \sum_{i=0}^{p} d_i L^i, \quad d_0 = d_p = 1, \quad p < n,$$

with an odd number of nonzero coefficients d_i. This will meliorate the correlation characteristics of the feedforward functions, without causing substantial drawbacks with regard to the other requirements listed above. For we have

Theorem 4 *The functions f_k defined by the sequences γ_k are balanced and the mapping $k \longmapsto f_k$ is injective. Moreover, if*

$$2^r + p \leq n,$$

then we shall have

$$\deg(f_k) \geq \lfloor \log_2(n - p) \rfloor.$$

We conclude the present paper by making the following remarks.

1. The lower bounds to $\deg(f_k)$ given in Theorem 2 and Theorem 4 are too modest as compared with concrete results computed for a large number of de Bruijn sequences. New ideas are needed for improving these bounds.

2. Since the functions f_k have been shown to be nonlinear, it is natural to ask, to what degree they can be approximated by linear functions $l(X) = l(x_0, x_1, \ldots, x_{n-1})$ over the space $V_n(GF(2))$. To state the problem more precisely, we require to make an estimation of the quantity

$$d = \max_{k} \max_{l(X)} \left\{ |\mathrm{Prob}(f_k(a_0, \ldots, a_{n-1}) = l(a_0, \ldots, a_{n-1})) - 1/2| \right\}.$$

This is in fact a problem concerning the I/O correlation immunity of the family of de Bruijn functions.

References

[1] Z.D.Dai, *On the Construction and Cryptographic Application of de Bruijn Sequences*, Submitted to Journal of Cryptology.

[2] Division of Algebra in the Institute of Mathematics and Department of USTC, *On Methods of Constructing Feedback Functions of M-Sequences*, ACTA MATHEMATICAE APPLICATAE SINICA, Nov.1977 (in Chinese).

[3] H. Fredricksen, *A Survey of Full Lenth Nonlinear Shift Register Cycle Algorithms*, SIAM REVIEW, Vol.24, No.2. 1982.

[4] R.H.Xiong, *The Methods of Feedback Functions of M-Sequences I*, ACTA MATHEMATICAE APPLICATAE SINICA, Vol.9, No.2. Apr.,1986.

[5] K.C.Zeng, *On the Key Entropy Leakage Phenomena in Cryptosystem*, Report to Symposium on Cryptography, Beijing, 1986.

NONLINEARITY CRITERIA FOR CRYPTOGRAPHIC FUNCTIONS

Willi Meier [1]   Othmar Staffelbach [2]

[1] HTL Brugg-Windisch
CH-5200 Windisch, Switzerland

[2] GRETAG Aktiengesellschaft
Althardstr. 70, CH-8105 Regensdorf
Switzerland

Abstract. Nonlinearity criteria for Boolean functions are classified in view of their suitability for cryptographic design. The classification is set up in terms of the largest transformation group leaving a criterion invariant. In this respect two criteria turn out to be of special interest, the distance to linear structures and the distance to affine functions, which are shown to be invariant under all affine transformations. With regard to these criteria an optimum class of functions is considered. These functions simultaneously have maximum distance to affine functions and maximum distance to linear structures, as well as minimum correlation to affine functions. The functions with these properties are proved to coincide with certain functions known in combinatorial theory, where they are called bent functions. They are shown to have practical applications for block ciphers as well as stream ciphers. In particular they give rise to a new solution of the correlation problem.

1. INTRODUCTION

For cryptographic systems the method of confusion and diffusion (as introduced by Shannon [8]) is used as a fundamental technique to achieve security. Confusion is reflected in nonlinearity of certain Boolean functions describing the cryptographic transformation. Nonlinearity is crucial since most linear systems are easily breakable. As a cipher which explicitly follows the principle of confusion and diffusion we mention DES. Likewise, this concept applies to other cryptosystems, block ciphers as well as stream ciphers.

In this context it is important to have criteria which are measures for nonlinearity. A variety of such criteria are known in cryptographic design. Our aim is to contribute to a general theory which classifies these criteria in view of their ability to measure nonlinearity. As a result of this investigation we are led to a class of nonlinear functions with many remarkable properties with regard to this theory.

Our considerations are based on the idea that a useful criterion should remain invariant under a certain group of transformations. This concept is fundamental in pure mathematics, e.g. in algebra. In cryptography it is motivated by the following point of view: A function is considered weak whenever it can be turned into a cryptographically weak function by means of simple (e.g. linear or affine) transformations. This is reminiscent to the situation, where Shannon introduced the notion of similar secrecy systems ([8], Chap. 8), R and S being similar if there is a transformation A such that R = AS. (Then by definition similar systems are cryptanalytically equivalent.)

To further illustrate our concept we consider the Boolean function $f(x_1, x_2, \ldots, x_n)$ whose algebraic normal form is obtained by summing up all possible product terms in

$x_1, x_2, \ldots, x_n$. At the first glance this seems to be a good nonlinear function, since it contains all nonlinear terms. However f can be written as the product $f(x_1, \ldots, x_n) = (1+x_1)(1+x_2) \cdots (1+x_n)$ which transforms into the monomial function $g(x_1, x_2, \ldots, x_n) = x_1 x_2 \cdots x_n$ by simply complementing all arguments. This turns f into a poor function with respect to the number of nonlinear terms. Thus, from the present point of view, a large number of nonlinear terms taken as a criterion is not suitable since it is not invariant under simple transformations.

It is desirable therefore that a nonlinearity criterion remains invariant under a larger group of transformations. For many applications this symmetry group should contain the group of all affine transformations. In Section 2 we develop a general method (Theorem 2.1) in order to show that several well known criteria satisfy this stronger requirement. Some of these criteria can be expressed in terms of a distance to appropriate sets of (cryptologically weak) functions. (The distance $d(f,S)$ of a function f to any subset S of Boolean functions is defined as the minimum of the Hamming distances of f to all members of S.) In particular the distance $\delta(f)$ to affine functions is defined as $\delta(f) = d(f,S)$, where S is the set of all affine functions (cf. also [7], p. 122). We show that δ is a nonlinearity criterion with the desired property since it remains invariant under the operation of the full affine group (Corollary 2.2).

Depending on the application different sets S of functions have to be considered as cryptographically weak. The set of affine functions may be replaced e.g. by functions of low nonlinear order, like quadratic functions. Therefore, as a design criterion for a Boolean function f, we may introduce its distance $\delta_k(f)$ to all functions with nonlinear order bounded by k. (Note that $\delta_1 = \delta$.) We show that the design criterion δ_k also remains invariant under affine transformations. This is proved as a consequence of the fact that similar invariance properties hold for the nonlinear order of Boolean functions (Theorem 2.4).

In certain applications the class of affine functions has to be extended to another class of cryptographically weak functions. The definition of these functions is motivated by the fact that for a linear (or affine) Boolean function f the values $f(\underline{x}+\underline{a})$ and $f(\underline{x})$, for every fixed $\underline{a}$, are either always equal or always different. Note however that many functions have this property without being linear or affine. The functions characterized by this condition appear to be important in the analysis and design of block ciphers, as has been pointed out by Chaum and Evertse in [1] and [3], where this property is termed a linear structure. We denote by $\sigma(f)$ the distance $d(f,S)$, where S is the set of all Boolean functions with linear structures. Then as for δ, the distance σ is invariant under the operation of the full affine group (Corollary 2.3).

With respect to linear structures, a function f has optimum nonlinearity if for every nonzero vector $\underline{a} \in GF(2)^n$ the values $f(\underline{x}+\underline{a})$ and $f(\underline{x})$ are equal for exactly half of the arguments $\underline{x} \in GF(2)^n$. If a function f satisfies this property we will call it perfect nonlinear with respect to linear structures, or briefly perfect nonlinear. In [3] Evertse has introduced a corresponding notion for DES-like S-boxes where he named it a 50%-linear structure. Furthermore he questioned whether S-boxes with this restrictive property do exist, a question which is settled in this paper.

In a different direction, this notion of perfect nonlinearity is closely related to another design criterion for S-boxes, namely the strict avalanche criterion (SAC). Basicly this is a diffusion criterion and has been investigated in [11] and [4]. Recall that a Boolean function satisfies SAC if the output changes with probability one half whenever a single input bit is complemented. This means that a function satisfies SAC if the condition stated in the definition of perfect nonlinearity merely holds for vectors **a** of weight 1. Therefore perfect nonlinearity effects diffusion, and it is in fact a much stronger requirement than SAC. It is remarkable that in this context diffusion can be linked with nonlinearity.

It turns out that perfect nonlinear functions correspond to certain functions known in combinatorial theory - in combinatorial theory Rothaus ([6]) has investigated a class of functions, which he called 'bent functions'. The coincidence of bent functions with perfect nonlinear functions is derived by using properties of the Walsh transform. The existence of these functions is established in [6] by giving explicit constructions. In particular, for n = 2m, the functions of the form $f(x_1, x_2, ..., x_n) = g(x_1, ... , x_m) + x_1 x_{m+1} + x_2 x_{m+2} + ... + x_m x_n$ are known to be perfect nonlinear, where $g(x_1, ... , x_m)$ is a completely arbitrary function. Moreover a systematic method allows to generate a large class of perfect nonlinear functions out of any existing one. However these constructions apply to even dimensions n only, whereas no perfect nonlinear functions exist in odd dimensions. It is furthermore known that the nonlinear order of bent functions is tightly bounded by n/2.

We show that perfect nonlinear functions are optimum with regard to the distances δ and σ. More precisely for an even number n of arguments the class $\pi(n)$ of perfect nonlinear functions simultaneously has maximum distance to all affine functions (Theorem 3.4) as well as maximum distance to linear structures (Theorem 3.2). These maximum values are shown to be $2^{n-1} - 2^{(n/2)-1}$ for δ, and 2^{n-2} for σ. Furthermore, perfect nonlinear functions have equal, and in fact minimum correlation to all affine functions (Theorem 3.5). The maximum distance δ to affine functions is of independent interest in coding theory, where it appears to be the covering radius of the Reed-Muller code of order 1 (cf. [2]). In the same context the maximum value of δ_k coincides with the covering radius of the k-th order Reed-Muller code.

These results allow for applications in several directions. In odd dimensions functions can be generated with properties close to those of perfect nonlinear functions (Section 3.3). Thus in every dimension we arrive at constructing Boolean functions with large distances δ and σ, i.e. functions with guaranteed lower bounds on δ and σ. For large n (e.g. n $\approx$ 64) an a priori verification of this property may be impossible since even the computation of Hamming distances between functions becomes infeasible.

Notably, our considerations have consequences for the design of block ciphers. Since perfect nonlinear functions are not exactly balanced the question as raised by Evertse can be answered: There are no DES-like S-boxes with maximum distance to linear structures (Corollary 3.6). A search for such S-boxes was motivated by the analysis of DES in [1] where linear structures of the S-boxes are considered.

The above example shows that in general perfect nonlinearity may not be compatible with other cryptographic design criteria, e.g. balance or highest nonlinear order. We indicate a method of finding functions which at the same time are nearly perfect nonlinear (i.e. with large δ and σ) and satisfy other criteria of cryptographic interest.

In particular this procedure is applied to propose functions which are useful in stream cipher design where one or more linear feedback shift registers are combined to produce the key stream. In this design there arise correlation problems, since any Boolean function f has correlation to some linear functions L. Such correlations are shown to lead to correlations to certain LFSR-sequences. For an individual L this correlation is reflected in a nonvanishing cross correlation coefficient c(f,L). The (normalized) correlations to all linear functions L are shown to satisfy

$$\sum_L c(f,L)^2 = 1,$$

which implies that correlations to linear functions do always exist whatever function f is used. However for perfect nonlinear functions the absolute values $|c(f,L)|$ turn out to be uniformly small. This motivates a general method to face the correlation problems in stream cipher design by choosing the combining functions to be (close to) a perfect nonlinear function (which in fact can be done in conjunction with other design criteria). By suitable design the remaining correlations may become as small as to defeat any kind of correlation attack.

This contrasts to the method of facing correlation by choosing correlation immune functions. A correlation immune function (of some order) has zero correlation to certain linear functions. However, as the above formula shows, the strongest correlations (to some other linear functions) are necessarily larger than for perfect (or nearly perfect) nonlinear functions.

2. NONLINEARITY CRITERIA FOR BOOLEAN FUNCTIONS

Cryptographic transformations are often described in terms of functions $GF(2)^n \longrightarrow GF(2)^m$. For small values of n and m these functions are usually given in form of tables. These tables can then be used as building blocks for generating functions in higher dimensions. As examples we mention the S-boxes $S: GF(2)^6 \longrightarrow GF(2)^4$ of DES, or the combining functions used in certain types of stream ciphers. The strength of the resulting algorithms heavily relies on the nonlinearity of these functions.

Most of the known nonlinearity criteria can be reduced to conditions imposed on Boolean functions $f: GF(2)^n \longrightarrow GF(2)$. This is illustrated by the notion of linear structures of S-boxes as introduced by Chaum and Evertse ([1],[3]): An S-box $S: GF(2)^n \longrightarrow GF(2)^m$ is said to have a linear structure if there is a nonzero vector $\underline{a} \in GF(2)^n$ together with a nontrivial linear mapping $L: GF(2)^m \longrightarrow GF(2)$ such that $LS(\underline{x}+\underline{a}) + LS(\underline{x})$ takes the same value (0 or 1) for all $\underline{x} \in GF(2)^n$. Thus linear structures of S can be expressed in terms of linear structures of the Boolean function LS. For this reason we concentrate hereafter on Boolean functions $f: GF(2)^n \longrightarrow GF(2)$. It is common to describe these functions in terms of their algebraic normal form

$$f(x_1,\ldots,x_n) = a_0 + \Sigma\, a_i x_i + \Sigma\, a_{ij} x_i x_j + \ldots + a_{12..n} x_1 x_2 \cdots x_n \qquad (2.1)$$

A function f is nonlinear (or non-affine) if its algebraic normal form contains terms of degree higher than one.

In this section we compare different criteria in view of their ability to measure nonlinearity of Boolean functions.

2.1. Distance to affine functions

The <u>distance to the nearest affine function</u> is defined as

$$\delta(f) = \min_{L \in A(n)} d(f,L) \qquad (2.2)$$

where $d(f,L)$ is the Hamming distance between f and L, and where the minimum is taken over the set $A(n)$ of all affine functions $L(x_1,\ldots,x_n) = a_0 + a_1 x_1 + \ldots + a_n x_n$. Thus $\delta(f)$ is the distance of f to the set $A(n)$. In order to investigate the properties of δ we introduce some additional notations.

Let $\Omega(n)$ denote the group of all invertible transformations of $GF(2)^n$, and let $AGL(n)$ denote the subgroup of all affine transformations. Recall that the elements of $AGL(n)$ can be described as functions $\alpha(\underline{x}) = A\underline{x} + \underline{a}$ where A is a regular $n \times n$ - matrix and $\underline{a} \in GF(2)^n$ is a vector. Moreover denote by $\Phi(n)$ the set of all Boolean functions $f: GF(2)^n \longrightarrow GF(2)$ of n arguments.

An operation of a group G on a set S means a mapping $G \times S \longrightarrow S$ which is compatible with group multiplication (cf. [5], Ch.I). For the image of a pair (g,s) $(g \in G$ and $s \in S)$ the notation $g \cdot s$ is commonly used. In these terms an operation of the group $\Omega(n)$ on the set $\Phi(n)$ is defined by

$$\alpha \cdot f(\underline{x}) = f(\alpha(\underline{x})), \quad \text{where } f \in \Phi(n) \text{ and } \alpha \in \Omega(n) \tag{2.3}$$

With this notion, we can now make some of our considerations in more general and more precise terms. Any design criterion is connected with a function D (valuation)

$$D: \Phi(n) \longrightarrow W \tag{2.4}$$

with values in a suitable set W, and a function f is considered to be "good" if the value $D(f)$ belongs to some well defined subset W_1 of W. It may be essential for a design criterion that the valuation D remains invariant under those transformations of $\Omega(n)$ which are considered "cryptographically weak". This guarantees that a good function cannot be made "worse" by means of weak transformations. For nonlinearity, weak transformations usually include affine transformations. To illustrate our terminology, the number of terms in the algebraic normal form (as exemplified in the introduction) is not invariant even under simple transformations like complementations of variables.

For any design criterion it is therefore of interest to introduce the largest subgroup $I(D)$ of $\Omega(n)$ which leaves D invariant, i.e.

$$I(D) = \{\alpha \in \Omega(n) \mid D(\alpha \cdot f) = D(f) \text{ for all } f \in \Phi(n)\} \tag{2.5}$$

Hereafter $I(D)$ will be called the symmetry group of D. In cryptography it may be essential that $I(D)$ is large. We therefore investigate various design criteria in view of their symmetry groups.

First we show that the distance δ to the nearest affine function remains in fact invariant under the operation of the whole affine group $AGL(n)$ (cf. Corollary 2.2 below.) It appears worthwhile to prove this result in a more general context, which allows to analyze other design criteria with regard to their symmetry group.

Let H be a subset of $\Phi(n)$, and for $f \in \Phi(n)$ let $d_H(f) = d(f,H)$ be the distance of f to the set H. (In applications, this subset will be the class of cryptographically weak functions, and for δ it will be the set $A(n)$ of all affine functions.) Moreover let

$$\Omega(n)^H = \{\alpha \in \Omega(n) \mid \alpha \cdot h \in H \text{ for all } h \in H\} \tag{2.6}$$

which will be called the symmetry group of the set H. This terminology is justified by the following result.

Theorem 2.1. For any subset H of $\Phi(n)$ the symmetry group of d_H coincides with the symmetry group of H

$$I(d_H) = \Omega(n)^H. \tag{2.7}$$

Proof. (a) $\Omega(n)^H$ is contained in $I(d_H)$: Suppose $\alpha \in \Omega(n)^H$ and $f \in \Phi(n)$. Let $h \in H$ such that $d_H(f) = d(f,h)$. Then $d_H(f) = d(f,h) = d(\alpha \cdot f, \alpha \cdot h) \geq d_H(\alpha \cdot f)$. Observe that the second equality is a consequence of the fact that the operation of $\Omega(n)$ on $\Phi(n)$ leaves the Hamming distance invariant. Moreover the last inequality holds as $\alpha \cdot h \in H$ by definition (2.6). Therefore

$$d_H(f) \geq d_H(\alpha \cdot f) \tag{2.8}$$

Since $\Omega(n)^H$ is a subgroup (2.8) may be applied with respect to the operation of α^{-1}. This yields $d_H(\alpha \cdot f) \geq d_H(\alpha^{-1} \cdot (\alpha \cdot f)) = d_H(f)$, and consequently $d_H(\alpha \cdot f) = d_H(f)$.

(b) $I(d_H)$ is contained in $\Omega(n)^H$: For any $\alpha \notin \Omega(n)^H$ there exists $h \in H$ such that $\alpha \cdot h$ is not in H. Hence $d_H(h) = 0$ but $d_H(\alpha \cdot h) \neq 0$. Therefore α is not in $I(d_H)$.

Corollary 2.2. The symmetry group $I(\delta)$ of δ is the affine group $AGL(n)$.

Proof. With regard to Theorem 2.1. it remains to show that $\Omega(n)^{A(n)} = AGL(n)$. Obviously $AGL(n)$ is contained in $\Omega(n)^{A(n)}$. In the other direction for any $\alpha \in \Omega(n) - AGL(n)$ there exists an index i such that the i-th component $\alpha_i(x_1, \ldots, x_n)$ of α is not affine. Then for $g(x_1, \ldots, x_n) = x_i$, the function $\alpha \cdot g(x_1, \ldots, x_n) = \alpha_i(x_1, \ldots, x_n)$ is not in $A(n)$, which implies that α is not in $\Omega(n)^{A(n)}$.

2.2. Distance to linear structures

According to the preliminary remarks to this section a general linear structure can be formulated in terms of linear structures of appropriate Boolean functions. Recall that a linear structure of a Boolean function $f: GF(2)^n \longrightarrow GF(2)$ can be identified with a vector $\underline{a} \in GF(2)^n$ such that the expression

$$f(\underline{x} + \underline{a}) + f(\underline{x}) \tag{2.9}$$

takes the same value (0 or 1) for all $\underline{x} \in GF(2)^n$ ([1],[3]). Let $LS(n)$ denote the subset of Boolean functions having linear structures. Observe that $LS(n)$ properly contains the set $A(n)$ of all affine functions.

For a Boolean function f the _distance to linear structures_ is defined as the distance of f to the set $LS(n)$:

$$\sigma(f) = d(f, LS(n)) = \min_{S \in LS(n)} d(f, S) \tag{2.10}$$

The distance to linear structures serves as a useful nonlinearity criterion as follows as a corollary to Theorem 2.1.

Corollary 2.3. The symmetry group $I(\sigma)$ of σ contains the affine group $AGL(n)$.

Proof. In order to apply Theorem 2.1 we show that $\Omega(n)^{LS(n)}$ contains $AGL(n)$. Let $f \in LS(n)$ and let $\underline{a} \in GF(2)^n$ be a linear structure of f, i.e. for all $\underline{x} \in GF(2)^n$ the equation $f(\underline{x}+\underline{a}) + f(\underline{x}) = c$ holds, where $c \in GF(2)$ is a constant. Then for $\alpha \in AGL(n)$

$$f(\alpha(\underline{x} + \alpha^{-1}(\underline{a}))) + f(\alpha(\underline{x})) = c \tag{2.11}$$

is satisfied for all $\underline{x} \in GF(2)^n$. This means that $\alpha^{-1}(\underline{a})$ is a linear structure of $\alpha \cdot f$. Hence $\alpha \in \Omega(n)^{LS(n)}$.

2.3. The nonlinear order

For a Boolean function $f \in \Phi(n)$ let $O(f)$ be the degree of the highest order terms in the algebraic normal form, which is called the <u>nonlinear order</u> of f. This defines another useful nonlinearity criterion $O: \Phi(n) \longrightarrow \{0,\ldots,n\}$ as is demonstrated by the following

Theorem 2.4. The symmetry group $I(O)$ of O is the affine group $AGL(n)$.

Proof. (a) $AGL(n)$ is contained in $I(O)$: Let $\alpha \in AGL(n)$ and $f \in \Phi(n)$ arbitrary. Compute the algebraic normal form of $\alpha \cdot f$ by formal reduction of $f(\alpha(\underline{x}))$. In this procedure existing nonlinear terms of $f(\underline{x})$ may disappear and new terms may be generated in $f(\alpha(\underline{x}))$. However terms of some degree k in $f(\underline{x})$ cannot create terms of degree higher than k in $f(\alpha(\underline{x}))$. This shows

$$O(\alpha \cdot f) \le O(f) \tag{2.12}$$

Formula (2.12) may be applied also with respect to the operation of α^{-1}. Hence

$$O(f) = O(\alpha^{-1} \cdot (\alpha \cdot f)) \le O(\alpha \cdot f)$$

and $O(\alpha \cdot f) = O(f)$. Therefore $\alpha \in I(O)$.

    (b) $I(O)$ is contained in $AGL(n)$: Suppose that α is not contained in $AGL(n)$. Then α has a nonlinear component $\alpha_i(x_1,\ldots,x_n)$. With $f(x_1,\ldots,x_n) = x_i$ we have $\alpha \cdot f = \alpha_i$. Thus $O(\alpha \cdot f) > 1$ whereas $O(f) = 1$. Therefore α is not contained in $I(O)$.

    Theorem 2.4 implies that other nonlinearity criteria, namely the distances δ_k to functions with nonlinear order bounded by k, also remain invariant under the operation of $AGL(n)$.

2.4. Correlation immunity

Refer to the notion of correlation immunity as introduced by Siegenthaler ([9]). It is known that correlation immunity is not a genuine nonlinearity criterion. Indeed the consideration of its symmetry group further illuminates this fact. In view of a later comparison to other design criteria, the study of correlation immunity in the context of symmetry groups is of independent interest. We start by defining a valuation $c: \Phi(n) \longrightarrow \{0,1, \ldots ,n-1\}$ by assigning to every function $f \in \Phi(n)$ its order $c(f)$ of correlation immunity.

Theorem 2.5. The symmetry group $I(c)$ is the group of permutations and complementations of variables, i.e. the group $P(n) = \{\alpha=(A,\underline{a}) \in AGL(n)$ where A is a permutation matrix$\}$.

Proof. First we show that $I(c)$ is contained in $AGL(n)$. Suppose that α is not contained in $AGL(n)$, and let $\alpha(\underline{x}) = (\alpha_1(\underline{x}),\ldots,\alpha_n(\underline{x}))$. For this we claim that there exists a sum of at least n-1 α_i's which is not linear. Suppose to the contrary that all the sums

$$\beta_0 = \alpha_1 + \alpha_2 + \ldots + \alpha_n \qquad \text{and} \qquad \beta_i = \sum_{j \neq i} \alpha_j$$

are linear. Then $\alpha_i = \beta_0 + \beta_i$ is linear for all i, in contradiction to the nonlinearity of α.

    In case β_0 is nonlinear take $f(\underline{x}) = x_1 + x_2 + \ldots + x_n$, and in case β_i is nonlinear for some $i > 0$ take $f(\underline{x}) = \sum_{j \neq i} x_j$. Then $\alpha \cdot f = \beta_i$ has nonlinear order at least 2.

Moreover $\alpha \cdot f$ is balanced as α is a permutation of $GF(2)^n$. Therefore by a result of Siegenthaler ([9]), $c(\alpha \cdot f) < n-2$ whereas $c(f) \geq n-2$. Hence α is not contained in $I(c)$.

In a second step we show that $I(c)$ is contained in $P(n)$. Suppose $\alpha \in AGL(n)$ is not contained in $P(n)$. Then there exists a component $\alpha_i(\underline{x}) = b_0 + b_1 x_1 + \ldots + b_n x_n$ with weight$(b_1,\ldots,b_n) = t > 1$. Take $f(\underline{x}) = x_i$. Then $c(\alpha \cdot f) = t-1 > 0$ and $c(f) = 0$. Therefore α is not contained in $I(c)$. Alltogether this shows that $I(c)$ is contained in $P(n)$. Since obviously $P(n)$ is contained in $I(c)$, this completes the proof of the theorem.

3. PERFECT NONLINEAR FUNCTIONS

In this section we investigate a class of functions whose definition is motivated by considering linear structures. With respect to linear structures (cf. (2.9)) a Boolean function f has optimum nonlinearity if $f(\underline{x}+\underline{a})$ coincides with $f(\underline{x})$ for exactly half of the arguments $\underline{x}$:

Definition 3.1. A Boolean function $f: GF(2)^n \longrightarrow GF(2)$ is called perfect nonlinear with respect to linear structures (or briefly perfect nonlinear) if for every nonzero vector $\underline{a} \in GF(2)^n$ the values $f(\underline{x}+\underline{a})$ and $f(\underline{x})$ are equal for exactly half of the arguments $\underline{x} \in GF(2)^n$.

The subset of $\Phi(n)$ consisting of the perfect nonlinear functions will be denoted by $\pi(n)$. We first show that these functions are optimum with respect to the distance σ to linear structures.

For an arbitrary function $f: GF(2)^n \longrightarrow GF(2)$ the distance to linear structures can be computed as follows. Let $\underline{a} \in GF(2)^n$ be a nonzero vector. Then the space $GF(2)^n$ can be exhausted by 2^{n-1} pairs $(\underline{x},\underline{x}+\underline{a})$. Denote by n_0 the number of elements in the set W_0 of pairs $(\underline{x},\underline{x}+\underline{a})$ for which $f(\underline{x})$ coincides with $f(\underline{x}+\underline{a})$. Similarly let n_1 be the number of elements in the set W_1 of pairs $(\underline{x},\underline{x}+\underline{a})$ for which $f(\underline{x})$ differs from $f(\underline{x}+\underline{a})$.

Furthermore any Boolean function can be derived from f by modifying an appropriate set of f-values. In this way f can be turned into a function with the linear structure $\underline{a}$ by changing the f-values of either $\underline{x}$ or $\underline{x}+\underline{a}$ for pairs in W_0, or by changing the f-values of either $\underline{x}$ or $\underline{x}+\underline{a}$ for the pairs in W_1. Thus n_0 values are to be changed to get a function g with the linear structure ($g(\underline{x}) \neq g(\underline{x}+\underline{a})$ for all $\underline{x}$), and n_1 values are to be changed to get a function g with the linear structure ($g(\underline{x}) = g(\underline{x}+\underline{a})$ for all $\underline{x}$). In order to generate any other function with these linear structures at least $\min(n_0,n_1)$ modifications are necessary. Therefore $n = \min(n_0,n_1)$ is the distance of f to the nearest functions with the linear structure $\underline{a}$. Observe that this n depends on the vector $\underline{a}$, i.e. $n = n_f(\underline{a}) = \min(n_0(\underline{a}),n_1(\underline{a}))$. Hence the distance of f to linear structures is given by

$$\sigma(f) = \min_{\underline{a} \neq 0} n_f(\underline{a}) \qquad (3.1)$$

Since $n_0(\underline{a}) + n_1(\underline{a}) = 2^{n-1}$ the derivation of formula (3.1) also proves that $n_f(\underline{a}) \leq 2^{n-2}$ for all $\underline{a} \neq 0$. This maximum distance is achieved by perfect nonlinear functions, as these functions are characterized by $n_0(\underline{a}) = n_1(\underline{a}) = 2^{n-2}$ for $\underline{a} \neq 0$, or equivalently by the property $\sigma(f) = 2^{n-2}$. This proves

Theorem 3.2. The class $\pi(n)$ of perfect nonlinear functions is the class of functions with maximum distance 2^{n-2} to linear structures.

3.1. Bent functions

We now establish a relationship between perfect nonlinear functions and the 'bent' functions which were introduced by Rothaus ([6]). The relation is expressed in terms of the Walsh transform.

Hereafter, in connection with Walsh transforms, all Boolean functions are considered with values +1 and -1 (i.e. $f(\underline{x})$ is replaced by $(-1)^{f(x)}$). Recall the definition of the Walsh transform

$$F(\underline{w}) = \sum_{\underline{x} \in GF(2)^n} f(\underline{x})(-1)^{\underline{x} \cdot \underline{w}} \tag{3.2}$$

where $\underline{w} \in GF(2)^n$ and $\underline{x} \cdot \underline{w}$ is the dot-product over $GF(2)$, and where the sum is evaluated over the reals.

For a Boolean function $f: GF(2)^n \longrightarrow \{+1,-1\}$ the condition of Definition 3.1. for given $\underline{a}$ reads as

$$\sum_{\underline{x} \in GF(2)^n} f(\underline{x})f(\underline{x}+\underline{a}) = 0 \tag{3.3}$$

The sum (3.3) equals $f*f(\underline{a})$ where $f*f$ denotes the convolution of f with itself. Thus a ± 1-valued function f is perfect nonlinear if and only if $f*f(\underline{a}) = 0$ for every nonzero vector $\underline{a} \in GF(2)^n$, i.e. if and only if $f*f$ is a δ-function. By the convolution theorem the function $f*f$ transforms into F^2, and a δ-function transforms into a constant. Therefore a ± 1-valued Boolean function f is perfect nonlinear if and only if $|F(\underline{w})|$ is constant for all $\underline{w}$. Since $f*f(\underline{0}) = 2^n$, this constant is

$$|F(\underline{w})| = 2^{n/2} \tag{3.4}$$

This property has been used by Rothaus to define the bent functions, which implies

Theorem 3.3. The class of perfect nonlinear functions coincides with the class of bent functions.

The following theorems A and B are the main results proved in [6] about bent functions.

Theorem A. Bent functions only exist for even numbers n of arguments, and their nonlinear order is always bounded by n/2.

Theorem B. For an even number n of arguments bent functions are constructed as follows.

(B1) Let $n = 2m$. Then the functions of the form $f(x_1,...,x_n) = g(x_1,...,x_m) + x_1 x_{m+1} + x_2 x_{m+2} + \cdots + x_m x_n$ are bent, where $g(x_1,...,x_m)$ is a completely arbitrary function of m variables.

(B2) Let $\underline{x} = (x_1,...,x_n)$ and let $a(\underline{x})$, $b(\underline{x})$ and $c(\underline{x})$ be bent functions such $a(\underline{x}) + b(\underline{x}) + c(\underline{x})$ is also bent. Then the function $f(\underline{x},x_{n+1},x_{n+2}) = a(\underline{x})b(\underline{x}) + b(\underline{x})c(\underline{x}) + c(\underline{x})a(\underline{x}) + [a(\underline{x})+b(\underline{x})]x_{n+1} + [a(\underline{x})+c(\underline{x})]x_{n+2} + x_{n+1}x_{n+2}$ is a bent function. The requirement that $a(\underline{x})+b(\underline{x})+c(\underline{x})$ is bent is readily met by taking $a(\underline{x})$, $b(\underline{x})$ and $c(\underline{x})$ from class B1, or by putting $a(\underline{x}) = b(\underline{x})$ or $b(\underline{x}) = c(\underline{x})$.

(B1) leads to an explicit construction of bent functions, whereas (B2) allows to generate new perfect nonlinear or bent functions out of any existing ones. This procedure can be combined with linear operations on given perfect nonlinear functions. In fact formula (2.11) implies that the class $\pi(n)$ of perfect nonlinear functions is invariant under the operation of the affine group AGL(n). Moreover addition of an arbitrary affine function does not affect perfect nonlinearity. Therefore assigning to $f \in \Phi(n)$ the function $\underline{x} \longrightarrow f(\alpha(\underline{x})) + L(\underline{x})$ defines an operation of AGL(n) x A(n) on $\Phi(n)$ which leaves $\pi(n)$ invariant. This leads to the following recursive construction of perfect nonlinear functions.

I. For n = 2 take the class C(2) consisting of all functions of nonlinear order 2.

II. For n > 2 take any functions a,b,c in C(n-2) such that their sum is also in C(n-2), and apply construction (B2). This defines a class C'(n) of perfect nonlinear functions. This class C'(n) is enlarged to a class C(n) by letting operate the whole group G = AGL(n) x A(n) on C'(n).

It can be shown that C(n) includes the functions obtained in (B1). It is not clear whether the class C(n) exhausts all functions in $\pi(n)$. But (B1) implies that there are at least $2^{2^{n/2}}$ perfect nonlinear functions among all 2^{2^n} Boolean functions. Thus only a very small fraction of all Boolean functions are perfect nonlinear. Already for n = 6 (i.e. in the input dimension of the DES S-boxes) it is virtually impossible to find perfect nonlinear functions by a pure random search.

3.2. Distance to affine functions and correlation

Let $L_W(\underline{x}) = \underline{w} \cdot \underline{x}$ denote an arbitrary linear function. Thus $(-1)^{\underline{w} \cdot \underline{x}}$ is the corresponding ±1-valued function which is also denoted by $L_W(\underline{x})$. Then the definition (3.2) of the Walsh transform implies

$$F(\underline{w}) = \#\{\underline{x} \mid f(\underline{x}) = L_W(\underline{x})\} - \#\{\underline{x} \mid f(\underline{x}) \neq L_W(\underline{x})\} = 2^n - 2d(f, L_W)$$

where d denotes the Hamming distance. Therefore

$$d(f, L_W) = 2^{n-1} - \frac{1}{2} F(\underline{w}) \tag{3.5}$$

For the corresponding affine function $L_W' = 1 + L_W$ the distance d is computed as $d(f, L_W') = 2^{n-1} + (1/2)F(\underline{w})$. Formula (3.5) can be used to find the best affine approximation to a given function by finding $\underline{w}$ such that $|F(\underline{w})|$ is maximum (cf. also Rueppel ([7], p. 122)), i.e.

$$\delta(f) = 2^{n-1} - \frac{1}{2} \max_w |F(\underline{w})| \tag{3.6}$$

Thus by property (3.4) the perfect nonlinear functions always have distance

$$\delta(f) = 2^{n-1} - 2^{n/2-1} \tag{3.7}$$

to the nearest affine functions. Suppose now that f is not perfect nonlinear. Then by Parseval's theorem

$$\sum_w F(\underline{w})^2 = 2^n \sum_x f(\underline{x})^2 = 2^{2n} \tag{3.8}$$

there exists a $\underline{w}$ with $|F(\underline{w})| > 2^{n/2}$. This implies $\delta(f) < 2^{n-1} - 2^{n/2-1}$ and therefore f is closer to the set of all affine functions than are perfect nonlinear functions.

This shows that the perfect nonlinear functions are not only optimum with respect to the distance to linear structures but also with respect to the distance to all affine functions.

Theorem 3.4. The class $\pi(n)$ of perfect nonlinear functions is the class of functions with maximum distance $2^{n-1} - 2^{(n/2)-1}$ to affine functions.

As formula (3.5) shows this result can be refined to the statement that the distance of a perfect nonlinear function f to any affine function is either $2^{n-1} + 2^{n/2-1}$ or $2^{n-1} - 2^{n/2-1}$. This fact can be expressed in terms of correlations of f to affine functions. In general the Hamming distance between two Boolean functions $f, g: GF(2)^n \longrightarrow \{+1, -1\}$ is tied up with the cross correlation between f and g which is defined as

$$c(f,g) = \frac{\#\{\underline{x} \mid f(\underline{x}) = g(\underline{x})\} - \#\{\underline{x} \mid f(\underline{x}) \neq g(\underline{x})\}}{2^n}$$

For $g = L_{\underline{w}}$ we have by definition of the Walsh transform (see also (3.5))

$$c(f, L_{\underline{w}}) = \frac{F(\underline{w})}{2^n} \tag{3.9}$$

Therefore the absolute value of the cross correlation between a perfect nonlinear function and any affine function is a constant equal to $2^{-n/2}$. Moreover for a function g which is not perfect nonlinear there is always an affine function L with cross correlation $c(g,L)$ larger than $2^{-n/2}$ in absolute value. This is summarized in the following

Theorem 3.5. The perfect nonlinear functions are the class of functions with minimum correlation to all affine functions.

This property contrasts to correlation immunity. Recall that a m-th order correlation immune function f satisfies $F(\underline{w}) = 0$ for all $\underline{w}$ with Hamming weight less or equal m (cf. [12]). Hence for these vectors $\underline{w}$ the cross correlation $c(f, L_{\underline{w}})$ vanishes. On the other hand Parseval's theorem implies

$$\sum_{\underline{w} \in GF(2)^n} c(f, L_{\underline{w}})^2 = 1 \tag{3.10}$$

for an arbitrary Boolean function f, which means that the "global correlation" to all linear (or affine) functions does not depend on the function f. Thus for correlation immune functions the vanishing of certain cross correlations necessarily leads to larger correlations to other affine functions.

The cross correlation $c(f, 0)$ to the all zero function measures the deviation from ± 1-balance of a Boolean function f. Therefore a perfect nonlinear function is never balanced. However its deviation from balance is given by $2^{-n/2}$ which rapidly tends to 0 as n grows larger. The same holds for the correlation to any other affine function. The fact that there exist no balanced perfect nonlinear functions answers a question raised by Evertse (cf. [3]).

Corollary 3.6. There are no DES-like S-boxes which are perfect nonlinear, or equivalently, S-boxes with maximum distance to linear structures.

3.3. Boolean functions with an odd number of arguments

Recall that there are no perfect nonlinear functions with an odd number of arguments. This relies on the fact that the absolute value of the Walsh transform of a perfect nonlinear function has to be constant (cf. formula (3.4)). However for odd dimensions we can construct functions with the property that the absolute value of their Walsh transform is two-valued. Such functions may be obtained by the following construction.

For $f \in \Phi(n)$, n odd, denote by f_0 the lower half of f, i.e. the function $f_0 \in \Phi(n-1)$ defined by $f_0(x_1, \ldots, x_{n-1}) = f(0,x_1,\ldots,x_{n-1})$, and by $f_1 \in \Phi(n-1)$ the upper half, $f_1(x_1, \ldots, x_{n-1}) = f(1,x_1,\ldots,x_{n-1})$. Similarly denote by F_0 and F_1 the lower and upper half of the Walsh transform F. Moreover let F_0' and F_1' be the Walsh transforms of f_0 and f_1, respectively. Then definition (3.2) implies

$$F_0 = F_0' + F_1'$$
$$F_1 = F_0' - F_1' \qquad (3.11)$$

Suppose now that f_0 and f_1 are perfect nonlinear. Then the values of F_0' and F_1' are $\pm 2^{(n-1)/2}$, which implies that the values of F_0 and F_1 are either 0 or $\pm 2^{(n+1)/2}$. Thus for any pair of perfect nonlinear functions $f_0, f_1 \in \pi(n-1)$ we can construct a function $f \in \Phi(n)$ such that the function $|F|$ takes two values (0 or $2^{(n+1)/2}$). More precisely, by Parseval's theorem (cf. (3.8)), half of the values of $|F|$ are 0 and the other half $2^{(n+1)/2}$.

For n odd denote by $\pi'(n)$ the class of all functions f such that $|F|$ takes the 2 values 0 and $2^{(n+1)/2}$. These classes π' of functions in odd dimensions are related to the classes π in even dimensions. This is reflected by similar properties of the two classes with regard to nonlinear order and distance to affine functions. In analogy to Theorem A it can be shown that the nonlinear order of a function $f \in \pi'(n)$ is always bounded by (n+1)/2. Moreover the distance of a function $f \in \pi'(n)$ to affine functions is obtained as

$$\delta(f) = 2^{n-1} - 2^{(n+1)/2-1}. \qquad (3.12)$$

This shows that in odd dimensions the elements of $\pi'(n)$ are nearly as far from affine functions as are the perfect nonlinear functions in even dimensions. Note however that it is possible to generate functions f in odd dimensions with larger distance $\delta(f)$. In general the maximum value of δ coincides with the covering radius of the Reed-Muller code R(1,n). This covering radius is unknown if n is an arbitrary odd number (cf. [2]).

4. CONCLUSIONS AND APPLICATIONS

The theory of perfect nonlinear (or bent) functions has interesting implications to the design of block ciphers as well as stream ciphers. We have already observed (cf. Corollary 3.6) that perfect nonlinearity may not be compatible with other cryptographic design criteria. For example perfect nonlinearity cannot be achieved in conjunction with balance or highest nonlinear order. However a reasonable strategy will be to find nearly perfect nonlinear functions which satisfy additional design criteria. This is illustrated by the following example of finding nearly perfect nonlinear functions which are balanced.

Recall that a function $f \in \pi(n)$ has distance $2^{n-1} \pm 2^{n/2-1}$ to each affine function. Suppose e.g. that f has distance $2^{n-1} + 2^{n/2-1}$ to the all zero function. Then complementing an arbitrary set of $2^{n/2-1}$ f-values 1 yields a balanced function f'. With regard to distance to affine functions this modified function f' still has desirable properties, since the triangle inequality implies

$$\delta(f') \geq 2^{n-1} - 2^{n/2}. \tag{4.1}$$

To illustrate this procedure take n = 8. In this case it is easy to generate balanced functions with distance δ at least 112 (compared to 120 for perfect nonlinear functions). Instead one could randomly try balanced functions until a function with δ = 112 has been found. However it has appeared (cf. Section 3) that perfect (or nearly perfect) nonlinear functions are very rare in the set of all Boolean functions. Therefore an exhaustive search in the set of balanced functions has virtually zero probability to succeed in reasonable time.

A similar method can be applied to other design criteria, e.g. nonlinear order or correlation immunity. This leads to the following general procedure, where we use a systematic approach to satisfy first those properties which cannot be achieved by a pure random search.

1. Generate a random perfect nonlinear function f using the recursive algorithm as described in Section 3.

2. Find a random function f' as close as possible to f which satisfies all the other desired criteria.

In this way we can construct functions which are useful in stream cipher design where one or more linear feedback shift registers (LFSRs) are combined to produce the key stream.

We start by considering the case where n different taps of one LFSR are nonlinearly combined by some Boolean function $f \in \Phi(n)$ (a situation which was originally treated in [10]). Denote by $\underline{a}_1, \underline{a}_2, \ldots, \underline{a}_n$ the output sequences of these taps. Now suppose that f is correlated to the linear function $L_{\underline{w}}, \underline{w} \in GF(2)^n$. Then the generator output sequence is correlated to the sum

$$\underline{x} = w_1\underline{a}_1 + w_2\underline{a}_2 + \ldots + w_n\underline{a}_n \tag{4.2}$$

which is a sequence (another phase) produced by the same LFSR. The corresponding cross correlation is obtained by (3.9)

$$c_f(\underline{w}) = \frac{F(\underline{w})}{2^n}$$

In this situation the use of correlation immune functions (of any order) is not adequate. To the contrary, correlation immunity of functions is equivalent to the vanishing of certain Walsh coefficients (or cross correlations to certain phases). But in this case Parseval's equality (cf. also (3.10))

$$\sum_{\underline{w} \in GF(2)^n} c_f(\underline{w})^2 = 1$$

implies that cross correlations to other phases are necessarily larger. In this context it is best to face the correlation problem by choosing f as close as possible to a perfect nonlinear function (where all cross correlations are minimum). This treatment also applies to the situation where taps of different LFSR's are combined.

Suppose that a Boolean function $f \in \Phi(n)$ combines a total number of n taps from k different LFSRs. Again, correlation will occur to sequences of the form (4.2) which is caused by correlation of f to the corresponding linear functions L_w. In this more general setting the sequence (4.2) can be expressed as

$$\underline{x} = \underline{b}_1 + \underline{b}_2 + \ldots + \underline{b}_k \tag{4.3}$$

by collecting terms coming from the same LFSR (i.e. $\underline{b}_i = \Sigma \, w_j \underline{a}_j$, summed over the set S_i of all indices j corresponding to tap positions belonging to LFSR i). It may happen that some of the $\underline{b}_i$'s in (4.3) are zero, in which case the generator is vulnerable to a divide and conquer attack by exploiting the correlation. Otherwise stated, if all summands $\underline{b}_i$ are nonzero, a divide and conquer correlation attack is not possible. To this aim maximum order correlation immunity has been postulated in [7]. In our terminology the generator is maximum order correlation immune if the combining function f satisfies the following condition MCI (expressed in terms of the Walsh transform):

$$\text{MCI:} \quad \begin{bmatrix} \text{For every } \underline{w} \\ \text{with } F(\underline{w}) \neq 0 \end{bmatrix} \implies \begin{bmatrix} \text{For every } i, \ 1 \leq i \leq k, \text{ there is at least} \\ \text{one index } \ j \in S_i \ \text{ such that } w_j = 1. \end{bmatrix}$$

In fact MCI is equivalent to the condition that all $\underline{b}_i$'s in (4.3) are nonzero.

In addition to MCI the combining function f may be designed such that the remaining correlations are uniformly small. This can be achieved e.g. by choosing f close to a perfect nonlinear function. By appropriate design these correlations may become as small as to defeat any kind of correlation attack.

Acknowledgement.

We wish to thank Bert den Boer for helpful discussions.

References

[1] D. Chaum, J.-H. Evertse, "Cryptanalysis of DES with a reduced number of rounds", Proceedings of Crypto'85, pp. 192-211.

[2] G.D. Cohen, M.G. Karpovsky, H.F. Mattson, J.R. Schatz, "Covering radius - Survey and recent results", IEEE Trans. Inform. Theory, Vol. IT-31, pp. 328-343, 1985.

[3] J.-H. Evertse, "Linear structures in block ciphers", Proceedings of Eurocrypt'87, pp. 249-266.

[4] R. Forré, "The strict avalanche criterion: Spectral properties of Boolean functions and an extended definition", Proceedings of Crypto'88.

[5] S. Lang, "Algebra", Addison-Wesley Publishing Company, 1971.

[6] O.S. Rothaus, "On bent functions", Journal of Combinatorial Theory (A), Vol. 20, pp. 300-305, 1976.

[7] R.A. Rueppel, "Analysis and design of stream ciphers", Springer-Verlag, 1986.

[8] C.E. Shannon, "Communications theory of secrecy systems", Bell Sys. Tech. Journal, Vol. 28, pp. 656-715, 1949.

[9] T. Siegenthaler, "Correlation-immunity of nonlinear combining functions for cryptographic applications", IEEE Trans. Inform. Theory, Vol. IT-30, pp. 776-780, 1984.

[10] T. Siegenthaler, "Cryptanalysts representation of nonlinearly filtered ML-sequences", Proceedings of Eurocrypt'85, pp. 103-110.

[11] A.F. Webster, S.E. Tavares, "On the design of S-boxes", Proceedings of Crypto'85, pp. 523-534.

[12] G.Z. Xiao, J.L. Massey, "A spectral characterization of correlation-immune combining functions", IEEE Trans. Inform. Theory, Vol IT-34, pp. 569-571, 1988.

On the Linear Complexity
of Feedback Registers

(extended abstract)

A. H. Chan   M. Goresky
A. Klapper

Northeastern University
College of Computer Science
360 Huntington Ave.
Boston, MA, 02115

ABSTRACT

In this paper, we study sequences generated by arbitrary feedback registers (not necessarily feedback shift registers) with arbitrary feedforward functions. We generalize the definition of linear complexity of a sequence to the notions of strong and weak linear complexity of feedback registers. A technique for finding upper bounds for the strong linear complexities of such registers is developed. This technique is applied to several classes of registers. We prove that a feedback shift register whose feedback function is of the form $x_1 + h(x_2, \ldots, x_n)$ can generate long periodic sequences with high linear complexites only if its linear and quadratic terms have certain forms.

I INTRODUCTION

Periodic sequences generated by feedback shift registers have many applications in modern communications systems because of their desirable properties, such as long period and balanced statistics. One measure of the strength (usefulness) of such a sequence is its linear complexity, as studied by various authors [1,2,4,6,7]. The *linear complexity* of a sequence is defined as the length of the shortest linear feedback shift register that

In general, however, these notions do not coincide. For example, the nonlinear feedback shift register $\mathcal{F}$ of length two with feedback function $f(x_1, x_2) = x_1 x_2$ generates the sequences $1111\ldots$, $0000\ldots$, $1000\ldots$, and $01000\ldots$ These sequences have linear complexities 1, 0, 2, and 2, respectively, so the weak linear complexity of $\mathcal{F}$ is two. The strong linear complexity of $\mathcal{F}$, however, is three since each of these sequences is generated by the linear feedback shift register of length three with feedback function x_3 and not by any shorter linear feedback shift register.

We also note that the strong linear complexity of a register $\mathcal{F}$ is equal to the degree of the least common multiple of the connection polynomials of the sequences generated by $\mathcal{F}$.

II UPPER BOUNDS

We derive a technique for computing bounds on the strong linear complexity of (linear and nonlinear) registers with arbitrary feedforward functions. The idea is to embed the given register into a linear register (of exponentially greater length, N). For such a register, the state transition function is considered to be a linear transformation on a vector space of dimension N. We then look for a supporting subspace of minimal dimension. The dimension of this subspace is an upper bound on the strong linear complexity of the original register.

Definition 1 Let $\mathcal{F} = (F, g)$ be a linear register of length n, and let W be a subspace of $GF(2)^n$. W supports F or is $\mathcal{F}$-supporting if there is a subspace U in $GF(2)^n$, complimentary to W, such that

1. $GF(2)^n = W + U$,

2. $F(U) \subseteq U$, and

3. If $w \in W$ and $u \in U$, then $g(w + u) = g(w)$.

Let w be in W and u be in U. For every i, $F^{(i)}$ is linear. By iterating condition 2, $F^{(i)}(u)$ is in U. It follows that $g \cdot F^{(i)}(w + u) = g(F^{(i)}(w) + F^{(i)}(u)) = g \cdot F^{(i)}(w)$. Thus the output from $\mathcal{F}$ can be completely determined from its action on W.

Lemma 1 Suppose $GF(2)^n$ *contains a* $\mathcal{F}$*-supporting subspace* W*. Then the strong linear complexity of* $\mathcal{F}$ *is less than or equal to the dimension of* W*.*

The strong linear complexity of a register is bounded from above by the length of any linear feedback register which can produce all the output sequences of the original register. For an arbitrary feedback register $\mathcal{F} = (F', g,)$ of length n, such a linear register $\mathcal{F}' = (F', g')$ of length $2^n - 1$ can be constructed as follows.

The Construction Let S be the set of nonempty subsets of $\{1, \ldots, n\}$. For every I in S, we construct a new variable x_I and identify it with the monomial $\prod_{i \in I} x_i$. Recall that every element a in $GF(2)$ satisfies $a^2 = a$, so all high degree terms such as $x_i^k, k \geq 1$ appear as x_i. S has cardinality $2^n - 1$, and is used as the index set for the $2^n - 1$ variables in $\mathcal{F}'$. For each I in S, let $F_I(x_1, \ldots, x_n) = \prod_{i \in I} F_i(x_1, \ldots, x_n)$, and let $F_I'(x_1, \ldots, x_{\{1, \ldots, n\}})$ be the linear function derived from F_I by replacing each monomial $\prod_{j \in J} x_j$ by the variable x_J, where J is in S. Then $F' = (F_{\{1\}}', \ldots, F_{\{1, \ldots, n\}}')$ defines a linear function from $V = GF(2)^{2^n - 1}$ to V. The feedforward function g' can be defined similarly as a linear combination of the monomials x_I, giving a linear function from V to $GF(2)$. $\mathcal{F}' = (F', g')$ defines a linear feedback register of length $2^n - 1$ with linear feedforward function.

To show that $\cdot \mathcal{F}'$ generates all the output sequences of $\mathcal{F}$, we consider the embedding $\theta : GF(2)^n \to V$ where the I-th coordinate of $\theta(x_1, \ldots, x_n)$ is $\prod_{i \in I} x_i$. We claim that $\theta \cdot F = F' \cdot \theta$ and $g = g' \cdot \theta$. In other words, the diagram in figure 1 commutes. To see this, note first that $(\theta \cdot F)_I(x_1, \ldots, x_n) = \prod_{i \in I} F_i(x_1, \ldots, x_n) = F_I(x_1, \ldots, x_n)$. On the other hand, $(F' \cdot \theta)_I(x_1, \ldots, x_n) = F_I'(\ldots, \prod_{j \in J} x_j, \ldots)$, i.e., is derived from F_I' by replacing x_J by $\prod_{j \in J}$. But F_I' was derived from F_I by doing the opposite, so $(F' \cdot \theta)_I = F_I = (\theta \cdot F)_I$, so $F' \cdot \theta = \theta \cdot F$. The second claim is proved similarly.

It follows that for any $\alpha \in GF(2)^n$ and any k, $g \cdot F^{(k)}(\alpha) = g' \cdot F'^{(k)}(\alpha)$. Thus the initial loading $\theta(\alpha)$ of $\mathcal{F}'$ gives the same output sequence as the initial loading α of $\mathcal{F}$.

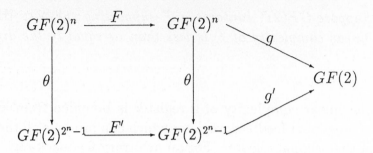

Figure 1: Linearizing a feedback register

Example Let $\mathcal{F} = (F, g)$ be a feedback shift register of length 4 with $g(x_1, x_2, x_3, x_4) = x_1$ and feedback function

$$f(x_1, x_2, x_3, x_4) = x_1 + x_2 x_4 + x_2 x_3 x_4.$$

Then

$$F'(x_1, x_2, x_3, x_4, x_{1,2}, x_{1,3}, x_{1,4}, x_{2,3}, x_{2,4}, x_{3,4}, x_{1,2,3}, x_{1,2,4}, x_{1,3,4}, x_{2,3,4}, x_{1,2,3,4})$$

$$= (x_2, x_3, x_4, x_1 + x_{2,4} + x_{2,3,4}, x_{2,3}, x_{2,4}, x_{1,2} + x_{2,4} + x_{2,3,4}, x_{3,4}, x_{1,3},$$

$$x_{1,4} + x_{2,4} + x_{2,3,4}, x_{2,3,4}, x_{1,2,3}, x_{2,4} + x_{1,2,4} + x_{2,3,4}, x_{1,3,4}, x_{1,2,3,4}).$$

The output sequence obtained from $\mathcal{F}$ with the initial loading $(1, 1, 0, 1)$ is obtained from $\mathcal{F}'$ with initial loading $(1, 1, 0, 1, 1, 0, 1, 0, 1, 0, 0, 1, 0, 0, 0)$.

From the construction above we observe that, if the sequence of polynomials $g'(\bar{x})$, $g' \circ F'(\bar{x})$, $g' \circ F' \circ F'(\bar{x}), \ldots$ contains only terms in $\{x_I | I \in Q\}$ for some $Q \subseteq S$, then we need only those monomials in $\mathcal{F}'$ indexed by elements of Q. Hence a linear feedback register of length $|Q|$ can be constructed that generates the same sequences as $\mathcal{F}$. This shows that the strong linear complexity of $\mathcal{F}$ can be bounded above by $|Q|$. The determination of Q is given by the following theorem.

Theorem 1 *Let $F(x_1, \ldots, x_n)$ be the state change function of a register of length n with feedforward function $g(x_1, \ldots, x_n)$. Let $T = \{I \in S : \Pi_{i \in I} x_i$ has a non-zero coefficient in $g\}$ and let Q be the smallest subset of S containing T such that if $I \in Q$ and the coefficient of x_J in F_I' is nonzero, then $J \in Q$. Then the strong linear complexity of (F, g) is bounded above by the cardinality of Q.*

Corollary 1 Let (F, g) be a feedback shift register with feedback function f. Let $T = \{I \in S : \prod_{i \in I} x_i \text{ has a non-zero coefficient in } g\}$, $R = \{I \in S : \prod_{i \in I} x_i \text{ has a non-zero coefficient in } f\}$. Let Q be the smallest subset of S containing T such that

1. If $I \in Q$ and $n \in I$, then for each $J \in R$, $J \cup \{i+1 \le n : i \in I\} \in Q$.

2. If $I \in Q$ and $n \notin I$, then $\{i+1 : i \in I\} \in Q$.

Then the strong linear complexity of (F, g) is bounded by the cardinality of Q.

We now treat the special case of a feedback shift register $\mathcal{F} = (F, g)$ of length n with feedback function $f(x_1, \ldots, x_n) = x_1 + h(x_2, \ldots, x_n)$ and standard feedforward function. Let T, R, and Q be as in corollary 1. Then $\{1\} \in T \subset Q$, so, by applying condition 2 repeatedly, $\{i\} \in Q$ for all i. In particular $\{n\} \in Q$. If J is the index set of a monomial that has a non-zero coefficient in $h(x_2, \ldots, x_n)$, then we can apply condition 1 with $I = \{n\}$, so $J \in Q$. Let I be any element of Q. Then applying either condition 1 with $J = \{1\}$ or condition 2 (only one condition is applicable to a given index set) $n - 1$ times, we get a sequence of elements of Q, $I = I_1, \ldots, I_n$. One more such application would give us I back again. Actually, we may return to I after a smaller number of applications of the conditions, but this number must divide n. If r is the cardinality of I, then r is the cardinality of each I_i and we call the set $\{I_1, \ldots, I_n\}$ a r-cycle, or simply a cycle if the cardinality is clear. For example, with $n = 4$, starting with $I = \{2, 3\}$ we get the 2-cycle $\{2, 3\}, \{3, 4\}, \{1, 4\}, \{1, 2\}$, whereas starting with $I = \{2, 4\}$, we get the 2-cycle $\{2, 4\}, \{1, 3\}$. These cycles are independent of $h(x_2, \ldots, x_n)$. The set S of all index sets decomposes into a disjoint union of such cycles, each cycle having cardinality dividing n (in fact, there is a relationship between this cycle decomposition and the decomposition of a finite field into cyclotomic cosets). If any one element of a cycle is in Q, then every element of that cycle must be in Q.

Recall again that each monomial in $x_1, \ldots, x_n$ corresponds to an index set, so $\mathcal{F}$ can have high linear complexity only if Q contains many index sets. As seen by the following theorem, this means that the feedback function must have many non-zero coefficients.

Theorem **2** *Let* $\mathcal{F} = (F, g)$ *be a feedback shift register of length* n *with feedback function* $f(x_1, \ldots, \dot{x}_n) = x_1 + h(x_2, \ldots, x_n)$ *and standard feedforward function. Let* r *be the smallest integer such that* $h(x_2, \ldots, x_n)$ *has a term of degree* r *with a non-zero coefficient. For any collection of* r-*cycles* $C_1, \ldots, C_k$, *each of whose corresponding monomials has a zero coefficient in* $h(x_1, \ldots, x_n)$, *the strong linear complexity of* $\mathcal{F}$ *is at most*

$$2^n - 2 - \sum_{i=2}^{r-1} \binom{n}{i} - \sum_{i=1}^{k} |C_i|.$$

This theorem makes precise the folklore belief that shift registers with only high degree terms are not good.

If the output sequence $(z_0, z_1, \ldots)$ from a register $\mathcal{F}$ of length n has maximal period $2^n - 1$, then any set of $2^n - 1$ consecutive bits contains 2^{n-1} ones and $2^{n-1} - 1$ zeros. Therefore the sequence satisfies the relation $z_i + z_{i+1} + \cdots + z_{i+2^n-2} = 0$ for every i. The linear complexity is thus at most $2^n - 2$, and there are registers of length n with linear complexity $2^n - 2$ (for example, the sequence consisting of $2^{n-1} - 1$ zeros followed by 2^{n-1} ones can be generated by such a register.) Note that in the case of a register that outputs a maximal period sequence, the strong and weak linear complexities of the register and the linear complexity of the output sequence all coincide.

In particular, if $\mathcal{F}$ and r are as in the previous theorem, then $\mathcal{F}$ cannot generate a maximal period, maximal linear complexity sequence unless h has quadratic terms and for every 2-cycle C there is an I in C whose corresponding monomial in $h(x_1, \ldots, x_n)$ has non-zero coefficient, or $h(x_1, \ldots, x_n)$ has linear terms.

Corollary **2** *Let* $\mathcal{F} = (F, g)$ *be a feedback shift register of length* n, *with feedback function* $x_1 + h(x_2, \ldots, x_n)$, *and standard feedforward function. If* $\mathcal{F}$ *generates a maximal period, maximal linear complexity sequence, then either* h *contains some linear terms or it has at least* $\lceil (n-1)/2 \rceil$ *quadratic terms.*

By a similar application of corollary 1, we generalize a theorem due to Key.

Proposition 1 (Key [4]) If every term of the feedback function of a feedback shift register with feedforward function has degree 1 (resp. ≤ 1), and every term of the feedforward function has degree $\leq k$, then the strong linear complexity of the register is bounded by $\sum_{i=1}^{k} \binom{n}{i}$ (resp. $\sum_{i=0}^{k} \binom{n}{i}$).

We also prove several similar results.

Proposition 2 If every term of the feedback and feedforward functions of a feedback shift register with feedforward function has degree greater than or equal to k, then the strong linear complexity of the register is bounded above by $\sum_{i=k}^{n} \binom{n}{i}$.

Proposition 3 If every term of the feedback function of a feedback shift register with feedforward function has degree $\geq k$, and the feedforward function is of the form $b_{m+1}x_{m+1} + \cdots + b_n x_n$ (resp. $a + b_{m+1}x_{m+1} + \cdots + b_n x_n$), then the strong linear complexity of the register is bounded above by $n - m + \sum_{i=k}^{n} \binom{n}{i}$ (resp. $1 + n - m + \sum_{i=k}^{n} \binom{n}{i}$).

Proposition 3 says that if the feedback function of a feedback register contains only high degree terms, then the linear complexity is low.

III   GENERALIZATION TO ARBITRARY FINITE FIELDS

The results of the previous section can be generalized to $GF(q)$, the finite field of q elements, where q is a power of an arbitrary prime. The definitions of feedback registers and their various special cases are the same, with 2 replaced by q. The only change is that now every element a of $GF(q)$ satisfies $a^q = a$, so that, when we consider functions as polynomials, we must include monomials in which each variable has degree up to $q - 1$. The remaining definitions (output sequence, weak and strong linear complexity, etc.) carry over verbatim. The counting techniques can then be generalized using multi-sets, and the main results are modified as follows: Theorem 2 holds with the upper bound

$$q^n - \sum_{j=2}^{r-1} \binom{n}{j} (q-1)^j - \sum_{i=1}^{k} |C_i|(q-1)^r - (q-1)^n$$

in the first case, and

$$q^n - 1 - \sum_{j=2}^{r-1} \binom{n}{j} (q-1)^j - \sum_{i=1}^{k} |C_i|(q-1)^r - (q-1)^n$$

in the second.

Let $\#(n, i)$ be the number of monomials of degree i in n variables in which each variable has degree at most $q - 1$. Proposition 1 then holds with $\binom{n}{i}$ replaced by $\#(n, i)$. In Proposition 2, we must require that each term of the feedback and feedforward functions contain at least k variables, and replace $\binom{n}{i}$ by $\#(n, i)$ in the conclusion. Similarly, in Proposition 3, we must require that each term of the feedback function contain at least k variables and replace $\binom{n}{i}$ by $\#(n, i)$ in the conclusion.

REFERENCES

[1] A.H. Chan, R.A.Games and E.L. Key. *On the complexity of deBruijn sequences.* Journal of Combinatorial Theory, Series A **33-3**, pp. 233-246, 1982.

[2] H. Fredricksen. *A Survey of Full Length Nonlinear Shift Register Cycle Algorithms.* SIAM Review **24**, pp. 195-221, 1982.

[3] S. Golomb, "Shift Register Sequences", Aegean Park Press, Laguna Hills, CA, 1982.

[4] E.L. Key. *An Analysis of the structure and complexity of nonlinear binary sequence generators.* IEEE Trans. Inform. Theory **IT-22** no. 6, pp. 732-736, Nov. 1976.

[5] J.L. Massey. *Shift Register Synthesis and BCH Decoding.* IEEE Trans. Inform. Theory **IT-15**, page 122-127, 1969.

[6] R.A. Rueppel. *New approaches to stream ciphers.* Ph.D. Thesis, Swiss Federal Institiute of Technology, Zurich, Switzerland. 1984.

[7] R.A. Rueppel and O.J. Staffelbach. *Products of Linear Recurring Sequences with Maximum Complexity* IEEE Trans. Inform. Theory **IT-33** no. 1, pp.124-131, 1987.

Linear Complexity Profiles and Continued Fractions

Muzhong Wang

Department of Electrical Engineering

University of Waterloo

Waterloo, Ontario, Canada N2L 3G1

Abstract

The linear complexity, $\mathcal{L}(s^n)$, of a sequence s^n is defined as the length of the shortest linear feedback shift-register (LFSR) that can generate the sequence. The linear complexity profile, $L_{s^n} = L_1 L_2 \ldots . L_n$, of s^n (where $L_i = \mathcal{L}(s^i)$, $1 \leq i \leq n$, denotes the linear complexity of first i digits of s^n) provides better insight into the complexity of an individual sequence. By the increment sequence $\Delta_{s^n} = \Delta_1 \Delta_2 \cdots \Delta_m$ in a linear complexity profile, $L_1 L_2 \cdots L_n$, we mean the subsequence of positive numbers in the sequence $L_1 (L_2 - L_1) \ldots (L_n - L_{n-1})$. For example, if $L_1 \cdots L_5 = 0\,2\,2\,2\,3$, its increment sequence is $\Delta_{s^5} = \Delta_1 \Delta_2 = 2\,1$. If we associate a sequence s^n over F with an element $S(z)$ in the field of Laurent series over F in the following way

$$s^n = s_1 s_2 \cdots s_n \iff S(z) = s_1 z^{-1} + s_2 z^{-2} + \cdots + s_n z^{-n},$$

$S(z)$ can then be written as

$$S(z) = a_0(z) + \cfrac{1}{a_1(z) + \cfrac{1}{a_2(z) + \cfrac{1}{\ddots + \cfrac{1}{a_k(z)}}}},$$

where $a_i(z) \in F[z]$, the ring of polynomials in z over F, for all $i \geq 0$. It will be shown that, for a sequence s^n, the increment sequence Δ_{s^n} of the linear complexity profile of s^n is as follows. (1) If $2 \cdot \sum_{i=1}^{k} \deg(a_i(z)) - \deg(a_k(z)) \leq n$, then $\Delta_{s^n} = \deg(a_1(z)) \deg(a_2(z)) \cdots \deg(a_k(z))$. (2) If $2 \cdot \sum_{i=1}^{k} \deg(a_i(z)) - \deg(a_k(z)) > n$, then $\Delta_{s^n} = \deg(a_1(z)) \deg(a_2(z)) \cdots \deg(a_{k'}(z))$, where $k' = \max\{j : 2 \cdot \sum_{i=1}^{j} \deg(a_i(z)) - \deg(a_j(z)) \leq n\}$.

1 Introduction

It has long been known that there is some sort of connection between linear complexity concepts and continued-fraction theory Recently, H. Niederreiter has done lots of works on the problem [NIED 87] [NIED 88] [NIED 89]. If sequences are associated with the elements in the field of Laurent series, the linear complexity profile of a sequence is totally specified by the degrees of partial quotients in the continued-fraction expansion of the corresponding Laurent series [NIED 87]. We will prove that the sequence of "jumps" in the linear complexity profile of a sequence is equal to the sequence of degrees of partial quotients in the continued-fraction expansion of the corresponding Laurent series. Therefore, sequences with desired linear complexity profiles can be constructed by choosing the degrees of partial quotients in the continued-fraction expansion.

We first give a short introduction to continued-fraction expansions in the field of Laurent series (Laurent series field). A *Laurent series* in the indeterminate z over the field F is an expression of the form

$$f_l(z) = \sum_{j=-\infty}^{+\infty} a_j z^j$$

for which $a_j \in F$, all j, and where $a_j = 0$ for $j > d$, where d is some integer. The *degree* of $f_l(z)$, denoted by $\deg f_l(z)$, is the largest j (if any) such that $a_j \neq 0$ and is, by way of convention, $-\infty$ when $a_j = 0$ for all j. For instance, the Laurent series $z + 1 + z^{-1} + z^{-2} + \ldots$ has degree 1 whereas the Laurent series $z^{-1} + z^{-2} + z^{-3} + \ldots$ has degree -1. Addition and multiplication of Laurent series is defined in the same way as for power series. The set of all Laurent series in z over the field F forms a field that we denote by $F(z^{-1})$. A *polynominal* is a Laurent series for which $a_j = 0$ for all $j < 0$. Note that the ring of polynomials in z over F, denoted by $F[z]$, is a subring of the field $F(z^{-1})$.

For a Laurent series $f_l(z)$, one defines its *valuation*, $\|f_l(z)\|$, by

$$f_l(z) = \begin{cases} 2^{\deg f_l(z)} & \text{if} \quad f_l(z) \neq 0 \\ 0 & \text{if} \quad f_l(z) = 0. \end{cases}$$

This is a *nonarchimedean valuation* because

$$\|f_l(z) + g_l(z)\| \leq \max\{\|f_l(z)\|, \|g_l(z)\|\},$$

which is stronger than the more usual "norm inequality" in which the right side is the sum of the two valuations.

For convenience, we summarize without proof some obvious properties of $\|.\|$.

Lemma 1 $\|.\|$ *has the following properties:*

P1. $\|f_l(z)\,g_l(z)\| = \|f_l(z)\|\,\|g_l(z)\|.$

P2. $\|f_l(z)\| \geq 0$ *with equality if and only if* $f_l(z) = 0.$

P3. $\|f_l(z) + g_l(z)\| \leq \max\{\|f_l(z)\|, \|g_l(z)\|\}$, *with equality if* $\|f_l(z)\| \neq \|g_l(z)\|.$

P4. $\|1\| = 1.$

P5. $\|f_l(z)^{-1}\| = \|f_l(z)\|^{-1}.$

P6. $\| - f_l(z)\| = \|f_l(z)\|.$

P7. *If* $f_l(z) \in F[z]$ *and* $f_l(z) \neq 0$, *then* $\|f_l(z)\| \geq 1.$

Euclid's division theorem for polynomials can be restated in terms of $\|.\|$ as follows.

Theorem 1 (**Euclid's Division Theorem for Polynomials**) *If* $f(z)$ *and* $g(z)$ *are in* $F[z]$ *with* $g(z) \neq 0$, *then there exists unique* $q(z)$ *and* $r(z)$ *in* $F[z]$ *such that*

$$f(z) = q(z)g(z) + r(z) \qquad and \quad \|r(z)\| < \|g(z)\|.$$

A *continued-fraction* in the indeterminate z over the field of F is an expression of the form

$$a_0(z) + \cfrac{1}{a_1(z) + \cfrac{1}{a_2(z) + \cfrac{}{\ddots + \cfrac{1}{a_i(z) + \cdots}}}} ,$$

where $a_i(z) \in F[z]$ for all $i \geq 0$ and either (1) $\deg a_i(z) \geq 1$ ($\|a_i(z)\| \geq 2$) for all $i \geq 1$ (in which case the continued-fraction is called to be *infinite*) or (2), for some positive integer N,

$\deg a_i(z) \geq 1$ for $1 \leq i \leq N$ and $a_i(z) = 0$ for all $i > N$ (in which case the continued-fraction is said to be *finite*). The polynomials $a_i(z)$ are called the *partial quotients* of the continued-fraction. There is a unique way in which the indicated divisions in a continued-fraction can be carried out to give a Laurent series, and we thus regard hereafter a continued-fraction as an element of $F(z^{-1})$.

Given any continued-fraction, let $[a_0(z); a_1(z), \ldots, a_n(z)]$ denote the finite continued-fraction obtained by setting $a_i(z) = 0$ for all $i \geq n$, i.e., the finite continued-fraction

$$a_0(z) + \cfrac{1}{a_1(z) + \cfrac{1}{a_2(z) + \cfrac{}{\ddots + \cfrac{1}{a_i(z)}}}} \ .$$

Every finite continued-fraction can, after clearing of denominators, be written as the ratio of two polynomials, i.e., as an element of $F(z)$, the field of rational functions over F. Thus one can write

$$[a_0(z); a_1(z), \cdots, a_n(z)] = \frac{p_n(z)}{q_n(z)}, \qquad n \geq 0,$$

where $p_n(z)$ and $q_n(z)$ are polynomials, defined recursively by

$$p_0(z) = a_0(z), \quad p_k(z) = a_k(z)p_{k-1}(z) + p_{k-2}(z) \ (k \geq 1), \tag{1}$$

$$q_0(z) = 1, \qquad q_k(z) = a_k(z)q_{k-1}(z) + q_{k-2}(z) \ (k \geq 1) \tag{2}$$

where, by way of convention, $p_{-1}(z) = 1$ and $q_{-1}(z) = 0$. The rational function $\frac{p_n(z)}{q_n(z)}$ is called the n-th *convergent* of the continued-fraction $[a_0(z); a_1(z), \cdots, a_i(z), \ldots]$.

The following lemma is proved in [LID-NIE 83, pp.235–239].

Lemma 2 *The convergents of* $[a_0(z); a_1(z), \ldots, a_i(z), \ldots]$ *have the following properties:*

$$p_k(z)q_{k-1}(z) - p_{k-1}(z)q_k(z) = (-1)^{k-1} \quad (k \geq 1) \tag{3}$$

or, equivalently,

$$\frac{p_k(z)}{q_k(z)} - \frac{p_{k-1}(z)}{q_{k-1}} = \frac{(-1)^{k-1}}{q_k(z)q_{k-1}(z)} \quad (k \geq 1).$$

Equation (3) implies that

$$\gcd\left(p_i(z), q_i(z)\right) = 1 \qquad \text{for } i \geq 1. \tag{4}$$

The following property of convergents, which appears to be new, will play an important role in the sequel.

Lemma 3 *The denominator $q_n(z)$ of the n-th convergent to $[a_0(z); a_1(z), \ldots, a_i(z), \ldots]$ satisfies*

$$\|q_0(z)\| = 1, \tag{5}$$
$$\|q_n(z)\| = \prod_{j=1}^{n} \|a_j(z)\|, \quad n \geq 1 \tag{6}$$

provided $a_n(z) \neq 0$.

Proof. Because $q_0(z) = 1$, we have $\|q_0(z)\| = 1$ as claimed. Because

$$q_1(z) = a_1(z),$$

(6) holds trivially for $n = 1$.

Suppose that (6) holds for $1 \leq n \leq N$. Because $\|a_j(z)\| > 1$ for $1 \leq j \leq N$, $\|q_{N-1}(z)\|$ is strictly smaller than $\|q_N(z)\|$. Thus

$$\begin{aligned}
\|q_{N+1}(z)\| &= \|a_{N+1}(z)q_N(z) + q_{N-1}(z)\| \\
&= \|a_{N+1}(z)q_N(z)\| \qquad \text{(P3 in Lemma 1)} \\
&= \|a_{N+1}(z)\| \cdot \|q_N(z)\| \qquad \text{(P1 in Lemma 1)} \\
&= \prod_{j=1}^{N+1} \|a_j(z)\|.
\end{aligned}$$

This completes the proof by induction.

The following theorem, which is proved in [WES-SCH 79, theorem 2], shows the sense in which the n-th convergent is the best approximation to

$$[a_0(z); a_1(z), \ldots, a_i(z), \ldots] = S(z).$$

Theorem 2 *The convergents to $[a_0(z); a_1(z), \ldots, a_i(z), \ldots]$ have the property that, for every n $(n \geq 0)$, if $q(z)$ is a polynomial with $\|q(z)\| < \|q_{n+1}(z)\|$, then, for any polynomial $p(z)$ such that*

$$\frac{p(z)}{q(z)} \neq \frac{p_n(z)}{q_n(z)},$$

it must hold that

$$\left\| \frac{p_n(z)}{q_n(z)} - S(z) \right\| < \left\| \frac{p(z)}{q(z)} - S(z) \right\|.$$

Let $g_l(z)$ be an element in $\boldsymbol{F}(z^{-1})$. If also $g_l(z) \in \boldsymbol{F}(z)$, then

$$g_l(z) = \frac{r_{-2}(z)}{r_{-1}(z)},$$

where $r_{-2}(z)$ and $r_{-1}(z)$ are polynomials with $\|r_{-1}(z)\| \geq 1$. There exist unique polynomials $a_0(z)$ and $r_0(z)$ such that $r_{-2}(z) = a_0(z) r_{-1}(z) + r_0(z)$ and $\|r_0(z)\| < \|r_{-1}(z)\|$. Equivalently,

$$g_l(z) = a_0(z) + \frac{r_0(z)}{r_{-1}(z)}.$$

If $\|r_0(z)\| \neq 0$, then by the same argument there exist unique polynomials $a_1(z)$ and $r_1(z)$ such that

$$\frac{r_{-1}(z)}{r_0(z)} = a_1(z) + \frac{r_1(z)}{r_0(z)}$$

and $\|r_1(z)\| < \|r_0(z)\| < \|r_{-1}(z)\|$. Continuing in this manner, we must eventually reach the case $r_N(z) = 0$ because the degrees of $r_{-1}(z), r_0(z), r_1(z), \cdots$ are strictly decreasing. Thus it follows that we can always write a rational function $\frac{r_{-2}(z)}{r_{-1}(z)}$ as a finite continued-fraction

$$\frac{r_{-2}(z)}{r_{-1}(z)} = [a_0(z); a_1(z), \cdots, a_N(z)].$$

The converse statement that every finite continued-fraction $[a_0(z); a_1(z), \ldots, a_N(z)]$ represents an element of $\boldsymbol{F}(z)$ was noted previously.

Example. $\quad r_2(z) = z^3 + z^2 + 1, \quad r_{-1}(z) = z^4,$

$\qquad\qquad\quad a_0(z) = 0,$

$\qquad\qquad\quad r_0(z) = z^3 + z^2 + 1,$

$\qquad\qquad\quad a_1(z) = z + 1, \quad r_1(z) = z^2 + z + 1,$

$$a_2(z) = z, \quad r_2(z) = z + 1,$$

$$a_3(z) = z, \quad r_3(z) = 1,$$

$$a_4(z) = z + 1, \quad r_4(z) = 0,$$

$$\frac{r_{-2}(z)}{r_{-1}(z)} = \frac{z^3 + z^2 + 1}{z^4}$$

$$= \cfrac{1}{z + 1 + \cfrac{1}{z + \cfrac{1}{z + \cfrac{1}{z + 1}}}}$$

$$= [0; z + 1, z, z, z + 1].$$

2 Relation between Linear Complexity Profile and Continued Fractions

The *linear complexity* $\mathcal{L}(s^n)$ of a sequence, $s^n = s_1 s_2 \cdots s_n$ where $s_1, s_2, \ldots, s_n$ are from a field F, can also be defined as the smallest nonnegative integer L such that there exist $c_0, c_1, \cdots, c_L$ in F satisfying

$$c_L s_{i+L} + c_{L-1} s_{i+L-1} + \cdots + c_0 s_i = 0, \ 1 \le i \le n - L, \tag{7}$$

where $c_L \ne 0$. The monic polynomial $c_L^{-1}(c_L D^L + \ldots + c_1 D + c_0)$ is called a *characteristic polynomial* of the sequence; we remark that the characteristic polynomial is unique if and only if $L \le n/2$.

The *linear complexity profile* L_{s^n} is defined as the sequence

$$L_{s^n} = L_1 L_2 \cdots L_n$$

where $L_i = \mathcal{L}(s^i)$. The definition of linear complexity implies

$$L_i \ge L_j \quad \text{for } i > j. \tag{8}$$

We associate a sequence s^n over F with an element $S(z)$ in the field of Laurent series over F in the following way

$$s^n = s_1 s_2 \cdots s_n \iff S(z) = s_1 z^{-1} + s_2 z^{-2} + \cdots + s_n z^{-n}. \tag{9}$$

We see immediately that the sequence s^n and the sequence $s^{n+m} = s^n 0^m$ (where $s^n 0^m$ denotes the concatenation of s^n and 0^m) are associated with the same element $\sum_{i=1}^n s_i z^{-i}$ in the field of Laurent series. Therefore, we can implicitly expand $s^n = s_1 s_2, \cdots s_n$ to a semi-infinite sequence s^∞ by concatenating s^n with infinitely many zeroes,

$$s^\infty = s_1 s_2 \cdots s_n 000 \cdots .$$

Suppose that $S(z) = s_1 z^{-1} + s_2 z^{-2} + \cdots$ is a Laurent series with $\|S(z)\| < 1$. Letting $q(z) = c_L z^L + \cdots + c_1 z + c_0$, $c_L \neq 0$, we see that the left side of (7) is the coefficient of z^{-i} in the product $S(z)q(z)$. Thus, if (7) holds, there is a unique polynomial $p(z)$ such that $\|p(z) - S(z)q(z))\| < 2^{n+L}$ and hence (by P1 and P5) such that

$$\|\frac{p(z)}{q(z)} - S(z)\| < 2^{-n}. \tag{10}$$

Moreover, $\|p(z)\| < \|q(z)\| = 2^L$.

Conversely, if (10) holds where $q(z) = c_L z^L + \cdots + c_1 z + c_0$ and $p(z)$ are polynomials with $\|q(z)\| = 2^L$, then (7) also holds and $\|p(z)\| < \|q(z)\|$. We have thus proved the following lemma.

Lemma 4 *The linear complexity of a sequence $s^n = s_1 s_2 \ldots s_n$ is equal to the minimum degree of polynomials $q(z)$ such that there exists a polynomial $p(z)$ satisfying*

$$\|\frac{p(z)}{q(z)} - S(z)\| < 2^{-n},$$

where $S(z) = s_1 z^{-1} + s_2 z^{-2} + \cdots + s_n z^{-n}$. Moreover, $c_L^{-1} q(z)$ is a characteristic polynomial of s^n, where c_L is the leading coefficient of $q(z)$.

Because $S(z)$ is in $F(z)$, $S(z)$ can be expressed as a finite continued-fraction

$$\begin{aligned}
S(z) &= a_0(z) + \cfrac{1}{a_1(z) + \cfrac{1}{a_2(z) + \cfrac{\ddots}{\qquad + \cfrac{1}{a_N(z)}}}} \\
&= [0; a_1(z), \cdots, a_N(z)]
\end{aligned}$$

for some N (where the polynomial part vanishes because $\|S(z)\| < 1$). Notice that $a_0(z)$ is always zero.

The following theorem is proved in [NIED 87]. For readers' convenience, we give an alternative proof here.

Theorem 3 *The linear complexity profile, $L_{\bullet\infty}$, of the sequence s^∞ is totally specified by the degrees of the partial quotients in the continued-fraction expansion of $S(z)$, in the following way:*

L1 if $\deg a_1(z) > 1$, $L_i = 0$, $1 \le i < \deg a_1(z)$;

L2 $L_i = \deg a_1(z)$, $\deg a_1(z) \le i < \deg a_2(z) + \deg a_1(z)$;
$L_i = \deg a_2(z) + \deg a_1(z)$, $\deg a_2(z) + 2\deg a_1(z) \le i < \deg a_3(z) + 2\sum_{i=1}^{2} \deg a_i(z)$;

$\vdots$

$L_i = \sum_{i=1}^{N-1} \deg a_i(z)$, $\deg a_{N-1}(z) + 2\sum_{i=1}^{N-2} \deg a_i(z) \le i < \deg a_N(z) + 2\sum_{i=1}^{N-1} \deg a_i(z)$;

$\vdots$.

Proof. L1 is obvious because $\deg a_1(z) > 1$ implies $s_1 = \ldots = s_{(\deg a_1(z)-1)} = 0$.

Consider the convergents

$$\frac{p_n(z)}{q_n(z)} = [0; a_1(z), a_2(z), \cdots, a_n(z)], \qquad n \ge 1.$$

We know from Lemma 2 and Lemma 3 that

$$\frac{p_{n+1}(z)}{q_{n+1}(z)} - \frac{p_n(z)}{q_n(z)} = \frac{(-1)^n}{q_{n+1}(z)p_{n+1}(z)}$$

and that $\xi_n = \deg[q_n(z)q_{n+1}(z)] = \deg a_{n+1}(z) + 2\sum_{i=1}^{n} \deg a_i(z)$. This implies that the coefficients of z^{-i} for $1 \le i < \xi_n$ in the Laurent series for $\frac{p_{n+1}(z)}{q_{n+1}(z)}$ and $\frac{p_n(z)}{q_n(z)}$ are the same but that the coefficients of z^{ξ_n} are different. Thus

$$\frac{p_n(z)}{q_n(z)} = s_1' z^{-1} + \cdots + s_{\xi_n-1}' z^{-(\xi_n-1)} + s_{\xi_n}' z^{-\xi_n} + \cdots$$

where

$$s_i' = s_i \quad \text{for } 1 \le i \le \xi_n - 1, \text{ and}$$

$$s_{\xi_n}' \ne s_{\xi_n}.$$

We have then

$$\left\| \frac{p_n(z)}{q_n(z)} - S(z) \right\| = 2^{-\xi_n}. \tag{11}$$

According to Lemma 4,

$$\mathcal{L}(s_1 s_2 \cdots s_{\xi_n}) \ge \deg q_n(z). \tag{12}$$

By the same argument that gives (11), we have

$$\left\| \frac{p_{n+1}(z)}{q_{n+1}(z)} - S(z) \right\| = 2^{-\xi_{(n+1)}}. \tag{13}$$

Theorem 4 shows that there exists no polynomials $p(z)$ and $q(z)$ with $\|q(z)\| < \|q_{n+1}(z)\|$ such that

$$\left\| \frac{p(z)}{q(z)} - S(z) \right\| < \left\| \frac{p_n(z)}{q_n(z)} - S(z) \right\|.$$

That is to say, $q_{n+1}(z)$ is the polynomial with minimum degree such that (13) holds.

It now follows from Theorem 2, (12) and (13) that

$$\mathcal{L}(s_1 s_2 \ldots s_i) = \deg q_{n+1}(z) \text{ for } \xi_n \le i < \xi_{n+1}.$$

This proves L2.

By the *increment sequence* $\Delta_1 \Delta_2 \cdots \Delta_m$ in a linear complexity profile, $L_1 L_2 \cdots L_n$, we mean the subsequence of positive numbers in the sequence $L_1 (L_2 - L_1) \ldots (L_n - L_{n-1})$. For example, if $L_1 \cdots L_5 = 0\,2\,2\,2\,3$, its increment sequence is $\Delta_1 \Delta_2 = 2\,1$.

Lemma 5 The linear complexity profile $L_1 L_2 \ldots L_n$ is uniquely determined by its increment sequence, and conversely.

Proof. The linear complexity profile trivially determines the increment sequence. The increment sequence uniquely determines the linear complexity after the k-th jump as $\Delta_1 + \Delta_2 + \cdots + \Delta_k$. Suppose this jump occurs at position $i+1$, i.e., $L_{i+1} = \Delta_1 + \Delta_2 + \cdots + \Delta_k > L - i =$

$\Delta_1 + \Delta_2 + \cdots + \Delta_{k-1}$. By the "Length-Change Property of LFSR's" proved in [MASS 69, theorem 2], $L_{i+1} \neq L_i$ implies $L_{i+1} = i + 1 - L_i$ for all $i > 0$ ($L_0 = 0$ by way of convention). Thus

$$i + 1 = L_{i+1} + L_i \tag{14}$$
$$= 2L_i + (L_{i+1} - L_i) \tag{15}$$
$$= 2(\Delta_1 + \Delta_2 + \cdots + \Delta_{k-1}) + \Delta_k. \tag{16}$$

Thus, the location $i + 1$ of the k-th jump ($L_{i+1} - L_i$) is also uniquely determined by the increment sequence. This proves the lemma.

For instance, suppose that the increment sequence of a linear complexity profile is 1 3 2, the linear complexity profile can only be

$$L_{\bullet\infty} = 1^4 4^4 6^\infty . \tag{17}$$

With the aid of Lemma 5, we now have our main result.

Corollary 1 to Theorem 3. If a semi-infinite sequence $s^\infty = s_1 s_2 \cdots$ over a field F is associated with the element $S(z) = \sum_{i=1}^\infty s_i z^{-i}$ in the field of Laurent series over F, then the increment sequence of the linear complexity profile of s^∞ is equal to the sequence of degrees of the partial quotients in the continued-fraction expansion of $S(z)$, i.e., $\Delta_k = \deg[a_k(z)]$.

Corollary 2 to Theorem 3. If a finite sequence $s^n = s_1 s_2 \cdots s_n$ over a field F is associated with the element $S(z) = \sum_{i=1}^n s_i z^{-i} = [0; a_1(z), a_2(z), \cdots, a_k(z)]$ in the field of Laurent series over F, then the increment sequence $\Delta_{\bullet^n}$ of the linear complexity profile of s^n is as follows.

1. *If $2 \cdot \sum_{i=1}^k \deg(a_i(z)) - \deg(a_k(z)) \leq n$, then $\Delta_{\bullet^n} = \deg(a_1(z)) \deg(a_2(z)) \cdots \deg(a_k(z))$.*

2. *If $2 \cdot \sum_{i=1}^k \deg(a_i(z)) - \deg(a_k(z)) > n$, then $\Delta_{\bullet^n} = \deg(a_1(z)) \deg(a_2(z)) \cdots \deg(a_{k'}(z))$, where $k' = max\{j : 2 \cdot \sum_{i=1}^j \deg(a_i(z) - \deg(a_j(z)) \leq n\}$.*

These corollaries tell us how to construct (finite and infinite) sequences with desired linear complexity profiles.

Example. Construct all sequences over F_2 that have the linear complexity profile $1^4 4^4 6^\infty$ of (17). The increment sequence of this linear complexity profile is 1 3 2 . According to the Corollary 1 to Theorem 3, a sequence with this increment sequence has the finite continued-fraction

$$S(z) = \cfrac{1}{a_1(z) + \cfrac{1}{a_2(z) + \cfrac{1}{a_3(z)}}},$$

where $\deg a_1(z) = 1$, $\deg a_2(z) = 3$, $\deg a_3(z) = 2$. There are 2^i ways to choose a polynomial over F_2 with degree i. There are thus $2^1 2^3 2^2 = 64$ different choices for $S(z)$, i.e., there are 64 semi-infinite binary sequences having the linear complexity profile of (17). For a specific such sequence, we choose

$$\begin{aligned} a_1(z) &= z, \\ a_2(z) &= z^3, \\ a_3(z) &= z^2. \end{aligned}$$

We have then

$$\begin{aligned} S(z) &= \cfrac{1}{z + \cfrac{1}{z^3 + \cfrac{1}{z^2}}} \\ &= \frac{z^5 + 1}{z^6 + z^2 + z}. \end{aligned}$$

By long division, we find

$$S(z) = z^{-1} + z^{-5} + z^{-9} + z^{-10} + \cdots .$$

The desired sequence is

$$s^\infty = 1(000100011001010111110)^\infty.$$

If a semi-infinite sequence s^∞ corresponds to the element $S(z)$ of the Laurent field in the manner (9) such that the continued-fraction expansion is infinite, i.e.,

$$S(z) = [0; a_1(z), \ldots, a_k(z), \ldots],$$

then by using the k-th convergent of $S(z)$ to approximate $S(z)$, we can see that L1 and L2 in Theorem 3 still hold. If k goes to infinity, the k-th convergent then approaches $S(z)$. Therefore, L1 and L2 in Theorem 3 and the Corollary 1 to Theorem 3 are also valid for the case that the continued-fraction expansion of $S(z)$ is infinite.

3 Remarks

In [NIED 86], Niederreiter showed the following result. If the continued-fraction expansion of $S(z)$ is infinite, which is the same as saying that $S(z)$ is irrational, the linear complexity profile satisfies

$$\frac{1}{2}(i+1-K(S)) \leq L_i \leq \frac{1}{2}(i+K(S)) \qquad \text{for all } i \geq 1, \tag{18}$$

where $L_i = \mathcal{L}(s_1 s_2 \cdots s_i)$ and $K(S) = \sup_{j \geq 1} \deg a_j(z)$. We now show that (18) is a simple consequence of L2 in Theorem 3 and the "length-change property of LFSR's" for the case that the continued-fraction expansion of $S(z)$ is infinite.

We restate (2) in Theorem 3 as follows.

For

$$\deg a_k(z) + 2 \sum_{j+1}^{k-1} \deg a_j(z) \leq i \leq \deg a_{k+1}(z) + 2 \sum_{j=1}^{k} \deg a_j(z) - 1, \tag{19}$$

where $k \geq 1$, we have

$$L_i = \sum_{j=1}^{k} \deg a_j(z) \tag{20}$$

$$= \frac{1}{2}\Big(\deg a_k(z) + 2 \sum_{j=1}^{k-1} \deg a_j(z) + \deg a_k(z) \Big) \tag{21}$$

$$\leq \frac{1}{2}(i + \deg a_k(z)) \tag{22}$$

with equality when $i = \deg a_k(z) + 2 \sum_{j=1}^{k-1}$, where the last step follows from the left inequality of (19).

Further,

$$L_i = \sum_{j=1}^{k} \deg a_j(z)$$

$$= \frac{1}{2}(a_{k+1}(z) + 2\sum_{j=1}^{k} \deg a_j(z) - a_{k+1}(z)).$$

It follows then from the right inequality of (19) that

$$L_i \geq \frac{1}{2}(i + 1 - \deg a_{k+1}(z)) \tag{23}$$

with equality when $i = \deg a_{k+1}(z) + 2\sum_{j=1}^{k} a_j(z) - 1$. Inequalities (22) and (23) immediately give (18).

Baum and Sweet [BAU-SWE 77] showed that all partial quotients of the continued-fraction expansion of $S(z)$ have degree one if and only if

$$S^2(z) + zS(z) + 1 = (1 + z)g^2(z) \tag{24}$$

for some polynomial $g(z)$. Their equation (24) is the same as

$$s_1 = 1, \quad \text{and}$$

$$s_{2i+1} = s_{2i} + s_i \quad \text{for } i \geq 1.$$

The Corollary 1 to Theorem 3 implies then that all sequences s^∞, for which $S(z)$ satisfies (24), have the linear complexity profile $1\,1\,2\,2\,3\,3\,\ldots$, defined as the *perfect linear complexity profile* (PLCP). This is consistent with the result proved in [WAN-MAS 86], namely, that s^∞ has a perfect linear complexity profile if and only if

$$s_1 = 1,$$

$$s_{2i+1} = s_{2i} + s_i \quad \text{for } i \geq 1.$$

Acknowledgements

I wish to thank Prof. James L. Massey for his guidance of this work. I further wish to thank Prof. Ian F. Blake for his encouragement and help.

References

[BAU-SWE 77] L. E. Baum and M. M. Sweet, "Badly approximable power series in characteristic 2" , *Ann. of Math.* , 105 (1977), pp.573–580.

[LID-NIE 83] R. Lidl and H. Niederreiter, *Finite Fields*, Addison-Wesley Publishing Company, Inc., 1983.

[MASS 69] J. L. Massey, "Shift-Register Synthesis and BCH Decoding", *IEEE Trans. on Info. Th.*, pp.122–127, IT-15, No.1, Jan. 1969.

[NIED 86] H. Niederreiter, "Continued Fractions for Formal Power Series, Pseudorandom Numbers, and Linear Complexity of Sequences", to appear in *Contributions to General Algebra 5* (Proc. Conf. Salzburg, 1986), Teubner, Stuttgart.

[NIED 87] H. Niederreiter, *"Sequences with almost perfect linear complexity profile"*, Proc. Eurocrypt'87, LNCS 304, 37-51 (1988).

[NIED 88] H. Niederreiter, *"The probabilistic theory of linear complexity"*, Proc. Eurocrypt'88, LNCS 330, 191-209, (1988).

[NIED 89] H. Niederreiter, *"Keystream sequence with good linear complexity profile for every starting point"*, paper presented at Eurocrypt'89, Hauthalen, Belgium.

[WAN-MAS 86] M. Z. Wang and J. L. Massey, "The Characterization of All Binary Sequences with Perfect Linear Complexity Profiles", paper presented at Eurocrypt'86, Linköping, Sweden.

[WES-SCH 79] L. R. Welch and R. A. Scholtz, "Continued Fractions and Berlekamp's Algorithm", *IEEE Trans. on Info. Th.*, pp.19–27, IT-25, No.1, January 1979.

A FAST CORRELATION ATTACK ON NONLINEARLY FEEDFORWARD FILTERED SHIFT–REGISTER SEQUENCES

Réjane Forré

Inst. for Communication Technology
Swiss Federal Institute of Technology
CH–8092 Zürich, Switzerland

ABSTRACT

An algorithm recently introduced by Meier and Staffelbach is modified to be applicable to stream–ciphers with running key generators (RKG) consisting of a single linear feedback shift–register (LFSR) with a (nonlinear) feedforward filter applied to it. It is shown that, under certain assumptions, this modified algorithm can be used by a cryptanalyst to determine an equivalent system –consisting of a couple of LFSR's together with a suitable combining function– which generates the same running key sequence. Finally, design criteria are given, which ensure that a RKG withstands the modified attack.

I. INTRODUCTION

A running key generator consisting of a maximum–length (ML) linear feedback shift–register and some nonlinear filtering function f is investigated (Fig. 1). Siegenthaler showed in [1] that the

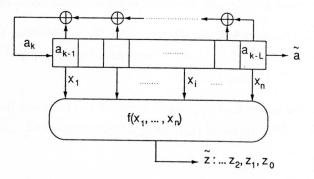

Figure 1: Structure of the investigated running key generator

output sequence $\tilde{z}$ and the ML–sequence $\tilde{a}$ of any RKG of the above type have a cross–correlation function (CCF) with a number of peaks (depending on the function f) whose magnitudes depend only on the Walsh–transform of f. He showed also by using a suitable set of linearly independent LFSR–sequences with initial states $S_1, S_2, \ldots, S_s$, which can be derived from these CCF–peaks, how it is possible to construct an equivalent system of the form as shown in Fig. 2, with s LFSR's with identical feedback connections and a nonlinear combining function g. However, the feasibility

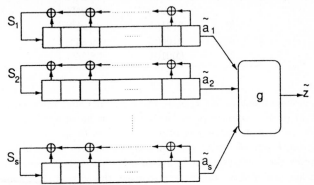

Figure 2: Cryptanalyst's equivalent system, with s identical LFSR's initially loaded with linearly independent states derived from the peaks of the CCF of $\tilde{a}$ and $\tilde{z}$

of this attack up to now was restricted to a RKG with a relatively short LFSR, because of the exponentially growing computational work needed to determine the peaks of the CCF of the sequences $\tilde{a}$ and $\tilde{z}$. In this paper, we show how to determine those peaks with a modified version of the correlation attack according to algorithm A as described by Meier and Staffelbach in [2,3], and therefore how to break a RKG of the above type with a long LFSR.

II. MODIFIED CORRELATION ATTACK

Meier and Staffelbach implicitely assumed in [2,3] a RGK built from a number s of LFSR's generating *cyclically different* (contrary to the situation as shown in Fig. 2) binary sequences $\tilde{a}_1, \tilde{a}_2, \ldots, \tilde{a}_s$ which are combined by some boolean function.

They considered the generated keystream $\tilde{z}$ as a noisy version of the sequence $\tilde{a}_i$, with the noise coming conceptually from the sequences $\tilde{a}_1, \tilde{a}_2, \ldots, \tilde{a}_j, \ldots, \tilde{a}_s$ for $j \neq i$. Their algorithms reconstruct from the sequence $\tilde{z}$ each of the sequences $\tilde{a}_i$ by using correlation properties between the sequences $\tilde{a}_i$ and $\tilde{z}$. The behaviour of their algorithms, however, is not clear for the case where the combined sequences are only cyclic shifts of each other. This is e.g. the case when the sequences are derived from the stages of a single LFSR as shown in Fig. 1 (or equivalently as shown in Fig. 2). Meier's and Staffelbach's algorithms may in this case not be able to converge to some defined result because there are instead of a single solution many (s) convergence points.

In this paper we investigate the corresponding problems and modify the algorithm to be applicable for RKG's as given in Fig. 1.
We first recall in this section the principles of the attack by Meier and Staffelbach.

Assume the cryptanalyst has observed N bits of a running key sequence $\tilde{z}$ known to be correlated to a ML–sequence $\tilde{a}$ produced by some MLLFSR of length k, having t feedback taps. The sequence $\tilde{z}$ may be viewed as a perturbation of $\tilde{a}$ by a binary asymmetric memoryless source (with $\mathrm{Prb}(0) = p_0 \neq 0.5$), and the purpose of the cryptanalyst is to reconstruct the LFSR–sequence $\tilde{a}$ from $\tilde{z}$. Every bit a_j of $\tilde{a}$ satisfies several linear relations (according to the basic feedback relation of the LFSR), each of them involving t other bits of $\tilde{a}$. The cryptanalyst checks how many of those

Table I: LFSR–initial states that yield ML–sequences highly correlated to the sequence $\tilde{z}$ of Fig. 3

Peak Nr.	Initial states	Correlation
1	101011	-69.84%
2	101010	68.25%
3	010100	68.25%
4	010000	-69.84%
5	000101	68.25%

relations hold for the corresponding bit z_j of $\tilde{z}$: the more they are, the higher is the probability for z_j to agree with a_j (if $p_0 > 0.5$), resp. to be the complement of a_j (if $p_0 < 0.5$). After having assigned to each bit z_j of $\tilde{z}$ a probability p_j of being equal to (resp. the complement of) a_j, the cryptanalyst selects the k bits of $\tilde{z}$ with the highest probabilities, uses these bits as a reference guess I_0 and computes the corresponding LFSR–initial state. Since some bits of the reference guess –usually with low probability– might be erroneous, the cryptanalyst sometimes has to test modifications of I_0 with Hamming–distances $1, 2, \ldots$ until he finds the correct initial state of the LFSR.

If this attack is applied to RKG's of the type of Fig. 1, there is no guarantee that it will succeed, since the high probable bits of the reference guess do not necessarily all correspond to the same CCF–peak, as the following example shows.

Example 1

The generator of Fig. 3 is investigated. The table I lists the initial states of the LFSR of Fig. 3

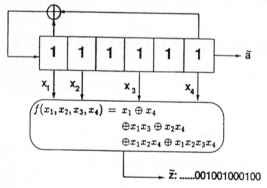

Figure 3: A running key generator with a MLLFSR of length 6 (feedback polynomial $x^6 + x + 1$)

that produce ML–sequences with a high correlation to $\tilde{z}$. A correlation of 70.00% in the second column of table I means that 70.00% of the bits of one period of $\tilde{z}$ coincide with the corresponding bits of the LFSR–sequence. The notation –70.00%, on the other hand, signifies that 70.00% of the bits of $\tilde{z}$ are the complements of the corresponding bits of the LFSR–sequence. Notice that the initial states and the correlations of Table I were computed without determining the full CCF of $\tilde{z}$ and $\tilde{a}$, by means of the Walsh–tranform of the function f [1].

We get a better insight into the situation by considering the table II. It contains the full period of 63 bits output by the generator of Fig. 3, the individual bit probabilities computed according to [2,3], and the five LFSR–sequences yielded by the five initial states of table I, where the sequences with negative correlations (Nr. 1 and 4) have been complemented.

Table II: Sequence output by the generator of Fig. 3, corresponding bit probabilities and correlated LFSR–sequences.

i	0	1	2	3	4	5	6	7	8	9	10	11
$Prb(z_i)$	0.493	0.621	0.493	0.621	0.887	0.621	0.366	0.887	0.734	0.366	0.887	0.887
z_i	0	0	1	0	0	0	1	0	0	1	0	0
$b_{1,i} = \overline{a_{1,i}}$	0	0	1	0	1	0	1	0	0	1	1	0
$b_{2,i} = a_{2,i}$	0	1	0	1	0	1	1	0	0	1	1	0
$b_{3,i} = a_{3,i}$	0	0	1	0	1	0	0	0	1	1	0	0
$b_{4,i} = \overline{a_{4,i}}$	1	1	1	1	0	1	1	1	1	1	0	0
$b_{5,i} = a_{5,i}$	1	0	1	0	0	0	1	1	0	0	0	0

i	12	13	14	15	16	17	18	19	20	21	22	23
$Prb(z_i)$	0.621	0.984	0.823	0.887	0.957	0.974	0.887	0.621	0.887	0.734	0.957	0.887
z_i	0	0	0	0	0	0	0	0	1	0	1	1
$b_{1,i} = \overline{a_{1,i}}$	0	1	0	0	0	1	0	0	1	0	1	1
$b_{2,i} = a_{2,i}$	1	1	1	0	1	1	0	1	0	0	1	0
$b_{3,i} = a_{3,i}$	0	0	1	0	0	0	0	0	1	1	1	1
$b_{4,i} = \overline{a_{4,i}}$	0	0	0	0	1	0	1	0	1	0	0	1
$b_{5,i} = a_{5,i}$	1	0	0	0	0	0	1	1	1	1	1	1

i	24	25	26	27	28	29	30	31	32	33	34	35
$Prb(z_i)$	0.929	0.957	0.823	0.957	0.957	0.493	0.734	0.734	0.887	0.621	0.734	0.366
z_i	0	1	0	1	0	0	0	1	0	0	1	0
$b_{1,i} = \overline{a_{1,i}}$	0	1	1	0	0	0	1	1	1	0	1	0
$b_{2,i} = a_{2,i}$	0	1	1	1	0	0	0	1	0	1	1	1
$b_{3,i} = a_{3,i}$	1	1	0	1	0	1	0	1	1	0	0	1
$b_{4,i} = \overline{a_{4,i}}$	1	0	0	1	0	0	0	1	0	0	1	0
$b_{5,i} = a_{5,i}$	0	1	0	1	0	1	1	0	0	1	1	0

i	36	37	38	39	40	41	42	43	44	45	46	47
$Prb(z_i)$	0.929	0.957	0.887	0.734	0.621	0.621	0.957	0.734	0.994	0.734	0.887	0.929
z_i	1	0	0	1	1	0	0	0	0	1	1	0
$b_{1,i} = \overline{a_{1,i}}$	0	0	0	1	1	0	1	0	1	1	1	0
$b_{2,i} = a_{2,i}$	1	0	0	1	0	1	0	0	0	1	1	0
$b_{3,i} = a_{3,i}$	1	0	1	1	1	0	1	1	0	1	0	0
$b_{4,i} = \overline{a_{4,i}}$	1	1	0	1	1	0	0	0	1	1	1	0
$b_{5,i} = a_{5,i}$	1	1	1	0	1	1	0	1	0	0	1	0

i	48	49	50	51	52	53	54	55	56	57	58	59
$Prb(z_i)$	0.991	0.991	0.887	0.957	0.957	0.887	0.991	0.957	0.984	0.887	0.929	0.823
z_i	0	0	0	1	0	0	0	0	0	1	1	1
$b_{1,i} = \overline{a_{1,i}}$	0	1	1	1	1	0	1	1	1	1	1	0
$b_{2,i} = a_{2,i}$	0	0	0	1	0	0	0	0	0	1	1	1
$b_{3,i} = a_{3,i}$	1	0	0	1	1	1	0	0	0	1	0	1
$b_{4,i} = \overline{a_{4,i}}$	1	0	0	0	0	1	1	0	1	0	1	1
$b_{5,i} = a_{5,i}$	0	1	1	1	0	0	0	1	0	1	1	1

i	60	61	62
$\mathrm{Prb}(z_i)$	0.734	0.493	0.493
z_i	1	0	0
$b_{1,i} = \overline{a_{1,i}}$	0	0	0
$b_{2,i} = a_{2,i}$	1	1	1
$b_{3,i} = a_{3,i}$	1	1	1
$b_{4,i} = \overline{a_{4,i}}$	1	0	0
$b_{5,i} = a_{5,i}$	1	0	0

The algorithm A of Meier and Staffelbach selects the most probable bits $(z_{13}, z_{44}, z_{48}, z_{49}, z_{54}, z_{56})$ as a reference guess I_0. Looking at the bits of the correlated LFSR–sequences at the same positions and introducing the notation $I_j = (b_{j,13}, b_{j,44}, b_{j,48}, b_{j,49}, b_{j,54}, b_{j,56})$, we see that I_0 has a Hamming–distance 1 of I_2, I_3 and I_5, but Hamming–distances 5 resp. 4 of I_1 resp. I_4.

Since the expected value of the number of errors in the reference guess is here one (computed according to [3]), modifications of I_0 with Hamming–distances up to one will be tried, as well as their complements (because we don't know in advance whether the sought initial state yields a positive or a negative correlation).

Thus, the states Nr. 1, or 2, or 3, or 5 can be discovered, but not the state Nr. 4. Note that the algorithm as described in [2,3] needs a special implementation (to be described later in this paper) to be able to find *all* of the above states (Nr. 1, 2, 3, 5). Note also that testing larger Hamming–distances –even if it seems at a first glance to be the best way to find state Nr. 4 in this small example– generally requires a very large amount of additional work in larger examples. Therefore, the algorithm needs another modification to deal with this situation (e.g. to eventually be able to find state Nr. 4). Instead of selecting the most probable bits as a reference guess, we propose to first select a set S of $M > 6$ high probable bits, and then randomly choose 6 bits out of this set. Several reference guesses leading to several initial states might be tested in that manner. Coming back to the above example, the set S of bits with probabilities ≥ 0.9 could be considered (21 bits). For $I_0 = (z_{13}, z_{17}, z_{27}, z_{28}, z_{42}, z_{49}) \subset S$ we can check that Hd $(I_0, I_4) = 0$, thus the initial state Nr. 4 will be detected. But if, for example, I_0 happens to be $(z_{13}, z_{16}, z_{37}, z_{49}, z_{52}, z_{58})$, we can check that Hd $(I_0, I_2) = $ Hd $(I_0, I_3) = $ Hd $(I_0, I_4) = $ Hd $(I_0, I_5) = 2$ and Hd $(I_0, I_1) = 3$ and neither of the five peaks will be discovered. Hereafter, we recapitulate the steps of the modified correlation attack.

1. Determine the average number m of linear relations per bit (according to [2,3]).

2. Determine, for each bit z_i of the observed running key sequence, the number of linear relations it fulfills, and compute the resulting probability p_i (again according to [2,3]).

3. *Select a set S consisting of M high probable bits z_i. The number M of bits in this set should be large enough to allow the selection of sufficiently many reference guesses I_0, but small enough to reduce the risk of enclosing erroneous bits.*

4. *Select randomly k bits in S (reference guess I_0) that form a non-singular linear system whose solution is the initial state of the LFSR leading to those particular bits at those particular positions.*

5. Test modifications of I_0 with Hamming–distances $0, 1, 2, \ldots, r$ by correlating the corresponding LFSR–sequences with the sequence $\tilde{z}$. Store the initial states that yield sufficiently high correlation values and go back to step 4, unless you have determined enough initial states.

6. Use a subset of linearly independent initial states (among those found in step 5) to construct an equivalent RKG according to the method described in [1].

In step 5, it is difficult to determine the upper limit r of the Hamming–distances to test. If we compute the expected number of erroneous bits in the reference guess I_0 according to [2,3], we obtain a value that sometimes lies far below the actual number of errors in the reference guess. This discordance ist due to the fact that, as already mentioned, the statistical model of [2,3] is not perfectly adequate for a running key generator of the type of Fig. 1. Instead of testing Hamming–distances up to an unknown upper bound, the cryptanalyst could just solve the linear system defined by the unmodified reference guess I_0, compute the LFSR–sequence associated to the obtained initial state and compare it with the running key sequence $\tilde{z}$. Then he keeps on selecting randomly new reference guesses until he has found enough initial states yielding high correlation values. We try now to answer the question, whether this alternative improves the efficiency of the modified attack or not.

Let M be the number of bits in the set $\mathcal{S}$, and assume that m bits $(0 \leq m \leq M)$ in $\mathcal{S}$ are erroneous with respect to some correlated LFSR–sequence produced by a LFSR of length k. In order to find at least one correct reference guess in $\mathcal{S}$, the inequality

$$m \leq M - k \tag{1}$$

must hold. There are $\binom{M}{k}$ different reference guesses that can be chosen in $\mathcal{S}$, and $\binom{M-m}{k}$ of them are "correct", i.e. they contain no erroneous bit with respect to the correlated LFSR–sequence. If we neglect the fact that some of these "correct" reference guesses yield singular linear systems and are useless in computing the searched initial state, we obtain for the probability P_0 of selecting (randomly and uniformly) a correct reference guess

$$P_0 = \binom{M-m}{k} \cdot \binom{M}{k}^{-1} \tag{2}$$

$$= \frac{M-m}{M} \cdot \frac{M-m-1}{M-1} \cdot \; \ldots \; \cdot \frac{M-m-k+1}{M-k+1}. \tag{3}$$

The probability that among N randomly and uniformly selected reference guesses exactly one is correct, is given by

$$P(N) = P_0 \cdot (1 - P_0)^{N-1}. \tag{4}$$

We are mainly interested in the expected number $E[N_1]$ of reference guesses to select in order to find a correct one:

$$E[N_1] = \sum_{n=1}^{n_{max}} n \cdot P(n), \quad \text{where } n_{max} = \binom{M}{k}, \tag{5}$$

$$= 1 \cdot P_0 + 2 \cdot P_0 \cdot (1 - P_0) + 3 \cdot P_0 \cdot (1 - P_0)^2 + \ldots + \binom{M}{k} \cdot P_0 \cdot (1 - P_0)^{\binom{M}{k} - 1}. \tag{6}$$

We now compute the expected number $E[N_2]$ of modifications of one randomly selected reference guess that are necessary to reconstruct the correct reference guess. We assume that the cryptanalyst begins by testing Hamming–distance 0, then 1, 2... and so on. The expected number m_0 of erroneous bits in a randomly selected reference guess is approximately given by (worst–case approximation)

$$m_0 = \left\lceil \frac{mk}{M} \right\rceil. \tag{7}$$

Therefore, the expected number $E[N_2]$ of modifications of a reference guess is given by

$$E[N_2] = \binom{k}{0} + \binom{k}{1} + \binom{k}{2} + \ldots + \binom{k}{m_0 - 1} + \frac{1}{2} \binom{k}{m_0}, \tag{8}$$

where we assumed that, on the average, half the possible tests with Hamming–distance m_0 have to be made for finding the correct reference guess. Fig. 1 shows a sample graph of the expected

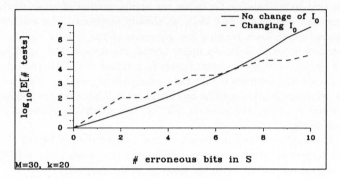

Figure 1: Expected number of tests to be made vs. the number of erroneous bits in the set $\mathcal{S}$, where $|\mathcal{S}| = M = 30$ and $k = 20$

number of tests to be made for both methods as a function of the number m of erroneous bits in the set $\mathcal{S}$. We observe that the first method (where reference guesses are picked up until a good one is found) is more efficient for the case where only few bits of $\mathcal{S}$ are erroneous. But when more bits in $\mathcal{S}$ are wrong, the method where Hamming–distances of some given reference guess are tested is to be preferred. This trend was confirmed by further sample curves. The two methods can of course be combined, for example in assigning a (small) value $r > 0$ to the maximal Hamming–distance to be tested and in picking up a new reference guess as soon as all possible tests have been made with the former one. This seems to be the most reasonable approach of the problem, since the cryptanalyst normally doesn't know the number m of erroneous bits in the set $\mathcal{S}$ but must necessarily choose a maximal Hamming–distance to test.

III.   LIMITS OF THE ATTACK

In order to judge the feasibility of the modified attack, two kinds of computer experiments were carried out for concrete examples of running key generators. The first series of experiments consisted of the full execution of the attack, as it would be done by an enemy cryptanalyst who can observe a limited amount of running key bits and knows nothing but the single LFSR used in the RKG. These experiments showed that the success of the attack depends on following factors.

- *The number of feedback taps of the LFSR:* the more taps there are, the more bits are involved in each linear relation and the less reliable is the assignment of probabilities in step 2 [2,3].

- *The (absolute and relative) heights of the correlation peaks between the running–key sequence and the LFSR–sequence.* Higher peaks are much easier detected by the algorithm than lower ones. If the CCF has one or a few high peaks and some lower peaks, the last ones are not easily discovered by the algorithm. In this case, it might be necessary to test modifications of I_0 with quite large Hamming–distances.

- *The number of bits in the set $\mathcal{S}$ in step 3.* It must be large enough to allow the cryptanalyst to extract a sufficient number of linearly independent sets of k bits.

The second series of experiments is described hereafter and the obtained results are then discussed. For a given RKG with an arbitrary initial state, we first determine the initial states $S_1, S_2, \ldots, S_s$

yielding the sequences $\tilde{a}_1, \tilde{a}_2, \ldots, \tilde{a}_s$ which lead to high cross–correlation values with the running–key sequence. We continue by assigning probabilities to the observed bits of the running–key sequence (as in step 3 above) and by selecting a set S of M high probable bits (as in step 2). We then compare the values of the bits in this set to those of the corresponding bits of the correlated sequences. If m_i bits of S coincide with the corresponding bits of the i–th ML–sequence ($1 \leq i \leq s$), the expected number of erroneous bits in a reference guess of k bits uniformly randomly chosen in S can be computed as

$$\epsilon_i = \left(1 - \frac{m_i}{M}\right) \cdot k. \tag{9}$$

Notice that "erroneous" means here "does not coincide with the corresponding bit of the sequence $\tilde{a}_i$". Under the assumption that the statistical model of [2,3] is suitable for the investigated type of RKG, the following two facts are expected to be observed experimentally.

1. The average numbers of errors ϵ_i ($1 \leq i \leq s$), should grow with the number M of bits in the set S, since including more bits in S implies including less reliable bits.

2. If the number M of bits in S gets very large, the number of errors ϵ_i should get closer and closer to the asymptotic value of $(1 - p_i) \cdot k$ (where p_i is the probability for any bit of the running key sequence to be equal to the corresponding bit of the i-th ML–sequence). This means that the bits of S could just as well have been chosen at random.

The first consideration was experimentally shown to hold more or less for RKG's with

- a small number of initial states $S_1, S_2, \ldots, S_s$ (the number s depends on properties of the feedforward function).

- cross–correlation peaks of sufficiently large amplitudes (70–75%),

- all the cross–correlation peaks having very similar amplitudes.

Fig. 2 shows the curves obtained for a RKG fulfilling the above conditions. We notice that

- The peaks Nr. 1 and Nr. 4 might be discovered by the attack if the set S contains for example 150 bits and if Hamming distances up to 17 are tested.

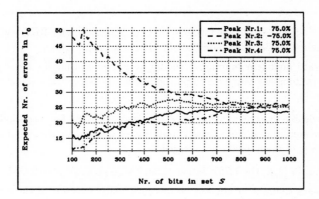

Figure 2: Expected number of errors in the reference guess vs. the number of bits in the set S, measured for a RKG with an LFSR of length 100 having 4 feedback taps; the first 1000 bits of the running key sequence were observed.

- Hamming distances up to 23 will have to be tested in order to detect the peak Nr. 3.

- The peak Nr. 2 is not detectable by the attack. For reasonably small numbers of bits in S, approximately half the bits of the reference guess are expected to be erroneous.

- As expected, taking more bits in S implies that the expected numbers of errors $\epsilon_1, \ldots, \epsilon_4$ tend to the asymptotic value of $(1 - 0.75) \cdot 100 = 25$.

- The average numbers of errors for the smallest set (of 100 bits) do not coincide with the theoretical value of $r = 14$ computed according to [2,3] (for $p_0 = 0.75$). This is due to the fact that the statistical model of [2,3] does not reproduce rigorously the situation where only cyclic shifts of *the same* LFSR–sequence are used.

Fig. 3 shows the error curves for a RKG with seven peaks of different amplitudes (one of 75%, one of 68.75% and five of 62.5%). Only the peak of 75% (lowest curve) is likely to be detected by the attack. In a way, this dominant peak "drowns" the effects of the lower peaks. For large sets S, the curves can be checked to converge towards the asymptotic values of $(1 - 0.75) \cdot 100 = 25$, $(1 - 0.6875) \cdot 100 = 31.25$ and resp. $(1 - 0.625) \cdot 100 = 37.5$ errors.

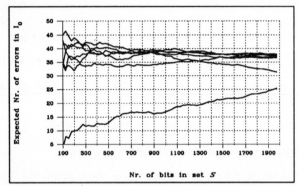

Figure 3: Expected number of errors in the reference guess vs. the number of bits in the set S, measured for a RKG with an LFSR of length 100 having 4 feedback taps; the first 2000 bits of the running key sequence were observed.

The above observations lead to the formulation of design criteria for RKG's with a single LFSR that are to withstand the described modified correlation attack.

1. The feedforward nonlinear function f should be chosen so that the cross–correlation peaks between the running key sequence $\tilde{z}$ and the LFSR–sequence $\tilde{a}$ take values of much less than 75% (cf. Fig. 1).

2. It is more advisable to have many cross–correlation peaks than few, especially when the peaks are of similar amplitudes, since the computation of bit probabilities tend to be less reliable when the effects of many peaks are merged.

3. As pointed out in [2,3], the LFSR in use should have no less than 10 feedback taps.

Finally, we remark that the attack can be more efficiently executed if the cryptanalyst knows the exact structure of the RKG. In that case, it is easy for him to plot the curves of Fig. 2 or 3 (for an arbitrarily chosen initial state of the LFSR), and he can test in priority Hamming distances corresponding to the expected numbers of errors for a given size of the set S. If he doesn't know the function f of Fig. 1, he has to systematically try out the Hamming distances $0, 1, \ldots$ up to some unknown upper bound. Indeed, we have seen that the expected number of errors r calculated according to [2,3] is not a reliable value for RKG's with a single LFSR.

Acknowledgements

The author is very grateful to Thomas Siegenthaler and Othmar Staffelbach for their helpful suggestions.

References

[1] Th. Siegenthaler, "Cryptanalysts Representation of Nonlinearly Filtered ML–Sequences", Advances in Cryptology, Eurocrypt'85, Springer–Verlag, pp. 103–110, 1986.

[2] W. Meier and O. Staffelbach, "Fast Correlation Attacks on Stream Ciphers", Advances in Cryptology, Eurocrypt'88, Springer–Verlag, pp. 301–314, 1988.

[3] W. Meier and O. Staffelbach, "Fast Correlation Attacks on Stream Ciphers", full paper, to appear in the Journal of Cryptology, Springer International.

Section 9

Algorithms

ON THE COMPLEXITY AND EFFICIENCY
OF A NEW KEY EXCHANGE SYSTEM

Johannes A. Buchmann[1]   Stephan Düllmann[1]
Hugh C. Williams[2]

[1] FB 10 Informatik
Universität des Saarlandes
D-6600 Saarbrücken
WEST GERMANY

[2] Department of Computer Science
University of Manitoba
Winnipeg, Manitoba
CANADA R3T 2N2

ABSTRACT

In [2] Buchmann and Williams presented a new public key exchange system based on imaginary quadratic fields. While in that paper the system was described theoretically and its security was discussed in some detail nothing much was said about the practical implementation.

In this paper we discuss the practical aspects of the new system, its efficiency and implementation. In particular we study the crucial point of the method: ideal reduction. We suggest a refinement of the well known reduction method which has been implemented on a computer. We present extensive running time statistics and a detailed complexity analysis of the methods involved.

The implementation of the reduction procedure on chips is subject of future research.

I. THE DIFFIE-HELLMAN SCHEME

In their paper "New Directions in Cryptography" [3] Diffie and Hellman introduced in 1976 the idea of public key exchange. By this method it is possible to communicate a secret key for some cryptosystem over a public insecure channel. We briefly review the idea of Diffie and Hellman.

Suppose that Alice (A) and Bob (B) wish to secretly exchange a key.

(1) A chooses a finite group G and – at random – an element $\lambda \in G$. Both G and λ are sent to B over the public channel.

(2) Both A and B now select – at random – an integer a and b, respectively. These integers are kept secret.

(3) A computes $\alpha = \lambda^a$ and transmits the result to B. In the same way B computes $\beta = \lambda^b$ and sends it to A. (Note that the group elements α and β are public.)

(4) A computes $\gamma = \beta^a$, B computes $\gamma' = \alpha^b$.

Because $\gamma = \beta^a = (\lambda^b)^a = (\lambda^a)^b = \alpha^b = \gamma'$ both A and B are in possession of the same key γ without this key having been sent over the public channel.

In order to be able to apply this algorithm in practice it is necessary to have an efficient multiplication routine in G i.e., if all the elements are represented by numbers in $\{0, 1, \ldots, |G| - 1\}$ the representation of the product of two elements of G should be computable in time polynomial in $\log |G|$.

Then, using the method of binary shifting (see [6, p. 441 ff.]) powers of group elements can be computed very efficiently even for large exponents d, namely in $O(\log d)$ elementary operations in G.

The scheme is secure if the key cannot be guessed easiliy and cannot be determined easily from the public information.

To avoid the key beeing guessed easily it is necessary to choose a group G of very high order, to pick a starting element λ of high order (close to $|G|$) and to select large exponents a and b.

To make sure that the choice of G is adequate one has to analyse its arithmetic properties carefully. A neccessary condition is that the determination of discrete logarithms in G is difficult. Note, however, that this condition is not sufficient for the security of the system because the determination of λ^{ab} from λ, λ^a and λ^b might be easier than the calculation of a and b from λ, λ^a and λ^b.

So far the following groups have been suggested:

- The group $G = GL_n(\mathbb{Z}/p\mathbb{Z})$ of invertible $n \times n$-matrices over the finite field $(\mathbb{Z}/p\mathbb{Z})$. ([14])

- The group of points on an elliptic curve over a finite field. ([11])

- Several groups associated with higher dimensional varieties. ([7])

- The group $G = (\mathbb{Z}/n\mathbb{Z})$ where n is the product of two large primes. ([10])

- The class group of an imaginary quadratic field. ([2])

In this paper we continue the discussion of the case where G is the class group of an imaginary quadratic field.

II. THE CLASS GROUP OF AN IMAGINARY QUADRATIC FIELD

First we summarize the main facts concerning imaginary quadratic fields. All these facts are well known and therefore will be given without proof. Proofs of the statements made here and more detailed descriptions can be found in standard texts like Hua [5] or Narkiewicz [12], [13].

Let $D < 0$ be a squarefree integer and let $K = \mathcal{Q}(\sqrt{D})$ be the quadratic field which is defined by adjoining $\sqrt{D}$ to the set of rational numbers $\mathcal{Q}$. If $\alpha \in K$ we denote by $\overline{\alpha}$ the complex conjugate of α, by $\mathrm{Tr}(\alpha)$ the *trace* of α, i.e. the value of $\alpha + \overline{\alpha}$, and by $N(\alpha)$ the *norm* of α, i.e. the value of $\alpha \cdot \overline{\alpha}$. Note that $N(\alpha) \geq 0$ for every $\alpha \in K$.

We define
$$r = \begin{cases} 1 & \text{if } D \equiv 2, 3 \bmod 4 \\ 2 & \text{if } D \equiv 1 \bmod 4. \end{cases}$$
Then the *discriminant* of K is given by
$$\Delta = \frac{4D}{r^2}.$$
For $\alpha, \beta \in K$ put $[\alpha, \beta] = \alpha\mathbb{Z} + \beta\mathbb{Z}$. Moreover, let
$$\omega = \frac{r - 1 + \sqrt{D}}{r}.$$
Then the *ring of algebraic integers* in K is given by
$$\mathcal{O}_K = [1, \omega].$$

The *fractional ideals* of $\mathcal{O}_\kappa$ form a multiplicative abelian group denoted by $\mathcal{F}$. By *1-ideal* we denote the neutral element in this group. The *principal ideals* of $\mathcal{O}_\kappa$ form a subgroup of $\mathcal{F}$ denoted by $\mathcal{P}$. The factor group $\mathcal{F}/\mathcal{P}$ is called the *class group* of $\mathcal{O}_\kappa$ denoted by $\mathcal{C}_\Delta$. The class group is finite, its order is called the *class number* of $\mathcal{O}_\kappa$ and is denoted by h. The elements of the class group are called the *ideal classes*. Two ideals of $\mathcal{O}_\kappa$ are said to be *equivalent* if they are in the same ideal class.

Every integral ideal a of $\mathcal{O}_\kappa$ can be written in the form

$$\mathbf{a} = [a, b + c\omega]$$

where $a, c \in \mathbb{Z}^1, b \in \mathbb{Z}$ and

$$c|a, \; c|b, \; ac|N(b + c\omega).$$

For a given ideal a the integers a and c are uniquely determined and b is unique modulo a. a is the least positive rational integer in a, denoted by $L(\mathbf{a})$. If $c = 1$ then a is called *primitive*.

Every integral primitive ideal a of $\mathcal{O}_\kappa$ can be uniquely presented in the form

$$\mathbf{a} = [L(\mathbf{a}), b + \omega]$$

with $b \in \mathbb{Z}$ and

$$-L(\mathbf{a}) < \mathrm{Tr}(b + \omega) \le L(\mathbf{a}).$$

This is called the *normal presentation* of a.

An integral primitive ideal a whose normal presentation is $\mathbf{a} = [L(\mathbf{a}), b + \omega]$ is called *reduced* if

$$|b + \omega| \ge L(\mathbf{a}),$$
$$\mathrm{Tr}(b + \omega) > 0 \quad \text{if } |b + \omega| = L(\mathbf{a}).$$

There are the following main facts concerning reduced ideals:

- For each reduced ideal we have $L(\mathbf{a}) < \sqrt{|\Delta|/3}$.

- Every ideal class of $\mathcal{O}_\kappa$ contains exactly one reduced ideal.

Each ideal class can be represented by its reduced ideal. So the arithmetic in the class group of imaginary quadratic fields can be reduced to ideal-arithmetic: In order to determine the product of two ideal classes multiply their reduced ideals and compute the reduced ideal equivalent to this product.

Since every reduced ideal is presented by a pair of integers which are less than $\sqrt{|\Delta|/3}$ in absolute value class groups can be used in the Diffie-Hellman scheme if multiplication and reduction of ideals can be carried out efficiently. The secret key must be one of the integers of the normal presentation of the reduced ideal computed in the scheme.

III. THE ALGORITHMS

In [2] it is pointed out that D has to be of order of magnitude 10^{200} to guarantee high security of the scheme. Also the exponents should be of this order of magnitude. So first of all we need a multi-precision integer-arithmetic for implementing the scheme. For details see [6] or [1].

After the initialization all computations have the following structure: For an integer n and an ideal class Γ given by its reduced ideal a_Γ compute the reduced ideal in Γ^n, i.e. the reduced ideal equivalent to $(a_\Gamma)^n$. This can be done by means of the well known fast exponentiation technique described in [6, p. 441].

Algorithm 3.1 (Exponentiation of ideal classes)
Input: *Reduced ideal* a, *exponent* $n \in \mathbb{Z}^{>0}$.
Output: *Reduced ideal* b *equivalent to* a^n.

(1) $N \leftarrow n$; b $\leftarrow$ 1-ideal; c $\leftarrow$ a;

(2) if N is even then goto (5);

(3) b $\leftarrow$ c $\cdot$ b;

(4) reduce b;

(5) $N \leftarrow \lfloor N/2 \rfloor$;

(6) if $N = 0$ terminate.

(7) c $\leftarrow$ c $\cdot$ c;

(8) reduce c and goto (2);

For practical purposes it is more convenient to represent an integral primitive normally presented ideal a $= [L(a), b+\omega]$ by $(A, B) \in \mathbb{Z}^2$ with

$A = rL(\mathbf{a})$ and $B = rb + r - 1$. Note that the new presentation is well defined and unique.

For the multiplication we use an algorithm based on Shanks [15] which computes a primitive ideal equivalent to the product of two primitive ideals:

Algorithm 3.2 (Multiplication of ideals)
Input: *Two primitive ideals* $(A_1, B_1), (A_2, B_2)$.
Output: *Primitive ideal* (A_3, B_3) *equivalent to the product of the input ideals.*

(1) $G' \leftarrow \gcd(A_1, A_2)$;
     compute the coefficient V' *in* $A_1 V' + A_2 W' = G'$;

(2) $B_3 \leftarrow V' A_1 (B_2 - B_1)$;

(3) $A_3 \leftarrow r A_1 A_2$;

(4) *if* $G' = 1$ *then goto* (8);

(5) $G \leftarrow \gcd(G', (B_1 + B_2))$;
     compute the coefficients U', U *in* $U'G' + U(B_1 + B_2) = G$;

(6) $B_3 \leftarrow [B_3 U' + U(D - B_1^2)]/G$;

(7) $A_3 \leftarrow A_3/G^2$;

(8) $B_3 \leftarrow B_1 + B_3$;

If the ideals to be multiplied are equal then step 1 of the multiplication algorithm can be replaced by

$$G = A_1, \ W' = 1, \ V' = 0$$

because $A_1 = A_2$. This simplification leads to the following algorithm for squaring an ideal:

Algorithm 3.3 (Squaring of an ideal)
Input: *Primitive ideal* (A_1, B_1).
Output: *Primitive ideal* (A_3, B_3) *equivalent to the square of the input ideal.*

(1) $G \leftarrow \gcd(A_1, 2B_1)$;
     compute the coefficient U *in* $U'A_1 + 2UB_1 = G$;

(2) $B_3 \leftarrow B_1 + [U(D - B_1^2)]/G;$

(3) $A_3 \leftarrow r(A_1/G)^2;$

(4) terminate.

Now we discuss ideal reduction. In order to decrease the size of the coefficients of the ideal representation whenever it is possible the ideals are reduced after each multiplication in Algorithm 3.1. So the reduction algorithm is applied very frequently and therefore it is necessary to have a fast reduction method. Our method is a refinement of the following well known algorithm (see [8] or [9]).

Algorithm 3.4 (Reduction of ideals, classical version)
Input: *Primitive ideal* (A', B').
Output: *Reduced ideal* (A, B) *(in normal presentation) equivalent to the input ideal.*

(1) $A \leftarrow A'; B \leftarrow B';$

(2) $B \leftarrow round(B/A) \cdot A - B;$

(3) $A_N \leftarrow (B^2 - D)/A;$

(4) if $A_N < A$ *then* $A \leftarrow A_N$ *and goto (2);*

(5) if $(2B < A)$ *and* $(B < 0$ *or* $A < A_N)$ *then* $B \leftarrow -B;$

In step 2 of the algorithms "round" means a multi precision integer function which computes the rounded quotient of two integers. This can be done by the following algorithm:

Algorithm 3.5 (Rounded quotient of two integers)
Input: *Integers* A *and* B, $B \neq 0$.
Output: *Integer* C *with* $C = round(A/B)$.

(1) $C \leftarrow A/B; R \leftarrow A \bmod B;$

(2) if $2|R| \geq |B|$ *then* $[$*if* $A > 0$ *then* $C \leftarrow C + 1$ *else* $C \leftarrow C - 1];$

Now we present the refinement of the algorithm which was theoretically described in [2].

Algorithm 3.6 (Reduction of ideals, optimized version)

Input: *Primitive ideal* (A', B').

Output: *Reduced ideal* (A, B) *(in normal presentation) equivalent to the input ideal.*

(1) $A \leftarrow A'$;
  if $B' < 0$ then $s \leftarrow -1$ else $s \leftarrow 1$;
  $B \leftarrow |B'|$;

(2) $Q \leftarrow B/A$; $R \leftarrow B$ mod A; $M \leftarrow A - 2R$;
  if $M \geq 0$ then $B \leftarrow R$ and $s \leftarrow -s$ else $B \leftarrow R + M$;
  $A_N \leftarrow (B^2 - D)/A$;

(3) if $A_N < A$ then $A_O \leftarrow A$ and $A \leftarrow A_N$ else goto *(6)*;

(4) $Q \leftarrow B/A$; $R \leftarrow B$ mod A; $M \leftarrow A - 2R$;

(5) $A_N \leftarrow A_O - (R + B) \cdot Q$;
  If $M \geq 0$ then $B \leftarrow R$ and $s \leftarrow -s$ else $B \leftarrow R + M$ and
  $A_N \leftarrow A_N + M$;
  goto *(3)*;

(6) if $s < 0$ then $B \leftarrow -B$;
  if $(2B < A)$ and $(B < 0$ or $A < A_N)$ then $B \leftarrow -B$;

Comparing the two versions of the algorithm we see that the number of iterations in both algorithms and also the sequences of the values for A and B computed in both algorithms are equal. There are two main differences between the two versions:

The computation of the new value of A in each iteration in the optimized version needs one division of multi precision integers less than in the first version because the division in step 3 is avoided. Instead of this the value of A from the preceeding iteration is used. Therefore this simplification cannot be made up in the first iteration. So step 2 of Algorithm 3.6 contains the first iteration of Algorithm 3.4.

The computation of the rounded quotient of A and B is avoided in the optimized version. Moreover the sign of B is stored in a seperate variable s. In this way one multiplication of multi precision integers in each iteration is avoided.

The following theorem makes sure that algorithm 3.6 indeed computes the reduced ideal equivalent to the input ideal in finitely many iterations. Moreover an estimation for the number of iterations is given. In [2] this theorem was proved for Algorithm 3.4 which requires one more iteration.

Theorem 3.1 *Let* (A, B) *represent a primitive ideal. Algorithm 3.4 finds the reduced ideal equivalent to* (A, B) *in at most* i *iterations of the steps 3,4 and 5 where*

$$i = \max \left\{ 0, \left\lfloor \frac{1}{2} \log \left(\frac{3A}{5\sqrt{|D|}} \right) \right\rfloor + 1 \right\}.$$

Finally we describe the initialization where we have to find an appropriate starting ideal class for a given value of D. We chose – at random – a prime number q ($q \equiv 3 \bmod 4$) such that D is quadratic residue modulo q. Then

$$A = r \cdot q$$
$$B = D^{\frac{q+1}{4}} \bmod rq$$

satisfy the condition

$$rA \mid B^2 - D.$$

So (A, B) represents a primitive ideal. Reduction of this ideal gives the starting ideal class.

IV. THE COMPLEXITY

First we compute the size of the integers occuring in the computations of our key exchange system for a fixed value of D.

Theorem 4.1

(a) If (A, B) *is a reduced ideal in normal presentation then*

$$A \leq 2\sqrt{\frac{|D|}{3}},$$
$$|B| \leq \sqrt{\frac{|D|}{3}}.$$

(b) If (A, B) is the primitive ideal computed by Algorithm 3.2 or 3.3 then

$$A \leq \frac{8|D|}{3},$$

$$|B| \leq \frac{16|D|}{3}\sqrt{\frac{|D|}{3}}$$

provided that the input ideals were reduced and given in normal presentation.

(c) The size of the integers occuring in the computations of our key exchange system is bounded by

$$M = 3|D|\sqrt{|D|}.$$

Proof:

(a) Let $a = [L(a), b+\omega] = (A, B)$. Then, according to the definition of A and B in section 4, we have $A = rL(a)$ and $B = rb+r-1$. Moreover

$$b + \omega = \frac{B+1}{r} - 1 + \frac{r-1+\sqrt{D}}{r} = \frac{B+\sqrt{D}}{r}$$

and therefore

$$\text{Tr}(b+\omega) = \frac{2B}{r} \leq 2B.$$

Since a is reduced we have

$$A = rL(a) < r\sqrt{\frac{|\Delta|}{3}} = r\sqrt{\frac{4|D|}{3r^2}} = 2\sqrt{\frac{|D|}{3}}.$$

Since a is in normal presentation we have $|\text{Tr}(b+\omega)| \leq L(a)$ and therefore

$$|B| \leq \frac{L(a)}{2} = \sqrt{\frac{|D|}{3}}.$$

(b) In Algorithm 3.2 the new value for A is computed in steps 3 and 7. From these steps and part (a) of this theorem we get

$$A \leq 2 \cdot \left(2\sqrt{\frac{|D|}{3}}\right)^2 = \frac{8|D|}{3}.$$

The new value for B is computed in steps 2, 4 and 8. The coefficients V', V, U' in step 2 and 5 can be choosen less than $2\sqrt{\frac{|D|}{3}}$. So

$$B \leq 2\sqrt{\frac{|D|}{3}} \cdot 2\sqrt{\frac{|D|}{3}} \cdot \sqrt{\frac{|D|}{3}}.$$

The same arguments apply to Algorithm 3.3.

(c) The input ideals for the multiplication and squaring algorithm are always reduced ideals in normal presentation. The input ideals for the reduction algorithm are always results of multiplication and squaring. Moreover the successive values of A in the reduction algorithm are strictly decreasing with the exeption of the last one and therefore the size of the input value of A is an upper bound for all values of A and B occuring in the algorithm. So by (a) and (b) we see the maximum integer occuring in the system is bounded by

$$\frac{16|D|}{3}\sqrt{\frac{|D|}{3}} = \frac{16\sqrt{3}}{9}|D|\sqrt{|D|} < 3|D|\sqrt{|D|}.$$

$\square$

Note that by part (c) of this theorem we also can estimate the number of computer words necessary to store the integers used in the algorithm. If for example D has 200 decimal digits then our multi precision arithmetic package must be able to handle numbers of 301 decimal digits.

Next we list the number of elementary operations (additions, multiplications and divisions) with multi precision integers for each of our algorithms. Multiplication by r is considered as an addition because r is either 1 or 2.

To analyze the multiplication and squaring algorithm we first need to know the number of elementary operations necessary to compute the gcd's. This computation can be done by the euclidian algorithm [6, p. 325]. If one has to compute a gcd and both coefficients of its representation the euclidian algorithm needs one division, three additions and three multiplications in each division step. This case will be called "gcd2". If one has to compute a gcd and only one of the coefficients of its representation one can avoid some computations. Then the euclidian algorithm only needs one division, two additions and two multiplications in each division step. This case will be called "gcd1".

Theorem 4.2

(a) The maximum number of division steps in the euclidian algorithm (see [6, p. 325]) used for the computation of the gcd's in the multiplication and squaring algorithm is

$$M' \leq \frac{3}{4}\log|D| + 1.$$

(b) In case $G' = 1$ Algorithm 3.2 requires at most $(3 + 2M')$ additions, $(3 + 2M')$ multiplications and M' divisions of multi precision integers.

(c) In case $G' > 1$ Algorithm 3.2 requires at most $(6 + 5M')$ additions, $(7 + 5M')$ multiplications and $(2 + 2M')$ divisions of multi precision integers.

(d) Algorithm 3.3 requires at most $(3 + 2M')$ additions, $(3 + 2M')$ multiplications and $(2 + M')$ divisions of multi precision integers.

Proof:

(a) From the bounds given in Theorem 4.1 (a) for the input values of the multiplication and squaring algorithm and by the listings of these algorithms in the previous section we see that the input values for the gcd's are bounded by $2\sqrt{\frac{|D|}{3}}$. By Corollary L in [6, p. 343] we can conclude that the number of division steps required when the euclidian algorithm is applied to numbers u and v with $0 \le u, v < N$ is at most $4.8 \log_{10} N + 0.68$. So in our special case we have

$$M' \le 4.8 \log_{10} \left(2\sqrt{\frac{|D|}{3}} \right) + 0.68$$

$$= 4.8 \log_{10} \frac{2}{\sqrt{3}} + \frac{4.8}{\log 10} \log \sqrt{|D|} + 0.68$$

$$\le \frac{3}{2} \log \sqrt{|D|} + 1$$

$$= \frac{3}{4} \log |D| + 1.$$

(b) For $G' = 1$ Algorithm 3.2 requires three additions, three multiplications and one gcd1.

(c) For $G' > 1$ Algorithm 3.2 requires three additions, seven multiplications, two divisions, one gcd1 and one gcd2.

(d) Algorithm 3.3 requires three additions, three multiplications, two divisions and one gcd1.
□

Now we turn to the reduction algorithm. Here we first have to compute the maximum number of iterations.

Theorem 4.3

(a) *The maximum number of iterations in the optimized version of the reduction algorithm is*

$$M'' \leq \frac{1}{4} \log |D| + 2.$$

(b) *Algorithm 3.4 requires $(1+4(M''+1))$ additions, $2(M''+1)$ multiplications and $2(M''+1)$ divisions of multi precision integers. Here the operations caused by the seperate " round"-algorithm are included.*

(c) *Algorithm 3.6 requires $(5+6M'')$ additions, $(1+M'')$ multiplications and $(2+M'')$ divisions of multi precision integers.*

Proof:

(a) Using the bound given in Theorem 4.1 (b) we have

$$A \leq \frac{8|D|}{3}$$

for the input ideal of the reduction algorithm. From Theorem 3.1 we conclude that the number of iterations in Algorithm 3.6 is bounded by

$$
\begin{aligned}
M'' &\leq \frac{1}{2} \log \left(\frac{3\frac{8|D|}{3}}{5\sqrt{|D|}} \right) + 1 \\
&= \frac{1}{2} \log \left(\frac{8}{5}\sqrt{|D|} \right) + 1 \\
&= \frac{1}{2} \log \frac{8}{5} + \frac{1}{4} \log |D| + 1 \\
&\leq \frac{1}{4} \log |D| + 2.
\end{aligned}
$$

(b) The number of iterations in Algorithm 3.4 is one more than in Algorithm 3.6. Each iteration (steps 2, 3 and 4) requires two additions, two multiplications, one division and one call of the round procedure in which there are two additions and one division. Step 5 of the reduction algorithm requires one additional addition.

(c) Each iteration (steps 3, 4 and 5) in Algorithm 3.6 requires six additions, one multiplication and one division. Additionally, steps 2 and 6 require five additions, one multiplication and two divisions.

□

Now we have to examine Algorithm 3.1 for the exponentiation of ideal classes.

Theorem 4.4 *If the exponents a and b are bounded by $a \cdot b < \sqrt{|D|}$ then the total number of iterations performed in all calls of the exponentiation algorithm in the key exchange system is at most*

$$M''' \leq \log|D| + 4.$$

Proof: The number of iterations in algorithm 3.1 for an arbitrary exponent n is lower than $\lfloor \log n \rfloor + 1$ (see [6, 443]). Each exponent is used twice in the scheme. Therefore the total number of iterations performed in the four calls of the exponentiation algorithm is bounded by

$$
\begin{aligned}
M''' &\leq 2(\log a + 1 + \log b + 1) \\
&\leq 2\log(a \cdot b) + 4 \\
&\leq 2\log\sqrt{|D|} + 4 \\
&= \log|D| + 4.
\end{aligned}
$$

□

Finally we give an upper bound for the number of elementary operations performed in the key exchange system which follows from Theorems 4.2, 4.3 and 4.4.

Theorem 4.5 *Using the optimized version of the reduction algorithm our key exchange system takes at most*

$$217 + \frac{169}{2}\log|D| + \frac{33}{4}\log^2|D| \quad \text{additions,}$$

$$95 + \frac{185}{4}\log|D| + \frac{23}{4}\log^2|D| \quad \text{multiplications,}$$

$$64 + \frac{105}{4}\log|D| + \frac{11}{4}\log^2|D| \quad \text{divisions}$$

of multi-precicison integers.

Now we give the asymptotical bit complexity for our algorithms and for the whole method. Clearly the complexity depends on the implementation of the multi precision routines. Using the so called "Classical Algorithms" (see [6, p. 250 ff.]) multiplication and divsion of n-bit numbers require $O(n^2)$ bit operations. Using the so-called "Fast Multiplication Techniques" these operations numbers require only $O(n \log n \log \log n)$ bit-operations (see [1, p. 272, 286]).

Theorem 4.6

(a) *The algorithms for squaring and multiplication of ideals and both versions of the reduction algorithm have bit complexity*

   (i) $O(\log^3 |D|)$ *if the classical algorithms are used,*

   (ii) $O(\log^2 |D| \log\log |D| \log\log\log |D|)$ *if fast multiplication technique is used.*

(b) *The method for public key exchange has bit complexity*

   (i) $O(\log^4 |D|)$ *if the classical algorithms are used,*

   (ii) $O(\log^3 |D| \log\log |D| \log\log\log |D|)$ *if fast multiplication technique is used.*

Proof: Both statements follow immediately from the previous results. Note that according to Theorem 4.1 (c) the binary length of the integers occuring in our scheme is $O(\log |D|)$. $\square$

The theorem shows that both versions of the reduction algorithm have the same asymptotical complexity. The improvement in the optimized version only affects the O-constant as can be seen from Theorem 4.3.

We see that using fast multiplication techniques the running time of our method for public key exchange is a cubic polynomial in the length of the input data. Therefore it is executable for big exponents and big discriminants.

V. RUNNING TIME STATISTICS

The method for public key exchange has been implemented on a SIEMENS 7580-S computer of the University of Düsseldorf [4]. The programming language was FORTRAN-77. For the multi-precision arithmetic we used the classical algorithms. These and some procedures to get the running time statistics presented below were taken from the number theoretic subroutine library KANT.

We computed many examples where the value of D was the product of two prime numbers each of size up to 10^{100}. This choice of D was done according to the remarks in [2] concerning the security of the method. Both cases $r = 1$ and $r = 2$ occured equally often.

In the table below we present the statistics of 32 of these examples. Here the size of D varies between 10^{120} and 10^{200}. For each size of D we took several exponents of different size. The columns of the table have the following contents:

(1) Number of the example

(2) Size of D (number of decimals)

(3) Size of the product of the exponents (number of decimals)

(4) Number of calls of the squaring algorithm

(5) Number of calls of the multiplication algorithm

(6) Number of calls of the reduction algorithm

(7) Average number of iterations in the reduction algorithm

(8) Maximum number of iterations in the reduction algorithm

(9) Theoretical bound for the number of iterations in the reduction algorithm

(10) Average running time of the squaring algorithm

(11) Average running time of the multiplication algorithm

(12) Average running time of the classical version of the reduction algorithm

(13) Average running time of the optimized version of the reduction algorithm

(14) Total running time of the program with the optimized version of the reduction algorithm

All times are given in seconds (CPU).

From the structure of the exponentiation algorithm we know the connection between the exponents and the number of calls of our algorithms: The number of squarings (col. 4) is exactly twice the sum of the logarithms of the two exponents. The number of multiplications (col. 5) is twice the number of One's in the binary representation of the exponents. The number of reductions (col. 6) minus 1 is exactly the sum of the columns 4 and 5 because there is a reduction after each squaring or multiplication and one reduction in the initialization.

The theoretical bound for the number of iterations in the reduction algorithm is computed from the formula in Theorem 4.3 (a) with a value of D rounded to the next power of ten.

Table (Running time statistics)

	Decimals		Subroutine Calls			Iterations in reduction			Average running time		Reduction		Total running time
	D	$a \cdot b$	Squ	Mul	Red	aver	max	theo	Squ	Mul	Cla	Opt	
(1)	(2)	(3)	(4)	(5)	(6)	(7)	(8)	(9)	(10)	(11)	(12)	(13)	(14)
1	198	10	62	30	93	57	80	166	0.23	0.23	1.60	0.27	47.11
2	198	30	192	88	281	64	79	166	0.26	0.28	1.80	0.30	159.83
3	197	50	324	162	487	65	77	165	0.27	0.28	1.81	0.30	280.79
4	197	71	468	236	705	66	82	165	0.27	0.28	1.82	0.30	408.24
5	197	97	640	294	935	66	80	165	0.27	0.27	1.84	0.30	540.30
6	182	11	66	32	99	51	71	153	0.20	0.21	1.25	0.22	42.54
7	182	30	190	82	273	59	79	153	0.23	0.24	1.44	0.26	133.59
8	182	48	310	140	451	60	75	153	0.23	0.24	1.48	0.27	225.44
9	182	71	466	238	705	61	74	153	0.24	0.24	1.50	0.27	356.13
10	182	93	608	286	895	62	74	153	0.24	0.24	1.51	0.27	454.34
11	178	19	118	62	181	54	72	148	0.21	0.21	1.28	0.23	80.75
12	177	40	246	132	379	57	74	148	0.22	0.22	1.34	0.25	177.06
13	177	60	394	202	597	59	75	148	0.23	0.23	1.39	0.25	288.64
14	177	79	518	238	757	59	75	148	0.23	0.23	1.40	0.25	363.59
15	167	10	58	26	85	48	72	140	0.18	0.18	1.03	0.20	32.25
16	167	40	256	122	379	55	69	140	0.20	0.21	1.18	0.22	162.92
17	166	60	394	202	597	55	70	139	0.20	0.20	1.20	0.22	255.87
18	166	79	518	238	757	55	70	139	0.20	0.21	1.22	0.23	326.39
19	157	10	62	40	103	46	64	132	0.17	0.17	0.89	0.18	46.50
20	157	30	196	98	295	51	66	132	0.18	0.19	0.99	0.20	113.45
21	152	51	334	182	517	50	63	128	0.18	0.18	0.94	0.19	193.32
22	158	71	466	228	695	53	67	133	0.19	0.19	1.04	0.21	276.74
23	146	10	62	30	93	42	59	123	0.14	0.16	0.72	0.16	28.40
24	147	30	190	84	275	47	60	124	0.16	0.16	0.83	0.18	93.22
25	136	50	328	188	517	46	59	115	0.15	0.15	0.72	0.16	161.26
26	142	71	466	238	705	48	62	119	0.16	0.17	0.81	0.18	240.87
27	127	20	126	44	171	40	51	107	0.13	0.13	0.58	0.14	45.33
28	133	39	246	132	379	44	56	112	0.14	0.14	0.67	0.15	111.96
29	127	60	394	202	597	43	53	107	0.13	0.14	0.61	0.14	167.23
30	123	20	128	54	183	40	54	104	0.12	0.13	0.54	0.13	47.35
31	129	41	258	140	399	43	55	109	0.13	0.14	0.62	0.15	113.18
32	124	59	380	176	557	42	52	104	0.13	0.13	0.58	0.14	149.86

The comparison of the values in column 8 and 9 shows that the maximum number of iterations in the reduction algorithms is about half of the theoretical bound.

The comparision of the average running time for the two versions of the reduction algorithm shows that there is indeed an important speeding up. The optimized version only requires 20% or less of the running time of the classical version. Most of this gain is caused by avoiding one multi precision division in each iteration. Note that the optimized version of the reduction algorithm is nearly as fast as the multiplication of ideals.

Note also that there is almost no difference in the average running time of the multiplication and the squaring algorithm. Here the theoretical improvement has no practical effect.

The examination of the total running time of the program confirms the complexity result given in Theorem 4.6.

Now we give the data computed by our program for the first example mentioned in the table above. This example has the highest values of D.

We list the information to be transmitted over the public channel (D, the starting ideal class Λ and the ideal classes $\Phi = \Lambda^a$ and $\Psi = \Lambda^b$) and also the secret information (the exponents a and b and the ideal class $\Gamma = \Lambda^{ab}$ which gives the secret key).

Note that even if we use low exponents like in the first example the secret key has a reasonable size of about $\sqrt{|D|}$. Nevertheless one should choose larger exponents like in the fifth example to have greater security.

Example

$$D = \begin{aligned} &-11980374469127450695419647802766644830542210 7338 \\ &66131993089788267603365417346323605263646114675501 \\ &72246998555004267772913478851528257403104704701886 \\ &44025223079332669700563976732928049691985604910619 \end{aligned}$$

Secret exponents: $a = 543210,$ $b = 7980$

Starting ideal class Λ: $A = 982,$ $B = 31$

Ideal class $\Phi = \Lambda^a$:

$$A = 9714296130363739860824457599172055225880660149\dot{4}9$$
$$1308628569800046625247791479368544638455492570475\dot{0}$$
$$B = 395993438464946485994639174385199545445240652683$$
$$753364678843011371468419191822334726511186284706\dot{4}1$$

Ideal class $\Psi = \Lambda^b$:

$$A = 324518422859147029485007758564193912232635385076\dot{8}$$
$$80241672632089462264174498544690212815603924965610$$
$$B = 134646776551626016985782284030044708699147398287\dot{4}$$
$$11630817948537340265799828623744461455733888148671$$

Ideal class $\Gamma = \Lambda^{ab}$:

$$A = 332643436351817816168967532848701108325101616841\dot{7}$$
$$44606450020063171247803446369585670411999866011434$$
$$B = -9757780944892653967990970142864073898882314417\dot{3}$$
$$532031955109450292182506204157234412540729934371\dot{0}5$$

References

[1] A.V. Aho, J.E. Hopcroft, J.D. Ullman, *The design and analysis of computer algorithms*, Addison-Wesley, Reading, Massachusetts, 1974.

[2] J. Buchmann and H.C. Williams, *A key exchange system based on imaginary quadratic fields*, Journal of Cryptology **1** (1989), to appear.

[3] W. Diffie and M. Hellman, *New directions in cryptography*, IEEE Transactions on Information theory **22** (1976), 472 – 492.

[4] S. Düllmann, *Ein neues Verfahren zum öffentlichen Schlüssel-austausch*, Staatsexamensarbeit, Universität Düsseldorf, 1988.

[5] Hua Loo Keng, *Introduction to number theory*, Springer Verlag, Berlin and New York, 1982.

[6] D.E. Knuth, *The art of computer programming*, vol. 2: *Seminumerical algorithms*, Addison-Wesley, Reading, Massachusetts, 2. Auflage, 1981.

[7] N. Koblitz, *A Family of Jacobians Suitable for Discrete Log Cryptosystems*, to appear in: Proceedings of Crypto'88, Lecture Notes of Computer Science, Springer-Verlag.

[8] J.C. Lagarias, *Worst-Case Complexity Bounds for Algorithms in the Theory of Integral Quadratic Forms*, Journal of Algorithms 1 (1980), 142 – 186.

[9] H.W Lenstra, Jr., *On the calculation of regulators and class numbers of quadratic fields*, London Math. Soc. Lecture Notes **56** (1982), 123 – 150.

[10] K.S. McCurley, *A Key Distribution System equivalent to Factoring*, preprint, 1987.

[11] V. Miller, *Use of Elliptic Curves in Cryptography*, Advances in cryptology (Proceedings of Crypto'85), Lecture Notes in Computer Science **218** (1986), Springer-Verlag, NY, 417 - 426.

[12] W. Narkiewicz, *Elementary and analytic theory of algebraic numbers*, Warszawa, 1974.

[13] W. Narkiewicz, *Number theory*, Warszawa, 1977, engl. edition 1983.

[14] R.W.K. Odoni, V. Varadharajan and P.W. Sanders, *Public Key Distribution in Matrix Rings*, Electronic Letters **20** (1984), 386 – 387.

[15] D. Shanks, *Class number, a theory of factorization and genera*, Proc. Symposia in Pure Mathematics **20** (1971), 415 – 440.

A New Multiple Key Cipher and an Improved Voting Scheme

Colin Boyd

Communications Research Group

Department of Electrical Engineering

University of Manchester, Manchester, M13 9PL, UK

1. Introduction

At Eurocrypt 88 [1] we introduced the notion of a multiple key cipher and illustrated it with an example based on RSA which we called "multiple key RSA". In this paper we consider another multiple key cipher also based on a well known cryptographic function, exponentiation in a prime field. The important difference from multiple key RSA is that this function does not possess the trapdoor property. At the end of [1] we speculated that such functions may have useful applications and here we give as one illustration a new voting scheme.

One of the applications of multiple key RSA given in [1] was a simple voting scheme. Although that scheme allowed voters to verify that their votes were counted while maintaining anonymity with respect to other voters, it did not maintain anonymity of voting from the "government" or vote-issuing authority. Indeed, we suggested, as had others [5], that the two properties that voters could only vote once, and that votes were anonymous to the authority, were incompatible. However, at the same conference Chaum [3] proved us wrong with a counterexample.

As an application of our new multiple key cipher we give an improved version of our voting scheme which also has the property that Chaum's has. It could equally be implemented with multiple key RSA.

2. Multiple Key Ciphers

In [1] we defined a Multiple Key Cipher (MKC) to be an abelian
group of transformations of some message space, M. In a
particular application a set of transformations (parametrised by
a set of keys) k1, k2, ...kn are chosen so that

k1 o k2 o ... o kn = identity(M).

The keys are distributed to the authorised users and then
messages of the form k1 o k2 o ... o kj(M) can be written by a
set of users possessing keys k1, k2, ... kj and read by a set of
users possessing k(j+1), k(j+2), ..., kn, or their product.

With the trapdoor property it is not feasible to calculate
inverses in the group of keys without knowledge of the trapdoor.
The applications described in [1] exploited this property of
multiple key RSA. Next we examine a MKC without this property.

3. The new MKC

The new MKC is simply that defined by the group of
exponentiation transformations in a prime field with exponent
prime to p-1. The message space is equal to the integers in the
same field. This function has received much attention in modern
cryptography starting with Diffie and Hellman's well known
public key distribution scheme [6]. This MKC can properly be
called a generalisaton of the cryptosystem proposed by Pohlig
and Hellman in [8]. The important difference between this
function and multiple key RSA, is the absence of a trapdoor.
Thus if a is a known key defining the transformation

M |-->   M**a mod p

then the inverse transformation with exponent b is easily calculated by solving

a.b = 1 mod p-1.

What can we say about the security of the multiple key cipher? Clearly a known plaintext attack is no harder than the discrete logarithm problem. In view of this we should certainly choose the prime p carefully as suggested by [8], for example p = 2p'+1 for some prime p'. According to the arguments from [8], a large number of plaintext/ciphertext pairs should not give an attacker any advantage.

Before going on to the main application we briefly mention that this MKC can also be used for schemes such as the selective distribution scheme from [1] if the keys are hidden from the users by tamperproofing.

In the selective distribution scheme each user has all keys except a single one which distinguishes the user. In order to distribute to a particular set of users the centre encrypts with exactly those keys that distinguish the users who are not to receive the information.

4. The improved Voting Scheme

As in Chaum's scheme [3] we assume the existence of a voting authority who will faithfully carry out elections and issue valid voting slips to every authorised voter exactly once. We assume the existence of some universally publicised large (enough) prime p so that it is universally accepted that the discrete logarithm problem in the field of integers modulo p is hard.

4.1 Choosing the parameters

In the first stage the voting authority selects three complementary keys a,b, and c. For example the first two can be chosen randomly (but coprime to p-1) and then the third selected so that

a.b.c = 1 mod p-1.

One of these keys, say a, is then made public for the purposes of this particular election. Note that there is a sufficient supply of numbers prime to p-1 as estimated in [8]; for example, when p = 2p'+1 the proportion of such numbers is about a half.

A primitive element e of the group of transformations prime to p-1 is also published. It may be a universal element for all elections as well as p.

4.2 Registration Phase

At the next stage each voter registers for the election by forming a block M consisting of a component of redundancy, a component of randomness and his voting intention. (Here is the crux of the difference between this scheme and our former scheme, in that there the block was formed by the authority.) The component of redundancy should be the same for all voters and could either be fixed for all elections or should be broadcast together with the key a during the first stage of the election. It should be large enough that random encryptions should have negligible chance of containing it.

The component of randomness should be chosen uniquely by the user for that particular vote. However it should also be chosen so that the whole voting block M is prime to p-1. As mentioned above this condition is easily satisfied and the randomness component needs to be large enough to ensure that many tries are possible until it can be satisfied. The voter's voting intention may be specified in a selected number of bits depending on the number of possible outcomes of the election.

The voter now performs two encryptions on M before sending it to the authority for registration. First the voter forms

B = e**M mod p.

Then the voter "blinds" B [2]. This may be done by splitting the public key a as follows. The voter selects a1 at random (but prime to p-1) and then calculates a2 such that

a1.a2 = a mod p-1.

B is encrypted with a1 by the voter and sent to the authority. Thus

e**(M.a1) = B**a1 mod p

is sent to the authority. The point of the double encryption is to guarantee anonymity since all votes will be an exponentiation of e by a random power prime to p-1. In order to reduce computation M.a1 mod p-1 can be calculated first and then a single exponentiation is required.

The authority will require authentication of the user's identity which may be done in several ways which we do not consider here. The authority records the fact that the voter has registered in order that no voter may vote twice, and encrypts the voting block with the secret key b and returns B**(a1.b) mod p to the voter.

4.3 Voting Phase

Each voter completes his vote by encrypting the returned block with a2 to form

$((B**a1)**b)**a2 \bmod p = B**ab \bmod p.$

and returning it anonymously to the authority together with the original block M. This may need to be done using a completely untraceable protocol such as that defined by Chaum in [4] in order to ensure that the vote is not traced.

The authority will encrypt the received anonymous block with c to recover the plaintext block. It will also check that $B = e**M \bmod p$. If the redundancy condition is satisfied then the authority will accept the vote. Note that in order to preserve anonymity all voters should register their votes before any voter returns an anonymous vote. The authority will publish all plain voting blocks with the result of the election. Each voter may then check that his random number is present to verify that his vote is counted. In order that voters may be able to distinguish their votes, the number of possible randomness components should be large in comparison with the number of voters. All identical votes are discarded by the authority to avoid repeat voting. If the randomness element is large enough this will result in disenfranchisment of any voters only with negligible probability.

5. Security of the Scheme

We consider the security of the scheme from two aspects. Firstly the difficulty of discovering the voting intentions of any voter, and secondly the difficulty to any voter of cheating to make more than one vote.

5.1 Anonymity of Votes

When considering the anonymity of voters we assume that during the voting phase there is a perfectly untraceable protocol which allows the voter to deliver the vote to the authority. Then the only information available to the authority or any other entity in order to discover voters' intentions is the set of messages passed in the registration phase and the set of published votes. It is impossible for any entity to associate any published vote with any registered vote since the registered votes are all random exponentiations of e by a number prime to p−1. Therefore any registered vote could take the value of any other registered vote if the public key a had been split by the voter in a different way. Thus anonymity of votes is unconditionally guaranteed.

5.2 Forgery of Votes

Attempts to forge votes may be made both by legitimate voters who want to vote more than once and by outsiders who wish to influence the outcome. Forgers may try to break the system completely by finding the authority's secret keys or they may try to forge votes without finding the keys. In addition they may want to forge votes without even knowing their values in order to simply disrupt the election.

In order to forge a vote it must be possible to convince the authority that the forged vote delivered in the final phase is a registered vote. This means that the redundancy condition must be satisfied. Thus the forger must be able to construct a pair (M,e**(M.ab)) with M satisfying the redundancy condition.

For an outsider to discover the secret authority key he must use the random (X,X**b) pairs exchanged in the registration phase to find b, or the (M,e**(M.a.b)) pairs sent in the voting phase. In other words he must solve the discrete logarithm problem for this instance. For an insider to find b he may also have a (X,X**b) pair with X chosen during the registration phase. This is obviously related to the discrete logarithm problem and is equivalent to finding the key for the Pohlig-Hellman cryptosystem with a chosen plaintext attack.

To forge a particular vote (including random number) means finding a M,e**(M.a.b) pair which is equivalent to breaking the Pohlig-Hellman cryptosystem without finding the secret key. However care must be taken to avoid attacks based on the multiplicative property of exponentiation. Thus for all integers k, if M,e**(M.a.b) is a valid pair then so is kM,e**(M.a.b)k. Therefore the redundancy condition should be chosen so that this is not possible.

6. Variations

As already mentioned the scheme could equally be implemented with multiple key RSA. In order to give further confidence in the difficulty of forging votes, one variation that might bear further investigation is to use multiple key RSA but with a modulus as defined by McCurley in [7] and then with 16 as the primitive element it follows that finding the key for an observer is equivalent to factoring the modulus as well as solving the discrete logarithm for the factors of the modulus.

A more radical variation is to dispense with the primitive element e and instead make each voting block M be a primitive element by suitable adjustment to the random part. (Checking this condition is harder than checking M is prime to p-1 but is straightforward if p-1 = 2p'.) Then M ** a1 mod p is sent by the voter to the authority who returns M ** a1.b mod p. Finally M ** ab mod p is delivered anonymously in the voting phase. This has the advantage that the vote does not have to be sent in cleartext since it is recovered completely by the authority. This variation may well be more secure against forgery since no plaintext/ciphertext pairs with the redundancy condition may be obtained.

7. Acknowledgement

I am grateful for the resources of the Data Security Laboratory, British Telecom, where I was when this work was started.

8. References

[1] C.A.Boyd, Some Applications of Multiple Key Ciphers, Proceedings of Eurocrypt 88, Springer-Verlag, 1988.

[2] D.L.Chaum, Untraceable Electronic Mail, Return Addresses, and Digital Pseudonyms, Comm.ACM, 24,2,(1981), 84-88.

[3] D.L.Chaum, Elections with Unconditionally-Secret Ballots and Disruption Equivalent to Breaking RSA, Proceedings of Eurocrpyt 88, Springer-Verlag, 1988.

[4] D.L.Chaum, The Dining Cryptographers Problem, Journal of Cryptology, 1,1,(1988), 65-75.

[5] J.D.Cohen & M.J.Fischer, A Robust and Verifiable Cryptographically Secure Election Scheme, Proceedings of IEEE Conference on Foundations of Computer Science, 1985.

[6] W.Diffie & M.Hellman, New Directions in Cryptography, IEEE Transactions on Information Theory, IT-22,6,1976.

[7] K.S.McCurley, A Key Distribution System Equivalent to Factoring, Journal of Cryptology, 1,2,(1988), 95-105.

[8] S.C.Pohlig & M.Hellman, An Improved Algorithm for Computing Logarithms over GF(p) and Its Cryptographic Significance, IEEE Transactions on Information Theory, IT-24,1,1978.

ATKIN'S TEST: NEWS FROM THE FRONT

François Morain *
Institut National de Recherche en Informatique et en Automatique
Domaine de Voluceau, B. P. 105
78153 LE CHESNAY CEDEX (France)

Département de Mathématiques
Université Claude Bernard
69622 Villeurbanne CEDEX (France)

Abstract

We make an attempt to compare the speed of some primality testing algorithms for certifying 100-digit prime numbers.

1. Introduction. The implementation of several public-key cryptosystems requires the ability to build *large* primes as fast as possible [13, 2]. Several authors [29, 12, 15, 22] have studied this problem and given some good algorithms, which give primes having special forms. Our purpose is to explain how to test a *random* integer for primality.

One possible solution to this problem is to use *probable* primes, recognized by a probabilistic primality testing algorithm, such as Miller-Rabin's [19]. This test is very fast and almost surely yields prime numbers. (For a philosophical interpretation of this test, see [4].)

Another way is to use *deterministic* primality testing algorithms which yield a proof for a number to be a prime. The first general purpose deterministic algorithm was introduced by Adleman, Rumely and Pomerance [1] and refined by H. Cohen, H. W. Lenstra (Jr.) and A. K. Lenstra [10, 11] (and more recently by Bosma and van der Hulst [6]). It gives good running times (on a huge computer). However, the proof given by their program is *yes* or *no* and the only way for someone else to verify the results is to rewrite and rerun the entire program. One of the most recent primality testing algorithm, due to Atkin [3], uses elliptic curves and generalizes the old theorems of Fermat on primality. The author used his own implementation of this algorithm to prove the primality of about fifty large numbers from Cunningham's tables [8] (thus finishing the list of probable primes that were waiting to be certified), the largest one being the 564-digit cofactor of F_{11} [24], and two other large primes, namely S_{1493} (572 digits) and S_{1901} (728 digits), where

$$S_p = \frac{(1 + \sqrt{2})^p + (1 - \sqrt{2})^p}{2}.$$

For this algorithm, the work needed to check the results is far less than that of establishing proofs.

The purpose of this paper, after a brief description of Atkin's test, is to attempt to compare these algorithms with respect to the following questions:

1. How long does it take to test a 100-digit number for primality?

2. How fast are these algorithms compared to the algorithm of Miller?

---

*On leave from the French Department of Defense, Délégation Générale pour l'Armement.

3. What kind of proof do we get? How long does it take to verify it?

It should be noted that we only describe the implementation of the algorithm that is needed to test 100-digit numbers. Many other strategies are used when dealing with larger numbers (see [24] or the forthcoming papers [23, 26]).

Notations. In the sequel, N will denote an odd integer to be tested for primality, **MR** Miller-Rabin's algorithm, **CL$_2$** Cohen-Lenstra's, and **ATK** Atkin's.

2. A brief description of Atkin's test.

(2.1) Elliptic curves. Let **K** be a field of characteristic prime to 6. An elliptic curve E over **K** is a non singular algebraic projective curve of genus 1. It can be shown [9, 33] that E is isomorphic to a curve with equation:

$$y^2 z = x^3 + axz^2 + bz^3, \tag{1}$$

with a and b in **K**. The *discriminant* of E is $\Delta = -16(4a^3 + 27b^2)$ and the *invariant* is

$$j = 2^8 3^3 \frac{a^3}{4a^3 + 27b^2}.$$

We write $E(\mathbf{K})$ for the set of points with coordinates $(x : y : z)$ which satisfy (1) with $z = 1$, together with the point at infinity: $O_E = (0 : 1 : 0)$. We will use the well-known *tangent-and-chord* addition law on a cubic [18] over a finite field $\mathbf{Z}/N\mathbf{Z}$ (see [16] for a justification).

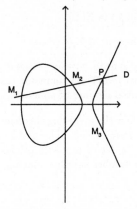

Figure 1: An elliptic curve over **R**

In order to add two points $M_1 = (x_1, y_1)$ and $M_2 = (x_2, y_2)$ on E resulting in $M_3 = (x_3, y_3)$, the equations are

$$\begin{cases} x_3 &= \lambda^2 - x_1 - x_2 \\ y_3 &= \lambda(x_1 - x_3) - y_1 \end{cases}$$

where

$$\lambda = \begin{cases} (y_2 - y_1)(x_2 - x_1)^{-1} & \text{if } x_2 \neq x_1 \\ \\ (3x_1^2 + a)(2y_1)^{-1} & \text{otherwise} \end{cases}$$

We can compute kP using the binary method (see also [11]) or addition-subtraction chains [28].

(2.2) Primality testing. Let us recall one of the converses of Fermat's theorem.

Theorem 1 *Let a be such that* $\gcd(a, N) = 1$, *q a prime divisor of $N - 1$. If*

$$a^{N-1} \equiv 1 \bmod N \text{ and } \gcd(a^{(N-1)/q} - 1, N) = 1$$

then each prime divisor p of N satisfies: $p \equiv 1 \bmod q$.

Corollary 1 *If $q > \sqrt{N}$ then N is prime.*

A similar theorem can be stated for elliptic curves.

Theorem 2 ([14, 21]) *Let N be an integer greater than 1 and prime to 6. Let E be an elliptic curve over $\mathbf{Z}/N\mathbf{Z}$, m and s two integers such that $s \mid m$. Suppose we have found a point P on E that satisfies $m\,P := O_E$, and that for each prime factor q of s, we have verified that $\frac{m}{q}P \neq O_E$. Then if p is a prime divisor of N, $\#E(Z/pZ) \equiv 0 \bmod s$.*

Corollary 2 *If $s > (\sqrt[4]{N}+1)^2$, then N is prime.*

In order to use the preceding theorem, we need to compute the number of points m. This process is far from trivial in general (see [32]). From a practical point of view, it is desirable to use deep properties of elliptic curves over finite fields. This involves the theory of complex multiplication and class fields and requires a lot of theory [24]. We can summarize the principal properties:

Theorem 3 *Every elliptic curve $E \bmod p$ has complex multiplication by the ring of integers of an imaginary quadratic field $K = \mathbf{Q}(\sqrt{-D})$.*

From a very down-to-earth point of view, this comes down to saying:

- p splits in K: $(p) = (\pi)\,(\pi')$ in K;

- $H_D(j(E)) \equiv 0 \bmod p$ for a fixed $H_D(X)$ in $\mathbf{Z}[X]$;

- $m = (\pi - 1)(\pi' - 1) = p + 1 - t$, where $|t| \leq 2\sqrt{p}$ (Hasse).

The computation of the polynomials H_D is dealt with in [24] and [25]: it requires some 1000 lines of MAPLE code. As a result, I have a list of 575 discriminants (those with $h \leq 10$ and some with $h = 12$), thus providing about 1158 potential number of points.

3. Atkin's algorithm. We now explain how the preceding theorems are used in a *factor and conquer* algorithm similar to the DOWNRUN process of [34]. The first phase of the algorithm consists in finding a sequence $N_0 = N > N_1 > \cdots > N_k$ of probable primes such that: N_{i+1} prime $\Longrightarrow N_i$ prime. The second then proves that each number is prime, starting from N_k.

Procedure SEARCHN

1. $i := 0$; $N_0 := N$;

2. find a fundamental discriminant $-D$ such that (N_i) splits as the product of two principal ideals in $\mathbf{Q}(\sqrt{-D})$;

3. for each solution of $(N_i) = (\pi)(\pi')$, find all factors of $m_\pi = (\pi - 1)(\pi' - 1)$ less than a given bound B and let N_π be the corresponding cofactor;

4. **if** one of the N_π is a probable prime **then** set $N_{i+1} := N_\pi$, store $\{N_i, D, \pi, m\}$ set $i := i + 1$, and go to step 2 **else** go to step 3.

5. **end.**

In Step 2, we use lattice reduction (see [24]). In Step 3, we use a sieve to find all factors less than 2^{15}, which is enough for our purpose (that is testing the primality of 100-digit integers). The sieving process is done as follows (this generalizes a trick described in [7, Section 7, Rem. 1] and [11]):

Procedure Sieve

1. **for** $i = 1..k$ **do** $RES[i] := (N + 1) \bmod p_i$;
2. $N \pm 1$ tests:
   for $i = 1..k$
   - **if** $RES[i] = 0$ **then** $p_i \mid N + 1$
   - **if** $RES[i] = 2$ **then** $p_i \mid N - 1$
3. $m_\pm = N + 1 \pm t$, $|t| \le 2\sqrt{N}$:
   for $i = 1..k$
   - $r := t \bmod p_i$
   - **if** $RES[i] = r$ **then** $p_i \mid m_-$
   - **if** $RES[i] = -r$ **then** $p_i \mid m_+$

If one of the steps of procedure SEARCHN cannot be achieved, this means that either one of our N_i is indeed composite or that this is a difficult number (see [26]).

The second phase consists in proving that the numbers N_i are indeed primes. This is done as follows:

Procedure PROOF

for $i = k..0$

1. compute a root j of $P_{D_i}(X) \bmod N_i$ using Berlekamp's algorithm if $\deg(P) > 4$, Shanks's if $\deg = 2$, Cailler-Williams' if $\deg = 3$, and Skolem's otherwise (see [27]);

2. let $k := j/(1728 - j) \bmod N_i$, $a := 3k$, $b := 2k$: $E(a, b)$ has for equation $y^2 \equiv x^3 + ax + b \bmod N_i$; choose a point P on $E(a, b)$ and compute $Q = m_i P$; if $Q \ne O_E$ then choose a non residue c and set: $a := ac^2$, $b := bc^3$.

3. verify the condition of theorem (2).

The same remarks can be made if we cannot complete our task. It has been observed that as soon as we can complete Phase 1, Phase 2 is no problem (apart from the execution time).

4. Implementation and empirical comparisons. I have implemented **ATK** on a SUN 3/60 (12 Mo) using the BigNum package described in [17]. This package includes about 700 lines of assembly code together with 1500 lines of Le-Lisp (or C). My program is written in Le-Lisp.

(4.1) A brief comparison with **CL$_2$**. In [11], the authors describe the implementation of **CL$_2$** on a CDC Cyber 170/750. They used a 47-bit arithmetic and they gave the time for doing elementary operations on *multiples* (i.e. 8 words of 47 bits) and *doubles* (16 words).

We can attempt to compare the speeds of these two arithmetics by measuring the time needed on a SUN to do the same operations on numbers having equal numbers of bits: a multiple consists of 12 words of 32 bits and a double of 24 words of 32 bits. We list below these times in milliseconds.

operation	CDC	SUN 3/60
multiple + multiple	0.014	0.260
multiple × multiple	0.070	1.270
double mod multiple	0.200	2.850

We can satisfy ourselves with the crude statement that our arithmetic is 15 times slower than that of Cohen and Lenstra.

Following the same line, we can compare the time needed to test 100-digit numbers for primality. We now describe the protocol we used (it is the protocole of [11] without the testing for small factors).

Protocol for measuring the time for algorithm $\mathcal{A}$

repeat 20 times

1. select a random odd number n;
2. if n passes four times **MR**
   then record the corresponding time;
   else $n := n + 2$; **go to** 2.
3. if n passes algorithm $\mathcal{A}$
   then ouput "Prime"; record the time;
   else output "Composite".

At the end of the process, we compute statistics. In the following table, we have listed the times for the fastest execution, the slowest, the mean running time and the standard deviation. There are two columns with "4× **MR** " which corresponds to four executions of the two versions of algorithm **MR**, one along with $\mathbf{CL_2}$, the other with **ATK**. Times are in seconds.

	$\mathbf{CL_2}$	4×**MR**	ratio	**ATK**	4×**MR**	ratio
minimum	26.031	0.544		350.4	4.0	
maximum	75.416	0.602		1042.3	5.0	
mean	50.442	0.567	88.92	661.0	4.4	150.23
standard deviation	15.203	0.015		197.0	0.4	

We now see that our **MR** is about 8 times slower, but **ATK** algorithm is less than 14 times slower than $\mathbf{CL_2}$. These arguments are not very strong, but gives some hints on the relative behaviors of $\mathbf{CL_2}$ compared to **ATK**.

For the sake of completeness, I list in the following table the corresponding times for the implementation of $\mathbf{CL_2}$ by A. K. Lenstra on a Cray I [20].

	$\mathbf{CL_2}$	4×**MR**
minimum	3.822	0.061
maximum	9.174	0.057
mean	7.047	0.058
standard deviation	1.549	0.001

We can also proceed to give in Table 1 the time needed to test a number of d words of 32 bits with my program, for $d = 2(2)20$. Time are in seconds. The first line is concerned with **ATK**, the second with the number of steps in procedure SEARCHN.

These data are reported in Figure 2. They suggest the following approximation for the average running time of our program (in seconds if d is the number of 32-byte words):

$$T_{ATK}(d) \approx 0.27 \times d^{3.41}.$$

Using the data in [11], we can compute a similar approximation for the running time of $\mathbf{CL_2}$ for numbers from 100 to 200 digits. This yields:

$$T_{CL}(d) \approx 0.024 \times d^{3.28}.$$

If we use the theoretical running time of $\mathbf{CL_2}$, we find

$$T_{th}(d) \approx 2 \times d^{1.65 \log \log d}.$$

We can draw some conclusions regarding this comparison between $\mathbf{CL_2}$ and **ATK**. It seems that $\mathbf{CL_2}$ is slightly faster for this range of numbers (100 to 200 digits). It is worth noting that the implementation of $\mathbf{CL_2}$ was very optimized for this range, as mine is not (at least for the time

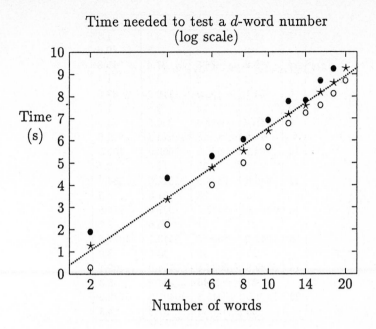

Time needed to test a d-word number
(log scale)

o: minimum time
•: maximum
⋆: mean

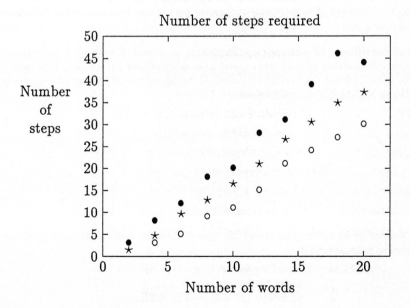

Number of steps required

d	min	max	mean	st. dev.
2	1.3	6.5	3.6	1.5
	0	3	1.5	0.9
4	9.0	74.8	28.4	15.8
	3	8	4.7	1.4
6	54.1	195.9	119.9	37.6
	5	12	9.6	1.7
8	146.1	413.4	250.6	73.9
	9	18	12.8	2.6
10	294.7	983.9	608.9	160.8
	11	20	16.5	2.7
12	846.6	2334.5	1312.0	384.0
	15	28	21.0	3.6
14	1369.4	2433.5	1937.0	276.5
	21	31	26.6	2.7
16	1945.1	5886.5	3513.1	1077.8
	24	39	30.5	3.9
18	3289.8	10009.5	5454.9	1469.1
	27	46	34.9	5.3
20	5871.0	17845.8	10405.6	2772.2
	30	44	37.3	4.3

Table 1: Time needed to test a d-word number for primality

being). My arithmetic is able to deal with arbitrary large numbers and I was able to certify numbers from 250 to 700 digits. (For these sizes, I use special algorithms, including Karatsuba's multiplication algorithm as implemented by P. Zimmermann at INRIA.) On the contrary, the size of integers used on the CDC must be fixed at compile time [20].

(4.2) Further remarks on my implementation. In a typical run of my program, three quarters of the time are needed to complete the first phase of the algorithm and one quarter for the second (again for a 100-digit number). The most time consuming part of the algorithm is the factorization of the number of points m.

5. Generalized certificate of primality. Certificates of primality have been introduced by Pratt [31]. Recently, Pomerance studied some certificates using elliptic curves [30]. It is very easy to generalize the work of Pratt. A certificate generated by algorithm **ATK** consists in blocks of numbers. Each block has the following structure:

$$n_i$$
$$type$$

where n_i is the number to be tested, *type* giving the type of theorem used to show the primality of n_i. This is an integer, chosen as follows:

-1 : use of the factors of $n_i - 1$,
1 : use of the factors of $n_i + 1$,
D : an integer ($D > 2$) used in **ATK**.

To each of the types corresponds a list of numbers used to complete the proof of n_i being prime, whenever the following block is valid. For instance, the format of a proof using elliptic curves is the following:

D	the discriminant used
m	the number of points on the curve
r_0	
$\cdots$	the factors of m
r_k	
0	
a	the curve is E: $y^2 = x^3 + ax + b$
b	E has complex multiplication by $\mathcal{O}(\sqrt{-D})$
x	the coordinates of a point P on the curve
y	
f_1	
$\cdots$	the factorization of the order of P on E
f_l	

For example, here is the first block of proof for a 50-digit prime:

```
n0=35090920174233837395447134480305116522935098213281
D=4
m=35090920174233837395447125326861763110277715724842
r1=937
    853
    13
    2
0
a=6
b=0
x=7778128793230599416235595023534938267181834875
y=18796494177062591514397495874290295720055455015935
f=16886233357214610359515335621502364304351 97
0
```

It is also possible to give a very short label to a certificate, which I call a *primality path*. It consists in some brief informations on the discriminants used and the actual value of m choosen (see [26]). For example, for the above mentioned prime, the path is: $4d1 - 1 + 1 + 3f1 + 1+$.

An independant verifier can check the proof by simply coding the necessary basic operations on elliptic curves.

The size of the whole program is about 230 kbytes (about 6500 lines of Le-Lisp) using 370 kbytes of data (the polynomials H_D and the primes below 2^{15}). The code needed for verifying a certificate is about 74 kbytes. Moreover, the time needed to verify a certificate for a 100-digit prime is about one fifth the time needed to build it.

6. Further directions of research. In the future, I intend to optimize my code a little further, namely using special techniques for gcd's, ... It will also be interesting to use the hardware multiplier built by J. Vuillemin and his team at INRIA and DEC-PRL [5] to speed up the modular operations.

Apart from this technological research, it is possible to give some hints concerning the analysis of the algorithm and also what a bad number is for **ATK**.

7. Conclusions. If we try to answer our original questions in brief, we see that $\mathbf{CL_2}$ gives a succinct proof in a short time on a huge computer, taking 90 times more time than four executions of **MR**; on the other hand, **ATK** gives a lengthy proof that can be verified in a small fraction of the time needed to establish it, and it works well on a small computer, achieving a ratio of 150 with $4\times$**MR**, which is not too bad compared to $\mathbf{CL_2}$, since the arithmetic is approximately fifteen times slower.

Acknowledgments. I wish to thank A. K. Lenstra for some useful comments on the first version of this article and S. S. Wagstaff, Jr. for his careful reading of this article.

References

[1] L. M. ADLEMAN, C. POMERANCE, AND R. S. RUMELY. On distinguishing prime numbers from composite numbers. *Annals of Math.*, 117:173–206, 1983.

[2] L. M. ADLEMAN, R. L. RIVEST, AND A. SHAMIR. A method for obtaining digital signatures and public-key cryptosystems. *Comm. ACM*, 21(2):120–126, 1978.

[3] A. O. L. ATKIN. Manuscript. August 1986.

[4] P. BEAUCHEMIN, G. BRASSARD, C. CRÉPEAU, C. GOUTIER, AND C. POMERANCE. The generation of random numbers that are probably prime. *J. Cryptology*, 1:53–64, 1988.

[5] P. BERTIN, D. RONCIN, AND J. VUILLEMIN. Introduction to programmable active memories. In *Proc. of the Internat. Conf. on Systolic Arrays*, 1989.

[6] W. BOSMA AND M.-P. VAN DER HULST. Faster primality testing. In *Proc. Eurocrypt '89*, 1989.

[7] J. BRILLHART, D. H. LEHMER, AND J. L. SELFRIDGE. New primality criteria and factorizations of $2^m \pm 1$. *Math. of Comp.*, 29(130):620–647, 1975.

[8] J. BRILLHART, D. H. LEHMER, J. L. SELFRIDGE, B. TUCKERMAN, AND S. S. WAGSTAFF (JR.). *Factorizations of $b^n \pm 1$, $b = 2, 3, 5, 6, 7, 10, 11, 12$ up to high powers*. Number 22 in Contemporary Mathematics. AMS, 1983.

[9] J. W. S. CASSELS. Diophantine equations with special references to elliptic curves. *J. London Math. Soc.*, 41:193–291, 1966.

[10] H. COHEN AND JR H. W. LENSTRA. Primality testing and Jacobi sums. *Math. of Comp.*, 42(165):297–330, 1984.

[11] H. COHEN AND A. K. LENSTRA. Implementation of a new primality test. *Math. of Comp.*, 48(177):103–121, 1987.

[12] C. COUVREUR AND J.J. QUISQUATER. An introduction to fast generation of large prime numbers. *Philips J. Research*, 37:231–264, 1982.

[13] W. DIFFIE AND M. E. HELLMAN. New directions in cryptography. *IEEE Trans. on Information Theory*, IT-22-6, nov 1976.

[14] S. GOLDWASSER AND J. KILIAN. Almost all primes can be quickly certified. In *Proc. 18th STOC*, pages 316—329, Berkeley, 1986.

[15] D. GORDON. Strong primes are easy to find. In *Proc. Eurocrypt '84*, pages 216–223. Springer, 1984.

[16] JR. H. W. LENSTRA. Factoring integers with elliptic curves. *Annals of Math.*, 126:649–673, 1987.

[17] J.-C. HERVÉ, F. MORAIN, D. SALESIN, B. SERPETTE, J. VUILLEMIN, AND P. ZIMMER-MANN. Bignum: A portable and efficient package for arbitrary precision arithmetic. Rapport de Recherche 1016, INRIA, avril 1989.

[18] D. HUSEMXEOLLER. *Elliptic curves*, volume 111 of *Graduate Texts in Mathematics*. Springer, 1987.

[19] D. E. KNUTH. *The Art of Computer Programming: Seminumerical Algorithms*. Addison-Wesley, 1981.

[20] A. K. LENSTRA. Data concerning the implementation of the Jacobi sums algorithm on a cray-1. Personnal Communication, April 1989.

[21] H. W. LENSTRA. Elliptic curves and number theoretic algorithms. Technical Report Report 86-19, Math. Inst., Univ. Amsterdam, 1986.

[22] U. M. MAURER. Fast generation of secure RSA-products with almost maximal diversity. In *Proc. Eurocrypt '89*, 1989.

[23] F. MORAIN. Distributed primality proving. In preparation.

[24] F. MORAIN. Implementation of the Atkin-Goldwasser-Kilian primality testing algorithm. Rapport de Recherche 911, INRIA, Octobre 1988.

[25] F. MORAIN. Construction of Hilbert class fields of imaginary quadratic fields and dihedral equations modulo p. Rapport de Recherche 1087, INRIA, Septembre 1989.

[26] F. MORAIN. Elliptic curves and primality proving. In preparation, 1989.

[27] F. MORAIN. Résolution d'équations de petit degré modulo de grands nombres premiers. Rapport de Recherche 1085, INRIA, Septembre 1989.

[28] F. MORAIN AND J. OLIVOS. Speeding up the computations on an elliptic curve using addition-subtraction chains. Rapport de Recherche INRIA 983, INRIA, Mars 1989.

[29] D. A. PLAISTED. Fast verification, testing and generation of large primes. *Theoretical Computre Science*, 9:1–16, 1979.

[30] C. POMERANCE. Very short primality proofs. *Math. of Comp.*, 48(177):315–322, 1987.

[31] V. R. PRATT. Every prime has a succint certificate. *SIAM J. Comput.*, 4:214–220, 1975.

[32] R. SCHOOF. Elliptic curves over finite fields and the computation of square roots mod p. *Math. of Comp.*, 44:483–494, 1985.

[33] J. T. TATE. The arithmetic of elliptic curves. *Inventiones Math.*, 23:179–206, 1974.

[34] M. C. WUNDERLICH. A performance analysis of a simple prime-testing algorithm. *Math. of Comp.*, 40(162):709–714, 1983.

Fast Generation of Secure RSA-Moduli with Almost Maximal Diversity

Ueli M. Maurer

Institute for Signal and Information Processing
Swiss Federal Institute of Technology
CH-8092 Zürich, Switzerland

Abstract

This paper describes a new method for generating primes together with a proof of their primality that is extremely efficient (for 100-digit primes the average running time is equal to the average time required for finding a "strong pseudoprime" of the same size that passes the Miller-Rabin test for only four bases), that yields primes that are nearly uniformly distributed over the set of all primes in a given interval, and that is easily modified to yield (with no additional computational effort) primes that are nearly uniformly distributed over the subset of these primes that satisfy certain security constraints for use in the RSA cryptosystem. This method is used to generate, for a given encryption exponent e, an RSA-modulus $m = pq$ that is nearly uniformly distributed over all secure RSA-moduli in a given interval I, i.e., over the set of all integers in I that are (1) the product of exactly two primes p and q none of which is smaller than a given limit L, where (2) $(p-1, e) = (q-1, e) = 1$ and (3) $p-1$ and $q-1$ each contain a prime factor greater than a given limit L', and where (4) for all but a provably (given) small fraction of plaintexts in Z_m^*, the minimum number of iterated encryptions with exponent e required to recover the plaintext, is provably greater than a given limit M. Our method exploits a well-known result due to Pocklington [20] that allows one to prove the primality of p when only a partial factorization of $p-1$ is known. These prime factors of $p-1$ are generated by recursive application of the prime generating procedure. Although the discussion is centered on the RSA system, our method can of course be used in other cryptographic systems, such as the Diffie-Hellman public key distribution system, that require large primes satisfying certain security constraints.

1. Introduction

The primes p and q in the RSA cryptosystem [26] must satisfy certain conditions in order to prevent that the system can easily be broken. Two kinds of attacks are usually considered: factorization of the modulus $m = pq$ and decryption by iterated encryption [25,27,29,31]. To prevent fast factorization of the modulus, p and q should be sufficiently large, and $p - 1$ (and similarly $q - 1$) should contain a large prime factor p' (q') to make the so-called $p - 1$ factoring method (see [24], p.172 and [21]) infeasible. Other conditions, for example that $p + 1$, $p^2 + 1$ or $p^2 \pm p + 1$ (and similarly for q) each contain a large prime factor can for a similar reason also be placed on the primes. However these additional conditions are less stringent and are usually not considered, with some exceptions for the (p+1)-condition [12,18,28]. The usual way to guarantee that the iterated encryption attack is infeasible for virtually all plaintexts is by requiring that p' (and similarly q') must contain a large prime factor p'' (q''). To satisfy these conditions [25,27] it is usually required that p and q are roughly of the same size (e.g. have equally many digits) and to choose p (and similarly q) of the special form $p = 2ap' + 1$ with $p' = 2bp'' + 1$ where p' and p'' are both primes and where a and b are small integers (often $a = b = 1$). This is a strong restriction on the set of allowed primes as will be shown.

Our approach differs in the way the security constraints are satisfied. We argue that the security requirements should be specified by the security parameters described in Section 3, but that the RSA-modulus should then be randomly chosen from the set of all RSA-moduli in a given interval that satisfy these constraints, in order to achieve maximal diversity and not to give the enemy any a priori information about the special form of the primes. For characterization of the iterated encryption attack, a new result (Theorem 2) is presented in Section 2 showing that the condition that $p' - 1$ has a large prime factor p'' is not necessary, although this is in practice virtually always the case. For one reasonable formulation of the security constraints, about 23% of all products of two sufficiently large primes are shown to be acceptable. This fraction is by orders of magnitude greater than that achieved by using the strong restrictions described above. We argue that there is little reason for such strong restrictions when the security requirements can also be met by a much looser restriction. Moreover, even if the general factorization problem were indeed difficult, it is at least conceivable that a specialized fast algorithm exists for factoring numbers that are the product of two primes of the described special form, although at present these instances of the factorization problem seem to be the most difficult ones. However, the security constraints in our method for generating primes can be specified flexibly, and by an appropriate choice the described strong restrictions can also be met if required by the security policy.

The method for generating RSA-moduli described in Section 3 is based on an analysis of the probability distribution of the size of the smaller prime factor of an integer randomly selected from the set of integers of a given size that are the product of exactly two primes, both of which are greater than a given limit. The underlying recursive method for generating primes is based on an analysis of the probability distribution of the size of the largest prime factor of an integer randomly selected from the set of integers n of a certain size, for which $2n + 1$ is prime. Other conditions on n, namely that $2an + 1$, and possibly also $2b(2an + 1) + 1$, $2c(2b(2an + 1) + 1) + 1$ etc., are primes, are also considered. These derivations, which are based on evident heuristic arguments similar to those used in [15] for estimating the number of primes p below n for which $(p - 1)/2$ is also prime, lead to number-theoretic conjectures about the asymptotic behaviour of the mentioned distributions, which are of independent interest.

There exist two approaches for generating a prime in a given interval: to choose odd integers at random and to apply a primality test until a prime is obtained, or to construct a prime (e.g. by the methods of [18], [19], [28] or [31]). The first approach is not suited for practical applications, because the most efficient presently known general primality tests, which are due to Morain [17], and to Adleman, Pomerance and Rumely [1] as well as Cohen and Lenstra [5,6], are not efficient enough to be suited for implementation on a small computer or on special purpose cryptographic hardware. Another related paper by Goldwasser and Killian [11] shows that almost all primes can be certified in time polynomial in their length, but this approach cannot be efficiently implemented either. This problem of bad efficiency is in practice usually remedied by using instead of a primality test a probabilistic compositeness test [23,30], which is efficient but does not yield provable primes. The construction approach on the other hand allows one to prove the primality of the generated integers, but has either had the disadvantage that the diversity of reachable primes is moderate [18,28,31], that the interval for the primes cannot be chosen flexibly enough [18], or was computationally not efficient enough ([19], method A).

2. Theoretical Results on the Decipherability by Multiple Encryption

In the following, we denote by $ord_n(x)$ the order of x in Z_n^*, the multiplicative group modulo n. A basic fact that will be used repeatedly is: $n|m \implies ord_n(x)|ord_m(x)$. The following lemma, which is the key to our method for constructing primes, is a special case of a well-known theorem due to Pocklington ([20], see also [24], p. 109), which is the basis of the contruction method presented in [28].

Lemma 1: Let $n - 1 = RF$ where $F = \prod_{j=1}^r q_j^{\beta_j}$ with $q_1, \ldots, q_r$ distinct primes. If there is an integer a, satisfying $a^{n-1} \equiv 1 \pmod{n}$ and $(a^{(n-1)/q_j} - 1, n) = 1$ for $j = 1, \ldots, r$, then each prime factor p of n has the form $p = Fm + 1$ for some integer $m \geq 1$. Moreover, if $F > \sqrt{n}$, then n is prime.

Proof: Let p be a prime divisor of n and let α be its multiplicity. We prove that $q_j^{\beta_j}|(p-1)$ for $j = 1, \ldots, r$. We have

$$a^{n-1} \equiv 1 \pmod{n} \implies ord_n(a)|(n-1) \implies ord_{p^\alpha}(a)|(n-1), \tag{1}$$

$$ord_{p^\alpha}(a)|\varphi(p^\alpha) \text{ where } \varphi(p^\alpha) = (p-1)p^{\alpha-1}, \tag{2}$$

$$p \nmid (n-1) \overset{by(1)}{\implies} p \nmid ord_{p^\alpha}(a) \overset{by(2)}{\implies} ord_{p^\alpha}(a)|(p-1), \text{ and} \tag{3}$$

$$(a^{(n-1)/q_j} - 1, n) = 1 \implies a^{(n-1)/q_j} \not\equiv 1 \pmod{p} \implies ord_{p^\alpha}(a) \nmid \frac{n-1}{q_j}. \tag{4}$$

Using (1) and (3) and the trivial fact that for a prime q, $q^\beta|y, z|y$ and $z \nmid (y/q)$ together imply $q^\beta|z$, finally yields

$$(4) \overset{by(1)}{\implies} q_j^{\beta_j}|ord_{p^\alpha}(a) \overset{by(3)}{\implies} q_j^{\beta_j}|(p-1) \text{ for } 1 \leq j \leq r \implies F|(p-1).$$

It is obvious that if every prime factor of a number is greater than its square root then the number itself must be prime. $\square$

More sophisticated conditions that guarantee the primality of a number p if the factored part of $p - 1$ is less than $\sqrt{p}$ are described in [4] and [7] but these conditions are computationally more costly to verify. Note that Lemma 1 is usually stated in a way that allows the base a to be different for each q_j [4,7,28]. We will only make use of the special case because the verification of the conditions is first computationally more efficient and secondly allows to prove, as will be demonstrated by Theorem 2, that iterated encryption is infeasible to recover the plaintext when a is chosen to be the encryption exponent.

A few basic facts about Euler's totient function $\varphi(.)$ are:

$$\varphi(ab) \geq \varphi(a)\varphi(b) \quad \text{with equality if and only if } (a,b) = 1, \tag{5}$$

$$\sum_{d|n} \varphi(d) = n, \quad \text{and} \tag{6}$$

$$p \text{ prime and } d|(p-1) \implies \#\left\{x \in Z_p^* : ord_p(x) = d\right\} = \varphi(d). \tag{7}$$

Lemma 2: *Let p be a prime and d a divisor of $p-1$. Then*

$$\#\left\{x \in Z_p^* : d|ord_p(x)\right\} \geq \frac{\varphi(d)}{d}(p-1)$$

with equality if and only if $(d, (p-1)/d) = 1$.

Proof: $\#\left\{x \in Z_p^* : d|ord_p(x)\right\} \overset{\text{by}(7)}{=} \sum_{d':d|d',d'|(p-1)} \varphi(d') = \sum_{k:k|\frac{p-1}{d}} \varphi(kd)$

$\overset{\text{by}(5)}{\geq} \sum_{k:k|\frac{p-1}{d}} \varphi(k)\varphi(d) = \varphi(d) \sum_{k:k|\frac{p-1}{d}} \varphi(k) \overset{\text{by}(6)}{=} \varphi(d)\frac{p-1}{d}.$

The inequality holds with equality if and only if $(d, (p-1)/d) = 1$. $\quad\square$

The following new result shows that virtually any a can be used in Lemma 1 if all q_j's are large.

Lemma 3: *Let $p = RF + 1$ be a prime with $F = \prod_{j=1}^r q_j^{\beta_j}$, $F > R$ and $(R, F) = 1$, where q_j, $1 \leq j \leq r$, are primes. Then the fraction of elements $a \in Z_p^*$ that are successful in proving the primality of p by Lemma 1 is $\varphi(F)/F \geq 1 - \sum_{j=1}^r 1/q_j$.*

Proof: $a^{p-1} \equiv 1 \ (mod \ p)$ is satisfied for every $a \in Z_p^*$. If $(R, F) = 1$ then $F|ord_p(a)$ and $a^{(p-1)/q_j} \not\equiv 1 \ (mod \ p)$ for $1 \leq j \leq r$ are equivalent statements. Application of Lemma 2 with $d = F$ completes the proof. $\quad\square$

The following theorem will be required to prove Theorem 2, the final result:

Theorem 1: *Let $m = pq$ be an RSA-modulus where $p - 1 = R_p F_p$ and $q - 1 = R_q F_q$ and where the prime factorizations of F_p and F_q are $F_p = \prod_{i=1}^r p_i'^{\alpha_i}$ and $F_q = \prod_{i=1}^s q_i'^{\beta_i}$, respectively. Then the fraction of plaintexts $x \in Z_m^*$ for which $ord_m(x)$ is not a multiple of $lcm(F_p, F_q)$ is upper bounded by $\sum_{i=1}^r 1/p_i' + \sum_{i=1}^s 1/q_i'$.*

Proof: By Lemma 2,

$$n_p \overset{\text{def}}{=} \#\left\{x \in Z_p^* : F_p|ord_p(x)\right\} \geq (p-1)\frac{\varphi(F_p)}{F_p} = (p-1)\prod_{i=1}^r (1 - \frac{1}{p_i'}), \quad \text{and}$$

$$n_q \overset{\text{def}}{=} \#\left\{x \in Z_q^* : F_q|ord_q(x)\right\} \geq (q-1)\frac{\varphi(F_q)}{F_q} = (q-1)\prod_{i=1}^s (1 - \frac{1}{q_i'}).$$

$F_p|ord_p(x)$ and $F_q|ord_q(x)$ for $x \in Z_m^*$ together imply $lcm(F_p, F_q)|ord_m(x)$. Since $Z_m^* = Z_p^* \times Z_q^*$ we have

$$\#\{x \in Z_m^* : lcm(F_p, F_q)|ord_m(x)\} \geq n_p n_q \geq (p-1)(q-1)\prod_{i=1}^r (1 - \frac{1}{p_i'})\prod_{i=1}^s (1 - \frac{1}{q_i'})$$

$$\geq |Z_m^*|\left(1 - \sum_{i=1}^r \frac{1}{p_i'} - \sum_{i=1}^s \frac{1}{q_i'}\right). \quad\square$$

Iterated t-fold encryption in an RSA cryptosystem reveals the plaintext x if and only if $x^{e^u} \equiv x \ (mod \ m)$ for some $u \le t$, i.e., if and only if $e^u \equiv 1 \ (mod \ ord_m(x))$ for some $u \le t$. Hence the minimal number of encryptions needed to recover the plaintext is $ord_{ord_m(x)}(e)$; it is required for security that this number be large for virtually all x.

Theorem 2: _Let $m = pq$ be an RSA-modulus where $p - 1 = R_p F_p$ and $q - 1 = R_q F_q$, and where the prime factorizations of F_p and F_q are $F_p = \prod_{i=1}^{r} p_i'^{\alpha_i}$ and $F_q = \prod_{i=1}^{s} q_i'^{\beta_i}$, respectively. Further let $p_i' - 1 = R_{p_i'} F_{p_i'}$ for $1 \le i \le r$ and $q_i' - 1 = R_{q_i'} F_{q_i'}$ for $1 \le i \le s$ where the prime factorizations of $F_{p_i'}$ and $F_{q_i'}$ are $F_{p_i'} = \prod_{j=1}^{r_i} p_{ij}''^{\alpha_{ij}}$ for $1 \le i \le r$ and $F_{q_i'} = \prod_{j=1}^{s_i} q_{ij}''^{\beta_{ij}}$ for $1 \le i \le s$. For every integer a relatively prime to $(p-1)(q-1)$ and satisfying_

$$\text{for } 1 \le i \le r: \quad a^{(p_i'-1)/p_{ij}''} \not\equiv 1 \ (mod \ p_i') \text{ for } 1 \le j \le r_i,$$
$$\text{and for } 1 \le i \le s: \quad a^{(q_i'-1)/q_{ij}''} \not\equiv 1 \ (mod \ q_i') \text{ for } 1 \le j \le s_i,$$

the fraction of plaintexts $x \in Z_m^*$ for which $ord_{ord_m(x)}(a)$ is not a multiple of $lcm(F_{p_1'}, \ldots, F_{p_r'}, F_{q_1'}, \ldots, F_{q_s'})$ is upper bounded by $\sum_{i=1}^{r} 1/p_i' + \sum_{i=1}^{s} 1/q_i'$.

Proof: Similar arguments as used in the proof of Lemma 1 allow one to show that $F_{p_i'} | ord_{p_i'}(a)$ and hence that $F_{p_i'} | ord_{p_i'^{\alpha_i}}(a)$ and also that $F_{p_i'} | ord_{F_p}(a)$ for $1 \le i \le r$. Thus $lcm(F_{p_1'}, \ldots, F_{p_r'}) | ord_{F_p}(a)$. Simililarly one obtains $lcm(F_{q_1'}, \ldots, F_{q_s'}) \mid ord_{F_q}(a)$ and hence that $lcm(F_{p_1'}, \ldots, F_{p_r'}, F_{q_1'}, \ldots, F_{q_s'})$ divides $ord_{lcm(F_p, F_q)}(a)$. From $lcm(F_p, F_q) \mid ord_m(x)$ it follows that $ord_{lcm(F_p, F_q)}(a) \mid ord_{ord_m(x)}(a)$ and hence application of Theorem 1 completes the proof. □

Theorem 2 illustrates that, in order to prevent decipherability by iterated encryption, the condition that $p' - 1$, where p' is the largest prime factor of $p - 1$, must again have a very large prime factor p'', is not necessary.

3. Recursive Algorithm for Generating Cryptographically Secure Primes and RSA-Moduli with Almost Maximal Diversity

Lemma 1 suggests a method for constructing large primes. One can form the product $F = \prod_{j=1}^{r} q_j^{\beta_j}$ of some known primes q_j, raised to some powers β_j, and repeatedly choose an integer $R < F$ at random until $n = 2RF + 1$ can be proved to be prime by an appropriate choice of the base a. Lemma 3 shows that if n is indeed prime and if all q_j's are large, then virtually every base a can be used for certifying the primality of n. On the other hand, if n is composite but does not contain a small prime factor that allows to detect this by trial division, then virtually every base a will satisfy $a^{n-1} \not\equiv 1 \ (mod \ n)$ and hence reveal the compositeness of n, unless n is of a very special form.

When this construction approach is used for generating RSA-primes p and q (see also [28]), the major problem is to avoid that the generated primes are of a special form, i.e., that the enemy has a priori information about the generated primes that could help him to factor $m = pq$. Therefore the prime factors of F must not be chosen from a small set of "base primes". Instead, in order to generate candidates n that are random odd integers, these prime factors should rather be selected according to their actual probability distribution. This can be achieved by selecting the sizes of the prime factors according to their probability distribution and then generating primes of the selected sizes. Note that by doing this, we are solving the original problem of generating a prime by reduction to itself, i.e., by recursion. It is somewhat surprising that this recursive construction turns out to be even faster than the generation of a strong pseudoprime that passes the Miller-Rabin test for an appropriate number of bases.

Let again $m = pq$ denote an RSA-modulus where $p < q$, $p'_1, \ldots p'_r$ are the r largest distinct prime factors of $p - 1$ in decreasing order and $q'_1, \ldots, q'_s$ are the s largest distinct prime factors of $q - 1$ in decreasing order. Our aim is to randomly generate secure RSA-moduli $m = pq$ for a given encryption exponent e, i.e., to randomly select with uniform distribution one of the integers that (1) lie in a given interval I centered at N, (2) are the product of exactly two primes p and q satisfying $(p-1, e) = (q-1, e) = 1$, and (3) satisfy certain security constraints. According to the results in Section 2 and recommendations in the literature, these conditions are (1) that p as well as q must be greater than a given limit $L = N^\gamma$ for some γ with $0 < \gamma < 1/2$ (this is to prevent factorization by a method specialized for finding small prime factors), (2a) that $p - 1$ and $q - 1$ must contain large prime factors p'_1 and q'_1, respectively, with $p'_1 \geq L'$ and $q'_1 \geq L'$ for a given limit $L' = N^{\gamma'}$ for some γ' with $0 < \gamma' < \gamma$ (this is to prevent factorization by the so-called $p - 1$ factoring method [24]), (2b) that the remaining $r - 1$ largest prime factors $p'_2 \ldots, p'_r$ of $p - 1$ and $s - 1$ largest prime factors $q'_2, \ldots, q'_s$ of $q - 1$, needed for the application of Lemma 1 and in order to satisfy condition (3) below, are each greater than a given limit $L'' = N^{\gamma''}$ for some γ'' with $0 < \gamma'' < \gamma'$ (this is to make the bound $\sum_{i=1}^r 1/p'_i + \sum_{i=1}^s 1/q'_i$ given in Theorem 2 sufficiently small), and (3) that $lcm(F_{p'_1}, \ldots, F_{p'_r}, F_{q'_1}, \ldots, F_{q'_s}) \geq M$ (see Theorem 2) for a given limit $M = N^\delta$ with $0 < \delta < 2\gamma$, where $F_{p'_1}, \ldots, F_{p'_r}$ and $F_{q'_1}, \ldots, F_{q'_s}$ are the factored parts of $p'_1 - 1, \ldots, p'_r - 1$ and $q'_1 - 1, \ldots, q'_s - 1$, respectively. Condition (3) can be satisfied by requiring that for some ρ with $0 < \rho < 1$, $F_{p'_i} \geq p''^\rho_i$ for $1 \leq i \leq r$ and $F_{q'_i} \geq q''^\rho_i$ for $1 \leq i \leq s$ as well as $F_p = \prod_{i=1}^r p'_i \geq p^\rho$ and $F_q = \prod_{i=1}^s q'_i \geq q^\rho$. If $p'_1, \ldots, p'_r$ and $q'_1, \ldots, q'_s$ are distinct, and $F_{p'_1}, \ldots, F_{p'_r}$ and $F_{q'_1}, \ldots, F_{q'_s}$ are pairwise relatively prime (as is virtually automatically the case and can be easily tested during the generation process), these conditions guarantee that $lcm(F_{p'_1}, \ldots, F_{p'_r}, F_{q'_1}, \ldots, F_{q'_s}) \geq N^{\rho^2}$, hence the choice $\rho = \sqrt{\delta}$ guarantees that condition (3) is satisfied. Note that $\rho \geq 1/2$ is required for the primality proof by Lemma 1 for the primes $p'_1, \ldots, p'_r, q'_1, \ldots, q'_s$, p and q, and in general the choice $\rho = 1/2$ guarantees a sufficiently large lower bound of $N^{1/4}$ on $lcm(F_{p'_1}, \ldots, F_{p'_r}, F_{q'_1}, \ldots, F_{q'_s})$, the guaranteed number (for virtually all plaintexts) of iterated encryptions required to recover the plaintext. However, other values for δ can easily be accommodated as well if necessary.

For practical applications where the RSA-moduli must be in a given interval $[M_1, M_2]$ centered at N (e.g. $N = 10^{200}$), we recommend to choose $\gamma \approx 0.4, \gamma' \approx 0.3, \rho = 1/2$ and $\delta = 1/4$. The choices $\gamma = 0.4$ and $\gamma' = 0.3$ guarantee that $p'_1 > \sqrt{p-1}$ and $q'_1 > \sqrt{q-1}$, and hence that if $\rho = 1/2$ then only one prime factor p'_1 of $p - 1$ and one prime factor q'_1 of $q - 1$ must be generated.

In the following we describe our algorithm for generating secure RSA-moduli with uniform distribution over a given interval $[M_1, M_2]$ centered at N, as well as the underlying recursive algorithm for generating primes in a given interval that satisfy certain conditions. The algorithm has been implemented on a VAX 8650 [8] which demonstrates its practicality and the correctness of the running time analysis. Many important and interesting implementation details can only be mentioned in this paper and are not discussed in detail. The algorithm is very well suited for implementation on a small computer like a PC or on special-purpose cryptographic hardware.

Assume the public encryption exponent e, an interval $[M_1, M_2]$ centered at N, and the security parameters $\gamma, \gamma', \gamma''$ and δ have been (arbitrarily) specified. The primes p and q will be generated to be compatible with e, i.e., to satisfy the contitions $(p - 1, e) = 1$ and $(q - 1, e) = 1$, and to satisfy the security constraints. In particular they satisfy the conditions given in Theorem 2 for a replaced by e. Note that e must be used as the base when Lemma 1 is applied for the primality proof of the largest prime factors of $p - 1$ and $q - 1$, i.e., of p'_1 and q'_1, in order to allow application of Theorem 2. The first step is to choose the relative size $\sigma_p = \log_N p$ of the smaller prime factor p according to its probability distribution $P[\sigma_p \geq \beta] \approx [\log(1 - \beta) - \log \beta]/[\log(1 - \gamma) - \log \gamma] \approx (1 - 2\beta)/(1 - 2\gamma)$ (see [16] for a derivation). Note that for $\gamma > 0.3$, σ_p is almost uniformly distributed over the interval $[\gamma, 1/2]$. The second step is to generate p at random in a small interval centered at N^{σ_p} by the recursive procedure

RANDOMPRIME described below. The last step is the generation of q at random in the interval $[M_1/p, M_2/p]$ by another application of the recursive procedure.

The probabilistic recursive procedure RANDOMPRIME generates a prime p at random with (virtually) uniform distribution over the interval $[P_1, P_2] = [P/c, cP]$, where c is a given interval span constant with $c \approx 1.05\ldots 2$. The interval is specified equivalently by P_1 and P_2 or by P and c. If the upper interval boundary P_2 is below a certain limit P_{lim} (e.g. $P_{lim} = 10^7$), then p is generated by repeatedly choosing integers from $[P_1, P_2]$ at random until one is shown to be prime by trial division by all primes below its square root, and the procedure is finished. If $P_2 > P_{lim}$ then the first step is to select the relative size $\sigma_{p_1'} = \log_{P/2} p_1'$ of the largest prime factor p_1' of $(p-1)/2$ according to the conditional probability distribution of the size of the largest prime factor of an integer n on the order of $P/2$, given that $2n + 1$ is prime. This distribution will be discussed in Section 4. In this first step, a lower bound L' on p_1' prescribed by security constraints can easily be accommodated by using the appropriate conditional distribution, given that $p_1' > L'$, which can easily be obtained from the unconditional distribution by a normalization step. The largest prime factor p_1' is then generated by a <u>recursive</u> call to RANDOMPRIME, where the interval for p_1' is $[P'/c', c'P']$ with $P' = (P/2)^{\sigma_{p_1'}}$ and where c' is a small interval span constant (e.g. $c' \approx 1.05...2$, preferably $c' = c$). If $p_1' > \sqrt{(P_2-1)/2}$ (i.e., roughly if $\sigma_{p_1'} > 1/2$), which happens with probability close to 70%, then an integer $R \in [(P_1-1)/2p_1', (P_2-1)/2p_1']$ is repeatedly chosen at random until $p = 2Rp_1' + 1$ can be proved to be prime by application of Lemma 1 for some base a. In this case the procedure is finished. If $p_1' < \sqrt{(P_2-1)/2}$, then the relative size $\sigma_{p_2'} = \log_{P/2} p_2'$ of the second largest prime factor p_2' of $(p-1)/2$ is selected according to the conditional probability distribution of the size of the second largest prime factor of an integer n on the order of $P/2$, given that the largest prime factor of n is p_1' and given that $2n + 1$ is prime. This conditional distribution is equal to the conditional probability distribution of the size of the largest prime factor of a number l on the order of $P/(2p_1')$, given that it is smaller than p_1' and given that $2p_1'l + 1$ is prime. This distribution is dicussed in [16]. Then p_2' is generated, also by a recursive call to RANDOMPRIME where the interval for p_2' is $[P''/c'', c''P'']$ with $P'' = (P/2)^{\sigma_{p_2'}}$ and where c'' is again a small interval span constant (e.g. $c'' = c$). This process continues until the product $x = \prod_{i=1}^{r} p_i'$ of the generated r distinct largest prime factors of $(p-1)/2$ satisfies $(P_2 - 1)/2x < p_r'$. This condition, which is stronger than the condition $x > \sqrt{(P_2-1)/2}$ suggested by the fact that the factored part of a number n in Lemma 1 must be greater than its square root, is necessary to ensure that a randomly chosen integer $R \in [(P_1 - 1)/2x, (P_2 - 1)/2x]$ cannot have a prime factor greater than p_r', a circumstance that would violate the condition that p_r' is the r-th largest prime factor, and thus would falsify the final distribution of the sizes of the prime factors of $(p-1)/2$. Note that in order to test $2Rx + 1$ for primality one first rules out all primes below a certain limit as divisors. (The optimal trial division limit is discussed in Section 5.) Instead of dividing $2Rx + 1$ by these small primes we suggest to divide the much smaller integer R by these small primes and checking whether R has a remainder modulo one of the small primes that would result in $2Rx + 1 \equiv 0$ modulo this prime. This gives an improvement over trial division of $2Rx + 1$ which contributes to the surprising efficiency of the method.

The size of the primes to be generated by RANDOMPRIME decreases with increasing level of recursion. The end of the recursion is reached as soon as the upper boundary P_2 of the interval for the prime is below P_{lim} (see above). The final result of RANDOMPRIME is hence a <u>tree</u> where the root is the generated prime $p \in [P_1, P_2]$, the intermediate and terminal nodes are the largest prime factors of the immediate predecessors in the tree reduced by 1, and where the terminal nodes are primes less than P_{lim}. This tree allows one to give a simple proof of the primality of all nodes, starting with the terminal nodes. Note that the obtained primality proof for p is an even more succinct certificate for p than the ones proposed by Pratt [22] and Plaisted [19]. The procedure RANDOMPRIME can also be modified to efficiently solve the problem of randomly generating an integer with uniform distribution over a given interval, such that its factorization, or partial factorization, is known. This is a problem that has been considered by Bach [2], whose results were applied by Blum and Micali [3] to show that, according to their definition, the set B of predicates related to the quadratic residuosity problem is accessible. Note that our method differs from Bach's method [2], which is of theoretical rather than of practical interest, in that it is computationally efficient but on the other hand sacrifices the rigorously provable uniform distribution.

As mentioned in Section 1, common security constraints placed on the primes p and q are that they be of about equal size (e.g. have equally many digits) and that $p = 2p' + 1$, $q = 2q' + 1$, $p' = 2p'' + 1$ and $q' = 2q'' + 1$, where p', q', p'' and q'' are primes. In the following we compare the restrictiveness of this choice with that of our method. One can show by convincing heuristic arguments [16] that the probability that a prime on the order of P be of this special form is $\Pr[p'$ is prime$|2p' + 1$ is prime$] \times \Pr[p''$ is prime$|2p'' + 1$ and $4p'' + 3$ are primes$] \approx C_2/\log(P/2) \times 3C_3/2\log(P/4)$ where $C_2 = \prod_{q\,\text{prime},q\geq 3} q(q-2)/(q-1)^2 \approx 0.66016$ and $C_3 = \prod_{q\,\text{prime},q\geq 5} q(q-3)/(q-1)(q-2) \approx 0.722$. Hence for example, among all numbers of size roughly N that are the product of exactly two roughly equally large primes, only a fraction $[C_2/\log(\sqrt{N}/2) \times 3C_3/2\log(\sqrt{N}/4)]^2$ has both prime factors of the special form. For $N = 10^{200}$, this fraction is one out of 5.4 billion. We argue that there is little reason for such a strong restriction, because the security requirements can also be met by a much looser restriction.

It can be shown again by convincing heuristic arguments (see [16]) that the fraction of numbers on the order of N that are the product of exactly two primes, none of which is less than N^α, is for $\alpha \in [0.03, 0.5]$ well approximated by $[\log(1-\alpha) - \log\alpha]/\log N$. We conjecture that for $0 < \alpha < 1/2$, $\lim_{N\to\infty} \#\{x \leq N : x = pq$ with p and q primes, $p \geq N^\alpha, q \geq N^\alpha\} \sim N(\log(1-\alpha) - \log\alpha)/\log N$. The fraction of numbers on the order of N that are the product of exactly two primes, p and q, both of which are greater than $L = N^\gamma$ and such that both $p - 1$ and $q - 1$ have a prime factor greater than $L' = N^{\gamma'}$ is approximately given by $\int_\gamma^{1/2}[1 + \log(\gamma'/x)][1 + \log(\gamma'/(1 - x))]/x(1 - x) \cdot dx/\log N$, which for $\gamma = 0.4$ and $\gamma' = 0.3$ becomes $0.094/\log N$. These numbers make up approximately 23% of all products on the order of N of exactly two primes both greater than $N^{0.4}$. (Note that the condition $lcm(F_{p'_1}, \ldots, F_{p'_t}, F_{q'_1}, \ldots, F_{q'_t}) \geq N^\delta$ is no essential additional restriction on the set of allowed RSA-moduli pq. It is only a restriction on the method for finding such primes.) Thus we can claim, unlike for the case where primes of a very special form are used, that factoring our RSA-moduli is about equivalent to solving the general problem of factoring the product of two large primes. Note that breaking the RSA system is not more difficult than the problem of factoring the product of two primes, which is not necessarily more and possibly much less, difficult than the problem of factoring general integers of comparable size. By allowing the RSA-modulus to be the product of more than 2 primes, i.e., a general integer, one could possibly obtain a cryptosystems which is as difficult to break as the general problem of factoring integers is difficult to solve. However this author does not suggest such a generalization for practical applications. To prove the equivalence of breaking the system and factoring would probably be even more difficult than for the original RSA-system.

4. On the Size of the Prime Factors of Certain Numbers

The basic probability distribution required by the procedure RANDOMPRIME is (for all relevant orders of magnitude, N) the probability distribution of the relative size $\sigma_{p'} = \log_N p'$ of the largest prime factor p' of an integer n on the order of N, given that n satisfies a certain condition. For the generation of the largest prime factor of $(p - 1)/2$ in the procedure RANDOMPRIME, this condition is simply that $2n + 1$ is prime. For the generation of the second and further largest prime factors of $(p - 1)/2$ (which is only necessary with probability 0.3, namely when the largest prime factor p' of $(p-1)/2$ is less than $\sqrt{p}$) this condition is that $2an + 1$ is prime, where a is the product of the prime factors generated so far. In the i-th level of the recursion, that is at depth i in the prime generation tree, the condition on n is of the form that $2a_1 + 1$, $2a_2(2a_1 + 1) + 1$, $2a_3(2a_2(2a_1 + 1) + 1) + 1, \ldots,$ $2a_{i-1}(2a_{i-2}(\ldots(2a_1 + 1)\ldots) + 1) + 1$ must all be prime.

All the arguments used in the derivation of these probability distributions are heuristic rather than mathematically rigorous, but they are empirically and numerically evident enough to be convincing. The arguments are similar to those used by Koblitz [15] for estimating the number $S(n)$ of primes p less than n for which $(p-1)/2$ is prime as $S(n) \sim C_2 n/(\log n)^2$, where $C_2 = \prod_{q \text{ prime}, q \geq 3} q(q-2)/(q-1)^2 \approx$ 0.66016. Similar heuristics can also be used to derive the conjecture stated by Hardy and Littlewood [13] that the number $P_2(n)$ of prime-pairs (primes p for which $p+2$ is prime) below n is asymptotically given as $P_2(n) \sim 2C_2 n/(\log n)^2$. Note that, as is also true for our analysis, the gap is wide between these precise and numerically evident conjectures and the best result that can be rigorously proved. It is namely not even proved yet that there exist infinitely many prime-pairs or infinitely many primes p for which $(p-1)/2$. The basic result of our analysis is that the distribution of the sizes of the prime factors of a number n on the order of N is virtually independent of the different conditions that can be placed on n, although for example the probabilities that n is prime, or that $n = kq$ with q prime for a specific k, can strongly depend on the condition.

In this paper we only briefly consider the probability distribution of the largest prime factor of a number n on the order of N, given that $2n + 1$ is prime. For further information we refer to a forthcoming paper [16]. Let $p_N(k)$ denote the probability that a randomly selected integer n on the order of N is of the form $n = kp'$ where p' is the largest prime factor of n, let $\overline{p}_N(k)$ denote the corresponding conditional probability, given that $2n + 1$ is prime, and let $\overline{F}_1^N(\alpha)$ and $F_1^N(\alpha)$ denote the probability, with and without the above condition, respectively, of the event that the largest prime factor p' of n is smaller or equal to N^α. Heuristic reasoning suggests for $k \leq \sqrt{N}$ that $p_N(k) \approx 1/k \cdot 1/\log(N/k)$, where $1/k$ is the probability that k divides n and $1/\log(N/k) = 1/(\log N - \log k)$ is the probability that a number on the order of N/k is prime. Similar heuristic reasoning as used in [15] (for the case $k = 1$) suggests that $\overline{p}_N(k) \approx A_k p_N(k)$ where $A_k = C_2 \prod_{q \text{ prime}, q \geq 3, q|k} (q-1)/(q-2)$ and where C_2 has been defined above. We conjecture from similar heuristic evidence that for every fixed k the number $S_k(n)$ of primes $p \leq n$ for which $(p-1)/2 = kq$ with q prime satisfies $S_k(n) \sim A_k n/k(\log n)^2$.

$\overline{F}_1^N(\alpha)$ (and similarly $F_1^N(\alpha)$) can be computed for $\alpha \geq 1/2$ as $\overline{F}_1^N(\alpha) = \sum_{k=1}^{N^{1-\alpha}} \overline{p}_N(k)$. It is interesting to note that $E[A_k] = 1$ for randomly chosen integers k. Because $\sum_{k=1}^{N^{1-\alpha}} \overline{p}_N(k) \approx \sum_{k=1}^{N^{1-\alpha}} A_k p_N(k) \approx E[A_k] \sum_{k=1}^{N^{1-\alpha}} p_N(k) = E[A_k] F_1^N(\alpha) = F_1^N(\alpha)$, it follows that $\overline{F}_1^N(\alpha) \approx F_1^N(\alpha)$. That is, the distribution of the size of the largest prime factor of a random integer n does not depend on the condition that $2n + 1$ is prime, although for example the probability that n is prime is 34% smaller when the condition "$2n + 1$ prime" is given. Similar results can be obtained if additional conditions are placed on n, for example that "$2a(2n+1)+1$ must also be prime" for a given constant a, "$2b(2a(2n+1)+1)+1$ must also be prime for a given constant b, etc.. Knuth and Trabb Pardo [14] proved that the limiting function $F_1(\alpha) = \lim_{N\to\infty} F_1^N(\alpha)$ exists, where n is assumed to be uniformly distributed in $[1, N]$. $F_1(\alpha)$ is defined by $F_1(\alpha) = 1 - \int_\alpha^1 F_1(x/(1-x))/x \cdot dx$ with $F_1(\alpha) = 1$ for $\alpha \geq 1$. In particular, $F_1(\alpha) = 1 + \log \alpha$ for $1/2 \leq \alpha \leq 1$. The probability that the largest prime factor of a randomly chosen large integer is greater than its square root is hence $\log 2 \approx 0.69$. We conjecture on the basis of substancial heuristic and empirical evidence that $\lim_{N\to\infty} \overline{F}_1^N(\alpha) = F_1(\alpha)$.

$\overline{F}_1^N(\alpha)$ can be very well approximated for finite N by $F_1(\alpha) - \epsilon/\log N$ for $1/2 \leq \alpha < 1$ where ϵ is a computable positive constant less than one, and by $(1 - \epsilon/(\log 2 \log N))F_1(\alpha)$ for $0 \leq \alpha \leq 1/2$.

For small values of k below a certain limit, the cases $(p-1)/2 = kq$ with q prime must be considered separately and a different type of recursion of RANDOMPRIME than the one described above must be used. k must be selected according to its discrete distribution $\bar{p}_N(k)$. Then p_1' must be generated by a recursive call to RANDOMPRIME with interval $[(P_1-1)/2k, (P_2-1)/2k]$. Note that this type of recursion, which occurs seldomly, requires the repeated generation of p_1' until $2kp_1' + 1$ is prime, as opposed to the reselection of a random integer R until $2Rp_1' + 1$ is prime. Hence the computation time for finding the desired prime p is considerably greater if this second type of recursion has to be used. However measures can be taken to keep the increase in computation time reasonably small. We recommend for the sake of speed to make use only of the fast type of recursion. Note that this slightly changes the a priori probabilities of those primes p in the given interval for which k is very small, i.e., $p-1$ has a very large prime factor, which is of no relevance for practical applications.

5. Comments on the Average Running Time and Conclusions

We can only present some main results of the running time analysis because the number of pages of this paper is limited. For more details we refer to a forthcoming paper to be submitted to the Journal of Cryptology. The aim of the running time analysis of the procedure RANDOMPRIME is to determine the ratios $c_{psp}(n) = T_{psp}(n)/T_{exp}(n)$ and $c_{pri}(n) = T_{pri}(n)/T_{exp}(n)$, where $T_{psp}(n)$, $T_{pri}(n)$ and $T_{exp}(n)$ are the expected times required for finding an n-bit strong pseudoprime that passes the Miller-Rabin test for one base, for generating one n-bit prime by our algorithm and for one full modular n-bit exponentiation, respectively. Since modular exponentiation is the most time consuming operation, $c_{psp}(n)$ and $c_{pri}(n)$ are almost implementation independent.

In order to find a strong pseudo-prime that passes the Miller-Rabin test for one base one can repeatedly select integers at random, rule out all primes below a certain limit as possible divisors by trial division, and if the trial division test is passed apply the Miller-Rabin test which costs one modular exponentiation. The optimal upper limit for trial division, i.e., the limit that minimizes the expected time for revealing the composite nature of an n-bit integer, can be shown to be $L_{opt}(n) = T_{exp}(n)/T_{div}(n)$, where $T_{div}(n)$ is the time for one division by a small integer (less than $L_{opt}(n)$). When this optimal trial division limit is used, $c_{psp}(n)$ can be shown to be a function growing as $n/\log n$; and, as an example, for $n = 332$ (100 decimal digits) it equals 14.5. Under the plausible simplifying assumption that in the procedure RANDOMPRIME the selected relative size $\sigma_{p_1'}$ of the largest prime factor is always equal to its average, i.e., $\sigma_{p_1'} = 0.624$, one can show that $c_{pri}(n) = 1.18 \cdot c_{psp}(n)$. In other words, the average running time for generating a prime is only roughly 20% greater than the time required for finding a strong pseudoprime that passes the Miller-Rabin test for only one base. For 100-digit integers, the expected time for generating a prime is equivalent to only 17.5 exponentiations (i.e., $c_{pri}(332) = 17.5$) compared to 14.5 for a pseudoprime. If the pseudoprime is tested for four different bases then the two methods are equally fast, but if more than four bases have to be tested (for a practical implementation 20 to 50 is reasonable in order to achieve a sufficient level of confidence), our method is considerably faster. The asymptotic running time is $n^4/\log n$.

Thus, our new method for generating primes not only offers the advantages of yielding provable primes that are virtually uniformly distributed over the set of primes in a given interval satisfying the flexibly specified RSA-security constraints, but moreover it is also faster than previous methods used to generate only "probable primes". Of course it is not restricted to the RSA cryptosystem, but it can be used in other cryptocraphic systems that require large primes satisfying certain security constraints, such as the Diffie-Hellman public key distribution system [9], the El-Gamal cryptosystem and signature scheme [10], the Blum-Micali pseudorandom sequence generator [3], etc..

Acknowledgement

I would like to thank Jim Massey for his valuable assistance during the preparation of this paper.

References

[1] L.M. Adleman, C. Pomerance, and R.S. Rumely, *On distinguishing prime numbers from composite numbers*, Annals of Mathematics, Vol. 117, 1983, pp.173-206.

[2] Eric Bach, *How to generate random integers with known factorization*, Proc. 15th annual ACM Symp. on Theory of Computing, 1983, pp. 184-188.

[3] M. Blum and S. Micali, *How to generate cryptographically strong sequences of pseudo-random bits*, SIAM J. Comput., Vol. 13, No. 4, Nov. 1984, pp. 850-864.

[4] J. Brillhart, D.H. Lehmer and J.L. Selfridge, *New primality criteria and factorizations of* $2^m \pm 1$, Math. Comp., Vol. 29, 1975, pp. 620-647.

[5] H. Cohen and W. Lenstra, Jr., *Primality testing and Jacobi sums*, Math. Comp., Vol. 42, 1984, pp. 297-330.

[6] H. Cohen and A.K. Lenstra, *Implementation of a new primality test*, Mathematics of Computation, Vol. 48, No. 177, Jan. 1987, pp. 103-121.

[7] C. Couvreur and J.J. Quisquater, *An introduction to fast generation of large prime numbers*, Philips Journal of Research, Vol. 37, 1982, pp. 231-264, (errata: id, Vol. 38., 1983, p. 77).

[8] R. De Moliner, *Effiziente Konstruktion zufälliger grosser Primzahlen*, Diploma Project, Institute for Signal and Information Processing, Swiss Federal Institute of Technology, Zurich, 1989.

[9] W. Diffie and M.E. Hellman, *New directions in cryptography*, IEEE Trans. Info. Th., Vol. IT-22, Nov. 1976, pp. 644-654.

[10] T. El-Gamal, *A public key cryptosystem and a signature scheme based on the discrete logarithm*, IEEE Trans. Info. Th., Vol. IT-31, No. 4, July 1985, pp. 469-472.

[11] S. Goldwasser and J. Kilian, *Almost all primes can be quickly certified*, Proc. of the Annual ACM Symp. on the Theory of Computing, 1986, pp.316-329.

[12] J. Gordon, *Strong primes are easy to find*, Proc. EUROCRYPT'84, Lecture Notes in Computer Science, Vol. 209, pp. 216-223, Springer Verlag.

[13] G.H. Hardy and J.E. Littlewood, *Some problems of 'partitio numerorum'; III: on the expression of a number as a sum of primes*, Acta Math., Vol. 44, 1923, pp. 1-70.

[14] D.E. Knuth and L. Trabb Pardo, *Analysis of a simple factorization algorithm*, Theoretical Computer Science, Vol. 3, 1976, pp. 321-348.

[15] N. Koblitz, *Primality of the number of points on an elliptic curve over a finite field*, Pacific J. of Mathematics, Vol. 131, No. 1, 1988, pp. 157-165.

[16] U.M. Maurer, *Some number-theoretic conjectures and their relation to the generation of cryptographic primes*, submitted to the 2nd IMA Symposium on Cryptography and Coding, Dec. 1989, Cirencester, UK.

[17] F. Morain, *Primality testing: news from the front*, these proceedings.

[18] M. Ogiwara, *A method for generating cryptographically strong primes*, Research Reports on Information Sciences, No. C-93, Dept. of Information Sciences, Tokyo Institute of Technology, April 1989.

[19] D.A. Plaisted, *Fast verification, testing, and generation of large primes*, Theoretical Computer Science, Vol. 9, 1979, pp. 1-16, (errata: id., Vol 14., 1981, p. 345).

[20] H.C. Pocklington, *The determination of the prime or composite nature of large numbers by Fermat's theorem*, Proc. Cambridge Philos. Soc., Vol. 18, 1914-1916, pp. 29-30.

[21] J.M. Pollard, *Theorems on factorization and primality testing*, Proc. Cambridge Philos. Soc., Vol. 76, 1974, pp. 521-528.

[22] V.R. Pratt, *Every prime has a succinct certificate*, SIAM J. Computing, Vol. 4, No. 3, Sept. 1975, pp.214-220.

[23] M.O. Rabin, *Probabilistic algorithm for testing primality*, Journal on Number Theory, Vol. 12, 1980, pp. 128-138.

[24] Hans Riesel, *Prime numbers and computer methods for factorization*, Boston, Basel, Stuttgart: Birkhäuser, 1985.

[25] R.L. Rivest, *Remarks on a proposed cryptanalytic attack on the M.I.T. public key cryptosystem*, Cryptologia, Vol. 2, No. 1, Jan. 1978, pp. 62-65.

[26] R.L. Rivest, A. Shamir, and L. Adleman, *A method for obtaining digital signatures and public-key cryptosystems*, Comm. ACM, Vol. 21, Feb. 1978, pp.120-126.

[27] C.P. Schnorr, *Zur Analyse des RSA-Schemas*, Preprint: Fachbereich Mathematik, Universität Frankfurt, Dec. 1981.

[28] J. Shawe-Taylor, *Generating strong primes*, Electronics Letters, Vol. 22, No. 16, July 1986, pp. 875-877.

[29] G. Simmons and M. Norris, *Preliminary comments on the M.I.T public key cryptosystem*, Cryptologia, Vol. 1, No. 4, Oct. 1977, pp. 406-414.

[30] R. Solovay and V. Strassen, *A fast Monte-Carlo test for primality*, SIAM J. Computing, Vol. 6, No.1, March 1977, pp. 84-85.

[31] H.C. Williams and B. Schmid, *Some remarks concerning the M.I.T. public-key cryptosystem*, BIT, Vol. 19, 1979, pp. 525-538.

Section 10

Old problems

Deciphering Bronze Age Scripts of Crete
The Case of Linear A

Yves Duhoux
Université Catholique de Louvain

0. Introducing linear A

Linear A is a script used in the Aegean area, mainly in the island of Crete. The texts discovered up to now can be dated between about 1850 B.C. and 1450 B.C. Linear A appears in two types of documents: accounting documents and the non accounting ones. More than 90 % of our texts are of the accounting type. The whole linear A corpus has 7.386 signs. This means about ten typewritten pages.

1. Decipherment of the script

Linear A has signs which are clearly ideographic: they represent objects, living creatures, or concepts.

Apart from these clear ideograms, there are about one hundred signs whose ideographic character cannot be proved. This number is just what we expect from a syllabary, and especially from an open syllable one, with signs for syllables like *ba, ca, da,* and so on.

Linear A is cognate to another script, linear B, which has been deciphered, so one can read most of its syllabograms. There are about seventy linear A and linear B syllabograms which have the same form. For various reasons, it seems quite probable that the creators of linear B have not drastically changed the phonetic values of linear A. We can, then, use the linear B values for the transliteration of the most frequent linear A signs.

2. Interpretation of linear A texts

The structure of the *linear A accounting texts* is rather easy to grasp, because they are full of ideograms, and especially of one kind of them: the numbers. Thanks to these numbers, we understand the only two words of the whole linear A corpus we are absolutely sure about: the names of the "total" and the "grand-total".

The *linear A non accounting documents* are extremely interesting, because they possess a very elaborate syntactic pattern and very developed morphological features. The difficulty is that we have not enough material to test our interpretations on extensive grounds.

3. Identification of the linear A language

The identification of the language of linear A is the most tantalizing task we are faced with. Linear A's language has been recognised as cognate to an impressive list of languages: Hittite, Luwian, Lycian, Sanskrit, Greek, Indo-european, Semitic, Carian, Basque, and so on. Unfortunately, there is no single one decipherment which fulfils our requirements.

4. What about the future?

We could work in three directions.

First, *find new texts*. Regularly, the archaeologists find new linear A inscriptions. These finds are vital for our studies, because they increase the number of our texts, which are for time being so few.

Second, *the internal study of linear A* should be systematically done. A full statistical study of the syllabograms and of their associations should be done - looking, for instance, for possible vocalic harmony. One should try to determine the probable status of every linear A word: is it a proper noun, or a term of the lexicon, and so on. Try to define the morphemes and describe their function. Create morphological and syntactical models, and test them on the largest possible number of texts. And so on.

Third: we should develop the *comparative study of linear A*. Other scripts than linear A were written in Crete. I think especially of the "hieroglyphic" Cretan script. Although we have very few texts, the comparison could be illuminating, and should be done.

Outside Crete, linear A should be compared with the languages of the Mediterranean world. Of course, in this area, the Semitic and the Indo-european languages are very important families, and one should pay enough attention to them. But there are many other languages which should deserve careful investigation. Languages which are neither Indo-european, nor Semitic - think for instance of languages such as Hattic, or Urartean, or Hurrian, and so on[1].

Place Pascal 1
B-1348 Louvain-la-Neuve
Belgium

[1] For more details on the whole subject, see Y. DUHOUX, Le linéaire A : problèmes de déchiffrement, in *Problems in Decipherment* (Y. DUHOUX, Th. G. PALAIMA, J. BENNET éd.), Louvain-la-Neuve, Peeters, 1989, pp. 59-119.

Section 11

Rump Session
(impromptu talks)

FASTER PRIMALITY TESTING

extended abstract

Wieb Bosma and Marc-Paul van der Hulst

Mathematisch Instituut
Universtiteit van Amsterdam
Roetersstraat 15
1018 WB Amsterdam
The Netherlands

Acknowledgement. Research was done while the authors were supported by the Nederlandse organisatie voor wetenschappelijk onderzoek NWO.

Abstract

Several major improvements to the Jacobi sum primality testing algorithm will speed it up in such a way that proving primality of primes of up to 500 digits will be a matter of routine. Primes of about 800 digits will take at most one night on a Cray.

Primality Testing and Factoring

Primality testing is one of two closely related classical problems in computational number theory, the other being that of *factoring integers*. Usually, if a positive integer n is composite, it is easy to find a proof for that. Since such a proof generally does not provide factors of n, for composite numbers the problem of factoring n remains. But if a number does not seem to be composite, one would like to find a *proof* for its primality; this is the object of primality testing and the subject of this paper. A *primality test* is an algorithm that gives a rigorous proof for the primality of prime numbers; one inputs an integer and the algorithm either yields a proof that n is prime, or it fails, indicating that n must be composite.

In this paper we describe several major improvements to the Jacobi sum primality test. As a consequence we will soon be able to prove the primality of prime numbers of up to many hundreds of digits routinely. Our estimates show that in the worst case proofs for 800 digit primes will take at most one night on a Cray. Furthermore, our implementation allows distribution on almost any number of processors; in this way we can achieve an m fold speedup by running the test on m identical machines.

There exist very fast *compositeness tests*, also called pseudo-prime tests, that on input n either give a proof for the fact that n is composite, or tell you that n is probably prime. A proof of compositeness usually consists of exhibiting an integer, called a *witness to the compositeness* of n, which has a special property (for instance concerning its order in the multiplicative group modulo n) that no integer can satisfy if n were prime. In particular, these proofs for compositeness do not give any clue as to the *divisors* of n; finding these is very hard in general, which is the raison d'être of the 'factoring industry'.

In the case of Rabin's compositeness test [R] the probability that a random integer is a witness to the compositeness of some composite number, is at least $\frac{3}{4}$. Since such a test can be repeated independently as many times as one would like, the probability that a composite integer is declared probably prime can be made arbitrarily small.

So the main problem in primality testing is that of proving the correctness of an answer that is easily obtained; in factoring it is just the other way around: there it is very hard to obtain the result (a factorization), but one checks its validity immediately (by multiplying out the factors).

But if one is willing to accept a small probability of giving the wrong answer, it is easy to answer the question whether a given integer n is a prime or a composite number. Thus the following paradoxical situation arises: for most practical purposes, one is perfectly happy with pseudo-prime tests, but this is mathematically unsatisfactory; on the other hand, it is easy to state sufficient conditions for primality, but it is much harder to make these criteria practical, i.e., to devise an efficient test! It should be remarked however, that primality testing is commonly regarded as "easier" than factoring, in terms of computational complexity.

To substantiate this belief, we mention here that under the assumption of some unproved hypotheses from analytic number theory, viz. sufficient generalized Riemann hypotheses, for every composite n there exists a witness a for the compositeness of n smaller than $2(\log n)^2$ (cf. [Ba]). Thus, under these hypotheses, primality testing is "polynomial time" [Mi].

Primality Tests

Two types of primality test ought to be distinguished: firstly, those tests that work only for primes with special arithmetic properties, and secondly, the general purpose type tests. Usually tests of the first type exploit divisibility properties of $n - 1$ or $n + 1$, more generally of $n^w - 1$ for small values of w. The classical examples give criteria for Fermat primes (i.e., primes of the form $2^{2^k} + 1$) and Mersenne primes (of the form $2^k - 1$) respectively. In both cases a property is used that is necessary as well as sufficient and that can be checked very quickly. It is these type of test that make headlines, because they are used to find gigantic primes, of up to tens of thousands of digits.

We will call these tests of *Lucas-Lehmer* type. In general they give a sufficient criterion for the primality of n that is applicable in case enough factors of $n - 1, n^2 - 1, \ldots$ can be found; here "enough factors" means that their product exceeds $\sqrt{n}$. Therefore these tests will only work for primes with very special arithmetic properties. Since they depend on the (hard) problem of factoring, there scope for general purposes is limited; in particular problems arose for certain primes of around 80 digits, for which not enough information on divisors could be gathered to complete a proof.

We turn to general purpose primality tests. The straightforward method of proving the primality of n by showing that no prime number smaller than $\sqrt{n}$ divides n, using trial division, rapidly becomes too time-consuming with increasing n, and a table of all primes up to $\sqrt{n}$ is needed. In fact, the trial division method can be employed quite efficiently for sieving out all composite integers up to a given bound, thus producing tables of primes; this is known as the sieve of Eratosthenes.

The first practical general purpose algorithm for primality proving was the *Jacobi sum test*. Based on observations made by Adleman, Pomerance and Rumely [APR], it was made practical by improvements of Cohen and H.W. Lenstra [CL]; in the implementation of A.K. Lenstra and Cohen [LC] it can routinely handle primes up to 212 digits and yields primality proofs for such numbers within 2 minutes on a Cray. Basically one restricts the possible divisors of the integer n to at most t different residue classes modulo s, for certain auxiliary integers t and s. If $s > \sqrt{n}$, then at least one divisor of n must be among these residues; if also their number t is not too large, one can thus prove the primality of n by showing that *none* of these residues does in fact divide n. It was proved that his gives rise to a subexponential algorithm. Below we will give a somewhat more detailed description of this algorithm.

Here we should also mention an important idea of H.W. Lenstra [L]. He proved that there can be at most 11 divisors of n in any given residue class modulo s if s exceeds $\sqrt[3]{n}$. Moreover there is an efficient algorithm for finding them. Therefore both the bound to be exceeded by the product of the factors found in the Lucas-Lehmer type tests, and the bound to be exceeded can be lowered to $\sqrt[3]{n}$, if one is willing to spend a little more time on checking the possibilities in each residue class. This idea was for instance used in proving the primality of the number all of whose 1031 decimal digits are equal to 1 (by means of Lucas-Lehmer type tests) [W].

In recent years, the theory of elliptic curves has been successfully applied to the problem of primality testing (as well as to factoring). Analogues of the Lucas-Lehmer type tests were devised using factorizations in the ring of integers of a quadratic number field that is the complex multiplication ring of an elliptic curve [Bo]. One should think of this as replacing the multiplicative group of integers modulo n by the group of points on certain elliptic curves .

An algorithm of Goldwasser and Kilian [GK] gives primality proofs for almost all primes in expected polynomial time. But it relies on an algorithm of Schoof [S] to compute the number of elements on an elliptic curve that – though polynomial – is considered to be too slow for practical purposes. A variant of this idea, working on hyperelliptic curves, by Adleman and Huang, [AH] yields primality proofs in expected polynomial time for all primes.

It seems that Atkin's method of using elliptic curves with complex multiplication for a general purpose test has proven to be practical (recently it was reported that it has been applied to primes of up to 564 digits [Mo]); in fact this formed a mayor incentive to improve the Jacobi sum algorithm as reported here, in order to let it maintain it's leading rôle. Very roughly, Atkin tries a list of elliptic curves until one is found for which the number of points on it, defined modulo n, is of the form kq, with k small and with q a number that is proven prime recursively. Although heuristically Atkin's algorithm is polynomial time, a rigorous analysis has not been given yet.

The Jacobi sum test
The Jacobi sum test can be roughly described as follows.

Select integers t and s such that $s = \prod q$ is the product of primes q with the property that $q - 1 | t$ and such that $s > \sqrt{n}$. For every pair (p^k, q), with q a prime divisor of s and p^k the highest power of the prime p dividing $q - 1$, perform a Jacobi sum test, which consists roughly of raising an element in $Z[\zeta_{p^k}]/nZ[\zeta_{p^k}]$ to the power n. Finally check that the integers $1 < r_i \le \sqrt{n}$ do not divide n, where $r_i \equiv n^i \bmod s$ for $i = 1, 2, \ldots, t$.

It can be shown that, in order to get $s > \sqrt{n}$, it suffices to take $t = (\log n)^{O(\log \log \log n)}$. For instance, for proving the primality of integers up to 212 digits, one could take t equal to (a divisor of) 55440.

We have made practical improvements on this algorithm in several directions. In the first place, the Jacobi sum test can be combined with the Lucas-Lehmer type tests; roughly speaking, this means that for every factor found in $n^w - 1$ the bound that the auxiliary number s for the Jacobi sum part of the combined test need to exceed, can be lowered by the same factor. Since the (modified) Lucas-Lehmer type tests are usually much cheaper than the Jacobi sum tests, this can be a tremendous gain. Of course one should compare this gain to the time needed to find more factors in $n^w - 1$, for small values of w. Using heuristics on the expected size of the factors that are to be found, a reasonable decision can be made here.

Secondly, it is possible to reduce the amount of work done in carrying out the Jacobi sum tests. Instead of doing the n-th powerings in the extension rings $Z[\zeta_{p^k}]/nZ[\zeta_{p^k}]$ of degree $\phi(p^k)$ over Z/nZ, where ϕ is Euler's function, it is possible to work in ring extensions of degree equal to $order(n \in Z/p^kZ)$, which is a divisor of $\phi(p^k)$, and which may be considerably smaller.

Thirdly, it is possible to combine several tests for pairs (p^k, q) into one larger test, provided that the primes p are different. The tests consist of n-th powerings of elements in a ring extension of degree u, which will be represented as polynomials of degree u on integer coordinates modulo n. Suppose that one test has to be done in an extension of degree u_1 and another in an extension of degree u_2; then they can be combined into one test in an extension of degree $\mathrm{lcm}(u_1, u_2)$. Of course that only makes sense if the combined test is cheaper; making the realistic assumption that multiplication is quadratic in the number of coordinates, combining the two tests is only advantageous if $\mathrm{lcm}^2(u_1, u_2) < u_1^2 + u_2^2$. But we are only able to deal with extensions of relatively small degree, and then it is easily seen that combining is only profitable if u_1 divides u_2 (or the other way around). In general, combinations should be made for degrees $u_1, u_2, \ldots, u_k$ with the property that every u_i divides $\max(u_1, u_2, \ldots, u_k)$. There is an easy, efficient procedure for finding the optimal combination, once the collection of all pairs (p^k, q) is known.

This combination method introduces another interesting optimalization problem: which choice of auxiliary numbers t and s leads to the least expensive collection of tests, i.e., of pairs (p^k, q)? Although some NP-complete parts of this problem prevented us from efficiently finding a solution that is guaranteed to be optimal, a procedure was found for generating a solution that is within a few percents of the optimal solution, in an amount of time that is negligible compared to the time saved in performing the rest of the algorithm using this solution.

Finally, Lenstra's idea of divisors in residue classes modulo $\sqrt[3]{n}$ applies here as well, which means that with some care the bound for s can be lowered to $\sqrt[3]{n}$.

Predictions

Although there has been no time yet to experiment extensively with the improved primality testing algorithm, some predictions can be made. We expect that in the very worst case, testing an integer of 800 digits for primality would take *one night* on a Cray. Here the worst case means that no factors are found for the Lucas-Lehmer part of the algorithm and moreover that the order of n is maximal modulo every divisor of the auxiliary number t. Also, the $\sqrt[3]{n}$ idea is not used here.

Experiments have shown that on the average, numbers of the same size will require about one third of the time to test the worst possible case; we expect that these experiments are reliable, even though they are only based on the optimalization part of the algorithm, since only the *size of* n and its residue class modulo the divisors of t determine the time needed for the Jacobi sum part of the test.

Taking into account that the algorithm is very well suited for parallelization (both of the time consuming steps, the Jacobi sum tests and the final trial divisions, can be performed in parallel), we predict that it will be possible, using this algorithm, to give primality proofs for random primes of up to 1000 digits in a few days, using either supercomputers or a network of small processors.

Thus the improved Jacobi sum test will once more prove to be the most powerful general purpose primality testing algorithm.

References

[AH] L.M. Adleman, M.A. Huang, *Recognizing primes in random polynomial time*, Proceedings of the nineteenth annual ACM symposium on the theory of computing (STOC), (1987), pp. 462-469.

[APR] L.M. Adleman, C. Pomerance and R. Rumely, *On distinguishing prime numbers from composite numbers*, Annals of Mathematics, **117** (1983), pp. 173-206.

[Ba] E. Bach, *Analytic methods in the analysis and design of number-theoretic algorithms*, MIT Press, (1985),

[Bo] W. Bosma, *Primality testing using elliptic curves*, Report 85-12, Universiteit van Amsterdam, (1985),

[CL] H. Cohen, H.W. Lenstra, Jr., *Primality testing and Jacobi sums*, Mathematics of Computation, **42** (1984), pp. 297-330.

[GK] S. Goldwasser, J. Kilian, *Almost all primes can be certified quickly*, Proceedings of the eighteenth annual ACM symposium on the theory of computing (STOC), (1986), pp. 316-329.

[LC] A.K. Lenstra, H. Cohen, *Implementation of a new primality test*, Mathematics of Computation, **48** (1987), pp. 103-121.

[L] H.W. Lenstra, Jr., *Divisors in residue classes*, Mathematics of Computation, **42** (1984), pp. 331-340.

[Mi] G.L. Miller, *Riemann's hypothesis and tests for primality*, J. Comp. Sys. Sci., **13** (1976), pp. 300-317.

[Mo] F. Morain, see update 2.2 to: *Factorizations to $b^n \pm 1$*, by Brillhart, Lehmer, Selfridge, Tuckerman and Wagstaff.

[R] M.O. Rabin, *Probabilistic algorithms for primality testing*, Journal of Number Theory, **12** (1980), pp. 128-138.

[W] H. Williams, H. Dubner, *The primality of R1031*, Mathematics of Computation, **47** (1986), pp. 703-711.

Extended Abstract

PRIVATE-KEY ALGEBRAIC-CODE CRYPTOSYSTEMS WITH HIGH INFORMATION RATES

Tzonelih Hwang

T.R.N. Rao

National Cheng Kung University
Institute of Information Engineering
Tainan, Taiwan, R.O.C.

University of Southwestern Louisiana
The Center for Advanced Computer Studies
Lafayette, Louisiana 70504

1. Introduction

Algebraic codes have been proven to be extremely powerful to combat errors in communications and computer systems. They can *reveal* information reliably in the presence of sustained interference due to channel noise or hardware failures [Lin 83, Rao 89]. Algebraic codes can also be used to *conceal* information from any unauthorized user. For examples, McEliece public-key cryptosystem applies the error-correcting capability of Goppa codes to provide data secrecy [McEliece 78]; Rao-Nam (RN) scheme applies simple algebraic codes to construct private-key cryptosystems [Rao 87]. These systems are called here *algebraic-code cryptosystems* (ACC).

Algebraic-code cryptosystems may encipher one plaintext in several different ways under one key, while retaining the independence of individual ciphertext. This property is also called *splitting* [Stinson 88], and these systems are referred here as *ACC with splitting*. ACC with splitting are important to prevent *ciphertext search attacks* and *pattern matching attacks* [Denning 82].

ACC with splitting require data expansion, a disadvantage of low information rate. To improve the information rate, we propose a new algebraic-code cryptosystem with splitting. The security of the system is investigated and appears to be as secure as the RN scheme modified by Struik and van Tilburg [Struik 87] which is called here the *ST scheme*. Although ACC could be constructed with nearly 100% information rates, but such systems do not provide splitting [Hwang 88].

A cipher is generally called *computationally secure* if it cannot be broken by systematic analysis with available resources [Denning 82]. By assuming that the ST scheme is computationally secure, the proposed scheme appears to be as secure as the ST scheme. Some comparisons will be given to show that the proposed scheme provides better information rate than the RN or ST schemes of the same block length.

1.1 RN and ST Scheme

Rao and Nam have proposed a private-key ACC using only simple codes (e.g., $d_{min} \approx 10$) which is called RN scheme here. RN scheme performs encryption by the following equation [Rao 87].

$$C = (MSG+Z)P,$$

where M is the plaintext of length k; C is the ciphertext of length n; S is a random non-singular matrix of rank k; G is the generator matrix of an (n, k) block code C which can correct t errors; P is a random permutation matrix of rank n; Z is an error vector randomly selected from the *syndrome-error table* that is constructed from the standard-array-decoding table of the code C.

Based on the linear structure of the system, Struik and van Tilburg proposed a chosen-plaintext attack to crack the RN scheme. They further modified the RN scheme to withstand similar chosen-plaintext attacks [Struik 87]. The ST scheme can be described as the following.

$$C = f(M, Z)G+Z,$$

where $f^{-1}(f(M, Z), Z) = M$. The details of enciphering and deciphering are simple and can be found in [Struik 87].

2. The Proposed Scheme

2.1 Encryption and Decryption

Let M be a plaintext block of $(n+k)$ bits to be enciphered into a ciphertext C of $2n$ bits (n and k are the parameters of the linear code $\mathbf{C}$ to be used here). M can be divided into two sub-blocks M_1 and M_2 where M_1 is an n-bit sequence and M_2 is a k-bit sequence. $\mathbf{P}$ is a random permutation matrix of rank $2n$. The syndrome-error table in the RN scheme is replaced by an arbitrary nonlinear function h for the purpose of saving storage space and also increasing the security level. The function h takes two parameters X and Y, where X is a k-bit sequence and Y is an n-bit sequence, and produces a k-bit block $h(X, Y)$.

Encryption. Let Ψ be an invertible, nonlinear (n-bit to n-bit) function. $\mathbf{G}$ is the generator matrix of an (n, k) linear code $\mathbf{C}$ that can correct t random errors. The encryption is performed by simple steps as follows.

    (a) Generate an n-bit random vector E of weight $\leq t$.
    (b) Compute an error vector Z by
$$Z = \tau + E,$$
      where $\tau = (h(M_2, E))\mathbf{G}$.
    (c) Obtain the ciphertext $C = (C_1, C_2)\cdot\mathbf{P}$, where

$$C_1 = \Psi(M_1 + \tau) \tag{1}$$

$$C_2 = M_1 + M_2\mathbf{G} + Z. \tag{2}$$

Since Z is a function of both M_2 and E, the total number (N) of Z's is 2^k if h is chosen carefully. Because the value N is not determined by the number of cosets in the standard array decoding table of the code, simple codes can be used in the system and still achieve a high level of security as can be seen from the discussion in Sec. 2.2.

Decryption. The decryption can be carried out easily by the following steps.

    (a) Compute $(C_1, C_2) = C \cdot \mathbf{P}^{-1}$.
    (b) Compute $\Psi^{-1}(C_1)$.
    (c) Obtain $M_2\mathbf{G} + E = \Psi^{-1}(C_1) + C_2$.
    (d) Decode the result of (c) by applying the decoding of the introduced
        code $\mathbf{C}$: Recover M_2 and obtain E.
    (e) Compute $\tau = h(M_2, E)$.
    (f) Recover $M_1 = \Psi^{-1}(C_1) + \tau$.

Note that any invertible, nonlinear function that can withstand ciphertext-only attacks can be used as the function Ψ. Therefore, Ψ can be very easy to implement. This will be shown in the next section.

2.2 Security of the Newly Developed Scheme

    The encryption and decryption steps given above are fairly simple and are easy to implement. Clearly, due to Step (a) of encryption, it is indeed an ACC with splitting. What remains to be studied is its security. The following discussion on the security of the new scheme is based on the assumption that the ST scheme is *computationally secure*. The lemmas and theorem are given here without proofs but these proofs will be given in the full paper.

First we show that the partial encryption specified by Equation (2) is at least as secure as the ST scheme that uses the same code. The method used to prove the lemma is to show that the ST scheme can be reduced to the partial encryption given in Equation (2). On the other hand, Equation (2) cannot be reduced to the ST scheme.

Lemma 1

The partial encryption given in Equation (2) is at least as secure as the ST scheme.

Next, we show how the proposed scheme can be secure by investigating the structure of the scheme. First, we investigate a simplified scheme obtained by removing both functions Ψ and $\mathbf{P}$ from the original scheme and show that the simplified scheme can be broken by a chosen-plaintext attack as follows.

Lemma 2

The encryption scheme
$$C = (M_1+\tau,\ M_1+M_2\mathbf{G}+Z)$$
can be broken by a chosen-plaintext attack in $O(kn^2)$ bit operations.

If a linear function $\Psi_{(L)}$ is introduced to scramble the first part $(M_1+\tau)$ of ciphertext C in Lemma 2, then the scheme is still insecure as shown by the following.

Lemma 3

The encryption scheme
$$C = (\Psi_{(L)}(M_1+\tau),\ M_1+M_2\mathbf{G}+Z)$$
can be partially broken by ST chosen-plaintext attacks in $O(n^2N^2\log N)$.

In what follows, we will show that if a nonlinear function $\Psi_{(N)}$, that is secure under ciphertext-only attacks while can be broken by a known-plaintext attack in polynomial time, is introduced, then the partial encryption given in Equation (1) is computationally secure.

Lemma 4

The partial encryption specified by Equation (1) is computationally secure if $\Psi_{(N)}$ is secure under ciphertext-only attacks.

So far we have shown that both Equations (1) and (2), i.e.,
$$C_1 = \Psi_{(N)}(M_1+\tau), \text{ and}$$
$$C_2 = M_1+M_2\mathbf{G}+Z$$
of the original scheme are computationally secure. However, it doesn't mean that the overall scheme is also computationally secure. For example, if $\Psi_{(N)}$ is specified by a key KEY_1 of length k_1 bits and the encryption of M_2 is specified by the key KEY_2 of k_2 bits, then an attack can crack the scheme in W operations, where $2^{k_2} \leq W \leq 2^{k_1}+2^{k_2}$, as follows.

(a) The cryptanalyst searches the key space of KEY_2 exhaustively to obtain M_2 and τ. It requires 2^{k_2} operations in the worst case.

(b) $C_1 = \Psi_{(N)}(M_1+\tau)$ can be broken by a known-plaintext attack in less than 2^{k_1} operations because we have assumed that $\Psi_{(N)}$ can withstand ciphertext-only attacks while can be broken by a known-plaintext attack in polynomial time.

Note that an exhaustive search on the key space of $\Psi_{(N)}$ will not be productive because it involves the search for key spaces of both KEY_1 and KEY_2. Based on this analysis, we also have the following.

Lemma 5

The security level of the encryption scheme
$$C = (\Psi_{(N)}(M_1+\tau),\ M_1+M_2\mathbf{G}+Z)$$
is $O(2^{k_1} + 2^{k_2})$.

Obviously, the cryptanalysis mentioned above highly depends on the correct partition of the ciphertext into C_1 and C_2. If the ciphertext is scrambled by a random permutation matrix P of rank $2n$, then the security level of the scheme will be highly increased. Based on the above discussion, the new scheme appears to be very secure. However, proving the security level of the scheme remains open and requires further research.

3. Comparison with the ST Scheme and Conclusion

First we note that the encryption block length of the proposed scheme is $2n$ where n is the block length of the error correcting code C. (The parameters of C are $(n, k, 2t+1)$.) The information rate of the code C is k/n, but for the encryption scheme, the rate is $(n+k)/2n$. To make a fair comparison, we choose an ST scheme (or RN scheme) of code of length $2n$ and t error correcting capability. If we choose BCH codes (or shortened codes) as examples for this comparison we arrive at Table 1.

Table 1. Comparison of information rates under the same ciphertext length $2n$.

			New Scheme		ST Scheme		
n	k	t	I.R. (%)	I.R. (%)	2n+1	k'	t
15	11	1	86.6	83.9	31	26	1
	7	2	73.3	67.7		21	2
	5	3	66.7	51.6		16	3
31	26	1	91.9	90.5	63	57	1
	21	2	83.9	81.0		51	2
	16	3	75.8	71.4		45	3
	11	5	67.7	57.1		36	5

If we compare the information rate of the two schemes under the condition that they are using the same $(n, k, 2t+1)$ base code, then Table 2 shows that our scheme also provides better information rates i.e., $R^* = \dfrac{n+k}{2n} > \dfrac{k}{n}$. Note that in this case, the ciphertext length of the new scheme is $2n$, while that of the ST scheme is n.

Table 2. Comparison of information rates under the same base code.

			New Scheme		ST Scheme		
n	k	t	I.R. (%)	I.R. (%)	2n+1	k'	t
15	11	1	86.6	73.3	15	11	1
	7	2	73.3	46.7		7	2
	5	3	66.7	33.3		5	3
31	26	1	91.9	83.9	31	26	1
	21	2	83.9	67.7		21	2
	16	3	75.8	51.6		16	3
	11	5	67.7	35.5		11	5

We have constructed a private-key algebraic-code encryption scheme with splitting property whose encryption/decryption can be carried out efficiently. We have investigated its security and show how the scheme can be secure. The new scheme provides higher information rate than the RN or ST schemes that use the same code or have the same ciphertext length and hence appears to be more practical.

References

[Denning 82] Dorothy E. Denning, *Cryptography and Data Security*, Addison Wesley, 1982.

[Denny 88] W.F. Denny and T.R.N. Rao, "Encryptions Using Linear and Nonlinear Codes: Implementation and Security Considerations", Ph.D Dissertation, Univ. of SW Louisiana, Spring 1988.

[Hwang 88] Tzonelih Hwang, *Secret Error-Correcting Codes and Algebraic-Code Cryptosystems*, Ph.D. Dissertation, Univ. of SW Louisiana, Summer, 1988.

[McEliece 78] R.J. McEliece, "A Public-Key Cryptosystem Based on Algebraic Coding Theory", *DSN Progress Report*, Jet Propulsion Laboratory, CA., Jan. & Feb. 1978, pp 42-44.

[NBS 77] "Data Encryption Standard", FIPS PUB 46, National Bureau of Standard, Washington, D.C., Jan. 1977.

[Rao 87] T.R.N. Rao and K.H. Nam "A Private-Key Algebraic-Code Cryptosystem," Advances in CRYPTO 86, editor A.M. Odlyzko, New York, Springler Verlag, pp. 35-48, 1987.

[Rao 89] T.R.N. Rao and E. Fujiwara, *Error Control Coding for Computer Systems*, Prentice Hall, 1989.

[Stinson 88] D.R. Stinson, "Some Construction and Bounds for Authentication Codes," *Journal of Cryptology*, Vol. 1, No. 1, 1988.

[Struik 87] R. Struik and van Tilburg J., "The Rao-Nam Scheme is Insecure Against a Chosen-plaintext Attack," CRYPTO '87.

[Tilburg 88] J. van Tilburg, "On the McEliece Public-key Cryptosystem," a paper presented in CRYPTO '88, to appear.

Zero-knowledge procedures for confidential access to medical records

Jean-Jacques Quisquater *   André Bouckaert **

*Philips Research Laboratory Belgium
Avenue Albert Einstein, 4
B–1348 Louvain-la-Neuve, Belgium
jjq@prlb.philips.be

** Université Catholique de Louvain
UMED - 72.25, Avenue E. Mounier
B-1200 Brussels, Belgium

(Extended summary)

During the seventies, several successful attempts were made to use computerized files for storing medical records, first by accessing to mainframes in large hospitals, and thereafter by use of microcomputers in decentralized medical services or general practitionners. The wide acceptance of computerized record keeping in general practice is witnessed *e.g.* by the fact there are more than fifty softwares available in the Belgian market for this purpose.

Such medical databases are vulnerable to spying, a hazard whose severity is increased by the traditional concept of medical secrecy (Mazen, 1988). According to this many-centuries-old concept, all medical data, or even non-medical data recorded during medical investigations, are not to be communicated by the doctor to anybody, except to another doctor if requested by the follow-up of the patient. Since paper medical records are very slowly and inefficiently retrieved, it could be thought that this ineffectiveness was a guarantee of confidentiality. This concept was not widely publicized but implicitly adopted and it explains the lack of security measures in most medical databases. There is no account of cases of piracy in medical databases although there is good reason to think that it has been increasingly practiced in the United States for litigation purposes.

Why not let the patients keep their own records? This question has been raised for the first time nine years ago (Metcalfe, 1980). Portable medical records were used episodicaly from the beginning of the eighties. Persons who advocated such portable records were usually public health doctors, motivated by the needs of the follow-up of the patients and, to a variable degree, by the feeling that it belongs to this patient to carry his own data or to be able to do so in a democratic society (Baldry & al., 1986; Gentilini & al., 1986; Bouckaert, 1984, 1988). The move towards portable records has been repeatedly criticized by medical professional unions and by their lobbies on the basis that such records would be vulnerable to blackmail.

In the European societies of the eighties, blackmailing a patient for his medical record is likely to be exercized by employers and by insurance companies. Other candidates to blackmail are police and security authorities, political parties, outlawed organizations. It should be understood that such a blackmail is always some kind of intermediate pressure, since the hard pressure is simply to ask for another medical examination. Nevertheless, it is not hard to imagine that an employed person could find it difficult to allow his employer to look at his record since the simple fact of refusing can be turned into a weapon against himself.

The concept of a portable computerized medical record stored on a smart card has evolved as a response to the dual challenge of confidentiality and portability. In the smart card, the access to the data is protected by access control rather than by enciphering although both methods can be used at the same time. In both cases, in order to access or/and to read, the usual approach is to request the production of a password by the patient. Such a patient password suffers from two weaknesses:

1. It exposes the patient to the hazard of blackmail.

2. It exposes the patient to the hazard of forgetting his own password.

At the same time, this policy achieves complete protection against the unauthorized use of stolen cards.

To avoid the blackmail, a possible solution is to use a two-key policy: both doctor and patient have a password and both use smart cards. Blackmail cannot be exercized against the patient without knowing the doctor's password. The main troubles are now:

1. The need for an external authority for password management, doctor habilitation, renewal procedures.

2. The possibility of monitoring the card readers to obtain the passwords.

3. The possibility of buffer spying.

4. The possibility of looking at the keyboard during password production.

It is possible to avoid these dangers and to reduce the managerial burden of the external authority by using a zero-knowledge procedure (Fiat and Shamir, 1987; Guillou and Quisquater, 1988). With such a procedure, the doctor does not know his own password and does not need to produce it. The password of the doctor issued by an external authority that does not keep trace of its own production. Hence nobody but the doctor's smart card itself will know the password. The patient's smart card can nevertheless check that the doctor's smart card contains the valid secret information.

In this situation, the possibility of copying passwords disappears since the password is never actually revealed during the process of interaction between smart cards. But this policy gives no protection for stolen smart cards. If the doctor's password can only be checked by his own patients however, the possibility of stealing the doctor's and patient's card is quite reduced. If no restriction is made when the passwords are issued, the zero-knowledge procedure should be supplemented by enciphering and a password-conditional deciphering procedure or by an additional password-controlled access.

References

[1] M Baldry, G. Chéal, B. Fisher and al., *Giving patients their own records in general practice: experience of patients and their staff*, Brit. Med. J., **292**, pp. 596–598, 1986.

[2] A. Bouckaert, *Security of transportable computerized files*, Advances in cryptology, Proceedings of EUROCRYPT '84, Springer-Verlag, pp. 416–425, 1985.

[3] A. Bouckaert, *Medical records, confidentiality, smart cards*, MIM News, 2, pp. 27-36, 1988.

[4] J. L. Gentilini, J. L. Samaille, M. Trotin and P. Marquis , *The PH-card: a portable and confidential medical file*, Medinfo '86, Proceedings, pp. 1010–1014, Amsterdam 1986.

[5] N. J. Mazen, *Le secret professionnel des praticiens de la santé*, Vigot, Paris, 1988.

[6] D. H. Metcalfe, *Why not let patients keep their own records?*, J. Roy. Coll. Gen. Pract., 30, p. 420, 1980.

FULL SECURE KEY EXCHANGE AND AUTHENTICATION WITH NO PREVIOUSLY SHARED SECRETS

JOSEP DOMINGO i FERRER
LLORENÇ HUGUET i ROTGER

Dept. Informàtica
UNIVERSITAT AUTONOMA DE BARCELONA
08193 BELLATERRA

When speaking about secure networks, the bootstrapping process is very often forgotten or at least ignored. Some of the methods used so far do not protect against impersonation (Diffie-Hellman exponential key exchange) or have an important computational complexity (public-key based methods). A new algorithm is presented which is able to achieve key exchange whilst ensuring secrecy and authentication with a reasonable amount of computation.

1. INITIALIZATION

Let us define the necessary conditions for our system.

q, α are a large prime (say 150 digits), with $q-1$ having at least one large prime factor, and a primitive element of $Z/(q)$, respectively, both publicly known. Each node i in the network has a <u>very probably unique name</u> (later we shall see what this means) in the form α^{t_i} mod q where t_i is chosen by i in the range $2..q-1$ and only known to i. The node name is also publicly known and together with the other node names and α, q is available on all communication media (radio, television, newspapers and so on, what has been called the <u>Merkle channel</u> from Merkle's proposal). This latter assumption may look a little bizarre, but it is the only way to ensure that everybody in the network has access to the public information without any distortion.

To generate the public information listed in the previous paragraph, we can proceed as follows. A particular node i (it is only important that it be a single node, no matter which one), generates α, q and sends them over the Merkle channel, so that everybody can reliably learn these numbers. Then each node i generates a number ti in the range 2..q-1 and keeps it secret; then node i computes its name as α^{ti} mod q and sends it over the Merkle channel. Now every node knows α, q and each other's name. The probability for all nodes to have different names is

P(q-3,n)/PR(q-3,n),

where n is the number of nodes, P denotes permutations and PR permutations with repetition. It is trivial to see that this quantity approaches 1 if q >> n^2.

2. THE METHOD. NORMAL MODE: KEY EXCHANGE PROTOCOL

In what we call "normal mode", the presented method works as a **key exchange protocol, providing secrecy and authentication**. After this mode has been used, the exchanged key can be used by this method in order either to encrypt and to decrypt messages, or to sign them, if some public parameters are added to those discussed in the previous section. Let us examine the "normal mode".

Node j first has α, q and knows all the node identifiers, in particular knows the identifier of node i, from which it wants to get a key, to be IDi = α^{ti} mod q. Of course, ti is only known to node i. Then node j and i proceed as follows:

STEP 1. NODE j: Compute X1, such that

gcd(X1,q-1) = 1   and   2 <= X1 <= q-2

Compute also Y1 = α^{X1} mod q .
Send Y1 to NODE i.

STEP 2. NODE i: Upon receiving Y1, pick X2, X4 such that

[1]   X2 + X4 = ti mod (q-1)   and   2 <= X2, X4 <= q-2

Compute the key K = α^{X2} mod q .
Compute Y2 = Y1^{X2} mod q.
Compute Y4 = Y1^{X4} mod q.
Send Y2 and Y4 to NODE j.

STEP 3. NODE j: Now NODE j computes the multiplicative inverse of X1 mod (q-1), X1^{-1} such that

[2]   X1^{-1} * X1 = 1 mod (q-1)

By computing

[3]   Y2$^{X1^{-1}}$ mod q = K

NODE j retrieves the key K it will, share with NODE i from now on. An additional exponentiation yields

[4]   Y4$^{X1^{-1}}$ mod q = α^{X4} mod q

In order to achieve authentication, NODE j performs the following test on K (if equation [5] does not hold, then the key K does not come from NODE i):

[5]   K * α^{X4} mod q = IDi = α^{ti} mod q

END.

Now NODE i and NODE j share the key K which can be used either straightforwardly as a DES key (taking for example the 8 lower bytes of it), or act as a key-encrypting key to transmit further keys, or be used as a key for the cryptosystem alternate mode of the presented method.

Note that some calculations **can be done in parallel** between NODE i and NODE j. For instance, NODE i can precalculate some (X2, X4) pairs for the next key transmissions to other nodes and can also precalculate the keys K it will send, in order to accelerate step 2. Also, while waiting for Y2, NODE j can compute $X1^{-1}$, so that step 3 is faster. A precalculation of some $(X1, X1^{-1}, Y1)$ triples is also possible at node j. With all these improvements, only <u>four interactive exponentiations</u> are necessary:

- To compute Y2 at node i.
- To compute Y4 at node i.
- To retrieve K at node j.
- To retrieve α^{X4} mod q at node j.

This is equivalent to the four exponentiations required to ensure secrecy and authentication with RSA (two at NODE i and two at NODE j to retrieve the key), but it avoids the strong prime calculation (two strong primes per node).

3. FINAL REMARKS

Proofs of correctness, secrecy and authentication, as well as some extensions to the basic algorithm will be given in the full paper. It will be shown that this algorithm can work in an **alternate mode**, thus operating as a **unidirectional conventional** or as a **signature scheme**.

4. REFERENCES

[DDD85] R. A. DeMillo, G. I. Davida, D. P. Dobkin et al., <u>Applied Cryptology, Cryptographic Protocols, and Computer</u>

Security Models, Proc. of Sympsia in Applied Math., Vol 29, American Mathematical Society (1985).

[Den82] D. E. Denning, Cryptography and Data Security, Addison-Wesley 1982.

[DH88] J. Domingo, L. Huguet, "Secure network bootstrapping: an algorithm for authentic key exchange and public-key encryption", IEEE Transactions on Information Theory (submitted).

[Diff88] W. Diffie, "The First Ten Years of Public-Key Cryptography", Proc. of the IEEE, Vol 76. No. 5, May 1988.

[FSh87] A. Fiat and A. Shamir, "How to prove yourself: Practical solutions to identification and signature problems", Crypto'86, Springer-Verlag (1987).

[GMR85] S. Goldwasser, S. Micali and C. Rackoff, "Knowledge complexity of interactive proofs", 17th Symposium on the Theory of Computing, 1985.

[HDP88] L. Huguet, J. Domingo and J. Ponsa, "Communications Cryptography: a DES-based System for PC's", Proc. of the MIMI'88, June 1988.

[RSA78] R. L. Rivest, A. Shamir and L. Adleman, "A Method for Obtaining Digital Signatures and Public-Key Cryptosystems", Communications of the ACM, Feb. 1978

[Sim88] G. J. Simmons, "A Survey on Information Authentication", Proc. of the IEEE, Vol. 76, No.5, May 1988.

Varying Feedback Shift Registers

Yves ROGGEMAN
Université Libre de Bruxelles
Laboratoire d'Informatique Théorique
Campus Plaine - CP212
boulevard du Triomphe
B-1050   Bruxelles
Belgium

1. Context

It is well known that a stream cipher system can be described in terms of a Vernam scheme using a Pseudo-Random Number Generator as key generator. Each character m_t of the plaintext (viewed as an integer) is enciphered by adding the corresponding pseudo-random key character s_t. Deciphering is obtained by subtracting the same value stream from the ciphertext (see Fig.1).

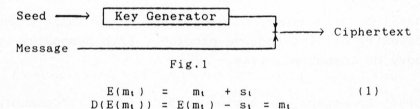

Fig.1

$$E(m_t) = m_t + s_t \qquad\qquad (1)$$
$$D(E(m_t)) = E(m_t) - s_t = m_t$$

A Lehmer Linear Congruential PRNG or a Linear Feedback Shift Register (LFSR) cannot be used in cryptographic systems because they can be cracked. In order to obtain Cryptographically Strong Number Generators, we can use Non-Linear Feedback Shift Registers. But a general model of such a NLFSR is difficult to implement and to study.

In another way, non-linearity is simulated in models involving more than one LFSR: product of sequences, cascade scheme, flip-flop, multiplexed LFSR, clock variation, a.s.o. But in most of these systems, every component can be isolated and/or the pseudo-random sequence is not always produced at a constant rate.

In this paper, we describe a new model based on FSR producing non-linear sequences, but which is easy to implement and can be used at a constant rate.

2. Classical theory

2.1. FSR

A k-satge FSR is a machine involving k memory cells X_0, X_1, ... X_{k-1} (see Fig.2). At each clock pulse, every value is shifted one position left, the leftmost value is output and the rightmost cells is filled with a value depending on the k previous ones.

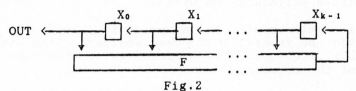

Fig.2

A solution (s_t) is an infinite sequence satifying

$$s_{t+k} = F(s_t, s_{t+1}, \ldots s_{t+k-1}) \qquad (2)$$

for some feedback function F. Such a solution is univokely determined by its initial state $[s_0, s_1, \ldots, s_{k-1}]$.

Classically, s_t belongs to a finite field GF(q) (GF(2) in most cases) and F is a rational function on GF(q). Such a register is noted $FSR^k(q)$.

2.2. Period and singularity

Each solution of a FSR is ultimately periodic. Its period and its singularity are the smallest integers π and σ satisfying

$$s_{t+\pi} = s_t \qquad \forall\, t \geq \sigma \geq 0, \pi > 0 \qquad (3)$$

A FSR is said to be non-singular if every solution is non-singular. This is achieved iff

$$F(\alpha, x_1, \ldots x_{k-1}) = F(\beta, x_1, \ldots x_{k-1}) \implies \alpha = \beta \qquad (4)$$

In cryptographic applications, registers have to be non-singular and with maximal period.

2.3. Linear FSR

A k-stage Linear Feedback Shift Register $LFSR^k(q)$ is defined by

$$s_{t+k} = \sum_{i=0}^{k-1} c_i\, s_{t+i} \qquad t \geq 0, \quad c_i \in GF(q) \qquad (5)$$

$$\text{or} \quad \sum_{i=0}^{k} c_i\, s_{t+i} = 0 \qquad \text{with } c_k = -1 \qquad (6)$$

Such a register is non-singular iff

$$c_0 \neq 0 \qquad (7)$$

The monic polynomial

$$f(x) = - \sum_{i=0}^{k} c_i \, x^i \qquad (8)$$

is called the characteristic polynomial associated with the LFSR^k(q).

Its maximal period is (q^k-1) which is reached iff $f(x)$ is a so-called primitive polynomial on GF(q).

The minimal polynomial of a periodic sequence is the characteristic polynomial of the smallest LFSR that can produced this sequence. Its degree is called the linear complexity of the sequence.

2.4. Transition matrix

The companion matrix C of $-f(x)$ is

$$C = \begin{bmatrix} 0 & \vdots & I \\ \text{-----} & \text{+} & \text{-------------} \\ c_0 & \vdots & c_1 \; \ldots \; c_{k-1} \end{bmatrix} \qquad (9)$$

Its characteristic polynomial is $(-1)^k f(x)$, and its determinant is $(-1)^{k-1}c_0$. C is called the transition matrix of the LFSR^k(q).

If we define the (transposed) state vector

$$\overline{s}_t{}' = [s_t, \, s_{t+1}, \, \ldots, \, s_{t+k-1}] \qquad (10)$$

the «'» indicating transposition, we have

$$\overline{s}_{t+1} = C \, \overline{s}_t \qquad (11)$$

The so-called generating functions for C^t and s_t are resp.

$$G(z) = \Sigma \, z^t \, C^t = (I - z \, C)^{-1} \qquad (12)$$

and

$$g(z) = \Sigma \, s_t \, z^t = \Phi(z)/\det(I-zC) \qquad (13)$$

where $\deg(\Phi) < k$.

3. Generalized LFSR

3.1. Non-degenerated solution

Generalizing Eq.11, we define a GLFSR^k(q) by

$$\overline{r}_{t+1} = M \ \overline{r}_t \qquad (14)$$

for any matrix M and $\overline{r}_t' = [r_{t,0}, \ \ldots \ r_{t,k-1}]$.

A GLFSR^k(q) is non-singular iff

$$\det(M) \neq 0 \qquad (15)$$

Let $R_0 = [\overline{r}_0, \ \overline{r}_1, \ \ldots \ \overline{r}_{k-1}]$ be the matrix whose columns are the first k states of a solution $(\overline{r}_t)$, this solution is called non-degenarated if $\det(R_0) \neq 0$.

We have the following property: if $(\overline{r}_t)$ is non-degenerated, it has the same minimal polynomial as M. Moreover, it is the characteristic polynomial of M.

3.2. Similar LFSR

Let C be the companion matrix of the monic characteristic polynomial of M. C is the transition matrix of a LFSR^k(q) similar to the GLFSR^k(q) defined by M. If $(\overline{r}_t)$ is a non-degenerated solution of M, let (s_t) be the solution of C corresponding to $\overline{s}_0' = [0, \ \ldots \ 0, 1]$ (the impulse), and let $S_0 = [\overline{s}_0, \ \ldots \ \overline{s}_{k-1}]$, we have

$$\begin{aligned} C &= S_0 \ R_0^{-1} \ M \ R_0 \ S_0^{-1} \\ \overline{r}_t &= R_0 \ S_0^{-1} \ \overline{s}_t \end{aligned} \qquad (16)$$

Thus, each non-degenerated GLFSR^k(q) is similar to the LFSR^k(q) corresponding to the same characteristic polynomial.

3.3. Affine LFSR

Let A be any lxk-matrix, $\overline{b}$ be any l-vector and (s_t) be any solution of a LFSR^k(q). The l-state sequence $(\overline{r}_t)$ defined by

$$\overline{r}_t = A \ \overline{s}_t + \overline{b} \qquad (17)$$

is the solution of a so-called Affine LFSR. It verifies

$$\sum_{i=0}^{k} c_i \ \overline{r}_{t+i} = \left(\sum_{i=0}^{l} c_i \right) \overline{b} = |c| \ \overline{b} = -f(1) \ \overline{b} \qquad (18)$$

If $R_t = [\overline{r}_t, \ \ldots \ \overline{r}_{t+k-1}]$ (a lxk-matrix),

$$R_{t+1} = R_t \ C' \qquad (19)$$

4. Varying FSR

4.1. Definitions

Modifying Eq.2 as

$$s_{t+k} = F_t(s_t, s_{t+1}, \ldots s_{t+k-1}) \qquad (20)$$

we define a FSR with varying feedback functions (F_t). If there exist σ and τ such that

$$F_t = F_{t+\tau} \qquad \forall\, t \geq \sigma \qquad (21)$$

every solution of Eq.20 (s_t) is ultimately periodic and the register is called a Periodic FSR.

If $\sigma = 0$ and F_t is a linear function for every t, the PFSR is called a τ-PLFSR^k(q) defined by

$$s_{t+k} = \sum c_{t,i}\, s_{t+i} = \overline{c_t}\,'.\,\overline{s_t} \qquad (22)$$
$$\overline{c_t} = \overline{c_{t+\tau}} \qquad \forall\, t$$

Such a τ-PLFSR^k(q) is equivalent to a classical LFSR^k(q) iff $\tau = 1$.

4.2. Generating function

Let C_t be companion matrix of $\overline{c_t}$, the generating function associated with the τ-PLFSR^k(q) is

$$G(z) = \sum z^t C_{t-1} \ldots C_1 C_0 \qquad (23)$$
$$= (I + z C_0 + \ldots + z^{\tau-1} C_{\tau-2} \ldots C_0)\, (I - z^\tau C)^{-1}$$

where $C = C_{\tau-1} \ldots C_1 C_0$.

Let $D_t = C_{t-1} \ldots C_1 C_0$, we have $C = D_\tau$,

$$\overline{s_{t+1}} = C_t\, \overline{s_t} = D_{t+1}\, \overline{s_0} \qquad (24)$$

and $G(z) = (\sum_{t=0}^{\tau-1} z^t D_t)\, (I - z^\tau C)^{-1} = \Gamma(z)\, G_c(z^\tau) \qquad (25)$

In Eq.25 we note $G_c(z)$ the generating function associated to the LFSR^k(q) with transition matrix C (see Eq.12).

If $m(x)$ is the minimal polynomial of (s_t), and $m_c(x)$ the minimal polynomial of C,

$$m(x) \mid m_c(x^\tau) \qquad (26)$$

and the linear complexity of any solution of a τ-PLFSR^k(q) is at most τk

4.3. Period and singularity

The state of a τ-PLFSR^k(q) does not only depend on the value of s_t. It includes the feedback index: t (mod τ). In this context, such a register is non-singular iff

$$c_{t,0} \neq 0 \qquad \forall\, t \qquad\qquad (27)$$

Let π be the actual period of any solution (s_t) of a τ-PLFSR^k(q), its state-period is $\mu = lcm(\pi, \tau)$.

It can be shown that any solution (s_t) of a τ-PLFSR^k(q) with actual period π can be produced by a δ-PLFSR^k(q) with $\delta = gcd(\pi, \tau)$. Thus, if τ is a prime, any solution can either be produced by a classical LFSR^k(q), or satisfies $\tau \mid \pi = \mu$.

5. Coupled LFSR

5.1. Definitions

In order to generate periodically varying feedback functions, we can use anaother LFSR. So we define a k,l-stage Coupled LFSR noted CLFSR^{k,l}(q) as a τ-PLFSR^l(q) where each c_t is the state of an Affine GLFSR^k(q) (see Fig.3).

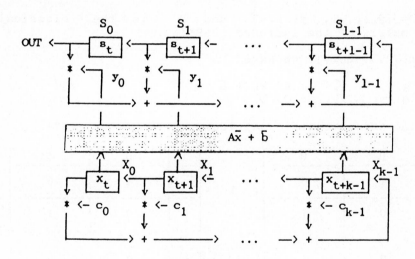

Fig.3

Such a model corresponds to equations

$$s_{t+l} = \sum_{j=0}^{l-1} y_{t,j}\, s_{t+j}$$

$$y_{t,j} = \sum_{i=0}^{k-1} a_{j,i}\, x_{t+i} + b_j \qquad j=0 \ldots l-1 \qquad\qquad (28)$$

$$x_{t+k} = \sum_{i=0}^{k-1} c_i\, x_{t+i}$$

It depends on $kl+k+1$ parameters: the matrix A and the vectors $\overline{c}$ and $\overline{b}$.

Such a CLFSR^{k,l}(q) is non-singular iff

$$y_{t,0} = b_0 \neq 0 \qquad \forall\, t \qquad (29)$$

i.e. $\qquad b_0 \neq 0$ and $a_{0,i} = 0, \qquad i=0...k-1.$

We shall only consider non-singular CLFSR.

5.2. Transition matrix

The state of a CLFSR^{k,l}(q) is given by the $(k+1)$-vector

$$\overline{v_t}' = [\ \overline{s_t}'\ |\ \overline{x_t}'\ ] = [s_t,\ ...\ s_{t+l-1}|x_t,\ ...\ x_{t+k-1}]\qquad (30)$$

So we define the associated transition matrix

$$
T_t \;=\;
\left[
\begin{array}{cc|c}
0 & I & \\
\hline
b_0 & \overline{y_t}\,{}^{\cdot\prime} & 0 \\
\hline
\multicolumn{2}{c|}{0} & C^{\cdot}
\end{array}
\right]
\qquad (31)
$$

where $\overline{y_t}\,{}^{\cdot\prime} = [y_{t,1},\ ...\ y_{t,l-1}]$ and $C^{\cdot}$ is the classical transition matrix of the included LFSR^k(q).

If $\overline{b}^{\cdot\prime} = [b_1,\ ...\ b_{l-1}]$, we have

$$
\begin{aligned}
\overline{y_t}^{\cdot} &= A^{\cdot}\ \overline{x_t} + \overline{b}^{\cdot} \\
\overline{x_{t+1}} &= C^{\cdot}\ \overline{x_t}
\end{aligned}
\qquad (32)
$$

We now define the invertible matrices $\qquad\qquad\qquad\qquad (33)$

$$
X_t =
\left[
\begin{array}{c|c|c}
1 & 0 & 0 \\
\hline
0 & I & 0 \\
\hline
\overline{x_t}^{\cdot} & 0 & I
\end{array}
\right],
\quad
Y_t =
\left[
\begin{array}{c|c|c}
1 & 0 & 0 \\
\hline
\overline{y_t}^{\cdot} & I & 0 \\
\hline
0 & 0 & I
\end{array}
\right]
$$

$$
A =
\left[
\begin{array}{c|c|c}
1 & 0 & 0 \\
\hline
0 & I & A^{\cdot} \\
\hline
0 & 0 & I
\end{array}
\right],
\quad
B =
\left[
\begin{array}{c|c|c}
b_0 & 0 & 0 \\
\hline
\overline{b}^{\cdot} & I & 0 \\
\hline
0 & 0 & I
\end{array}
\right]
$$

$$
C =
\left[
\begin{array}{c|c|c}
1 & 0 & 0 \\
\hline
0 & I & 0 \\
\hline
0 & 0 & C^{\cdot}
\end{array}
\right],
\quad
P =
\left[
\begin{array}{c|c|c}
0 & 1 & 0 \\
\hline
I & 0 & 0 \\
\hline
0 & 0 & I
\end{array}
\right]
$$

Using the commutator $[X_t, A^{-1}] = X_t^{-1} A X_t A^{-1}$, we have

$$T_t' = C' Y_t P = C' B [X_t, A^{-1}] P \qquad (34)$$
$$= C' B [C^t X_0 C^{-t}, A^{-1}] P = C' B [X_0, C^{-t} A^{-1} C^t] P$$

In these formulae, we have the following properties:

- C' can be placed anywhere;

- B & $[X_t, A^{-1}]$ commute iff $b_0 = 1$ or $A^* = 0$;

- B & P commute iff $b_0 = 1$ and $\overline{b}^* = 0$ (i.e. $B = I$);

- P & $[X_t, A^{-1}]$ commute only if $\overline{x}_t \in Ker(A^*)$.

This last property assure that if (x_t) is a solution of a primitive $LFSR^k(q)$, and if $A^* = 0$, T'_{t+1} can never be expressed as a linear function of T'_t.

5.3. Statistical properties

In order to obtain the best statistical properties for the solution (s_t) of a $CLFSR^{k,l}(q)$, we choose (x_t) as a solution of a primitive $LFSR^k(q)$ which has period $\tau = q^k - 1$.

For coupled registers, it can be proved that the period of any non-degenerated solution (s_t) is divisible by τ. The maximal period is then $(q^k - 1)(q^l - 1)$ which can be reached only if $(q-1) \mid l$.

There exist sufficient conditions to assure this maximal period, but they are not easy to verify. In practical applications however, $q = 2$ and we can choose k and l such that $2^k - 1$ and $2^l - 1$ are Mersenne primes. In this case, most solutions are maximal.

In a maximal solution, the distribution of multigrams $[s_t, \ldots s_{t+\mu}]$ satisfies:

- if $\mu < l$, the null multigram occurs $(q^{l-\mu-1}-1)(q^k-1)$ times
        and the other ones   occur $q^{l-\mu-1} (q^k-1)$ times.

- if $\mu = l$, there exist

$(q^l - q^{l-r})$ multigrams occurring	$(q^{k-1}-1)$ times
$(q-1)(q^l - q^{l-r})$	q^{k-1}
$(q^{l-r}-1)$	(q^k-1)
$(q-1)(q^{l-r}-1)+q$	0

where $0 \leq r \leq l-1$ is the rank of A^*. Bigger is r, more uniform is the multigrams distribution.

Moreover, the X^2 of the cross distribution of s_t and $s_{t+\mu}$ is

$$X^2(\beta) = \frac{(q-1)(q^l-1)}{(q^k-1)(q^l-q)^2} \{(q-1)(q^k-1)^2 - 2 \beta (q^k-1)q^{l-2}(q^l-1)$$
$$+ \beta^2 [((q^l-1)-(q-2))^2+(q-2)]\}$$

$$(35)$$

where β eventually depends on the rank of $M = [\overline{b}^*|A^*]$.

If $\mu < 1$, then $\beta = 0$ and $X^2(0)$ has the same constant value as in a classical primitive $LFSR^l(q)$.

If $\mu = 1$,
 if rank(M) = r+1, then $\beta = 0$ as in the previous case;
 if rank(M) = r, but $\overline{b}^* \neq 0$, then $\beta = q^{k-r}$;
 if $\overline{b}^* = 0$, then $\beta = (q^{k-r}-1)$.

In the binary case, Eq.35 corresponds to the auto-correlation

$$P(\beta) = \frac{-1}{2^l-2} [1 - \beta \frac{2^l-1}{2^k-1}] \qquad (36)$$

Thus, the classical Golomb's theorems are locally satisfied for maximal solutions of a CLFSR.

6. Conclusion

Coupled Linear Feedback Shift Registers are simple designs involving LFSR producing non-linear pseudo-random sequences. They seem to be cryptographically strong enough for stream cipher systems, because the behaviour of one component is not independent from the other one. Nevertheless, a CLFSR can easy be implemented as a piece of hardware or software in a very efficient way.

7. Bibliography

GOLOMB, S.W., *Shift Register Sequences*, Holden-Day, San Fransisco, Calif., 1967.

MEYER, C.H. & TUCHMAN, W.L., "Pseudorandom Codes can be Cracked", *Elec. Des*, 23, Nov. 1972, p74-76.

REEDS, J.A., "«Cracking» a Random Number Generator", *Cryptologia*, 1, Jan. 1977, p20-26.

ROGGEMAN, Y., "Remarks on the Auto-Correlation Function of Binary Periodic Sequences", *Cryptologia*, 10 (2), April 1986, p96-100.

ROGGEMAN, Y., *Quelques classes de registres à décalage et leurs applications en cryptographie*, Ph. D. Thesis, Univ. Libre de Bruxelles, 1986.

RUEPPEL, R.A., *Analysis and Design of Stream Ciphers*, Springer Verlag, Berlin, 1986.

A Cryptanalysis of Step$_{k,m}$-Cascades

Dieter Gollmann[1] William G.Chambers[2]

[1] Fakultät für Informatik, Universität Karlsruhe
D-7500 Karlsruhe, Germany

[2] Department of Electronic and Electrical Engineering
King's College London
London WC2R 2LS, United Kingdom

We examine cascades of clock-controlled shift registers where registers are clocked by more general schemes than simply "stop-and-go". In particular, we consider the relation between the stepping function and the number of keys of such a cascade.

1   Introduction

The history of stop-and-go generators can be traced back to mechanical devices where a rotor is stepped if and only if a pin in a controlling wheel has been set. In an electronic device clock-control can be easily generalized to arbitrary stepping functions and it seems plausible that the security of a generator can be improved by choosing a stepping function other than stop-and-go. Indeed, several attacks on "clock-controlled" rotor machines rely on the fact that rotors do not step on input 0. We extend results for "stop-and-go"-registers to registers with more general stepping functions, in particular, we examine the existence of equivalent states in such cascades. Clock controlled shift registers of length 3 will demonstrate the influence of the stepping function on the cryptographic merits of the resulting cascade.

2 Cascades of clock-controlled shift registers

We consider cascades of clock-controlled cyclic shift registers over $GF(2)$ (see Fig.1). A $step_{k,m}$-register steps k times on input 0 and m times on input 1. Stop-and-go is thus $step_{0,1}$. The output of a register and the input to its clock are added modulo 2 to give the input to the clock of the next register (or the output of the cascade respectively). No register shall be loaded with $0\ldots0$ or $1\ldots1$. Properties of $step_{0,1}$-cascades have been reported e.g. in [2,3] and are stated in the following theorem. $prob_n(w)$ will denote the probability to observe a binary word w in the output sequence generated by a cascade of n registers when the input to the cascade is constantly 1.

Theorem 1 *$Step_{0,1}$-cascades of n clock-controlled shift registers of length p, $p \geq 3$ prime, generate sequences with*

- *period p^n*
- *linear complexity greater or equal $d\frac{p^n-1}{p-1}$ where d is the degree of the irreducible polynomials with period p and $p^2 \nmid 2^{p-1} - 1$*
- *$\lim_{n\to\infty} prob_n(w) = \dfrac{1}{2^{|w|}}$*
- *different output sequences for all $(2^p - 2)^n$ legal initial states.*

$Step_{0,1}$-cascades are invertible and the inverse cascade can be synchronized [2]. If the initial states of all registers are known but for their rotations these can be deduced with high probability by feeding the output sequence to the inverse cascade. The security of the cascade should thus not be related to $(2^p - 2)^n$, the number of legal initial states, but to $((2^p - 2)/p)^n$, the number of equivalence classes modulo rotation of registers.

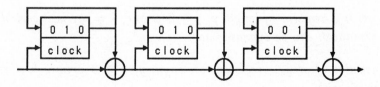

Figure 1: A cascade of clock-controlled shift registers of length 3

3   Properties of step$_{k,m}$-cascades

Most of the properties of step$_{0,1}$-cascades also hold for step$_{k,m}$-cascades. We first state a simple but fundamental lemma on the period of the output sequences.

Lemma 1 *A cascade of n step$_{k,m}$-registers of length p, $p \geq 3$ prime, $k \neq m$, generates sequences with period p^n.*

Proof. One may show by induction that sequences generated by a cascade of length n have period p^n and contain a number of 1's which is coprime to p. More details can be taken from the proof for step$_{0,1}$-cascades given in [3]. □

Given the period of the sequences we can state a lower bound for their linear complexity.

Lemma 2 *A cascade of n step$_{k,m}$-registers of length p, $p \geq 3$ prime, $p^2 \nmid 2^{p-1} - 1$, $k \neq m$, generates sequences with linear complexity greater or equal $d(p^n - 1)/(p - 1)$ where d is defined as in Theorem 1.*

Proof. The last register in the cascade contributes at least one polynomial $f(x^{p^{n-1}})$ to the generating polynomial of some given output sequence where $f(x)$ is an irreducible polynomial of period p. With [1], Theorem 6.23, we find that $f(x^{p^{n-1}})$ decomposes into irreducible factors of period p^n. [1], Theorem 6.52 and $p^2 \nmid 2^{p-1} - 1$ imply that $f(x^{p^{n-1}})$ is irreducible and has degree $d \cdot p^{n-1}$. Again, arguments from [4] can be used to construct a detailed proof. □

The results from [2] also can be extended to step$_{k,m}$-cascades. The impact on the security has already been stated above for step$_{0,1}$-cascades.

Lemma 3 *A step$_{k,m}$-register of length p, $p \geq 3$ prime, $k \neq m$, is invertible. The inverse automaton can be synchronized.*

Lemma 4 *Cascades of step$_{k,m}$-registers of length p, $p \geq 3$ prime, $k \neq m$, generate sequences with*
$$\lim_{n \to \infty} \text{prob}_n(w) = \frac{1}{2^{|w|}}.$$

Proof. We adapt the corresponding proof from [3]. We have to prove that for any initial state $\underline{q}$ of a cascade of length n, any input $\underline{x}$ and output $\underline{y}$ of length n there exists a state $\underline{q}'$ so that

- there exists an input sequence that sends the cascade from $\underline{q}$ to $\underline{q'}$

- input $\underline{x}$ applied to the cascade in state $\underline{q'}$ generates output $\underline{y}$.

Assume that this proposition holds up to some given n. Prefix $\underline{x}$ and $\underline{y}$ with new bits x_0 and y_0 respectively and consider a cascade of length $n + 1$ with initial state $(\underline{q}, q_{n+1})$. Without loss of generality we choose q_{n+1}, the state of the last register, so that if the last register is turned back k steps its output is y_0 and if it is turned back m steps its output is $\bar{y}_0$.

As step$_{k,m}$-registers are invertible, there exists a word $\underline{z}$ of length n so that $\underline{x}$ generates the internal signal $\underline{z}$ after n stages of the cascade and output $\underline{y}$. Furthermore, there exists a state $\underline{q_0}$ so that x_0 sends the first n registers of the cascade from $\underline{q_0}$ to $\underline{q}$, generating some output bit c. If $c = 0$ we start the last register with q_{n+1} turned back k steps, i.e. with y_0 in the output position, thus the overall output is also y_0. If $c = 1$ we start the last register with q_{n+1} turned back m steps, i.e. with $\bar{y}_0$ in the output position, thus the overall output is $1 \oplus \bar{y}_0 = y_0$.

As a next step, for any $s \geq 1$ and any cascade of length n, consider $2^s \times 2^s$-matrices where the entry in position x, y , $1 \leq x, y \leq 2^s$ gives the number of states where input $\underline{x}$ produces output $\underline{y}$, $\underline{x}$ and $\underline{y}$ are the binary representations of x, y. Divide all entries by $1/p^n$. The matrix corresponding to a cascade is the product of the matrices of the individual stages, these matrices are contraction operators with the equidistribution of s-tupels as fixed point (see [3]). Thus the output distribution of s-tupels will converge to equidistribution as the length of the cascade goes to infinity. □

The rate of convergence that follows from the above proof is in general much slower than the rate observed in practice. However, in the particular case of step$_{1,2}$-cascades of registers of length 3 the output sequences can be proven to have almost ideal statistical properties. The distribution of s-tupels in the output is nearly flat for all $s \geq 1$. Words of the same length appear with probabilities differing at most by $1/3^n$. We have

Remark 1 *Step$_{1,2}$-cascades of n clock-controlled shift registers of length 3 generate sequences with*

$$\frac{1}{3^n} \left\lceil \frac{3^n}{2^{|w|}} \right\rceil \leq \text{prob}_n(w) \leq \frac{1}{3^n} \left\lfloor \frac{3^n}{2^{|w|}} \right\rfloor , \text{ for all } w \in \{0,1\}^*.$$

Proof. Consider a register with initial state 001. We will see later (Lemma 6) that this is no undue restriction. Let $x_1 x_2$ denote an input string and $y_1 y_2$ the corresponding output. Map the states $\{100, 001, 010\}$ to $\{0, 1, 2\}$

as follows, $100 \to 0, 001 \to 1, 010 \to 2$. Let $q(0)$ be the initial state of the register and $q(i)$ the state after processing $x_1 \ldots x_i$, $i \geq 1$. Tabulate $x_1 \bar{x}_2$ and $q(2)$ in dependence of $y_1 \bar{y}_2$ and $q(0)$.

<table>
<tr><td colspan="4" align="center">$q(2)$</td><td></td><td colspan="4" align="center">$x_1 \bar{x}_2$</td></tr>
<tr><td>$y_1 \bar{y}_2$</td><td>$q(0) = 0$</td><td>1</td><td>2</td><td></td><td>$y_1 \bar{y}_2$</td><td>$q(0) = 0$</td><td>1</td><td>2</td></tr>
<tr><td>00</td><td>1</td><td>1</td><td>1</td><td></td><td>00</td><td>00</td><td>10</td><td>01</td></tr>
<tr><td>01</td><td>0</td><td>0</td><td>2</td><td></td><td>01</td><td>01</td><td>11</td><td>00</td></tr>
<tr><td>10</td><td>1</td><td>0</td><td>0</td><td></td><td>10</td><td>11</td><td>00</td><td>10</td></tr>
<tr><td>11</td><td>2</td><td>2</td><td>2</td><td></td><td>11</td><td>10</td><td>01</td><td>11</td></tr>
</table>

On inspection we see that the above tables define the computation

$$\tilde{x} = 3 \cdot \tilde{y} + q(0) \quad (\bmod 4)$$
$$\tilde{y} = \lfloor (3 \cdot \tilde{y} + q(0))/4 \rfloor$$

where $\tilde{x} = 2\bar{x}_2 + x_1$, $\tilde{y} = 2\bar{y}_2 + y_1$. The final state $q(2)$ serves as a carry that is handed on to the next inputs and outputs of length 2, say $x_3 x_4$ and $y_3 y_4$. Repeating the above argument we get for $x_1 \ldots x_m$, $y_1 \ldots y_m$, m even

$$\tilde{x} \equiv 3 \cdot \tilde{y} + q(0) \quad (\bmod 2^m)$$
$$\tilde{y} \equiv \lfloor (3 \cdot \tilde{y} + q(0))/2^m \rfloor$$

with

$$\tilde{x} = \bar{x}_m 2^m + x_{m-1} 2^{m-1} + \cdots + \bar{x}_2 2 + x_1 ,$$
$$\tilde{y} = \bar{y}_m 2^m + y_{m-1} 2^{m-1} + \cdots + \bar{y}_2 2 + y_1 .$$

For odd m we find the same relation between inputs, outputs, and the states of the register. However, in this case we have to invert the odd bits in input and output. Now consider a cascade of length n. We get

$$\tilde{x} \equiv 3^n \cdot \tilde{y} + \tilde{q} \quad (\bmod 2^m)$$

with

$$\tilde{q} = \sum_{j=1}^{n} 3^{j-1} q_j(0)$$

where $q_j(0)$ is the initial state of the jth register. As the cascade has period 3^n, $\tilde{q}$ will take on all values in $[0, 3^n - 1]$ exactly once. Hence

$$\left\lceil \frac{3^n}{2^{|w|}} \right\rceil \leq \tilde{q} \ (\bmod 2^m) \leq \left\lfloor \frac{3^n}{2^{|w|}} \right\rfloor$$

and the same holds for the outputs $\tilde{y}$ if the input $\tilde{x}$ is fixed. $\qquad \square$

4   Equivalent states in $step_{k,m}$-cascades

Two states of a finite automaton are called equivalent when, for any input, both will produce the same output. A finite automaton is called minimal if equivalence implies equality. (For more details see e.g.[5]).

We consider $step_{k,m}$-cascades with constant input 1. In the context of cryptographic applications, equivalent states are seeds that produce the same pseudo random sequence. The set of keys is thus not the set of seeds but the set of equivalence classes. Different seeds generate different sequences if and only if the cascade is minimal. In a first step we examine the structure of internal signals in equivalent states of a $step_{k,m}$-cascade.

Lemma 5 *If two two states $\underline{q}$, $\underline{q}'$ of a $step_{k,m}$-cascade are equivalent then corresponding internal signals are either identical or bitwise complemented.*

Proof. More generally, consider two states $\underline{q}$, $\underline{q}'$ of a cascade of length n producing outputs that are either identical or bitwise complemented and assume that the proposition does not hold for the input to the last stage. Without loss of generality assume that there exists a time frame $(i, i+1)$, where inputs 01 and 11 respectively are fed to the last stage. We know from Lemma 1 that the cascade has period p^n. If we observe the cascade at instances $(i + \lambda p^{n-1}, i + 1 + \lambda p^{n-1})$, $0 \le \lambda < p$, then the inputs to the last stage will be repeated but the last register will be in a different rotation every time. If $\underline{q}$ and $\underline{q}'$ produce the same output then the observations at times $i + \lambda p^{n-1}$ imply that the number of 1's in the last stage of $\underline{q}$ is the complement (mod p) of the number of 1's in the last stage of $\underline{q}'$. From the observations at times $i + 1 + \lambda p^{n-1}$ we find that both numbers should be the same. This is impossible as p is odd. We get the same contradiction when the outputs are complemented. By induction, the lemma can be shown to hold for all internal signals.   □

Lemma 6 *A $step_{k,m}$-cascade is minimal if and only if $k \ne p - m$.*

Proof. Consider two equivalent states $\underline{q}$, $\underline{q}'$ of a cascade of length $n > 1$. If all internal signals were the same then obviously the states would be the same. So we may assume that the inputs to the last stage are bitwise complemented.

Let $(q_n(1), \ldots, q_n(p))$ and $(q_n'(1), \ldots, q_n'(p))$ be the respective states of the last stages. We know that there must be times i, $i + j$, $p \nmid j$, so that the cascade starting in $\underline{q}$ is in the same position, otherwise the last register would rotate with period p. Denote the number of 1's in the interval $[i, i + j - 1]$ by t. Without loss of generality assume that the output of the last stage is

$q_n(p)$ and that the outputs of the last stage for the initial state $\underline{q}'$ are $q'_n(r)$ and $q'_n(r')$. Let σ_n denote the number of 1's in the input to the last stage. Observing the cascade at instances $(i + \lambda p^{n-1}, i + j + \lambda p^{n-1})$, $0 \leq \lambda < p$, we get

$$q'_n(r - \lambda \sigma_n(k - m)) = \bar{q}_n(p - \lambda \sigma_n(m - k)) = q'_n(r' - \lambda \sigma_n(k - m)) \; ,$$

thus the shifts of $(q'_n(1), \ldots, q'_n(p))$ corresponding to r and r' are equal. Because p is prime we have $r = r'$. This implies

$$\begin{aligned} t \cdot m + (j - t) \cdot k \equiv 0 \quad (\text{mod } p) \\ t \cdot k + (j - t) \cdot m \equiv 0 \quad (\text{mod } p) \end{aligned}$$

hence

$$(m + k) \cdot j \equiv 0 \quad (\text{mod } p)$$

and finally

$$m + k \equiv 0 \quad (\text{mod } p) \; .$$

$\square$

5 Conclusion

We collect our results in

Theorem 2 *A* $step_{k,m}$*-register of length* p, $p \geq 3$ *prime,* $k \neq m$, *is invertible. The inverse automaton can be synchronized. A cascade of* n $step_{k,m}$*-registers of length* p, $p \geq 3$ *prime,* $k \neq m$, *generates sequences with*

- *period* p^n
- *linear complexity* $\geq d\frac{p^n - 1}{p - 1}$ *where* d *is defined as in Theorem 1 and* $p^2 \nmid 2^{p-1} - 1$
- $\lim\limits_{n \to \infty} \text{prob}_n(w) = \dfrac{1}{2^{|w|}}$
- *different output sequences for all* $(2^p - 2)^n$ *legal initial states if and only if* $k \neq -m$.

The minimality of $step_{k,m}$-cascades depends only on the choice of the stepping function. For $step_{k,-k}$ the number of useful initial states is reduced to $((2^p - 2)/2p)^n$ as we may replace registers which contain more than $p/2$ 1's by registers with less than $p/2$ 1's. This will invert the internal signal after the stage where this replacement had taken place but there also exits a modification for the state of the next stage so that we can recover the original signal after that next stage.

It is interesting to note that sequences from $step_{1,2}$-cascades of registers of length 3 are at the same time totally insecure and almost perfect with respect to standard criteria like linear complexity or statistical distribution.

References

[1] E.Berlekamp, *Algebraic Coding Theory*, McGraw-Hill, New York, 1968

[2] W.G.Chambers, D.Gollmann, *Lock-in Effect in Cascades of Clock-Controlled Shift-Registers*, Proc.Eurocrypt88, Springer LNCS 330, pp.331-342, 1988

[3] D.Gollmann, *Pseudo Random Properties of Cascade Connections of Clock Controlled Shift Registers*, Proc.Eurocrypt84, Springer LNCS 209, pp.93-98, 1985

[4] D.Gollmann, *Linear Recursions of Cascaded Sequences*, Contributions to General Algebra 3, Hölder-Pichler-Tempsky, Wien, Teubner, Stuttgart, 1985

[5] Z.Kohavi, *Switching and Finite Automata Theory*, McGraw-Hill, New York, 1970

Efficient Identification and Signatures
for Smart Cards[1]

C.P. Schnorr

Universität Frankfurt

Abstract[2]

We present an efficient interactive identification scheme and a related signature scheme that are based on discrete logarithms and which are particularly suited for smart cards. Previous cryptosystems, based on the discrete logarithm, have been proposed by El Gamal (1985), Chaum, Evertse, van de Graaf (1988), Beth (1988) and Günther (1989). The new scheme comprises the following novel features.

1. We propose an efficient algorithm to preprocess the exponentiation of random numbers. This preprocessing makes signature generation very fast. It also improves the efficiency of the other discrete log-cryptosystems. The preprocessing algorithm is based on two fundamental principles *local randomization* and *internal randomization*.

2. We use a prime modulus p such that $p-1$ has a prime factor q of appropriate size (e.g. 140 bits long) and we use a base α for the discrete logarithm such that $\alpha^q = 1 (\mathrm{mod} p)$. All logarithms are calculated modulo q. The length of the signatures is about 212 bits, i.e. it is less than half the length of RSA and Fiat-Shamir signatures. The number of communication bits of the identification scheme is less than half that of other schemes.

---

[1]European patent application 89103290.6 from 24.2.1989.

[2]Extended abstract: C.P. Schnorr, "Efficient Identification and Signatures for Smart Cards", *Advances in Cryptology: Proceedings of CRYPTO '89 (Lecture Notes in Computer Science; 435)*, G. Brassard, Ed., Springer Verlag, 1990, pp. 239–252.

The new scheme minimizes the work to be done by the smart card for generating a signature or for proving its identity. This is important since the power of current processors for smart cards is rather limited. Previous signature schemes require many modular multiplications for signature generation. In the new scheme signature generation costs about 12 modular multiplications, and these multiplications do not depend on the message/identification, i.e. they can be done in preprocessing mode during idle time of the processor.

The security of the scheme relies on the one-way property of the exponentiation $y \longrightarrow \alpha^y (\bmod p)$, i.e. we assume that discrete logarithms with base α are difficult to compute. The security of the preprocessing is established by information theoretic arguments.

THE DINING CRYPTOGRAPHERS IN THE DISCO:

UNCONDITIONAL SENDER AND RECIPIENT UNTRACEABILITY WITH COMPUTATIONALLY SECURE SERVICEABILITY

Michael Waidner Birgit Pfitzmann

Institut für Rechnerentwurf und Fehlertoleranz, Universität Karlsruhe
Postfach 6980, D-7500 Karlsruhe 1, F.R. Germany

Abstract

In Journal of Cryptology 1/1 (1988) 65–75, CHAUM describes a beautiful technique, the *DC-net*, which should allow participants to send and receive messages anonymously in an arbitrary network. The untraceability of the senders is proved to be unconditional, but that of the recipients implicitly assumes a *reliable* broadcast network. This assumption is unrealistic in some networks, but it can be removed completely by using the fail-stop key generation schemes by WAIDNER (these proceedings). In both cases, however, each participant can untraceably and permanently disrupt the entire DC-net.

We present a protocol which guarantees *unconditional untraceability*, the original goal of the DC-net, on the *inseparability assumption* (i.e. the attacker must be unable to prevent honest participants from communicating, which is considerably less than reliable broadcast), and *computationally secure serviceability*: Computationally restricted disrupters can be identified and removed from the DC-net.

On the one hand, our solution is based on the lovely idea by CHAUM of setting traps for disrupters. He suggests a scheme to guarantee unconditional untraceability and computationally secure serviceability, too, but on the reliable broadcast assumption. The same scheme seems to be used by BOS and DEN BOER (these proceedings). We show that this scheme needs some changes and refinements before being secure, even on the reliable broadcast assumption.

On the other hand, our solution is based on the idea of *digital signatures whose forgery by an unexpectedly powerful attacker is provable*, which might be of independent interest. We propose such a (one-time) signature scheme based on claw-free permutation pairs; the forgery of signatures is equivalent to finding claws, thus in a special case to the factoring problem. In particular, with such signatures we can, for the first time, realize *fail-stop Byzantine Agreement*, and also *adaptive Byzantine Agreement*, i.e. Byzantine Agreement which can only be disrupted by an attacker who controls at least a third of all participants *and* who can forge signatures.

We also sketch applications of these signatures to a payment system, solving disputes about shared secrets, and signatures which cannot be shown round.

SOME CONDITIONS ON
THE LINEAR COMPLEXITY PROFILES
OF CERTAIN BINARY SEQUENCES

Glyn Carter

Racal Comsec Ltd.
Milford Industrial Estate
Tollgate Road
Salisbury
Wiltshire, SP1 2JG
ENGLAND

ABSTRACT

In this paper we consider the binary sequences
whose bits satisfy any set of linear equations from
a wide class of sets, of which the equations in the
perfect profile characterization theorem are
typical. We show that the linear complexity
profile any such sequence will be restricted in the
sense that it will have no jumps of a certain
parity above a certain height.

1. INTRODUCTION

The *linear complexity* of a binary sequence is the length of the
shortest linear feedback shift register (LFSR) on which the
sequence can be generated. There are two forms of linear
complexity; *global linear complexity*, which applies to infinite
periodic binary sequences, and *local linear complexity*, which
applies to binary sequences of finite length. In this paper we
will be interested in the latter.

Consider an n-bit sequence $s_0 s_1 \ldots s_{n-1}$. The local linear
complexity $L(n)$ of $s_0 s_1 \ldots s_{n-1}$ and the connection polynomial

$C_n(x)$ of an L(n)-stage LFSR on which the sequence can be
generated can be computed using the Berlekamp-Massey algorithm
[3]. The algorithm also yields, in the course of computing L(n),
the local linear complexities L(1), L(2), ..., L(n-1) of the
subsequences s_0, $s_0 s_1$, ..., $s_0 s_1 \ldots s_{n-2}$ of $s_0 s_1 \ldots s_{n-1}$. We call
the n-vector (L(1),L(2),...,L(n-1),L(n)) the *linear complexity
profile* of $s_0 s_1 \ldots s_{n-1}$, and we say that the profile *jumps* with
s_{k-1} if L(k)-L(k-1) > 0. L(k)-L(k-1) is known as the *height* of
the jump. From the Berlekamp-Massey algorithm it can be seen
that a profile can only jump with s_{k-1} if 2.L(k-1) ⩽ k-1, and
that if it does jump then L(k) must be equal to k-L(k-1).

A sequence is said to have the *perfect linear complexity profile*
if all the jumps in its linear complexity profile have height 1.
In 1984 Rueppel conjectured in his thesis [5] that the sequence
110100010000000100... (i.e. the binary sequence such that s_i = 1
if and only if i = 2^r-1 for some integer r ⩾ 0) has the perfect
linear complexity profile. This conjecture was later proved to
be true by Dai [2]. In 1986 Wang and Massey extended this result
by characterizing the set of binary sequences having the perfect
linear complexity profile; in [6] they showed that an n-bit
sequence $s_0 s_1 \ldots s_{n-1}$ has the perfect profile if and only if
s_0 = 1 and $s_{2i} = s_{2i-1} + s_{i-1}$ for 1 ⩽ i ⩽ $\frac{n-1}{2}$. We will refer
to this result as the *perfect profile characterization theorem*.

In this paper we will consider the binary sequences whose bits
satisfy any set of linear equations from a wide class of sets, of
which the equations in the perfect profile characterization
theorem are typical. We will show that the linear complexity
profile any such sequence will be constrained in some way.

2. MAIN RESULTS

We now move on to the main results of this paper. In the
sequences in the perfect profile characterization theorem, every
other bit is the sum of the preceding bit and a bit approximately
"half way back". The results in this section involve sequences
in which, roughly speaking, every other bit is the sum of a
number of the preceding few bits and a number of bits

approximately "half way back". We can show that the linear complexity profile of any sequence of this type is restricted in the sense that it can have no jumps of a certain parity above a certain height. The proofs of these results are rather long, and unfortunately there is no room to include them in this paper. The interested reader is referred to [1].

We deal with the sequences in two groups, according to whether their "fixed" bits (i.e. the ones which can be expressed as a sum of previous bits) are the ones with odd or even indices. The results in the two cases are very similar; we separate them for clarity only. We begin by considering the sequences whose fixed bits have odd indices :-

Theorem 2.1.

Let $s_0 s_1 \ldots s_{n-1}$ be an n-bit sequence with

$$
\begin{aligned}
s_{2i+1-2w} = {} & s_{2i+1-2x(1)} + s_{2i+1-2x(2)} + \ldots + s_{2i+1-2x(a)} \\
& + s_{2i-2y(1)} + s_{2i-2y(2)} + \ldots + s_{2i-2y(b)} \\
& + s_{i-z(1)} + s_{i-z(2)} + \ldots + s_{i-z(c)}
\end{aligned}
$$

for $\min(w, z(1)) \leqslant i \leqslant \min(\frac{n}{2}+w-1, n+z(1)-1)$,

where $s_\ell := 0$ for $\ell < 0$

$(w < x(1) < x(2) < \ldots < x(a), \quad w \leqslant y(1) < y(2) < \ldots < y(b),$
$\qquad z(1) < z(2) < \ldots < z(c), \quad a \geqslant 0, \ b \geqslant 0, \ c > 0)$.

Then the height j of any jump in the linear complexity profile of $s_0 s_1 \ldots s_{n-1}$ must satisfy either (i) or (ii) below :-

    (i)   j odd

    (ii) $j \leqslant \max(2z(c)-2w, 2y(b)-2w+1, 2x(a)-2w)$

Proof

See [1].

We now deal with the sequences whose fixed bits have even indices :-

Theorem 2.2.

Let $s_0 s_1 \ldots s_{n-1}$ be an n-bit sequence with

$$s_{2i-2w} = s_{2i+1-2x(1)} + s_{2i+1-2x(2)} + \ldots + s_{2i+1-2x(a)}$$
$$+ s_{2i-2y(1)} + s_{2i-2y(2)} + \ldots + s_{2i-2y(b)}$$
$$+ s_{i-z(1)} + s_{i-z(2)} + \ldots + s_{i-z(c)}$$

$$\text{for } \min(w,z(1)) \leqslant i \leqslant \min(\tfrac{n-1}{2}+w, n+z(1)-1),$$

where $s_\ell := 0$ for $\ell < 0$

$$(w < x(1) < x(2) < \ldots < x(a), \quad w < y(1) < y(2) < \ldots < y(b),$$
$$z(1) < z(2) < \ldots < z(c), \quad a \geqslant 0, \ b \geqslant 0, \ c > 0).$$

Then the height j of any jump in the linear complexity profile of $s_0 s_1 \ldots s_{n-1}$ must satisfy either (i) or (ii) below :-

    (i) j even
    (ii) $j \leqslant \max(2z(c)-2w-1, 2y(b)-2w, 2x(a)-2w-1)$

Proof

See [1].

$\square$

As an example of how these results can be applied to a sequence whose bits satisfy a particular set of linear equations, consider an n-bit sequence $s_0 s_1 \ldots s_{n-1}$ whose bits satisfy the following equations :-

$$s_1 = 0$$
$$s_3 = s_2 + s_1$$
$$s_{2i+1} = s_{2i} + s_{2i-2} + s_i + s_{i-1} \quad \text{for } 2 \leqslant i \leqslant \tfrac{n}{2}-1$$

By Theorem 2.1, the linear complexity profile of $s_0 s_1 \ldots s_{n-1}$ can have no jumps of even height greater than 2.

The above theory yields a large class of sequences in which non-randomness in the sequences is reflected in their linear complexity profiles. Empirical tests suggest that, in many cases, such non-randomness would not be identified by established statistical tests; it would in most cases, however, be identified by statistical tests based on linear complexity profiles (see [1] and [4]).

REFERENCES

[1] **Carter, G.D.**, *'Aspects of Local Linear Complexity'*, Ph.D. Thesis, University of London, (1988).

[2] **Dai, Z.D.**, 'Proof of Rueppel's linear complexity conjecture'. To appear.

[3] **Massey, J.L.**, 'Shift register synthesis and BCH decoding', *IEEE Transactions on Information Theory*, **IT-15**, (1969), pp 122-127.

[4] **Niederreiter, H.**, 'The probabilistic theory of linear complexity', *Advances in Cryptology: Proceedings of Eurocrypt 88*, Springer-Verlag, Berlin, (1988), pp 191-209.

[5] **Rueppel, R.A.**, *'New Approaches to Stream Ciphers'*, D.Sc. Dissertation, Swiss Federal Institute of Technology, Zurich, (1984).

[6] **Wang, M.Z. and Massey, J.L.**, 'The characterization of all binary sequences with perfect linear complexity profiles'. Presented at Eurocrypt 86.

ON THE DESIGN OF PERMUTATION P IN DES TYPE CRYPTOSYSTEMS

Lawrence Brown   *Jennifer Seberry*

Department of Computer Science
University College, UNSW, Australian Defence Force Academy
Canberra ACT 2600. Australia

Abstract

This paper reviews some possible design criteria for the permutation P in a DES style cryptosystem. These permutations provide the diffusion component in a substitution-permutation network. Some empirical rules which seem to account for the derivation of the permutation used in the DES are first presented. Then it is noted that these permutations may be regarded as latin-squares which link the outputs of S-boxes to their inputs at the next stage. A subset of these with a regular structure, and which perform well in a dependency analysis are then presented. Some design rules are then derived, and it is suggested these be used to design permutations in future schemes for an extended version of the DES.

1. Introduction

The Data Encryption Standard (DES) [NBS77] is currently the only certified encryption standard. It has achieved wide utilization, particularly in the banking and electronic funds transfer areas, and is an Australian standard [ASA85] among others.

However at the time of its introduction, there was a considerable amount of controversy due both to the design criteria being classified, and to the choice of a 56-bit key that was considered too small [DiHe77], [Hell79]. However, on all other accounts DES appears to be an excellent cryptosystem. No short-cuts have been found to aid in the cryptanalysis of DES other than by exhaustive key-space search. With the current significant use of DES (especially in banking), there is interest in designing and building a DES-type cryptosystem with an extended key length.

In Brown [Brow88], the design criteria used in the DES, both those reported in the literature, and those noted by the author, were discussed. This paper further considers the design of the permutation box P, which provides the diffusion component of a **substitution-permutation** network in DES type cryptosystems. Such systems were original devised by Shannon [Shan49], who termed this arrangement a **mixing transformation**. The S-boxes provided **confusion** of the input, and the P-boxes provided **diffusion** of the S-box outputs across the inputs to the next stage. Two key properties of such S-P networks are the **avalanche** property, identified by Feistel [Feis73]; and the **completeness** property, identified by Kam and Davida [KaDa79]. They ensure that every output bit becomes a function of each input bit in as few rounds as possible. Meyer [Meye78] (also in [MeMa82]) has quantified this property for the DES, by showing that after 5 rounds, every output bit is a function of all input bits. We have used this form of analysis as a measure of effectiveness in the design of permutation P. It is intended that as a result of this work, the design of the permutation P in an extended DES style scheme may be performed on a sounder theoretical basis.

2. Empirical P-box Design Criteria

The central component of the DES cryptosystem is the function g. This is implemented as a composition of an expansion function E which provides an autoclave function, eight substitution boxes (S-boxes) S which **confuse** the input bits, and then a permutation P which **diffuses** the outputs from each S-box to the inputs of a number of S-boxes in the next stage. A more detailed description of these functions may be found in [NBS77], [ASA85] or [SePi88]. For the purposes of this analysis, it is convenient to regard the DES as a mixing function, as detailed in Davio et al. [DDFG83], which emphasizes the analysis of the functional composition $P.S.E$ (see Fig 1.).

In Brown [Brow88], the following analysis of, and design rules for, permutation P were derived. This was done by analysing the $S.P.E$ functional composition, which forms one round of the S-P network in the DES:

$$R(i) = L(i-1) \oplus P(S(E(R(i-1)) \oplus K(i))).$$

The input of permutation P may be divided into 4-bit outputs from the eight S-boxes, and the output from E is divided into 6-bit inputs to the eight S-boxes at the next stage. Instead of expressing this permutation in terms of bit positions, it may be written in terms of which S-box output is connected by $P.E$ to each input bit of the S-boxes at the next stage (see Table 1). Provided the S-boxes fulfil their design requirements, it may be assumed that all S-box outputs are equivalent, and that the inputs may be considered as three pairs ab, cd and ef. Note that due to the expansion function E, columns ab are identical to columns ef for the previous S-box.

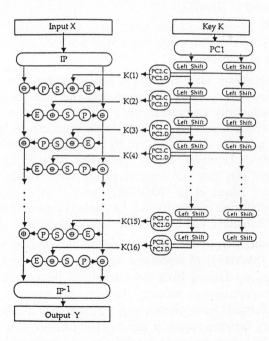

Fig 1. DES as a Mixing Function

Table 1 - P.E Permutation							
S-box	Inputs from S-boxes						Excluded S-box
	a	b	c	d	e	f	
1	7	4	2	5	6	8	3
2	6	8	3	7	5	1	4
3	5	1	4	6	7	2	8
4	7	2	5	8	3	1	6
5	3	1	2	6	4	8	7
6	4	8	7	1	3	5	2
7	3	5	4	8	2	6	1
8	2	6	3	1	7	4	5

From this Brown [Brow88] derived the following set of empirical rules for designing permutation P:

R1   Each of the S-box input bits ab cd ef come from the outputs of different S-boxes.

R2   None of the input bits ab cd ef to a given S-box S(i) comes from the output of that same S-box S(i).

R3   An output from S(i-1) goes to one of the ef input bits of S(i), and hence via E an output from S(i-2) goes to one of the ab input bits.

R4   An output from S(i+1) goes to one of the cd input bits of S(i).

R5   For each S-box output, two bits go to ab or ef input bits, the other two go to cd input bits as noted in [Davi82].

These rules all appear consistent with the implementation of the avalanche and completeness effects by ensuring that every output bit becomes a function of each input bit in as few rounds as possible.

Permutations satisfying these rules have been generated. A total of 178 permutations were found, from 96 possible exclusion sets (the mandatory wirings required by rules 3 and 4 were arbitrarily assigned to bits c and e/a respectively). Each of these permutations could generate a number of possible permutations by swapping the order of bits in each of the pairs ab, cd, and ef; or by rearranging the order of the four output bits from each S-box. However, Brown [Brow88] believes these variations are not significant, and notes that they do not change the result of the dependency analysis. Davies [Davi82] and Davio et al. [DDFG83] have also observed that a functionly equivalent $S.P.E$ combination can be formed by rearranging the order of the S-box output bits (by rearranging columns), and by then altering permutation P to compensate.

To provide a measure of the effectiveness of the derived permutations, Meyer's analysis [Meye78], [MeMa82] of ciphertext dependence on plaintext bits was extended for the 178 permutations derived from these empirical rules. Briefly, following Meyer, this analysis may be described as follows. To provide a measure of this dependency, a $64*64$ array $G_{a,b}$ is formed. Each element $G_{a,b}(i,j)$ specifies a dependency of output bit $X(j)$ on input bit $X(i)$, between rounds a and b. The number of marked elements in $G_{0,r}$ indicates the degree to which complete dependence was achieved by round r. Details of the derivation of this matrix, and the means by which entries are propagated, may be found in [MeMa82]. The extended analysis [Brow88] for the current DES, the 178 empirically generated permutations, and the worst case P (see Table 2), gave results as shown in columns 2, 3, and 7 of Table 3.

Table 2 - Worst Case DES P
1 1 1 1 2 2 2 2 3 3 3 3 4 4 4 4 5 5 5 5 6 6 6 6 7 7 7 7 8 8 8 8

3. Further Analysis of the Design of Permutation P

On closer examination, it was noted that these empirically derived rules for the design of permutation P, when written as in Table 1, actually result in a latin square. A **latin square** is a square of numbers in which each number occurs exactly once in each row and each column (see [DeKe74]). Note that the permutation P used in the current DES is not a latin square, however it may be transformed into one by selectively swapping some $ablef$ (these being equivalent due to E) and cd pairs to align the highlighted values in Table 1 into the same columns. This transformation makes no difference to the dependency result obtained. Such a result cannot be accidental, and must be related to the desired properties of permutation P, namely to provide maximal diffusion among the S-boxes.

In order to explore this further, some possible permutations which could be used in a DES type system, and which form a latin-square when written as specified above, were generated. A sample space of 657096 such permutations were generated. These were analysed for effectiveness using Meyer's ciphertext dependence on plaintext, with results as summarised in column 4 of Table 3.

Table 3 - Dependency of Ciphertext on Plaintext Bits for various DES permutations P by Round						
Round	Current P	Empirical P	Sample Latin-Square P	Best Latin-Square P	Regular Form P	Worst P
1	6.25	6.25	6.25	6.25	6.25	6.25
2	32.06	32.03-32.10	30.49-32.20	32.02-32.23	30.47-32.23	21.09
3	73.49	73.44-73.58	70.36-73.78	73.44-73.83	70.31-73.83	43.75
4	96.90	96.87-96.95	95.34-97.05	96.88-97.07	95.31-97.07	68.75
5	100.00	100.00	100.00	100.00	100.00	89.06
6	100.00	100.00	100.00	100.00	100.00	98.44
7	100.00	100.00	100.00	100.00	100.00	100.00
8	100.00	100.00	100.00	100.00	100.00	100.00

The P-box used in the current DES has a propagation profile that falls fairly close to to the median of the 178 permutations generated by the empirical rules. It thus appears to be a fairly representative example of them. In turn, these also fall within the range found for the permutations derived from latin-squares, which obviously encompass the criteria needed to design them. In conjunction with the substantially inferior profile of the worst case P-box, this provides a strong indication that the design rules identified are comprehensive, at least as far as this aspect of the P-box design is concerned.

4. Analysis of the Best Latin-Square Permutations P

From the sample space of 657096 permutations, those permutations resulting in the highest percentage values in each round of the dependency analysis were extracted. A total of 20 such permutations were found. When examined, half of the columns (namely the $ablef$ columns) of these permutations were found to be identical, with values as given in Table 4. These columns provide inputs to two S-boxes due to expansion function E. Further, they were found to be nearly identical to the values generated by applying a difference function of

$$[-2 +1 \ c \ d \ -1 \ +2] \qquad c, d \ \text{arbitary} \tag{1}$$

to the input S-box number, as shown in full in Table 4.

Table 4 - Fixed Columns in Sample Permutations P
Replicated pattern in the best sample permutations
3 . . 7 8 . . 1 4 . . 5 2 . . 3 6 . . 4 7 . . 8 5 . . 6 1 . . 2
Pattern generated by a [-2 +1 c d -1 +2] difference function
2 . . 8 3 . . 1 4 . . 2 5 . . 3 6 . . 4 7 . . 5 8 . . 6 1 . . 7

From this, we suggested that permutations with a fixed *ablef* column structure whose values are those generated by difference function (1) may perform well. Permutations of this form were generated, a total of 264 being found. The results of the dependency analysis on these permutations is given in column 5 of Table 3. Those providing the best results in this analysis were extracted, 100 such permutations being found. These were found to be grouped into pairs, with only columns c and d interchanged. These pairs in turn were grouped by exclusion set (the set of values not used as inputs to each S-box). The only difference between them is the swapping of some of the cd bits, a transformation which makes no difference to the dependency results. A total of 18 exclusion sets were found among the best permutations, and a sample from each is given in Table 5.

Table 5 - Best Latin-Square Permutations P		
Sample Permutation	Exclusion Set	Number
2 5 6 8 3 6 7 1 4 7 8 2 5 8 1 3 6 1 2 4 7 2 3 5 8 3 4 6 1 4 5 7	4 5 6 7 8 1 2 3	2
2 5 6 8 3 7 5 1 4 6 8 2 5 1 7 3 6 8 2 4 7 2 1 5 8 3 4 6 1 4 3 7	4 6 7 8 1 3 2 5	4
2 5 6 8 3 7 5 1 4 6 7 2 5 1 8 3 6 8 2 4 7 2 1 5 8 3 4 6 1 4 3 7	4 6 8 7 1 3 2 5	8
2 5 6 8 3 7 5 1 4 6 7 2 5 1 8 3 6 8 1 4 7 2 3 5 8 4 2 6 1 3 4 7	4 6 8 7 2 1 3 5	8
2 5 6 8 3 6 5 1 4 7 8 2 5 1 7 3 6 8 2 4 7 3 1 5 8 2 4 6 1 4 3 7	4 7 6 8 1 2 3 5	4
2 5 6 8 3 6 5 1 4 7 8 2 5 1 7 3 6 8 1 4 7 2 3 5 8 4 2 6 1 3 4 7	4 7 6 8 2 1 3 5	8
2 4 6 8 3 7 5 1 4 6 8 2 5 8 7 3 6 1 2 4 7 2 1 5 8 3 4 6 1 5 3 7	5 6 7 1 8 3 2 4	4
2 4 6 8 3 7 5 1 4 6 8 2 5 1 7 3 6 8 1 4 7 2 3 5 8 3 2 6 1 5 4 7	5 6 7 8 2 1 4 3	4
2 4 6 8 3 7 5 1 4 6 7 2 5 1 8 3 6 8 1 4 7 3 2 5 8 2 4 6 1 5 3 7	5 6 8 7 2 1 3 4	4
2 4 6 8 3 6 5 1 4 7 8 2 5 8 7 3 6 1 2 4 7 3 1 5 8 2 4 6 1 5 3 7	5 7 6 1 8 2 3 4	4
2 4 6 8 3 6 5 1 4 7 8 2 5 8 7 3 6 1 2 4 7 3 1 5 8 2 3 6 1 5 4 7	5 7 6 1 8 2 4 3	8
2 4 6 8 3 6 5 1 4 7 8 2 5 1 7 3 6 8 1 4 7 2 3 5 8 3 2 6 1 5 4 7	5 7 6 8 2 1 4 3	8
2 4 5 8 3 7 6 1 4 6 8 2 5 8 7 3 6 1 2 4 7 3 1 5 8 2 3 6 1 5 4 7	6 5 7 1 8 2 4 3	8
2 4 5 8 3 7 6 1 4 6 8 2 5 8 7 3 6 1 2 4 7 2 1 5 8 3 4 6 1 5 3 7	6 5 7 1 8 3 2 4	8
2 4 5 8 3 7 6 1 4 6 8 2 5 1 7 3 6 8 2 4 7 3 1 5 8 2 3 6 1 5 4 7	6 5 7 8 1 2 4 3	4
2 4 5 8 3 7 6 1 4 6 7 2 5 1 8 3 6 8 2 4 7 3 1 5 8 2 4 6 1 5 3 7	6 5 8 7 1 2 3 4	4
2 4 5 8 3 7 6 1 4 6 7 2 5 1 8 3 6 8 2 4 7 2 1 5 8 3 4 6 1 5 3 7	6 5 8 7 1 3 2 4	8
2 4 5 8 3 5 6 1 4 6 7 2 5 7 8 3 6 8 1 4 7 1 2 5 8 2 3 6 1 3 4 7	6 7 8 1 2 3 4 5	2

We noted that among these permutations are two which may be generated with difference functions

$$[-2 +1 +4 -3 -1 +2], \quad \text{and} \tag{2}$$

$$[-2 +1 +3 +4 -1 +2] \tag{3}$$

being the first and last entries in Table 5 respectively.

5. A Regular Form for Permutation P

As further confirmation of this result, all permutations P which may be generated by a regular difference function on the target S-box were constructed, 120 being found. There cipher-plaintext bit dependency propagation is given in Table 3. The 8 best of these are indeed the regular permutations formed using the difference functions (2) and (3), and their equivalents formed by swapping *ab/ef* or *cd* columns. They are listed in Table 6. These permutations would thus seem to be excellent candidates for any reimplementation of the DES using 64-bit blocks. However, some problems are noted when the dependency results are examined further, as indicated below.

Table 6 - Best Regular Form Permutations P	
Permutation	Difference Fn
7 4 5 3 8 5 6 4 1 6 7 5 2 7 8 6 3 8 1 7 4 1 2 8 5 2 3 1 6 3 4 2	-2 +3 +4 +2
7 5 4 3 8 6 5 4 1 7 6 5 2 8 7 6 3 1 8 7 4 2 1 8 5 3 2 1 6 4 3 2	-2 +4 +3 +2
7 5 6 3 8 6 7 4 1 7 8 5 2 8 1 6 3 1 2 7 4 2 3 8 5 3 4 1 6 4 5 2	-2 +4 -3 +2
7 6 5 3 8 7 6 4 1 8 7 5 2 1 8 6 3 2 1 7 4 3 2 8 5 4 3 1 6 5 4 2	-2 -3 +4 +2
2 4 5 8 3 5 6 1 4 6 7 2 5 7 8 3 6 8 1 4 7 1 2 5 8 2 3 6 1 3 4 7	+1 +3 +4 -1
2 5 4 8 3 6 5 1 4 7 6 2 5 8 7 3 6 1 8 4 7 2 1 5 8 3 2 6 1 4 3 7	+1 +4 +3 -1
2 5 6 8 3 6 7 1 4 7 8 2 5 8 1 3 6 1 2 4 7 2 3 5 8 3 4 6 1 4 5 7	+1 +4 -3 -1
2 6 5 8 3 7 6 1 4 8 7 2 5 1 8 3 6 2 1 4 7 3 2 5 8 4 3 6 1 5 4 7	+1 -3 +4 -1

6. A Further Criterion for Best Permutations

In the analysis so far, the measure of effectiveness being used is the rate of growth of dependence of output bits on input bits, using Meyer's analysis. However this dependence may be further characterized, by a dependence through message inputs (bits *bcde*) only, autoclave inputs (bits *af*) only, or through both. If, as a measure of effectiveness, we examine the rate of growth of dependence on input bits through both autoclave and message bits, this demonstrates that a strictly regular permutation is not necessarily the most effective (see Table 7). Rather, the best latin-square permutations, that is with only two of four columns fixed, performs better by this criterion. Also, the original DES performs better still, but at a price of lower growth in overall dependence. There thus, seems to be some tradeoff between regularity, and best performance.

In the extended scheme being developed, the choice of some of the best latin square permutations, would thus seem to be indicated. In subsequent work on permutation PC2 and the key schedule, a subset of these, which also perform best using a measure of dependence of output bits on key bits have been found. These are listed in Table 8, and were used to produce the results in Table 7.

Round	Current P				Best Latin Square P				Best Regular P			
	Msg	Auto	Both	Total	Msg	Auto	Both	Total	Msg	Auto	Both	Total
1	192	64	0	6.25	192	64	0	6.25	192	64	0	6.25
2	641	499	173	32.06	656	512	152	32.23	616	512	192	32.23
3	834	806	1370	73.49	864	864	1296	73.83	784	960	1280	73.83
4	353	371	3245	96.90	368	448	3160	97.07	328	640	3008	97.07
5	0	0	4096	100.00	0	32	4064	100.00	0	128	3968	100.00
6	0	0	4096	100.00	0	0	4096	100.00	0	0	4096	100.00

Table 7 - Dependency of Ciphertext bits on
Plaintext bits via Message, Autoclave and Both Inputs

Table 8 - Best Latin Square Permutations P
by both Ciphertext-Plaintext and Ciphertext-Key Dependence

```
2 5 6 8 3 6 5 1 4 7 8 2 5 1 7 3 6 8 1 4 7 2 3 5 8 4 2 6 1 3 4 7
2 5 6 8 3 6 5 1 4 7 8 2 5 1 7 3 6 8 1 4 7 3 2 5 8 2 4 6 1 4 3 7
2 5 6 8 3 6 5 1 4 8 7 2 5 7 1 3 6 1 8 4 7 2 3 5 8 4 2 6 1 3 4 7
2 5 6 8 3 6 5 1 4 8 7 2 5 7 1 3 6 1 8 4 7 3 2 5 8 2 4 6 1 4 3 7
2 6 5 8 3 5 6 1 4 7 8 2 5 1 7 3 6 8 1 4 7 2 3 5 8 4 2 6 1 3 4 7
2 6 5 8 3 5 6 1 4 7 8 2 5 1 7 3 6 8 1 4 7 3 2 5 8 2 4 6 1 4 3 7
2 6 5 8 3 5 6 1 4 8 7 2 5 7 1 3 6 1 8 4 7 2 3 5 8 4 2 6 1 3 4 7
2 6 5 8 3 5 6 1 4 8 7 2 5 7 1 3 6 1 8 4 7 3 2 5 8 2 4 6 1 4 3 7
```

7. Design Rules for Permutation P

In the light of these experiments, the current set of rules for designing permutation P in a DES like scheme are:

R1 Each of the input bits $a\ b\ c\ d\ e\ f$ for S-box S(i) must come from the outputs of different S-boxes in the previous round.

R2 None of the input bits $a\ b\ c\ d\ e\ f$ for S-box S(i) may come from the output of that same S-box S(i) in the previous round.

R3 For each input bit $a, b, c, d, e,$ or f over all of the S-boxes, the S-boxes permuted to that bit must differ (ie each column forms a permutation of integers 1 to 8).

R4 An output from S(i-2) goes to one of the ab input bits of S(i), and from S(i+1) to the other. Hence via E an output from S(i-1) goes to one of the ef input bits, and from S(i+2) to other.

R5 Outputs from two of S(i-3), S(i+3), and S(i+4) go to cd input bits of S(i).

R6 For each S-box output, two bits go to ab or ef input bits, the other two go to cd input bits as noted in [Davi82].

Rules R1, R2 and R3 constrain the permutation to be a latin-rectangle, whilst the remaining rules further constrain the permutation to select outputs which maximize the growth of bit dependence. Having used these rules to construct a permutation, the dependency analysis may then be used to verify its effectiveness.

8. Regular Permutations P in An Extended DES

Permutations with strictly regular form were generated for an extended DES using 128-bit blocks. The 4 best of these were produced using a difference function

$$[-2 +1 -7 +7 -1 +2] \tag{4}$$

and its equivalents by column swapping, as shown in Table 9. The cipher-plaintext dependency results are given in Table 10 (along with those for a worst case, and sample permutations given in Brown [Brow88]). These best permutations alone of all those generated, resulted in complete dependency of ciphertext bits on all plaintext bits in 5 rounds, the same as for the current size scheme. Work is continuing to derive the best permutations for use in the extended scheme when dependence on both message and autoclave inputs is considered.

Table 9 - Best Regular Form Extended Permutations P
2 10 8 16 3 11 9 1 4 12 10 2 5 13 11 3 6 14 12 4 7 15 13 5 8 16 14 6 9 1 15 7 10 2 16 8 11 3 1 9 12 4 2 10 13 5 3 11 14 6 4 12 15 7 5 13 16 8 6 14 1 9 7 15
2 8 10 16 3 9 11 1 4 10 12 2 5 11 13 3 6 12 14 4 7 13 15 5 8 14 16 6 9 15 1 7 10 16 2 8 11 1 3 9 12 2 4 10 13 3 5 11 14 4 6 12 15 5 7 13 16 6 8 14 1 7 9 15
15 10 8 3 16 11 9 4 1 12 10 5 2 13 11 6 3 14 12 7 4 15 13 8 5 16 14 9 6 1 15 10 7 2 16 11 8 3 1 12 9 4 2 13 10 5 3 14 11 6 4 15 12 7 5 16 13 8 6 1 14 9 7 2
15 8 10 3 16 9 11 4 1 10 12 5 2 11 13 6 3 12 14 7 4 13 15 8 5 14 16 9 6 15 1 10 7 16 2 11 8 1 3 12 9 2 4 13 10 3 5 14 11 4 6 15 12 5 7 16 13 6 8 1 14 7 9 2

Table 10 - Dependency of Ciphertext bits on Plaintext bits in Extended (128-bit) DES			
Round	Best Regular Form P	Sample XDES P	Worst XDES P
1	3.13	3.125	3.125
2	17.58	17.68	10.55
3	52.34	50.98	21.88
4	87.50	84.47	34.38
5	100.00	98.44	46.88
6	100.00	100.00	59.38
7	100.00	100.00	71.88
8	100.00	100.00	84.38
9	100.00	100.00	94.53
10	100.00	100.00	99.21
11	100.00	100.00	100.00

9. Conclusion

This paper reviews some possible design criteria for the permutation P in a DES style cryptosystem. These permutations provide the diffusion component in a substitution-permutation network. Some empirical rules which seem to account for the derivation of the permutation used in the DES are first presented. Then it is noted that these permutations may be regarded as latin-squares which link the outputs of S-boxes to their inputs at the next stage. When a sample of these latin-square permutations were generated, they were found to span those generated using the empirical rules. The best of these permutations suggested that permutations having a fixed column structure as generated by difference function (1) would perform well. Permutations with this structure were generated, and the best of these analysed. They were found to include the permutations generated using difference functions (2) and (3), which were strictly regular. Whilst these performed well in growth of overall bit dependence, when dependence on both message and autoclave bits were examined, these were found to be slightly less than optimal. From these a set of design rules have been developed for permutation P. An extension of these results to an extended DES was then analysed, and 4 permutations formed by difference function (4) were found to result in complete ciphertext dependency on plaintext bits after 5 rounds, a result that is the same as with the current size scheme.

Acknowledgements

To the following members of the Centre for Computer Security Research: Leisa Condie, Thomas Hardjono, Arthur Lagos, Mike Newberry, Cathy Newberry, Josef Pieprzyk, and Jennifer Seberry; and to: Dr. George Gerrity, Dr. Andrez Goscinski, and Dr. Charles Newton; for their comments on, suggestions about, and critiques of this paper. Thankyou.

References

[ASA85] ASA, *"Electronics Funds Transfer - Requirements for Interfaces, Part 5, Data Encryption Algorithm,"* AS2805.5-1985, Standards Association of Australia, Sydney, Australia, 1985.

[Brow88] L. Brown, "A Proposed Design for an Extended DES," in *Proc. Fifth International Conference and Exhibition on Computer Security*, IFIP, Gold Coast, Queensland, Australia, 19-21 May, 1988.

[Davi82] D. W. Davies, "Some Regular Properties of the Data Encryption Standard," in *Advances in Cryptology - Proc. of Crypto 82*, D. Chaum, R. L. Rivest and A. T. Sherman (editors), pp. 89-96, Plenum Press, New York, Aug. 23-25, 1982.

[DDFG83] M. Davio, Y. Desmedt, M. Fosseprez, R. Govaerts, J. Hulsbosch, P. Neutjens, P. Piret, J. Quisquater, J. Vanderwalle and P. Wouters, "Analytical Characteristics of the DES," in *Advances in Cryptology - Proc. of Crypto 83*, D. Chaum, R. L. Rivest and A. T. Sherman (editors), pp. 171-202, Plenum Press, New York, Aug. 22-24, 1983.

[DeKe74] J. Denes and A. D. Keedwell, *Latin Squares and their Applications*, English Universities Press Limited, London UK, 1974.

[DiHe77] W. Diffie and M. E. Hellman, "Exhaustive Cryptanalysis of the NBS Data Encryption Standard," *Computer*, vol. 10, no. 6, pp. 74-84, June 1977.

[Feis73] H. Feistel, "Cryptography and Computer Privacy," *Scientific American*, vol. 228, no. 5, pp. 15-23, May 1973.

[Hell79] M. E. Hellman, "DES will be totally insecure within ten years," *SPECTRUM*, vol. 16, no. 7, pp. 31-41, July 1979. With rebuttals from George I. Davida, Walter Tuchman, & Dennis Branstad.

[KaDa79] J. B. Kam and G. I. Davida, "Structured Design of Substitution-Permutation Encryption Networks," *IEEE Trans. on Computers*, vol. C-28, no. 10, pp. 747-753, Oct. 1979.

[Meye78]   C. H. Meyer, "Ciphertext/plaintext and ciphertext/key dependence vs number of rounds for the data encryption standard," in *AFIPS Conf. Proc. 47*, pp. 1119-1126, AFIPS Press, Montvale NJ, USA, June 1978.

[MeMa82]   C. H. Meyer and S. M. Matyas, *Cryptography: A New Dimension in Data Security*, John Wiley & Sons, New York, 1982.

[NBS77]   NBS, *"Data Encryption Standard (DES),"* FIPS PUB 46, US National Bureau of Standards, Washington, DC, Jan. 1977.

[SePi88]   J. Seberry and J. Pieprzyk, *Cryptography: An Introduction to Computer Security*, Prentice Hall, Englewood Cliffs, NJ, 1988.

[Shan49]   C. E. Shannon, "Communication Theory of Secrecy Systems," *Bell System Technical Journal*, vol. 28, no. 10, pp. 656-715, Oct. 1949.

A Fast Elliptic Curve Cryptosystem

G.B. Agnew R.C. Mullin S.A. Vanstone
University of Waterloo

Introduction

In the fall of 1986, the authors developed a prototype of a fast GF(2^{593}) multiplier/exponentiator. This device was based on the discovery of optimal normal basis structures in fields of characteristic two [1][2]. In the ensuing years, much effort has gone into fabricating this structure as a VLSI device. In the early months of 1989, the first functioning VLSI devices were fabricated [3]. These devices, which implement a cryptographic system based on discrete exponentiation [4][5], have throughput rates of up to 300 Kbps.. Many cryptographic applications based on discrete exponentiation have been implemented or proposed [6][7], and several new applications of the normal basis multiplier have recently been considered.

In 1985, Miller [8], presented a method of implementing a cryptographic system based on elliptic curves. The advantage of such cryptosystems is that, unlike RSA or discrete exponentiation, no sub-exponential method is known for attacking elliptic curves [9]. Thus, smaller block sizes could be used to implement a computationally secure system.

The problem with elliptic curve implementations is the complexity (and thus computation time) of calculating points on the curve. In this paper, we will examine an implementation (not optimized) of an elliptic curve system using the fast normal basis multiplier structure.

Elliptic Curve Calculations

Koblitz [10], lists several forms of elliptic curves which form groups. For curves of characteristic two, the curve

$$y^2 + y = x^3$$

was chosen. In this system, the calculation of a new point KP from a point P on the curve and value K ($0 < K \le 2^n - 1$) can be realized as

$$KP = k_0 P + k_1 2P + k_2 4P + \ldots + k_{n-1} 2^{n-1} P$$

or

$$KP = (\ldots((k_0 P + k_1 2P) + k_2 4P) + \ldots + k_{n-1} 2^{n-1} P).$$